OXFORD GRADUATE TEXTS IN MA

Series Editors

OXFORD GRADUATE TEXTS IN MATHEMATICS

Books in the series

1. Keith Hannabuss: *An Introduction to Quantum Theory*
2. Reinhold Meise and Dietmar Vogt: *Introduction to Functional Analysis*
3. James G. Oxley: *Matroid Theory*
4. N.J. Hitchin, G.B. Segal, and R.S. Ward: *Integrable Systems: Twistors, Loop Groups, and Riemann Surfaces*
5. Wulf Rossmann: *Lie Groups: An Introduction through Linear Groups*
6. Qing Liu: *Algebraic Geometry and Arithmetic Curves*
7. Martin R. Bridson and Simon M. Salamon (eds): *Invitations to Geometry and Topology*
8. Shmuel Kantorovitz: *Introduction to Modern Analysis*
9. Terry Lawson: *Topology: A Geometric Approach*
10. Meinolf Geck: *An Introduction to Algebraic Geometry and Algebraic Groups*
11. Alastair Fletcher and Vladimir Markovic: *Quasiconformal Maps and Teichmüller Theory*
12. Dominic Joyce: *Riemannian Holonomy Groups and Calibrated Geometry*
13. Fernando Villegas: *Experimental Number Theory*
14. Péter Medvegyev: *Stochastic Integration Theory*
15. Martin A. Guest: *From Quantum Cohomology to Integrable Systems*
16. Alan D. Rendall: *Partial Differential Equations in General Relativity*
17. Yves Félix, John Oprea, and Daniel Tanré: *Algebraic Models in Geometry*
18. Jie Xiong: *Introduction to Stochastic Filtering Theory*
19. Maciej Dunajski: *Solitons, Instantons, and Twistors*
20. Graham R. Allan: *Introduction to Banach Spaces and Algebras*
21. James Oxley: *Matroid Theory, Second Edition*
22. Simon Donaldson: *Riemann Surfaces*
23. Clifford Henry Taubes: *Differential Geometry: Bundles, Connections, Metrics and Curvature*
24. Gopinath Kallianpur and P. Sundar: *Stochastic Analysis and Diffusion Processes*
25. Selman Akbulut: *4-Manifolds*
26. Fon-Che Liu: : *Real Analysis*
27. Dusa Mcduff and Dietmar Salamon: *Introduction to Symplectic Topology, Third Edition*

Introduction to Symplectic Topology

THIRD EDITION

DUSA MCDUFF

Professor of Mathematics, Barnard College, Columbia University

DIETMAR SALAMON

Professor of Mathematics, ETH Zürich

Great Clarendon Street, Oxford, OX2 6DP,
United Kingdom

Oxford University Press is a department of the University of Oxford.
It furthers the University's objective of excellence in research, scholarship,
and education by publishing worldwide. Oxford is a registered trade mark of
Oxford University Press in the UK and in certain other countries

First Edition published in 1995

Third Edition published in 2017

Impression: 2

Published in the United States of America by Oxford University Press
198 Madison Avenue, New York, NY 10016, United States of America

British Library Cataloguing in Publication Data
Data available

Library of Congress Control Number: 2016946840

ISBN 978–0–19–879489–9 (hbk.)
ISBN 978–0–19–879490–5 (pbk.)

Printed and bound by
CPI Group (UK) Ltd, Croydon, CR0 4YY

PREFACE

This book had its beginnings in lecture courses which the authors gave at their respective universities in 1989 and 1990. Since then many people have made various useful comments on the text. We would like to thank them all, and in particular the students Eleonora Ciriza, Brian Kasper, Gang Liu, Wladyslaw Lorek, Alan McRae, Simon Richard, and Lisa Traynor at Stony Brook as well as Mohan Bhupal, Thomas Davies, Shaun Martin, and Marcin Pozniak at Warwick. We are grateful to Simon Donaldson and Peter Kronheimer for explaining to us their recent results which are discussed in Chapter 4. We also owe a special debt of gratitude to Santiago Simanca who drew all the illustrations, to Phil Boyland who provided Fig. 8.3 and to Scott Sutherland who helped with sundry queries about TeX on the SUNs.

Stony Brook D.M.
Warwick D.S.
1994

PREFACE TO THE SECOND EDITION

The second edition is a revised and enlarged version of the original book. The most significant additions are: a section on J-holomorphic curves in Chapter 4, new material on Hamiltonian fibrations in Chapter 6, a section on the topology of symplectomorphism groups in Chapter 10, a generating function proof of the Arnold conjecture for cotangent bundles and a section on Floer homology in Chapter 11, and a more extensive discussion of Taubes and Donaldson's recent contributions in Chapter 13. Chapter 12 on symplectic capacities has been revised to make the proofs more transparent, and some infinite-dimensional examples of moment maps have been added to Chapter 5. We have reorganized Chapters 6 and 7. Chapter 6 is now exclusively devoted to symplectic fibrations and the new Chapter 7 contains the remaining material of the old Chapter 6 (blowing up and down and fibre connected sums) as well as Gromov flexibility. The remaining sections from the old Chapter 7 now appear as part of the new Chapter 13. We have added references and sometimes brief discussions of a number of interesting new developments of the past 4 years. Finally, we took the opportunity to make some needed corrections. We wish to thank everyone who told us of errors and typos in the original text, including Paul Biran, Peter Kahn, Yael Karshon, Yuri Rudyak, David Théret and the students Meike Akveld, Silvia Anjos, Olguta Buse, Leonor Godinho, Haydee Guzman-Herrera, Jenn Slimowitz, and Luisa Stelling.

Stony Brook D.M.
Warwick D.S.
1998

This reprint of the second edition corrects some typos as well as some significant errors, notably the statement of Theorem 6.36 about Hamiltonian bundles and the discussion preceding Exercise 10.28.

Stony Brook D.M.
Zürich D.S.
2005

PREFACE TO THE THIRD EDITION

Our goal in the third edition of this book is two-fold. On the one hand we have corrected mistakes and improved the exposition at various places in the hope that this will make the text more accessible for graduate students first encountering the subject of symplectic topology. On the other hand we have updated the book at several places to take account of the many new developments in the field, which has grown into a vast area of research over the past quarter of a century and has many connections to other subjects. We hope that this will make the book also useful for advanced researchers in the field. In particular, there are several places where we give overviews of some important developments in symplectic topology, without providing a full expository account, and where we only summarize some of the central results. These parts will certainly be much harder to digest for beginning graduate students. They are Section 4.5 on J-holomorphic curves, Section 5.7 on geometric invariant theory, Section 7.4 on Donaldson hypersurfaces, Section 10.4 on the topology of symplectomorphism groups, Section 11.4 on Floer homology, Section 13.3 on Taubes–Seiberg–Witten theory, and Section 13.4 on symplectic four-manifolds. We have also included in the third edition a new Chapter 14, where we list some open questions in symplectic topology and discuss some of the known results about these questions (without any attempt at completeness).

In comparison to the second edition of this book, the main changes, apart from the aforementioned Chapter 14, include the following. Section 7.1 contains an 'expanded and much more detailed discussion of the construction of symplectic forms on blow-ups; some technical parts of the construction are relegated to Appendix A and to a new Section 3.3 on isotopy extensions. We have also expanded the material on Donaldson hypersurfaces, previously contained in Chapter 13, and moved it to Chapter 7 as an additional Section 7.4. Chapter 13 now starts with a new section on existence and uniqueness problems in symplectic topology. It includes a much more detailed section on Taubes–Seiberg–Witten theory and its applications as well as a new section on symplectic four-manifolds. In the earlier chapters, Section 2.5 on linear complex structures and Section 3.5 on contact structures have been significantly expanded, Section 4.5 on J-holomorphic curves has been rewritten and updated, and Section 5.7, containing an overview of GIT, has been added. In the later chapters, Section 10.4 on the topology of symplectomorphism groups, Section 11.4 on Floer theory, and Section 12.3 on Hofer geometry have been updated and in parts rewritten.

We would like to thank the many mathematicians who have sent us comments and suggestions to improve this current edition; and especially Janko Latschev and Andrew McInerney for help with Section 3.5, Rahul Pandharipande for help with Section 4.3, Yael Karshon for perceptive comments on Chapter 5, Catherine

Cannizzo for pointing out many small inaccuracies, and Mohammed Abouzaid, Denis Auroux, Paul Biran, Simon Donaldson, Yasha Eliashberg, Mark McLean, Leonid Polterovich, and Felix Schlenk for suggesting problems for Chapter 14.

New York — D.M.
Zürich — D.S.
2016

CONTENTS

INTRODUCTION

Symplectic topology has a long history. It has its roots in classical mechanics and geometric optics and in its modern guise has many connections to other fields of mathematics and theoretical physics ranging from dynamical systems, low-dimensional topology, algebraic and complex geometry, representation theory, and homological algebra, to classical and quantum mechanics, string theory, and mirror symmetry. One of the origins of the subject is the study of the equations of motion arising from the Euler–Lagrange equations of a one-dimensional variational problem. The Hamiltonian formalism arising from a Legendre transformation leads to the notion of a *canonical transformation* which preserves Hamilton's form of the equations of motion (see Chapter 1). From this vantage point one can define a symplectic manifold as a smooth manifold equipped with an atlas whose transition maps are canonical transformations. Symplectic manifolds are then spaces with a global intrinsic structure on which Hamilton's differential equations can be defined and take the classical form in each coordinate chart. Since canonical transformations preserve a standard 2-form associated to Hamilton's equations, a symplectic manifold is equipped with a closed nondegenerate 2-form. It is called a **symplectic form** and will typically be denoted by ω. A foundational result in the subject is *Darboux's Theorem* which asserts that the existence of a symplectic form is equivalent to the existence of an atlas whose transition maps are canonical transformations, and that this atlas (if chosen maximal) and the symplectic form determine one another. This is the starting point for the field of symplectic topology that we chose in this book.

This discussion illustrates the passage from the *local* study of Hamiltonian dynamics to *global* phenomena. The emphasis on global properties is significant for the entire subject of symplectic topology. Darboux's theorem asserts that every symplectic manifold is locally isomorphic to the Euclidean space of the appropriate dimension with its standard symplectic structure, and thus there cannot be any local invariants that distinguish symplectic manifolds. This is in sharp contrast to Riemannian geometry where the curvature is a local invariant. Second, every manifold admits a Riemannian metric while there is a very interesting (and rather little understood) question in symplectic topology of which manifolds admit symplectic structures. Third, a Riemannian manifold has only a finite-dimensional group of isometries, while a symplectic manifold has a very rich infinite-dimensional group of diffeomorphisms that preserve the symplectic structure. They are called **symplectomorphisms**. The counterpart of this third observation is that there is an infinite-dimensional family of non-isometric Riemannian metrics on any manifold, while by *Moser isotopy* (another foundational principle in the subject) there is only a finite-dimensional family of nonisomorphic symplectic structures on a closed manifold.

Symplectomorphisms generated by global Hamiltonian differential equations are called **Hamiltonian symplectomorphisms**. Their study can be viewed as a branch of the field of dynamical systems and has had a profound influence on the development of the subject of symplectic topology, and vice versa. One important topic, with its origins in celestial mechanics, has been the search for periodic solutions of Hamiltonian differential equations; many mathematicians, including Poincaré, Weierstrass, Arnold, Siegel, and Moser, have contributed to this subject. Such periodic orbits can be viewed as fixed points of Hamiltonian symplectomorphisms. In the search for fixed points one again encounters a local versus global dichotomy. It is fairly easy to show that a Hamiltonian symplectomorphism of a closed symplectic manifold must have a fixed point when it is C^1 close to the identity (the local case), while it has been a deep and for a long time unsolved problem to show that every Hamiltonian symplectomorphism of a closed symplectic manifold must have a fixed point regardless of how far away it is from the identity (the global case). A strengthened form of this problem, which gives a lower bound on the number of fixed points, is the *Arnold conjecture* (see Chapter 11). Other versions of this question are also interesting in Euclidean space with its standard linear symplectic structure.

Aside from the concepts of a *symplectic form* and a *symplectomorphism*, that of a *Lagrangian submanifold* is the third fundamental notion in the field of symplectic topology. It has no counterpart in Riemannian geometry. A symplectic manifold M carries a symplectic form ω, and it follows from the nondegeneracy and skew-symmetry of ω that M must have even dimension $2n$. Moreover, for every nondegenerate skew-symmetric bilinear form on a vector space of dimension $2n$, there exist subspaces on which this 2-form vanishes, and the maximal dimension of such a subspace is n. A **Lagrangian submanifold** of a $2n$-dimensional symplectic manifold (M, ω) is an n-dimensional submanifold $L \subset M$ such that the symplectic form ω vanishes on each tangent space of L. The study of Lagrangian submanifolds is a central topic in symplectic topology that can tell us a great deal about the symplectic manifold (M, ω). Examples of Lagrangian submanifolds are the graph of a symplectomorphism, the zero section of a cotangent bundle (which admits a *canonical symplectic structure*), the real part of a smooth projective variety, any embedded circle in a two-dimensional symplectic manifold, or any product of Lagrangian submanifolds in a product symplectic manifold. In the words of Alan Weinstein '*everything is a Lagrangian submanifold*'. There are many interesting questions one can ask about Lagrangian submanifolds, some of which are formulated in (vi) below. For many symplectic manifolds these problems are wide open.

Here is a list of some vaguely worded questions, including some mentioned above, that are of central importance to the subject of symplectic topology.

(i) Which manifolds support a symplectic structure? What symplectic invariants distinguish one from another?

(ii) Are there any special distinguishing features of Hamiltonian flows on arbitrary compact hypersurfaces in Euclidean space? For example, must they always have a periodic orbit? (These surfaces are known as *compact energy surfaces* since they appear as level sets of the Hamiltonian energy function.)
(iii) Must a symplectomorphism always have 'a lot' of fixed points?
(iv) What can be said about the shape of a symplectic image of a ball? Can it be long and thin, or must it always be in some sense round?
(v) Is there a geometric way to understand the fact that a symplectic structure makes 2-*dimensional* measurements? (A symplectic structure is a special kind of 2-form, but what geometric meaning does that have?)
(vi) Which homology classes can be represented by Lagrangian submanifolds? Are there any restrictions on the topology of Lagrangian submanifolds? Can a given Lagrangian submanifold *be disjoined* from itself by a Hamiltonian isotopy? Are any two Lagrangian submanifolds in the same homology class smoothly (or symplectically or Hamiltonian) isotopic?

Classically, very little was known about such questions. In dynamical systems there is *Poincaré's last geometric theorem*, which asserts that an area- and orientation-preserving twist map of the annulus must have at least two different fixed points. It was proved by Birkhoff in the 1920s. In two dimensions a symplectic structure is just an area form and hence the Poincaré–Birkhoff theorem can be considered as the first global theorem in symplectic topology. It was quickly realized that the right generalization of the Poincaré–Birkhoff theorem was not to volume-preserving diffeomorphisms but to symplectomorphisms. But it was not until the 1960s that some of these questions began to be formulated more precisely. For example, Arnold's conjectures (1965) give a precise formulation of (iii). And it was not until the late 1970s that some of these questions began to be answered. For example, the work of Weinstein established the existence of periodic orbits on convex hypersurfaces in Euclidean space. Another important milestone was Moser's proof in 1965 of the stability of symplectic structures on compact manifolds. His theorem asserts that any cohomologous deformation of a symplectic form on a compact manifold is diffeomorphic to the original form. This is another expression of the Darboux principle that symplectic forms have no local invariants, and makes the search for global structure harder but more interesting. It also inspired an important series of papers by Weinstein, who worked out the basic structure theorems of global symplectic and Poisson geometry. He showed, for example, that every Lagrangian submanifold admits a neighbourhood which is symplectomorphic to a neighbourhood of the zero section in the cotangent bundle of the submanifold. Some of these results are described in his influential set of lecture notes [671].

Several developments have led to a flowering of the subject since the 1980s. The variational methods pioneered by Rabinowitz and Weinstein in the late 1970s led to the Conley–Zehnder proof of Arnold's conjectures for the standard torus in 1983. They were taken up by Viterbo, who in 1987 confirmed the Weinstein conjecture about the existence of periodic orbits on a special kind of compact energy surface, namely those of contact type. Then Hofer, working in collaboration with Ekeland and Zehnder, developed the Ekeland–Hofer variational theory of capacities, and another related capacity called the Hofer–Zehnder capacity, which cast light on questions (iv) and (v) above. Hofer also found a very interesting new metric on the (infinite-dimensional) group of symplectomorphisms which he called the energy of a symplectomorphism. These methods work best when the manifold has some underlying linear structure, for example in Euclidean space or on cotangent bundles.

Quite different lines of development were initiated by Gromov. Already in the late 1960s he had shown in his thesis that symplectic structures on open manifolds can be classified by homotopy-theoretic data. This is what he called **flexibility**. Later he and Eliashberg investigated the limits of this flexibility, i.e. at what point would rigidity appear through the existence of some kind of global structure? Gromov discovered a fundamental **hard versus soft** alternative: either the group of symplectomorphisms is closed in the group of all diffeomorphisms with respect to the C^0-topology (hardness or rigidity) or it is C^0-dense in the group of volume-preserving diffeomorphisms (softness). An important signpost was Eliashberg's proof in the late 1970s of the rigid alternative via his analysis of wave fronts [174, 176].

In 1985, Gromov found a completely different way in which symplectic rigidity manifested itself. He brought elliptic and geometric methods into play by considering geometric properties of the family of almost complex structures associated to a symplectic form. More precisely, he examined the properties of the moduli space of J-holomorphic curves in a symplectic manifold.[1] One of his most celebrated results is the nonsqueezing theorem, which states that a symplectic image of a ball cannot be 'long and thin'. He also established the existence of exotic symplectic structures on Euclidean space, which cannot be embedded into the standard structure. These elliptic methods are very powerful and work well on arbitrary manifolds. They have many deep connections with the Yang–Mills theory of connections, and they inspired Floer to develop his theory of Floer homology, which applies in both contexts. In the symplectic context, Floer homology has been used to establish the Arnold conjectures for many manifolds. It has also been combined with the Ekeland–Hofer variational theory of capacities to produce the Floer–Hofer theory of symplectic homology, which gives rise to powerful geometric invariants of open symplectic manifolds.

A third strand in modern symplectic topology, which is closely related to the variational techniques discussed above, is a globalization of the classical idea of a generating function, stemming from the work of Chaperon, Laudenbach, Sikorav, and Viterbo. This gives another basic approach to the foundational theorems of symplectic topology. For example, Gromov's nonsqueezing theorem in Euclidean space can now be proved either by Gromov's original elliptic methods, by the

[1] Given a compatible almost complex structure J on the symplectic manifold M, a J-holomorphic curve is a map of a Riemann surface to M whose differential is complex linear with respect to J. Hence in the case of an embedding, its image is a complex submanifold.

Ekeland–Hofer theory of capacities, or by Viterbo's development of the theory of generating functions. Tamarkin [605] pointed out that it can even be proved by sheaf-theoretic methods, a micro-local version of generating functions; for details see Guillermou–Kashiwara–Schapira [300].

A variety of purely topological methods have been developed. Notable among these are Gompf's construction of fibre connected sums which he used to produce a huge number of new symplectic four-manifolds, and the Lalonde–McDuff construction of symplectic embeddings of balls which has been used to prove the nonsqueezing theorem for arbitrary symplectic manifolds.

The mid 1990s brought two exciting breakthroughs in symplectic topology. First, Donaldson proved a remarkable theorem about the existence of symplectic submanifolds of codimension two. This result can be viewed as an analogue of the construction, in Kähler geometry, of divisors as zero sets of holomorphic sections of ample line bundles. Donaldson adapts these techniques to the almost complex case and also proves that every symplectic four-manifold, after blow-up at finitely many points, admits the structure of a topological Lefschetz fibration. We shall discuss these results in more detail in Section 7.4 and show, in Remark 7.4.3, how they can be combined with results by Kronheimer and Mrowka about the minimal genus of embedded surfaces to obtain obstructions to the existence of symplectic structures in a given homology class.

The second event, happening as the first edition of this book went to press, was Taubes' calculation of the Seiberg–Witten invariants for symplectic four-manifolds. This allowed him to show that certain almost complex four-manifolds support no symplectic structure at all. Taubes proved subsequently that the Seiberg–Witten invariants are equal to the Gromov invariants and, as a result, that the canonical class $K = -c_1(TM, J)$ of every symplectic four-manifold with $b^+ > 1$ can be represented by a symplectic submanifold. Thus Donaldson's theorem can be interpreted as a symplectic version of the *ample divisor* and Taubes' theorem as a symplectic version of the *canonical divisor*. Taubes' theorem has many important consequences for our understanding of symplectic four-manifolds and we shall summarize some of these in Section 13.3.

In this book we attempt to give an introduction to all of these ideas. However, the details of the more far reaching developments, such as Gromov's theory of J-holomorphic curves (Section 4.5), Floer homology (Section 11.4), Taubes' work on the Seiberg–Witten invariants (Section 13.3), and Donaldson's theory of symplectic submanifolds and topological Lefschetz fibrations (Section 7.4), go beyond the scope of this text; each of them would require a book to itself.[2] Here we only give brief overviews of these four subjects. The deepest proofs

[2] Good references for J-holomorphic curves are the collection of articles *Holomorphic Curves in Symplectic Geometry* [34] that describes the theory from a fairly elementary, differential-geometric point of view and gives many applications, the Bern seminar [13] that follows Gromov's original approach to compactness questions, and the book *J-holomorphic Curves and Symplectic Topology* [470] that takes a more analytic approach. A comprehensive introduction to the Seiberg–Witten invariants and their applications is given in [556].

in our book use a finite-dimensional variational technique which in spirit lies midway between the infinite-dimensional universal variational methods in Hofer and Zehnder's book [325] and Viterbo's finite-dimensional generating function techniques in [646].

The book does not assume any familiarity with symplectic geometry, and so Part I is taken up with developing the basic facts about symplectic forms and Hamiltonian flows, and so on. Part II is concerned with global symplectic topology on manifolds, group actions, and methods of construction. Part III discusses properties of symplectomorphisms, with special emphasis on generating functions. Part IV is really the heart of the book. Here we give full proofs of the simplest versions of important theorems, such as the nonsqueezing theorem in Euclidean space, Arnold's conjectures for the torus and cotangent bundles, and the energy-capacity inequality for symplectomorphisms of Euclidean space. Readers should then be in a position to pursue later developments in the original literature, to which we give copious references.

It goes without saying that much is omitted from this book. In particular, we say almost nothing about classical mechanics, quantization, and applications of symplectic geometry to representation theory. Comprehensive books about these subjects include Arnold [23], Siegel–Moser [584], Marsden [437], Woodhouse [689], Bates–Weinstein [59], and Guillemin–Lerman–Sternberg [292].

Here is a brief description of the contents of the book, chapter by chapter. Part I develops the basic background material. Chapter 1 begins by discussing Hamiltonian systems in Euclidean space from a classical viewpoint and then describes some of the problems which have motivated the development of modern symplectic topology. Chapter 2 develops linear symplectic geometry, the theory of symplectic vector spaces and vector bundles. It includes a proof of the existence of the first Chern class. Chapter 3 is a foundational chapter which treats symplectic forms on arbitrary manifolds. It establishes Darboux's theorem and Moser's stability theorem, and also contains an introduction to contact geometry, the odd-dimensional analogue of symplectic geometry. Chapter 4, on almost complex structures, first discusses the difference between symplectic and Kähler manifolds, describes the classification of Kähler surfaces, and then outlines the theory of J-holomorphic curves and explains some of its applications.

Part II discusses examples of symplectic manifolds, and questions about their structure. Chapter 5 is concerned with the important topic of symplectic reduction. This has its roots in the study of mechanical systems with symmetry, and leads to many significant modern applications of symplectic topology, for example in the theory of group representations and in the study of moduli spaces. This chapter includes proofs of the Atiyah–Guillemin–Sternberg convexity theorem for moment maps and of the Duistermaat–Heckman localization formula. Chapter 6 is devoted to symplectic fibrations. It explains symplectic connections and curvature, gives a proof of the Guillemin–Lerman–Sternberg construction of the coupling form, and characterizes symplectic connections with Hamiltonian holonomy. Chapter 7 describes different ways of constructing symplectic manifolds, by symplectic blowing up and down, fibre connected sums, and Gromov's telescope construction of symplectic forms on open almost complex manifolds. We include a proof of Gompf's result that the fundamental group of a compact symplectic four-manifold may be any finitely presented group. Chapter 7 also includes a brief introduction to Donaldson submanifolds, Weinstein domains, and symplectic Lefschetz fibrations. Here the proofs of the main theorems go beyond the scope this book and we content ourselves with giving an overview and pointing the reader towards the relevant references.

Part III is concerned with symplectomorphisms. We begin in Chapter 8 with a classical example, giving an essentially complete proof of the Poincaré–Birkhoff theorem, which states that an area-preserving twist map of the annulus has two distinct fixed points. An important special case is that of strongly monotone twist maps, which admit a simple kind of generating function. In Chapter 9 we explore generating functions in more detail, in both their modern and classical guise, and show how they lead to discrete-time variational problems. Chapter 10 develops basic results on the structure of the group of symplectomorphisms, paying particular attention to properties of the subgroup of Hamiltonian (or exact) symplectomorphisms. It includes an introduction to the flux homomorphism and to the Calabi homomorphism, as well as a brief overview of some known results about the topology of symplectomorphism groups.

In Part IV we use the finite-dimensional variational methods described in Chapter 9 to establish some of the central results in the subject. Chapter 11 returns to the study of the Arnold conjectures on the existence of fixed points of Hamiltonian symplectomorphisms, and proves these conjectures for the standard torus. The necessary critical point theory (for example the Conley index and Ljusternik–Schnirelmann theory) is developed along the way. The chapter also contains a generating function proof of the Arnold conjecture for Lagrangian intersections in cotangent bundles. Chapter 12 gives a variational proof of the existence of capacities, and applies it to establish the nonsqueezing theorem, symplectic rigidity, and the Weinstein conjecture for hypersurfaces of Euclidean space. It also introduces the Hofer metric on the group of Hamiltonian symplectomorphisms and gives a geometric proof of the energy–capacity inequality for symplectomorphisms in Euclidean space. Chapter 13 first discusses the fundamental existence and uniqueness problems for symplectic structures, and then surveys a variety of different examples of symplectic manifolds and symplectic invariants. It proceeds with an overview of Taubes–Seiberg–Witten theory and some of its consequences for symplectic four-manifolds. Here the main theorems are carefully stated. However, their proofs go much beyond the scope of this book and we content ourselves with deriving their consequences rigorously. The final Chapter 14 discusses various open problems and conjectures in symplectic topology, as well as some of the known results about these questions.

Because symplectic topology is a meeting point for many different subjects it is impossible to cover all the relevant background material. We have tried to make the book accessible to graduate students who have some familiarity with topology, differential equations, differential manifolds, and differential forms. (The first chapter of Bott and Tu [78] would be good preparation.) Some parts require more. For example, a knowledge of Morse theory would be useful in Section 7.2, Morse theory and dynamical systems come up in Section 11.2, and more differential topology and geometry (including complex geometry) would be useful in Part II. However, we use only techniques from finite-dimensional analysis.

The first three chapters are foundational, and we recommend that readers familiarize themselves with their basic contents before proceeding. The later chapters are mostly independent of each other and can be read in more or less any order. For example, readers who want to understand the Arnold conjecture might read Chapter 8, and then Chapter 11, referring back to Section 9.2 as necessary. To understand capacities, a reader could go directly to Chapter 12 and read relevant parts of Chapter 9 as it becomes necessary. We have provided ample cross-references throughout, together with a comprehensive index, to guide readers who are interested in only parts of the book. Also, there is some repetition to accommodate those who do not read in linear order.

PART I

FOUNDATIONS

1

FROM CLASSICAL TO MODERN

Symplectic structures first arose in the study of classical mechanical systems, such as the planetary system, and classical work focused on understanding how these systems behave. Over the years methods have been developed that give insight into the global features of symplectic geometry, leading to the birth of a new field — symplectic topology.

There are many possible ways in which to approach symplectic topology. Because its variational roots colour the whole modern development of the theory, we have chosen to begin by showing how the basic elements of symplectic geometry emerge as one looks at a variational problem. Thus we show how the equations that describe this kind of problem lead to a system of Hamiltonian differential equations, and hence to the notion of a symplectic structure. On the way, we discuss some of the other concepts which arise in this context, such as Poisson structures, Lagrangian submanifolds, and symplectomorphisms. Throughout our approach is very explicit. Readers who are unfamiliar with the kind of analysis used here should attempt to understand the general picture without getting hung up on the details. We shall see in Chapter 3 that most (but not all) of the concepts discussed here may be defined in a coordinate-free way, and so generalize easily to arbitrary manifolds.

In the second section we illustrate the transition from the classical local theory to the modern global theory by formulating four foundational problems which have been focal points of development since the 1970s: the question of existence of periodic orbits for Hamiltonian flows (the Weinstein conjecture); the topological behaviour of symplectomorphisms (Gromov's nonsqueezing theorem); the question of rigidity (or what is the C^0-closure of the group of symplectomorphisms); and the question of counting fixed points of symplectomorphisms (Arnold's conjecture). We will return to these problems in the last part of the book, and there we will develop variational tools which are powerful enough to provide at least partial solutions.

1.1 Hamiltonian mechanics

Legendre transformations

The equations of motion in classical mechanics arise as solutions of variational problems. The most immediate example is **Fermat's principle of least time**, which states that light (considered as a stream of particles) moves from one point to another by a path which takes the shortest possible amount of time. This implies that light travels in straight lines through a homogeneous medium such as air, and allows one to calculate how much the rays bend when they pass an interface, say between air and water. Similarly, all systems that possess kinetic but not potential energy[3] move along geodesics, which are paths minimizing energy. A general mechanical system such as a pendulum or top possesses both kinetic and potential energy, and in this case the quantity which is minimized is the mean value of kinetic *minus* potential energy. This is a less intuitive quantity, and is often called the **action**. As we proceed, we will find that the action crops up in many apparently unrelated contexts, for example when we consider the algebraic structure of the group of symplectomorphisms in Chapter 10. However, this should really be no surprise, since the principle of least action underlies all symplectic motion.

Consider a system whose configurations are described by points x in Euclidean space $\mathbb{R}^n$ which move along trajectories $t \mapsto x(t)$. As we shall see in Lemma 1.1.1 below, the assumption that these paths minimize some action functional gives rise to a system of n second-order differential equations called the **Euler–Lagrange equations** of the variational problem. We will then show how these equations can be transformed into a Hamiltonian system of $2n$ first-order equations. As we investigate the structure of this larger system, we will see the concepts of symplectic geometry slowly emerge.

Let $L = L(t, x, v)$ be a twice continuously differentiable function of $2n + 1$ variables $(t, x_1, \ldots, x_n, v_1, \ldots, v_n)$, where we think of $v \in \mathbb{R}^n = T_x\mathbb{R}^n$ as a tangent vector at the point $x \in \mathbb{R}^n$ which represents the velocity $\dot{x}$. Consider the problem of minimizing the **action integral**

$$I(x) = \int_{t_0}^{t_1} L(t, x, \dot{x})dt \tag{1.1.1}$$

over the set of paths $x \in C^1([t_0, t_1], \mathbb{R}^n)$ which satisfy the boundary condition

$$x(t_0) = x_0, \qquad x(t_1) = x_1.$$

The function L is called the **Lagrangian** of this variational problem. A continuously differentiable path $x : [t_0, t_1] \to \mathbb{R}^n$ is called **minimal** if $I(x) \le I(x + \xi)$ for every variation $\xi \in C^1([t_0, t_1], \mathbb{R}^n)$ with $\xi(t_0) = \xi(t_1) = 0$. Thus we vary x keeping its endpoints fixed. (In any variational problem it is important to know exactly what conditions the objects which are varying satisfy, since the form of the resulting Euler–Lagrange equations will depend on these conditions.)

[3] Kinetic energy is energy of motion, while potential energy is energy determined by position. For example, the kinetic energy of a particle of mass m moving with speed v is $\frac{1}{2}mv^2$. If the particle is moving freely under gravity its potential energy has the form mgh, where h is the height above some reference level and g is the gravitational constant. Good references for further reading about all the notions discussed in this section are the books by Abraham and Marsden [7] and by Arnold [23].

Lemma 1.1.1 *A minimal path $x : [t_0, t_1] \to \mathbb{R}^n$ is a solution of the* **Euler–Lagrange equations**

$$\frac{d}{dt}\frac{\partial L}{\partial v} = \frac{\partial L}{\partial x}. \tag{1.1.2}$$

Here we use the vector notation $\partial L/\partial v = (\partial L/\partial v_1, \ldots, \partial L/\partial v_n) \in \mathbb{R}^n$, and similarly for $\partial L/\partial x$.

Proof: If x is minimal, all directional derivatives of I at x must vanish. Hence

$$\begin{aligned} 0 &= \left.\frac{d}{d\varepsilon}\right|_{\varepsilon=0} I(x + \varepsilon\xi) \\ &= \int_{t_0}^{t_1} \left(\sum_{j=1}^n \frac{\partial L}{\partial x_j}(t, x, \dot{x})\, \xi_j + \sum_{j=1}^n \frac{\partial L}{\partial v_j}(t, x, \dot{x})\, \dot{\xi}_j \right) dt \\ &= \int_{t_0}^{t_1} \sum_{j=1}^n \left(\frac{\partial L}{\partial x_j}(t, x, \dot{x}) - \frac{d}{dt}\frac{\partial L}{\partial v_j}(t, x, \dot{x}) \right) \xi_j \, dt \end{aligned}$$

for every $\xi \in C^1([t_0, t_1], \mathbb{R}^n)$ with $\xi(t_0) = \xi(t_1) = 0$. The proof of the last equality uses integration by parts and the boundary conditions. □

Remark 1.1.2 As always in a variational problem, the Euler–Lagrange equations correspond to critical points of the functional under consideration rather than to actual minima or maxima. This is often glossed over when talking about principles of least action, but in some contexts it is important to consider critical points which are not minima. For example, sometimes there are no obvious minima, and all one can find are saddle points. □

Under the **Legendre condition**

$$\det\left(\frac{\partial^2 L}{\partial v_j \partial v_k}\right) \neq 0, \tag{1.1.3}$$

equation (1.1.2) defines a regular system of second-order ordinary differential equations in the n variables $x_1, \ldots, x_n$. The **Legendre transformation** produces a system of first-order differential equations in $2n$ variables. It is convenient to introduce the new variables

$$y_k := \frac{\partial L}{\partial v_k}(x, v), \qquad k = 1, \ldots, n. \tag{1.1.4}$$

Then

$$\dot{y}_k = \frac{d}{dt}\frac{\partial L}{\partial v_k} = \frac{\partial L}{\partial x_k}$$

whenever x is a solution of (1.1.2). Now it follows from condition (1.1.3) and the implicit function theorem that v can be locally expressed as a function of t, x, and y, which we denote by

$$v_k = G_k(t, x, y).$$

Define the **Hamiltonian**

$$H := \sum_{j=1}^{n} y_j v_j - L \tag{1.1.5}$$

and consider H as a function of t, x, and y. (Namely, equation (1.1.5) is shorthand notation for $H(t,x,y) := \sum_j y_j G_j(t,x,y) - L(t,x,G(t,x,y))$.) Then

$$\frac{\partial H}{\partial x_k} = -\frac{\partial L}{\partial x_k}, \qquad \frac{\partial H}{\partial y_k} = G_k.$$

Hence the Euler–Lagrange equations (1.1.2) transform into the **Hamiltonian differential equations**

$$\dot{x} = \frac{\partial H}{\partial y}, \qquad \dot{y} = -\frac{\partial H}{\partial x}. \tag{1.1.6}$$

This proves the following result.

Lemma 1.1.3 *Let $x : [t_0, t_1] \to \mathbb{R}^n$ be a continuously differentiable path and let $y : [t_0, t_1] \to \mathbb{R}^n$ be the new variables given by* (1.1.4). *Then x is a solution of the Euler–Lagrange equations* (1.1.2) *if and only if the functions x and y satisfy the Hamiltonian system* (1.1.6).

Remark 1.1.4 The above process can be reversed. Under the condition

$$\det\left(\frac{\partial^2 H}{\partial y_j \partial y_k}\right) \neq 0, \tag{1.1.7}$$

the Hamiltonian system (1.1.6) can be transformed into the Euler–Lagrange equations (1.1.2) of an associated variational problem. □

Exercise 1.1.5 Carry out the inverse Legendre transformation. □

Remark 1.1.6 If the matrix (1.1.3) is positive definite, then every solution of (1.1.2) minimizes the functional $I(x)$ on sufficiently small time intervals. If in addition L satisfies suitable quadratic growth conditions at infinity, then it follows from the direct methods in the calculus of variations that for any two points $x_0, x_1 \in \mathbb{R}^n$ there exists a minimal solution $x \in C^2([t_0, t_1], \mathbb{R}^n)$ of (1.1.2) with $x(t_0) = x_0$ and $x(t_1) = x_1$. □

Example 1.1.7 (Kepler problem) Consider the differential equation

$$\frac{d^2 x_k}{dt^2} = -\frac{\partial V}{\partial x_k}, \qquad k = 1, \ldots, n.$$

Then $y = \dot{x} = v$ and the Hamiltonian

$$H = \tfrac{1}{2}|y|^2 + V(x)$$

is the sum of the kinetic and potential energies, while the Lagrangian

$$L = \frac{1}{2}|v|^2 - V(x)$$

is their difference. (Check this.) The **Kepler problem** is the special case $n = 3$ with the potential

$$V(x) = -\frac{1}{|x|}.$$

This potential corresponds to a force towards the origin which varies with the inverse square of the distance, and so represents forces such as gravity or an electrostatic charge. Thus the solutions of this equation describe the paths of planets orbiting around the sun, or of a (classical) point charge orbiting around a fixed centre. □

So far, this has been a purely formal calculation. However, the new variables y_j have a natural physical interpretation as the coordinates of momentum, and we will see that the Hamiltonian system of equations has a rich internal structure which is crystallized in the concept of a symplectic form.

Symplectic action

We have seen that the solutions of a Hamiltonian differential equation obey a variational principle whenever the nondegeneracy condition (1.1.7) is satisfied. We shall now give an alternative formulation of this principle which does not rely on the condition (1.1.7) and hence is valid for all Hamiltonian systems.

Assume that the Lagrangian L satisfies the Legendre condition (1.1.3) and let $H(t, x, y)$ be the corresponding Hamiltonian function defined by (1.1.5). Given a twice differentiable curve $x : [t_0, t_1] \to \mathbb{R}^n$, let $y : [t_0, t_1] \to \mathbb{R}^n$ be given by (1.1.4) with $v = \dot{x}$ and denote $z(t) = (x(t), y(t))$. Then the integral $I(x)$ defined by (1.1.1) agrees with the **action integral** $\mathcal{A}_H(z)$ defined by

$$\mathcal{A}_H(z) := \int_{t_0}^{t_1} \Big(\langle y, \dot{x} \rangle - H(t, x, y) \Big)\, dt, \tag{1.1.8}$$

where $\langle \cdot, \cdot \rangle$ denotes the standard inner product on $\mathbb{R}^n$. This is the integral of the **action form**

$$\lambda_H := \sum_{j=1}^{n} y_j dx_j - H\, dt$$

along the curve $z : [t_0, t_1] \to \mathbb{R}^{2n}$. It agrees with $I(x)$ whenever y is given by (1.1.4). However, $\mathcal{A}_H(z)$ is well-defined for any path z in the phase space $\mathbb{R}^{2n}$ and for any Hamiltonian function H even if the nondegeneracy condition (1.1.7) does not hold. The next lemma shows that the Euler–Lagrange equations of the action integral (1.1.8) are Hamilton's equations (1.1.6). Thus Hamilton's equations can be directly expressed in terms of the **principle of least action**.

Lemma 1.1.8 *A curve $z : [t_0, t_1] \to \mathbb{R}^{2n}$ is a critical point of $\mathcal{A}_H$ (with respect to variations with fixed endpoints $x(t_0) = x_0$ and $x(t_1) = x_1$) if and only if it satisfies Hamilton's equations* (1.1.6).

Proof: Let $z_s = (x_s, y_s) : [t_0, t_1] \to \mathbb{R}^{2n}$ be a 1-parameter family of curves satisfying $x_s(t_0) = x_0$ and $x_s(t_1) = x_1$ for all s and $z_0 = z$. Denote the partial derivatives with respect to s at $s = 0$ by

$$\widehat{x}(t) := \frac{\partial}{\partial s} x_s(t)\bigg|_{s=0}, \qquad \widehat{y}(t) := \frac{\partial}{\partial s} y_s(t)\bigg|_{s=0}, \qquad \widehat{z}(t) := \frac{\partial}{\partial s} z_s(t)\bigg|_{s=0}.$$

Differentiate under the integral sign to obtain

$$\begin{aligned} d\mathcal{A}_H(z)\widehat{z} &= \frac{d}{ds}\mathcal{A}_H(z_s)\bigg|_{s=0} \\ &= \int_{t_0}^{t_1} \frac{\partial}{\partial s}\bigg|_{s=0} \Big(\langle y_s(t), \partial_t x_s(t)\rangle - H(t, x_s(t), y_s(t))\Big)\, dt \\ &= \int_{t_0}^{t_1} \Big(\langle \widehat{y}, \partial_t x\rangle + \langle y, \partial_t \widehat{x}\rangle - \langle \partial_x H, \widehat{x}\rangle - \langle \partial_y H, \widehat{y}\rangle\Big)\, dt \\ &= \int_{t_0}^{t_1} \langle \widehat{y}, \dot{x} - \partial_y H\rangle\, dt - \int_{t_0}^{t_1} \langle \dot{y} + \partial_x H, \widehat{x}\rangle\, dt. \end{aligned} \tag{1.1.9}$$

The last equation follows from integration by parts and uses the fact that $\widehat{x}(t_0) = \widehat{x}(t_1) = 0$. Hence z is a critical point of $\mathcal{A}_H$ with respect to the boundary conditions $x(t_0) = x_0$ and $x(t_1) = x_1$ if and only if the last two integrals on the right in (1.1.9) vanish for all $\widehat{x}$ and $\widehat{y}$ satisfying $\widehat{x}(t_0) = \widehat{x}(t_1) = 0$. This holds if and only if z satisfies Hamilton's equations (1.1.6). □

Hamiltonian flows

Assume the time-independent case where the Hamiltonian function $H : \mathbb{R}^{2n} \to \mathbb{R}$ in the differential equation (1.1.6) is independent of t. Then, in the coordinates $z = (x_1, \dots x_n, y_1, \dots, y_n) \in \mathbb{R}^{2n}$, the Hamiltonian system (1.1.6) has the form

$$\dot{z} = -J_0 \nabla H(z).$$

Here ∇H denotes the gradient of H and J_0 denotes the $2n \times 2n$-matrix

$$J_0 := \begin{pmatrix} 0 & -\mathbb{1} \\ \mathbb{1} & 0 \end{pmatrix}. \tag{1.1.10}$$

This matrix represents a rotation through the right angle $\pi/2$ and it squares to $J_0^2 = -\mathbb{1}$. The vector field

$$X_H := -J_0 \nabla H = \begin{pmatrix} \partial H/\partial y \\ -\partial H/\partial x \end{pmatrix} : \mathbb{R}^{2n} \to \mathbb{R}^{2n} \tag{1.1.11}$$

is called the **Hamiltonian vector field** associated to the Hamiltonian function H, or the **symplectic gradient** of H. (See Fig. 1.1.)

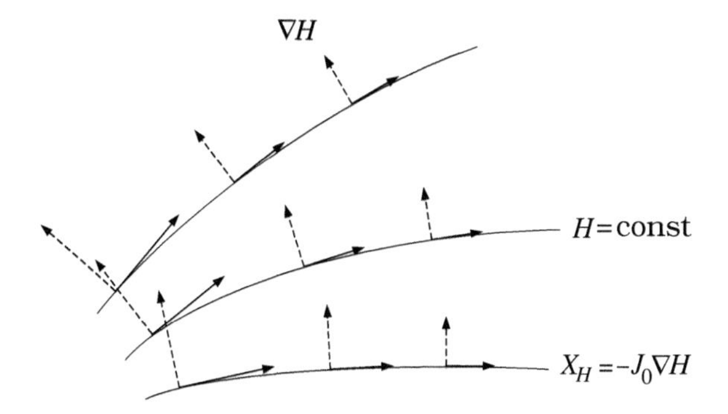

FIG. 1.1. X_H is the symplectic gradient of H.

Given a smooth time-dependent Hamiltonian function

$$\mathbb{R} \times \mathbb{R}^{2n} \to \mathbb{R} : (t, x, y) \mapsto H_t(x, y),$$

denote by ϕ_H^{t,t_0} the solution operator of (1.1.6). It is defined by $\phi_H^{t,t_0}(z_0) = z(t)$, where $z(t)$ is the unique solution of (1.1.6) with initial condition $z(t_0) = z_0$. The domain of definition of ϕ_H^{t,t_0} is the open set $\Omega_{t_0,t}$ of all $z_0 \in \mathbb{R}^{2n}$ for which the solution exists on the time interval $[t_0, t]$ (respectively $[t, t_0]$ when $t < t_0$). The diffeomorphisms $\phi_H^{t,t_0} : \Omega_{t_0,t} \to \Omega_{t,t_0}$ satisfy

$$\frac{\partial}{\partial t}\phi_H^{t,t_0} = X_{H_t} \circ \phi_H^{t,t_0}$$

and

$$\phi_H^{t_2,t_1} \circ \phi_H^{t_1,t_0} = \phi_H^{t_2,t_0}, \qquad \phi_H^{t_0,t_0} = \mathrm{id}.$$

In the time-independent case $H_t \equiv H$, one normalizes by setting $t_0 = 0$ and calls the resulting family $\phi_H^t = \phi_H^{t,0}$ of diffeomorphisms the **Hamiltonian flow** generated by H.

Symplectomorphisms

A diffeomorphism $\psi : \mathbb{R}^{2n} \to \mathbb{R}^{2n}$ is called a **symplectomorphism** if

$$d\psi(\zeta)^{\mathrm{T}} J_0 \, d\psi(\zeta) = J_0 \tag{1.1.12}$$

for every $\zeta \in \mathbb{R}^{2n}$. (Here we write A^{T} for the transpose of the matrix A.) As the name suggests, it will turn out that these are the diffeomorphisms which preserve the symplectic structure. However, at the moment all we have to work with is the notion of a Hamiltonian system of differential equations. Therefore, we will begin by showing that symplectomorphisms preserve the class of Hamiltonian differential equations. In the classical literature on Hamiltonian mechanics, symplectomorphisms are sometimes called **canonical transformations**.

Lemma 1.1.9 *Let $\psi : \mathbb{R}^{2n} \to \mathbb{R}^{2n}$ be a symplectomorphism and let $H : \mathbb{R}^{2n} \to \mathbb{R}$ be a smooth Hamiltonian function. Suppose that $\zeta(t)$ is a solution of the Hamiltonian differential equation*

$$\dot{\zeta} = X_{H\circ\psi}(\zeta). \tag{1.1.13}$$

Then $z(t) := \psi(\zeta(t))$ is a solution of (1.1.6). In other words

$$X_{H\circ\psi} = \psi^* X_H. \tag{1.1.14}$$

Proof: An easy calculation shows that, for any diffeomorphism ψ of $\mathbb{R}^{2n}$ and any function H, the gradient ∇H transforms by

$$\nabla(H \circ \psi)(p) = d\psi(p)^{\mathrm{T}}\, \nabla H(\psi(p)).$$

Hence, if $\zeta : I \to \mathbb{R}^{2n}$ is a solution of equation (1.1.13) on an interval I, then

$$\begin{aligned} d\psi(\zeta)^{\mathrm{T}} \nabla H(\psi(\zeta)) &= \nabla(H \circ \psi)(\zeta) \\ &= J_0 \dot{\zeta} \\ &= d\psi(\zeta)^{\mathrm{T}} J_0\, d\psi(\zeta)\dot{\zeta} \\ &= d\psi(\zeta)^{\mathrm{T}} J_0 \dot{z}. \end{aligned}$$

Multiplying by $(d\psi(\zeta)^{\mathrm{T}})^{-1}$ we find that $\dot{z} = -J_0 \nabla H(z) = X_H(z)$. This proves Lemma 1.1.9. □

We next show that the diffeomorphisms generated by a Hamiltonian system of differential equations are all symplectomorphisms.

Lemma 1.1.10 *Wherever defined, ϕ_H^{t,t_0} is a symplectomorphism.*

Proof: Let $z_0 \in \mathbb{R}^{2n}$ and define $z(t) := \phi_H^{t,t_0}(z_0)$ and

$$\Phi(t) := d\phi_H^{t,t_0}(z_0) \in \mathbb{R}^{2n\times 2n}.$$

Then for every $\zeta_0 \in \mathbb{R}^{2n}$ the function $\zeta(t) := \Phi(t)\zeta_0$ is the solution of the linearized differential equation

$$\dot{\zeta} = dX_{H_t}(z)\zeta.$$

Since $J_0 X_{H_t} = \nabla H_t$, it follows that

$$J_0 \dot{\Phi}(t) = S(t)\Phi(t), \qquad \Phi(t_0) = \mathbb{1},$$

where $S(t) \in \mathbb{R}^{2n\times 2n}$ denotes the symmetric matrix with entries

$$S_{jk}(t) = \frac{\partial^2 H_t}{\partial z_j \partial z_k}\left(\phi_H^{t,t_0}(z_0)\right).$$

We must show that $\Phi^{\mathrm{T}} J_0 \Phi = J_0$. This is clearly satisfied for $t = t_0$. Moreover

$$\begin{aligned} \frac{d}{dt}\Phi^{\mathrm{T}} J_0 \Phi &= \Phi^{\mathrm{T}} J_0 \dot{\Phi} + \dot{\Phi}^{\mathrm{T}} J_0 \Phi \\ &= \Phi^{\mathrm{T}}(S - S^{\mathrm{T}})\Phi \\ &= 0. \end{aligned}$$

This proves Lemma 1.1.10. □

Symplectic matrices

The most basic class of symplectomorphisms consists of the linear ones. Since we work in Euclidean space with its canonical basis we shall identify a linear map with the matrix $\Psi \in \mathbb{R}^{2n\times 2n}$ which represents it. Such a matrix is called **symplectic** if it satisfies

$$\Psi^{\mathrm{T}} J_0 \Psi = J_0, \tag{1.1.15}$$

where J_0 is as in (1.1.6). It follows from equation (1.1.15) that $\det(\Psi)^2 = 1$ and so every symplectic matrix is invertible. In fact, we shall see below that every symplectic matrix has determinant one (Lemma 1.1.15). In the terminology introduced above, a matrix $\Psi \in \mathbb{R}^{2n\times 2n}$ is symplectic if and only if the corresponding linear transformation $\Psi : \mathbb{R}^{2n} \to \mathbb{R}^{2n}$ is a symplectomorphism. More generally, a diffeomorphism $\psi : \mathbb{R}^{2n} \to \mathbb{R}^{2n}$ is a symplectomorphism if and only if the Jacobian $d\psi(z)$ is a symplectic matrix for every $z \in \mathbb{R}^{2n}$ (see page 17). The set of symplectic matrices will be denoted by

$$\mathrm{Sp}(2n) := \mathrm{Sp}(2n, \mathbb{R}) := \left\{\Psi \in \mathbb{R}^{2n\times 2n} \,|\, \Psi^{\mathrm{T}} J_0 \Psi = J_0\right\}. \tag{1.1.16}$$

The next two lemmas show that this is a Lie group.

Lemma 1.1.11 *If Φ and Ψ are symplectic matrices then so are $\Phi\Psi$, Ψ^{-1}, Ψ^{T}.*

Proof: Multiply the identity $\Psi^{\mathrm{T}} J_0 \Psi = J_0$ by Ψ^{-1} from the right and by $(\Psi^{\mathrm{T}})^{-1}$ from the left to obtain $J_0 = (\Psi^{\mathrm{T}})^{-1} J_0 \Psi^{-1}$. So Ψ^{-1} is a symplectic matrix. Now invert both sides of the latter identity to get $J_0^{-1} = \Psi J_0^{-1} \Psi^{\mathrm{T}}$ and hence $J_0 = \Psi J_0 \Psi^{\mathrm{T}}$. So Ψ^{T} is a symplectic matrix. □

Lemma 1.1.12 $\mathrm{Sp}(2n)$ *is a Lie group with Lie algebra*

$$\mathfrak{sp}(2n) := \mathrm{Lie}(\mathrm{Sp}(2n)) = \left\{-J_0 S \,|\, S \in \mathbb{R}^{2n\times 2n},\ S^T = S\right\}. \tag{1.1.17}$$

Proof: Denote by $\mathfrak{so}(2n) \subset \mathbb{R}^{2n\times 2n}$ the linear subspace of skew-symmetric $2n \times 2n$ matrices and define the map $f : \mathbb{R}^{2n\times 2n} \to \mathfrak{so}(2n)$ by

$$f(\Psi) := \Psi^T J_0 \Psi$$

for $\Psi \in \mathbb{R}^{2n\times 2n}$. Its differential at $\Psi \in \mathbb{R}^{2n\times 2n}$ is given by

$$df(\Psi)A = A^T J_0 \Psi + \Psi^T J_0 A$$

for $A \in \mathbb{R}^{2n\times 2n}$. If $\Psi \in \mathrm{Sp}(2n)$ and $B \in \mathfrak{so}(2n)$ then the matrix $A := -\frac{1}{2}\Psi J_0 B$ satisfies $df(\Psi)A = B$. Hence J_0 is a regular value of f and so $\mathrm{Sp}(2n) = f^{-1}(J_0)$ is a smooth submanifold of $\mathrm{GL}(2n, \mathbb{R})$. Hence it is a Lie group by Lemma 1.1.11 and its Lie algebra is

$$\mathfrak{sp}(2n) = T_{\mathbb{1}}\mathrm{Sp}(2n) = \ker\, df(\mathbb{1}) = \left\{A \in \mathbb{R}^{2n\times 2n} \,|\, A^T J_0 + J_0 A = 0\right\}.$$

Since $S := J_0 A$ is symmetric when $A \in \mathfrak{sp}(2n)$, this proves Lemma 1.1.12. □

Exercise 1.1.13 Consider the matrix

$$\Psi = \begin{pmatrix} A & B \\ C & D \end{pmatrix}, \tag{1.1.18}$$

where A, B, C, and D are real $n \times n$ matrices. Prove that Ψ is symplectic if and only if its inverse is of the form

$$\Psi^{-1} = \begin{pmatrix} D^{\mathrm{T}} & -B^{\mathrm{T}} \\ -C^{\mathrm{T}} & A^{\mathrm{T}} \end{pmatrix}.$$

More explicitly this means that

$$A^{\mathrm{T}}C = C^{\mathrm{T}}A, \qquad B^{\mathrm{T}}D = D^{\mathrm{T}}B, \qquad A^{\mathrm{T}}D - C^{\mathrm{T}}B = \mathbb{1},$$

or equivalently

$$AB^{\mathrm{T}} = BA^{\mathrm{T}}, \qquad CD^{\mathrm{T}} = DC^{\mathrm{T}}, \qquad AD^{\mathrm{T}} - BC^{\mathrm{T}} = \mathbb{1}.$$

Deduce that a 2×2 matrix is symplectic if and only if its determinant is equal to 1. □

Exercise 1.1.14 Show that the set

$$\mathrm{Symp}(\mathbb{R}^{2n}) := \left\{ \psi \in \mathrm{Diff}(\mathbb{R}^{2n}) \,\middle|\, d\psi(z) \in \mathrm{Sp}(2n) \text{ for all } z \in \mathbb{R}^{2n} \right\} \tag{1.1.19}$$

of symplectomorphisms of $\mathbb{R}^{2n}$ is a group. □

As noted above, the square of the determinant of a symplectic matrix is obviously 1. It is somewhat nontrivial to prove that the determinant itself is 1, and we will accomplish this by expressing the condition for a matrix to be symplectic in terms of a differential form. This result is known as Liouville's theorem. Another proof may be found in Arnold [23].

The symplectic form

Consider the 2-form

$$\omega_0 := \sum_{j=1}^{n} dx_j \wedge dy_j \tag{1.1.20}$$

on $\mathbb{R}^{2n}$. We can think of this either as a differential form on $\mathbb{R}^{2n}$ with constant coefficients, or as a nondegenerate skew-symmetric bilinear form

$$\omega_0 : \mathbb{R}^{2n} \times \mathbb{R}^{2n} \to \mathbb{R}$$

on the vector space $\mathbb{R}^{2n}$. (The two notions coincide when the vector space $\mathbb{R}^{2n}$ is viewed as the tangent space of the manifold $\mathbb{R}^{2n}$ at a point z.) The value of ω_0 on a pair of vectors $\zeta = (\xi, \eta)$ and $\zeta' = (\xi', \eta')$ with $\xi, \eta, \xi', \eta' \in \mathbb{R}^n$ is given by

$$\omega_0(\zeta, \zeta') = \sum_{j=1}^{n} (\xi_j \eta'_j - \eta_j \xi'_j) = \langle J_0 \zeta, \zeta' \rangle = -\zeta^{\mathrm{T}} J_0 \zeta'. \tag{1.1.21}$$

Lemma 1.1.15 *Every symplectic matrix has determinant* 1.

Proof: Let $\Psi \in \mathrm{Sp}(2n)$. Then $\Psi^T J_0 \Psi = J_0$ and hence it follows from equation (1.1.21) that $\omega_0(\Psi\zeta, \Psi\zeta') = \omega_0(\zeta, \zeta')$ for all $\zeta, \zeta' \in \mathbb{R}^{2n}$. Hence $\Psi^*\omega_0 = \omega_0$ and therefore Ψ also preserves the top exterior power ω_0^n. Since

$$\frac{\omega_0^n}{n!} = \frac{\omega_0 \wedge \cdots \wedge \omega_0}{n!} = dx_1 \wedge dy_1 \wedge \cdots \wedge dx_n \wedge dy_n =: \mathrm{dvol}_{\mathbb{R}^{2n}}, \qquad (1.1.22)$$

it follows that $\Psi^*\mathrm{dvol}_{\mathbb{R}^{2n}} = \mathrm{dvol}_{\mathbb{R}^{2n}}$ or, equivalently, $\det(\Psi) = 1$. This proves Lemma 1.1.15. □

All the concepts introduced so far can be expressed in terms of the 2-form ω_0. For example, a Hamiltonian vector field X_H is determined by the identity

$$\iota(X_H)\omega_0 = dH,$$

and a diffeomorphism ψ of $\mathbb{R}^{2n}$ is a canonical transformation if and only if

$$\psi^*\omega_0 = \omega_0.$$

Based on the observation that canonical transformations preserve the class of Hamiltonian differential equations (Lemma 1.1.9), it is natural to introduce the notion of a manifold equipped with an atlas whose transition maps are canonical transformations. Such a manifold is called **symplectic**. The above discussion shows that a symplectic manifold is equipped with a 2-form ω that agrees with ω_0 on each coordinate chart. This 2-form is clearly nondegenerate and closed. Conversely, the theorem of Darboux (Theorem 3.2.2) asserts that every closed nondegenerate 2-form ω is locally isomorphic to ω_0 and hence gives rise to an atlas whose transition maps are canonical transformations. Thus in the modern theory a symplectic structure is defined to be a closed nondegenerate 2-form and all the concepts are developed from that basis. This will be the subject of Chapter 3.

The proof of Lemma 1.1.15 shows that every symplectomorphism ψ of $\mathbb{R}^{2n}$ preserves the standard volume form or, equivalently, satisfies $\det(d\psi(z)) = 1$ for all z. A 2×2 matrix is symplectic if and only if it has determinant one. So in the case $n = 1$ the symplectomorphisms are precisely the area-preserving transformations of $\mathbb{R}^2$. In general the volume-preserving condition is only necessary. For $n \geq 2$ it is easy to construct volume-preserving diffeomorphisms of $\mathbb{R}^{2n}$ which are not symplectic, and hence $\mathrm{Symp}(\mathbb{R}^{2n})$ is a proper subgroup of $\mathrm{Diff}_{\mathrm{vol}}(\mathbb{R}^{2n})$.

Exercise 1.1.16 Find an element of the special linear group $\mathrm{SL}(4, \mathbb{R})$ which does not belong to $\mathrm{Sp}(4)$. **Hint:** If you find this hard, read Section 2.2 to get more information on the properties of symplectic matrices. □

The difference between volume-preserving and symplectic diffeomorphisms only started to be better understood in the 1980s. We will return to this question in the next section, when we discuss symplectic rigidity.

The Poisson bracket

Let $z(t) = (x(t), y(t))$ be a solution of Hamilton's equations (1.1.6) in the case where $H_t \equiv H : \mathbb{R}^{2n} \to \mathbb{R}$ is independent of time. Then

$$\frac{d}{dt} H \circ z = \sum_{j=1}^{n} \frac{\partial H}{\partial x_j} \dot{x}_j + \sum_{j=1}^{n} \frac{\partial H}{\partial y_j} \dot{y}_j = 0.$$

This means that the Hamiltonian function H is constant along the solutions of (1.1.6). This is the familiar law of **conservation of energy**. Geometrically, it says that the level sets of H are invariant under the Hamiltonian flow.

Example 1.1.17 (Hamiltonian flow on the sphere) The level sets of the function

$$H = \tfrac{1}{2} \sum_{j=1}^{n} (x_j^2 + y_j^2)$$

are spheres centred at the origin. The symplectic gradient is the vector field $X_H(z) = -J_0 z$ and the Hamiltonian differential equation has the form

$$\dot{x}_j = y_j, \qquad \dot{y}_j = -x_j.$$

The easiest way to understand the corresponding flow is to identify $\mathbb{R}^{2n}$ with complex n-space $\mathbb{C}^n$ via $z = (z_1, \ldots, z_n) = (x_1 + iy_1, \ldots, x_n + iy_n)$. Then J_0 acts by multiplication with i and the Hamiltonian vector field is $X_H(z) = -iz$. Hence the orbit of $z = (z_1, \ldots, z_n)$ under the Hamiltonian flow is given by

$$z(t) = (e^{-it} z_1, \ldots, e^{-it} z_n).$$

Geometrically, the vector field $X_H(z) = -iz$ is tangent to the circle formed by the intersection of the complex line $\mathbb{C}z$ with the sphere S^{2n-1}, and the flow of the Hamiltonian vector field X_H rotates points around this circle with uniform speed. This flow is considered again at the beginning of Section 5.1 as an example of a symplectic circle action. □

A smooth function $F : \mathbb{R}^{2n} \to \mathbb{R}$ is called an **integral** of the differential equation (1.1.6), with a time-independent Hamiltonian function $H_t = H$, if F is constant along the solutions of (1.1.6). In particular, the Hamiltonian function itself is an integral. Now if $z(t)$ is a solution of (1.1.6) then

$$\frac{d}{dt} F \circ z = (\nabla F)^{\mathrm{T}} \dot{z} = -(\nabla F)^{\mathrm{T}} J_0 \nabla H = \sum_{j=1}^{n} \left(\frac{\partial F}{\partial x_j} \frac{\partial H}{\partial y_j} - \frac{\partial F}{\partial y_j} \frac{\partial H}{\partial x_j} \right).$$

This expression is called the **Poisson bracket** of F and H and is denoted by

$$\{F, H\} := -(\nabla F)^{\mathrm{T}} J_0 \nabla H. \tag{1.1.23}$$

Thus F is an integral for (1.1.6) if and only if $\{F, H\} = 0$, and this holds if and only if H is an integral of the Hamiltonian differential equation associated to F. An important fact is that the space $C^\infty(\mathbb{R}^{2n})$ of all smooth functions on $\mathbb{R}^{2n}$ becomes a Lie algebra under the Poisson bracket.

Lemma 1.1.18 *The Poisson bracket defines a Lie algebra structure on the vector space $C^\infty(\mathbb{R}^{2n})$, i.e. it is skew-symmetric and satisfies the* **Jacobi identity**

$$\{\{F,G\},H\} + \{\{G,H\},F\} + \{\{H,F\},G\} = 0. \tag{1.1.24}$$

Moreover,

$$\{FG,H\} = F\{G,H\} + G\{F,H\} \tag{1.1.25}$$

and

$$[X_F, X_G] = X_{\{F,G\}} \tag{1.1.26}$$

for all $F, G, H \in C^\infty(\mathbb{R}^{2n})$. Thus the Hamiltonian vector fields form a Lie subalgebra of the Lie algebra of vector fields on $\mathbb{R}^{2n}$.

An essential ingredient in the proof of Lemma 1.1.18 is the correct choice of signs. Our conventions are fully explained in Remark 3.1.6 below. Here it suffices to note that we define the Lie bracket of two vector fields X and Y on $\mathbb{R}^{2n}$ by

$$[X,Y](z) := dX(z)\,Y(z) \; - \; dY(z)\,X(z),$$

where $X(z)$ and $Y(z)$ are column vectors for each $z \in \mathbb{R}^{2n}$ and $dX(z)$ and $dY(z)$ are matrices representing the derivatives at z of the vector fields X and Y, understood as smooth maps from $\mathbb{R}^{2n}$ to $\mathbb{R}^{2n}$.[4]

Proof of Lemma 1.1.18: Denote by $d^2F(z) = d\nabla F(z)$ the Hessian of F at z, understood as the symmetric $2n \times 2n$-matrix of second partial derivatives of F. Then

$$\nabla\{F,G\} = -\nabla\left((\nabla F)^T J_0 \nabla G\right) = (d^2G)J_0\nabla F - (d^2F)J_0\nabla G$$

and hence

$$\begin{aligned} X_{\{F,G\}} &= -J_0\nabla\{F,G\} \\ &= J_0(d^2F)J_0\nabla G - J_0(d^2G)J_0\nabla F \\ &= dX_F \cdot X_G - dX_G \cdot X_F \\ &= [X_F, X_G]. \end{aligned}$$

This proves equation (1.1.26). It follows from (1.1.26) and the Jacobi identity for vector fields that the function $\{\{F,G\},H\} + \{\{G,H\},F\} + \{\{H,F\},G\}$ is

[4] In this notation our choice of sign is natural. The reverse sign, which is often used in the literature, has its origin in the identification of vector fields with derivations. Namely, if $z_1, \ldots, z_{2n}$ are the coordinates on $\mathbb{R}^{2n}$ and the vector fields are written in the form

$$X = \sum_{j=1}^{2n} a_j \frac{\partial}{\partial z_j}, \qquad Y = \sum_{j=1}^{2n} b_j \frac{\partial}{\partial z_j},$$

then the Lie bracket (with our choice of sign) is given by

$$[X,Y] = YX - XY = \sum_j \sum_i \left(b_i \frac{\partial a_j}{\partial z_i} - a_i \frac{\partial b_j}{\partial z_i} \right) \frac{\partial}{\partial z_j}.$$

This point is explained more fully in Remark 3.1.6.

constant. The constant vanishes when the functions F, G, H have compact support. To prove that it always vanishes, multiply F, G, H by a smooth cutoff function with compact support that is equal to one on some open set and note that the constant cannot change. Thus we have proved equation (1.1.24). Equation (1.1.25) follows from the Leibniz rule and this proves Lemma 1.1.18. □

The next lemma shows that symplectomorphisms are characterized among all diffeomorphisms by the property that they preserve the Poisson bracket.

Lemma 1.1.19 *A diffeomorphism $\psi : \mathbb{R}^{2n} \to \mathbb{R}^{2n}$ is a symplectomorphism if and only if*

$$\{F, G\} \circ \psi = \{F \circ \psi, G \circ \psi\}$$

for all $F, G \in C^\infty(\mathbb{R}^{2n})$.

Proof: Let $\psi : \mathbb{R}^{2n} \to \mathbb{R}^{2n}$ be a diffeomorphism. Then, by (1.1.23),

$$\{F, G\} \circ \psi = -(\nabla F \circ \psi)^T J_0 (\nabla G \circ \psi),$$

$$\{F \circ \psi, G \circ \psi\} = -\nabla(F \circ \psi)^T J_0 \nabla(G \circ \psi) = -(\nabla F \circ \psi)^T d\psi J_0 (d\psi)^T (\nabla G \circ \psi).$$

for all $F, G \in C^\infty(\mathbb{R}^{2n})$. This shows that ψ preserves the Poisson bracket if and only if $d\psi(z) J_0 d\psi(z)^T = J_0$ for every z. By Lemma 1.1.11 this is equivalent to the condition $d\psi(z)^T J_0 d\psi(z) = J_0$ for all z and hence to the condition that ψ is a symplectomorphism. This proves Lemma 1.1.19. □

Exercise 1.1.20 Prove that the Poisson bracket is given by

$$\{F, G\} = \omega_0(X_F, X_G).$$

Deduce that the Lie bracket $[X, Y]$ of two symplectic vector fields X, Y on an open subset of $\mathbb{R}^{2n}$ is the Hamiltonian vector field generated by $H := \omega_0(X, Y)$. □

The Hamiltonian system (1.1.6) is called **integrable** in the domain $\Omega \subset \mathbb{R}^{2n}$ if there exist n independent **Poisson commuting** integrals $F_1, \ldots, F_n$ in Ω. This means that the vectors $\nabla F_1(z), \ldots, \nabla F_n(z)$ are linearly independent for $z \in \Omega$ and that

$$\{F_j, F_k\} = 0 \qquad \text{for } j, k = 1, \ldots, n.$$

Poisson commuting integrals are said to be in **involution**.

Integrable systems have a particularly simple structure, and one can choose coordinates (called **action–angle variables**) in which there are explicit formulae for their solutions. Geometrically the action–angle variables can be described as follows. Because the functions F_i are constants of the motion, the level sets

$$T_c := \{z \in \Omega \,|\, F_j(z) = c_j\}$$

form n-dimensional invariant submanifolds of the Hamiltonian flow of H. These submanifolds are also invariant under the Hamiltonian flows of the F_j. Since these flows commute, one can show that if the level set T_c is compact and connected, it is diffeomorphic to the n-torus. Moreover, in a neighbourhood of every such invariant torus one can find coordinates

$$z = \psi(\xi, \eta),$$

where $\xi \in \mathbb{T}^n = \mathbb{R}^n/\mathbb{Z}^n$ and $\eta \in \mathbb{R}^n$ is a function of the variables c_j, such that the Hamiltonian flow in the variables $\zeta = (\xi, \eta)$ is given by

$$\dot{\xi} = \frac{\partial K}{\partial \eta}, \qquad \dot{\eta} = 0.$$

In other words, the Hamiltonian function $K = H \circ \psi$ depends only on η and the space is foliated by invariant tori on which the Hamiltonian flow is given by straight lines. The coordinates ξ and η are the action–angle variables and they show that the dynamics of integrable Hamiltonian systems are extremely simple.

The invariant level sets T_c are also interesting from a symplectic point of view, because the symplectic form ω_0 vanishes on them. Such submanifolds are called **Lagrangian**. They are an important element of symplectic topology, and will be discussed more fully in Chapter 3.

The type of dynamical behaviour exhibited by an integrable system is highly exceptional, and an arbitrarily small perturbation may destroy many of these invariant tori. On the other hand, if the frequency vector $\omega = \partial K/\partial \eta$ has rationally independent coordinates which satisfy suitable Diophantine inequalities and if $\partial^2 K/\partial \eta^2$ is nonsingular, then the corresponding invariant torus survives under sufficiently small perturbations which are sufficiently smooth. This is the content of the **Kolmogorov–Arnold–Moser theorem** for which we refer to Arnold [23], Moser [491], Salamon–Zehnder [562], Salamon [557], and the references therein. In spite of the fact that integrable Hamiltonian systems are very rare, there is a vast literature on this subject and over the years many new and interesting integrable Hamiltonian systems have been discovered. They are related to a variety of phenomena in classical and quantum physics, in dynamical systems, spectral theory, algebraic geometry, and many other areas, including symplectic field theory and enumerative geometry. We shall not pursue this direction here but instead refer the interested reader to Arnold [23], Moser [492], Eliashberg–Givental–Hofer [186], Okounkov–Pandharipande [512], Dubrovin [167], and the references therein.

It is also worth mentioning that in many situations one is interested in dynamical systems on manifolds which are more general than the finite-dimensional symplectic manifolds which are the subject of this book. One very useful idea is the concept of a Poisson manifold. These are finite- (or infinite-)dimensional manifolds in which the basic structure is a Poisson bracket on the space of functions rather than a 2-form. The properties of such manifolds have been studied for example by Weinstein in [676]. See also Ratiu–Weinstein–Zung [537] for further developments in this subject.

Examples and exercises

Exercise 1.1.21 Check that in the Kepler problem (Example 1.1.7) the three components of the angular momentum $x \times \dot{x}$ are integrals of the motion which are not in involution. (It turns out that the Kepler problem is in fact integrable; however, it is not so easy to see this.) □

Exercise 1.1.22 (Harmonic oscillator) Consider the Hamiltonian

$$H = \sum_{j=1}^{n} a_j \left(x_j^2 + y_j^2\right)$$

with $a_j > 0$. Find the solutions of the corresponding Hamiltonian differential equation. Prove that this system is integrable. Find all periodic solutions on the energy surface $H = c$ for $c > 0$. This exercise generalizes Example 1.1.17. We will see in Lemma 2.4.6 that every quadratic, positive definite Hamiltonian function is symplectically equivalent to one of the form considered here. □

Example 1.1.23 (Geodesic flow) Geodesics on a Riemannian manifold are defined as paths which locally minimize length. We now show that they are given by a Hamiltonian system of equations. This fact has been exploited in some work on the behaviour of geodesics, particularly in the problem of counting the number of closed geodesics on compact manifolds: see [239] for example.

More precisely, consider the variational problem with the Lagrangian

$$L(x, v) = \frac{1}{2} v^{\mathrm{T}} g(x) v = \frac{1}{2} \sum_{i,j=1}^{n} g_{ij}(x) v_i v_j,$$

where the matrix $g_{ij}(x) = g_{ji}(x)$ is positive definite. Then

$$y_k = \frac{\partial L}{\partial v_k} = \sum_{j=1}^{n} g_{kj}(x) \dot{x}_j$$

and hence the Euler–Lagrange equations are

$$\begin{aligned} 0 &= \dot{y}_k - \frac{\partial L}{\partial x_k} \\ &= \sum_{j=1}^{n} g_{kj} \ddot{x}_j + \sum_{i,j=1}^{n} \frac{\partial g_{kj}}{\partial x_i} \dot{x}_i \dot{x}_j - \frac{1}{2} \sum_{i,j=1}^{n} \frac{\partial g_{ij}}{\partial x_k} \dot{x}_i \dot{x}_j \\ &= \sum_{j=1}^{n} g_{kj} \ddot{x}_j + \frac{1}{2} \sum_{i,j=1}^{n} \left(\frac{\partial g_{kj}}{\partial x_i} + \frac{\partial g_{ki}}{\partial x_j} - \frac{\partial g_{ij}}{\partial x_k} \right) \dot{x}_i \dot{x}_j. \end{aligned}$$

These equations can be rewritten in the form

$$\ddot{x}_k = - \sum_{i,j=1}^{n} \Gamma_{ij}^k(x) \dot{x}_i \dot{x}_j. \tag{1.1.27}$$

The coefficients Γ_{ij}^k are called the **Christoffel symbols**. They are given by

$$\Gamma_{ij}^k := \frac{1}{2}\sum_{\ell=1}^n g^{k\ell}\left(\frac{\partial g_{\ell j}}{\partial x_i} + \frac{\partial g_{\ell i}}{\partial x_j} - \frac{\partial g_{ij}}{\partial x_\ell}\right),$$

where g^{jk} denotes the inverse matrix of g_{ij}, i.e. $\sum_{j=1}^n g_{ij}g^{jk} = \delta_{ik}$. The solutions of (1.1.27) are called **geodesics**. They minimize the energy

$$E(x) := \int_{t_0}^{t_1} \langle \dot{x}, g(x)\dot{x}\rangle dt$$

on small time intervals. Minimizing energy turns out to be equivalent to minimizing length but it yields nicer equations. □

Exercise 1.1.24 Carry out the Legendre transformation for the geodesic flow. Prove that the g-norm of the velocity

$$|\dot{x}|_g = \sqrt{\langle \dot{x}, g(x)\dot{x}\rangle}$$

is constant along every geodesic. □

Exercise 1.1.25 (Exponential map) Assume $g(x) = \mathbb{1}$ for large x so that the solutions $x(t)$ of equation (1.1.27) exist for all time. The solution with initial conditions $x(0) = x$ and $\dot{x}(0) = \xi$ is called the **geodesic through** (x, ξ). Define the **exponential map**

$$E : \mathbb{R}^n \times \mathbb{R}^n \to \mathbb{R}^n, \qquad E(x, \xi) = x(1),$$

where $x(t)$ is the geodesic through (x, ξ). Prove that this geodesic is given by $x(t) = E(x, t\xi)$. Prove that there exists a constant $c > 0$ such that

$$|E(x, \xi) - x - \xi| \le c|\xi|^2$$

and deduce that

$$\frac{\partial E_j}{\partial x_k}(x, 0) = \frac{\partial E_j}{\partial \xi_k}(x, 0) = \delta_{jk}, \qquad \frac{\partial^2 E_j}{\partial x_k \partial \xi_\ell}(x, 0) = 0.$$

□

Exercise 1.1.26 Suppose that $\phi : \mathbb{R}^n \to \mathbb{R}^n$ is a diffeomorphism and

$$g(x) = \phi^* h(x) = d\phi(x)^{\mathrm{T}} h(\phi(x)) d\phi(x).$$

Prove that every geodesic $x(t)$ for g is mapped under ϕ to a geodesic $y(t) = \phi(x(t))$ for h. Deduce that the concept of the exponential map extends to manifolds. □

Exercise 1.1.27 The covariant derivative of a vector field $\xi(s) \in \mathbb{R}^n$ along a curve $x(s) \in \mathbb{R}^n$ is defined by

$$(\nabla\xi)_k = \dot{\xi}_k + \sum_{i,j=1}^n \Gamma_{ij}^k(x)\dot{x}_i\xi_j.$$

A submanifold $L \subset \mathbb{R}^n$ is called **totally geodesic** if $\nabla\dot{x}(s) \in T_{x(s)}L$ for every smooth curve $x(s) \in L$. Prove that L is totally geodesic if and only if TL is invariant under the geodesic flow. □

1.2 The symplectic topology of Euclidean space

In this section we discuss some fundamental questions about the simplest symplectic manifold, namely Euclidean space with the standard symplectic form

$$\omega_0 := \sum_{j=1}^{n} dx_j \wedge dy_j,$$

introduced in (1.1.20). These questions have been very influential in the development of the subject, and will be treated in much more detail in later chapters.

Euclidean space is the most basic symplectic manifold, not only because it is easy to describe, but also because it provides a local model for *every* symplectic manifold. (This is the content of Darboux's theorem; cf. the discussion after Lemma 1.1.15.) An important fact about this structure is that it is essentially unchanged by scalar multiplication. In other words, for $\lambda \in \mathbb{R}$ the map

$$\phi_\lambda(z) := e^\lambda z$$

acts on the form ω_0 by a simple rescaling, i.e. $\phi_\lambda^*\omega_0 = e^{2\lambda}\omega_0$. All the global structure of interest to us is invariant under rescaling. Thus, all such structure that is carried by compact pieces of Euclidean space is reflected inside every symplectic manifold. Note also that the derivative at the origin of a diffeomorphism ϕ which fixes the origin is the limit of rescalings; in other words

$$d\phi(0)v = \lim_{t\to 0} \frac{\phi(tv)}{t}$$

for $v \in \mathbb{R}^{2n}$. Thus one can think of the derivative as being given geometrically by looking at what happens on smaller and smaller pieces of the manifold. In this way, the local (or linear) theory may be thought of as the limit of the global, nonlinear theory.

It is probably most accurate to think of Euclidean space as a 'semi-global' (or perhaps 'semi-local') symplectic manifold. It is a manifold, and therefore one can state global nonlinear problems on it. On the other hand, its linear structure at infinity gives it special properties which make available variational and topological methods which are not applicable in general. We will exploit this in Part IV. Thus we know more about the symplectic topology of Euclidean space than of other manifolds. In fact, much research in symplectic topology has been devoted to extending results known for $\mathbb{R}^{2n}$ to as large a class of manifolds as possible. The most widely applicable method is the technique of J-holomorphic curves in its various forms, which is based on the close connection between symplectic and complex geometry. This theory has many important applications that are not accessible to other methods. It is briefly described in Section 4.5. Readers who wish to learn more about this theory could consult [34, 470, 684].

We will begin by considering the Weinstein conjecture, which can be thought of as a nonlinear analogue of the harmonic oscillator.

The Weinstein conjecture

Let $c \in \mathbb{R}$ be a regular value of a smooth Hamiltonian $H : \mathbb{R}^{2n} \to \mathbb{R}$ and suppose that the energy surface

$$S := H^{-1}(c)$$

is compact and nonempty. Then S is invariant under the flow of the Hamiltonian differential equation (1.1.6). This flow is generated by the vector field X_H in (1.1.11) that has no zeros on S. A basic question is whether this flow has a periodic orbit or, equivalently, whether any of its orbits close up to form a circle. The answer is independent of the choice of the Hamiltonian function H and depends only on the surface S. This may be proved by explicit calculation, as in Exercise 1.2.2 below. Here is a more geometric proof.

Observe first that if $S \subset \mathbb{R}^{2n}$ is a hypersurface, then for every $z \in S$ there is a natural 1-dimensional subspace

$$L_z = \{J_0 v \,|\, v \perp T_z S\} \subset T_z S$$

of the tangent space $T_z S$, where $\perp$ denotes the orthogonal complement with respect to the standard metric and J_0 is as in (1.1.11).[5] The integral curves of this distribution determine a foliation of S called the **characteristic foliation**. If H is a Hamiltonian function with S as a regular level surface, it is easy to see that

$$X_H(z) \in L_z$$

for every $z \in S$. Hence the solutions of the Hamiltonian differential equation (1.1.6) on S are precisely the leaves of the characteristic foliation, or **characteristics**, up to a reparametrization of time. Thus the above question is equivalent to asking whether any nonempty compact hypersurface of $\mathbb{R}^{2n}$ admits a closed characteristic. In this generality the question has been answered in the negative by Herman [307] and Ginzburg [263, 264, 265]. In dimensions $2n \geq 6$ they found smooth hypersurfaces without closed characteristics that are diffeomorphic to spheres. A C^2 counterexample in $\mathbb{R}^4$ was constructed by Ginzburg and Gurel [266]. However, as we shall see in Chapter 12, there are large classes of hypersurfaces which do admit closed characteristics.

In some cases, of course, the periodic orbits are easy to compute. For example, it follows from Example 1.1.17 and Exercise 1.1.22 on the harmonic oscillator that all the characteristics on the sphere are closed, and that there are at least n distinct closed characteristics on each ellipsoid in $\mathbb{R}^{2n}$. But as soon as one moves away from these simple examples, which one can think of as the linear case, the question becomes much more difficult. For example, even the case when S is convex is highly nontrivial.

Weinstein's conjecture refers to a special class of hypersurfaces S called **hypersurfaces of contact type**. The precise definition of these hypersurfaces is given in Chapter 3 (see Proposition 3.5.31). They can be characterized by

[5] One can also describe L_z as the set $\{v \,|\, \omega_0(v, T_z S) = 0\}$.

the existence of a vector field X near S which is transverse to S and satisfies $\mathcal{L}_X\omega_0 = \omega_0$ (i.e. X is a **Liouville vector field**). An example is a hypersurface which is star-shaped about the origin, since such a hypersurface is transverse to the radial vector field which is Liouville. In 1979 Weinstein conjectured that any hypersurface of contact type must have a closed characteristic (cf. [673]). This was based on two earlier existence theorems by Weinstein in the convex case and by Rabinowitz for star-shaped hypersurfaces (cf. [536]). In 1987 the Weinstein conjecture was proved for all contact hypersurfaces of $\mathbb{R}^{2n}$ by Viterbo in his pioneering paper [643], which inspired many further developments by Hofer and his collaborators.

Theorem 1.2.1. (Viterbo) *Every hypersurface of contact type in $\mathbb{R}^{2n}$ has a closed characteristic.*

Proof: See Theorem 12.4.6 on page 486. □

In Chapter 12 we will give a version of the proof which is due to Hofer and Zehnder. The aforementioned examples of Herman [307] and Ginzburg [263, 264, 265] show that there are smooth hypersurfaces in $\mathbb{R}^{2n}$ for $n \geq 3$ that do not admit any closed characteristics. Thus the contact condition in Theorem 1.2.1 cannot be removed. We also point out that in [673] Weinstein actually formulated his conjecture for Reeb flows (defined in equation (3.5.2)) on general closed contact manifolds, whether or not they can be embedded as contact hypersurfaces in some Euclidean space. In dimension three this was confirmed by Hofer [317] for the 3-sphere and by Taubes [612] in general. In higher dimensions the general Weinstein conjecture is still open.

Exercise 1.2.2 Let $S = H^{-1}(c) = (H')^{-1}(c')$ be a regular energy surface for two Hamiltonian functions $H, H' : \mathbb{R}^{2n} \to \mathbb{R}$, and choose the function $\lambda : S \to \mathbb{R}$ such that $X_{H'}(z) = \lambda(z)X_H(z)$ for $z \in S$. Suppose that $z(t) = z(t+T) \in S$ is a periodic solution of (1.1.6), and consider the reparametrization $t \mapsto \tau(t)$ which satisfies the conditions $\dot{\tau} = \lambda(z(\tau))$ and $\tau(0) = 0$. Show that there is a $T' \neq 0$ such that $\tau(kT') = kT$ for all $k \in \mathbb{Z}$. Deduce that the function $z'(t) = z \circ \tau(t) = z'(t+T')$ is a periodic solution of the Hamiltonian system (1.1.6), with H replaced by H'. □

The nonsqueezing theorem and the symplectic camel

If one wants to understand what a map looks like geometrically, the first thing one might try to understand is how it transforms basic geometric subsets such as balls, ellipsoids, and cylinders. The behaviour of volume-preserving diffeomorphisms in this regard is well understood, namely there is a volume-preserving diffeomorphism from a ball B onto a subset $U \subset \mathbb{R}^{2n}$ if and only if the two sets are diffeomorphic and have the same volume. As we remarked above, every symplectomorphism preserves volume. It had long been suspected that the group of symplectomorphisms is significantly smaller than that of volume-preserving diffeomorphisms, and that its elements are not as malleable as those of the larger group. However, there was no result which pinpointed a difference until Gromov proved his celebrated nonsqueezing theorem in 1985. This says that a standard symplectic ball cannot be symplectically embedded into a thin cylinder.

More precisely, denote by

$$B^{2n}(r) := \{z \in \mathbb{R}^{2n} \mid |z| \le r\}$$

the closed Euclidean ball in $\mathbb{R}^{2n}$ with centre 0 and radius r, and by

$$Z^{2n}(r) := B^2(r) \times \mathbb{R}^{2n-2} := \left\{z \in \mathbb{R}^{2n} \mid \sqrt{x_1^2 + y_1^2} \le r\right\}$$

the symplectic cylinder. (Observe that the disc $B^2(r)$ here is a symplectic disc, i.e. its coordinates are (x_1, y_1) and not (x_1, x_2). We also assume $n \ge 2$.) We say that $\phi : U \to V$ is a symplectic embedding of one subset $U \subset \mathbb{R}^{2n}$ into another subset $V \subset \mathbb{R}^{2n}$ if ϕ is a smooth embedding of U into V which preserves the symplectic form, that is, $\phi^*\omega_0 = \omega_0$. (See Fig. 1.2.)

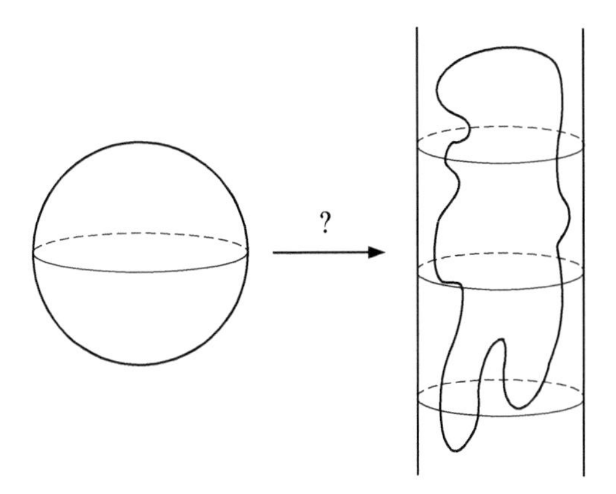

FIG. 1.2. The nonsqueezing theorem.

Theorem 1.2.3 (Nonsqueezing theorem) *If there exists a symplectic embedding $B^{2n}(r) \hookrightarrow Z^{2n}(R)$, then $r \le R$.*

Proof: The result is restated in Theorem 12.1.1 and proved on page 484. □

This is a foundational theorem in symplectic topology, and each fundamental approach to the theory (variational methods, generating functions, J-holomorphic curves) provides a proof. It shows that there is a basic property of the ball and cylinder that is preserved by symplectomorphisms. Intuitively, one can think of this as their 'symplectic width' or 'symplectic area'. This idea was crystallized in Ekeland and Hofer's definition of a symplectic **capacity**. This, and related developments, will be discussed further in Chapter 12. However, as we shall see below, just the fact that this invariant exists immediately led Eliashberg and Ekeland–Hofer to a proof of symplectic rigidity.

Another problem with a similar flavour is that of the symplectic camel. Here one asks whether it is possible to isotop a ball B symplectically through a small 'hole' in a wall. We may take the wall W to be the hyperplane $\{y_1 = 0\}$, and the hole H_ε to be the set

$$H_\varepsilon := \left\{ z \in W \,\middle|\, |z| < \varepsilon \right\}.$$

Then the question is whether one can find a continuous family ϕ_t, $0 \le t \le 1$, of symplectic embeddings of the closed unit ball B into $\mathbb{R}^{2n}$ which go through the hole, i.e.

$$\phi_t(B) \subset \mathbb{R}^{2n} \setminus (W \setminus H_\varepsilon) = (\mathbb{R}^{2n} \setminus W) \cup H_\varepsilon,$$

and begin and end at different sides of the wall, i.e.

$$\phi_0(B) \subset \{y_1 < 0\}, \qquad \phi_1(B) \subset \{y_1 > 0\}.$$

(See Fig. 1.3.) Clearly, this is possible when the radius ε of the hole is greater than 1, so the real problem occurs when $\varepsilon \le 1$.

FIG. 1.3. The symplectic camel.

The methods that Gromov developed to prove the nonsqueezing theorem suffice to answer this question in the negative. An exposition of his argument was given by McDuff and Traynor in [472]. Viterbo gave a completely different proof in [646] that uses generating functions.

Rigidity

This question also concerns the relative size of the subgroup $\mathrm{Symp}(\mathbb{R}^{2n})$ in the group $\mathrm{Diff}_{\mathrm{vol}}(\mathbb{R}^{2n})$ of all volume-preserving diffeomorphisms. Suppose that you have a sequence of symplectomorphisms which converge uniformly on some compact set to a diffeomorphism ϕ_∞. Clearly, the limit ϕ_∞ must preserve volume, but does it preserve the symplectic structure? This is not at all obvious, because the usual characterization of a symplectomorphism involves the first derivative while the C^0, or uniform, limit depends only on how far points are moved. Put another way, is it possible to characterize symplectic diffeomorphisms in terms of a condition which is continuous with respect to the C^0-topology? This question lies at the heart of symplectic topology because it involves the very nature of what it means to be symplectic.

Gromov formulated this question in the 1960s in terms of the contrast between **flexibility** versus **rigidity** (see his thesis [286] and its expanded version [288]), and in the 1980s [289] as **softness** versus **hardness**. In the present context, hardness (or rigidity) means that $\mathrm{Symp}(\mathbb{R}^{2n})$ is C^0-closed in $\mathrm{Diff}(\mathbb{R}^{2n})$, and implies that there are robust symplectic invariants which persist under C^0-perturbations. Softness (or flexibility) means that the C^0-closure of $\mathrm{Symp}(\mathbb{R}^{2n})$ in $\mathrm{Diff}(\mathbb{R}^{2n})$ is $\mathrm{Diff}_{\mathrm{vol}}(\mathbb{R}^{2n})$. The soft option would imply that there cannot be any nontrivial symplectic invariants of a topological nature, and thus that there is no symplectic topology.

From the above description it should be clear that in the situation which we are now considering, rigidity prevails. In fact, the nonsqueezing theorem is the key to establishing rigidity. As we shall see in Chapter 12, it can be used to define a C^0-continuous symplectic invariant, and symplectomorphisms can then be characterized as the diffeomorphisms that preserve this invariant. Readers interested in the details of this argument will find them in Sections 2.4, 12.1, and 12.2. The discussion there is easily accessible. It just uses a few facts about symplectic linear algebra from Section 2.1.

The characterization of what it means to be symplectic is still somewhat mysterious, and many questions remain without answers. In this connection the nonsqueezing theorem may be taken to be a geometric expression of what a symplectic structure is. Another global (albeit rather indirect) way in which the nature of what it means to be symplectic manifests itself is through the topological properties of those manifolds which admit symplectic structures. Even now, not much is understood about the distinction between manifolds which have symplectic structures and those which do not, although the 1990s saw spectacular progress on this problem by Donaldson and Taubes, which is discussed in Sections 7.4 and 13.3. In Part II we describe some general constructions that work well in the symplectic context. In particular, it is easy to construct symplectic forms on connected manifolds with nonempty boundary, which shows that the real geometric obstruction to the existence of symplectic structures occurs in the top dimension. So far, there is little hint of what this obstruction might be.

Fixed point theorems

The final question which we will consider here is that of counting the number of fixed points of time-1 maps of time-dependent Hamiltonian flows. This question makes the best sense when considered on a compact manifold, and we will illustrate the problem by working on the 2-torus

$$\mathbb{T}^2 = \mathbb{R}^2/\mathbb{Z}^2.$$

The points of the 2-torus are equivalence classes of pairs (x, y) of real numbers, where $(x, y) \equiv (x+m, y+n)$ iff $(m, n) \in \mathbb{Z}^2$. Since the torus inherits its symplectic structure from $\mathbb{R}^2$, we may equivalently consider doubly periodic Hamiltonian flows on $\mathbb{R}^2$ and count the number of fixed points which are nonequivalent under the above equivalence relation $\equiv$. Thus we have a family H_t, $0 \le t \le 1$, of Hamiltonian functions on $\mathbb{R}^2$ which are periodic in the sense that

$$H_t(x+m, y+n) = H_t(x, y)$$

for $m, n \in \mathbb{Z}$. If $\phi : \mathbb{R}^2 \to \mathbb{R}^2$ is the time-1 map of the associated flow, we want to estimate the number of nonequivalent solutions (x, y) to the equation $\phi(x, y) = (x+m, y+n)$. If H_t is independent of time, then the critical points of H are fixed points of the motion. Indeed, at such a point the Hamiltonian equations

$$\dot{x} = \frac{\partial H}{\partial y}, \qquad \dot{y} = -\frac{\partial H}{\partial x}$$

reduce to the system $\dot{x} = \dot{y} = 0$, and so their solution is constant. If the C^2-norm of H is sufficiently small then these will be the only fixed points (see Lemma 12.4.1). In other words, for small Hamiltonian functions, the number of fixed points of ϕ is equal to the number of critical points of H.

The minimal number of critical points of a function on a closed (i.e. compact and without boundary) manifold M is a topological invariant of that manifold, as is the minimal number of critical points of a Morse function (where all the critical points are nondegenerate). We denote these invariants by

$$\mathrm{Crit}(M) := \min_{f:M\to\mathbb{R}} \#\mathrm{Crit}(f), \qquad \mathrm{Crit}_{\mathrm{Morse}}(M) := \min_{\substack{f:M\to\mathbb{R} \\ \text{is Morse}}} \#\mathrm{Crit}(f).$$

Here the minimum is taken over all smooth functions (respectively over all Morse functions) $f : M \to \mathbb{R}$ and $\mathrm{Crit}(f) \subset M$ denotes the set of critical points of f. It is obvious that $\mathrm{Crit}(M) \ge 2$ for every manifold M containing at least two points, because every nonconstant function must have a distinct maximum and minimum. A more precise estimate of the number $\mathrm{Crit}(M)$ is given by Ljusternik–Schnirelmann theory (see Chapter 11). Moreover, it follows from the Morse inequalities that $\mathrm{Crit}_{\mathrm{Morse}}(M) \ge \dim H_*(M)$ for every coefficient field. Note that the number $\mathrm{Crit}_{\mathrm{Morse}}(M)$ may be strictly bigger than $\dim H_*(M)$, as is shown by the example of a homology 3-sphere.

The numbers for the 2-torus are $\mathrm{Crit}(\mathbb{T}^2) = 3$ and $\mathrm{Crit}_{\mathrm{Morse}}(\mathbb{T}^2) = 4$. Thus, in the case of a time-independent Hamiltonian on $M = \mathbb{T}^2$, the time-1 map has at least three distinct fixed points, and at least four if they are all nondegenerate. In fact, it is easy to extend this argument to C^2-small time-dependent Hamiltonian functions. However the extension to arbitrarily large flows presents completely new problems. Arnold conjectured in the 1960s that analogous estimates remain valid for arbitrary Hamiltonian symplectomorphisms on arbitrary closed symplectic manifolds.

Conjecture 1.2.4 (Arnold) *Let ϕ be the time-1 map of a time-dependent Hamiltonian differential equation on a closed symplectic manifold M. Then the number of fixed points of ϕ is bounded below by the minimal number of critical points of a function on M, i.e.*

$$\#\mathrm{Fix}(\phi) \geq \mathrm{Crit}(M).$$

If the fixed points of ϕ are all nondegenerate, then $\#\mathrm{Fix}(\phi) \geq \mathrm{Crit}_{\mathrm{Morse}}(M)$.

As we will see in Chapter 11, much progress has been made with understanding this problem. For closed symplectic 2-manifolds it was proved by Eliashberg [174] and for the $2n$-torus it was proved by Conley and Zehnder in their celebrated paper [124]. The breakthrough came with the work of Floer [226, 230] who developed a new elliptic approach to Morse theory in suitable infinite-dimensional settings which is now called *Floer homology*. In particular, his theory is applicable to the symplectic action functional $\mathcal{A}_H$ and this led to his proof of the Arnold conjecture, in the nondegenerate case and with the homological lower bound, for a large class of closed symplectic manifolds. Floer's approach was inspired by the Conley–Zehnder proof of the Arnold conjecture for the torus, as well as by Gromov's theory of pseudoholomorphic curves and by the Morse–Thom–Smale–Milnor–Witten complex in Morse theory. All subsequent work on the Arnold conjecture has built on the ideas of Floer.

The lower bounds in Conjecture 1.2.4 are, in general, larger than those predicted by the Lefschetz fixed point theorem. In the nondegenerate case, the latter implies that the number of fixed points of a map $\phi : M \to M$ that is homotopic to the identity is bounded below by the number of zeros of a vector field on M (counted with multiplicities). This is the Euler characteristic, i.e. is the alternating sum of the Betti numbers. This number vanishes for the torus, and so Lefschetz theory gives no information in this case. This is appropriate, since the torus admits diffeomorphisms such as rotations with no fixed points. As we shall see in Chapter 10 the condition in Arnold's conjecture that ϕ be Hamiltonian restricts consideration to a subgroup of finite codimension in the symplectic group consisting of maps that, loosely speaking, are 'irrotational'. In particular it excludes rotations of the torus. In exchange, this restricted class of symplectomorphisms is conjectured to have significantly more fixed points than a general diffeomorphism on manifolds whose odd Betti numbers do not all vanish.

Arnold's conjecture for the 2-torus can also be viewed as a generalization of the Poincaré–Birkhoff theorem on the existence of two fixed points for area-preserving twist maps of the annulus. We give a complete proof for the existence of one fixed point in Section 8.2. The argument is elementary and is accessible to readers without any background in symplectic geometry.

The fact that the Arnold conjecture gives sharper estimates for the number of fixed points than the Lefschetz fixed point theorem is a global version of the observation that a function on a manifold must have more critical points than a vector field (or 1-form) must have zeros. The latter statement has a direct interpretation in symplectic geometry since the differential of a function is an exact Lagrangian submanifold of the cotangent bundle, and the critical points are the intersections with the zero section. Another version of the Arnold conjecture extends this statement about Lagrangian intersections to arbitrary exact Lagrangian submanifolds of a cotangent bundle, whether or not they are graphs. This is yet another indication of how local symplectic structure is reflected in large-scale, global phenomena. The aim of symplectic topology is to map out and understand the manifestations of this principle.

Comments

The symplectic action functional $\mathcal{A}_H$ introduced in (1.1.8) plays a central role in symplectic topology. We showed in Lemma 1.1.8 that its critical points are the solutions of Hamilton's equations. In [226, 227, 228, 229, 230] Floer developed an infinite-dimensional version of Morse theory for the symplectic action functional, called **(symplectic) Floer homology**. Floer theory has many facets and has become a very powerful tool in symplectic topology and other areas of mathematics. We give a brief outline of symplectic Floer homology in Section 11.4. Because it treats $\mathcal{A}_H$ as a Morse function on an infinite-dimensional space, the arguments needed to set it up rigorously are heavily analytic and beyond the scope of this book; cf. Audin–Damian [33]. However, in Chapter 9 we develop the theory of generating functions, that includes a theory of *finite-dimensional approximations* to the symplectic action as one important special case. These functions will be used in Chapter 11 to prove a version of the Arnold conjecture for tori and in Chapter 12 to define the Hofer–Zehnder capacity and prove nontriviality of the Hofer metric, the Weinstein conjecture, and Gromov's nonsqueezing theorem.

Lemma 1.1.19 shows that the Poisson bracket is another fundamental structure in the symplectic world. This also exhibits interesting and unexpected rigidity phenomena; cf. Cardin–Viterbo [93], Entov–Polterovich's work on symplectic function theory [199, 200, 201, 202], Buhovsky's 2/3 estimate [82], and the invariants of Entov–Polterovich–Zapolsky [204] and Buhovsky–Entov–Polterovich [86]. The book [532] by Polterovich and Rosen explains this work in context.

Our book concentrates on finite-dimensional symplectic topology. However, foundational results such as the nonsqueezing theorem are beginning to find infinite-dimensional analogues and applications; see for example Abbondandolo–Majer [1] and the references therein.

2

LINEAR SYMPLECTIC GEOMETRY

The geometry of a bilinear skew form is very different from that of a symmetric form and its main features, for example compatible complex structures and Lagrangian subspaces, appear later as significant elements of the global theory. Moreover, it is increasingly apparent that the linear theory is a profound paradigm for the nonlinear theory. For example, the fact that all symplectic vector spaces of the same dimension are isomorphic translates into the statement that all symplectic manifolds are locally diffeomorphic (Darboux's theorem in Chapter 3). A more startling example is Gromov's nonsqueezing theorem which implies that a ball can be symplectically embedded into a cylinder if and only if it has a smaller radius. We shall discuss the linear version of this result in Section 2.4.

The theory is enriched by the interplay between symmetric and skew-symmetric forms. This appears in many guises: either directly, as in the theorem that a nondegenerate symmetric bilinear form and a nondegenerate skew-symmetric form have a common diagonalization, or indirectly, for example in the theorem that the eigenvalues of a symplectic matrix occur in pairs of the form $\lambda, 1/\lambda$. In the presence of a skew-symmetric form, a symmetric form gives rise to other related geometric objects such as Lagrangian subspaces and almost complex structures, and we shall also explore the elementary theory of these objects. Throughout we focus on the properties of real symplectic vector spaces, not venturing into the more complicated world of complex symplectic geometry.

We begin this chapter by discussing symplectic vector spaces in Section 2.1, linear symplectomorphisms in Section 2.2, and Lagrangian subspaces in Section 2.3. These can be thought of as the three fundamental notions in linear symplectic geometry. In Section 2.4 we prove the linear nonsqueezing theorem and discuss its consequences for the action of the symplectic linear group on ellipsoids. We prove that a linear transformation is symplectic (or anti-symplectic) if and only if it preserves the *linear symplectic width* of ellipsoids. Section 2.5 discusses linear complex structures on symplectic vector spaces and Section 2.6 deals with symplectic vector bundles. In Section 2.7 we develop the theory of the first Chern class from scratch.

The material of the first three sections in this chapter is essential for the study of symplectic manifolds in Chapter 3. The linear nonsqueezing theorem will play an important role in the proof of symplectic rigidity in Chapter 12.

However, the various discussions of the Maslov index and also the subsections on trivializations and the first Chern class may be omitted at first reading.

2.1 Symplectic vector spaces

The archetypal example of a symplectic vector space is the Euclidean space $\mathbb{R}^{2n}$ with the skew-symmetric form

$$\omega_0 = \sum_{j=1}^{n} dx_j \wedge dy_j.$$

Another basic example is the direct sum $E \oplus E^*$ of a real vector space E with its dual, equipped with the skew form

$$\omega\big((v, v^*), (w, w^*)\big) = w^*(v) - v^*(w).$$

More generally, a **symplectic vector space** is a pair (V, ω) consisting of a finite-dimensional real vector space V and a nondegenerate skew-symmetric bilinear form $\omega : V \times V \to \mathbb{R}$. This means that the following conditions are satisfied:

(skew-symmetry) For all $v, w \in V$

$$\omega(v, w) = -\omega(w, v).$$

(nondegeneracy) For every $v \in V$

$$\omega(v, w) = 0 \quad \forall\, w \in V \qquad \Longrightarrow \qquad v = 0.$$

The vector space V is necessarily of even dimension since a real skew-symmetric matrix of odd dimension must have a kernel. (See Exercise 2.1.12 below.) When V is an even-dimensional real vector space, the set of nondegenerate skew-symmetric bilinear forms on V will be denoted by $\Omega(V)$.

A **linear symplectomorphism** of the symplectic vector space (V, ω) is a vector space isomorphism $\Psi : V \to V$ that preserves the symplectic structure in the sense that $\Psi^*\omega = \omega$, where $(\Psi^*\omega)(v, w) := \omega(\Psi v, \Psi w)$ for $v, w \in V$. The linear symplectomorphisms of (V, ω) form a group which we denote by $\mathrm{Sp}(V, \omega)$. In the case of the standard symplectic structure on Euclidean space we use the notation $\mathrm{Sp}(2n) = \mathrm{Sp}(\mathbb{R}^{2n}, \omega_0)$.

The **symplectic complement** of a linear subspace $W \subset V$ is defined as the subspace

$$W^\omega = \{v \in V \,|\, \omega(v, w) = 0 \quad \forall\, w \in W\}. \tag{2.1.1}$$

The symplectic complement need not be transversal to W. A subspace W is called

isotropic if $W \subset W^\omega$,
coisotropic if $W^\omega \subset W$,
symplectic if $W \cap W^\omega = \{0\}$,

Lagrangian if $W = W^{\omega}$.

Note that W is isotropic if and only if ω vanishes on W, and W is symplectic if and only if $\omega|_W$ is nondegenerate.

Lemma 2.1.1 *For any subspace $W \subset V$,*

$$\dim W + \dim W^{\omega} = \dim V, \qquad W^{\omega\omega} = W.$$

Thus

- *W is symplectic if and only if W^{ω} is symplectic;*
- *W is isotropic if and only if W^{ω} is coisotropic;*
- *W is Lagrangian if and only if it is isotropic and has half the dimension of V.*

Proof: Define a map ι_{ω} from V to the dual space V^* by setting

$$\iota_{\omega}(v)(w) = \omega(v, w).$$

Since ω is nondegenerate ι_{ω} is an isomorphism. It identifies W^{ω} with the annihilator $W^{\perp}$ of W in V^*. But for any subspace W of any vector space V we have

$$\dim W + \dim W^{\perp} = \dim V.$$

This proves the first statement. The further assertions follow by combining this with the special properties of the pair W, W^{ω} when W is symplectic, Lagrangian, or (co)isotropic. □

The following exercise shows that Lagrangian subspaces are closely related to linear symplectomorphisms.

Exercise 2.1.2 Let (V, ω) be a symplectic vector space and $\Psi : V \to V$ be a linear map. Prove that Ψ is a linear symplectomorphism if and only if its graph

$$\Gamma_{\Psi} = \{(v, \Psi v) \,|\, v \in V\}$$

is a Lagrangian subspace of $V \times V$ with symplectic form

$$(-\omega) \oplus \omega := \mathrm{pr}_2^*\omega - \mathrm{pr}_1^*\omega.$$

(The symplectic form is standard in the target and reversed in the source.) □

The following result is the main theorem of this section. It implies that all symplectic vector spaces of the same dimension are linearly symplectomorphic.

Theorem 2.1.3 *Let (V, ω) be a symplectic vector space of dimension $2n$. Then there exists a basis $u_1, \dots, u_n, v_1, \dots, v_n$ such that*

$$\omega(u_j, u_k) = \omega(v_j, v_k) = 0, \qquad \omega(u_j, v_k) = \delta_{jk}.$$

Such a basis is called a **symplectic basis**. *Moreover, there exists a vector space isomorphism $\Psi : \mathbb{R}^{2n} \to V$ such that $\Psi^*\omega = \omega_0$.*

Proof: The proof is by induction over n. Since ω is nondegenerate there exist vectors $u_1, v_1 \in V$ such that $\omega(u_1, v_1) = 1$. Hence the subspace spanned by u_1 and v_1 is symplectic. Let W denote its symplectic complement. Then (W, ω) is a symplectic vector space of dimension $2n - 2$. By the induction hypothesis, there exists a symplectic basis $u_2, \dots, u_n, v_2, \dots, v_n$ of W. Hence the vectors $u_1, \dots, u_n, v_1, \dots, v_n$ form a symplectic basis of V. The linear map $\Psi : \mathbb{R}^{2n} \to V$, defined by

$$\Psi z = \sum_{j=1}^{n} x_j u_j + \sum_{j=1}^{n} y_j v_j$$

for $z = (x, y) \in \mathbb{R}^{2n}$, satisfies $\Psi^* \omega = \omega_0$ as required. □

A symplectic basis is sometimes called an **ω-standard basis.**

Corollary 2.1.4 *Let V be a $2n$-dimensional real vector space and let ω be a skew-symmetric bilinear form on V. Then ω is nondegenerate if and only if its n-fold exterior power is nonzero, i.e. $\omega^n = \omega \wedge \cdots \wedge \omega \neq 0$.*

Proof: Assume first that ω is degenerate. Let $v \neq 0$ such that $\omega(v, w) = 0$ for all $w \in V$. Now choose a basis $v_1, \dots, v_{2n}$ of V such that $v_1 = v$. Then $\omega^n(v_1, \dots, v_{2n}) = 0$. Conversely, suppose that ω is nondegenerate. Then, since ω_0^n is a volume form, it follows from Theorem 2.1.3 that $\omega^n \neq 0$. □

Lemma 2.1.5 *Every isotropic subspace of V is contained in a Lagrangian subspace. Moreover, every basis $u_1, \dots, u_n$ of a Lagrangian subspace Λ can be extended to a symplectic basis of (V, ω).*

Proof: Let W be an isotropic subspace, i.e. $W \subset W^\omega$. If the subspace W_1 is obtained by adjoining some vector $v \in W^\omega \backslash W$ to W, then ω vanishes on W_1. Hence a maximal isotropic subspace must satisfy $W = W^\omega$ and hence be Lagrangian. Because $\Lambda \subset W$ implies $W^\omega \subset \Lambda^\omega$, no subspace W that is strictly larger than a Lagrangian space Λ can be isotropic. Hence the Lagrangian subspaces are precisely the maximal isotropic subspaces. This proves the first statement.

To prove the second statement it suffices to consider the case $V = \mathbb{R}^{2n}$ with the standard symplectic structure ω_0 and complex structure J_0. Given a Lagrangian subspace Λ the subspace $\Lambda' = J_0 \Lambda$ is also Lagrangian and can be identified with the dual space Λ^* via the isomorphism $\iota_{\omega_0} : \mathbb{R}^{2n} \to (\mathbb{R}^{2n})^*$ of Lemma 2.1.1. Hence we may choose $\{v_1, \dots, v_n\} \subset \Lambda'$ to be the basis dual to $u_1, \dots, u_n$. This proves Lemma 2.1.5. □

Example 2.1.6 Consider the direct sum $E \oplus E^*$ of a vector space E with its dual, provided with the form

$$\omega\big((v, v^*), (w, w^*)\big) = w^*(v) - v^*(w).$$

Then ω is nondegenerate and has standard basis given by $e_1, \dots, e_n, e_1^*, \dots, e_n^*$, where $e_1, \dots, e_n$ is a basis of E with dual basis $e_1^*, \dots, e_n^*$. □

Linear symplectic reduction

Every coisotropic subspace $W \subset V$ gives rise to a new symplectic vector space obtained by dividing W by its symplectic complement. This construction of a subquotient is called symplectic reduction.

Lemma 2.1.7 *Let (V, ω) be a symplectic vector space and let $W \subset V$ be a coisotropic subspace. Then the following holds.*
(i) *The quotient*

$$\overline{W} := W/W^\omega$$

carries a unique symplectic structure $\overline{\omega}$ such that the restriction $\omega|_W$ agrees with the pullback of $\overline{\omega}$ under the projection $W \to \overline{W}$.
(ii) *If $\Lambda \subset V$ is a Lagrangian subspace then*

$$\overline{\Lambda} := \big((\Lambda \cap W) + W^\omega\big)/W^\omega$$

is a Lagrangian subspace of $(\overline{W}, \overline{\omega})$.

Proof: Denote $[w] := w + W^\omega \in \overline{W}$ for $w \in W$. By the definition of coisotropic, W^ω is an isotropic subspace of W and $\omega(v, w) = 0$ for $v \in W^\omega$ and $w \in W$. Hence $\omega(w_1, w_2)$ depends only on the equivalence classes $[w_1]$ and $[w_2]$ in $\overline{W} = W/W^\omega$. Hence ω descends to a 2-form $\overline{\omega}$ on $\overline{W}$. Moreover, if $w \in W$ and $\omega(v, w) = 0$ for all $v \in W$, then $w \in W^\omega$ and hence $\overline{\omega}$ is nondegenerate. This proves (i).

To prove (ii) we first show that the subspace

$$\widetilde{\Lambda} := (\Lambda \cap W) + W^\omega$$

is a Lagrangian subspace of V, namely

$$\widetilde{\Lambda}^\omega = (\Lambda \cap W)^\omega \cap W = (\Lambda + W^\omega) \cap W = (\Lambda \cap W) + W^\omega = \widetilde{\Lambda}.$$

Let $w \in W$ such that $\overline{\omega}([w], [v]) = 0$ for all $[v] \in \overline{\Lambda} = \widetilde{\Lambda}/W^\omega$. Then $\omega(w, v) = 0$ for all $v \in \widetilde{\Lambda}$ and so $w \in \widetilde{\Lambda}^\omega = \widetilde{\Lambda}$. Thus $\overline{\Lambda} \in \mathcal{L}(\overline{W}, \overline{\omega})$ and this proves (ii). □

Example 2.1.8 Let (V_0, ω_0) and (V_1, ω_1) be symplectic vector spaces of the same dimension, $\Lambda_0 \subset V_0$ be a Lagrangian subspace, and $\Psi : V_0 \to V_1$ be a linear symplectomorphism. Consider the symplectic vector space (V, ω) defined by $V := V_0 \times V_0 \times V_1$ and $\omega := \omega_0 \oplus (-\omega_0) \oplus \omega_1$. Then

$$W := \Delta \times V_1 \subset V$$

is a coisotropic subspace with isotropic complement $W^\omega \simeq \Delta \times \{0\}$ and quotient

$$\overline{W} := W/W^\omega \cong V_1.$$

The subspace $\Lambda := \Lambda_0 \times \Gamma_\Psi \subset V$ is Lagrangian and intersects W transversally. The reduced Lagrangian subspace is isomorphic to $\Lambda_1 := \Psi\Lambda_0$. □

The above example shows that the action of a symplectic transformation Φ on Lagrangian subspaces may be described in terms of symplectic reduction. The next exercise describes composition of matrices in a similar way. This idea can be greatly generalized, leading to a more general kind of morphism in the category of symplectic vector spaces, called a 'Lagrangian correspondence' (see Weinstein et al [429, 669, 680, 681] and the references therein).

Exercises

Exercise 2.1.9 Identify a matrix with its graph as in Exercise 2.1.2 and use a construction similar to that in Example 2.1.8 to interpret the composition of symplectic matrices in terms of symplectic reduction. □

Exercise 2.1.10 Let (V, ω) be a symplectic vector space and $W \subset V$ be any subspace. Prove that the quotient $V' = W/W \cap W^\omega$ carries a natural symplectic structure. □

Exercise 2.1.11 Let $A = -A^{\mathrm{T}} \in \mathbb{R}^{2n\times 2n}$ be a nondegenerate skew-symmetric matrix and define $\omega(z, w) = \langle Az, w\rangle$. Prove that a symplectic basis for $(\mathbb{R}^{2n}, \omega)$ can be constructed from the eigenvectors $u_j + iv_j$ of A. **Hint:** Use the fact that the matrix $iA \in \mathbb{C}^{2n\times 2n}$ is self-adjoint and therefore can be diagonalized. For more details see the proof of Lemma 2.4.5 below. □

Exercise 2.1.12 Show that if β is any skew-symmetric bilinear form on the vector space W, there is a basis $u_1, \dots, u_n, v_1, \dots, v_n, w_1, \dots, w_p$ of W such that $\beta(u_j, v_k) = \delta_{jk}$ and all other pairings $\beta(b_1, b_2)$ vanish. A basis with this property is called a **standard basis** for (W, β), and the integer $2n$ is the **rank** of β. □

Exercise 2.1.13 Show that if W is an isotropic, coisotropic or symplectic subspace of a symplectic vector space (V, ω), then any standard basis for (W, ω) extends to a symplectic basis for (V, ω). □

Exercise 2.1.14 Show that any hyperplane W in a $2n$-dimensional symplectic vector space (V, ω) is coisotropic. Thus $W^\omega \subset W$ and $\omega|_W$ has rank $2(n-1)$. **Hint:** By Exercise 2.1.12 the 2-form $\omega|_W$ has even rank. Hence there is some nonzero vector $w \in W$ such that $\omega(w, x) = 0$ for all $x \in W$. Show that this vector w spans W^ω. □

Exercise 2.1.15 Recall that $\Omega(V)$ denotes the space of all symplectic forms on the vector space V. Consider the (covariant) action of the general linear group $\mathrm{GL}(2n, \mathbb{R})$ on $\Omega(V)$ via $\mathrm{GL}(2n, \mathbb{R}) \times \Omega(V) \to \Omega(V) : (\Psi, \omega) \mapsto (\Psi^{-1})^*\omega$ and show that $\Omega(V)$ is homeomorphic to the homogeneous space $\mathrm{GL}(2n, \mathbb{R})/\mathrm{Sp}(2n)$. □

Exercise 2.1.16 (The Gelfand–Robbin quotient) It has been noted by physicists for a long time that symplectic structures often arise from boundary value problems. The underlying abstract principle can be formulated as follows. Let H be an infinite-dimensional Hilbert space and let $D : \mathrm{dom}(D) \to H$ be a symmetric linear operator with a closed graph and a dense domain $\mathrm{dom}(D) \subset H$. Prove that the quotient

$$V := \mathrm{dom}(D^*)/\mathrm{dom}(D)$$

is a symplectic vector space with symplectic structure

$$\omega([x], [y]) := \langle x, D^*y\rangle - \langle D^*x, y\rangle \qquad \text{for } x, y \in \mathrm{dom}(D^*).$$

Here $[x] \in \mathrm{dom}(D^*)/\mathrm{dom}(D)$ denotes the equivalence class of $x \in \mathrm{dom}(D^*)$. Show that self-adjoint extensions of D are in one-to-one correspondence to Lagrangian subspaces $\Lambda \subset V$. If D has a closed range, show that the kernel of D^* determines a Lagrangian subspace

$$\Lambda_0 := \{[x] \mid x \in \mathrm{dom}(D^*),\ D^*x = 0\}.$$

In applications D is a differential operator on a manifold with boundary, V is a suitable space of boundary data, and the symplectic form can, via Stokes' theorem, be expressed as an integral over the boundary. **Hints:** The symmetry condition asserts that $\langle x, Dy\rangle = \langle Dx, y\rangle$ for all $x, y \in \mathrm{dom}(D)$. Recall that the domain of the operator $D^* : \mathrm{dom}(D^*) \to H$ is defined as the set of all vectors $y \in H$ such that the linear functional $\mathrm{dom}(D) \to \mathbb{R} : x \mapsto \langle y, Dx\rangle$ extends to a bounded linear functional on H. Thus for $y \in \mathrm{dom}(D^*)$ there exists a unique vector $z \in H$ such that $\langle z, x\rangle = \langle y, Dx\rangle$ for all $x \in \mathrm{dom}(D)$ and one defines $D^*y = z$. The symmetry condition is equivalent to $\mathrm{dom}(D) \subset \mathrm{dom}(D^*)$ and $D^*y = Dy$ for $y \in \mathrm{dom}(D)$. A symmetric operator is called **self-adjoint** if $D^* = D$. Show that a linear operator $D : \mathrm{dom}(D) \to H$ is self-adjoint if and only if $\mathrm{graph}(D) \subset H \times H$ is a Lagrangian subspace with respect to the standard symplectic structure of Example 2.1.6. Interpret the exercise as linear symplectic reduction in an infinite-dimensional setting. □

Exercise 2.1.17 Let $J_0 \in \mathbb{R}^{2n\times 2n}$ be the matrix (1.1.10) and consider the linear operator $D := J_0 \frac{\partial}{\partial t}$ on the Hilbert space

$$H = L^2([0,1], \mathbb{R}^{2n})$$

with

$$\mathrm{dom}(D) = W_0^{1,2}([0,1], \mathbb{R}^{2n})$$

(the Sobolev space of absolutely continuous functions which vanish on the boundary and whose first derivative is square integrable). Show that in this case the Gelfand–Robbin quotient is given by $V = \mathbb{R}^{2n} \times \mathbb{R}^{2n}$ with symplectic form $(-\omega_0) \oplus \omega_0$. □

2.2 The symplectic linear group

In this section we shall examine the group

$$\mathrm{Sp}(V, \omega) := \{\Psi \in \mathrm{GL}(V) \mid \Psi^*\omega = \omega\} \tag{2.2.1}$$

of linear symplectomorphisms in more detail. By Theorem 2.1.3 all symplectic vector spaces of the same dimension are isomorphic. Hence Lemma 1.1.12 asserts that $\mathrm{Sp}(V, \omega)$ is a Lie group and that its Lie algebra is given by

$$\mathfrak{sp}(V, \omega) := \mathrm{Lie}(\mathrm{Sp}(V, \omega)) = \{A \in \mathrm{End}(V) \mid \omega(A\cdot, \cdot) + \omega(\cdot, A\cdot) = 0\}. \tag{2.2.2}$$

To understand this group it suffices to consider the case $V = \mathbb{R}^{2n}$ with the standard symplectic form ω_0. We can think of the elements of $\mathrm{Sp}(2n) = \mathrm{Sp}(\mathbb{R}^{2n}, \omega_0)$ as real $2n \times 2n$ matrices Ψ that satisfy equation (1.1.15), i.e. $\Psi^{\mathrm{T}} J_0 \Psi = J_0$, or equivalently $\Psi^*\omega_0 = \omega_0$. Matrices which satisfy this condition are called **symplectic** and have determinant one by Lemma 1.1.15.

We identify $\mathbb{R}^{2n}$ with $\mathbb{C}^n$ in the usual way with the real vector $z = (x, y)$ corresponding to the complex vector $x+iy$ for $x, y \in \mathbb{R}^n$. Then multiplication by the matrix J_0 in (1.1.10) in $\mathbb{R}^{2n}$ corresponds to multiplication by i in $\mathbb{C}^n$. With this identification the complex linear group $\mathrm{GL}(n, \mathbb{C})$ is a subgroup of $\mathrm{GL}(2n, \mathbb{R})$ and $\mathrm{U}(n)$ is a subgroup of $\mathrm{Sp}(2n)$.[6]

The following lemma demonstrates the close connection between symplectic and complex linear maps. It is the first step on the way to proving the important Proposition 2.2.4, which states that the unitary group $\mathrm{U}(n)$ is a maximal compact subgroup of $\mathrm{Sp}(2n)$. As usual, we denote the orthogonal group by $\mathrm{O}(2n)$.

Lemma 2.2.1

$$\mathrm{Sp}(2n) \cap \mathrm{O}(2n) = \mathrm{Sp}(2n) \cap \mathrm{GL}(n, \mathbb{C}) = \mathrm{O}(2n) \cap \mathrm{GL}(n, \mathbb{C}) = \mathrm{U}(n).$$

Proof: A matrix $\Psi \in \mathrm{GL}(2n, \mathbb{R})$ satisfies

$$\begin{aligned} \Psi \in \mathrm{GL}(n, \mathbb{C}) \quad &\Longleftrightarrow \quad \Psi J_0 = J_0 \Psi, \\ \Psi \in \mathrm{Sp}(2n) \quad &\Longleftrightarrow \quad \Psi^{\mathrm{T}} J_0 \Psi = J_0, \\ \Psi \in \mathrm{O}(2n) \quad &\Longleftrightarrow \quad \Psi^{\mathrm{T}} \Psi = \mathbb{1}. \end{aligned}$$

Any two of these conditions imply the third. By Exercise 1.1.13 the subgroup $\mathrm{Sp}(2n) \cap \mathrm{O}(2n)$ consists of those matrices

$$\Psi = \begin{pmatrix} X & -Y \\ Y & X \end{pmatrix} \in \mathrm{GL}(2n, \mathbb{R})$$

which satisfy $X^{\mathrm{T}} Y = Y^{\mathrm{T}} X$ and $X^{\mathrm{T}} X + Y^{\mathrm{T}} Y = \mathbb{1}$. This is precisely the condition on $U := X + iY$ to be unitary. □

Lemma 2.2.2 *Let $\Psi \in \mathrm{Sp}(2n)$. Then*

$$\lambda \in \sigma(\Psi) \quad \Longleftrightarrow \quad \lambda^{-1} \in \sigma(\Psi)$$

and the multiplicities of λ and λ^{-1} agree. If ± 1 is an eigenvalue of Ψ then it occurs with even multiplicity. Moreover,

$$\Psi z = \lambda z, \quad \Psi z' = \lambda' z', \quad \lambda \lambda' \neq 1 \qquad \Longrightarrow \qquad \omega_0(z, z') = 0.$$

Proof: The first statement follows from the fact that Ψ^{T} is similar to Ψ^{-1}:

$$\Psi^{\mathrm{T}} = J_0 \Psi^{-1} J_0^{-1}.$$

Hence the total multiplicity of all eigenvalues not equal to 1 or -1 is even. Since the determinant is the product of all eigenvalues it follows from Lemma 1.1.15

[6] In the present context it is convenient to identify $\mathrm{GL}(n, \mathbb{C})$ and $\mathrm{U}(n)$ with subgroups of $\mathrm{GL}(2n, \mathbb{R})$. However, in other cases it will be important to keep the distinction between an orthogonal symplectic $2n \times 2n$-matrix and the corresponding unitary $n \times n$-matrix.

that if -1 is an eigenvalue then it occurs with even multiplicity. Hence the eigenvalue 1 occurs with even multiplicity as well. The last statement follows from the identity

$$\lambda\lambda'\langle z', J_0 z\rangle = \langle \Psi z', J_0 \Psi z\rangle = \langle z', J_0 z\rangle.$$

This proves Lemma 2.2.2. □

Lemma 2.2.3 *If $P = P^{\mathrm{T}} \in \mathrm{Sp}(2n)$ is a symmetric, positive definite symplectic matrix then $P^\alpha \in \mathrm{Sp}(2n)$ for every real number $\alpha \geq 0$.*

Proof: Every positive definite symmetric matrix has positive eigenvalues and an orthonormal basis of eigenvectors. Hence there is an orthogonal decomposition

$$\mathbb{R}^{2n} = \bigoplus_{i=1}^{k} E_i, \qquad E_i = \ker(\lambda_i \mathbb{1} - P),$$

where $\lambda_i > 0$, $i = 1, \dots, k$, are the eigenvalues of P. Choose two vectors

$$z = \sum_{i=1}^{k} z_i, \qquad w = \sum_{i=1}^{k} w_i$$

in $\mathbb{R}^{2n}$, where $z_i, w_i \in E_i$ for $i = 1, \dots, k$. Since P is a symplectic matrix,

$$\omega_0(z_i, w_j) = \omega_0(Pz_i, Pw_j) = \lambda_i\lambda_j\omega_0(z_i, w_j)$$

for all i, j, and hence either $\lambda_i\lambda_j = 1$ or $\omega_0(z_i, z_j) = 0$. This implies

$$\omega_0(P^\alpha z, P^\alpha w) = \sum_{i,j=1}^{k} (\lambda_i\lambda_j)^\alpha \omega_0(z_i, w_j) = \sum_{i,j=1}^{k} \omega_0(z_i, w_j) = \omega_0(z, w)$$

for every $\alpha \geq 0$. Since this holds for all $z, w \in \mathbb{R}^{2n}$ it follows that P^α is a symplectic matrix for every $\alpha \geq 0$. This proves Lemma 2.2.3. □

Proposition 2.2.4 **(i)** *The unitary group* $\mathrm{U}(n)$ *is a maximal compact subgroup of the symplectic linear group* $\mathrm{Sp}(2n)$.

(ii) *The inclusion of* $\mathrm{U}(n)$ *into* $\mathrm{Sp}(2n)$ *is a homotopy equivalence. In particular,* $\mathrm{Sp}(2n)$ *is connected.*

Proof: We prove (ii). Define the map $f : [0,1] \times \mathrm{Sp}(2n) \to \mathrm{Sp}(2n)$ by

$$f(t, \Psi) := f_t(\Psi) := \Psi(\Psi^T\Psi)^{-t/2}.$$

Since $\Psi^T\Psi$ is a symmetric, positive definite symplectic matrix, so is its inverse and hence, so is $(\Psi^T\Psi)^{-t/2}$, by Lemma 2.2.3. Thus $f_t(\Psi) \in \mathrm{Sp}(2n)$ for every $t \geq 0$ and every $\Psi \in \mathrm{Sp}(2n)$. Moreover, f is continuous, $f_0 = \mathrm{id}$, $f_t|_{\mathrm{U}(n)} = \mathrm{id}$ for every t, and $f_1(\mathrm{Sp}(2n)) = \mathrm{U}(n)$ because $f_1(\Psi)$ is both symplectic and orthogonal. Hence $f_1 : \mathrm{Sp}(2n) \to \mathrm{U}(n)$ is a homotopy inverse of the inclusion of $\mathrm{U}(n)$ into $\mathrm{Sp}(2n)$. This proves (ii).

We prove (i). Suppose G is a compact subgroup of $\mathrm{Sp}(2n)$ that contains $\mathrm{U}(n)$ but is not equal to $\mathrm{U}(n)$. Then there is an element $\Psi \in \mathrm{G} \setminus \mathrm{U}(n)$. We saw above that $f_1(\Psi) = \Psi\left(\Psi^T\Psi\right)^{-1/2} \in \mathrm{U}(n) \subset \mathrm{G}$ and hence the symmetric, positive definite symplectic matrix $P := \left(\Psi^T\Psi\right)^{1/2}$ belongs to $\mathrm{G} \setminus \mathrm{U}(n)$. Hence it has an eigenvalue $\lambda > 1$. Hence the sequence $P, P^2, P^3, \ldots$ in G has no convergent subsequence, in contradiction to our assumption that G is compact. This proves (i) and Proposition 2.2.4. □

Remark 2.2.5 The above argument shows that there is a **symplectic polar decomposition** for elements in $\mathrm{Sp}(2n)$. Namely each $\Psi \in \mathrm{Sp}(2n)$ can be written uniquely as a product

$$\Psi = UP,$$

where

$$U := \Psi\left(\Psi^T\Psi\right)^{-1/2} \in \mathrm{U}(n), \qquad \text{and} \qquad P := \left(\Psi^T\Psi\right)^{1/2}$$

is a symmetric, positive definite, symplectic matrix. □

Proposition 2.2.6 *The fundamental group of* $\mathrm{U}(n)$ *is isomorphic to the integers. The complex determinant map* $\det_{\mathbb{C}} : \mathrm{U}(n) \to S^1$ *induces an isomorphism of fundamental groups.*

Proof: The determinant map $\det_{\mathbb{C}} : \mathrm{U}(n) \to S^1$ is a fibration with fibre $\mathrm{SU}(n)$. Hence the homotopy exact sequence

$$\pi_1(\mathrm{SU}(n)) \to \pi_1(\mathrm{U}(n)) \to \pi_1(S^1) \to \pi_0(\mathrm{SU}(n))$$

shows that $\pi_1(\mathrm{U}(n)) \cong \pi_1(S^1) \cong \mathbb{Z}$.

Here we have used the fact that $\mathrm{SU}(n)$ is simply connected. This is best seen by an induction argument. It obviously holds for $n = 1$. So suppose $n \geq 2$ and consider the map $\mathrm{SU}(n) \to S^{2n-1}$ that sends a matrix $U \in \mathrm{SU}(n)$ to its first column. This is a fibration with fibre $\mathrm{SU}(n-1)$. Hence there is an exact sequence

$$\pi_2(S^{2n-1}) \to \pi_1(\mathrm{SU}(n-1)) \to \pi_1(\mathrm{SU}(n)) \to \pi_1(S^{2n-1})$$

and this shows that if $\mathrm{SU}(n-1)$ is simply connected then so is $\mathrm{SU}(n)$. □

With these results we have obtained a great deal of information about symplectic matrices. The following exercises take these ideas somewhat further. Proposition 2.2.4 is the first sign of the intimate relation between symplectic forms and complex structures that we discuss in detail in Section 2.5. In particular, we show in Proposition 2.5.8 below that every compact subgroup of $\mathrm{Sp}(2n)$ is conjugate to a subgroup of $\mathrm{U}(n)$. More information about the structure of symplectic matrices and its implications for the stability of Hamiltonian flows may be found in the articles by Arnold–Givental [25] and Lalonde–McDuff [395] and the books by Ekeland [170] and Long [433].

Exercises

Exercise 2.2.7 (i) Show that if $\Psi \in \mathrm{Sp}(2n)$ is diagonalizable, it can be diagonalized by a symplectic matrix.
(ii) Deduce from Lemma 2.2.2 that the eigenvalues of $\Psi \in \mathrm{Sp}(2n)$ occur either in pairs $\lambda, 1/\lambda \in \mathbb{R}$, $\lambda, \bar{\lambda} \in S^1$ or in complex quadruplets

$$\lambda, \frac{1}{\lambda}, \bar{\lambda}, \frac{1}{\bar{\lambda}}.$$

(iii) Find all conjugacy classes of matrices in Sp(2) and Sp(4): see [25] and [395]. □

Exercise 2.2.8 Show that the exponential map $\exp : \mathfrak{sp}(2n) \to \mathrm{Sp}(2n)$ is not surjective. **Hint:** Consider the diagonal matrix $\Psi := \mathrm{diag}(-2, -1/2) \in \mathrm{Sp}(2)$. □

Exercise 2.2.9 Use the argument of Proposition 2.2.4 to prove that the inclusion

$$\mathrm{O}(2n)/\mathrm{U}(n) \hookrightarrow \mathrm{GL}(2n,\mathbb{R})/\mathrm{GL}(n,\mathbb{C})$$

of homogeneous spaces is a homotopy equivalence. Prove similarly that the inclusion

$$\mathrm{O}(2n)/\mathrm{U}(n) \hookrightarrow \mathrm{GL}(2n,\mathbb{R})/\mathrm{Sp}(2n)$$

is a homotopy equivalence. We saw in Exercise 2.1.15 that $\mathrm{GL}(2n,\mathbb{R})/\mathrm{Sp}(2n)$ can be identified with the space of symplectic structures on $\mathbb{R}^{2n}$. Similarly, by Proposition 2.5.2, the homogeneous space $\mathrm{GL}(2n,\mathbb{R})/\mathrm{GL}(n,\mathbb{C})$ can be identified with the space of complex structures on $\mathbb{R}^{2n}$. (For explicit constructions of these homotopy equivalences in an intrinsic setting see Proposition 2.5.6 and Remark 2.5.10 below.) □

Exercise 2.2.10 Let $\mathrm{SP}(n,\mathbb{H})$ denote the group of quaternionic matrices $W \in \mathbb{H}^{n\times n}$ such that $W^*W = \mathbb{1}$. Prove that $\mathrm{SP}(n,\mathbb{H})$ is a maximal compact subgroup of the compexified linear symplectic group $\mathrm{Sp}(2n,\mathbb{C})$ and that the quotient $\mathrm{Sp}(2n,\mathbb{C})/\mathrm{SP}(n,\mathbb{H})$ is contractible. □

Exercise 2.2.11 Let

$$\Psi = \begin{pmatrix} X & -Y \\ Y & X \end{pmatrix} \in \mathrm{GL}(2n,\mathbb{R}).$$

What is the relationship between $\det(\Psi) \in \mathbb{R}$ and $\det_{\mathbb{C}}(X + iY) \in \mathbb{C}$? Compare the proof of Theorem 2.2.12. □

The Maslov index for loops of symplectic matrices

It follows from Proposition 2.2.4 and Proposition 2.2.6 that the fundamental group of $\mathrm{Sp}(2n)$ is isomorphic to the integers. An explicit isomorphism from $\pi_1(\mathrm{Sp}(2n))$ to the integers is given by the Maslov index.[7]

[7] **Warning:** In the following we shall use the notation $\Psi : \mathbb{R}/\mathbb{Z} \to \mathrm{Sp}(2n)$ for a loop $\Psi(t) = \Psi(t+1)$ of symplectic matrices. Sometimes we shall also use the same letter to denote an individual symplectic matrix. In each case it should be clear from the context which notation is being used.

Theorem 2.2.12 *There exists a unique function μ, called the* **Maslov index**, *which assigns an integer $\mu(\Psi)$ to every loop*

$$\Psi : \mathbb{R}/\mathbb{Z} \to \mathrm{Sp}(2n)$$

of symplectic matrices and satisfies the following axioms:

(homotopy) *Two loops in $\mathrm{Sp}(2n)$ are homotopic if and only if they have the same Maslov index.*

(product) *For any two loops $\Psi_1, \Psi_2 : \mathbb{R}/\mathbb{Z} \to \mathrm{Sp}(2n)$ we have*

$$\mu(\Psi_1\Psi_2) = \mu(\Psi_1) + \mu(\Psi_2).$$

In particular, the constant loop $\Psi(t) \equiv \mathbb{1}$ has Maslov index 0.

(direct sum) *If $n = n' + n''$ identify $\mathrm{Sp}(2n') \oplus \mathrm{Sp}(2n'')$ in the obvious way with a subgroup of $\mathrm{Sp}(2n)$. Then*

$$\mu(\Psi' \oplus \Psi'') = \mu(\Psi') + \mu(\Psi'').$$

(normalization) *The loop $\Psi : \mathbb{R}/\mathbb{Z} \to \mathrm{Sp}(2)$, defined by*

$$\Psi(t) := \begin{pmatrix} \cos(2\pi t) & -\sin(2\pi t) \\ \sin(2\pi t) & \cos(2\pi t) \end{pmatrix},$$

has Maslov index 1.

Proof: Define the map $\rho : \mathrm{Sp}(2n) \to S^1$ by

$$\rho(\Psi) := \det{}_{\mathbb{C}}(X + iY), \qquad \begin{pmatrix} X & -Y \\ Y & X \end{pmatrix} := \Psi\left(\Psi^{\mathrm{T}}\Psi\right)^{-1/2} \in \mathrm{Sp}(2n) \cap \mathrm{O}(2n),$$

for $\Psi \in \mathrm{Sp}(2n)$. Here Ψ is an individual matrix, not a loop, and the matrix

$$U := \Psi(\Psi^{\mathrm{T}}\Psi)^{-1/2}$$

is the orthogonal part of Ψ in the polar decomposition $\Psi = UP$ in Remark 2.2.5. The **Maslov index** of a loop $\Psi(t) = \Psi(t+1) \in \mathrm{Sp}(2n)$ can be defined as the degree of the composition $\rho \circ \Psi : \mathbb{R}/\mathbb{Z} \to S^1$; thus

$$\mu(\Psi) := \deg(\rho \circ \Psi),$$

or equivalently

$$\mu(\Psi) := \alpha(1) - \alpha(0),$$

where $\alpha : \mathbb{R} \to \mathbb{R}$ is a continuous lift of the loop $\rho \circ \Psi$, i.e. it satisfies

$$e^{2\pi i\,\alpha(t)} = \det{}_{\mathbb{C}}(X(t) + iY(t)), \qquad \begin{pmatrix} X(t) & -Y(t) \\ Y(t) & X(t) \end{pmatrix} := \Psi(t)\left(\Psi(t)^{\mathrm{T}}\Psi(t)\right)^{-1/2},$$

for all $t \in \mathbb{R}$.

With this definition the Maslov index is obviously an integer and depends only on the homotopy class of the loop Ψ. By Propositions 2.2.4 and 2.2.6 the map $\rho : \mathrm{Sp}(2n) \to S^1$ induces an isomorphism of fundamental groups. This proves the homotopy axiom. The product axiom is obvious for loops of unitary matrices. Hence, because every symplectic loop is homotopic to a unitary loop, it follows from the homotopy axiom. The direct sum and normalization axioms are obvious. This proves the existence part of Theorem 2.2.12. The proof of uniqueness is left to the reader. □

We now sketch an alternative interpretation of the Maslov index as the intersection number of a loop in $\mathrm{Sp}(2n)$ with the **Maslov cycle** $\overline{\mathrm{Sp}}_1(2n)$ of all symplectic matrices Ψ which satisfy $\det(B) = 0$ in the decomposition of Exercise 1.1.13. This set is a singular hypersurface of codimension one which admits a natural coorientation (i.e. an orientation of the normal bundle). It is stratified by the rank of the matrix B and a generic loop will intersect each stratum transversely and thus will only intersect the top stratum, where the rank of B is $n-1$. More precisely, let

$$\mathbb{R}/\mathbb{Z} \to \mathrm{Sp}(2n) : t \mapsto \Psi(t) = \begin{pmatrix} A(t) & B(t) \\ C(t) & D(t) \end{pmatrix}$$

be a loop of symplectic matrices. A **crossing** is a number $t \in \mathbb{R}/\mathbb{Z}$ such that $\Psi(t) \in \overline{\mathrm{Sp}}_1(2n)$. A crossing t is called **regular** if the **crossing form**

$$\Gamma(\Psi, t) : \ker B(t) \to \mathbb{R},$$

defined by

$$\Gamma(\Psi, t)(y) = -\langle \dot{B}(t)y, D(t)y \rangle, \qquad y \in \ker B(t),$$

is nonsingular. It is called **simple** if it is regular and $\dim(\ker B(t)) = 1$. At a regular crossing the **crossing index** is the signature (the number of positive minus the number of negative eigenvalues) of the crossing form. By the Kato selection theorem of [542, Theorem 4.28], regular crossings are isolated. The Maslov index of a loop $\Psi(t)$ with only regular crossings can now be defined by

$$\mu(\Psi) := \frac{1}{2} \sum_t \operatorname{sign} \Gamma(\Psi, t),$$

where the sum runs over all crossings. Since the kernel of $B(t)$ can have dimension greater than one at a regular crossing, the crossings on such a loop need not occur on the top stratum of $\overline{\mathrm{Sp}}_1(2n)$. However, again by the Kato selection theorem, there is a nearby path with only simple crossings and with the same sum of the crossing indices. It then follows from standard arguments in differential topology (e.g. [484]) that this number is a homotopy invariant, is therefore well-defined for all loops, and satisfies the first three axioms of Theorem 2.2.12. The reader can check by direct calculation that the last (normalization) axiom also holds. It follows that both our definitions of the Maslov index agree. For more details the reader may consult Robbin and Salamon [541, 542].

2.3 Lagrangian subspaces

In this section we shall discuss Lagrangian subspaces in more detail. We denote by $\mathcal{L}(V, \omega)$ the set of Lagrangian subspaces of (V, ω) and abbreviate

$$\mathcal{L}(n) = \mathcal{L}(\mathbb{R}^{2n}, \omega_0).$$

In more explicit terms Lagrangian subspaces of $\mathbb{R}^{2n}$ are characterized as follows.

Lemma 2.3.1 *Let X and Y be real $n \times n$ matrices and define $\Lambda \subset \mathbb{R}^{2n}$ by*

$$\Lambda = \operatorname{im} Z, \qquad Z = \begin{pmatrix} X \\ Y \end{pmatrix}. \tag{2.3.1}$$

Then $\Lambda \in \mathcal{L}(n)$ if and only if the matrix Z has rank n and

$$X^{\mathrm{T}} Y = Y^{\mathrm{T}} X.$$

In particular, the graph $\Lambda = \{(x, Ax) \,|\, x \in \mathbb{R}^n\}$ of a matrix $A \in \mathbb{R}^{n \times n}$ is Lagrangian if and only if A is symmetric.

Proof: Given two vectors $z = (Xu, Yu)$ and $z' = (Xu', Yu')$ in Λ, we have by (1.1.21) that $\omega_0(z, z') = u^{\mathrm{T}}(X^{\mathrm{T}}Y - Y^{\mathrm{T}}X)u'$. This proves the first assertion. The second assertion is the special case $X = \mathbb{1}$, $Y = A$. □

A matrix $Z \in \mathbb{R}^{2n \times n}$ of the form (2.3.1) which satisfies $X^{\mathrm{T}}Y = Y^{\mathrm{T}}X$ and has rank n is called a **Lagrangian frame**. If Z is a Lagrangian frame then its columns form an orthonormal basis of Λ if and only if the matrix

$$U := X + iY$$

is unitary. In this case Z is called a **unitary Lagrangian frame**. One consequence of Lemma 2.3.1 is that $\mathcal{L}(n)$ is a manifold of dimension $n(n+1)/2$. To see this, note first that the space of symmetric $n \times n$ matrices can be identified with an open neighbourhood in $\mathcal{L}(n)$ of the horizontal Lagrangian

$$\Lambda_{\mathrm{hor}} = \left\{ z = (x, y) \in \mathbb{R}^{2n} \,|\, y = 0 \right\}.$$

Second, use the next lemma to see that any Lagrangian plane can be identified with Λ_{hor} via a linear symplectomorphism.

Lemma 2.3.2 **(i)** *If $\Lambda \in \mathcal{L}(n)$ and $\Psi \in \mathrm{Sp}(2n)$ then $\Psi\Lambda \in \mathcal{L}(n)$.*
(ii) *For any two Lagrangian subspaces $\Lambda, \Lambda' \in \mathcal{L}(n)$ there exists a symplectic matrix $\Psi \in \mathrm{U}(n)$ such that $\Lambda' = \Psi\Lambda$.*
(iii) *There is a natural diffeomorphism $\mathcal{L}(n) \cong \mathrm{U}(n)/\mathrm{O}(n)$.*

Proof: Statement (i) is obvious. To prove (ii) fix a Lagrangian subspace $\Lambda \subset \mathbb{R}^{2n}$ and choose a unitary frame of the form (2.3.1). Define the matrix

$$\Psi := \begin{pmatrix} X & -Y \\ Y & X \end{pmatrix}.$$

Then $\Psi \in \mathrm{Sp}(2n) \cap \mathrm{O}(2n)$ and $\Psi\Lambda_{\mathrm{hor}} = \Lambda$. This proves (ii). Statement (iii) holds because the unitary matrix $U = X + iY \in \mathrm{U}(n)$ determined by a unitary

Lagrangian frame is uniquely determined by Λ up to right multiplication by a matrix in $\mathrm{O}(n)$. □

Exercises

Exercise 2.3.3 Prove that the orthogonal complement of a Lagrangian subspace

$$\Lambda \subset \mathbb{R}^{2n}$$

with respect to the standard metric is $\Lambda^\perp = J_0\Lambda$. Deduce that if $u_1, \ldots, u_n$ is an orthonormal basis of Λ then the vectors $u_1, \ldots, u_n, J_0u_1, \ldots, J_0u_n$ form a basis for $\mathbb{R}^{2n}$ which is both symplectically standard and orthogonal. Relate this to Lemma 2.3.2 and Lemma 2.4.5. □

Exercise 2.3.4 State and prove the analogue of Lemma 2.3.2 for isotropic, symplectic and coisotropic subspaces. □

Exercise 2.3.5 Consider the vertical Lagrangian

$$\Lambda_{\mathrm{vert}} = \left\{ z = (x, y) \in \mathbb{R}^{2n} \,|\, x = 0 \right\}.$$

Use Lemma 2.3.1 to show that $\mathcal{L}(n)$ is the disjoint union

$$\mathcal{L}(n) = \mathcal{L}_0(n) \cup \Sigma(n),$$

where $\mathcal{L}_0(n)$ can be identified with the affine space of symmetric $n \times n$ matrices and $\Sigma(n)$ consists of all Lagrangian subspaces which do not intersect Λ_{vert} transversally. The set $\Sigma(n)$ is called the **Maslov cycle** and is discussed further below. □

Exercise 2.3.6 For $U = X + iY \in \mathrm{U}(n)$ denote by $\Lambda = \Lambda_U \in \mathcal{L}(n)$ the Lagrangian subspace defined by (2.3.1).

(i) Let $U, V \in \mathrm{U}(n)$. Prove that $\Lambda_U = \Lambda_V$ if and only if $UU^T = VV^T$.

(ii) Denote by $\mathcal{S}(n) \subset \mathbb{C}^{n\times n}$ the space of symmetric complex $(n \times n)$-matrices. Prove that $\mathrm{U}(n) \cap \mathcal{S}(n)$ is a submanifold of $\mathrm{U}(n)$ and that the map

$$\mathcal{L}(n) \to \mathrm{U}(n) \cap \mathcal{S}(n) : \Lambda = \Lambda_U \mapsto A_\Lambda := UU^T$$

is a diffeomorphism. **Hint:** For $n = 1$ this is the diffeomorphism

$$\mathcal{L}(1) \cong \mathrm{U}(1)/\{\pm 1\} \to \mathrm{U}(1) : [x + iy] \mapsto x^2 - y^2 + 2ixy.$$

If $A + iB \in \mathrm{U}(n) \cap \mathcal{S}(n)$ then A, B have a common orthonormal basis of eigenvectors. Thus the proof of surjectivity can be reduced to the case $n = 1$.

(iii) Let $U = X + iY \in \mathrm{U}(n)$ and $\Lambda := \Lambda_U \in \mathcal{L}(n)$. Define

$$R_\Lambda := \begin{pmatrix} A_1 & A_2 \\ A_2 & -A_1 \end{pmatrix}, \quad A_1 + iA_2 := A_\Lambda = XX^T - YY^T + i(XY^T + YX^T).$$

Show that

$$R_\Lambda^*\omega_0 = -\omega_0, \qquad R_\Lambda^2 = \mathbb{1}, \qquad \Lambda = \ker(\mathbb{1} - R_\Lambda).$$

Deduce that R_Λ is the unique anti-symplectic involution with fixed point set Λ. □

The Maslov index for loops of Lagrangian subspaces

Lemma 2.3.2 implies that the fundamental group of $\mathcal{L}(n)$ is isomorphic to the integers. An explicit homomorphism $\pi_1(\mathcal{L}(n)) \to \mathbb{Z}$ is given by the Maslov index.[8]

Theorem 2.3.7 *There exists a function μ, called the* **Maslov index**, *which assigns an integer $\mu(\Lambda)$ to every loop $\Lambda : \mathbb{R}/\mathbb{Z} \to \mathcal{L}(n)$ of Lagrangian subspaces and satisfies the following axioms:*

(homotopy) *Two loops in $\mathcal{L}(n)$ are homotopic if and only if they have the same Maslov index.*

(product) *For any two loops $\Lambda : \mathbb{R}/\mathbb{Z} \to \mathcal{L}(n)$ and $\Psi : \mathbb{R}/\mathbb{Z} \to \mathrm{Sp}(2n)$ we have*

$$\mu(\Psi\Lambda) = \mu(\Lambda) + 2\mu(\Psi).$$

(direct sum) *If $n = n' + n''$ identify $\mathcal{L}(n') \oplus \mathcal{L}(n'')$ in the obvious way with a submanifold of $\mathcal{L}(n)$. Then*

$$\mu(\Lambda' \oplus \Lambda'') = \mu(\Lambda') + \mu(\Lambda'').$$

(zero) *A constant loop $\Lambda(t) \equiv \Lambda_0$ has Maslov index zero.*

(normalization) *The loop $\Lambda : \mathbb{R}/\mathbb{Z} \to \mathcal{L}(1)$, defined by*

$$\Lambda(t) := e^{\pi i t}\mathbb{R} \subset \mathbb{C} = \mathbb{R}^2,$$

has Maslov index one.

The Maslov index is uniquely determined by the homotopy, product, direct sum, and zero axioms.

Proof: Define the map $\rho : \mathcal{L}(n) \to S^1$ by[9]

$$\rho(\Lambda) := \det_{\mathbb{C}}(U^2), \qquad \mathrm{im}\begin{pmatrix} X \\ Y \end{pmatrix} = \Lambda, \qquad U := X + iY \in \mathrm{U}(n),$$

for $\Lambda \in \mathcal{L}(n)$. Here Λ is an individual Lagrangian subspace, not a loop, and $\det_{\mathbb{C}}$ is the complex determinant. The **Maslov index** of a loop $\Lambda(t) = \Lambda(t+1) \in \mathcal{L}(n)$ can be defined as the degree of the composition $\rho \circ \Lambda : \mathbb{R}/\mathbb{Z} \to S^1$; thus

$$\mu(\Lambda) := \deg(\rho \circ \Lambda),$$

or equivalently

$$\mu(\Lambda) := \alpha(1) - \alpha(0),$$

where $\alpha : \mathbb{R} \to \mathbb{R}$ is a continuous lift of $\rho \circ \Lambda$, i.e. it satisfies

$$e^{\pi i\,\alpha(t)} = \det_{\mathbb{C}}(X(t) + iY(t)), \quad \mathrm{im}\begin{pmatrix} X(t) \\ Y(t) \end{pmatrix} = \Lambda(t), \quad X(t) + iY(t) \in \mathrm{U}(n),$$

for all $t \in \mathbb{R}$.

[8] **Warning:** As before we shall use the notation $\Lambda : \mathbb{R}/\mathbb{Z} \to \mathrm{Sp}(2n)$ for a loop $\Lambda(t) = \Lambda(t+1)$ of Lagrangian subspaces. Sometimes we shall also use the same letter Λ to denote an individual Lagrangian subspace. Again it should be clear from the context which notation is being used.

[9] Note the square in this formula. It is needed because we consider unoriented Lagrangian subspaces. Compare with the proof of Theorem 2.2.12 and Exercise 2.2.11.

With this definition the Maslov index is obviously an integer and depends only on the homotopy class of the loop Λ. Conversely, assume that $\Lambda_0(t) = \Lambda_0(t+1)$ and $\Lambda_1(t) = \Lambda_1(t+1)$ are loops of Lagrangian subspaces with the same Maslov index $\mu(\Lambda_0) = \mu(\Lambda_1)$. By Lemma 2.3.2 we may assume without loss of generality that $\Lambda_j(0) = \Lambda_j(1) = \mathbb{R}^n \times \{0\}$. Choose lifts $U_j(t) = X_j(t) + iY_j(t) \in \mathrm{U}(n)$ as above such that $U_0(0) = U_1(0) = \mathbb{1}$. If necessary, we may alter $U_j(t)$ by right multiplication with a path of orthogonal matrices to obtain

$$U_0(1) = U_1(1) = \begin{pmatrix} \pm 1 & 0 & \cdots & 0 \\ 0 & 1 & \ddots & \vdots \\ \vdots & \ddots & \ddots & 0 \\ 0 & \cdots & 0 & 1 \end{pmatrix}.$$

Hence the unitary matrices $U(t) := U_1(t)U_0(t)^{-1}$ form a loop. Moreover, the loop $\det_{\mathbb{C}}(U) : S^1 \to S^1$ is contractible because $\mu(\Lambda_0) = \mu(\Lambda_1)$. By Proposition 2.2.6, the path U_0 is homotopic to U_1 with fixed endpoints, and hence Λ_0 is homotopic to Λ_1.

Thus we have proved that our Maslov index, as defined above, satisfies the homotopy axiom. That it also satisfies the product, direct sum, zero, and normalization axioms is obvious. This proves the existence statement of Theorem 2.3.7.

We prove that the normalization axiom follows from the direct sum, product, and zero axioms. The loop $\Lambda(t) := e^{\pi i t}\mathbb{R} \subset \mathbb{C}$ satisfies

$$\Lambda(t) \oplus \Lambda(t) = \Psi(t)\Lambda_0,$$

where $\Lambda_0 = \mathbb{R}^2 \subset \mathbb{C}^2$ is the constant loop and $\Psi : \mathbb{R}/\mathbb{Z} \to \mathrm{U}(2)$ is the loop of unitary matrices given by

$$\Psi(t) := e^{\pi i t}X(t) := e^{\pi i t}\begin{pmatrix} \cos(\pi t) & -\sin(\pi t) \\ \sin(\pi t) & \cos(\pi t) \end{pmatrix}, \qquad 0 \le t \le 1.$$

Note here that path of real matrices $X(t)$ fixes Λ_0 and has constant determinant $\det_{\mathbb{C}}(X(t)) = 1$, so that the loop Ψ has Maslov index $\mu(\Psi) = 1$. Hence it follows from the direct sum, product, and zero axioms that

$$2\mu(\Lambda) = \mu(\Lambda \oplus \Lambda) = \mu(\Lambda_0) + 2\mu(\Psi) = 2.$$

Hence $\mu(\Lambda) = 1$ as claimed. The proof that all five axioms uniquely determine the Maslov index is left to the reader. □

Alternatively the Maslov index can be defined as the intersection number of the loop $\Lambda(t)$ with the **Maslov cycle** $\Sigma(n)$ of all Lagrangian subspaces Λ which intersect the vertical $\{0\} \times \mathbb{R}^n$ nontransversally. This set is a singular hypersurface of $\mathcal{L}(n)$ of codimension 1 which admits a natural coorientation. It is stratified by the dimension of the intersection $\Lambda \cap \Lambda_{\mathrm{vert}}$. A generic loop will intersect only the highest stratum (where the intersection is 1-dimensional) and all the intersections will be transverse. More explicitly, let $\Lambda(t)$ be a path of Lagrangian planes represented by a lift $X(t)+iY(t) \in \mathrm{U}(n)$ of unitary Lagrangian frames. A **crossing** is a number t such that $\det(X(t)) = 0$. A crossing t is called **regular** if the **crossing form** $\Gamma(\Lambda, t) : \ker X(t) \to \mathbb{R}$, defined by

$$\Gamma(\Lambda, t)(u) = -\langle \dot{X}(t)u, Y(t)u \rangle,$$

is nonsingular. (See Fig. 2.1.)

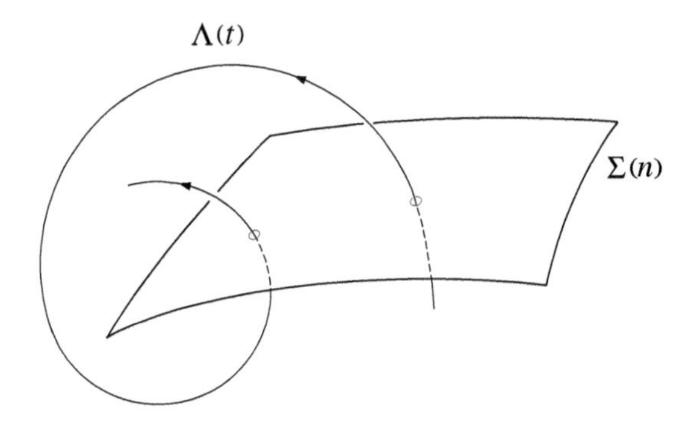

FIG. 2.1. The Maslov cycle.

At a regular crossing the **crossing index** is the signature (the number of positive minus the number of negative eigenvalues) of the crossing form. The Maslov index of a loop $\Lambda(t) = \Lambda(t+1)$ with only regular crossings can be defined by

$$\mu(\Lambda) = \sum_t \operatorname{sign} \Gamma(\Lambda, t),$$

where the sum runs over all crossings. As in the case of symplectic matrices this definition satisfies the axioms of Theorem 2.3.7. For more details see Robbin–Salamon [541].

Exercise 2.3.8 The Maslov index of a loop $\Lambda : \mathbb{R}/\mathbb{Z} \to \mathcal{L}(V, \omega)$ of Lagrangian subspaces in a general symplectic vector space is defined as the Maslov index of the loop $t \mapsto \Psi^{-1}\Lambda(t) \in \mathcal{L}(n)$, where $\Psi : (\mathbb{R}^{2n}, \omega_0) \to (V, \omega)$ is a linear symplectomorphism. Show that this definition is independent of Ψ. Show that if one reverses the sign of ω the sign of the Maslov index reverses also. □

Exercise 2.3.9 Let $\Psi : \mathbb{R}/\mathbb{Z} \to \mathrm{Sp}(V, \omega)$ be a loop of linear symplectomorphisms. Prove that the corresponding loop

$$\Gamma_\Psi : \mathbb{R}/\mathbb{Z} \to \mathcal{L}(V \times V, (-\omega) \times \omega)$$

of Lagrangian graphs has twice the Maslov index

$$\mu(\Gamma_\Psi) = 2\mu(\Psi).$$

(Γ_Ψ is defined in Exercise 2.1.2.) □

2.4 The affine nonsqueezing theorem

An **affine symplectomorphism** of $\mathbb{R}^{2n}$ is a map $\psi : \mathbb{R}^{2n} \to \mathbb{R}^{2n}$ of the form

$$\psi(z) = \Psi z + z_0,$$

where $\Psi \in \mathrm{Sp}(2n)$ and $z_0 \in \mathbb{R}^{2n}$. We denote by $\mathrm{ASp}(2n)$ the group of affine symplectomorphisms. The affine nonsqueezing theorem asserts that a ball in $\mathbb{R}^{2n}$ can only be embedded into a symplectic cylinder by an affine symplectomorphism if the ball has smaller radius than the cylinder. Let $e_1, \dots, e_n, f_1, \dots, f_n$ be the standard basis of $\mathbb{R}^{2n}$. As in Section 1.2, we denote the symplectic cylinder of radius $R > 0$ by

$$Z^{2n}(R) := B^2(R) \times \mathbb{R}^{2n-2} = \left\{ z \in \mathbb{R}^{2n} \,|\, \langle z, e_1 \rangle^2 + \langle z, f_1 \rangle^2 \le R^2 \right\} \subset \mathbb{R}^{2n}.$$

Here the splitting $\mathbb{R}^{2n} = \mathbb{R}^2 \times \mathbb{R}^{2n-2}$ is a symplectic one, i.e. the ball $B^2(R)$ corresponds to the coordinates (x_1, y_1).

Theorem 2.4.1 (Affine nonsqueezing) *Let $\psi \in \mathrm{ASp}(2n)$ be an affine symplectomorphism and assume that $\psi(B^{2n}(r)) \subset Z^{2n}(R)$. Then $r \le R$.*

Proof: Assume without loss of generality that $r = 1$ and write $\psi(z) = \Psi z + z_0$. Define

$$u := \Psi^{\mathrm{T}} e_1, \quad v := \Psi^{\mathrm{T}} f_1, \quad a := \langle e_1, z_0 \rangle, \quad b := \langle f_1, z_0 \rangle.$$

Then, since $\Psi, \Psi^{\mathrm{T}} \in \mathrm{Sp}(2n)$, we have $\omega_0(u, v) = 1$ and hence the condition $\psi(B^{2n}(1)) \subset Z^{2n}(R)$ can be restated in the form

$$\sup_{|z|=1} \left(\langle e_1, \psi(z) \rangle^2 + \langle e_2, \psi(z) \rangle^2 \right) = \sup_{|z|=1} \left((\langle u, z \rangle + a)^2 + (\langle v, z \rangle + b)^2 \right) \le R^2.$$

But $1 = \omega_0(u, v) \le |u| \cdot |v|$ and hence either u or v has length greater than or equal to 1. Assume without loss of generality that $|u| \ge 1$ and choose $z := \pm u/|u|$ (with sign depending on a) to obtain $R \ge 1$. This proves Theorem 2.4.1. □

The nonsqueezing property can be formulated in a symplectically invariant way. A set $B \subset \mathbb{R}^{2n}$ is called a **linear** (resp. **affine**) **symplectic ball** of **radius** r if it is linearly (resp. affinely) symplectomorphic to $B^{2n}(r)$. Similarly, a subset $Z \subset \mathbb{R}^{2n}$ is called a **linear** (resp. **affine**) **symplectic cylinder** if there exists an element $\Psi \in \mathrm{Sp}(2n)$ (resp. $\Psi \in \mathrm{ASp}(2n)$) and a number $R > 0$ such that $Z = \Psi Z^{2n}(R)$. It follows from Theorem 2.4.1 that for any such set Z the number $R > 0$ is a linear symplectic invariant. It is called the **radius** of Z. A matrix $\Psi \in \mathbb{R}^{2n \times 2n}$ is said to have the **linear nonsqueezing property** if, for every linear symplectic ball B of radius r and every linear symplectic cylinder Z of radius R, we have

$$\Psi B \subset Z \quad \Longrightarrow \quad r \le R.$$

The following theorem shows that linear symplectomorphisms are characterized by the nonsqueezing property. More precisely, we must also include the case of **anti-symplectic** matrices Ψ which satisfy $\Psi^* \omega_0 = -\omega_0$.

Theorem 2.4.2 (Affine rigidity) *Let $\Psi \in \mathbb{R}^{2n\times 2n}$ be a nonsingular matrix such that Ψ and Ψ^{-1} have the linear nonsqueezing property. Then Ψ is either symplectic or anti-symplectic.*

Proof: Assume that Ψ is neither symplectic nor anti-symplectic. Then neither is Ψ^{T} and so there exist vectors $u, v \in \mathbb{R}^{2n}$ such that

$$\omega_0(\Psi^{\mathrm{T}}u, \Psi^{\mathrm{T}}v) \neq \pm\omega_0(u, v).$$

Perturbing u and v slightly, and using the fact that Ψ is nonsingular, we may assume that $\omega_0(u, v) \neq 0$ and $\omega_0(\Psi^{\mathrm{T}}u, \Psi^{\mathrm{T}}v) \neq 0$. Moreover, replacing Ψ by Ψ^{-1} if necessary, we may assume that $|\omega_0(\Psi^{\mathrm{T}}u, \Psi^{\mathrm{T}}v)| < |\omega_0(u, v)|$. Now, by rescaling u if necessary, we obtain

$$0 < \lambda^2 = |\omega_0(\Psi^{\mathrm{T}}u, \Psi^{\mathrm{T}}v)| < \omega_0(u, v) = 1.$$

Hence there exist symplectic bases $u_1, v_1 \ldots, u_n, v_n$ and $u_1', v_1' \ldots, u_n', v_n'$ of $\mathbb{R}^{2n}$ such that

$$u_1 = u, \qquad v_1 = v, \qquad u_1' = \lambda^{-1}\Psi^{\mathrm{T}}u, \qquad v_1' = \pm\lambda^{-1}\Psi^{\mathrm{T}}v.$$

Denote by $\Phi \in \mathrm{Sp}(2n)$ the matrix which maps the standard basis $e_1, \ldots, f_n$ to $u_1, \ldots, v_n$ and by $\Phi' \in \mathrm{Sp}(2n)$ the matrix which maps $e_1, \ldots, f_n$ to $u_1', \ldots, v_n'$. Then the matrix $A := {\Phi'}^{-1}\Psi^{\mathrm{T}}\Phi$ satisfies

$$Ae_1 = \lambda e_1, \qquad Af_1 = \pm\lambda f_1.$$

This implies that the transposed matrix A^{T} maps the unit ball $B^{2n}(1)$ to the cylinder $Z^{2n}(\lambda)$. Since $\lambda < 1$ this means that Ψ does not have the nonsqueezing property in contradiction to our assumption. This proves Theorem 2.4.2. □

The affine nonsqueezing theorem gives rise to the notion of the **linear symplectic width** of an arbitrary subset $A \subset \mathbb{R}^{2n}$, defined by

$$w_L(A) = \sup\left\{\pi r^2 \,|\, \psi(B^{2n}(r)) \subset A \ \text{ for some } \ \psi \in \mathrm{ASp}(\mathbb{R}^{2n})\right\}.$$

It follows from Theorem 2.4.1 that the linear symplectic width has the following properties:

(monotonicity) If $\psi \in \mathrm{ASp}(2n)$ and $\psi(A) \subset B$ then $w_L(A) \leq w_L(B)$;
(conformality) $w_L(\lambda A) = \lambda^2 w_L(A)$;
(nontriviality) $w_L(B^{2n}(r)) = w_L(Z^{2n}(r)) = \pi r^2$.

The nontriviality axiom implies that w_L is a two-dimensional invariant. It is obvious from the monotonicity property that affine symplectomorphisms preserve the linear symplectic width. We shall prove that this property in fact characterizes symplectic and anti-symplectic linear maps.

Exercise 2.4.3 Prove that every anti-symplectic linear map has determinant $(-1)^n$. Prove that every anti-symplectic linear map preserves the linear symplectic width of subsets of $\mathbb{R}^{2n}$. **Hint:** The linear symplectic width $w_L(A)$ equals the maximal radius of a ball which can be mapped into A by an anti-symplectic affine transformation. □

Although the nonsqueezing property refers only to the images of balls, it is natural to state the following property in terms of ellipsoids centred at zero, since this class of subsets is invariant under linear isomorphisms. Recall that an ellipsoid $E \subset \mathbb{R}^{2n}$, centred at zero, is a compact set of the form $\{z \in \mathbb{R}^{2n} \,|\, Q(z) \leq 1\}$, where Q is any positive definite quadratic form.

Theorem 2.4.4 *Let $\Psi : \mathbb{R}^{2n} \to \mathbb{R}^{2n}$ be a linear map. Then the following are equivalent.*
(i) *Ψ preserves the linear symplectic width of ellipsoids centred at zero.*
(ii) *The matrix Ψ is either symplectic or anti-symplectic, i.e. $\Psi^*\omega_0 = \pm\omega_0$.*

Proof: It follows from Theorem 2.4.1 that (ii) implies (i). Hence assume (i). We prove that Ψ has the nonsqueezing property. To see this let B be a linear symplectic ball of radius r and Z be a linear symplectic cylinder of radius R such that

$$\Psi B \subset Z.$$

Then $w_L(B) = w_L(\Psi B)$ by hypothesis, so that by the monotonicity property of the linear symplectic width, we have

$$\pi r^2 = w_L(B) = w_L(\Psi B) \leq w_L(Z) = \pi R^2,$$

and hence $r \leq R$. It also follows from (i) that Ψ must be nonsingular because otherwise the image of the unit ball under Ψ would have linear symplectic width zero. Moreover, Ψ^{-1} also satisfies (i) because $w_L(\Psi^{-1}E) = w_L(\Psi\Psi^{-1}E) = w_L(E)$ for every ellipsoid E which is centred at zero. Thus we have proved that both Ψ and Ψ^{-1} have the nonsqueezing property, and in view of Theorem 2.4.2 this implies that Ψ is either symplectic or anti-symplectic. This proves Theorem 2.4.4. □

We shall now study in more detail the action of the symplectic linear group on ellipsoids. The main tools for understanding this action are the affine nonsqueezing theorem and the following lemma about the simultaneous normalization of a symplectic form and an inner product. Note that this lemma provides an alternative proof of Theorem 2.1.3.

Lemma 2.4.5 *Let (V, ω) be a symplectic vector space and $g : V \times V \to \mathbb{R}$ be an inner product. Then there exists a basis $u_1, \dots, u_n, v_1, \dots, v_n$ of V which is both g-orthogonal and ω-standard. Moreover, this basis can be chosen such that*

$$g(u_j, u_j) = g(v_j, v_j)$$

for all j.

Proof: Consider the vector space $V = \mathbb{R}^{2n}$ with the standard inner product $g = \langle \cdot, \cdot \rangle$ and assume that

$$\omega(z, w) = \langle z, Aw \rangle$$

is a nondegenerate skew form. Then A is nondegenerate and $A^{\mathrm{T}} = -A$. Hence $iA \in \mathbb{C}^{2n \times 2n}$ is a Hermitian matrix, and so the spectrum of A is purely imaginary and there exists an orthonormal basis of eigenvectors. Because $\overline{iA} = -iA$, the eigenvalues occur in pairs $\pm i\alpha_j$ for $j = 1, \ldots, n$ with $\alpha_j > 0$. Choose eigenvectors $z_j = u_j + iv_j \in \mathbb{C}^{2n}$ such that

$$Az_j = i\alpha_j z_j, \qquad \bar{z}_j^{\mathrm{T}} z_k = \delta_{jk}.$$

Then we have $A\bar{z}_j = -i\alpha_j \bar{z}_j$, and by the orthogonality of the eigenvectors $\bar{z}_j$ to the z_k, we have

$$z_j^{\mathrm{T}} z_k = 0.$$

With $z_j = u_j + iv_j$ we obtain

$$Au_j = -\alpha_j v_j, \qquad Av_j = \alpha_j u_j$$

and

$$u_j^{\mathrm{T}} v_k = u_j^{\mathrm{T}} u_k = v_j^{\mathrm{T}} v_k = 0, \qquad j \neq k.$$

This implies

$$\omega(u_j, v_j) = u_j^{\mathrm{T}} A v_j = \alpha_j |u_j|^2 > 0$$

and similarly $\omega(u_j, v_k) = \omega(u_j, u_k) = \omega(v_j, v_k) = 0$ for $j \neq k$. Rescaling the vectors u_j and v_j, if necessary, we obtain the required symplectic orthogonal basis of $\mathbb{R}^{2n}$. This proves Lemma 2.4.5. □

The next lemma is a geometric interpretation of this result. Given an n-tuple $r = (r_1, \ldots, r_n)$ with $0 < r_1 \leq \cdots \leq r_n$, consider the closed ellipsoid

$$E(r) = \left\{ z \in \mathbb{C}^n \,\middle|\, \sum_{i=1}^{n} \left| \frac{z_i}{r_i} \right|^2 \leq 1 \right\}.$$

Lemma 2.4.6 *Given any compact ellipsoid*

$$E = \left\{ w \in \mathbb{R}^{2n} \,\middle|\, \sum_{i,j=1}^{2n} a_{ij} w_i w_j \leq 1 \right\},$$

there is a symplectic linear transformation $\Psi \in \mathrm{Sp}(2n)$ such that $\Psi E = E(r)$ for some n-tuple $r = (r_1, \ldots, r_n)$ with $0 < r_1 \leq \cdots \leq r_n$. Moreover, the numbers r_j are uniquely determined by E.

Proof: Consider the inner product

$$g(v, w) = \sum_{i,j=1}^{2n} a_{ij} v_i w_j$$

on $\mathbb{R}^{2n}$. Then the ellipsoid E is given by

$$E = \{w \in \mathbb{R}^{2n} \mid g(w, w) \leq 1\}.$$

By Lemma 2.4.5 there is a basis $u_1, \ldots, u_n, v_1, \ldots, v_n$ of $\mathbb{R}^{2n}$ which is both symplectically standard and orthogonal for g. Moreover, we may assume that

$$g(u_j, u_j) = g(v_j, v_j) = \frac{1}{{r_j}^2}.$$

Let $\Psi : \mathbb{R}^{2n} \to \mathbb{R}^{2n}$ be the symplectic linear transformation which takes the standard basis of $\mathbb{R}^{2n}$ to this new basis, i.e.

$$\Psi z = \sum_{j=1}^{n} (x_j u_j + y_j v_j)$$

for $z = (x_1, \ldots, x_n, y_1, \ldots, y_n)$. Then

$$g(\Psi z, \Psi z) = \sum_{j=1}^{n} \frac{{x_j}^2 + {y_j}^2}{{r_j}^2}$$

and hence $\Psi^{-1} E = E(r)$.

To prove uniqueness of the n-tuple $r_1 \leq \cdots \leq r_n$, consider the diagonal matrix

$$\Delta(r) = \mathrm{diag}(1/r_1^2, \ldots, 1/r_n^2, 1/r_1^2, \ldots, 1/r_n^2).$$

We must show that if there is a symplectic matrix Ψ such that

$$\Psi^{\mathrm{T}} \Delta(r) \Psi = \Delta(r'),$$

then $r = r'$. Since $J_0 \Psi^{\mathrm{T}} = \Psi^{-1} J_0$, the above identity is equivalent to

$$\Psi^{-1} J_0 \Delta(r) \Psi = J_0 \Delta(r').$$

Hence $J_0 \Delta(r)$ and $J_0 \Delta(r')$ must have the same eigenvalues. But it is easy to check that the eigenvalues of $J_0 \Delta(r)$ are $\pm i/{r_1}^2, \ldots, \pm i/{r_n}^2$, and this proves Lemma 2.4.6. □

Remark 2.4.7 In the case $n = 1$ the existence statement of Lemma 2.4.6 asserts that every ellipse in $\mathbb{R}^2$ can be mapped into a circle by an area-preserving linear transformation. □

In view of Lemma 2.4.6 we define the **symplectic spectrum** of an ellipsoid E to be the unique n-tuple $r = (r_1, \ldots, r_n)$ with $0 < r_1 \leq \cdots \leq r_n$ such that E is linearly symplectomorphic to $E(r) = E(r_1, \ldots, r_n)$. The spectrum is invariant under linear symplectic transformations and, in fact, two ellipsoids in $\mathbb{R}^{2n}$, which are centred at 0, are linearly symplectomorphic if and only if they have the same spectrum. Moreover, the volume of an ellipsoid $E \subset \mathbb{R}^{2n}$ is given by

$$\operatorname{Vol} E = \int_E \frac{\omega_0^n}{n!} = \frac{\pi^n}{n!} \prod_{j=1}^n r_j^2.$$

Ellipsoids with spectrum $(r, \ldots, r)$ are linear symplectic balls. The following theorem characterizes the linear symplectic width of an ellipsoid in terms of the spectrum.

Theorem 2.4.8 *Let $E \subset \mathbb{R}^{2n}$ be an ellipsoid centred at 0. Then*

$$w_L(E) = \sup_{B \subset E} w_L(B) = \inf_{Z \supset E} w_L(Z),$$

where the supremum runs over all affine symplectic balls contained in E and the infimum runs over all affine symplectic cylinders containing E. If E has symplectic spectrum $0 < r_1 \leq \cdots \leq r_n$, then

$$w_L(E) = \pi {r_1}^2.$$

Proof: Assume that E has symplectic spectrum $0 < r_1 \leq r_2 \leq \cdots \leq r_n$. Then there exists a symplectic matrix $\Psi \in \operatorname{Sp}(2n)$ such that

$$\Psi E = E(r_1, \ldots, r_n).$$

Hence

$$\Psi^{-1} B^{2n}(r_1) \subset E \subset \Psi^{-1} Z^{2n}(r_1)$$

and so

$$\inf_{Z \supset E} w_L(Z) \leq \pi {r_1}^2 \leq \sup_{B \subset E} w_L(B).$$

Now suppose that B is an affine symplectic ball of radius r contained in E. Then $\Psi B \subset \Psi E \subset Z^{2n}(r_1)$ and so $r \leq r_1$ by Theorem 2.4.1. Similarly, if $Z = \psi(Z^{2n}(R))$ is an affine symplectic cylinder containing E, then the inclusion

$$B^{2n}(r_1) \subset \Psi E \subset \Psi Z = \Psi\big(\psi(Z^{2n}(R))\big)$$

implies that the cylinder $Z^{2n}(R)$ contains an affine ball of radius r_1 so that $r_1 \leq R$. Hence

$$\sup_{B \subset E} w_L(B) \leq \pi {r_1}^2 \leq \inf_{Z \supset E} w_L(Z).$$

Since $w_L(E) = \sup_{B \subset E} w_L(B)$ this proves Theorem 2.4.8. □

Exercise 2.4.9 Let $E \subset \mathbb{R}^{2n}$ be an ellipsoid and define the dual ellipsoid by

$$E^* = \{v \in \mathbb{R}^{2n} \mid \langle v, e\rangle \leq 1 \; \forall e \in E\},$$

where $\langle \cdot, \cdot \rangle$ is the standard inner product on $\mathbb{R}^{2n}$. Prove that

$$E^{**} = E, \qquad (\Psi E)^* = (\Psi^{\mathrm{T}})^{-1} E^*,$$

for $\Psi \in \mathrm{Sp}(2n)$. Prove that the symplectic spectrum of E^* is given by $(1/r_n, \ldots, 1/r_1)$, where $(r_1, \ldots, r_n)$ is the symplectic spectrum of E. Deduce that the dual of a linear symplectic ball is again a linear symplectic ball. □

This discussion of symplectic capacities and nonsqueezing is continued in Chapter 12, while open problems concerning invariants of ellipsoids are discussed further in Section 14.8.

2.5 Linear complex structures

A **(linear) complex structure** on a real vector space V is an automorphism

$$J : V \to V$$

such that

$$J^2 = -\mathbb{1}.$$

With such a structure V becomes a complex vector space with multiplication by $i = \sqrt{-1}$ corresponding to J. Thus the scalar multiplication is given by

$$\mathbb{C} \times V \to V : (s + it, v) \mapsto sv + tJv.$$

In particular, V is necessarily of even dimension over the reals. We denote the space of linear complex structures on V by $\mathcal{J}(V)$. The basic example is the automorphism of $\mathbb{R}^{2n}$ induced by the matrix

$$J_0 := \begin{pmatrix} 0 & -\mathbb{1} \\ \mathbb{1} & 0 \end{pmatrix}.$$

If we identify $\mathbb{R}^{2n}$ with $\mathbb{C}^n$ via the isomorphism $(x, y) \mapsto x + iy$ for $x, y \in \mathbb{R}^n$, then the matrix J_0 corresponds to multiplication by i.

In this section we first discuss the properties of the set of complex structures on a vector space and then discuss complex structures that are compatible with a symplectic form. Finally we consider the larger set of complex structures that are tamed by a symplectic form. The close interconnection between symplectic and complex geometry is shown by the power of the ideas of mirror symmetry. Also the intriguing notion of a **generalized complex structure**, initially proposed by Hitchin, interpolates between complex and symplectic structures. The paper by Gualtieri [291] develops both its linear and nonlinear aspects.

The next proposition shows that every linear complex structure is isomorphic to the standard complex structure J_0. We give three proofs, highlighting different approaches that are developed later.

Proposition 2.5.1 *Let V be a $2n$-dimensional real vector space and let J be a linear complex structure on V. Then there exists a vector space isomorphism $\Phi : \mathbb{R}^{2n} \to V$ such that $J\Phi = \Phi J_0$.*

Proof 1: Every complex vector space has a complex basis. For (V, J) this means that there exist vectors $v_1, \ldots, v_n \in V$ such that the vectors

$$v_1, Jv_1, \ldots, v_n, Jv_n$$

form a real basis of V. The required transformation $\Phi : \mathbb{R}^{2n} \to V$ is given by

$$\Phi\zeta = \sum_{j=1}^{n} (\xi_j v_j + \eta_j J v_j) \tag{2.5.1}$$

for $\zeta = (\xi_1, \ldots, \xi_n, \eta_1, \ldots, \eta_n)$. □

Proof 2: Let V^{c} denote the complexification of V and denote by

$$E^{\pm} := \ker(\mathbb{1} \pm iJ) = \operatorname{im}(\mathbb{1} \mp iJ)$$

the eigenspaces of J. Then $V^{\mathrm{c}} = E^+ \oplus E^-$ and hence $\dim E^{\pm} = n$. Choose a basis $v_j - iw_j$, $j = 1, \ldots, n$, of E^+. Then the vectors $v_1, \ldots, v_n, w_1, \ldots, w_n$ form a basis of V and

$$Jv_j = w_j, \qquad Jw_j = -v_j.$$

Hence the linear map $\Phi : \mathbb{R}^{2n} \to V$ in (2.5.1) satisfies $\Phi J_0 = J\Phi$. □

Proof 3: There exists an inner product $g : V \times V \to \mathbb{R}$ such that $J^*g = g$. (Choose any inner product h on V and define $g := \frac{1}{2}(h + J^*h)$.) If $n > 0$ choose a unit vector v_1. Then the unit vectors v_1, Jv_1 are orthogonal. If $n > 1$, choose a unit vector v_2, orthogonal to v_1, Jv_1. Then the unit vectors v_1, Jv_1, v_2, Jv_2 are pairwise orthogonal and hence linearly independent. Continue by induction to obtain a g-orthonormal basis of the form $v_1, Jv_1, \ldots, v_n, Jv_n$, and define the map $\Phi : \mathbb{R}^{2n} \to V$ by (2.5.1). □

Proposition 2.5.2 (i) *The space $\mathcal{J}(\mathbb{R}^{2n})$ is diffeomorphic to the homogeneous space $\mathrm{GL}(2n, \mathbb{R})/\mathrm{GL}(n, \mathbb{C})$, and so has two connected components.*

(ii) *The connected component $\mathcal{J}^+(\mathbb{R}^{2n})$ containing J_0 is the space of all complex structures on $\mathbb{R}^{2n}$ that are compatible with the standard orientation. It is diffeomorphic to the homogeneous space $\mathrm{GL}^+(2n, \mathbb{R})/\mathrm{GL}(n, \mathbb{C})$ and is homotopy equivalent to the submanifold*

$$\mathcal{J}^+(\mathbb{R}^{2n}) \cap \mathrm{SO}(2n) \cong \mathrm{SO}(2n)/\mathrm{U}(n)$$

of orthogonal complex structures that are compatible with the standard orientation.

(iii) *$\mathcal{J}^+(\mathbb{R}^4)$ is homotopy equivalent to S^2.*

Proof: Consider the map

$$\mathrm{GL}(2n,\mathbb{R}) \to \mathcal{J}(\mathbb{R}^{2n}) : \Psi \mapsto \Psi J_0 \Psi^{-1}.$$

By Proposition 2.5.1 this map is surjective and it descends to the quotient space $\mathrm{GL}(2n,\mathbb{R})/\mathrm{GL}(n,\mathbb{C})$. (Here we identify $\mathrm{GL}(n,\mathbb{C})$ with a subgroup of $\mathrm{GL}(2n,\mathbb{R})$ as in Section 2.2.) Thus we have identified $\mathcal{J}(\mathbb{R}^{2n})$ with the homogeneous space $\mathrm{GL}(2n,\mathbb{R})/\mathrm{GL}(n,\mathbb{C})$. This space has two connected components distinguished by the determinant. This proves (i).

To prove (ii), note that by Exercise 2.2.9, the connected component

$$\mathrm{GL}^+(2n,\mathbb{R})/\mathrm{GL}(n,\mathbb{C}) \cong \mathcal{J}^+(\mathbb{R}^{2n})$$

is homotopy equivalent to the quotient $\mathrm{SO}(2n)/\mathrm{U}(n) \cong \mathcal{J}^+(\mathbb{R}^{2n}) \cap \mathrm{SO}(2n)$. Likewise, $\mathcal{J}(\mathbb{R}^{2n})$ is homotopy equivalent to $\mathcal{J}(\mathbb{R}^{2n}) \cap \mathrm{O}(2n)$. (For an explicit construction of these homotopy equivalences see Proposition 2.5.6 below.)

Now assume $n = 2$. Denote a vector in $\mathbb{R}^4$ by $x = (x_0, x_1, x_2, x_3)$ and let e_0, e_1, e_2, e_3 be the standard basis of $\mathbb{R}^4$. We claim that a matrix

$$J \in \mathcal{J}(\mathbb{R}^4) \cap \mathrm{SO}(4)$$

is completely determined by the unit vector

$$Je_0 \in \left\{ x = (x_0, x_1, x_2, x_3) \in \mathbb{R}^4 \,|\, x_0 = 0 \right\}.$$

To see this note that the vectors e_0, Je_0 form an orthonormal basis of a 2-plane $E \subset \mathbb{R}^4$. The matrix J is then completely determined by the fact that for any unit vector $v \in E^\perp$ the vectors v, Jv form an orthonormal basis of $E^\perp$, which is oriented so that the vectors e_0, Je_0, v, Jv form a positively oriented basis for $\mathbb{R}^4$. This proves Proposition 2.5.2. □

Compatible complex structures

Let (V, ω) be a symplectic vector space. A complex structure $J \in \mathcal{J}(V)$ is said to be **compatible with** ω if

$$\omega(Jv, Jw) = \omega(v, w) \tag{2.5.2}$$

for all $v, w \in V$ and

$$\omega(v, Jv) > 0 \tag{2.5.3}$$

for all nonzero $v \in V$. If J is a compatible complex structure then

$$g_J(v, w) := \omega(v, Jw) \tag{2.5.4}$$

defines an inner product on V and J is skew-adjoint with respect to g_J, i.e. $g_J(v, Jw) + g_J(Jv, w) = 0$ for all $v, w \in V$. We denote the space of compatible complex structures on (V, ω) by

$$\mathcal{J}(V, \omega) := \{J \in \mathcal{J}(V) \,|\, J \text{ satisfies (2.5.2) and (2.5.3)}\}.$$

Exercise 2.5.3 Let (V, ω) be a symplectic vector space and J be a complex structure on V. Prove that the following are equivalent.

(i) J is compatible with ω.

(ii) The bilinear form $g_J : V \times V \to \mathbb{R}$ defined by $g_J(v, w) := \omega(v, Jw)$ is symmetric, positive definite, and J-invariant.

(iii) The real bilinear map $H : V \times V \to \mathbb{C}$ defined by

$$H(v, w) := \omega(v, Jw) + i\,\omega(v, w)$$

is complex linear in w, complex anti-linear in v, satisfies $H(w, v) = \overline{H(v, w)}$, and has a positive definite real part. Such a form is called a **Hermitian inner product** on (V, J). Note that here V is understood as a complex vector space with the complex structure given by J. Thus one must show that $H(v, Jw) = iH(v, w)$, etc. □

Proposition 2.5.4 *Let (V, ω) be a symplectic vector space and J be a linear complex structure on V. Then the following are equivalent.*

(i) *J is compatible with ω.*

(ii) *(V, ω) has a symplectic basis of the form*

$$v_1, \ldots, v_n, Jv_1, \ldots, Jv_n.$$

(iii) *There exists a vector space isomorphism $\Psi : \mathbb{R}^{2n} \to V$ such that*

$$\Psi^*\omega = \omega_0, \qquad \Psi^*J = J_0.$$

(iv) *J satisfies (2.5.3) and*

$$\Lambda \in \mathcal{L}(V, \omega) \qquad \Longrightarrow \qquad J\Lambda \in \mathcal{L}(V, \omega). \tag{2.5.5}$$

Proof: We prove that (i) implies (ii). By (i) the formula (2.5.4) defines an inner product $g_J : V \times V \to \mathbb{R}$. By Theorem 2.1.3 there exists a Lagrangian subspace $\Lambda \subset V$. Choose an orthonormal basis $v_1, \ldots, v_n$ of Λ with respect to g_J. Then

$$\omega(v_i, Jv_j) = g_J(v_i, v_j) = \delta_{ij}$$

and, by (2.5.2),

$$\omega(Jv_i, Jv_j) = \omega(v_i, v_j) = 0$$

for all i and j. Hence $v_1, \ldots, v_n, Jv_1, \ldots, Jv_n$ is a symplectic basis of (V, ω). Thus we have proved that (i) implies (ii).

That (ii) implies (iii) follows by taking $\Psi : \mathbb{R}^{2n} \to V$ to be the isomorphism, defined by

$$\Psi z := \sum_{i=1}^{n} (x_i v_i + y_i J v_i)$$

for

$$z := (x_1, \ldots, x_n, y_1, \ldots, y_n) \in \mathbb{R}^{2n}.$$

That (iii) implies (i) is obvious. Thus we have proved that assertions (i), (ii), and (iii) are equivalent.

We prove that (i) implies (iv). If $J \in \mathcal{J}(V,\omega)$ then J satisfies (2.5.3) and $J^*\omega = \omega$. The last condition implies (2.5.5), by definition of a Lagrangian subspace, and thus J satisfies (iv).

We prove that (iv) implies (i). Define the bilinear form $g_J : V \times V \to \mathbb{R}$ by (2.5.4). We prove that g_J is symmetric. Assume, by contradiction, that g_J is not symmetric. Then there exist vectors $u, v \in V$ such that

$$\omega(v, Ju) \neq \omega(u, Jv).$$

Hence v is nonzero and hence $\omega(v, Jv) > 0$ by (2.5.3). Define

$$w := u - \frac{\omega(v, Ju)}{\omega(v, Jv)} v.$$

Then

$$\omega(v, Jw) = 0, \qquad \omega(w, Jv) = \omega(u, Jv) - \omega(v, Ju) \neq 0.$$

This shows that the vectors w, Jv are linearly independent and so are the vectors v, Jw. Since

$$\omega(v, Jw) = 0,$$

it follows from Lemma 2.1.5 that there is a Lagrangian subspace $\Lambda \subset V$ containing v, Jw. It follows that $Jv, w \in J\Lambda$ and, since

$$\omega(Jv, w) \neq 0,$$

$J\Lambda$ is not a Lagrangian subspace, in contradiction to (iv). Hence our assumption that g_J is not symmetric must have been wrong. Thus g_J is symmetric, so J satisfies (2.5.2) (and (2.5.3) by assumption), and hence $J \in \mathcal{J}(V,\omega)$. This proves Proposition 2.5.4. □

Our next aim is to show that the space $\mathcal{J}(V,\omega)$ is contractible. We give three proofs of this important fact.

- The first proof identifies $\mathcal{J}(V,\omega)$ with the space of symmetric, positive definite, and symplectic matrices (Lemma 2.5.5).
- The second proof constructs an explicit homotopy equivalence between $\mathcal{J}(V,\omega)$ and the space $\mathfrak{Met}(V)$ of all inner products on V (Corollary 2.5.7). This construction is based on Proposition 2.5.6, which shows that the space of linear complex structures on an even-dimensional real vector space V is homotopy equivalent to the space of nondegenerate skew-symmetric bilinear forms on V. A consequence of the second proof is the observation that every compact subgroup of $\mathrm{Sp}(2n)$ is conjugate to a subgroup of $\mathrm{U}(n)$ (Proposition 2.5.8).
- The third proof identifies the space $\mathcal{J}(V,\omega)$ with the Siegel upper half space (Lemma 2.5.12). A geometric interpretation of the third proof in relation to Lagrangian subspaces is described in Exercise 2.5.16.

Lemma 2.5.5 *Let (V, ω) be a symplectic vector space of dimension $2n$. Then the space $\mathcal{J}(V, \omega)$ of ω-compatible complex structures on V is diffeomorphic to the space $\mathcal{P}$ of symmetric positive definite symplectic $2n \times 2n$-matrices. In particular, $\mathcal{J}(V, \omega)$ is contractible.*

Proof: By Theorem 2.1.3 we may assume $V = \mathbb{R}^{2n}$ and $\omega = \omega_0$. A matrix $J \in \mathbb{R}^{2n \times 2n}$ is a compatible complex structure if and only if

$$J^2 = -\mathbb{1}, \qquad J^{\mathrm{T}} J_0 J = J_0, \qquad \langle v, -J_0 J v \rangle > 0 \ \forall v \neq 0.$$

The first two identities imply that

$$(J_0 J)^{\mathrm{T}} = -J^{\mathrm{T}} J_0 = J^{\mathrm{T}} J_0 J^2 = J_0 J.$$

Hence the matrix

$$P := -J_0 J$$

is symmetric, positive definite and symplectic. Conversely, if a matrix P has these properties, then $J := -J_0^{-1} P \in \mathcal{J}(\mathbb{R}^{2n}, \omega_0)$. Now it follows from Lemma 2.2.3 that $\mathcal{J}(V, \omega)$ is contractible. This proves Lemma 2.5.5. □

As preparation for the second proof, we show that the space of linear complex structures on V is homotopy equivalent to the space of symplectic bilinear forms on V. The homotopy equivalences are given by explicit formulas, depending on the choice of an inner product on V, and the construction carries over to vector bundles. Recall that the space of symplectic bilinear forms on V is denoted by

$$\Omega(V) := \left\{ \omega : V \times V \to \mathbb{R} \,\middle|\, \begin{array}{l} \omega \text{ is bilinear, skew-symmetric} \\ \text{and nondegenerate} \end{array} \right\} \tag{2.5.6}$$

and that the space of inner products on V is denoted by

$$\mathfrak{Met}(V) := \left\{ g : V \times V \to \mathbb{R} \,\middle|\, \begin{array}{l} g \text{ is bilinear, symmetric} \\ \text{and positive definite} \end{array} \right\}. \tag{2.5.7}$$

Thus $\mathfrak{Met}(V)$ is a convex open subset of the vector space of symmetric bilinear forms on V and so is contractible. For $g \in \mathfrak{Met}(V)$ define

$$\mathcal{J}(V, g) := \{ J \in \mathcal{J}(V) \,|\, J^* g = g \}, \qquad \Omega(V, g) := \{ g(J \cdot, \cdot) \,|\, J \in \mathcal{J}(V, g) \}.$$

The condition $J \in \mathcal{J}(V, g)$ is equivalent to the assertion that the bilinear form $g(J \cdot, \cdot)$ is skew-symmetric. Hence $\Omega(V, g) \subset \Omega(V)$. An element $J \in \mathcal{J}(V)$, respectively $\omega \in \Omega(V)$, is called **compatible with the inner product** g if it belongs to the set $\mathcal{J}(V, g)$, respectively $\Omega(V, g)$.

Consider the map

$$\mathfrak{Met}(V) \times \mathcal{J}(V) \to \Omega(V) : (g, J) \mapsto \omega_{g,J} := \tfrac{1}{2}\big(g(J\cdot,\cdot) - g(\cdot, J\cdot)\big). \qquad (2.5.8)$$

For every $g \in \mathfrak{Met}(V)$ and every $J \in \mathcal{J}(V)$, the skew-symmetric bilinear form $\omega_{g,J}$ in (2.5.8) is nondegenerate and compatible with J because

$$\omega_{g,J}(\cdot, J\cdot) = \tfrac{1}{2}\big(g + J^*g\big),$$

and hence $J \in \mathcal{J}(V, \omega_{g,J})$. Moreover, the map (2.5.8) is smooth and $\mathrm{GL}(V)$-equivariant. Fixing an inner product $g \in \mathfrak{Met}(V)$, we obtain a smooth map $f_g : \mathcal{J}(V) \to \Omega(V)$ given by $f_g(J) := \omega_{g,J}$. This map restricts to the diffeomorphism $f_g : \mathcal{J}(V, g) \to \Omega(V, g)$ given by $f_g(J) = g(J\cdot, \cdot)$. Hence the following diagram commutes

$$\begin{array}{ccc} \mathcal{J}(V,g) & \xrightarrow{f_g} & \Omega(V,g) \\ \downarrow & & \downarrow \\ \mathcal{J}(V) & \xrightarrow{f_g} & \Omega(V) \end{array}.$$

Here the vertical maps are the obvious inclusions. The next proposition shows that all four maps in this diagram are homotopy equivalences.

Proposition 2.5.6 *Let V be an even-dimensional real vector space. There exists a $\mathrm{GL}(V)$-equivariant smooth map*

$$\mathfrak{Met}(V) \times \Omega(V) \to \mathcal{J}(V) : (g, \omega) \mapsto J_{g,\omega} \qquad (2.5.9)$$

with the following properties.

(i) *For all $g \in \mathfrak{Met}(V)$, $\omega \in \Omega(V)$, and $J \in \mathcal{J}(V)$, we have*

$$J_{g,\omega} \in \mathcal{J}(V, \omega) \cap \mathcal{J}(V, g), \qquad J \in \mathcal{J}(V, \omega_{g,J}), \qquad (2.5.10)$$

$$g = \omega(\cdot, J\cdot) \quad \Longrightarrow \quad J_{g,\omega} = J \quad \text{and} \quad \omega_{g,J} = \omega. \qquad (2.5.11)$$

In particular, for every $g \in \mathfrak{Met}(V)$, the map $\mathcal{J}(V,g) \to \Omega(V,g) : J \mapsto \omega_{g,J}$ is a diffeomorphism with inverse $\Omega(V,g) \to \mathcal{J}(V,g) : \omega \mapsto J_{g,\omega}$.

(ii) *For every $g \in \mathfrak{Met}(V)$ the map*

$$\Omega(V) \to \mathcal{J}(V) : \omega \mapsto J_{g,\omega}$$

is a homotopy equivalence with homotopy inverse

$$\mathcal{J}(V) \to \Omega(V) : J \mapsto \omega_{g,J}.$$

(iii) *For every $g \in \mathfrak{Met}(V)$ the inclusion $\mathcal{J}(V,g) \hookrightarrow \mathcal{J}(V)$ is a homotopy equivalence with homotopy inverse*

$$\mathcal{J}(V) \to \mathcal{J}(V,g) : J \mapsto J_{g,\omega_{g,J}},$$

and the inclusion $\Omega(V,g) \hookrightarrow \Omega(V)$ is a homotopy equivalence with homotopy inverse

$$\Omega(V) \to \Omega(V,g) : \omega \mapsto \omega_{g,J_{g,\omega}}.$$

Proof: The proof has three steps. The first step explains the construction of the map (2.5.9) satisfying condition (i).

Step 1. *Let $g \in \mathfrak{Met}(V)$ and $\omega \in \Omega(V)$. Define $A_{g,\omega} \in \mathrm{GL}(V)$ by*

$$\omega(v, w) = g(A_{g,\omega}v, w). \tag{2.5.12}$$

Then there exists a unique g-self-adjoint and g-positive definite automorphism $Q_{g,\omega} : V \to V$ that satisfies the equation

$$Q_{g,\omega}^2 = -A_{g,\omega}^2. \tag{2.5.13}$$

The automorphism $Q_{g,\omega}$ commutes with $A_{g,\omega}$ and

$$J_{g,\omega} := Q_{g,\omega}^{-1}A_{g,\omega} \in \mathcal{J}(V, \omega) \cap \mathcal{J}(V, g). \tag{2.5.14}$$

The map $(g, \omega) \mapsto J_{g,\omega}$ is smooth and $\mathrm{GL}(V)$-equivariant.

Since ω is skew-symmetric and nondegenerate, it follows that the linear map

$$A := A_{g,\omega} : V \to V$$

is a g-skew-adjoint automorphism, i.e. $g(Av, w) = -g(v, Aw)$. Hence, writing A^* for the g-adjoint of A, we find that the automorphism $P := A^*A = -A^2$ is g-self-adjoint and g-positive definite. Hence there exists a unique automorphism $Q = Q_{g,\omega} : V \to V$ which is g-self-adjoint, g-positive definite, and satisfies equation (2.5.13). (To see this, use the fact that P can be represented as a diagonal matrix with positive diagonal entries in a suitable g-orthonormal basis of V. Details are given in Exercise 2.5.15 below.) Since $Q^2 = -A^2$ commutes with A so does Q. Hence $J := Q^{-1}A$ is a linear complex structure on V. It satisfies

$$J^* = A^*Q^{-1} = -AQ^{-1} = -Q^{-1}A = -J$$

and hence $J \in \mathcal{J}(V, g)$. Moreover, the bilinear map

$$\omega(\cdot, J\cdot) = g(A\cdot, J\cdot) = g(QJ\cdot, J\cdot)$$

is an inner product and hence $J \in \mathcal{J}(V, \omega)$. If ω is compatible with g then $A \in \mathcal{J}(V)$ and $g = \omega(\cdot, A\cdot)$, hence $-A^2 = \mathbb{1}$, hence $Q = \mathbb{1}$ and hence $J = A$. This completes the construction of a map (2.5.9) that satisfies (i).

We prove that the map (2.5.9) is $\mathrm{GL}(V)$-equivariant. Let $\Phi \in \mathrm{GL}(V)$ and $g \in \mathfrak{Met}(V)$ and $\omega \in \Omega(V)$ be given and abbreviate $A := A_{g,\omega}$ and $Q := Q_{g,\omega}$. If g is replaced by $\Phi^*g = g(\Phi\cdot, \Phi\cdot)$ and ω is replaced by $\Phi^*\omega$, then

$$(\Phi^*g)(\Phi^{-1}A\Phi v, w) = g(A\Phi v, \Phi w) = \omega(\Phi v, \Phi w) = (\Phi^*\omega)(v, w)$$

for $v, w \in V$. Hence $A_{\Phi^*g,\Phi^*\omega} = \Phi^{-1}A\Phi$, hence $Q_{\Phi^*g,\Phi^*\omega} = \Phi^{-1}Q\Phi$, and hence $J_{\Phi^*g,\Phi^*\omega} = \Phi^{-1}Q^{-1}A\Phi = \Phi^{-1}J_{g,\omega}\Phi$. Thus the map (2.5.9) is $\mathrm{GL}(V)$-equivariant. That this map is smooth is the content of Exercise 2.5.15 below. This completes the proof of Step 1.

Step 2. *We prove (ii).*

The composition of the maps $\omega \mapsto J_{g,\omega}$ and $J \mapsto \omega_{g,J}$ is the map

$$\phi : \Omega(V) \to \Omega(V), \qquad \phi(\omega) := \omega_{g,J_{g,\omega}}. \tag{2.5.15}$$

It is homotopic to the identity via $\phi_t : \Omega(V) \to \Omega(V)$, defined by

$$\phi_t(\omega) := (1-t)\omega + t\omega_{g,J_{g,\omega}}, \qquad 0 \le t \le 1. \tag{2.5.16}$$

To see this, note that $J_{g,\omega} \in \mathcal{J}(V,\omega) \cap \mathcal{J}(V,\omega_{g,J_{g,\omega}})$ by (2.5.10). Thus both ω and $\omega_{g,J_{g,\omega}}$ are compatible with the linear complex structure $J_{g,\omega}$ and hence, so is each of their convex combinations. Hence ϕ_t indeed takes values in $\Omega(V)$ and is a homotopy from $\phi_0 = \mathrm{id}$ to $\phi_1 = \phi$.

The composition of the maps $J \mapsto \omega_{g,J}$ and $\omega \mapsto J_{g,\omega}$ is the map

$$\psi : \mathcal{J}(V) \to \mathcal{J}(V), \qquad \psi(J) := J_{g,\omega_{g,J}}. \tag{2.5.17}$$

It is homotopic to the identity via $\psi_t : \mathcal{J}(V) \to \mathcal{J}(V)$, defined by

$$\psi_t(J) := J_{g_t,\omega_{g_t,J}}, \qquad g_t := \tfrac{1+t}{2}\, g + \tfrac{1-t}{2}\, J^*g, \qquad 0 \le t \le 1. \tag{2.5.18}$$

In fact, the inner product $g_0 = \frac{1}{2}(g + J^*g)$ is compatible with J and hence also with the symplectic form

$$\omega_0 := \omega_{g_0,J} = g_0(J\cdot,\cdot) = \tfrac{1}{2}\big(g(J\cdot,\cdot) - g(\cdot,J\cdot)\big) = \omega_{g,J}.$$

Hence $g_0 = \omega_0(\cdot, J\cdot)$ and so it follows from (i) that

$$\psi_0(J) = J_{g_0,\omega_0} = J.$$

Moreover, $g_1 = g$ and hence $\psi_1 = \psi$ as required. This completes Step 2.

Step 3. *We prove (iii).*

If $\omega \in \Omega(V,g)$ then there is a $J \in \mathcal{J}(V)$ such that $g = \omega(\cdot, J\cdot)$. Hence

$$J_{g,\omega} = J, \qquad \omega_{g,J_{g,\omega}} = \omega_{g,J} = \tfrac{1}{2}\big(g(J\cdot,\cdot) - g(\cdot,J\cdot)\big) = g(J\cdot,\cdot) = \omega.$$

Thus the composition of the inclusion $\Omega(V,g) \to \Omega(V)$ with the projection $\Omega(V) \to \Omega(V,g) : \omega \mapsto \omega_{g,J_{g,\omega}}$ is the identity. The converse composition is the map ϕ in (2.5.15) and hence is homotopic to the identity via (2.5.16).

If $J \in \Omega(V,g)$ then $\omega_{g,J} = g(J\cdot,\cdot)$ and hence $J_{g,\omega_{g,J}} = J$, by (i). This shows that the composition of the inclusion $\mathcal{J}(V,g) \to \mathcal{J}(V)$ with the projection $\mathcal{J}(V) \to \mathcal{J}(V,g) : J \mapsto J_{g,\omega_{g,J}}$ is the identity. The converse composition is the map ψ in (2.5.17) and hence is homotopic to the identity via (2.5.18). This completes the proof of Step 3 and Proposition 2.5.6. □

Corollary 2.5.7 *For every $\omega \in \Omega(V)$ the map*

$$\mathcal{J}(V,\omega) \to \mathfrak{Met}(V) : J \mapsto g_J := \omega(\cdot, J\cdot)$$

is a homotopy equivalence with homotopy inverse

$$\mathfrak{Met}(V) \to \mathcal{J}(V,\omega) : g \mapsto J_{g,\omega}.$$

Hence $\mathcal{J}(V,\omega)$ is contractible.

Proof: The composition of the map $\mathcal{J}(V,\omega) \to \mathfrak{Met}(V) : J \mapsto g_J$ with the map $\mathfrak{Met}(V) \to \mathcal{J}(V,\omega) : g \mapsto J_{g,\omega}$ is the identity, by Proposition 2.5.6 (i). The converse composition is the map $\mathfrak{Met}(V) \to \mathfrak{Met}(V) : g \mapsto g_{J_{g,\omega}}$. It is homotopic to the identity because the manifold $\mathfrak{Met}(V)$ is a convex open subset of a vector space. □

We saw in Proposition 2.2.4 that $\mathrm{U}(n)$ is a maximal compact subgroup of $\mathrm{Sp}(2n)$. The next result shows that more is true.

Proposition 2.5.8 *Every compact subgroup $\mathrm{G} \subset \mathrm{Sp}(2n)$ is conjugate to a subgroup of $\mathrm{U}(n)$.*

Proof: The proof has three steps.

Step 1. *There exists a symmetric positive definite matrix $S = S^T \in \mathbb{R}^{2n\times 2n}$ such that $\Psi^T S \Psi = S$ for every $\Psi \in \mathrm{G}$.*

The Haar measure on a compact topological group G is the unique bounded linear functional $C(\mathrm{G},\mathbb{R}) \to \mathbb{R} : f \mapsto \int_{\mathrm{G}} f(\Psi)\, d\mu$ on the space of continuous functions $f : \mathrm{G} \to \mathbb{R}$, which is invariant under left and right translations, and maps the constant function $f \equiv 1$ to the real number one. Define

$$S := \int_{\mathrm{G}} \Psi^T \Psi \, d\mu.$$

This matrix is evidently symmetric, and it is positive definite because the integral of a positive continuous function on G with respect to the Haar measure is positive. Moreover, it follows from the invariance of the Haar measure that $\Psi^T S \Psi = S$ for every $\Psi \in \mathrm{G}$.

Step 2. *There is an ω_0-compatible complex structure $J \in \mathcal{J}(\mathbb{R}^{2n}, \omega_0)$ that commutes with every element of* G.

Let S be as in Step 1 and let $g := \langle \cdot, S\cdot \rangle \in \mathfrak{Met}(\mathbb{R}^{2n})$ be the inner product determined by S. Then g and ω_0 are G-invariant and hence, so is the linear complex structure $J := J_{g,\omega_0} \in \mathcal{J}(\mathbb{R}^{2n}, \omega_0)$ in Proposition 2.5.6.

Step 3. *There is a matrix $\Phi \in \mathrm{Sp}(2n)$ such that $\Phi^{-1}\mathrm{G}\Phi \subset \mathrm{U}(n)$.*

By Proposition 2.5.4, there is a vector space isomorphism $\Phi : \mathbb{R}^{2n} \to \mathbb{R}^{2n}$ such that $\Phi^*\omega_0 = \omega_0$ and $\Phi^* J = J_0$. Hence $\Phi \in \mathrm{Sp}(2n)$ and $\Phi^{-1}J\Phi = J_0$. This implies that for every $\Psi \in \mathrm{G}$, the matrix $\Phi^{-1}\Psi\Phi \in \mathrm{Sp}(2n)$ commutes with J_0 and hence belongs to $\mathrm{U}(n)$. This proves the Proposition 2.5.8. □

Remark 2.5.9 The construction of Proposition 2.5.6 also implies the following. Let V be an even-dimensional real vector space and denote by

$$\mathcal{C}(V) := \{(\omega, J) \,|\, \omega \in \Omega(V),\, J \in \mathcal{J}(V, \omega)\}$$

the space of compatible pairs, consisting of a nondegenerate skew-symmetric bilinear form ω and an ω-compatible linear complex structure J. Proposition 2.5.6 implies that the obvious projections $\pi_\Omega : \mathcal{C}(V) \to \Omega(V)$ and $\pi_\mathcal{J} : \mathcal{C}(V) \to \mathcal{J}(V)$ are homotopy equivalences. For $g \in \mathfrak{Met}(V)$ the map

$$\iota_\Omega : \Omega(V) \to \mathcal{C}(V), \qquad \iota_\Omega(\omega) := (J_{g,\omega}, \omega),$$

is a homotopy inverse of π_Ω, and the map

$$\iota_\mathcal{J} : \mathcal{J}(V) \to \mathcal{C}(V), \qquad \iota_\mathcal{J}(J) := (J, \omega_{g,J}),$$

is a homotopy inverse of $\pi_\mathcal{J}$. The composition $\iota_\Omega \circ \pi_\Omega : \mathcal{C}(V) \to \mathcal{C}(V)$ sends (J, ω) to $(J_{g,\omega}, \omega)$ and is homotopic to the identity via the homotopy

$$[0,1] \times \mathcal{C}(V) \to \mathcal{C}(V) : (t, J, \omega) \mapsto (J_{tg+(1-t)\omega(\cdot, J\cdot),\omega}, \omega).$$

The composition $\iota_\mathcal{J} \circ \pi_\mathcal{J} : \mathcal{C}(V) \to \mathcal{C}(V)$ sends (J, ω) to $(J, \omega_{g,J})$ and is homotopic to the identity via the homotopy

$$[0,1] \times \mathcal{C}(V) \to \mathcal{C}(V) : (t, J, \omega) \mapsto (J, t\omega_{g,J} + (1-t)\omega).$$

□

Remark 2.5.10 In the case $V = \mathbb{R}^{2n}$ the homotopy equivalences in Proposition 2.5.6 and Remark 2.5.9 can be understood in terms of homogeneous spaces as follows. There is a commutative diagram

$$\begin{array}{ccccccccc}
\frac{\mathrm{O}(2n)}{\mathrm{U}(n)} & \longrightarrow & \frac{\mathrm{GL}(2n,\mathbb{R})}{\mathrm{GL}(n,\mathbb{C})} & \longleftarrow & \frac{\mathrm{GL}(2n,\mathbb{R})}{\mathrm{U}(n)} & \longrightarrow & \frac{\mathrm{GL}(2n,\mathbb{R})}{\mathrm{Sp}(2n)} & \longleftarrow & \frac{\mathrm{O}(2n)}{\mathrm{U}(n)} \\
\downarrow \cong & & \downarrow \cong & & \downarrow \cong & & \downarrow \cong & & \downarrow \cong \\
\mathcal{J}(V,g) & \longrightarrow & \mathcal{J}(V) & \xleftarrow{\pi_\mathcal{J}} & \mathcal{C}(V) & \xrightarrow{\pi_\Omega} & \Omega(V) & \longleftarrow & \Omega(V,g)
\end{array}$$

where the horizontal maps are homotopy equivalences and the vertical maps are diffeomorphisms. The fact that the outer vertical maps are diffeomorphisms follows from Theorem 2.1.3, Lemma 2.2.1 and Propositions 2.5.1 and 2.5.2. To understand the central map, recall from Proposition 2.5.4 that a linear complex structure $J \in \mathcal{J}(V)$ is compatible with a symplectic bilinear form $\omega \in \Omega(V)$ if and only if (V, ω) has a symplectic basis of the form

$$v_1, \dots, v_n, Jv_1, \dots, Jv_n.$$

It follows that the map

$$\mathrm{GL}(2n, \mathbb{R}) \to \mathcal{C}(\mathbb{R}^{2n}) : \Psi \mapsto \Psi_*(\omega_0, J_0) := \big((\Psi^{-1})^*\omega_0, \Psi J_0 \Psi^{-1}\big)$$

is onto and descends to a diffeomorphism $\mathrm{GL}(2n, \mathbb{R})/\mathrm{U}(n) \to \mathcal{C}(\mathbb{R}^{2n})$. □

The Siegel upper half space

The **Siegel upper half space** $\mathcal{S}_n$ is the space of complex symmetric matrices $Z = X+iY \in \mathbb{C}^{n\times n}$ with positive definite imaginary part Y. It carries a transitive action of the symplectic linear group $\mathrm{Sp}(2n)$ and the isotropy subgroup of the element $i\mathbb{1} \in \mathcal{S}_n$ is the subgroup $\mathrm{U}(n)$ of unitary matrices under our usual identification of complex $n\times n$-matrices with real $2n\times 2n$-matrices. The action of $\mathrm{Sp}(2n)$ on $\mathcal{S}_n$ is a natural generalization of the action of the group of fractional linear transformations on the upper half plane in the case $n=1$. In the notation of Exercise 1.1.13 a symplectic matrix $\Psi \in \mathrm{Sp}(2n)$ determines a diffeomorphism $\Psi_* : \mathcal{S}_n \to \mathcal{S}_n$, defined by

$$\Psi_* Z := (AZ+B)(CZ+D)^{-1}, \qquad \Psi = \begin{pmatrix} A & B \\ C & D \end{pmatrix}. \tag{2.5.19}$$

This group action was discovered by Siegel [583].

Exercise 2.5.11 Prove that (2.5.19) defines a holomorphic group action of $\mathrm{Sp}(2n)$ on $\mathcal{S}_n$ and that

$$\Psi_*(i\mathbb{1}) = i\mathbb{1} \qquad \Longleftrightarrow \qquad \Psi \in \mathrm{U}(n).$$

Deduce that the map $\mathrm{Sp}(2n) \to \mathcal{S}_n : \Psi \mapsto \Psi_*(i\mathbb{1})$ descends to a diffeomorphism from the homogeneous space $\mathrm{Sp}(2n)/\mathrm{U}(n)$ to the Siegel upper half space $\mathcal{S}_n$. Thus $\mathrm{Sp}(2n)/\mathrm{U}(n)$ inherits the complex structure of $\mathcal{S}_n$.

Hint 1. Prove that for each $Z \in \mathcal{S}_n$ the matrix $CZ+D$ represents a surjective linear transformation from $\mathbb{C}^n$ to itself. The proof uses the fact that the symmetric matrix $(CX+D)Y^{-1}(CX+D)^T + CYC^T$ is positive definite.

Hint 2. Prove that the matrix $\widetilde{Z} = \widetilde{X} + i\widetilde{Y} := \Psi_* Z$ is given by

$$\begin{aligned} \widetilde{Y}^{-1} &= (CX+D)Y^{-1}(CX+D)^T + CYC^T, \\ \widetilde{X}\widetilde{Y}^{-1} &= (AX+B)Y^{-1}(CX+D)^T + AYC^T, \\ \widetilde{Y}^{-1}\widetilde{X} &= (CX+D)Y^{-1}(AX+B)^T + CYA^T. \end{aligned} \tag{2.5.20}$$

Hint 3. Prove that

$$\Psi_*\Phi_* Z = (\Psi\Phi)_* Z$$

for $\Phi, \Psi \in \mathrm{Sp}(2n)$ and $Z \in \mathcal{S}_n$.

Hint 4. To prove that the action is transitive, take $Z = i\mathbb{1}$ and use the diffeomorphism $\mathcal{S}_n \to \mathcal{J}(\mathbb{R}^{2n}, \omega_0)$ constructed in Lemma 2.5.12 below. □

The next lemma shows that $\mathcal{S}_n$ is diffeomorphic to the space $\mathcal{J}(\mathbb{R}^{2n}, \omega_0)$ of ω_0-compatible complex structures on $\mathbb{R}^{2n}$. This is our third proof of the contractibility of the space $\mathcal{J}(\mathbb{R}^{2n}, \omega_0)$. When $n=1$ a complex structure $J \in \mathcal{J}(\mathbb{R}^2)$ is compatible with ω_0 if and only if it induces the same orientation (see Exercise 2.5.19). Hence in this case we recover the standard identification of the space of positively oriented complex structures on $\mathbb{R}^2$ with the upper half plane in $\mathbb{C}$.

Lemma 2.5.12 *There is a unique bijection $\mathcal{S}_n \to \mathcal{J}(\mathbb{R}^{2n}, \omega_0) : Z \mapsto J(Z)$ such that for all $Z \in \mathcal{S}_n$ and $\Psi \in \mathrm{Sp}(2n)$,*

$$J(i\mathbb{1}) = J_0, \qquad J(\Psi_* Z) = \Psi J(Z) \Psi^{-1}. \tag{2.5.21}$$

This bijection is a diffeomorphism and is given by the formula

$$J(Z) = \begin{pmatrix} XY^{-1} & -Y - XY^{-1}X \\ Y^{-1} & -Y^{-1}X \end{pmatrix}, \qquad Z =: X + iY \in \mathcal{S}_n. \tag{2.5.22}$$

(See Exercise 2.5.16 below for a geometric interpretation of this formula.)

Proof: It follows from Proposition 2.5.4 that the group $\mathrm{Sp}(2n)$ acts transitively on $\mathcal{J}(\mathbb{R}^{2n}, \omega_0)$ and it follows from Lemma 2.2.1 that the stabilizer subgroup of J_0 is $\mathrm{Sp}(2n) \cap \mathrm{GL}(n, \mathbb{C}) = \mathrm{U}(n)$. Hence the existence and uniqueness of a bijection $\mathcal{S}_n \to \mathcal{J}(\mathbb{R}^{2n}, \omega_0) : Z \mapsto J(Z)$ that satisfies (2.5.21) follows from Exercise 2.5.11. Now define the map $\mathcal{S}_n \to \mathcal{J}(\mathbb{R}^{2n}, \omega_0) : Z \mapsto J(Z)$ by (2.5.22). This map clearly satisfies $J(i\mathbb{1}) = J_0$. Moreover, it follows by direct calculation from Exercise 1.1.13 and the formula (2.5.20) in Exercise 2.5.11 that the map (2.5.22) is $\mathrm{Sp}(2n)$-equivariant. The formula shows that it is a diffeomorphism. This proves Lemma 2.5.12. □

Tame complex structures

In many situations we do not need to work with compatible complex structures. For example, in order to ensure that the compactness theorems hold for J-holomorphic curves, it is enough that J satisfies only the positivity condition (2.5.3), i.e.

$$\omega(v, Jv) > 0$$

for every nonzero $v \in V$. A linear complex structure $J \in \mathcal{J}(V)$ that satisfies this condition is called **ω-tame**. We denote the space of all ω-tame linear complex structures on V by $\mathcal{J}_\tau(V, \omega)$. This is an open subset of the space of all complex structures on V. It follows directly from the definition that the formula

$$g_{\omega,J}(v, w) := \tfrac{1}{2}\big(\omega(v, Jw) - \omega(Jv, w)\big) \tag{2.5.23}$$

defines an inner product on V for every $J \in \mathcal{J}_\tau(V, \omega)$.

Proposition 2.5.13 *The space $\mathcal{J}_\tau(V, \omega)$ is contractible.*

We again give three proofs of this result. The first appeared in Gromov [287] and requires some homotopy theory. The second proof gives explicit formulas for the homotopy equivalences in Gromov's argument. The third proof is by direct calculation and is due to Sévennec [34, Chapter II,§1.1].

Proof 1: Consider the manifolds

$$\begin{aligned} \mathcal{C}_\tau(V) &:= \{(\omega, J)\,|\,\omega \in \Omega(V),\ J \in \mathcal{J}_\tau(V,\omega)\}, \\ \mathcal{C}(V) &:= \{(\omega, J)\,|\,\omega \in \Omega(V),\ J \in \mathcal{J}(V,\omega)\}. \end{aligned}$$

Given $(\omega, J) \in \mathcal{C}_\tau(V)$, define

$$\pi_1(\omega, J) := \tfrac{1}{2}\big(\omega + J^*\omega\big). \tag{2.5.24}$$

The identity

$$\tfrac{1}{2}\big(\omega + J^*\omega\big)(\cdot, J\cdot) = \tfrac{1}{2}\big(\omega(\cdot, J\cdot) - \omega(J\cdot, \cdot)\big) = g_{\omega,J}$$

shows that the form $\pi_1(\omega, J)$ is nondegenerate and hence symplectic. Further, $\pi_1(\omega, J)$ is compatible with J and determines the same inner product as ω, namely $g_{\omega,J}$. The fibre of the projection

$$\mathrm{pr} : \mathcal{C}_\tau(V) \to \mathcal{C}(V) : (\omega, J) \mapsto \big(\tfrac{1}{2}(\omega + J^*\omega), J\big) \tag{2.5.25}$$

over a compatible pair $(\omega, J) \in \mathcal{C}(V)$ is the space of all pairs $(\omega + \tau, J)$, where $\tau \in \Lambda^2 V^*$ satisfies $\tau + J^*\tau = 0$ (all such pairs are tame and determine the same inner product). The condition $\tau + J^*\tau = 0$ means that τ is the real part of a $(2,0)$-form. Thus $\mathcal{C}_\tau(V)$ is a vector bundle over $\mathcal{C}(V)$. Hence the projection (2.5.25) is a homotopy equivalence and so is the projection $\pi_\Omega : \mathcal{C}(V) \to \Omega(V)$, by Remark 2.5.9. Their composition is the projection $\pi_1 : \mathcal{C}_\tau(V) \to \Omega(V)$ in (2.5.24). If we define

$$\pi_0 : \mathcal{C}_\tau(V) \to \Omega(V), \qquad \pi_0(\omega, J) := \omega,$$

to be the standard projection, we therefore have a diagram

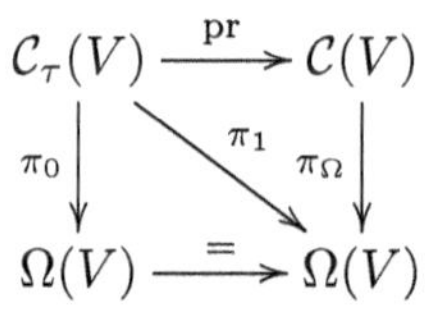

$$\begin{array}{ccc} \mathcal{C}_\tau(V) & \xrightarrow{\ \mathrm{pr}\ } & \mathcal{C}(V) \\ {\scriptstyle \pi_0}\downarrow & \searrow^{\pi_1} & \downarrow{\scriptstyle \pi_\Omega} \\ \Omega(V) & \xrightarrow{\ =\ } & \Omega(V) \end{array}$$

in which the top triangle commutes but the bottom one does not. However, π_1 is homotopic to π_0 via the homotopy $\pi_t := (1-t)\pi_0 + t\pi_1$. Hence $\pi_0 : \mathcal{C}_\tau(V) \to \Omega(V)$ is a homotopy equivalence. Since it is also a fibration, it follows that the fibres $\pi_0^{-1}(\omega) \cong \mathcal{J}_\tau(V,\omega)$ are contractible. This proves Proposition 2.5.13. □

Proof 2: We prove that the inclusion of $\mathcal{J}(V,\omega)$ into $\mathcal{J}_\tau(V,\omega)$ is a homotopy equivalence. Our proof constructs an explicit homotopy inverse, depending on the choice of an inner product $g \in \mathfrak{Met}(V)$. The construction is equivariant under the action of $\mathrm{GL}(V)$ and carries over to vector bundles. It requires three steps. The first step constructs a connection, depending on g, on the tautological symplectic vector bundle over $\Omega(V)$ whose fibre over ω is the vector space V equipped with the symplectic bilinear form ω. Since the bundle is trivial, the connection can be written as an endomorphism valued 1-form on $\Omega(V)$; see Section 6.3 for a related discussion. The second step shows how the connection identifies the fibre over ω with the fibre over $\frac{1}{2}(\omega + J^*\omega)$ along the straight line $\omega_t := \omega + \frac{t}{2}(J^*\omega - \omega)$, $0 \le t \le 1$. The resulting linear symplectomorphism $\Psi : (V, \omega) \to (V, \frac{1}{2}(\omega + J^*\omega))$ will depend on ω (source), J (determining the target), and g (determining the connection).

Step 1. *There is a collection of endomorphism valued* 1*-forms*

$$\Omega(V) \times \Lambda^2 V^* \to \mathrm{End}(V) : (\omega, \widehat{\omega}) \mapsto A_g(\omega, \widehat{\omega}),$$

one for each even-dimensional vector space V and each $g \in \mathfrak{Met}(V)$, such that for every $g \in \mathfrak{Met}(V)$, every $\omega \in \Omega(V)$, every $\widehat{\omega} \in T_\omega \Omega(V) = \Lambda^2 V^*$, and every vector space isomorphism $\Phi : V' \to V$, we have

$$\widehat{\omega} + \omega(A_g(\omega, \widehat{\omega})\cdot, \cdot) + \omega(\cdot, A_g(\omega, \widehat{\omega})\cdot) = 0. \tag{2.5.26}$$

$$B \in \mathrm{End}(V),\ \omega(B\cdot, \cdot) + \omega(\cdot, B\cdot) = 0 \implies \langle A_g(\omega, \widehat{\omega})^{*_g}, B \rangle_g = 0, \tag{2.5.27}$$

$$A_{\Phi^* g}(\Phi^*\omega, \Phi^*\widehat{\omega}) = \Phi^{-1} A_g(\omega, \widehat{\omega}) \Phi. \tag{2.5.28}$$

For each $g \in \mathfrak{Met}(V)$ the 1*-form $A_g \in \Omega^1(\Omega(V), \mathrm{End}(V))$ is uniquely determined by the conditions* (2.5.26) *and* (2.5.27). *For each V the map*

$$\mathfrak{Met}(V) \times \Omega(V) \times \Lambda^2 V^* \to \mathrm{End}(V) : (g, \omega, \widehat{\omega}) \mapsto A_g(\omega, \widehat{\omega})$$

is smooth.

That the 1-form $A_g \in \Omega^1(\Omega(V), \mathrm{End}(V))$ exists and is uniquely determined by equations (2.5.26) and (2.5.27) follows from the fact that for every $\omega \in \Omega(V)$, the linear map $\mathrm{End}(V) \to \Lambda^2 V^* : A \mapsto \omega(A\cdot, \cdot) + \omega(\cdot, A\cdot)$ is surjective. Equation (2.5.27) asserts that $A_g(\omega, \widehat{\omega})$ is g-orthogonal to the kernel of this linear map, and with this constraint is therefore uniquely determined by equation (2.5.26). The proof of smoothness and the verification of (2.5.28) are left to the reader.

Step 2. *There exists a collection of smooth maps*

$$[0, 1] \times \mathcal{C}_\tau(V) \to \mathrm{GL}(V) : (t, \omega, J) \mapsto \Psi_g(t, \omega, J), \tag{2.5.29}$$

one for each even-dimensional vector space V and each $g \in \mathfrak{Met}(V)$, such that for every $g \in \mathfrak{Met}(V)$, every $t \in [0, 1]$, every $(\omega, J) \in \mathcal{C}_\tau(V)$, and every vector space isomorphism $\Phi : V' \to V$, we have

$$\Psi_g(0, \omega, J) = 1\!\!1_V, \tag{2.5.30}$$

$$\omega = J^*\omega \qquad \implies \qquad \Psi_g(t, \omega, J) = 1\!\!1_V, \tag{2.5.31}$$

$$\Psi_g(t, \omega, J)^* \left(\omega + \tfrac{t}{2}\left(J^*\omega - \omega\right)\right) = \omega, \tag{2.5.32}$$

$$\Psi_{\Phi^* g}(t, \Phi^*\omega, \Phi^* J) = \Phi^{-1} \Psi_V(t, g, \omega, J) \Phi. \tag{2.5.33}$$

For each V the map

$$\mathfrak{Met}(V) \times [0, 1] \times \mathcal{C}_\tau(V) \to \mathrm{GL}(V) : (g, t, \omega, J) \mapsto \Psi_g(t, \omega, J)$$

is smooth.

Let $g \in \mathfrak{Met}(V)$, $\omega \in \Omega(V)$, and $J \in \mathcal{J}_\tau(V,\omega)$ be given. Define the smooth path $t \mapsto \Psi_g(t,\omega,J)$ as the unique solution of the ordinary differential equation

$$\frac{d}{dt}\Psi_g(t,\omega,J) = A_g\left(\omega + \tfrac{t}{2}(J^*\omega - \omega), \tfrac{1}{2}(J^*\omega - \omega)\right)\Psi_g(t,\omega,J) \tag{2.5.34}$$

with the initial condition $\Psi_g(0,\omega,J) = 1\!\!1$ as in (2.5.30). Abbreviate

$$\begin{aligned} \omega_t &:= \omega + \tfrac{t}{2}(J^*\omega - \omega), \\ \widehat{\omega} &:= \tfrac{1}{2}(J^*\omega - \omega), \\ \Psi_t &:= \Psi_g(t,\omega,J), \\ A_t &:= A_g(\omega_t, \widehat{\omega}). \end{aligned}$$

Then, by (2.5.34) and (2.5.30), we have

$$\partial_t\Psi_t = A_t\Psi_t, \qquad \Psi_0 = 1\!\!1, \qquad \partial_t\omega_t = \widehat{\omega}, \tag{2.5.35}$$

and hence

$$\begin{aligned} \frac{d}{dt}\Psi_t^*\omega_t &= \Psi_t^*\partial_t\omega_t + \omega_t(\partial_t\Psi_t\cdot, \Psi_t\cdot) + \omega_t(\Psi_t\cdot, \partial_t\Psi_t\cdot) \\ &= \Psi_t^*\widehat{\omega} + \omega_t(A_t\Psi_t\cdot, \Psi_t\cdot) + \omega_t(\Psi_t\cdot, A_t\Psi_t\cdot) \\ &= \Psi_t^*\left(\widehat{\omega} + \omega_t(A_g(\omega_t,\widehat{\omega})\cdot,\cdot) + \omega_t(\cdot, A_g(\omega_t,\widehat{\omega})\cdot)\right) \\ &= 0. \end{aligned} \tag{2.5.36}$$

Here the third equation follows from the formula $A_t = A_g(\omega_t,\widehat{\omega})$ and the last equation follows from equation (2.5.26) in Step 1. This shows that $\Psi_t^*\omega_t = \omega_0 = \omega$ for all t and hence Ψ_g satisfies (2.5.32). That it satisfies (2.5.30) is obvious from the definition. That Ψ_g satisfies (2.5.31) follows from the fact that when J is compatible with ω, we have $J^*\omega = \omega$, hence $\widehat{\omega} = 0$ in the above notation, hence $A_t = 0$ for all t, and hence $\Psi_t = 1\!\!1$ for all t. The equivariance condition (2.5.33) for Ψ_g follows from the analogous condition (2.5.28) for A_g, and smoothness follows from the smooth dependence of solutions of linear differential equations on the coefficients and the smoothness of A in Step 1.

Step 3. *Let $\omega \in \Omega(V)$ and $g \in \mathfrak{Met}(V)$. Then the inclusion of $\mathcal{J}(V,\omega)$ into $\mathcal{J}_\tau(V,\omega)$ is a homotopy equivalence, with homotopy inverse given by*

$$\mathcal{J}_\tau(V,\omega) \to \mathcal{J}(V,\omega) : J \mapsto \Psi_g(1,\omega,J)^{-1}J\Psi_g(1,\omega,J). \tag{2.5.37}$$

Here Ψ_g is as in Step 2.

For $0 \le t \le 1$ and $J \in \mathcal{J}_\tau(V,\omega)$ define

$$f_t(J) := \Psi_g(t,\omega,J)^{-1}J\Psi_g(t,\omega,J), \qquad \omega_t := \omega + \tfrac{t}{2}(J^*\omega - \omega).$$

Since $\Psi_g(t,\omega,J)^*\omega_t = \omega$ by (2.5.32) and J is tamed by the symplectic bilinear form ω_t for every t, we have

$$f_t(J) = \Psi_g(t, \omega, J)^* J \in \mathcal{J}_\tau(V, \omega)$$

for all t. If $t = 1$ then J is compatible with ω_1 and so $f_1(J) \in \mathcal{J}(V, \omega)$. If $J \in \mathcal{J}(V, \omega)$ then $\Psi_g(t, \omega, J) = \mathbb{1}$ and hence $f_t(J) = J$ for all t. Thus the map $[0,1] \times \mathcal{J}_\tau(V, \omega) \to \mathcal{J}_\tau(V, \omega) : (t, J) \mapsto f_t(J)$ is smooth and satisfies

$$f_0 = \mathrm{id}, \qquad f_t|_{\mathcal{J}(V,\omega)} = \mathrm{id}, \qquad f_1(\mathcal{J}_\tau(V, \omega)) = \mathcal{J}(V, \omega).$$

Hence the inclusion of $\mathcal{J}(V, \omega)$ into $\mathcal{J}_\tau(V, \omega)$ is a homotopy equivalence with homotopy inverse f_1. This completes the second proof of Proposition 2.5.13. □

Proof 3: We may assume that $(V, \omega) = (\mathbb{R}^{2n}, \omega_0)$. Then $J \in \mathcal{J}(\mathbb{R}^{2n})$ is tamed by ω_0 if and only if the matrix $Z := -J_0 J$ satisfies

$$Z > 0, \quad Z^{-1} = J_0^{-1} Z J_0.$$

Here we write $Z > 0$ to mean that $\langle v, Zv \rangle > 0$ when $v \neq 0$ but the matrix Z is not required to be symmetric. To transform the second identity into something which is manageable we use the Cayley transform[10] $z \mapsto (1-z)(1+z)^{-1} = w$. As a map of the Riemann sphere this takes the positive half space $\{z \in \mathbb{C} \,|\, \mathrm{Re}\, z > 0\}$ into the unit disc, interchanges the points $\pm i$, and transforms the involution $z \mapsto 1/z$ into the involution $w \mapsto -w$. Similarly, if we think of this as a map on matrices

$$W := F(Z) := (\mathbb{1} - Z)(\mathbb{1} + Z)^{-1},$$

it is easy to check that it is well-defined on the half space $Z > 0$ and takes it onto the open unit disc $\{W \,|\, \|W\| < 1\}$. Moreover, it interchanges $\pm J_0$, and transforms the identity $Z^{-1} = J_0^{-1} Z J_0$ into

$$-W = J_0^{-1} W J_0.$$

But the set of such matrices W is convex and hence contractible. This completes the third proof of Proposition 2.5.13. □

We note that the calculation in Proof 3 above can also be applied to show that $\mathcal{J}(V, \omega)$ is contractible. The only difference in this case is that Z is now symmetric, and this property is preserved by F.

Corollary 2.5.14 *Let ω_t be a smooth family of nondegenerate skew-symmetric bilinear forms on V depending on a real parameter t. Then there exists a smooth family of isomorphisms $\Psi_t : V \to V$ such that $\Psi_0 = \mathbb{1}$ and $\Psi_t^* \omega_t = \omega_0$ for all t.*

Proof: Fix an inner product $g \in \mathfrak{Met}(V)$ and let $A_g : \Omega(V) \times \Lambda^2 V^* \to \mathrm{End}(V)$ be the map constructed in Step 1 of the second proof of Proposition 2.5.13. Let Ψ_t be the solution of the differential equation (2.5.35) with $\widehat{\omega} = \frac{d}{dt}\omega_t$ now depending on t. Then it follows from equation (2.5.36) that Ψ_t satisfies the requirements of Corollary 2.5.14. □

[10] This also appears in [25, I.4.2].

Exercises

The following exercises explore further the interconnections between symplectic forms, inner products, and almost complex structures.

Exercise 2.5.15 **(i)** Prove that the map $(g, \omega) \mapsto J_{g,\omega}$ in Proposition 2.5.6 is smooth. **Hint:** Assume $V = \mathbb{R}^{2n}$, let $\omega \in \Omega(\mathbb{R}^{2n})$ be represented by a skew-symmetric matrix

$$B = -B^T \in \mathbb{R}^{2n \times 2n}, \qquad \omega(v, w) = (Bv)^T w,$$

and let $g \in \mathfrak{Met}(\mathbb{R}^{2n})$ be represented by a positive definite symmetric matrix

$$S = S^T \in \mathbb{R}^{2n \times 2n}, \qquad g(v, w) = v^T S w.$$

The formula $\omega(v, w) = g(Av, w)$ determines the matrix $A = S^{-1}B$. Prove that the g-adjoint of A is represented by the matrix $A^* = S^{-1}A^T S = -A$. Prove that the g-square root Q of the matrix

$$P := A^*A = -A^2 = S^{-1}B^T S^{-1}B$$

is given by

$$Q = S^{-1/2}\left(S^{-1/2}B^T S^{-1}BS^{-1/2}\right)^{1/2} S^{1/2}.$$

Deduce that the map $(S, B) \mapsto J := Q^{-1}S^{-1}B$ is smooth.

(ii) The algebra here is a reformulation of that in the proof of Lemma 2.4.5. Use the current methods to give an alternative proof of this result. **Hint:** Find a symplectic basis which is orthogonal with respect to both g and g_J, where $J := J_{g,\omega}$. □

Exercise 2.5.16 Here is another proof of the contractibility of $\mathcal{J}(V, \omega)$ taken from [34, Chapter II]. This proof illustrates in a clear geometric way the relationship between Lagrangian subspaces, complex structures, and inner products. Given a Lagrangian subspace $\Lambda_0 \in \mathcal{L}(V, \omega)$ there is a natural bijection

$$\mathcal{J}(V, \omega) \to \mathcal{L}_0(V, \omega, \Lambda_0) \times \mathcal{S}(\Lambda_0),$$

where $\mathcal{L}_0(V, \omega, \Lambda_0)$ is the space of all Lagrangian subspaces which intersect Λ_0 transversally and $\mathcal{S}(\Lambda_0)$ is the space of all positive definite quadratic forms on Λ_0. Note that by Lemma 2.3.1, the space $\mathcal{L}_0(V, \omega, \Lambda_0)$ is contractible. The above correspondence is given by the map

$$J \mapsto (J\Lambda_0, g_J|_{\Lambda_0}),$$

where $g_J(v, w) = \omega(v, Jw)$ as above. Show that this is a bijection. Compare this geometric construction in the case $(V, \omega, \Lambda_0) = (\mathbb{R}^{2n}, \omega_0, \mathbb{R}^n \times \{0\})$ with the formula (2.5.22) in Lemma 2.5.12. □

Exercise 2.5.17 Let ω and g be given. Show that that each basis for V that is both g-orthogonal and ω-standard gives rise to a Lagrangian subspace Λ whose g-orthogonal complement $\Lambda^\perp$ is also Lagrangian. Conversely, given such a Lagrangian, construct a g-orthogonal and ω-standard basis. □

Exercise 2.5.18 Suppose that J_t is a smooth family of complex structures on V depending on a parameter t. Prove that there exists a smooth family of isomorphisms $\Phi_t : \mathbb{R}^{2n} \to V$ such that $J_t\Phi_t = \Phi_t J_0$ for every t. **Hint:** Either carry out the third proof of Proposition 2.5.1 with an extra parameter t, or construct a connection on a suitable bundle. □

Exercise 2.5.19 Prove that the real 2×2 matrix

$$J = \begin{pmatrix} a & b \\ c & d \end{pmatrix}$$

satisfies $J^2 = -\mathbb{1}$ if and only if

$$ad - bc = 1, \qquad a = -d.$$

Deduce that J_0 and $-J_0$ lie in different connected components of $\mathcal{J}(\mathbb{R}^2)$. Prove that each connected component of $\mathcal{J}(\mathbb{R}^2)$ is contractible. □

Exercise 2.5.20 Let V be a two-dimensional real vector space. Fix a nondegenerate skew-symmetric bilinear form $\omega \in \Omega(V)$ and a linear complex structure $J \in \mathcal{J}(V)$. Prove that ω and J are compatible if and only if they induce the same orientation on V. □

Exercise 2.5.21 A linear subspace $W \subset V$ is called **totally real** if it is of dimension n and

$$JW \cap W = \{0\}.$$

If $W \subset V$ is a totally real subspace show that the space of nondegenerate skew forms $\omega : V \times V \to \mathbb{R}$ which are compatible with J and satisfy

$$W \in \mathcal{L}(V, \omega)$$

is naturally isomorphic to the space of inner products on W and hence is convex. □

2.6 Symplectic vector bundles

In this section we discuss the basic properties of symplectic vector bundles. For convenience, we will always assume that the base space M of the bundle is a smooth manifold, perhaps with boundary ∂M, but the theory can of course be developed in more generality. The main result is Theorem 2.6.3 which says that symplectic vector bundles are essentially the same as complex vector bundles. After sketching how this follows from the theory of classifying spaces, we give an explicit and elementary proof.

A **symplectic vector bundle** over a manifold M is a pair (E, ω) consisting of a real vector bundle $\pi : E \to M$ and a family of symplectic bilinear forms $\omega_q : E_q \times E_q \to \mathbb{R}$ on the fibres $E_q = \pi^{-1}(q)$ of the vector bundle that vary smoothly with $q \in M$. These skew forms ω_q fit together to give a smooth section ω of the exterior power $E^* \wedge E^*$ of the dual bundle $E^* = \mathrm{Hom}(E, \mathbb{R})$. This section ω is a nondegenerate skew-symmetric bilinear form on E which we will call a **symplectic bilinear form**. Two symplectic vector bundles (E_1, ω_1) and (E_2, ω_2) are called **isomorphic** if there exists a vector bundle isomorphism $\Psi : E_1 \to E_2$ such that $\Psi^*\omega_2 = \omega_1$.

Example 2.6.1 If $E \to M$ is any vector bundle, the sum $E \oplus E^*$ is a symplectic bundle with form Ω_{can} given by

$$\Omega_{\mathrm{can}}(v_0 \oplus v_1^*, w_0 \oplus w_1^*) = w_1^*(v_0) - v_1^*(w_0).$$

Compare Example 2.1.6. □

Exercise 2.6.2 Let (E, ω) be a symplectic vector bundle of rank $2n$ over a manifold M. Prove that E admits a system of local trivializations such that the transition maps take values in $\mathrm{Sp}(2n) \subset \mathrm{GL}(2n, \mathbb{R})$. More precisely, show that every point $q \in M$ has a neighbourhood $U \subset M$ such that the symplectic vector bundle $(\pi^{-1}(U), \omega)$ is isomorphic to the trivial bundle $(U \times \mathbb{R}^{2n}, \omega_0)$ over U. **Hint:** Construct sections which form a symplectic basis in each fibre, for example by using the method in the proof of Theorem 2.1.3. □

It follows from Exercise 2.6.2 that the structure group of a symplectic vector bundle (E, ω) can be reduced from $\mathrm{GL}(2n; \mathbb{R})$ to $\mathrm{Sp}(2n)$. This means that E can be constructed by patching together disjoint pieces $U_\alpha \times \mathbb{R}^{2n}$ using transition functions

$$U_\alpha \cap U_\beta \to \mathrm{Sp}(2n).$$

According to the theory of classifying spaces (see, for example, [488] and [335]) isomorphism classes of symplectic bundles over the base M correspond to homotopy classes of maps from M to the classifying space $B\mathrm{Sp}(2n)$. But $B\mathrm{Sp}(2n)$ is homotopy equivalent to $B\mathrm{U}(n)$ by Proposition 2.2.4. Hence each symplectic vector bundle has a complex structure which is well-defined up to homotopy. Moreover, this complex structure characterizes the isomorphism class of the bundle in the following sense.

Theorem 2.6.3 *For $i = 1, 2$ let (E_i, ω_i) be a symplectic vector bundle over a manifold M and let J_i be an ω_i-tame complex structure on E_i. Then the symplectic vector bundles (E_1, ω_1) and (E_2, ω_2) are isomorphic if and only if the complex vector bundles (E_1, J_1) and (E_2, J_2) are isomorphic.*

Proof: See page 82 below for a proof that does not rely on the theory of classifying spaces. □

As in the case of vector spaces, a **complex structure** on a vector bundle $E \to M$ is an automorphism J of E such that $J^2 = -\mathbb{1}$ (think of J as multiplication by i). This complex structure is said to be **compatible** with ω if the induced complex structure J_q on the fibre E_q is compatible with ω_q for all $q \in M$. For any such compatible pair the bilinear form $g_J : E \times E \to \mathbb{R}$, defined by

$$g_J(v, w) := \omega(v, Jw),$$

is symmetric and positive definite. A triple (ω, J, g) with these properties is called a **Hermitian structure** on E. The following proposition asserts that every symplectic vector bundle admits a Hermitian structure.

Proposition 2.6.4 *Let $E \to M$ be a $2n$-dimensional vector bundle.*

(i) *For every symplectic bilinear form ω on E the space $\mathcal{J}(E,\omega)$ of ω-compatible complex structures on E and the space $\mathcal{J}_\tau(E,\omega)$ of ω-tame complex structures on E are nonempty and contractible.*

(ii) *Let J be a complex structure on E. Then the space of symplectic bilinear forms ω on E that are compatible with J (respectively that tame J) is nonempty and contractible.*

(iii) *Let g be an inner product on E. Then the map $\omega \mapsto J_{g,\omega}$ in Proposition 2.5.6 induces a homotopy equivalence from the space of symplectic bilinear forms on E to the space of complex structures on E.*

Proof: We prove (i). That $\mathcal{J}(E,\omega)$ is contractible follows directly from the proof of Corollary 2.5.7. Namely the map $J \mapsto g_J := \omega(\cdot, J\cdot)$ defines a homotopy equivalence from $\mathcal{J}(E,\omega)$ to the space of inner products on E whose homotopy inverse is the map $g \mapsto J_{g,\omega}$ of Proposition 2.5.6. To prove that $\mathcal{J}_\tau(E,\omega)$ is contractible we use Sévennec's method in the third proof of Proposition 2.5.13. Choose a complex structure $J_0 \in \mathcal{J}(E,\omega)$, denote by

$$\langle\cdot,\cdot\rangle_0 := \omega(\cdot, J_0\cdot)$$

the associated inner product on E, let

$$|v|_0 := \sqrt{\langle v,v\rangle_0}$$

be the norm of $v \in E_p$, and let

$$\|A\|_0 := \sup_{0\neq v\in E_p} \frac{|Av|_0}{|v|_0}$$

be the operator norm of an endomorphism $A : E_p \to E_p$. Define

$$\mathcal{A} := \left\{ A \in \Omega^0(M, \mathrm{End}(E)) \,\middle|\, AJ_0 + J_0A = 0,\ \|A(p)\|_0 < 1\ \forall\, p \in M \right\}.$$

Then the formula

$$\mathcal{F}(J) := (\mathbb{1} + J_0J)(\mathbb{1} - J_0J)^{-1}$$

defines a homeomorphism $\mathcal{F} : \mathcal{J}_\tau(M,\omega) \to \mathcal{A}$ with inverse

$$\mathcal{F}^{-1}(A) = J_0(\mathbb{1} + A)^{-1}(\mathbb{1} - A).$$

Since $\mathcal{A}$ is convex, this proves (i).

The existence of a compatible symplectic bilinear form ω on E in (ii) is equivalent to the existence of an inner product with respect to which J is skew-adjoint. This follows from the existence of any inner product g by taking $\frac{1}{2}(g + J^*g)$. In (ii) contractibility follows from the fact that the relevant spaces are convex. Assertion (iii) follows directly from Proposition 2.5.6 (ii). This proves Proposition 2.6.4. □

Proof of Theorem 2.6.3: Assume first that there exists a bundle isomorphism $\Psi : E_1 \to E_2$ such that $\Psi^*\omega_2 = \omega_1$. Then the complex structures J_1 and $\Psi^* J_2$ are both compatible with ω_1. Hence it follows from Proposition 2.6.4 (i) that there exists a smooth family of complex structures $J_t : E_1 \to E_1$ which are compatible with ω_1 and connect $J_0 = \Psi^* J_2$ to J_1. By Exercise 2.5.18 there exists a smooth family of bundle isomorphisms $\Phi_t : E_1 \to E_1$ such that $\Phi_t^* J_t = J_1$. Thus the bundle isomorphism $\Psi \circ \Phi_0 : E_1 \to E_2$ intertwines J_1 and J_2.

Conversely, suppose that there is a bundle ismorphism $\Psi : E_1 \to E_2$ such that $\Psi^* J_2 = J_1$. Then $\Psi^*\omega_2$ and ω_1 are nondegenerate 2-forms on E_1 that are compatible with (or tame) the same complex structure J_1. Hence so is

$$\omega_t := t\omega_1 + (1-t)\Psi^*\omega_2.$$

Hence $\Psi^*\omega_2$ and ω_1 are connected by a smooth family of nondegenerate skew-symmetric bilinear forms on the fibres of E_1. By Corollary 2.5.14 this implies that there exists a vector bundle isomorphism $\Phi : E_1 \to E_1$ such that $\Phi^*\Psi^*\omega_2 = \omega_1$. Hence $\Psi \circ \Phi$ is a symplectic vector bundle isomorphism from (E_1, ω_1) to (E_2, ω_2). This proves Theorem 2.6.3. □

Similar results hold for pairs (E, F), where E is a symplectic bundle and F is a subbundle with fibres which are symplectic or Lagrangian subspaces. The next exercise formulates this precisely in the case when F is a Lagrangian subbundle over the boundary ∂M of M.

Exercise 2.6.5 Let $E \to M$ be a $2n$-dimensional vector bundle with complex structure J and $F \to \partial M$ be an n-dimensional totally real subbundle. This means

$$J_q F_q \cap F_q = \{0\}$$

for all $q \in \partial M$. Prove that there exists a symplectic bilinear form ω which is compatible with J and satisfies $F_q \in \mathcal{L}(E_q, \omega_q)$ for $q \in \partial M$. Prove that the space of such forms is contractible. □

Trivializations

A trivialization of a bundle E is an isomorphism from E to the trivial bundle which preserves the structure under consideration. For example, a symplectic trivialization preserves the symplectic structure, and a complex trivialization is an isomorphism of complex vector bundles. It follows from Theorem 2.6.3 that these notions are essentially the same. For example, a symplectic bundle (E, ω) is symplectically trivial if and only if, given any compatible complex structure J on E, the complex vector bundle (E, J) is trivial as a complex bundle. It is therefore convenient to combine these notions and consider unitary trivializations. The results we develop now will be used to define the first Chern class in the next section.

A **unitary trivialization** of a Hermitian vector bundle E is a smooth map $M \times \mathbb{R}^{2n} \to E : (q, \zeta) \mapsto \Phi(q)\zeta$ which transforms ω, J and g to the standard structures on $\mathbb{R}^{2n}$:

$$\Phi^* J = J_0, \qquad \Phi^* \omega = \omega_0, \qquad \Phi^* g = g_0,$$

where $g_0(\xi, \eta) = \langle \xi, \eta \rangle$ for $\xi, \eta \in \mathbb{R}^{2n}$. A **unitary trivialization along a curve** $\gamma : [0,1] \to M$ is a unitary trivialization of the pullback bundle $\gamma^* E$.

Lemma 2.6.6 *Let $E \to M$ be a vector bundle with Hermitian structure (ω, J, g). Let $\gamma : [0,1] \to M$ be a smooth curve and let $\Phi_0 : \mathbb{R}^{2n} \to E_{\gamma(0)}$, $\Phi_1 : \mathbb{R}^{2n} \to E_{\gamma(1)}$ be unitary isomorphisms at the endpoints. Then there exists a unitary trivialization $\Phi(t) : \mathbb{R}^{2n} \to E_{\gamma(t)}$ of $\gamma^* E$ such that $\Phi(0) = \Phi_0$ and $\Phi(1) = \Phi_1$.*

Proof: We first prove that such a trivialization exists on some interval $0 \le t < \varepsilon$. Choose $s_{j0} \in E_{\gamma(0)}$ such that

$$\Phi_0 \zeta = \sum_j s_{j0} \zeta_j$$

for $\zeta \in \mathbb{R}^{2n}$. We must construct $2n$ sections $s_j(t) \in E_{\gamma(t)}$ which satisfy

$$\langle s_j, s_k \rangle = \delta_{jk}, \qquad \omega(s_j, s_{j+n}) = 1$$

and $\omega(s_j, s_k) = 0$ for all other values of j and k. Here $\langle s_j, s_k \rangle = g(s_j, s_k)$ denotes the inner product. To construct these sections choose a Riemannian connection ∇ on E (see Kobayashi and Nomizu [371] or Donaldson and Kronheimer [160]) and choose $\widetilde{s}_j(t) \in E_{\gamma(t)}$ to be parallel:

$$\nabla \widetilde{s}_j = 0, \qquad \widetilde{s}_j(0) = s_{j0}.$$

Then for small t the first n vectors $\widetilde{s}_1(t), \ldots, \widetilde{s}_n(t)$ are linearly independent over $\mathbb{C}$. Now use Gram–Schmidt over the complex numbers to obtain a unitary basis s_k such that

$$s_{k+n} = J s_k, \qquad k = 1, \ldots, n.$$

More precisely, the vectors s_k are defined inductively by $s_1 := \widetilde{s}_k / |\widetilde{s}_1|$ and, after $s_1, \ldots, s_{k-1}$ have been constructed, by

$$s_k := \frac{\widetilde{s}_k}{|\widetilde{s}_k|} - \sum_{j=1}^{k-1} \frac{\langle s_j, \widetilde{s}_k \rangle}{|\widetilde{s}_k|} s_j - \sum_{j=1}^{k-1} \frac{\omega(s_j, \widetilde{s}_k)}{|\widetilde{s}_k|} J s_j.$$

This works for small time intervals. Now cover the interval $[0,1]$ by finitely many intervals over which a unitary trivialization exists and use a patching argument on overlaps. More precisely, given two trivializations Φ_1 and Φ_2 over an interval (a,b) choose a smooth path $\Psi : (a,b) \to \mathrm{U}(n) \subset \mathrm{Sp}(2n)$ such that $\Psi(t) = \mathbb{1}$ for t near a and $\Psi(t) = \Phi_1(t)^{-1} \Phi_2(t)$ for t near b. Then the path $\Phi(t) := \Phi_1(t) \Psi(t)$ agrees with Φ_1 near a and with Φ_2 near b. This proves Lemma 2.6.6. □

Proposition 2.6.7 *A Hermitian vector bundle $E \to \Sigma$ over a compact oriented 2-manifold Σ with nonempty boundary $\partial\Sigma$ admits a unitary trivialization.*

Proof: The proof of the proposition is by induction over the number

$$k(\Sigma) = 2g(\Sigma) + \ell(\Sigma),$$

where $\ell(\Sigma) \geq 1$ is the number of boundary components and $g(\Sigma) \geq 0$ is the genus. If $k(\Sigma) = 1$ then Σ is diffeomorphic to the unit disc and in this case the statement is a parametrized version of Lemma 2.6.6: trivialize along rays starting at the origin.

Suppose the statement has been proved for $k(\Sigma) \leq m$ and let $k(\Sigma) = m + 1$. Then Σ can be decomposed as $\Sigma = \Sigma_1 \cup_C \Sigma_2$ such that $k(\Sigma_1) = m$ and Σ_2 is diffeomorphic to the unit disc with two holes. (See Fig. 2.2.)

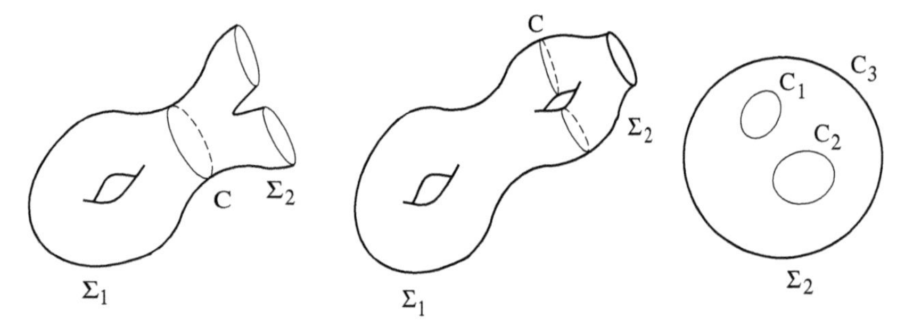

FIG. 2.2. Decomposing Σ.

There are two cases. Either C has one component and

$$g(\Sigma_1) = g(\Sigma), \qquad \ell(\Sigma_1) = \ell(\Sigma) - 1$$

or C has two components and

$$g(\Sigma_1) = g(\Sigma) - 1, \qquad \ell(\Sigma_1) = \ell(\Sigma) + 1.$$

In either case $k(\Sigma_1) = k(\Sigma) - 1$.

By the induction hypothesis the bundle E admits a trivialization over Σ_1. We must prove that any trivialization of E over C extends to a trivialization over Σ_2. Denote $\partial\Sigma_2 =: C_1 \cup C_2 \cup C_3$, where C_1 and C_2 are the (interior) boundaries of the two holes and C_3 is the (exterior) boundary of the disc. If $C = C_1 \cup C_2$ connect the two circles C_1 and C_2 by a straight line. By Lemma 2.6.6 extend the trivialization over C to that line. Then use Lemma 2.6.6 along rays to extend the trivialization to Σ. If $C = C_1$ choose first any trivialization over C_2 and then proceed as above. This proves Proposition 2.6.7. □

Exercise 2.6.8 Define the notion 'symplectic trivialization'. Show that a Hermitian bundle has a unitary trivialization if and only if its underlying symplectic bundle has a symplectic trivialization. □

Exercise 2.6.9 Prove that the space of paths $\Psi : [0,1] \to \mathrm{Sp}(2n)$ of symplectic matrices satisfying

$$\Psi(1) = \Psi(0)^{-1}$$

has two connected components. Deduce that up to isomorphism in each even dimension there are precisely two symplectic vector bundles over the real projective plane $\mathbb{R}\mathrm{P}^2$. **Hint:** Think of $\mathbb{R}\mathrm{P}^2$ as the 2-disc with opposite points on the boundary identified. □

2.7 First Chern class

Since the set of isomorphism classes of symplectic vector bundles coincides with the set of isomorphism classes of complex vector bundles, symplectic vector bundles have the same characteristic classes as complex vector bundles, namely the Chern classes. In this section we establish the existence of the first Chern class. This is an element of the integral 2-dimensional cohomology of the base manifold. For bundles over closed (i.e. compact without boundary) oriented 2-dimensional bases, the first Chern class c_1 is completely described by the first Chern number, which is the value taken by c_1 on the fundamental 2-cycle of the base. Therefore we will begin by describing this integer invariant for symplectic vector bundles over closed oriented 2-manifolds. It can be defined axiomatically as follows.

Theorem 2.7.1 *There exists a unique functor c_1, called the* **first Chern number**, *that assigns an integer*

$$c_1(E) \in \mathbb{Z}$$

to every symplectic vector bundle E over a closed oriented 2-manifold Σ and satisfies the following axioms.

(naturality) *Two symplectic vector bundles E and E' over a closed oriented 2-manifold Σ are isomorphic if and only if they have the same rank and the same first Chern number.*

(functoriality) *For any smooth map $\phi : \Sigma' \to \Sigma$ of closed oriented 2-manifolds and any symplectic vector bundle $E \to \Sigma$*

$$c_1(\phi^* E) = \deg(\phi) \cdot c_1(E).$$

(additivity) *For any two symplectic vector bundles E and E' over a closed oriented 2-manifold Σ*

$$c_1(E \oplus E') = c_1(E) + c_1(E').$$

(normalization) *The Chern number of the tangent bundle of Σ is*

$$c_1(T\Sigma) = 2 - 2g,$$

where g is the genus.

Proof: See page 87 below. □

Remark 2.7.2 **(i)** It follows from the axioms that the first Chern number vanishes if and only if the bundle is trivial. Hence the first Chern number $c_1(E)$ can be viewed as an obstruction for the bundle E to admit a symplectic trivialization.

(ii) If E is a symplectic vector bundle over any manifold M then the first Chern number assigns an integer $c_1(f^*E)$ to every smooth map $f : \Sigma \to M$ from a compact oriented 2-manifold without boundary to M. We will see in Exercise 2.7.10 that this integer depends only on the homology class of f. Thus the first Chern number generalizes to a homomorphism $H_2(M;\mathbb{Z}) \to \mathbb{Z}$. This gives rise to a cohomology class

$$c_1(E) \in H^2(M;\mathbb{Z})/\text{torsion}.$$

There is in fact a natural choice of a lift of this class to $H^2(M;\mathbb{Z})$, also denoted by $c_1(E)$, which is called the **first Chern class**. We shall not discuss this lift in detail, but only remark that, in the case of a complex line bundle $L \to M$ over a closed oriented manifold M, the first Chern class $c_1(L) \in H^2(M;\mathbb{Z})$ is Poincaré dual to the homology class determined by the zero set of a generic section and that it determines the isomorphism class of the complex line bundle L uniquely.

(iii) It is customary to define the Chern class as an invariant for *complex* vector bundles. It follows from Theorem 2.6.3 that our definition agrees with the usual one. □

We will prove Theorem 2.7.1 by giving an explicit definition of the first Chern number and checking that the axioms are satisfied. Given a compact oriented 2-manifold Σ without boundary, choose a splitting

$$\Sigma = \Sigma_1 \cup_C \Sigma_2, \qquad \partial\Sigma_1 = \partial\Sigma_2 = \Sigma_1 \cap \Sigma_2 =: C.$$

Orient the 1-manifold C as the boundary of Σ_1: a vector $v \in T_qC$ is positively oriented if $\{\nu(q), v\}$ is a positively oriented basis of $T_q\Sigma$, where $\nu : C \to T\Sigma$ is a normal vector field along C which points out of Σ_1.

Now let E be a symplectic vector bundle over Σ and choose symplectic trivializations

$$\Sigma_k \times \mathbb{R}^{2n} \to E : (q, \zeta) \mapsto \Phi_k(q)\zeta$$

of E over Σ_1 and Σ_2. The **overlap map** $\Psi : C \to \mathrm{Sp}(2n)$ is defined by

$$\Psi(q) = \Phi_1(q)^{-1}\Phi_2(q)$$

for $q \in C$. Consider the map $\rho : \mathrm{Sp}(2n) \to S^1$ defined by

$$\rho(\Psi) := \det_{\mathbb{C}}(X + iY), \qquad \begin{pmatrix} X & -Y \\ Y & X \end{pmatrix} := (\Psi\Psi^{\mathrm{T}})^{-1/2}\Psi,$$

as in Theorem 2.2.12. We shall prove that the first Chern number of E is the degree of the composition $\rho \circ \Psi : C \to S^1$:

$$c_1(E) = \deg(\rho \circ \Psi).$$

In other words, the first Chern number is the sum of the Maslov indices of the loops $\Psi \circ \gamma_j : \mathbb{R}/\mathbb{Z} \to \mathrm{Sp}(2n)$, i.e.

$$c_1(E) = \sum_{j=1}^{\ell} \mu(\Psi \circ \gamma_j),$$

where ℓ is the number of components of C and each component is parametrized by a loop $\gamma_j : \mathbb{R}/\mathbb{Z} \to C$ such that $\dot{\gamma}_j(t)$ is positively oriented. A direct application of this definition is given in Example 2.7.6 below. The next lemma is required to prove that with this definition the first Chern number is independent of the choices, i.e. that the degree of $\rho \circ \Psi$ is independent of the splitting and of the trivializations.

Lemma 2.7.3 *Let Σ be a compact connected oriented 2-manifold with nonempty boundary. A smooth map $\Psi : \partial\Sigma \to \mathrm{Sp}(2n)$ extends to Σ if and only if*

$$\deg(\rho \circ \Psi) = 0.$$

Proof: First assume that Ψ extends to a smooth map $\Sigma \to \mathrm{Sp}(2n)$. Then the composition $\rho \circ \Psi : \partial\Sigma \to S^1$ extends to a smooth map $\Sigma \to S^1$ and hence must have degree zero.

Now assume that $\partial\Sigma = C_0 \cup C_1$, where both C_0 and C_1 are nonempty. We prove that any smooth map $\Psi : C_0 \to \mathrm{Sp}(2n)$ extends over Σ. This is obvious in the case of the cylinder $\Sigma = S^1 \times [0,1]$. The case where Σ is the disc with two holes and C_1 is the outer boundary can be reduced to that of the cylinder by extending the map $\Psi : C_0 \to \mathrm{Sp}(2n)$ over a line which connects the two holes. The general case is proved by decomposing Σ as in the proof of Proposition 2.6.7.

Conversely, assume $\deg(\rho \circ \Psi) = 0$. By what we have just proved, the map $\Psi : \partial\Sigma \to \mathrm{Sp}(2n)$ extends over $\Sigma \setminus B$, where B is a disc. By the first part of the proof the resulting loop $\partial B \to \mathrm{Sp}(2n)$ has Maslov index zero. By Theorem 2.2.12 this loop is contractible. This proves Lemma 2.7.3. □

Proof of Theorem 2.7.1: We prove that the degree of $\rho \circ \Psi$ as defined above is independent of the choice of the splitting and the trivialization, and satisfies the axioms of Theorem 2.7.1. By Lemma 2.7.3 the degree of $\rho \circ \Psi$ is independent of the choice of the trivialization. We prove that it is independent of the choice of the splitting. Consider a threefold splitting

$$\Sigma = \Sigma_{32} \cup_{C_2} \Sigma_{21} \cup_{C_1} \Sigma_{10}$$

and choose trivializations Φ_{31} and Φ_{20} of E over

$$\Sigma_{31} = \Sigma_{32} \cup_{C_2} \Sigma_{21}, \qquad \Sigma_{20} = \Sigma_{21} \cup_{C_1} \Sigma_{10}.$$

(See Fig. 2.3.)

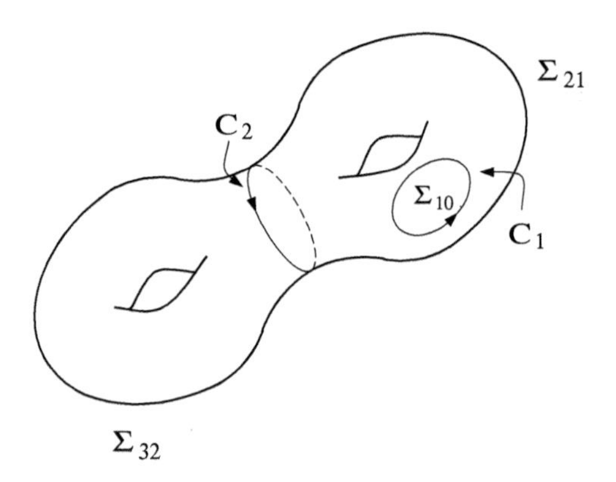

FIG. 2.3. A threefold splitting.

The corresponding overlap map is

$$\Psi_{21}(q) := \Phi_{31}(q)^{-1}\Phi_{20}(q)$$

for $q \in \Sigma_{21}$. The restriction $\Psi_1 := \Psi_{21}|C_1$ corresponds to the splitting

$$\Sigma = \Sigma_{31} \cup_{C_1} \Sigma_{10}$$

and $\Psi_2 := \Psi_{21}|C_2$ to the splitting

$$\Sigma = \Sigma_{32} \cup_{C_2} \Sigma_{20}.$$

By Lemma 2.7.3 the loops $\Psi_1 : C_1 \to \mathrm{Sp}(2n)$ and $\Psi_2 : C_2 \to \mathrm{Sp}(2n)$ with the induced orientations of $C_1 = \partial\Sigma_{10}$ and of $C_2 = \partial\Sigma_{20}$ have the same Maslov index. This proves the statement for two splittings along disjoint curves C_1 and C_2. The general case can be reduced to this one by choosing a third splitting.

We prove naturality. Fix a splitting $\Sigma = \Sigma_1 \cup_C \Sigma_2$, where C has only one connected component. Choose trivializations of E and E' over Σ_1 and Σ_2 and let $\Psi : C \to \mathrm{Sp}(2n)$ and $\Psi' : C \to \mathrm{Sp}(2n)$ be the corresponding loops of symplectic matrices. If Ψ and Ψ' have the same Maslov index, it follows that the loop $\Psi'\Psi^{-1} : C \to \mathrm{Sp}(2n)$ extends to a smooth map $\Sigma_1 \to \mathrm{Sp}(2n)$. Use this map to construct the required symplectic isomorphism $E \to E'$.

We prove functoriality. Choose a regular value $q \in \Sigma$ of $\phi : \Sigma' \to \Sigma$ and cut out a small neighbourhood $B \subset \Sigma$ of q. Then $\phi^{-1}(B)$ consists of $d = \deg(\phi)$ discs $B'_1, \dots, B'_d \subset \Sigma'$. Let $\Psi'_j : \partial B'_j \to \mathrm{Sp}(2n)$ be the corresponding parametrized loops of symplectic matrices arising from trivializations of ϕ^*E over B'_j and $\Sigma' - \cup_j B'_j$. These loops all have the same Maslov index $\deg \rho \circ \Psi'_j = c_1(E)$. Hence $c_1(\phi^*E) = d \cdot c_1(E)$.

Additivity follows from the identity $\det_{\mathbb{C}}(U \oplus U') = \det_{\mathbb{C}}(U)\det_{\mathbb{C}}(U')$ for two unitary matrices $U \in \mathrm{U}(n)$ and $U' \in \mathrm{U}(n')$. The axiom of normalization is left as an exercise and so is the proof of uniqueness. This proves Theorem 2.7.1. □

Chern–Weil theory

The first Chern class can also expressed as a curvature integral. We explain this in the case of a line bundle $L \to \Sigma$ over a compact oriented Riemann surface Σ. Let L be equipped with a Hermitian structure and consider the circle bundle (or unitary frame bundle) $\pi : P \to \Sigma$ of all vectors in L of length 1:

$$P = \{(z, v) \mid z \in \Sigma,\, v \in L_z,\, |v| = 1\}.$$

The circle $S^1 = \{\lambda \in \mathbb{C} \mid |\lambda| = 1\}$ acts on this bundle in the obvious way. We denote this action by $P \times S^1 \to P : (p, \lambda) \mapsto p \cdot \lambda$. For $i\theta$ in the Lie algebra $i\mathbb{R} = \mathrm{Lie}(S^1)$ we denote the induced vertical vector field on P by

$$p \cdot i\theta = \left.\frac{d}{dt}\right|_{t=0} p \cdot e^{i\theta t}$$

for $p \in P$. A **connection** on P is a 1-form $A \in \Omega^1(P, i\mathbb{R})$ with values in $i\mathbb{R} = \mathrm{Lie}(S^1)$ which is invariant under the action of S^1 and canonical in the vertical direction, namely

$$A_{p\cdot\lambda}(v \cdot \lambda) = A_p(v), \qquad A_p(p \cdot i) = i$$

for $p \in P$, $v \in T_pP$, and $\lambda \in S^1$. Since the group is abelian, the curvature of a connection 1-form A agrees with the differential dA. This differential is invariant and horizontal in the sense that $dA_p(p \cdot i, v) = 0$ for all $v \in T_pP$. Any such form descends to the base Σ, i.e.

$$dA = \pi^* F_A, \qquad F_A \in \Omega^2(\Sigma, i\mathbb{R}).$$

The 2-form F_A is called the **curvature of** A. The following result is a simple case of Chern–Weil theory.

Theorem 2.7.4 *Every connection* 1*-form* A *on* P *satisfies*

$$c_1(P) = \frac{i}{2\pi} \int_\Sigma F_A.$$

Moreover, for every 2*-form* $\tau \in \Omega^2(\Sigma, i\mathbb{R})$ *with* $\int_\Sigma \tau = -2\pi i c_1(P)$ *there exists a connection* 1*-form* A *on* P *with curvature* $F_A = \tau$.

Proof: Decompose $\Sigma = \Sigma_1 \cup_C \Sigma_2$ and choose sections $s_j : \Sigma_j \to P$. These give rise to trivializations $\Sigma_j \times \mathbb{C} \to L : (z, \zeta) \mapsto s_j(z)\zeta$ and, by the discussion after Theorem 2.7.1, the first Chern number $c_1(L) = c_1(P) \in \mathbb{Z}$ is the degree of the loop $\gamma : C \to S^1$ defined by

$$s_1(z)\gamma(z) = s_2(z), \qquad z \in C,$$

where C is oriented as the boundary of Σ_1. Now the connection A pulls back to 1-forms $\alpha_j = s_j^* A \in \Omega^1(\Sigma_j, i\mathbb{R})$ and we have

$$\begin{aligned}\int_\Sigma F_A &= \int_{\Sigma_1} d\alpha_1 + \int_{\Sigma_2} d\alpha_2 \\ &= \int_C (\alpha_1 - \alpha_2) \\ &= -\int_C \gamma^{-1} d\gamma \\ &= -2\pi i \deg(\gamma).\end{aligned}$$

The last identity is a simple exercise. Since $c_1(L) = c_1(P) = \deg(\gamma)$ this proves the first statement of the theorem. To prove the second statement, let $\tau \in \Omega^2(\Sigma)$ be any 2-form such that

$$\int_\Sigma \tau = -2\pi i c_1(P).$$

Then $\tau - F_A$ is exact and hence $\tau - F_A = d\alpha$ for some 1-form $\alpha \in \Omega^1(\Sigma, i\mathbb{R})$. It follows that $A + \pi^*\alpha$ is a connection 1-form with

$$d(A + \pi^*\alpha) = \pi^*(F_A + d\alpha) = \pi^*\tau.$$

Hence $F_{A+\pi^*\alpha} = \tau$. This proves Theorem 2.7.4. □

Chern class and Euler class

Let $L \to \Sigma$ be a complex line bundle (or symplectic bundle of rank 2) over an oriented Riemann surface Σ. Then the first Chern number can also be interpreted as the self-intersection number of the zero section of L. More precisely, let $s : \Sigma \to L$ be a section of L which is transverse to the zero section. Then s has finitely many zeros and at each zero z the intersection index $\iota(z, s)$ is defined to be ± 1 according to whether or not the linearized map

$$Ds(z) := \mathrm{pr} \circ ds(z) : T_z\Sigma \to L_z$$

is orientation-preserving or orientation-reversing. (Here $\mathrm{pr} : T_{(z,0)}L \to L_z$ denotes the projection associated to the canonical splitting $T_{(z,0)}L = T_z\Sigma \oplus L_z$.) The following theorem shows that the sum of these indices agrees with the first Chern number of L. In other words $c_1(L)$ is the obstruction to the existence of a nonvanishing section of L and so coincides with the **Euler number** of L.

Theorem 2.7.5 *If the section $s : \Sigma \to L$ is transverse to the zero section then the first Chern number of L is given by*

$$c_1(L) = \sum_{s(z)=0} \iota(z, s). \tag{2.7.1}$$

Proof: The proof is an exercise with hint. Cut out a small neighbourhood U of the zero set of s and use s to trivialize the bundle L over the complement $\Sigma' := \Sigma \setminus U$. Use a different method to trivialize Σ over U and then compare the two trivializations over the boundary of U (which consists of finitely many circles). □

Examples and Exercises

We begin by working out the first Chern class of the normal bundle $\nu_{\mathbb{C}P^1}$ to the line $\mathbb{C}P^1$ in the complex projective plane $\mathbb{C}P^2$. We will use the notation of Example 4.3.3 and, for simplicity, will work in the complex context, choosing complex trivializations rather than symplectic ones. By Theorem 2.6.3, this will make no difference to the final result.

Example 2.7.6 Consider the line $\{[z_0 : z_1 : 0]\}$ in $\mathbb{C}P^2$. It is covered by the two discs

$$\Sigma_1 := \{[1 : z_1 : 0] \,|\, |z_1| \le 1\}, \qquad \Sigma_2 := \{[z_0 : 1 : 0] \,|\, |z_0| \le 1\},$$

which intersect in the circle

$$C := \{[1 : e^{2\pi i\theta} : 0]\} = \Sigma_1 \cap \Sigma_2.$$

Note that θ is a positively oriented coordinate for C. The fibre of the normal bundle $\nu_{\mathbb{C}P^1}$ over the point $[1 : z_1 : 0] \in \Sigma_1$ embeds into $\mathbb{C}P^2$ as the submanifold $\{[1 : z_1 : w] \,|\, w \in \mathbb{C}\}$. Therefore we may choose the complex trivializations

$$\begin{aligned}
\Phi_1 &: \Sigma_1 \times \mathbb{C} \to \nu_{\mathbb{C}P^1}, \quad \Phi_1([1 : z_1 : 0], w) = [1 : z_1 : w], \\
\Phi_2 &: \Sigma_2 \times \mathbb{C} \to \nu_{\mathbb{C}P^1}, \quad \Phi_2([z_0 : 1 : 0], w) = [z_0 : 1 : w].
\end{aligned}$$

The map $\Phi_1^{-1}\Phi_2(z) : \mathbb{C} \to \mathbb{C}$ is the composite

$$w \mapsto [1/z : 1 : w] = [1 : z : zw] \mapsto zw,$$

and so induces the map $C \to U(1) : \theta \mapsto e^{2\pi i\theta}$. This shows that $c_1(\nu_{\mathbb{C}P^1}) = 1$. □

Exercise 2.7.7 Use the formula (2.7.1) to calculate the first Chern number $c_1(\nu_{\mathbb{C}P^1})$ of the normal bundle $\nu_{\mathbb{C}P^1}$ of $\mathbb{C}P^1$ in $\mathbb{C}P^2$. □

Exercise 2.7.8 Let $L \subset \mathbb{C}^n \times \mathbb{C}P^{n-1}$ be the incidence relation:

$$\begin{aligned}
L &= \{(z, \ell) \,|\, z \in \ell\} \\
&= \{(z_1, \dots, z_n; [w_1 : \dots : w_n]) \,|\, w_j z_k = w_k z_j \; \forall\, j, k\}.
\end{aligned}$$

The projection $\mathrm{pr} : L \to \mathbb{C}P^{n-1}$ gives L the structure of a complex line bundle over $\mathbb{C}P^{n-1}$. Show that when $n = 2$ the first Chern number of the restriction $L|_{\mathbb{C}P^1}$ is -1, and hence calculate $c_1(L)$ for arbitrary n. Another approach to this calculation is given in Lemma 7.1.1. This bundle is called either the **tautological line bundle** over $\mathbb{C}\mathrm{P}^n$ or the **universal line bundle**, where the second name derives from the fact that for $n > k$ it classifies line bundles over k-dimensional CW-complexes; cf. [488]. □

Exercise 2.7.9 Prove that every symplectic vector bundle over a Riemann surface decomposes as a direct sum of 2-dimensional symplectic vector bundles. You can either prove this directly, by constructing suitable nonvanishing sections, or use the naturality axiom in Theorem 2.7.1. □

Exercise 2.7.10 **(i)** Suppose that $E \to \Sigma$ is a symplectic vector bundle over a closed oriented 2-manifold Σ that extends over a compact oriented 3-manifold Y with boundary $\partial Y = \Sigma$. Prove that the restriction $E|_\Sigma$ has first Chern class zero. **Hint:** This argument relies on the nontrivial fact that every complex (or equivalently symplectic) vector bundle over a 3-dimensional manifold splits as the sum of a complex line bundle with a trivial bundle. (By obstruction theory, this holds essentially because $\pi_2(\mathrm{U}(n)) = 0$.) Thus it suffices to consider line bundles. Argue further using either the curvature of a connection as in Theorem 2.7.4 or the zero set of a section as in Theorem 2.7.5.

(ii) Use (i) above and Exercise 2.7.9 to substantiate the claim made in Remark 2.7.2 that the Chern class $c_1(f^*E)$ depends only on the homology class of f. Here the main problem is that when $f_*([\Sigma])$ is null-homologous the 3-chain C that bounds it need not be representable by a 3-manifold. However, because we can assume that we are working in a manifold of dimension at least 4, its singularities can be assumed to have codimension 2 in C and so the proof of (i) goes through. □

Exercise 2.7.11 Prove that every symplectic vector bundle $E \to \Sigma$ that admits a Lagrangian subbundle can be symplectically trivialized. **Hint:** Use the proof of Theorem 2.7.1 to show that $c_1(E) = 0$. □

Exercise 2.7.12 Let E and E' be two complex vector bundles of ranks n and n', respectively, over a closed oriented two-manifold Σ. Prove that the first Chern number of their complex tensor product $E \otimes E'$ is given by

$$c_1(E \otimes E') = n'c_1(E) + nc_1(E').$$

Hint: Use the formula

$$\det_{\mathbb{C}}(U \otimes U') = \left(\det_{\mathbb{C}}(U)\right)^{n'} \left(\det_{\mathbb{C}}(U')\right)^{n}$$

for two unitary matrices $U \in \mathrm{U}(n)$ and $U' \in \mathrm{U}(n')$. □

Exercise 2.7.13 An isomorphism class of complex line bundles over the torus

$$\mathbb{T}^m = \mathbb{R}^m/\mathbb{Z}^m$$

can be described as an equivalence class of **cocycles**

$$\mathbb{Z}^m \to \mathrm{Map}(\mathbb{R}^m, S^1) : k \mapsto \phi_k$$

which satisfy the **cocycle condition**

$$\phi_{k+\ell}(x) = \phi_\ell(x+k)\phi_k(x)$$

for $x \in \mathbb{R}^m$ and $k, \ell \in \mathbb{Z}^m$. Two such cocycles ϕ and ψ are called **equivalent** if there is a function $g : \mathbb{R}^m \to S^1$ such that $\psi_k(x) = g(x+k)^{-1}\phi_k(x)g(x)$ for $x \in \mathbb{R}^m$ and $k \in \mathbb{Z}^m$. The complex line bundle associated to a cocycle ϕ is the quotient

$$L := L(\phi) := \frac{\mathbb{R}^m \times \mathbb{C}}{\mathbb{Z}^m}, \qquad k \cdot (x, z) = (x+k, \phi_k(x)z).$$

A section of L is a function $s : \mathbb{R}^m \to \mathbb{C}$ that satisfies $s(x+k) = \phi_k(x)s(x)$ for $x \in \mathbb{R}^m$ and $k \in \mathbb{Z}^m$.

(i) Prove that $L(\phi)$ is isomorphic to $L(\psi)$ if and only if ϕ is equivalent to ψ.

(ii) Let $A = \sum_{\nu=1}^m A_\nu dx_\nu \in \Omega^1(\mathbb{R}^m, i\mathbb{R})$, where the $A_\nu : \mathbb{R}^m \to i\mathbb{R}$ satisfy

$$A_\nu(x+k) - A_\nu(x) = -\phi_k(x)^{-1}\frac{\partial \phi_k}{\partial x_\nu}(x), \qquad \nu = 1, \dots, m. \tag{2.7.2}$$

Prove that $\nabla_A := d + A$ is a unitary connection on $L(\phi)$.

(iii) Let $B \in \mathbb{Z}^{m\times m}$ and consider the cocycle

$$\phi_{B,k}(x) := \exp(2\pi i k^T Bx). \tag{2.7.3}$$

Prove that the 1-form

$$A := -2\pi i \sum_{\nu,\mu=1}^m x_\nu B_{\nu\mu} dx_\mu$$

satisfies (2.7.2) and has the curvature $F_A = -2\pi i \sum_{\nu<\mu} (B_{\nu\mu} - B_{\mu\nu})\, dx_\nu \wedge dx_\mu$. Deduce that

$$c_1(L(\phi_B)) = \left[\sum_{\nu<\mu}^m C_{\nu\mu} dx_\nu \wedge dx_\mu\right], \qquad C := B - B^T. \tag{2.7.4}$$

(See Remark 2.7.2 and Theorem 2.7.4.)

(iv) Prove that the complex line bundle in (iii) admits a trivialization whenever B is symmetric and admits a square root whenever B is skew-symmetric.

(v) Prove that another cocycle with first Chern class (2.7.4) is given by

$$\phi_k(x) := \varepsilon(k)\exp(\pi i k^T Cx),$$

where the numbers $\varepsilon(k) = \pm 1$ are chosen such that

$$\varepsilon(k+\ell) = \varepsilon(k)\varepsilon(\ell)\exp(\pi i k^T C\ell).$$

If $C = B - B^T$ then the numbers $\varepsilon(k) = \exp(\pi i k^T Bk)$ satisfy this condition.

(vi) Prove that every cocycle is equivalent to one of the form (2.7.3). **Hint:** Choose two unitary connections A and A' with constant curvature form for two cocycles ϕ and ϕ' with the same first Chern class. Show that $A' - A = d\xi$ for some function $\xi : \mathbb{R}^m \to i\mathbb{R}$ and deduce that $g = \exp(\xi) : \mathbb{R}^m \to S^1$ transforms ϕ into ϕ'. □

3

SYMPLECTIC MANIFOLDS

This is a foundational chapter, and everything in it (except perhaps Section 3.5 on contact structures) is needed to understand later chapters. The first section contains elementary definitions and first examples of symplectic manifolds. The second section is devoted to Darboux's theorem. As Gromov notes in [289], symplectic geometry is a curious mixture of the 'hard' and the 'soft'. Some situations are flexible and there are no nontrivial invariants, while other situations are rigid. In this chapter we deal with 'soft' phenomena, the fact that in symplectic geometry there are no local invariants. The classical formulation of this principle is known as Darboux's theorem: all symplectic forms are locally diffeomorphic. However there are many other related results, such as Moser's stability theorem and various versions of the symplectic neighbourhood theorem. We prove these in Sections 3.2 and 3.4 using Moser's homotopy method. Section 3.3 establishes some elementary isotopy extension theorems. The chapter ends with a discussion of contact geometry. This is the odd-dimensional analogue of symplectic geometry, and there are many connections between the two subjects.

3.1 Basic concepts

Throughout we will assume that M is a connected C^∞-smooth manifold, which (unless specific mention is made to the contrary) has no boundary. Very often, M will also be compact. A **closed** manifold is a compact manifold without boundary.

A **symplectic structure** on a smooth manifold M is a nondegenerate closed 2-form $\omega \in \Omega^2(M)$. Nondegeneracy means that each tangent space (T_qM, ω_q) is a symplectic vector space. The manifold M is necessarily of even dimension $2n$ and, by Corollary 2.1.4, the n-fold wedge product

$$\omega^n = \omega \wedge \cdots \wedge \omega$$

never vanishes. Thus M is oriented.

Example 3.1.1 The first example of a symplectic manifold is $\mathbb{R}^{2n}$ itself with the standard symplectic form

$$\omega_0 = \sum_{j=1}^{n} dx_j \wedge dy_j$$

which was defined in Section 1.1. □

Example 3.1.2 Another basic example is the 2-sphere with its standard area form. If we think of S^2 as the unit sphere

$$S^2 = \left\{ (x_1, x_2, x_3) \in \mathbb{R}^3 \,\middle|\, x_1^2 + x_2^2 + x_3^2 = 1 \right\},$$

then the induced area form is given by

$$\omega_x(\xi, \eta) = \langle x, \xi \times \eta \rangle$$

for $\xi, \eta \in T_x S^2$ and the total area of S^2 is 4π. □

Example 3.1.3 The construction in Example 3.1.2 extends to every 2-dimensional oriented submanifold $\Sigma \subset \mathbb{R}^3$, equipped with a normal vector field $\nu : \Sigma \to S^2$ so that $\nu(x) \perp T_x\Sigma$ for all $x \in \Sigma$. An example of a symplectic form on Σ is the standard area form $\omega \in \Omega^2(\Sigma)$ associated to the normal vector field ν. It is given by

$$\omega_x(\xi, \eta) := \langle \nu(x), \xi \times \eta \rangle = \det(\nu(x), \xi, \eta)$$

for $x \in \Sigma$ and $\xi, \eta \in T_x\Sigma = \nu(x)^\perp$. □

The next exercise goes back to Archimedes.

Exercise 3.1.4 Consider cylindrical polar coordinates (θ, x_3) on the sphere minus its poles $S^2 \setminus \{(0, 0, \pm 1)\}$, where $0 \leq \theta < 2\pi$ and $-1 < x_3 < 1$. Show that the area form induced by the Euclidean metric is precisely the form $\omega = d\theta \wedge dx_3$. In other words, the horizontal projection from the cylinder to the sphere preserves the surface area. (See Fig. 3.1.) Here the sphere is oriented as the boundary of the unit ball: given a point $x \in S^2$, a positively oriented basis for $T_x S^2$ is $\{\partial_\theta, \partial_3\}$ because $\det(x, \partial_\theta, \partial_3) > 0$. □

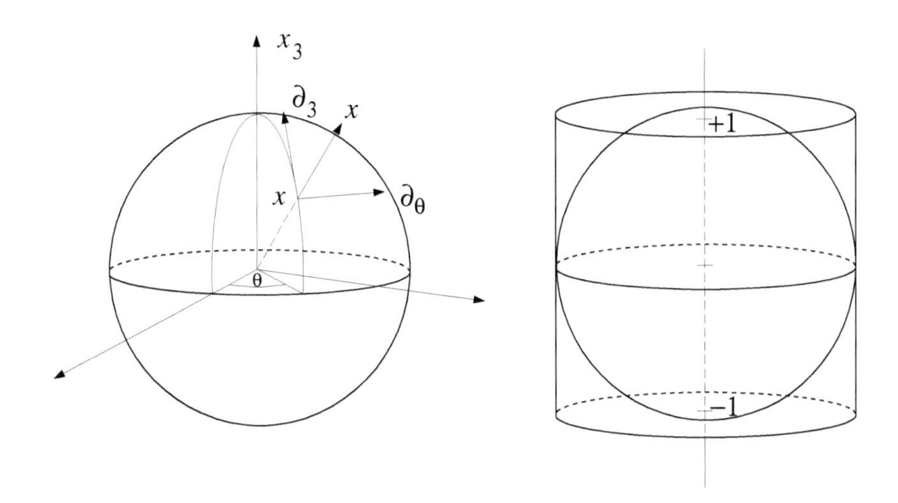

FIG. 3.1. The sphere and cylinder.

Observe that a symplectic form satisfies two conditions. The first (nondegeneracy) is purely algebraic and means that there is a canonical isomorphism between the tangent and cotangent bundles, namely

$$TM \to T^*M \,:\, X \mapsto \iota(X)\omega = \omega(X, \cdot). \tag{3.1.1}$$

The second (closedness) is geometric. Both conditions come together in Proposition 3.1.10 below, which shows how a function on a symplectic manifold generates a family of diffeomorphisms which preserve the symplectic structure. This is a key feature of symplectic geometry which sharply distinguishes it from Riemannian geometry.

One immediate consequence of closedness is that the symplectic form ω represents a cohomology class

$$a = [\omega] \in H^2(M; \mathbb{R}).$$

If M is closed (i.e. compact and without boundary) then the cohomology class $a^n \in H^{2n}(M; \mathbb{R})$ is represented by the volume form $\omega^n \in \Omega^{2n}(M)$ and the integral of this form over M does not vanish. Hence the symplectic form ω cannot be exact or, equivalently, the n-fold cup product of a is nonzero:

$$a^n = a \cup \cdots \cup a \neq 0.$$

It follows that there are orientable even-dimensional manifolds which do not admit a symplectic structure, for example any sphere S^{2n} with $n \geq 2$. It is a natural question whether every closed oriented $2n$-dimensional manifold M, equipped with a nondegenerate 2-form $\tau \in \Omega^2(M)$ compatible with the orientation and a cohomology class $a \in H^2(M; \mathbb{R})$ satisfying $a^n > 0$, admits a symplectic form (in the cohomology class a and homotopic to τ through nondegenerate 2-forms). This is the **existence problem** for symplectic structures. In dimension two the answer is positive and orientability is the only condition. In dimension four additional necessary conditions, arising from Donaldson's symplectic submanifolds (see [146]) and from Seiberg–Witten theory (see Taubes [606, 607]), were discovered in 1994. These conditions will be discussed further in Chapter 14. In dimensions $2n \geq 6$ the existence problem is still open at the time of writing.

A **symplectomorphism** of a symplectic manifold (M, ω) is a diffeomorphism $\psi \in \mathrm{Diff}(M)$ which preserves the symplectic form

$$\omega = \psi^*\omega.$$

We denote the group of symplectomorphisms of (M, ω) by

$$\mathrm{Symp}(M, \omega) := \{\phi \in \mathrm{Diff}(M) \,|\, \phi^*\omega = \omega\}, \tag{3.1.2}$$

or sometimes by $\mathrm{Symp}(M)$. Since ω is nondegenerate the homomorphism (3.1.1) is bijective. Thus, just as in Riemannian geometry, there is a one-to-one correspondence between vector fields and 1-forms via $\mathcal{X}(M) \to \Omega^1(M) : X \mapsto \iota(X)\omega$.

A vector field $X \in \mathcal{X}(M)$ is called **symplectic** if $\iota(X)\omega$ is closed. We denote the space of symplectic vector fields by

$$\mathcal{X}(M,\omega) := \{X \in \mathcal{X}(M) \mid \mathcal{L}_X\omega = d\iota(X)\omega = 0\}. \tag{3.1.3}$$

The next result shows that when M is closed, $\mathcal{X}(M,\omega)$ is the Lie algebra of the group $\mathrm{Symp}(M,\omega)$.

Proposition 3.1.5 *Let M be a closed manifold. If $t \mapsto \psi_t \in \mathrm{Diff}(M)$ is a smooth family of diffeomorphisms generated by a family of vector fields $X_t \in \mathcal{X}(M)$ via*

$$\frac{d}{dt}\psi_t = X_t \circ \psi_t, \qquad \psi_0 = \mathrm{id},$$

then $\psi_t \in \mathrm{Symp}(M,\omega)$ for every t if and only if $X_t \in \mathcal{X}(M,\omega)$ for every t. Moreover, if $X, Y \in \mathcal{X}(M,\omega)$ then $[X,Y] \in \mathcal{X}(M,\omega)$ and

$$\iota([X,Y])\omega = dH, \quad \textit{where} \quad H = \omega(X,Y).$$

Proof: The first statement follows from the identity

$$\frac{d}{dt}{\psi_t}^*\omega = {\psi_t}^*\left(\mathcal{L}_{X_t}\omega\right) = {\psi_t}^*\left(\iota(X_t)d\omega + d(\iota(X_t)\omega)\right) = {\psi_t}^*d(\iota(X_t)\omega),$$

which, in turn, follows from Cartan's formula for the Lie derivative

$$\mathcal{L}_X\omega = \iota(X)d\omega + d(\iota(X)\omega)$$

and the fact that ω is closed. Note, moreover, that X is a symplectic vector field if and only if $\mathcal{L}_X\omega = 0$.

Now let X and Y be symplectic vector fields with corresponding flows ϕ_t and ψ_t. According to the sign conventions in Remark 3.1.6 below we have

$$[X,Y] = \mathcal{L}_Y X = \left.\frac{d}{dt}\right|_{t=0} {\psi_t}^* X$$

and hence

$$\begin{aligned}
\iota([X,Y])\omega &= \left.\frac{d}{dt}\right|_{t=0} \iota(\psi_t^* X)\omega \\
&= \mathcal{L}_Y(\iota(X)\omega) \\
&= d\iota(Y)\iota(X)\omega \\
&= d(\omega(X,Y)).
\end{aligned}$$

Here the second equality follows from the fact that $\psi_t^*\omega = \omega$ for all t and hence $\iota({\psi_t}^*X)\omega = {\psi_t}^*(\iota(X)\omega)$. The third equality follows from Cartan's formula and the fact that $\iota(X)\omega$ is closed. This proves Proposition 3.1.5. □

Remark 3.1.6 (Sign conventions) There are various mutually inconsistent sign conventions in common use, and unfortunately it is impossible to make consistent choices that yield all the identities which one would wish. We have chosen to arrange the signs so as to obtain the formula $[X_F, X_H] = X_{\{F,H\}}$ in Lemma 1.1.18 above and Proposition 3.1.10 below, and as a result a somewhat awkward minus sign appears somewhere else. Many authors deal with this by defining X_H by $\iota(X_H)\omega_0 = -dH$, but this gives the wrong sign for the Hamiltonian equations, unless the sign of the symplectic form ω_0 is reversed as well, which is inconsistent with complex geometry. Another possibility is to reverse the sign in the definition of the Lie bracket of vector fields. This is contrary to customary usage but is more natural.[11] In other words, we define

$$[X, Y] = -\mathcal{L}_X Y = -\left.\frac{d}{dt}\right|_{t=0} \phi_t^* Y,$$

where $\phi^* Y(q) = d\phi(q)^{-1} Y(\phi(q))$ denotes the pullback of the vector field Y under the diffeomorphism ϕ and ϕ_t denotes the flow of X.

To put this in context, recall that a vector field X induces a derivation $\mathcal{L}_X$ on the space of smooth functions $C^\infty(M)$ which is given by differentiating a function f in the direction X, namely

$$\mathcal{L}_X f = df \circ X = \left.\frac{d}{dt}\right|_{t=0} f \circ \phi_t.$$

The derivations form a Lie algebra $\mathrm{Der}(M)$ of operators on $C^\infty(M)$ with the Lie bracket given by the commutator

$$[\mathcal{L}_X, \mathcal{L}_Y] = \mathcal{L}_X \mathcal{L}_Y - \mathcal{L}_Y \mathcal{L}_X.$$

With our definition of the Lie bracket of vector fields we have

$$\mathcal{L}_{[X,Y]} = -[\mathcal{L}_X, \mathcal{L}_Y].$$

This means that the operator

$$\mathcal{X}(M) \to \mathrm{Der}(M) : X \mapsto \mathcal{L}_X$$

is a *Lie algebra anti-homomorphism*. However, this operator is the differential of the map $\mathrm{Diff}(M) \to \mathrm{Aut}(C^\infty(M)) : \phi \mapsto \phi^*$ given by $\phi^* f = f \circ \phi$. Since

$$(\phi \circ \psi)^* = \psi^* \phi^*$$

this map is a *Lie group anti-homomorphism* with respect to the natural group operations. Hence our sign convention for the Lie bracket of vector fields is consistent with the usual action of the diffeomorphism group on $C^\infty(M)$. It is also in accordance with standard conventions in the theory of Lie groups and algebras. □

[11] Arnold [23] also uses this definition of the Lie bracket of vector fields, although his sign conventions differ from ours in other places.

Hamiltonian flows

We now extend the concepts introduced in Chapter 1 to general symplectic manifolds. For any smooth function $H : M \to \mathbb{R}$, the vector field $X_H : M \to TM$ determined by the identity

$$\iota(X_H)\omega = dH$$

is called the **Hamiltonian vector field** associated to the **Hamiltonian function** H. If M is closed, the vector field X_H generates a smooth 1-parameter group of diffeomorphisms $\phi_H^t \in \mathrm{Diff}(M)$ satisfying

$$\frac{d}{dt}\phi_H^t = X_H \circ \phi_H^t, \qquad \phi_H^0 = \mathrm{id}.$$

This is called the **Hamiltonian flow** associated to H. The identity

$$dH(X_H) = (\iota(X_H)\omega)(X_H) = \omega(X_H, X_H) = 0$$

shows that the vector field X_H is tangent to the level sets $H = \mathit{const}$ of H.

Example 3.1.7 As a simple example, take $H : S^2 \to \mathbb{R}$ to be the height function x_3 on the 2-sphere in Exercise 3.1.4. The level sets are circles at constant height, and the Hamiltonian flow ϕ_H^t rotates each circle at constant speed. In fact, in cylindrical polar coordinates, X_H is simply the vector field $\partial/\partial\theta$. Thus ϕ_H^t is the rotation of the sphere about its vertical axis through the angle t. (See Fig. 3.2.) This flow is considered again at the beginning of Chapter 5 as an example of a circle action. □

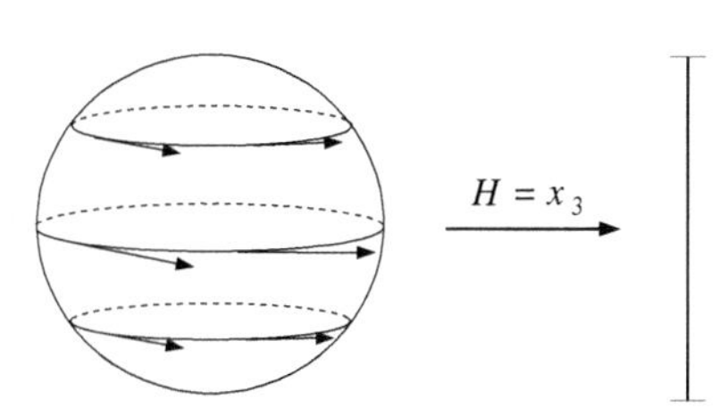

FIG. 3.2. Rotating the sphere.

As in Section 1.1, a smooth function $F \in C^\infty(M)$ is constant along the orbits of the flow of H if and only if the **Poisson bracket**

$$\{F, H\} = \omega(X_F, X_H) = dF(X_H)$$

vanishes. Because ω is closed, the Poisson bracket defines a Lie algebra structure on the space of smooth functions on M. The easiest way to prove this is to use Lemma 1.1.18 together with Darboux's theorem which is discussed in Section 3.2 below. Alternatively, one can use the following exercise.

Exercise 3.1.8 Assume that τ is a nondegenerate 2-form on M which is not necessarily closed. (Such a pair (M,τ) is sometimes called an **almost symplectic manifold**.) In this case Hamiltonian vector fields and Poisson brackets can be defined as above. Show that

$$\{\{F,G\},H\}+\{\{G,H\},F\}+\{\{H,F\},G\}=\tfrac{1}{2}d\tau(X_F,X_G,X_H) \tag{3.1.4}$$

for any three functions $F,G,H\in C^\infty(M)$. (**Note:** Some relevant formulae can be found in the proof of Lemma 4.1.14 below.) □

Exercise 3.1.9 Let (M,ω) be a $2n$-dimensional symplectic manifold. Prove that the Poisson bracket of two functions $F,G:M\to\mathbb{R}$ satisfies

$$\{F,G\}\frac{\omega^n}{n!}=dF\wedge dG\wedge\frac{\omega^{n-1}}{(n-1)!}. \tag{3.1.5}$$

If one of the functions F,G has compact support, deduce that their Poisson bracket has mean value zero (i.e. $\int_M\{F,G\}\omega^n=0$). □

Proposition 3.1.10 *Let (M,ω) be a symplectic manifold.*

(i) *Wherever defined, the Hamiltonian flow ϕ_H^t is a symplectomorphism which is tangent to the level surfaces of H.*

(ii) *For every Hamiltonian function $H:M\to\mathbb{R}$ and every symplectomorphism $\psi\in\mathrm{Symp}(M,\omega)$ we have $X_{H\circ\psi}=\psi^*X_H$.*

(iii) *The Lie bracket of two Hamiltonian vector fields X_F and X_G is the Hamiltonian vector field $[X_F,X_G]=X_{\{F,G\}}$.*

Proof: Statement (i) was proved in Proposition 3.1.5. Statement (ii) is the identity

$$\iota(X_{H\circ\psi})\omega=d(H\circ\psi)=\psi^*dH=\psi^*\iota(X_H)\omega=\iota(\psi^*X_H)\omega.$$

To prove statement (iii), note first that since $\phi_F^t\in\mathrm{Symp}(M,\omega)$, we have

$$[X_F,X_G]=-\left.\frac{d}{dt}\right|_{t=0}(\phi_F^t)^*X_G=-\left.\frac{d}{dt}\right|_{t=0}X_{G\circ\phi_F^t}.$$

Hence

$$\begin{aligned}\iota([X_F,X_G])\omega&=-\left.\frac{d}{dt}\right|_{t=0}d(G\circ\phi_F^t)\\&=-d\left.\frac{d}{dt}\right|_{t=0}G\circ\phi_F^t\\&=-d(dG(X_F))\\&=-d\{G,F\}\\&=d\{F,G\}\end{aligned}$$

as required. □

Proposition 3.1.10 shows that the Hamiltonian vector fields form a Lie subalgebra of the Lie algebra of symplectic vector fields. The map $H \mapsto X_H$ is a surjective Lie algebra homomorphism from the Lie algebra of smooth functions on M with the Poisson bracket to the Lie algebra of Hamiltonian vector fields. The kernel of this homomorphism consists of the constant functions.

Remark 3.1.11 Let (M, ω) be a compact connected symplectic manifold. Then the space of smooth functions $F : M \to \mathbb{R}$ that have mean value zero with respect to ω (i.e. $\int_M F\omega^n = 0$) is closed under the Poisson bracket by Exercise 3.1.9. Hence it follows from Proposition 3.1.10 that the map $H \mapsto X_H$ is a Lie algebra isomorphism from the space of smooth functions on M with mean value zero under the Poisson bracket to the space of Hamiltonian vector fields on M under the Lie bracket. □

By Proposition 3.1.10 the Hamiltonian function H is constant along the flow lines of the associated Hamiltonian vector field X_H and hence every level set of H is an invariant submanifold of M. Conversely, let $S \subset M$ be any compact orientable hypersurface (that is, a submanifold of codimension 1) of a symplectic manifold (M, ω). By Exercise 2.1.14 every such hypersurface is a coisotropic submanifold. Hence the vector space

$$L_q = (T_qS)^\omega = \{v \in T_qM \,|\, \omega(v, w) = 0 \text{ for } w \in T_qS\}$$

is a 1-dimensional subspace of T_qS for every $q \in S$ and hence defines a real line bundle L over S. It integrates to give a 1-dimensional foliation of S called the **characteristic foliation**. As we have seen in the beginning of Section 1.2, the leaves of this foliation are the integral curves of any Hamiltonian vector field X_H for which S is a regular level surface of the function H (or a component of such a surface).

Exercise 3.1.12 Let S be a compact orientable hypersurface in the symplectic manifold (M, ω). Prove that there exists a smooth function $H : M \to \mathbb{R}$ such that 0 is a regular value of H and $S \subset H^{-1}(0)$. Prove that $X_H(q) \in L_q$ for $q \in S$. (Note the example of a noncontractible circle S in $M = \mathbb{T}^2$. In this case there is no smooth function $H : M \to \mathbb{R}$ such that $S = H^{-1}(0)$. Such a function can only exist if the complement of S is disconnected.) □

Hamiltonian isotopies

Let (M, ω) be a symplectic manifold without boundary. A **symplectic isotopy** of (M, ω) is a smooth map $[0, 1] \times M \to M : (t, q) \mapsto \psi_t(q)$ such that ψ_t is a symplectomorphism for every t and ψ_0 is the identity. Any such isotopy is generated by a smooth family of vector fields $X_t : M \to TM$ via

$$\frac{d}{dt}\psi_t = X_t \circ \psi_t, \qquad \psi_0 = \mathrm{id}. \tag{3.1.6}$$

Since ψ_t is a symplectomorphism for every t, the vector fields X_t are symplectic, i.e. $d\,\iota(X_t)\omega = 0$ (see Proposition 3.1.5). A symplectic isotopy $\{\psi_t\}_{0 \le t \le 1}$ is called a **Hamiltonian isotopy** if the 1-form $\iota(X_t)\omega$ is exact and so X_t is a Hamiltonian vector field for every t. In this case there is a smooth function $H : [0,1] \times M \to \mathbb{R}$ such that, for each t, the function $H_t := H(t,\cdot)$ generates the vector field X_t via

$$\iota(X_t)\omega = dH_t. \tag{3.1.7}$$

The function H is called a **time-dependent Hamiltonian**. It is determined by the Hamiltonian isotopy only up to an additive function $c : [0,1] \to \mathbb{R}$. If M is simply connected then every symplectic isotopy is a Hamiltonian isotopy.

Definition 3.1.13 *A symplectomorphism $\psi \in \mathrm{Symp}(M,\omega)$ is called* **Hamiltonian** *if there exists a Hamiltonian isotopy $\psi_t \in \mathrm{Symp}(M,\omega)$ from $\psi_0 = \mathrm{id}$ to $\psi_1 = \psi$. We denote the space of Hamiltonian symplectomorphisms by*

$$\mathrm{Ham}(M,\omega) := \left\{ \psi \in \mathrm{Symp}(M,\omega) \;\middle|\; \begin{array}{l} \exists\, [0,1] \to C^\infty(M) : t \mapsto H_t \\ \exists\, [0,1] \to \mathrm{Diff}(M) : t \mapsto \psi_t \\ \textit{with } (3.1.6),\ (3.1.7),\ \psi_1 = \psi \end{array} \right\}. \tag{3.1.8}$$

Every compactly supported Hamiltonian function $H : [0,1] \times M \to \mathbb{R}$ determines a compactly supported Hamiltonian isotopy $\{\psi_t\}_{0 \le t \le 1}$ via (3.1.6) *and* (3.1.7), *and its time-*1 *map will be denoted by*

$$\phi_H := \psi_1.$$

Every such Hamiltonian symplectomorphism is called **compactly supported** *and we denote*

$$\mathrm{Ham_c}(M,\omega) := \{\phi_H \,|\, H \in C_0^\infty([0,1] \times M)\}. \tag{3.1.9}$$

Thus $\mathrm{Ham_c}(M,\omega) = \mathrm{Ham}(M,\omega)$ *whenever M is closed. It is sometimes convenient to abbreviate*

$$\mathrm{Ham}(M) := \mathrm{Ham}(M,\omega), \qquad \mathrm{Ham_c}(M) := \mathrm{Ham_c}(M,\omega).$$

It turns out that $\mathrm{Ham}(M,\omega)$ is a normal subgroup of $\mathrm{Symp}(M,\omega)$ and that its Lie algebra is the algebra of all Hamiltonian vector fields. The last statement requires some proof because it is not clear from the definition that an arbitrary path in $\mathrm{Ham}(M,\omega)$ is a Hamiltonian isotopy. We will prove this and other related results in Chapter 10; see also Section 9.3. For now we observe that the existence of this rich infinite-dimensional group of symplectomorphisms sharply distinguishes symplectic geometry from Riemannian geometry where the group of isometries is always finite-dimensional.

Exercise 3.1.14 Let ϕ_t be a Hamiltonian isotopy generated by F_t, let ψ_t be a Hamiltonian isotopy generated by G_t, and let χ be a symplectomorphism.

(i) Prove that $\phi_t \circ \psi_t$ is the Hamiltonian isotopy generated by

$$(F\#G)_t := F_t + G_t \circ \phi_t^{-1}. \tag{3.1.10}$$

Hint: Use the fact that $\partial_t(\phi_t \circ \psi_t) = (\partial_t\phi_t) \circ \psi_t + d\phi_t \circ \partial_t\psi_t$.

(ii) Prove that ϕ_t^{-1} is the Hamiltonian isotopy generated by $K_t := -F_t \circ \phi_t$.

(iii) Prove that $\chi^{-1} \circ \phi_t \circ \chi$ is the Hamiltonian isotopy generated by $F_t \circ \chi$.

(iv) Deduce that $\mathrm{Ham}(M, \omega)$ is a normal subgroup of $\mathrm{Symp}(M, \omega)$. □

Remark 3.1.15 We have formulated Definition 3.1.13 so that it applies to all symplectic manifolds without boundary, whether or not they are compact. For closed manifolds there is a one-to-one correspondence between flows and vector fields, respectively between isotopies and time-dependent vector fields. The same applies to closed symplectic manifolds and symplectic (or Hamiltonian) isotopies and vector fields. However, if (M, ω) is noncompact the situation is somewhat more delicate, because the solutions of a differential equation may not exist for all time. If they do, a vector field (or time-dependent family of vector fields) determines a global flow (or global isotopy). This is the situation used in the above discussion. A particular case where the existence of solutions is guaranteed is when the time-dependent vector fields have uniform compact support. This condition is of course vacuous whenever M is a closed manifold. □

The next exercise establishes the existence of Hamiltonian symplectomorphisms that cannot be generated by a time-independent Hamiltonian function.

Exercise 3.1.16 Let (M, ω) be a symplectic manifold and let $\phi \in \mathrm{Ham}(M, \omega)$ be generated by a time-independent Hamiltonian function $H : M \to \mathbb{R}$.

(i) Prove that every isolated fixed point $p \in M$ of ϕ is a critical point of H.

(ii) Let p be a critical point of H. Prove that

$$d\phi(p) = \exp(A), \qquad A := DX_H(p) \in \mathfrak{sp}(T_pM, \omega_p).$$

Here $DX(p) : T_pM \to T_pM$ denotes the intrinsic derivative of a vector field $X \in \mathcal{X}(M)$ at a point $p \in M$ where the vector field vanishes and $\mathfrak{sp}$ denotes the symplectic Lie algebra as in (2.2.2).

(iii) Let $\psi : M \to M$ be a Hamiltonian symplectomorhism and let $p \in M$ be an isolated fixed point of ψ such that the differential $d\psi(p) \in \mathrm{Sp}(T_pM, \omega_p)$ does not belong to the image of the exponential map $\exp : \mathfrak{sp}(T_pM, \omega_p) \to \mathrm{Sp}(T_pM, \omega_p)$. Show that ψ cannot be generated by a time-independent Hamiltonian function.

(iv) Prove that the map $\mathbb{R}^2 \to \mathbb{R}^2 : (x, y) \mapsto (-2x, -\frac{1}{2}y)$ is a Hamiltonian symplectomorphism that cannot be generated by a time-independent Hamiltonian function. **Hint:** See Exercise 2.2.8.

(v) Define the Hamiltonian function $H : S^2 \to \mathbb{R}$ by

$$H(x) := 2x_1^2 + \tfrac{1}{2}x_2^2 + x_3^2$$

for $x = (x_1, x_2, x_3) \in S^2$ and define the map $\psi : S^2 \to S^2$ by

$$\psi(x_1, x_2, x_3) := \phi_H(-x_1, -x_2, x_3).$$

Prove that ψ is a Hamiltonian symplectomorphism that cannot be generated by a time-independent Hamiltonian function. **Hint:** Show that the north pole $p := (0, 0, 1)$ is an isolated fixed point of ψ. □

Some basic examples of symplectic manifolds

Every oriented Riemann surface with its area form is a symplectic manifold. Another example is the $2n$-dimensional torus $\mathbb{T}^{2n} := \mathbb{R}^{2n}/\mathbb{Z}^{2n}$ with its standard form

$$\omega_0 = \sum_{j=1}^{n} dx_j \wedge dy_j.$$

The product of two symplectic manifolds $M_1 \times M_2$ is a symplectic manifold with the symplectic form $\omega_1 \oplus \omega_2 := \mathrm{pr}_1^*\omega_1 + \mathrm{pr}_2^*\omega_2$. Furthermore, the Kähler form on a Kähler manifold is symplectic (see Section 4.3 below). An interesting class of 4-dimensional symplectic manifolds consists of surface bundles over a Riemann surface. The following example of a torus bundle over a torus was known to Kodaira in the 1950s as an example of a complex nonKähler manifold and rediscovered in the 1970s by Thurston [624] as the first example of a compact symplectic manifold with no Kähler structure. It turns out that these two structures are related by mirror symmetry; cf. Abouzaid [4].

Example 3.1.17 (A 4-dimensional symplectic manifold) Consider the group $\Gamma = \mathbb{Z}^2 \times \mathbb{Z}^2$ with the noncommutative group operation

$$(j', k') \circ (j, k) = (j + j', A_{j'}k + k'), \qquad A_j := \begin{pmatrix} 1 & j_1 \\ 0 & 1 \end{pmatrix},$$

where $j = (j_1, j_2) \in \mathbb{Z}^2$, and similarly for k. The group Γ acts on $\mathbb{R}^4$ via

$$\Gamma \times \mathbb{R}^4 \to \mathbb{R}^4 : ((j,k),(x,y)) \mapsto (x + j, A_j y + k).$$

This action preserves the symplectic form $\omega := dx_1 \wedge dx_2 + dy_1 \wedge dy_2$. Hence the quotient $M = \mathbb{R}^4/\Gamma$ is a compact symplectic manifold. Its fundamental group is $\pi_1(M) = \Gamma$ and hence

$$H_1(M;\mathbb{Z}) = \Gamma/[\Gamma,\Gamma] \cong \mathbb{Z} \oplus \mathbb{Z} \oplus \mathbb{Z}.$$

Here $[\Gamma,\Gamma] = 0 \oplus 0 \oplus \mathbb{Z} \oplus 0$ is the subgroup of Γ which is generated by the elements of the form $aba^{-1}b^{-1}$ for $a, b \in \Gamma$. Since the odd-dimensional Betti numbers of a Kähler manifold must be even, M does not admit a Kähler structure.

The manifold $M = \mathbb{R}^4/\Gamma$ can also be thought of as a torus bundle over a torus, that is $M = \mathbb{R}^2 \times_{\mathbb{Z}^2} \mathbb{T}^2$. Here $\mathbb{Z}^2$ acts on $\mathbb{T}^2$ via $j \mapsto A_j$, and so $M = [0,1] \times S^1 \times \mathbb{T}^2/\sim$, where $(0, x_2, y_1, y_2) \sim (1, x_2, y_1 + y_2, y_2)$. A whole class of similar examples is discussed by Gray and his co-workers in [127, 212]. For example, it is shown in [212] (see also Exercise 6.5.2) that the Kodaira–Thurston manifold has a complex structure which admits an indefinite Kähler metric, while other examples admit no complex structure at all; see also Geiges [259]. These examples all have a nontrivial fundamental group. The first simply connected nonKähler symplectic manifold was constructed by McDuff in [444], and is described in Exercise 7.1.33 below. Examples in all dimensions have been constructed by Gompf [278] using the fibre connected sum operation of Section 7.2. □

Cotangent bundles

Cotangent bundles form a fundamental class of symplectic manifolds because they are the phase spaces of classical mechanics, with coordinates q and p corresponding to position and momentum. We show in Exercise 3.1.22 below how to generalize the Legendre transformation between Lagrangian and Hamiltonian mechanics to this context.

Our goal here is to explain the symplectic structure of these spaces. A cotangent bundle T^*L is the vector bundle whose sections are 1-forms on L and so it carries a universal (or canonical) 1-form $\lambda_{\mathrm{can}} \in \Omega^1(T^*L)$. Its symplectic form is then given by

$$\omega_{\mathrm{can}} = -d\lambda_{\mathrm{can}}. \tag{3.1.11}$$

In standard local coordinates (x, y), where $x \in \mathbb{R}^n$ is the coordinate on L and $y \in \mathbb{R}^n$ is the coordinate on the fibre T_xL, the canonical 1-form and its differential are given by the formulae[12]

$$\lambda_{\mathrm{can}} = y\,dx, \qquad \omega_{\mathrm{can}} = dx \wedge dy.$$

To explain this more fully, let $x : U \to \mathbb{R}^n$ be a local coordinate chart on L. Think of the coordinates $x_1, \ldots, x_n$ of x as real valued functions on U. Then their differentials at $q \in U$ are linear maps $dx_j(q) : T_qL \to \mathbb{R}$. These form a basis of the dual space T_q^*L and so any vector $v^* \in T_q^*L$ can be written in the form

$$v^* = \sum_{j=1}^{n} y_j dx_j.$$

The coordinates y_j in this formula are uniquely determined by q and v^* and determine coordinate functions $T^*U \to \mathbb{R}^n : (q, v^*) \mapsto y(q, v^*)$. In summary, we have a coordinate chart[13] $T^*U \to \mathbb{R}^n \times \mathbb{R}^n : (q, v^*) \mapsto (x(q), y(q, v^*))$. In these coordinates the canonical 1-form is given by

$$\lambda_{\mathrm{can}} := \sum_{j=1}^{n} y_j dx_j.$$

This formula for λ_{can} is independent of the choice of local coordinate x. One way to prove this is simply to work out what happens to the form $\sum_j y_j dx_j$ under a change of coordinates. Another way is to give a coordinate-free definition of the form λ_{can}. Here we adopt the second approach, and leave the first as an exercise for the reader.

[12] We inserted the minus sign in the formula $\omega_{\mathrm{can}} = -d\lambda_{\mathrm{can}}$ in order to get $dx \wedge dy$ rather than $-dx \wedge dy$. For further discussion of signs, see Remark 3.5.35.

[13] Strictly speaking the first component of this chart is the composition $x \circ \pi : T^*U \to \mathbb{R}^n$, where $\pi : T^*U \to U$ is the projection. But we shall not distinguish notationally between x and $x \circ \pi$. Moreover, it is often convenient to think of coordinates as independent variables in an open set of $\mathbb{R}^n$ rather than functions, defined on an open set in a manifold, and then the above distinction becomes irrelevant.

Consider the projection

$$\pi : T^*L \to L, \quad (q, v^*) \mapsto q.$$

The differential of π is a linear map

$$d\pi(q, v^*) : T_{(q,v^*)}T^*L \to T_qL.$$

Now v^* itself is a linear functional on T_qL, and the value of the canonical 1-form at the point (q, v^*) is the composition

$$\lambda_{\mathrm{can}\,(q,v^*)} = v^* \circ d\pi(q, v^*) : T_{(q,v^*)}T^*L \to \mathbb{R}, \tag{3.1.12}$$

or, in other words, the pullback of v^* to $T_{(q,v^*)}T^*L$ under the projection $d\pi$. Speaking loosely, the canonical 1-form at (x, v^*) is v^* itself. This agrees with our previous definition because, in the local coordinates (x, y) on T^*L, the linear map $d\pi(q, v^*)$ is given by $(\xi, \eta) \mapsto \xi$ and so $v^* \circ d\pi(q, v^*)$ is the linear functional $(\xi, \eta) \mapsto \langle y, \xi\rangle$ which can be written as $\sum_j y_j dx_j$. Equivalently, we can write

$$d\pi(q, v^*)\frac{\partial}{\partial x_j} = \frac{\partial}{\partial x_j}, \qquad d\pi(q, v^*)\frac{\partial}{\partial y_j} = 0.$$

It is obvious from the expression for λ_{can} in local coordinates that the associated 2-form $\omega_{\mathrm{can}} = -d\lambda_{\mathrm{can}}$ is nondegenerate. Hence $(T^*L, \omega_{\mathrm{can}})$ is a symplectic manifold.

Proposition 3.1.18 *The* 1*-form* $\lambda_{\mathrm{can}} \in \Omega^1(T^*L)$ *is uniquely characterized by the property that*

$$\sigma^*\lambda_{\mathrm{can}} = \sigma$$

for every 1*-form* $\sigma : L \to T^*L$.

Proof: In the local coordinates x on L, a 1-form σ on L can be written as

$$\sigma = \sum_{j=1}^{n} a_j(x)dx_j.$$

When regarded as a map from L to T^*L, this 1-form is represented in the local coordinates (x, y) by $x \mapsto (x_1, \ldots, x_n, a_1(x), \ldots, a_n(x))$. The pullback of λ_{can} by this map is exactly the form σ. □

Exercise 3.1.19 Show that there is an isomorphism

$$T_{(q,0)}T^*L \cong T_qL \oplus T_q^*L$$

and

$$-d\lambda_{\mathrm{can}\,(q,0)}(v, w) = w_1^*(v_0) - v_1^*(w_0)$$

for $v = (v_0, v_1^*) \in T_qL \oplus T_q^*L$ and $w = (w_0, w_1^*) \in T_qL \oplus T_q^*L$. □

The previous exercise shows that the restriction of the tangent bundle $T(T^*L)$ to the zero section is naturally isomorphic to the bundle $TL \oplus T^*L$ over L, and that the bilinear form induced on this bundle by $\omega_{\mathrm{can}} = -d\lambda_{\mathrm{can}}$ is just the canonical bilinear form Ω_{can} defined in Example 2.6.1. This result can be extended to the whole of T^*L.

Exercise 3.1.20 Prove that there is a bundle isomorphism $\Phi : TL \oplus T^*L \to T(T^*L)$ which identifies the summand T^*L with the vertical vectors. Prove that Φ can be chosen such that the composite $d\pi \circ \Phi$ restricts to the identity on the summand TL and $\Phi^*\omega_{\mathrm{can}} = \Omega_{\mathrm{can}}$. (For example, Φ could be induced by some connection on T^*L.) □

Exercise 3.1.21 **(i)** Any diffeomorphism $\psi : L \to L$ lifts to a diffeomorphism

$$\Psi : T^*L \to T^*L$$

by the formula

$$\Psi(q, v^*) = \left(\psi(q), \left(d\psi(q)^{-1}\right)^* v^*\right).$$

Prove that $\Psi^*\lambda_{\mathrm{can}} = \lambda_{\mathrm{can}}$ and hence Ψ is a symplectomorphism of T^*L.
(ii) Let $Y : L \to TL$ be a vector field on L which integrates to the 1-parameter group ψ_t of diffeomorphisms of L. Let $X : T^*L \to TT^*L$ generate the corresponding group of symplectomorphisms Ψ_t of $(T^*L, \omega_{\mathrm{can}})$. Show that $X = X_H$ is the Hamiltonian vector field of the function $H : T^*L \to \mathbb{R}$ given by

$$H(q, v^*) = v^*(Y(q)).$$

Hint: It follows from (i) that $\mathcal{L}_X\lambda_{\mathrm{can}} = \iota(X)d\lambda_{\mathrm{can}} + d(\iota(X)\lambda_{\mathrm{can}}) = 0.$ □

Exercise 3.1.22 **(i)** A Lagrangian for a variational problem on a manifold M is a function $L : TM \to \mathbb{R}$. Formulate an appropriate version of the nondegeneracy condition which permits the Legendre transformation. What is the corresponding Hamiltonian function H on $(T^*M, \omega_{\mathrm{can}})$? Check that the Euler equations for L on TM and the corresponding Hamiltonian equations on T^*M are invariant under coordinate transformations.
(ii) Consider the Lagrangian

$$L(x, v) := \tfrac{1}{2}|v|^2$$

for some Riemannian metric $|\cdot|$ on M. Then the solutions of the Euler–Lagrange equations are the geodesics on M (see Example 1.1.23). Describe explicitly the energy levels $H = c$ of the corresponding Hamiltonian function $H : T^*M \to \mathbb{R}$. □

The examples of symplectic manifolds discussed so far are all of a very simple form. They are Euclidean space (Example 3.1.1), surfaces (Example 3.1.3), tori, the Kodaira–Thurston manifold (Example 3.1.17), and cotangent bundles. Throughout the book we will encounter a wealth of other examples that illustrate the rich world of symplectic topology. One interesting method of constructing symplectic manifolds is Luttinger surgery, explained in Example 3.4.16. We will encounter various examples of Kähler manifolds while examining almost complex structures in Chapter 4. In Chapter 5 we shall study Hamiltonian group actions and show how, in favourable cases, they give rise to symplectic quotient manifolds, including homogeneous spaces, coadjoint orbits, and toric manifolds. Chapter 6 constructs symplectic forms on certain fibrations, while Chapter 7 explains various other constructions, including symplectic blowing up and down, Gompf's construction of symplectic forms on fibre connected sums (leading to symplectic 4-manifolds with arbitrary finitely generated fundamental groups), and Gromov's construction of symplectic forms on almost complex open manifolds via the h-principle. Before explaining any of these sophisticated methods, we return in the present chapter to the basic foundations of symplectic topology.

3.2 Moser isotopy and Darboux's theorem

We will begin by describing Moser's homotopy method for constructing isotopies [490]. This has many important consequences, which have to do with the fact that symplectic structures have no local invariants. The word *local* here has many possible interpretations: one can localize at a point and get Darboux's theorem, one can localize near a symplectic or Lagrangian submanifold, or one can localize in the space of symplectic forms and consider properties of smooth deformations of a symplectic structure.

Roughly speaking, Moser's argument shows that, for every family of symplectic forms $\omega_t \in \Omega^2(M)$ with an exact derivative

$$\frac{d}{dt}\omega_t = d\sigma_t, \tag{3.2.1}$$

there exists a family of diffeomorphisms $\psi_t \in \mathrm{Diff}(M)$ such that

$$\psi_t^*\omega_t = \omega_0. \tag{3.2.2}$$

The key idea is to determine the diffeomorphisms ψ_t by representing them as the flow of a family of vector fields X_t on M. Thus we suppose that

$$\frac{d}{dt}\psi_t = X_t \circ \psi_t, \qquad \psi_0 = \mathrm{id}. \tag{3.2.3}$$

The vector fields X_t have to be constructed so that (3.2.2) is satisfied. Differentiating this gives

$$0 = \frac{d}{dt}\psi_t^*\omega_t = \psi_t^*\left(\frac{d}{dt}\omega_t + \iota(X_t)d\omega_t + d\iota(X_t)\omega_t\right).$$

Since ω_t is closed for every t and its derivative is exact, this identity will be satisfied if

$$\sigma_t + \iota(X_t)\omega_t = 0. \tag{3.2.4}$$

Since ω_t is nondegenerate, there exists a unique family of vector fields X_t satisfying (3.2.4). Hence the diffeomorphisms ψ_t determined by (3.2.3) satisfy (3.2.2) as required.

This argument is perfectly rigorous in the case where M is compact. In general, one has to verify that the solutions of the differential equation (3.2.3) exist on the required time interval. The following case is of interest to us.

Lemma 3.2.1 (Moser isotopy) *Let M be a $2n$-dimensional smooth manifold and $Q \subset M$ be a compact submanifold. Suppose that $\omega_0, \omega_1 \in \Omega^2(M)$ are closed 2-forms such that at each point q of Q the forms ω_0 and ω_1 are equal and non-degenerate on T_qM. Then there exist open neighbourhoods $\mathcal{N}_0$ and $\mathcal{N}_1$ of Q and a diffeomorphism $\psi : \mathcal{N}_0 \to \mathcal{N}_1$ such that*

$$\psi|_Q = \mathrm{id}, \qquad \psi^*\omega_1 = \omega_0.$$

Proof: In view of Moser's argument it is enough to prove that there exists a 1-form $\sigma \in \Omega^1(\mathcal{N}_0)$ such that

$$\sigma|_{T_QM} = 0, \qquad d\sigma = \omega_1 - \omega_0. \tag{3.2.5}$$

Then we may consider the family of closed forms

$$\omega_t = \omega_0 + t(\omega_1 - \omega_0) = \omega_0 + td\sigma$$

on $\mathcal{N}_0$. Shrinking $\mathcal{N}_0$ if necessary we may assume that ω_t is nondegenerate in $\mathcal{N}_0$ for every t. Thus we can solve equation (3.2.4) and the resulting vector fields X_t vanish on Q. Shrinking $\mathcal{N}_0$ again we obtain that in $\mathcal{N}_0$ the solutions of (3.2.2) exist on the required time interval $0 \le t \le 1$.

To prove (3.2.5), we use the method of 'integration over the fibre' as in [78, Chapter 1,§6]. Consider the restriction of the exponential map to the normal bundle $TQ^\perp$ of the submanifold Q with respect to any Riemannian metric on M. We denote this restriction by

$$\exp : TQ^\perp \to M.$$

Consider the neighbourhood of the zero section:

$$U_\varepsilon := \left\{(q, v) \in TM \mid q \in Q, v \in T_qQ^\perp, |v| < \varepsilon\right\}.$$

Then the restriction of the exponential map to U_ε is a diffeomorphism onto

$$\mathcal{N}_0 := \exp(U_\varepsilon)$$

for $\varepsilon > 0$ sufficiently small. Now define $\phi_t : \mathcal{N}_0 \to \mathcal{N}_0$ for $0 \le t \le 1$ by

$$\phi_t(\exp(q, v)) := \exp(q, tv).$$

Then ϕ_t is a diffeomorphism for $t > 0$ and we have $\phi_0(\mathcal{N}_0) \subset Q$, $\phi_1 = \mathrm{id}$, and $\phi_t|_Q = \mathrm{id}$. This implies

$$\phi_0^*\tau = 0, \qquad \phi_1^*\tau = \tau,$$

where $\tau := \omega_1 - \omega_0$. Since ϕ_t is a diffeomorphism for $t > 0$ we may define the vector field

$$X_t := \left(\frac{d}{dt}\phi_t\right) \circ \phi_t{}^{-1} \qquad \text{for } t > 0.$$

Note that X_t becomes singular at $t = 0$. However, we obtain

$$\frac{d}{dt}\phi_t^*\tau = \phi_t^*(\mathcal{L}_{X_t}\tau) = d\big(\phi_t^*(\iota(X_t)\tau)\big) = d\sigma_t,$$

where the family of 1-forms $\sigma_t = \phi_t^*(\iota(X_t)\tau)$ given for $w \in T_x\mathcal{N}_0$ by

$$\sigma_t(x; w) := \tau\left(\phi_t(x); X_t(\phi_t(x)), d\phi_t(x)(w)\right)$$

is smooth at $t = 0$ and vanishes on Q. Hence

$$\tau = \phi_1^*\tau - \phi_0^*\tau = \int_0^1 \frac{d}{dt}(\phi_t^*\tau)\, dt = d\sigma, \qquad \sigma := \int_0^1 \sigma_t\, dt.$$

This proves (3.2.5). □

Our first application of this lemma is to prove Darboux's theorem. We saw in Theorem 2.1.3 that all symplectic structures on a manifold M are isomorphic when restricted to a point of M. Darboux's theorem says that even their germs at a point are isomorphic. This is a great contrast to Riemannian geometry where the curvature is a local invariant.

Theorem 3.2.2 (Darboux) *Every symplectic form ω on M is locally diffeomorphic to the standard form ω_0 on $\mathbb{R}^{2n}$.*

Proof: Apply Lemma 3.2.1 to the case where Q is a single point and use Theorem 2.1.3. □

This proof of Darboux's theorem is due to Moser in [490]. It was used by Weinstein in [669] to prove an equivariant version of Darboux's theorem and the symplectic neighbourhood theorems of the next section. Arnold and Givental use the same method in [25], and call it the homotopy method. In [23, Section 43B] Arnold gives a completely different and much more geometric proof of Darboux's theorem.

Corollary 3.2.3 *Let (M, ω) be a symplectic manifold of dimension $2n$. Then there exists an open cover $\{U_\alpha\}_\alpha$ of M and charts $\alpha : U_\alpha \to \alpha(U_\alpha) \subset \mathbb{R}^{2n}$ such that $\alpha^*\omega_0 = \omega$. This atlas has symplectic transition matrices*

$$d(\beta \circ \alpha^{-1})(x) \in \mathrm{Sp}(2n)$$

for $x \in \alpha(U_\alpha \cap U_\beta)$. Charts with these properties are called **Darboux charts**.

Theorem 3.2.2 shows that symplectic manifolds, in contrast to Riemannian manifolds, have no local invariants. They have some obvious global invariants such as the cohomology class $[\omega]$ of the symplectic form in de Rham cohomology and the first Chern class of the tangent bundle in $H^2(M;\mathbb{Z})$ (See Chapter 4). Other global invariants (such as symplectic capacities) will be discussed later (see Chapters 11 and 12). The following **Moser stability theorem for symplectic structures** shows that one cannot vary the diffeomorphism type of a symplectic form by perturbing it within the same cohomology class.

Theorem 3.2.4 (Moser stability) *Let M be a closed manifold (i.e. a compact manifold without boundary) and suppose that ω_t is a smooth family of cohomologous symplectic forms on M. Then there is a family of diffeomorphisms ψ_t of M such that*

$$\psi_0 = \mathrm{id}, \qquad \psi_t^* \omega_t = \omega_0.$$

Proof: To apply Moser's argument we must find a smooth family of 1-forms σ_t such that

$$d\sigma_t = \frac{d}{dt}\omega_t.$$

Since the ω_t are cohomologous, each form $\omega_t - \omega_0$ is exact and so are the forms

$$\tau_t = \frac{d}{dt}\omega_t.$$

Therefore, for each t there is a 1-form σ_t such that $d\sigma_t = \tau_t$, and the only problem is to choose these forms smoothly with respect to t. One can prove that this is possible by induction over the number of sets in a good cover of M as in Bott and Tu [78]. One proves it first for compactly supported forms in a coordinate chart by direct calculation, using integration over the fibre as in Lemma 3.2.1 above. The inductive step is achieved by using the Mayer–Vietoris sequence. We leave further details of this approach to the reader.

Another way to construct the σ_t is to use an intrinsically global method, such as Hodge theory. To do this, choose a Riemannian metric on M and denote by $d^* : \Omega^2 \to \Omega^1$ the L^2-adjoint operator of d with respect to the induced metric on forms. By Hodge theory, d restricts to an isomorphism from the image of d^* to the exact 2-forms on M. Thus, for each t, there is a unique 1-form $\sigma_t \in \operatorname{im} d^*$ so that $d\sigma_t = \tau_t$. By elliptic regularity, σ_t depends smoothly on t and this proves Theorem 3.2.4. □

Definition 3.2.5 *Two symplectic forms ω_0 and ω_1 on a manifold M are said to be* **isotopic** *if they can be joined by a smooth family ω_t of cohomologous symplectic forms on M. They are said to be* **strongly isotopic** *if there is an isotopy ψ_t of M such that $\psi_1^*\omega_1 = \omega_0$. Theorem 3.2.4 shows that these two notions agree when M is closed. This is not the case for open manifolds (see Section 13.1).*

Exercise 3.2.6 This exercise establishes a relative form of Moser's theorem that is often useful. Let M be a compact manifold with boundary. Suppose that ω_t, $0 \le t \le 1$, is a smooth family of symplectic forms that agree on T_xM for every $x \in \partial M$ and satisfy, for every compact 2-manifold Σ and every smooth map $u : \Sigma \to M$,

$$u(\partial\Sigma) \subset \partial M \qquad \Longrightarrow \qquad \frac{d}{dt}\int_\Sigma u^*\omega_t = 0. \tag{3.2.6}$$

Prove that there exists a smooth isotopy $\psi_t : M \to M$ such that

$$\psi_0 = \mathrm{id}, \qquad \psi_t|_{\partial M} = \mathrm{id}, \qquad \psi_t^*\omega_t = \omega_0.$$

If $\omega_t = \omega_0$ in some neighbourhood of ∂M, prove that ψ_t can be chosen equal to the identity in a (possibly smaller) neighbourhood of ∂M. **Hint:** The area forms $\omega_t - \omega_0$ vanish on $T_{\partial M}M$ and hence define a family $[\omega_t - \omega_0]$ of elements of the relative group $H^2(M, \partial M; \mathbb{R})$. Condition (3.2.6) asserts that these cohomology classes vanish. Deduce that there is a smooth family of 1-forms $\sigma_t \in \Omega^1(M)$ such that σ_t vanishes on $T_{\partial M}M$ for all t and $d\sigma_t = \partial_t \omega_t$. □

Exercise 3.2.7 Suppose that ω_t and τ_t are two families of symplectic forms on a closed manifold M such that $\omega_0 = \tau_0$ and ω_t is cohomologous to τ_t for all $t \in [0,1]$. Prove that for some $\varepsilon > 0$ there is an isotopy ψ_t such that $\psi_t^* \tau_t = \omega_t$ for $0 \le t \le \varepsilon$. Examples are known where one cannot take $\varepsilon = 1$ (see Example 13.2.9 in Chapter 13). □

Exercise 3.2.8 **(i)** Prove Darboux's theorem in the 2-dimensional case, using the fact that every nonvanishing 1-form on a surface can be written locally as $f\,dg$ for suitable functions f and g.

(ii) Let Σ be a closed 2-manifold. Prove that a symplectic (or area) form on Σ is determined up to strong isotopy by its cohomology class. **Hint:** Use Moser's stability Theorem 3.2.4 and the fact that the cohomologous symplectic forms on Σ form a convex set.

(iii) Prove a similar result for volume forms on closed manifolds. □

3.3 Isotopy extension theorems

Let $Q \subseteq M$ be a submanifold. A family of embeddings $\psi_t : Q \to M$ which starts at the inclusion is called an **isotopy** of Q in M. If $\psi_t = \psi_0$ on some subset $Y \subset Q$ then ψ_t is called an **isotopy rel** Y. If ψ_t preserves a symplectic form it is called a **symplectic isotopy**.

We now discuss the question of when a symplectic isotopy $\psi_t : Q \to M$ can be extended symplectically, either to a neighbourhood of $Q \subset M$ or to the whole of M. As a warmup we prove that symplectic embeddings of closed balls extend to symplectic embeddings of slightly larger balls and that this extension result carries over to isotopies. This result will be needed in Section 7.1 for the construction of symplectic forms on blowups. In the following $B(r)$ denotes the closed ball in $\mathbb{R}^{2n}$ of centre 0 and radius r, with the standard symplectic form.

Theorem 3.3.1 (Symplectic ball extension theorem) *Let (M, ω) be a connected symplectic manifold without boundary.*

(i) *Let $\lambda > 0$. Then every symplectic embedding $\psi : B(\lambda) \to M$ extends to a symplectic embedding of $B(r)$ into M for some $r > \lambda$.*

(ii) *Let $0 \le \lambda < r$ and let $\psi_0, \psi_1 : B(r) \to M$ be symplectic embeddings that are joined by a smooth family of symplectic embeddings $\psi_t : B(\lambda) \to M$, $0 \le t \le 1$, of the smaller ball $B(\lambda)$. Then there exists a Hamiltonian isotopy $\{\phi_t\}_{0 \le t \le 1}$ of M and a constant $\lambda < \rho < r$ such that*

$$\phi_0 = \mathrm{id}, \qquad \phi_1 \circ \psi_0|_{B(\rho)} = \psi_1|_{B(\rho)}, \qquad \phi_t \circ \psi_0|_{B(\lambda)} = \psi_t$$

for all t. Thus the isotopy from ψ_0 to ψ_1 on $B(\lambda)$ extends to $B(\rho)$. The extension is given by $\phi_t \circ \psi_0 : B(\rho) \to M$ for $0 \le t \le 1$.

Proof: The proof has four steps.

Step 1. *Let $[0,1]\times B(\lambda)\to M:(t,z)\mapsto\psi_t(z)$ be a smooth map such that $\psi_t:B(\lambda)\to M$ is a symplectic embedding for every t. Then there exists a Hamiltonian isotopy $[0,1]\times M\to M:(t,p)\mapsto\phi_t(p)$ such that*

$$\phi_0=\mathrm{id},\qquad \phi_t\circ\psi_0=\psi_t \tag{3.3.1}$$

for all t.

Differentiating the equation $\psi_t^*\omega=\omega_0$ we find that the 1-form

$$\alpha_t:=\omega(\partial_t\psi_t,d\psi_t\cdot)\in\Omega^1(B(\lambda)) \tag{3.3.2}$$

is closed for each t. Hence there is a smooth family of functions $h_t:B(\lambda)\to\mathbb{R}$ such that

$$dh_t=\alpha_t,\qquad h_t(0)=0. \tag{3.3.3}$$

By Theorem A.2.1 and Lemma A.2.2, there exists a smooth family of functions $H_t:M\to\mathbb{R}$ such that

$$H_t\circ\psi_t=h_t. \tag{3.3.4}$$

(First extend ψ_t to a smooth isotopy of embeddings of $\mathbb{R}^{2n}$ by Theorem A.2.1 and then extend the Hamiltonian functions by Lemma A.2.2.) Consider the Hamiltonian isotopy generated by H_t via

$$\partial_t\phi_t=X_{H_t}\circ\phi_t,\qquad \phi_0=\mathrm{id},\qquad \iota(X_{H_t})\omega=dH_t.$$

By (3.3.2), (3.3.3), and (3.3.4) this isotopy satisfies (3.3.1). This proves Step 1.

Step 2. *We prove (i).*

By part (i) of Theorem A.2.1 the symplectic embedding $\psi:B(\lambda)\to M$ extends to a smooth embedding of $B(\lambda+\varepsilon)$ into M, still denoted by ψ. The symplectic form

$$\psi^*\omega\in\Omega^2(B(\lambda+\varepsilon))$$

agrees with ω_0 on $B(\lambda)$. Hence it follows from Poincaré's lemma that there exists a 1-form $\alpha\in\Omega^1(B(\lambda+\varepsilon))$ such that

$$\alpha|_{B(\lambda)}=0,\qquad d\alpha=\psi^*\omega-\omega_0.$$

Shrinking ε, if necessary, we may assume that $\omega_s:=\omega_0+sd\alpha$ is nondegenerate on $B(\lambda+\varepsilon)$ for every $s\in[0,1]$. Then it follows from Moser isotopy that there exists a $\delta>0$ and a smooth family of embeddings

$$\chi_s:B(\lambda+\delta)\to B(\lambda+\varepsilon)$$

such that

$$\chi_0=\mathrm{id},\qquad \chi_s|_{B(\lambda)}=\mathrm{id},\qquad \chi_s^*(\omega_0+sd\alpha)=\omega_0$$

for all s. Hence $\chi_1^*\psi^*\omega=\omega_0$ and hence $\psi\circ\chi_1:B(\lambda+\delta)\to M$ is the desired extension of ψ. This proves Step 2.

Step 3. *We prove (ii) for $\lambda > 0$.*

This follows from a parametrized version of the argument in the proof of Step 2. By part (ii) of Theorem A.2.1, there exists a smooth family of smooth embeddings $\Psi_t : \mathbb{R}^{2n} \to M$, $0 \le t \le 1$, such that

$$\Psi_0|_{B(r)} = \psi_0, \qquad \Psi_1|_{B(r)} = \psi_1, \qquad \Psi_t|_{B(\lambda)} = \psi_t$$

for all t. Then the 2-form $\Psi_t^*\omega \in \Omega^2(\mathbb{R}^{2n})$ agrees with ω_0 on $B(\lambda)$ for all t. By Poincaré's lemma there is a smooth family of 1-forms $\alpha_t \in \Omega^1(\mathbb{R}^{2n})$ such that

$$\alpha_t|_{B(\lambda)} = 0, \qquad d\alpha_t = \Psi_t^*\omega - \omega_0, \qquad \alpha_0 = \alpha_1 = 0.$$

An explicit formula for α_t is

$$\alpha_t(z;\widehat{z}) := \int_0^1 (\Psi_t^*\omega - \omega_0)_{\theta z}(z;\theta\widehat{z})\,d\theta. \tag{3.3.5}$$

Choose $\varepsilon > 0$ so small that $\omega_{s,t} := \omega_0 + s d\alpha_t$ is nondegenerate on $B(\lambda+\varepsilon)$ for all $s,t \in [0,1]$. Then it follows from Moser isotopy that there exists a $\delta > 0$ and a smooth family of embeddings $\chi_{s,t} : B(\lambda+\delta) \to B(\lambda+\varepsilon)$ such that

$$\chi_{s,t}^*(\omega_0 + s d\alpha_t) = \omega_0, \qquad \chi_{s,t}|_{B(\lambda)} = \mathrm{id}, \qquad \chi_{0,t} = \chi_{s,0} = \chi_{s,1} = \mathrm{id}$$

for all $s,t \in [0,1]$. Hence

$$(\Psi_t \circ \chi_{1,t})^*\omega = \chi_{1,t}^*\Psi_t^*\omega = \chi_{1,t}^*(\omega_0 + d\alpha_t) = \omega_0 \tag{3.3.6}$$

for all t, and so $\Psi_t \circ \chi_{1,t} : B(\lambda+\delta) \to M$ is a smooth family of symplectic embeddings connecting $\psi_0|_{B(\lambda+\delta)} = \Psi_0 \circ \chi_{1,0}$ to $\psi_1|_{B(\lambda+\delta)} = \Psi_1 \circ \chi_{1,1}$ and satisfying $\Psi_t \circ \chi_{1,t}|_{B(\lambda)} = \psi_t$ for all t. This proves Step 3.

Step 4. *We prove (ii) for $\lambda = 0$.*

Let $\psi_0, \psi_1 : B(r) \to M$ be symplectic embeddings and let $\gamma : [0,1] \to M$ be a smooth path such that $\gamma(0) = \psi_0(0)$ and $\gamma(1) = \psi_1(0)$. By part (iii) of Theorem A.2.1, there exists a smooth family of embeddings $\Psi_t : \mathbb{R}^{2n} \to M$ for $0 \le t \le 1$, such that

$$\Psi_t(0) = \gamma(t), \qquad \Psi_i|_{B(r)} = \psi_i$$

for $i = 0,1$ and all t. Consider the smooth path of nondegenerate skew-symmetric bilinear forms $\omega_{t,0} : \mathbb{R}^{2n} \times \mathbb{R}^{2n} \to \mathbb{R}$ obtained by evaluating the pullback symplectic form $\Psi_t^*\omega \in \Omega^2(M)$ at the origin. Then $\omega_{0,0} = \omega_{1,0} = \omega_0$ is the standard symplectic form. By Corollary 2.5.14 there exists a smooth path

$$[0,1] \to \mathrm{GL}(2n,\mathbb{R}) : t \mapsto \Phi_t$$

such that

$$\Phi_0 = 1\!\!1, \qquad \Phi_t^*\omega_{t,0} = \omega_0$$

for all t. In particular, $\Phi_1 \in \mathrm{Sp}(2n)$. Since $\mathrm{Sp}(2n)$ is connected (see Proposition 2.2.4), we can modify the path $t \mapsto \Psi_t$ (by composition with a smooth path in Sp$(2n)$ connecting the identity matrix to Φ_1^{-1}) to obtain $\Phi_1 = \mathbb{1}$. Now replace the embeddings $\Psi_t : \mathbb{R}^{2n} \to M$ by $\Psi_t \circ \Phi_t$. Then Ψ_0, Ψ_1, and $\Psi_t(0) = \gamma(t)$ remain unchanged, and the pullback symplectic forms $\Psi_t^*\omega$ now agree with ω_0 at the origin. The remainder of the proof is verbatim the same as that of Step 3. In particular, the 1-forms $\alpha_t \in \Omega^1(\mathbb{R}^{2n})$ defined by (3.3.5) vanish at the origin and so determine a family of smooth embeddings $\chi_{1,t} : B(\delta) \to \mathbb{R}^{2n}$ such that

$$\chi_{1,0} = \chi_{1,1} = \mathrm{id}, \qquad \chi_{1,t}(0) = 0, \qquad (\Psi_t \circ \chi_{1,t})^*\omega = \omega_0$$

for all t. This proves Step 4 and Theorem 3.3.1. □

We close this section by proving a version of the symplectic isotopy extension theorem. It is an elementary fact in differential topology that any isotopy of a compact submanifold $Q \subset M$ can be extended to an isotopy of M (see [522]; it follows by arguments similar to those in Appendix A), and the question is whether this remains true in the symplectic category. The problem may be divided into two parts: one can try to extend the isotopy first to a neighbourhood of Q and then to the whole of M. As we shall see in Exercise 3.4.20 the first extension is always possible provided that Q is nicely related to the symplectic form ω, for example if Q is symplectic or Lagrangian. In contrast, there are cohomological obstructions to the extension over M. An example of such an obstruction is given in the following result, which is due to Banyaga [50].

Theorem 3.3.2 (Symplectic isotopy extension theorem) *Let (M, ω) be a compact symplectic manifold and let $Q \subset M$ be a compact subset that is a neighbourhood deformation retract and satisfies $H^2(M, Q; \mathbb{R}) = 0$. Let $\phi_t : U \to M$ be a symplectic isotopy of an open neighbourhood U of Q. Then there exists a neighbourhood $\mathcal{N} \subset U$ of Q and a symplectic isotopy $\psi_t : M \to M$ such that*

$$\psi_t|_{\mathcal{N}} = \phi_t|_{\mathcal{N}} \tag{3.3.7}$$

for every t. Moreover, if $H^1(Q; \mathbb{R}) = 0$, we may construct the extension ψ_t to be Hamiltonian.

Proof: Choose a neighbourhood $\mathcal{N} \subset U$ of Q which retracts onto Q. Then $H^*(\mathcal{N}, Q; \mathbb{R}) = 0$ and hence it follows from the cohomology exact sequence of the triple $(Q, \mathcal{N}, M)$ that $H^2(M, \mathcal{N}; \mathbb{R}) = 0$. Now choose any extension ρ_t of $\phi_t|_{\mathcal{N}}$ to M. Define $\omega_t = \rho_t^*\omega$ and note that ω_t agrees with ω in $\mathcal{N}$. Hence the forms $\tau_t = \frac{d}{dt}\omega_t$ vanish in $\mathcal{N}$ and so represent relative cohomology classes

$$a_t = [\tau_t] \in H^2(M, \mathcal{N}; \mathbb{R}) = 0.$$

Applying relative Hodge theory to the compact manifold $M \setminus \mathrm{int}(\mathcal{N})$ we can find 1-forms σ_t on M which vanish on $\mathcal{N}$ and satisfy $d\sigma_t = \tau_t$. By Moser isotopy these 1-forms σ_t give rise to diffeomorphisms χ_t of M which restrict to the identity on $\mathcal{N}$ and satisfy $\chi_t^*\omega_t = \omega_0 = \omega$ for all t. Thus $\rho_t \circ \chi_t$ is the desired extension. This proves the first statement in Theorem 3.3.2.

To prove the second claim, differentiate the equation $\psi_t^*\omega|_{\mathcal{N}} = \omega_0|_{\mathcal{N}}$ to obtain that the 1-form

$$\alpha_t := \omega(\partial_t\psi_t, d\psi_t\cdot) \in \Omega^1(\mathcal{N}) \tag{3.3.8}$$

is closed for each t. Since $H^1(\mathcal{N};\mathbb{R}) = H^1(Q;\mathbb{R}) = 0$, there is a smooth family of functions $h_t : \mathcal{N} \to \mathbb{R}$ such that

$$dh_t = \alpha_t, \qquad h_t(0) = 0. \tag{3.3.9}$$

After slightly shrinking $\mathcal{N}$ if necessary, we can find a smooth family of functions $H_t : M \to \mathbb{R}$ such that

$$H_t \circ \psi_t = h_t. \tag{3.3.10}$$

Consider the Hamiltonian isotopy generated by H_t via

$$\partial_t\phi_t = X_{H_t} \circ \phi_t, \qquad \phi_0 = \mathrm{id}, \qquad \iota(X_{H_t})\omega = dH_t.$$

By (3.3.8), (3.3.9), and (3.3.10) this isotopy satisfies (3.3.7). □

3.4 Submanifolds of symplectic manifolds

Let (M^{2n}, ω) be a symplectic manifold with submanifold $Q \subset M$. We understand this to mean that every point $q \in Q$ has an open neighbourhood U in M with a coordinate chart $\phi : U \to \mathbb{R}^{2n}$ such that $\phi(U \cap Q)$ is an open set in a d-dimensional linear subspace of $\mathbb{R}^{2n}$. A submanifold $Q \subset M$ is called **symplectic** (or **isotropic**, **coisotropic**, **Lagrangian**) if for every $q \in Q$ the subspace T_qQ of the symplectic vector space (T_qM, ω_q) is symplectic (or isotropic, coisotropic, Lagrangian).

Example 3.4.1 (i) The zero section and the fibres of the cotangent bundle T^*L with its canonical symplectic structure are Lagrangian because the canonical 1-form λ_{can} vanishes on these submanifolds (see Proposition 3.1.18). More generally, given a smooth submanifold $Q \subset L$, the annihilator (often called the conormal)

$$TQ^\perp = \left\{(q, v^*) \in T^*L \,\middle|\, q \in Q,\, v^*|_{T_qQ} = 0\right\}$$

is Lagrangian.

(ii) Given a symplectic manifold (M, ω) the product $M \times M$ admits a natural symplectic structure $(-\omega) \oplus \omega$. The submanifolds $M \times \mathrm{pt}$ and $\mathrm{pt} \times M$ are symplectic whereas the diagonal is Lagrangian.

(iii) Any hypersurface S is coisotropic. The 1-dimensional symplectic complement $T_qS^\omega \subset T_qS$ is tangent to the characteristic foliation of S which was defined in Section 1.2. (See also Exercise 3.1.12.) □

Proposition 3.4.2 *The graph $\Gamma_\sigma \subset T^*L$ of a 1-form σ on L is Lagrangian if and only if σ is closed.*

Proof: The graph Γ_σ is the image of the embedding $\sigma : L \to T^*L$. Hence the symplectic form $-d\lambda_{\mathrm{can}}$ vanishes on Γ_σ if and only if

$$0 = \sigma^*(d\lambda_{\mathrm{can}}) = d(\sigma^*\lambda_{\mathrm{can}}) = d\sigma.$$

Here we have used Proposition 3.1.18. □

Proposition 3.4.3 *Let (M, ω) be a symplectic manifold and let $\psi : M \to M$ be a diffeomorphism. Then ψ is a symplectomorphism if and only if its graph*

$$\mathrm{graph}(\psi) = \{(q, \psi(q)) \,|\, q \in M\} \subset M \times M$$

is a Lagrangian submanifold of $(M \times M, (-\omega) \oplus \omega)$.

Proof: Exercise. □

This proposition shows that Lagrangian submanifolds are very important in symplectic geometry. Indeed, Weinstein [675], with the motto 'everything is a Lagrangian submanifold!', created a dictionary in which he interpreted many of the concepts of symplectic geometry in terms of Lagrangian submanifolds. (For a much more ambitious version of this idea see [680] and the references therein.) A significant example of this is discussed in Chapter 11, where we will see that Arnold's conjectures about fixed points of symplectic maps may be generalized to conjectures about intersections of Lagrangian submanifolds. The central role played by these submanifolds in the modern theory is also illustrated by Gromov's brilliant argument which uses the properties of Lagrangian submanifolds in $\mathbb{R}^{2n}$ to construct an exotic symplectic structure on Euclidean space (see Theorem 13.2.3). An interesting survey article about Lagrangian submanifolds was written by Audin–Lalonde–Polterovich [35] in the early 1990s. Much progress has been made since then, some of which will be discussed in Chapters 13 and 14.

Definition 3.4.4 (Maslov class and monotonicity) *Let L be a Lagrangian submanifold of $(\mathbb{R}^{2n}, \omega_0)$. Then every loop $\gamma : \mathbb{R}/\mathbb{Z} \to L$ determines a loop of Lagrangian subspaces*

$$\mathbb{R}/\mathbb{Z} \to \mathcal{L}(n) : t \mapsto T_{\gamma(t)}L =: \Lambda_\gamma(t).$$

The Maslov index of Λ_γ depends only on the homotopy class of γ (see Theorem 2.3.7) and hence gives rise to a homomorphism (called the **Maslov class***)*

$$\mu_L : \pi_1(L) \to \mathbb{Z}, \qquad \mu_L([\gamma]) := \mu(\Lambda_\gamma).$$

The symplectic action $\mathcal{A}_0(\gamma)$ in (1.1.8) with Hamiltonian $H = 0$ also depends only on the homotopy class of γ. (Prove this!) Hence it gives rise to another homomorphism (called the **symplectic area class***)*

$$a_L : \pi_1(L) \to \mathbb{R}, \qquad a_L([\gamma]) := -\mathcal{A}_0(\gamma) := -\int_\gamma \lambda_0, \qquad \lambda_0 := \sum_{j=1}^n y_j dx_j.$$

The Lagrangian submanifold L is called **monotone** *if there is a constant $c > 0$ such that $a_L = c\mu_L : \pi_1(L) \to \mathbb{R}$. The number c is called the* **monotonicity**

factor. *When there is a loop with nontrivial Maslov index, the* **minimal Maslov number** *of L is defined as the unique integer $N > 0$ such that $\mu_L(\pi_1(L)) = N\mathbb{Z}$. When there is no such loop the* **minimal Maslov number** *is $N := \infty$.* □

Exercise 3.4.5 **(i)** Prove that the Maslov class and the symplectic area class are invariant under Hamiltonian isotopy (of Lagrangian embeddings).

(ii) Extend the definition of Maslov class and symplectic area class to Lagrangians in the cotangent bundle $T^*\mathbb{T}^n = \mathbb{T}^n \times \mathbb{R}^n$. Prove that they vanish for the zero section.

(iii) Prove that an embedded circle $\Gamma \subset \mathbb{C}$ enclosing a disc of area a is a monotone Lagrangian submanifold with factor $a/2$ and minimal Maslov number 2.

(iv) Prove that a product of embedded circles $L := \Gamma_1 \times \cdots \times \Gamma_n$ in $(\mathbb{C}^n \cong \mathbb{R}^{2n}, \omega_0)$ is monotone if and only if the areas enclosed by the Γ_i agree.

(v) Prove that any two embedded circles in $\mathbb{C}$ which enclose the same area are Hamiltonian isotopic. **Hint:** Use Moser isotopy and a theorem of D.B.A. Epstein [206] which asserts that every embedded circle is the boundary of an embedded closed disc. (See Akveld–Salamon [18, Proposition A.1].) □

Remark 3.4.6 (Audin conjecture) In [29] Audin conjectured that every Lagrangian torus in $(\mathbb{R}^{2n}, \omega_0)$ has minimal Maslov number two. This was confirmed by Viterbo [645] for $n = 2$ and by Buhovski [81] and Fukaya–Oh–Ohta–Ono [249, Theorem 6.4.35] for monotone Lagrangian tori (see also Damian [131]). In 2014 the Audin conjecture was settled by Cieliebak–Mohnke [117]. Their work extends the result to all orientable Lagrangian submanifolds of $\mathbb{C}\mathrm{P}^n$ (see Example 4.3.3) that admit metrics of nonpositive sectional curvature. They also prove that nonorientable Lagrangian submanifolds of $\mathbb{C}\mathrm{P}^n$ with metrics of nonpositive sectional curvature have minimal Maslov number one or two. A different approach to these theorems and related results and conjectures was outlined by Fukaya in [247]. □

Example 3.4.7 (Chekanov torus) The set

$$L_{\mathrm{Ch}} := \left\{ \begin{pmatrix} (e^s + ie^{-s}t)\cos(\theta) \\ (e^s + ie^{-s}t)\sin(\theta) \end{pmatrix} \,\middle|\, \begin{array}{l} \theta, s, t \in \mathbb{R}, \\ s^2 + t^2 = 1 \end{array} \right\} \tag{3.4.1}$$

is a monotone Lagrangian torus in $(\mathbb{C}^2 \cong \mathbb{R}^4, \omega_0)$, called the **Chekanov torus**. To see this, denote by $\theta \in \mathbb{R}/2\pi\mathbb{Z}$ the coordinate on the unit circle $S^1 \subset \mathbb{C}$ and consider the diffeomorphism $S^1 \times \mathbb{R} \to \mathbb{R}^2 \setminus \{0\} : (\theta, s) \mapsto (e^s\cos(\theta), e^s\sin(\theta))$. Lift it to a symplectomorphism of the cotangent bundles

$$\psi : T^*S^1 \times T^*\mathbb{R} \to \mathbb{C}^2 \setminus i\mathbb{R}^2 \cong T^*(\mathbb{R}^2 \setminus \{0\})$$

via Exercise 3.1.21. Denote the coordinates on $T^*S^1 = S^1 \times \mathbb{R}$ by (θ, τ) and the coordinates on $T^*\mathbb{R} = \mathbb{R}^2$ by (s, t). Then ψ is given by

$$\psi(\theta, \tau, s, t) = \begin{pmatrix} (e^s + ie^{-s}t)\cos(\theta) - i\tau e^{-s}\sin(\theta) \\ (e^s + ie^{-s}t)\sin(\theta) + i\tau e^{-s}\cos(\theta) \end{pmatrix}.$$

The Chekanov torus L_{Ch} is the image under ψ of the product of the zero section in T^*S^1 with the unit circle in $\mathbb{R}^2$. Since $\psi^*\lambda_0 = \lambda_{\mathrm{can}}$, it is monotone with factor $\pi/2$, as is the **Clifford torus** $L_{\mathrm{Cl}} := S^1 \times S^1$ (see Exercise 3.4.5). In [99] Chekanov proved that L_{Ch} is Lagrangian isotopic, but not Hamiltonian isotopic, to the Clifford torus. □

Another monotone Lagrangian torus in $\mathbb{C}^2$ with factor $\pi/2$ is given by

$$L_{\mathrm{EP}} := \left\{ (e^{i\theta} z, e^{-i\theta} z) \in \mathbb{C}^2 \,|\, \theta \in \mathbb{R},\, z \in \Gamma \right\}, \tag{3.4.2}$$

where

$$\Gamma \subset \{ z \in \mathbb{C} \,|\, \mathrm{Re}(z) > 0 \}$$

is the boundary of an embedded disc of area $\pi/2$. This construction is due to Eliashberg and Polterovich [196]. They also found many examples of isotopy classes of (knotted) embedded tori in $\mathbb{C}^2$ that do not contain any Lagrangian submanifolds. Their proof is based on Luttinger surgery (see Example 3.4.16 below). It was shown by Gadbled [256] that L_{Ch} and L_{EP} are Hamiltonian isotopic; see also Oakley–Usher [505]. So-called special Lagrangian tori in the Hamiltonian isotopy classes of L_{Cl} and L_{Ch} were studied by Auroux [38, 39] in the context of mirror symmetry. For more information see also the work of Chekanov–Schlenk [103] and Biran–Cornea [71], and the discussion in Section 14.3.

Exercise 3.4.8 Give examples of symplectic, isotropic, coisotropic, and Lagrangian submanifolds of the symplectic manifold $\mathbb{R}^4/\Gamma$ of Example 3.1.17. (If you find this difficult, first do the same problem for the torus $\mathbb{R}^4/\mathbb{Z}^4$ with the symplectic form induced by ω_0 on $\mathbb{R}^4$ by investigating the properties of the subtori obtained by fixing some of the coordinates.) □

Exercise 3.4.9 If Q is a coisotropic submanifold in (M, ω), show that the complementary distribution $TQ^\omega \subset TQ$ is integrable. Since ω vanishes on TQ^ω the leaves of the corresponding foliation of Q are isotropic. This generalizes the characteristic foliation on a hypersurface, discussed in Section 1.2. **Hint:** Suppose that Q has codimension k, and choose functions $H_1, \ldots, H_k$ near $p \in Q$ which locally define Q. Show that the vector fields $X_{H_1}, \ldots, X_{H_k}$ span the distribution T_xQ^ω for all $x \in Q \cap U$. Check that the functions $H_1, \ldots, H_k$ are in involution along Q, i.e. the Poisson bracket $\{H_i, H_j\} = dH_i \cdot X_{H_j}$ vanishes on Q for all i and j. Use this to show that the vector fields X_{H_j} span an integrable distribution in TQ. □

Neighbourhoods

We now use Moser's homotopy method to show that when Q is one of these special submanifolds, a neighbourhood of Q in M is determined, up to a symplectomorphism which is the identity on Q, by the restriction of ω to Q and some suitable data on the normal bundle ν_Q of Q in M. Throughout, we will denote any, suitably small, neighbourhood of Q by $\mathcal{N}(Q)$. (Think of $\mathcal{N}(Q)$ as a germ of M at Q.)

The symplectic case

We first consider the case when Q is symplectic. In this case the normal bundle ν_Q may be identified with the complementary symplectic bundle TQ^ω. Since the restriction of ω is a symplectic form on TQ^{ω}, we may consider ν_Q to be a symplectic vector bundle. Recall from Theorem 2.6.3 that its isomorphism class as a symplectic vector bundle is determined by the isomorphism class of its underlying complex bundle. The next result, due to Weinstein [669], shows that a neighbourhood of Q is completely determined by the restriction of ω to Q together with the isomorphism class of the symplectic bundle (ν_Q, ω).

Theorem 3.4.10 (Symplectic neighbourhood theorem) *For $j = 0, 1$ let (M_j, ω_j) be a symplectic manifold with compact symplectic submanifold Q_j. Suppose that there is an isomorphism*

$$\Phi : \nu_{Q_0} \to \nu_{Q_1}$$

of the symplectic normal bundles to Q_0 and Q_1 which covers a symplectomorphism $\phi : (Q_0, \omega_0) \to (Q_1, \omega_1)$. Then ϕ extends to a symplectomorphism $\psi : (\mathcal{N}(Q_0), \omega_0) \to (\mathcal{N}(Q_1), \omega_1)$ such that $d\psi$ induces the map Φ on $\nu_{Q_0} = (TQ_0)^{\omega}$.

Proof: First observe that ϕ extends to a diffeomorphism $\phi' : \mathcal{N}(Q_0) \to \mathcal{N}(Q_1)$ such that $d\phi'$ induces the map Φ on $\nu_{Q_0} = TQ_0^{\omega}$. Here we may take

$$\phi' = \exp_1 \circ \Phi \circ {\exp_0}^{-1},$$

where $\exp_j$ is an exponential map which takes a neighbourhood of the zero section in the normal bundle to Q_j onto some neighbourhood of Q_j in M_j. Thus we may assume that ω_0 and ω_1 are two symplectic forms on M whose restrictions to $T_Q M$ agree. The result now follows from Moser isotopy (see Lemma 3.2.1). □

Exercise 3.4.11 Let Q be a 2-dimensional compact symplectic submanifold of a symplectic 4-manifold (M, ω). Prove that a neighbourhood of Q is determined up to symplectomorphism by the self-intersection number $Q \cdot Q$ and the integral $\int_Q \omega$. Use Theorem 2.7.5 and Exercise 3.2.8 (ii). □

Exercise 3.4.12 Suppose that the normal bundles ν_{Q_0} and ν_{Q_1} are trivial as symplectic (or, equivalently, complex) bundles, and fix a symplectic isomorphism from ν_{Q_0} to the trivial symplectic bundle $Q_0 \times \mathbb{R}^{2k}$. Then, choosing the isomorphism Φ in the preceding theorem is equivalent to choosing a symplectic framing of ν_{Q_1}, and so there may well be several nonisotopic choices.

Here is an explicit example. For $i = 0, 1$ let $(M_i, Q_i) = (\mathbb{T}^2 \times \mathbb{C}, \mathbb{T}^2 \times \{0\})$ with the usual product form and let $\phi = \mathrm{id}$. Take the obvious identifications $\nu_{Q_0} = \nu_{Q_1} = \mathbb{T}^2 \times \mathbb{C}$, and define Φ by $\Phi(s, t, v) := (s, t, e^{2\pi i t} v)$, where $(s, t) \in \mathbb{T}^2 = \mathbb{R}^2/\mathbb{Z}^2$. Show that Φ is an isomorphism of the symplectic vector bundle $\nu_{\mathbb{T}^2}$, and find a formula for the symplectomorphism $\psi : \mathcal{N}(Q_0) \to \mathcal{N}(Q_1)$. This example is taken from Gompf [277], who used it to establish the existence of a nonKähler and simply connected symplectic 4-manifold. For a further discussion, see Section 7.2. □

The Lagrangian case

Next let us consider a Lagrangian submanifold L in (M, ω). The following important result, again due to Weinstein [669], asserts that the the symplectomorphism class of a sufficiently small neighbourhood of L is completely determined by the diffeomorphism class of L itself.

Theorem 3.4.13 (Lagrangian neighbourhood theorem) *Let (M,ω) be a symplectic manifold and $L \subset M$ be a compact Lagrangian submanifold. Then there exists a neighbourhood $\mathcal{N}(L_0) \subset T^*L$ of the zero section, a neighbourhood $V \subset M$ of L, and a diffeomorphism $\phi : \mathcal{N}(L_0) \to V$ such that*

$$\phi^*\omega = -d\lambda, \qquad \phi|_L = \mathrm{id},$$

where λ is the canonical 1*-form on T^*L.*

Proof: The proof rests on the fact that the normal bundle of L in M is isomorphic to the tangent bundle. To define an explicit isomorphism, one may use a compatible complex structure J on the tangent bundle TM, which exists by Proposition 2.6.4. (Such a complex structure is called an almost complex structure on M: see Chapter 4.) By Proposition 2.5.4 the subspace $J_qT_qL \subset T_qM$ is the orthogonal complement of T_qL with respect to the metric g_J induced by J, and is a Lagrangian subspace of (T_qM,ω). Let

$$\Phi_q : T_q^*L \to T_qL$$

be the isomorphism induced by the metric g_J, i.e.

$$g_J(\Phi_q(v^*), v) := v^*(v), \qquad v \in T_qL.$$

Now consider the map $\phi : T^*L \to M$ given by the exponential map of the Riemannian metric g_J:

$$\phi(q, v^*) := \exp_q(J_q\Phi_q(v^*)).$$

Then for $v = (v_0, v_1^*) \in T_qL \oplus T_q^*L = T_{(q,0)}T^*L$ we have

$$d\phi_{(q,0)}(v) = v_0 + J_q\Phi_q(v_1^*),$$

and hence

$$\begin{aligned}
\phi^*\omega_{(q,0)}(v,w) &= \omega_q(v_0 + J_q\Phi_q(v_1^*), w_0 + J_q\Phi_q(w_1^*)) \\
&= \omega_q(v_0, J_q\Phi_q(w_1^*)) - \omega_q(w_0, J_q\Phi_q(v_1^*)) \\
&= g_J(v_0, \Phi_q(w_1^*)) - g_J(w_0, \Phi_q(v_1^*)) \\
&= w_1^*(v_0) - v_1^*(w_0) \\
&= -d\lambda_{(q,0)}(v,w).
\end{aligned}$$

This shows that the 2-form $\phi^*\omega \in \Omega^2(T^*L)$ agrees with the canonical 1-form $-d\lambda_{\mathrm{can}}$ on the zero section. Hence the assertion of Theorem 3.4.13 follows from Moser isotopy (see Lemma 3.2.1). □

As an example, consider the diagonal Δ in the product $(M \times M, (-\omega) \oplus \omega)$. The previous proposition implies that a neighbourhood $\mathcal{N}(\Delta)$ of Δ is symplectomorphic to a neighbourhood $\mathcal{N}(M_0)$ of the zero section in T^*M. It follows that a neighbourhood of the identity in the group $\mathrm{Symp}(M)$ of all symplectomorphisms of M may be identified with a neighbourhood of zero in the vector space of closed 1-forms on M.

Proposition 3.4.14 *Let (M,ω) be a closed symplectic manifold. Then a neighbourhood of the identity in* $\mathrm{Symp}(M)$ *can be identified with a neighbourhood of zero in the vector space of closed* 1*-forms on* M.

Proof: By Theorem 3.4.13, there exist a neighbourhood $\mathcal{N}(\Delta) \subset M \times M$ of the diagonal Δ, a convex neighbourhood $\mathcal{N}(M_0) \subset T^*M$ of the zero section M_0, and a diffeomorphism $\Psi : \mathcal{N}(\Delta) \to \mathcal{N}(M_0)$ such that

$$\Psi(q,q) = (q,0)$$

for $q \in M$ and

$$\Psi^*(-d\lambda_{\mathrm{can}}) = (-\omega) \oplus \omega.$$

(See Fig. 3.3.) Now suppose that $\psi \in \mathrm{Symp}(M,\omega)$ is sufficiently C^1-close to the identity. By Proposition 3.4.3 the graph of ψ is a Lagrangian submanifold of $\mathcal{N}(\Delta)$. Hence

$$L := \Psi(\mathrm{graph}(\psi))$$

is a Lagrangian submanifold of T^*M. Moreover, L is the image of a smooth map

$$M \to T^*M : q \mapsto \Psi(q, \psi(q))$$

which is C^1-close to the canonical embedding of the zero section. Hence L is the graph of a 1-form σ on M (see Exercise 3.4.15 below), which is closed by Proposition 3.4.2. This proves Proposition 3.4.14. □

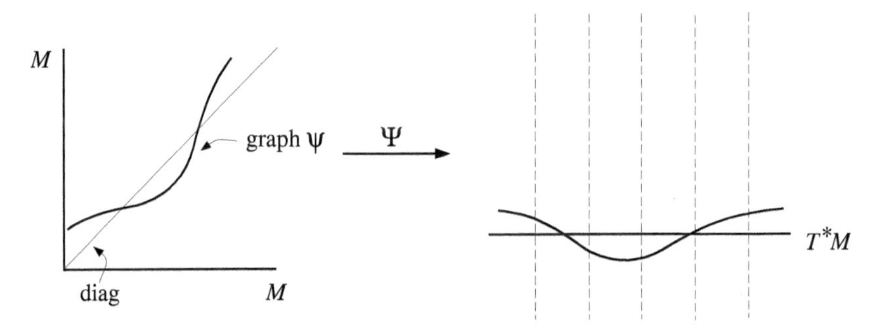

FIG. 3.3. The graph of ψ.

Exercise 3.4.15 **(i)** Let $g : M \to T^*M$ be an embedding which is sufficiently close to the canonical embedding of the zero section in the C^1-topology. Prove that the image of g is the graph of a 1-form.
(ii) Let $g : M \to M \times M$ be an embedding which is sufficiently close to the canonical embedding of the diagonal in the C^1-topology. Prove that the image of g is the graph of a diffeomorphism. **Hint:** A smooth map $f : M \to M$ which is sufficiently close to the identity in the C^1-topology is a diffeomorphism. □

Example 3.4.16 (Luttinger surgery) Let (M, ω) be a symplectic 4-manifold without boundary and let $\iota : \mathbb{T}^2 \to M$ be a Lagrangian embedding of the standard 2-torus with coordinates $x = (x_1, x_2) \in \mathbb{T}^2 = \mathbb{R}^2/\mathbb{Z}^2$. **Luttinger surgery** assigns a new symplectic manifold $M(\iota, k)$ to the Lagrangian embedding and an integer k. The manifold $M(\iota, k)$ is obtained by cutting out a neighbourhood of the Lagrangian torus (using the Lagrangian Neighbourhood Theorem) and gluing it back in with a suitable diffeomorphism of the boundary. Here are the details of the construction (following Auroux–Donaldson–Katzarkov [44]).

The cotangent bundle of the torus is $T^*\mathbb{T}^2 = \mathbb{T}^2 \times \mathbb{R}^2$ with coordinates (x_1, x_2, y_1, y_2) and symplectic form $\omega_{\mathrm{can}} = dx_1 \wedge dy_1 + dx_2 \wedge dy_2$. The Lagrangian Neighbourhood Theorem 3.4.13 asserts that the original Lagrangian embedding extends to a symplectic embedding

$$\iota : \mathbb{T}^2 \times [-r, r]^2 \to M, \qquad \iota^*\omega = \omega_{\mathrm{can}},$$

(denoted by the same letter) of a compact neighbourhood of the zero section in $T^*\mathbb{T}^2$. Choose a smooth cutoff function $\beta : [-r, r] \to [0, 1]$ such that

$$\beta(t) = \begin{cases} 0, & \text{for } t \leq -r/3, \\ 1, & \text{for } t \geq r/3, \end{cases} \qquad \int_{-r}^{r} t\beta'(t)\, dt = 0.$$

For $0 < \varepsilon \leq r$ define

$$U_\varepsilon := \mathbb{T}^2 \times [-\varepsilon, \varepsilon]^2.$$

Given an integer $k \in \mathbb{Z}$ define the symplectomorphism

$$\phi_k : U_r \setminus U_{r/2} \to U_r \setminus U_{r/2}$$

by

$$\phi_k(x_1, x_2, y_1, y_2) := \begin{cases} (x_1, x_2, y_1, y_2), & \text{for } y_2 < r/2, \\ (x_1 + k\beta(y_1), x_2, y_1, y_2), & \text{for } y_2 \geq r/2. \end{cases}$$

Thus ϕ_k acts by a k-fold Dehn twist in the (x_1, y_1) coordinates in the region $y_2 \geq r/2$. The manifold $M(\iota, k)$ is defined by

$$M(\iota, k) := (M \setminus \iota(U_{r/2})) \cup_{\phi_k} U_r := \big((M \setminus \iota(U_{r/2})) \sqcup U_r\big)/\sim,$$

where the equivalence relation identifies an element $\iota(x_1, x_2, y_1, y_2) \in \iota(U_r \setminus U_{r/2})$ in the first summand $M \setminus \iota(U_{r/2})$ with the element $\phi_k(x_1, x_2, y_1, y_2) \in U_r \setminus U_{r/2}$ in the second summand U_r. The resulting manifold $M(\iota, k)$ still contains a Lagrangian torus (the zero section in U_r). Note that $M(\iota, 0) = M$.

For more information and interesting applications of Luttinger surgery, see the original paper by Luttinger [435], and the papers by Eliashberg–Polterovich [196] and Auroux–Donaldson–Katzarkov [44]. For a generalization to coisotropic tori, see Baldridge–Kirk [48]. □

Exact Lagrangian submanifolds and their generating functions are discussed in detail in Section 9.4. There has also been a great deal of interest in **Lagrangian fibrations**, i.e. fibrations $\mathrm{pr} : (M, \omega) \to B$ of a symplectic manifold with Lagrangian fibres, since according to the program of Strominger–Yau–Zaslow [600] such fibrations can be used to give a geometric construction of mirror manifolds.

The (co)isotropic case

Exercise 3.4.17 (Hypersurfaces) Let ω_0 and ω_1 be symplectic forms on M which agree on a compact orientable hypersurface S. (They are not required to agree on $T_S M$.) Prove that the obvious inclusion $\iota : S \to M$ extends to an embedding ϕ of a neighbourhood U of S into M such that $\phi^*\omega_1 = \omega_0$. Deduce that a neighbourhood of S is symplectomorphic to the product $S \times (-\varepsilon, \varepsilon)$ with the symplectic form

$$\omega = \iota^*\omega_0 + d(t\alpha).$$

Here t is the coordinate on $(-\varepsilon, \varepsilon)$ and α is any 1-form on S that does not vanish on any nonzero vector in the line bundle $L \subset TS$ with fibres $L_q = (T_qS)^{\omega_0} = (T_qS)^{\omega_1}$. **Hint:** Since S is orientable, the normal bundle ν_S admits a nonzero section $q \mapsto \nu(q)$. Let $S \to L : q \mapsto \xi(q)$ be a nonzero section of the characteristic line bundle L. Prove that $\omega_i(\nu(q), \xi(q)) \neq 0$ for $i = 0, 1$ and all $q \in S$. Define the locally constant function $\varepsilon : S \to \{-1, +1\}$ by $\varepsilon(q) := \mathrm{sign}\big(\omega_0(\nu(q), \xi(q))/\omega_1(\nu(q), \xi(q))\big)$ for $q \in S$. Next choose a diffeomorphism $\psi : U \to U$ of a neighbourhood U of S such that $\psi(q) = q$ and $d\psi(q)\nu(q) = \varepsilon(q)\nu(q)$ for all $q \in S$. Prove that the 2-forms $\omega_t := \omega_0 + t(\psi^*\omega_1 - \omega_0)$ are nondegenerate near S for $0 \le t \le 1$. Prove that $\tau := \psi^*\omega_1 - \omega_0$ is exact near S. Now use Moser isotopy. □

Exercise 3.4.18 State and prove the analogues of Theorem 3.4.10 and Theorem 3.4.13 for isotropic and coisotropic submanifolds. □

Further discussion of the structure of coisotropic submanifolds and their role in the process of symplectic reduction may be found in Section 5.4.

Further exercises and remarks

Exercise 3.4.19 Show that any point q of a symplectic $2k$-dimensional submanifold Q of M has a Darboux chart such that Q is given by the equations $x_i = 0$, for $i > 2k$. State and prove similar theorems for Lagrangian, isotropic, and coisotropic submanifolds. In the coisotropic case the chart should be adapted to the isotropic foliation of Exercise 3.4.9. □

Exercise 3.4.20 Let $Q \subset (M, \omega)$ be a symplectic, Lagrangian, isotropic, or coisotropic submanifold, and let $\psi_t : Q \to M$ be a smooth family of embeddings starting at the inclusion such that $\psi_t^*\omega = \psi_0^*\omega$ for all t. Show that ψ_t extends symplectically over a neighbourhood of Q. **Hint:** First adjust the forms $\psi_t^*\omega$ to be constant on $T_Q M$ by using results such as Corollary 2.5.14. □

We have stated the results in this section for compact manifolds, but they remain true in the noncompact case and in the presence of a group action. There are also corresponding results for any submanifold on which the restriction of ω has constant rank (see Exercise 2.1.12).

3.5 Contact structures

Contact geometry is the odd-dimensional analogue of symplectic geometry. Here we will just briefly sketch some of the highlights of the theory. For more information and further references see Arnold [23], the survey paper by Eliashberg [182] from the early 1990s, Eliashberg–Gromov [187], Eliashberg–Thurston [197], Arnold–Givental [25], Geiges [261], Cieliebak–Eliashberg [111].

A **contact structure** on a manifold M of dimension $2n+1$ is a field of hyperplanes $\xi \subset TM$ which is as far as possible from being integrable. Recall that the Frobenius integrability theorem asserts that ξ is integrable if and only if the sections of ξ are closed under the Lie bracket.

For simplicity, we assume throughout that ξ is transversally orientable, i.e. that it can be described as the kernel of some 1-form α.

Thus, a vector field X is a section of ξ if and only if $\alpha(X) = 0$, and so ξ is integrable if and only if $\alpha([X,Y]) = 0$ whenever $\alpha(X) = \alpha(Y) = 0$. In view of the identity[14]

$$d\alpha(X,Y) = \mathcal{L}_X(\alpha(Y)) - \mathcal{L}_Y(\alpha(X)) + \alpha([X,Y]) \tag{3.5.1}$$

this amounts to requiring that $d\alpha$ be zero when restricted to the vectors in ξ or, equivalently, that $\alpha \wedge d\alpha = 0$. The contact condition is as far from this as possible. It requires that $d\alpha$ restricts to a nondegenerate form on ξ. Thus ξ can be thought of as a maximally nonintegrable distribution. The next proposition shows that every contact manifold is orientable.

Proposition 3.5.1 *Let M be a manifold of dimension $2n+1$ and $\xi \subset TM$ be a transversally orientable hyperplane field.*

(i) *Let α be a 1-form with $\xi = \ker\alpha$. Then $d\alpha$ is nondegenerate on ξ if and only if*

$$\alpha \wedge (d\alpha)^n \neq 0.$$

In this case ξ is called a **contact structure** *and α a* **contact form** *for ξ.*

(ii) *Let α and α' be 1-forms with $\xi = \ker\alpha = \ker\alpha'$. Then α is a contact form if and only if α' is.*

(iii) *If ξ is a contact structure then the symplectic bilinear form on ξ induced by $d\alpha$ is independent of the choice of the contact form α up to a positive scaling function.*

Proof: By Corollary 2.1.4, the 2-form $d\alpha$ is nondegenerate on $\xi = \ker\alpha$ if and only if $(d\alpha)^n$ is nonzero on ξ. This is equivalent to $\alpha \wedge (d\alpha)^n \neq 0$. If α and α' are as in (ii), then there exists a nonzero function $f : M \to \mathbb{R}$ such that

$$\alpha' = f\alpha.$$

Hence $d\alpha' = f d\alpha + df \wedge \alpha$ and $d\alpha'$ agrees with $f d\alpha$ on ξ. This proves statements (ii) and (iii). □

[14] Recall the sign conventions in Remark 3.1.6.

Proposition 3.5.2 *Let (M^{2n+1}, ξ) be a contact manifold, $\alpha \in \Omega^1(M)$ a contact form for ξ, and $L \subset M$ an integral submanifold of ξ. Then T_qL is an isotropic subspace of the symplectic vector space $(\xi_q, d\alpha_q)$ for every $q \in L$. In particular,*

$$\dim L \leq n.$$

If $\dim L = n$ *then L is called* **Legendrian**.

Proof: If $X, Y \in \mathcal{X}(L)$ then $\iota(X)\alpha = \iota(Y)\alpha = \iota([X,Y])\alpha = 0$ and equation (3.5.1) shows that $d\alpha(X,Y) = 0$. Hence $d\alpha$ vanishes on TL. □

We emphasize that the contact structure is the hyperplane field ξ and not the contact form α. A contact structure can be viewed as an equivalence class of 1-forms $\alpha \in \Omega^1(M)$ which satisfy $\alpha \wedge (d\alpha)^n \neq 0$, where two such 1-forms are equivalent iff they differ by a nonzero function on M. Often one fixes a transverse orientation of ξ. This corresponds to the equivalence relation $\alpha \sim \alpha'$ iff there exists a positive function f such that $\alpha' = f\alpha$.

Thus, specifying a particular contact form is equivalent to specifying a function. Now a function on a symplectic manifold generates a flow. Analogously, there is a canonically defined flow associated to any contact form. In fact, given a contact form α there exists a unique vector field $Y = Y_\alpha : M \to TM$ such that

$$\iota(Y)d\alpha = 0, \qquad \alpha(Y) = 1. \tag{3.5.2}$$

This vector field is called the **Reeb vector field** determined by α. The first condition says that Y points along the unique null direction of the form $d\alpha$ and the second condition normalizes Y. Because

$$\mathcal{L}_Y\alpha = d\,\iota(Y)\alpha + \iota(Y)\,d\alpha = 0,$$

the flow of Y preserves the contact form α and hence the contact structure ξ. Note that if one chooses a different contact form $f\alpha$, the corresponding vector field $Y_{f\alpha}$ is different from Y_α, and its flow may have quite different properties. If, however, α and $f\alpha$ are related by a contactomorphism[15] then so are the Reeb vector fields Y_α and $Y_{f\alpha}$.

Examples

The basic example is the contact structure on $\mathbb{R}^{2n+1}$ given by the 1-form

$$\alpha_0 := \sum_j y_j dx_j - dz, \tag{3.5.3}$$

where we use the coordinates $x_1, \ldots, x_n, y_1, \ldots, y_n, z$ on $\mathbb{R}^{2n+1}$. This is the **standard contact structure** on $\mathbb{R}^{2n+1}$. Its Reeb vector field is $Y = -\frac{\partial}{\partial z}$. The standard contact structure is invariant under translation in the x_i and z directions,

[15] A contactomorphism of (M, ξ) is a diffeomorphism of M that preserves the transversally oriented hyperplane field ξ: see the discussion before Lemma 3.5.14 below.

and the planes parallel to the y-axis are Legendrian submanifolds. In particular, in dimension 3, the lines parallel to the y-axis are Legendrian curves, and as one moves along them the contact plane turns slowly, becoming more and more vertical as $|y|$ increases.

Exercise 3.5.3 **(i)** Show that the 1-form

$$\alpha_1 := \tfrac{1}{2}\sum_j (y_j dx_j - x_j dy_j) - dz \tag{3.5.4}$$

is another contact form on $\mathbb{R}^{2n+1}$ that is diffeomorphic to α_0. **Hint:** Look for a diffeomorphism of the form $(x, y, z) \mapsto (x, y, z + F(x, y))$. (A general method for understanding problems of this kind is given in the discussion of contact isotopy, just before Exercise 3.5.20. Note that here we can choose ψ_t so that $f_t \equiv 1$.)

(ii) Show that the contact form α_1 in (3.5.4) is invariant under the product of rotations in the (x_j, y_j) planes. In dimension 3, show that all horizontal rays through the z-axis are Legendrian. Again, as one moves out along each such ray, the contact plane turns slowly, but never becomes vertical. What are the images of these families of rotations and rays in $(\mathbb{R}^{2n+1}, \alpha_0)$?

(iii) Denote the coordinates on $\mathbb{R}^3$ by θ, x, y and consider the 1-form

$$\alpha_2 := \sin\left(\frac{x^2+y^2}{2}\right)\frac{ydx - xdy}{x^2+y^2} - \cos\left(\frac{x^2+y^2}{2}\right)d\theta. \tag{3.5.5}$$

This 1-form descends to $S^1 \times \mathbb{R}^2$, where $S^1 := \mathbb{R}/2\pi\mathbb{Z}$. Show that it is a contact form with $\alpha_2 \wedge d\alpha_2 = d\theta \wedge dx \wedge dy$. Show that the rays $\theta = \text{const}$, $x/y = \text{const}$ and the circles $\theta = \text{const}$, $(x^2 + y^2)/2 = k\pi$ (with k a positive integer) are Legendrian. □

Definition 3.5.4 *An embedded disc $D \subset M$ in a contact 3-manifold (M, ξ) is called* **overtwisted** *if its boundary is a Legendrian circle and its tangent planes at interior points are transverse to the contact distribution except at one point. A contact 3-manifold is called* **overtwisted** *if it contains an overtwisted disc, and is called* **tight** *otherwise.*

Example 3.5.5 The contact structure on $\mathbb{R}^3$ determined by the contact form α_2 in (3.5.5) contains an overtwisted disc $D := \{\theta = 0,\ (x^2 + y^2)/2 \le \pi\} \subset \mathbb{R}^3$. More generally, let $\beta : [0,\infty) \to [0,\infty)$ be any smooth function such that $\beta' > 0$ and $\beta(s) = s$ for s near 0. Then the 1-form

$$\alpha_\beta := \sin\bigl(\beta\bigl(\tfrac{x^2+y^2}{2}\bigr)\bigr)\frac{ydx - xdy}{x^2+y^2} - \cos\bigl(\beta\bigl(\tfrac{x^2+y^2}{2}\bigr)\bigr)\, d\theta \tag{3.5.6}$$

is a contact form on $\mathbb{R}^3$ with a Legendrian circle $\theta = \text{const}$, $(x^2 + y^2)/2 = s > 0$ whenever $\beta(s) \in \pi\mathbb{N}$. If $\beta(s) = \pi$ then the disc $D := \{\theta = \text{const},\ (x^2+y^2)/2 \le s\}$ is overtwisted. **Exercise:** Show that the volume form of α_β is given by

$$\alpha_\beta \wedge d\alpha_\beta = \beta'\bigl(\tfrac{x^2+y^2}{2}\bigr)\, d\theta \wedge dx \wedge dy.$$

□

One of the first nontrivial results in modern contact topology was Bennequin's proof [60] that the standard contact structure $\xi_0 = \ker \alpha_0$ on $\mathbb{R}^3$ does not contain an overtwisted disc. It follows that the contact structure defined on $\mathbb{R}^3$ by α_2 is not standard. On the other hand, Eliashberg [177, 182] later showed that, up to diffeomorphism, there is only one tight and one overtwisted contact structure on $\mathbb{R}^3$. Therefore the set of forms α_β divides into two classes; if $\sup_s \beta(s) \leq \pi$ the corresponding contact structures are all diffeomorphic to the standard one, while if $\sup_s \beta(s) > \pi$ they are diffeomorphic to the overtwisted structure.

Example 3.5.6 Let L be any compact manifold. Then the 1-jet bundle

$$J^1L := T^*L \times \mathbb{R}$$

is a contact manifold with the contact form

$$\alpha := \lambda_{\mathrm{can}} - dz$$

and the Reeb vector field $Y = -\partial/\partial z$, where z is the real parameter. For any function $S : L \to \mathbb{R}$ the submanifold

$$L_S := \{(x, dS(x), S(x)) \,|\, x \in L\} \subset J^1L$$

is Legendrian. This is the analogue of the construction of Lagrangian submanifolds of T^*L as the graphs of exact 1-forms. Any compact Legendrian submanifold L of any contact manifold M has a neighbourhood which is contactomorphic to a neighbourhood of the zero section in J^1L. □

The previous example is a 1-dimensional extension of the symplectic manifold T^*L. In contrast, the next example describes a natural contact structure on a codimension-1 reduction of a cotangent bundle.

Example 3.5.7 Given a Riemannian manifold L, let $M = S(T^*L)$ be the unit sphere bundle in T^*L. Then the restriction of the canonical 1-form

$$\alpha := \lambda_{\mathrm{can}}|_{S(T^*L)}$$

is a contact form. The corresponding Reeb vector field Y is the Hamiltonian vector field dual to the geodesic flow on TL. In the case $L = \mathbb{R}^{n+1}$ this construction yields a contact structure on $\mathbb{R}^{n+1} \times S^n$ which extends the above contact structure on $\mathbb{R}^{2n+1}$.

Neglecting the Riemannian structure we may think of $M = S(T^*L)$ as the manifold formed by all oriented hyperplanes in TL. Then an element of $S(T^*L)$ is an oriented line in T^*L, so that $S(T^*L)$ is the oriented projectivization[16] of the cotangent bundle T^*L. The contact structure ξ is given by

[16] Here we take the oriented rather than the unoriented projectivization because we only consider transversely orientable contact structures. Arnold [23] calls the unoriented projectivization the 'manifold of contact elements' and regards it as the archetypal example of a contact manifold.

$$\xi_{(q,[v^*])} := \ker v^* \circ d\pi_{(q,[v^*])}$$

for $q \in L$ and $v^* \in T_q^*L$. Here $[v^*]$ is the line $\{rv^* \,|\, r > 0\}$ and $\pi : S(T^*L) \to L$ denotes the natural projection.

Any Lagrangian submanifold of T^*L which is transverse to $S(T^*L)$ intersects it in a Legendrian submanifold. In particular, if Q is a proper submanifold of L, then $S(TQ^\perp)$ is a Legendrian submanifold of $S(T^*L)$ (see Example 3.4.1). □

Example 3.5.8 A special case of Example 3.5.7 is the torus $L = \mathbb{T}^n$. Its unit cotangent bundle

$$M := \mathbb{T}^n \times S^{n-1}$$

is a contact manifold with contact form

$$\alpha := \sum_{j=1}^{n} y_j dx_j,$$

where $x = (x_1, \ldots, x_n) \in \mathbb{T}^n = \mathbb{R}^n/\mathbb{Z}^n$ and $y = (y_1, \ldots, y_n) \in S^{n-1}$. The Reeb flow is given by $\dot{x} = y$ and $\dot{y} = 0$. This is the restriction to $\mathbb{T}^n \times S^{n-1}$ of the Hamiltonian flow on $T^*\mathbb{T}^n = \mathbb{T}^n \times \mathbb{R}^n$ generated by the Hamiltonian function

$$H(x, y) := \tfrac{1}{2}|y|^2.$$

Let $y_0 = (y_{01}, \ldots, y_{0n}) \in S^{n-1}$ be any frequency vector with rationally dependent components, i.e. all the ratios y_{0i}/y_{0j} are rational. Then the manifold

$$L := \{(x + j, y_0) \,|\, \langle x, y_0 \rangle = 0,\ j \in \mathbb{Z}^n\} / \mathbb{Z}^n$$

is a Legendrian submanifold of $\mathbb{T}^n \times S^{n-1}$. Note that this is one of the two components of a Legendrian submanifold of the form $S(TQ^\perp)$.

For $n = 2$ this example furnishes a tight contact structure on the 3-torus. There are many other (i.e. noncontactomorphic) tight contact structures on the 3-torus. (See Giroux [270, 271], Kanda [353], and Honda [329, 330]). □

Example 3.5.9 Let $\Sigma := S^{2n+1}$ be the unit sphere in $\mathbb{R}^{2n+2} = \mathbb{C}^{n+1}$. At each point $q \in \Sigma$, the complex part

$$\xi_q := T_q\Sigma \cap J_0 T_q\Sigma$$

of its tangent space has codimension 1. These hyperplanes form a contact structure on Σ with the contact form

$$\lambda_0 := \tfrac{1}{2} \sum_{j=0}^{n} (y_j dx_j - x_j dy_j),$$

and the Reeb flow is given by $\dot{x} = 2y$ and $\dot{y} = -2x$. □

The contact structure in Example 3.5.9 is called the **standard contact structure on S^{2n+1}**. The next exercise shows that the standard contact structure on the complement of a point in S^{2n+1} is diffeomorphic to the standard structure on $\mathbb{R}^{2n+1}$. A different proof, based on properties of the stereographic projection, may be found in [261, Proposition 2.1.8].

Exercise 3.5.10 **(i)** Let D_{n+1} denote the **Siegel domain**

$$D_{n+1} = \{(z, w) \in \mathbb{C}^n \times \mathbb{C} \,|\, \operatorname{Im} w > |z|^2\},$$

and consider the map $f : \mathbb{C}^{n+1} \setminus (\mathbb{C}^n \times \{-1\}) \to \mathbb{C}^{n+1} \setminus (\mathbb{C}^n \times \{-i\})$ defined by

$$f(z, w) = \left(\frac{z}{w+1}, -i\frac{w-1}{w+1} \right).$$

Show that f maps the interior of the unit ball biholomorphically onto D_{n+1}.
(ii) It follows from (i) that the boundary $Q := \partial D_{n+1}$ has a canonical contact structure ξ defined as in Example 3.5.9. Namely, at each point $q \in Q$, the contact hyperplane is defined to be the complex part

$$\xi_q := T_qQ \cap J_0 T_qQ$$

of the tangent space T_qQ. Prove this by direct calculation, and check that the contact structure so obtained is contactomorphic to the standard structure on $\mathbb{R}^{2n+1}$.
Note: The function $\phi := \operatorname{Im} w - |z|^2$ is weakly plurisubharmonic, i.e. $\omega_\phi := -d(d\phi \circ J)$ satisfies $\omega_\phi(v, Jv) \geq 0$ with strict inequality if $v \in \xi$ is nonzero, but one can easily change it in the direction normal to $Q := F^{-1}(0)$ to a function F' with $Q := F'^{-1}(-0)$ that is plurisubharmonic in the sense of Definition 7.4.7. In particular, the hypersurface $Q = \partial D_{n+1} \subset \mathbb{C}^{n+1}$ has contact type in the sense of Definition 3.5.32 below.
(iii) Find an explicit contactomorphism $S^{2n+1} \setminus \{\mathrm{pt}\} \to \mathbb{R}^{2n+1}$. □

Example 3.5.11 Let (V, ω) be a symplectic manifold and assume that the cohomology class of ω admits an integral lift in $H^2(V; \mathbb{Z})$. Then there exists a circle bundle $\pi : P \to V$ with Euler class $-[\omega]$. Denote the circle action on P by

$$P \times S^1 \to P : (p, e^{i\theta}) \mapsto pe^{i\theta}, \qquad \theta \in \mathbb{R}/2\pi\mathbb{Z}.$$

The action is generated by the vector field $X \in \mathcal{X}(P)$, defined by

$$X(p) := \left.\frac{d}{d\theta}\right|_{\theta=0} pe^{i\theta}.$$

A connection on P is an imaginary valued 1-form $A \in \Omega^1(P, i\mathbb{R})$ that is S^1-invariant and satisfies $A(X) = i$. Since the Euler class of P is $-[\omega]$, the connection can be chosen such that its curvature is

$$F_A = dA = 2\pi i\, \pi^*\omega.$$

(See Theorem 2.7.4.) In the language of Bott and Tu [78], A is a global angular form. The 1-form

$$\alpha := \frac{i}{2\pi} A$$

is a contact form on P with $d\alpha = -\pi^*\omega$. It is S^1-invariant and $\alpha(X) = -\frac{1}{2\pi}$. Hence the Reeb vector field of α is

$$Y := -2\pi X,$$

and all Reeb orbits have period one. This contact manifold is often called the **prequantization bundle** of (V, ω), and its properties are studied in relation to those of (V, ω). In particular, the group of diffeomorphisms of P that preserve the contact form α, sometimes called the **quantomorphism group**, is closely related to the group of symplectomorphisms of the base (V, ω). (See Souriau [596] for an extensive discussion of this topic and its connections to physics and dynamical systems.) □

Example 3.5.12 The contact structure on the sphere in Example 3.5.9 is a special case of Example 3.5.11. It is obtained by considering the Hopf fibration

$$S^{2n+1} \to \mathbb{C}\mathrm{P}^n.$$

(Compare with Example 4.3.3 below.) In that case the circle action is generated by the vector field

$$X := \sum_{j=0}^{n} \left(x_j \frac{\partial}{\partial y_j} - y_j \frac{\partial}{\partial x_j} \right).$$

The 2-form

$$\tau := \frac{1}{\pi} \sum_{j=0}^{n} dx_j \wedge dy_j$$

on S^{2n+1} descends to a symplectic form $\omega := \frac{1}{\pi}\omega_{\mathrm{FS}}$ on $\mathbb{C}\mathrm{P}^n$ which represents an integral cohomology class. The Euler class of the Hopf fibration is $-[\omega]$, the connection 1-form is

$$A := i \sum_{j=0}^{n} (x_j dy_j - y_j dx_j)$$

with

$$\frac{i}{2\pi} F_A = -\tau,$$

the contact form is

$$\alpha := \frac{i}{2\pi} A = \frac{1}{2\pi} \sum_{j=0}^{n} (y_j dx_j - x_j dy_j)$$

with $d\alpha = -\tau$, and the Reeb vector field is $Y = -2\pi X$. □

Example 3.5.13 Another special case of Example 3.5.11 is where the symplectic form on V is exact and P is the trivial circle bundle over V, i.e.

$$\omega = -d\lambda, \qquad P = V \times S^1.$$

In this case

$$A = -2\pi i\lambda + id\theta,$$

where $e^{i\theta}$ is the coordinate on S^1. Thus the contact form is

$$\alpha = \lambda - dt,$$

where $t = \theta/2\pi \in \mathbb{R}/\mathbb{Z}$. □

Contactomorphisms

Let (M, ξ) be a contact manifold and $\alpha \in \Omega^1(M)$ be a contact form for ξ. A diffeomorphism $\psi : M \to M$ is called a **contactomorphism** if ψ preserves the oriented hyperplane field ξ. This is equivalent to the condition

$$\psi^*\alpha = e^h\alpha$$

for some function $h : M \to \mathbb{R}$. A **contact isotopy** is a smooth family of contactomorphisms $\psi_t : M \to M$ such that

$$\psi_t^*\alpha = e^{h_t}\alpha, \qquad \psi_0 = \mathrm{id}.$$

Suppose these contactomorphisms are generated by a smooth family of vector fields $X_t : M \to TM$ via the differential equation

$$\frac{d}{dt}\psi_t = X_t \circ \psi_t.$$

Then the formula

$$\psi_t^*\mathcal{L}_{X_t}\alpha = \frac{d}{dt}\psi_t^*\alpha = \left(\frac{d}{dt}h_t\right)e^{h_t}\alpha = \left(\frac{d}{dt}h_t\right)\psi_t^*\alpha = \psi_t^*(g_t\alpha),$$

with $g_t := (\frac{d}{dt}h_t) \circ {\psi_t}^{-1}$, shows that

$$\mathcal{L}_{X_t}\alpha = g_t\alpha.$$

Conversely, if X_t satisfies this condition then the diffeomorphisms $\psi_t : M \to M$ generated by X_t as above determine a contact isotopy with $h_t = \int_0^t g_s \circ \psi_s\, ds$. A vector field $X : M \to TM$ such that $\mathcal{L}_X\alpha = g\alpha$ for some function $g : M \to \mathbb{R}$ is called a **contact vector field**. The above argument shows that the Lie algebra of the group of contactomorphisms is the space of contact vector fields.

Lemma 3.5.14 *Let Y be the Reeb field of the contact form α.*

(i) *The vector field $X : M \to TM$ is a contact vector field if and only if there exists a function $H : M \to \mathbb{R}$ such that*

$$\iota(X)\alpha = H, \qquad \iota(X)d\alpha = (\iota(Y)dH)\,\alpha - dH. \tag{3.5.7}$$

(ii) *For every function $H : M \to \mathbb{R}$ there exists a unique contact vector field $X = X_H : M \to TM$ which satisfies* (3.5.7).

Proof: If (3.5.7) holds then $\mathcal{L}_X\alpha = g\alpha$ with $g = \iota(Y)dH$. Conversely, assume $\mathcal{L}_X\alpha = g\alpha$ and define $H := \iota(X)\alpha$. Then

$$\iota(X)d\alpha = \mathcal{L}_X\alpha - d\iota(X)\alpha = g\alpha - dH.$$

Evaluate this 1-form on Y to obtain $g = \iota(Y)dH$. This proves (i).

To prove (ii) let $H : M \to \mathbb{R}$ be given. Then there exists a unique vector field $Z : M \to TM$ such that Z is a section of ξ and

$$-\iota(Z)d\alpha|_\xi = dH|_\xi.$$

Then the vector field

$$X_H := HY + Z$$

satisfies $\alpha(X_H) = H\alpha(Y) = H$ and

$$\iota(X_H)d\alpha = \iota(Z)d\alpha = (\iota(Y)dH)\,\alpha - dH.$$

The last equation holds because the 1-forms on both sides of the equation vanish on Y and agree on tangent vectors in ξ. This proves Lemma 3.5.14. □

Lemma 3.5.14 shows that for each fixed contact form α there is a one-to-one correspondence between contact vector fields $X : M \to TM$ and smooth functions $H : M \to \mathbb{R}$. For example, the Reeb vector field Y of α is generated by the function $H(q) \equiv 1$. The proof also shows that the only contact vector field which is everywhere tangent to the contact distribution ξ is the zero vector field. In other words, there is no nontrivial contact flow which is everywhere tangent to ξ. For example in $\mathbb{R}^3$ with contact structure given by $ydx - dz$, the vector field $\partial/\partial y$ is everywhere tangent to the contact distribution, but as one moves along a ray parallel to the y-axis the contact planes are not preserved, but rather slowly turn. This may be thought of as another expression of the nonintegrability of ξ.

The above discussion shows that there are many contactomorphisms that change the contact form by a nonconstant scalar function. It is an interesting question as to which functions can occur here; it was already noted by Gray and Bennequin that not all functions are possible. The next exercise gives an example from Bennequin [60], and also shows that such functions do not exist in symplectic geometry except in dimension two.

Exercise 3.5.15 **(i)** Show that the 1-form $\alpha = r^2 d\theta + dz$ is diffeomorphic to the contact form $\alpha_1 = -\frac{1}{2}r^2 d\theta - dz$ of Exercise 3.5.3, where (r, θ) are polar coordinates on the x, y-plane. But there is no diffeomorphism ψ such that $\psi^*\alpha = (1+r^2+z^2)^{-2}\alpha =: \beta$. **Hint:** If $\psi^*\alpha = \beta$, then $\psi^* d\alpha = d\beta$ so that $d\psi$ pushes the Reeb field of β forward to that of α. But the Reeb field of β (which lies in $\ker d\beta$) has a periodic orbit, namely $r = 1, z = 0$.

(ii) Show that if ω is a symplectic form on a closed connected manifold of dimension $2n \geq 4$, the only functions f such that $f\omega$ is closed are the constant functions. Because of this, one gets nothing new by defining a conformal-symplectic diffeomorphism ψ to be one such that $\psi^*\omega = f\omega$. □

Exercise 3.5.16 Let $H = H(x_1, \ldots, x_n, y_1, \ldots, y_n, z)$ be a smooth function on $\mathbb{R}^{2n+1}$. Prove that the contact vector field generated by H with respect to the standard form $\alpha_0 = \sum_j y_j dx_j - dz$ is given by the differential equation

$$\dot{x}_j = \frac{\partial H}{\partial y_j}, \quad \dot{y}_j = -\frac{\partial H}{\partial x_j} - y_j \frac{\partial H}{\partial z}, \quad \dot{z} = \sum_j y_j \dot{x}_j - H. \tag{3.5.8}$$

Note, in particular, that if H is independent of z then $z(t_1) - z(t_0)$ is given by the symplectic action of the solution curve $(x(t), y(t))$ of the Hamiltonian flow on the time interval $t_0 \leq t \leq t_1$ as defined in equation (1.1.8). □

Exercise 3.5.17 Prove that the solutions of (3.5.8) are the characteristics of the **Hamilton–Jacobi equation**

$$\partial_t S + H(x, \partial_x S, S) = 0 \tag{3.5.9}$$

for a function $S = S(t, x)$ on $\mathbb{R}^{n+1}$.

More precisely, if S is a solution of (3.5.9) and $x(t)$ is a solution of the ordinary differential equation

$$\dot{x} = \partial_y H(x, \partial_x S, S),$$

prove that the functions

$$x(t), \qquad y(t) := \partial_x S(t, x(t)), \qquad z(t) := S(t, x(t))$$

satisfy equation (3.5.8). Conversely, given an initial function $S(0, x) = S_0(x)$, use the solutions of the contact differential equation (3.5.8) with initial conditions of the form $x(0) = x_0$, $y(0) = \partial_x S_0(x_0)$, $z(0) = S_0(x_0)$, to construct a solution of the Hamilton–Jacobi equation (3.5.9) for small t. Moreover, prove that a function $S = S(t, x)$ satisfies (3.5.9) if and only if the corresponding Legendrian submanifolds

$$L_t := \{(x, \partial_x S(t, x), S(t, x)) \,|\, x \in \mathbb{R}^n\}$$

are related by $L_t = \psi_t(L_0)$, where $\psi_t : \mathbb{R}^{2n+1} \to \mathbb{R}^{2n+1}$ is the flow of the differential equation (3.5.8). Compare this with Example 3.5.6. For a further discussion of the Hamilton–Jacobi equation see Section 9.1. □

Remark 3.5.18 The analogue of the Poisson bracket in contact geometry is

$$\begin{aligned}\{F, G\} &:= \alpha([X_F, X_G]) \\ &= dF(X_G) - dG(X_F) + d\alpha(X_F, X_G) \\ &= dF(X_G) - dG(Y) \cdot F\end{aligned} \tag{3.5.10}$$

for $F, G : M \to \mathbb{R}$. (Recall the sign conventions in Remark 3.1.6.) Here $Y = Y_\alpha$ denotes the Reeb vector field of α in (3.5.2). The second equation in (3.5.10) follows from (3.5.1) and the third equation follows from (3.5.7). The contact vector fields form a Lie algebra with

$$[X_F, X_G] = X_{\{F,G\}}$$

for $F, G : M \to \mathbb{R}$. Hence the map $(F, G) \mapsto \{F, G\}$ determines a Lie algebra structure on $C^\infty(M)$. □

Lemma 3.5.14 shows that the group $\mathrm{Cont}(M, \xi)$ of contactomorphisms of (M, ξ) is infinite-dimensional and Remark 3.5.18 shows that its Lie algebra is the space of functions on M with the Poisson bracket in (3.5.10). To date, the group $\mathrm{Cont}(M, \xi)$ has been less studied than the group of symplectomorphisms, though there are some very interesting results, for example the work of Eliashberg–Kim–Polterovich [189] and Sandon [564] on nonsqueezing and orderability.

Exercise 3.5.19 **(i)** Not every contact vector field is the Reeb field of some contact form. Show that X is the Reeb field of some contact form which defines ξ if and only if X is transverse to ξ, i.e. $\alpha(X) \neq 0$, for any defining form α. (If ξ is transversally oriented, then the condition becomes $\alpha(X) > 0$.)

(ii) Let (M, ξ) be a contact manifold with contact form α and corresponding Reeb field Y. If β is any 1-form such that $\beta(Y) = 0$ prove that there is a unique vector field X which is tangent to $\ker \alpha$ and such that $\beta = \iota(X) d\alpha$. □

Contact isotopy

Appropriate versions of Darboux's theorem and Moser's stability theorem hold in contact geometry. Roughly speaking, we want to show that given a family α_t of contact forms on M, there is a family of diffeomorphisms ψ_t such that

$$\psi_t^* \alpha_t = f_t \alpha_0 \tag{3.5.11}$$

for some nonvanishing functions f_t. Again, we determine the diffeomorphisms ψ_t by representing them as the flow of a family of vector fields X_t on M:

$$\frac{d}{dt}\psi_t = X_t \circ \psi_t, \qquad \psi_0 = \mathrm{id}.$$

Differentiating (3.5.11) gives

$$\psi_t^* \left(\frac{d}{dt}\alpha_t + \mathcal{L}_{X_t}\alpha_t \right) = g_t \psi_t^* \alpha_t, \qquad g_t = \frac{1}{f_t}\frac{d}{dt} f_t.$$

Hence we must find X_t and $h_t = g_t \circ \psi_t^{-1}$ such that

$$\frac{d}{dt}\alpha_t + \mathcal{L}_{X_t}\alpha_t = h_t\alpha_t. \tag{3.5.12}$$

To solve equation (3.5.12) let Y_t be the Reeb vector field of α_t and define

$$h_t := \iota(Y_t)\frac{d}{dt}\alpha_t.$$

Then $h_t\alpha_t - \frac{d}{dt}\alpha_t$ vanishes on Y_t. By Exercise 3.5.19 there exists a unique one-parameter family of vector fields X_t such that

$$\iota(X_t)d\alpha_t = h_t\alpha_t - \frac{d}{dt}\alpha_t, \qquad \alpha_t(X_t) = 0.$$

Since $d(\iota(X_t)\alpha_t) = 0$, equation (3.5.12) is satisfied. Thus, when one wants to change the contact form one can flow along a vector field that is everywhere tangent to the contact distribution.

Exercise 3.5.20 (Darboux's theorem) Prove that every contact structure is locally diffeomorphic to the standard structure on $\mathbb{R}^{2n+1}$. □

Exercise 3.5.21 (Gray's stability theorem) Prove that every one-parameter family of contact forms α_t on a closed manifold M has the form

$$\alpha_t = \psi_t^*(f_t\alpha_0)$$

for some family of positive functions f_t and some isotopy ψ_t such that $\psi_0 = \mathrm{id}$ and $f_0 = 1$. □

Contact neighbourhood theorems

As in the symplectic world, the neighbourhood of a submanifold that is suitably placed with respect to the contact structure is determined by normal bundle data along the submanifold. The following exercise gives some explicit examples.

Exercise 3.5.22 (i) Let L be a Legendrian submanifold of a transversally oriented contact manifold (M, ξ). Show that it has a neighbourhood contactomorphic to a neighbourhood of the zero section in the 1-jet bundle $J^1L = T^*L \times \mathbb{R}$.

(ii) Let (M, ξ) be a transversely oriented contact 3-manifold and let $C \subset M$ be a transverse embedded circle, i.e. one that is never tangent to ξ. Show that C is the closed orbit of some Reeb flow on M. Show further that a neighbourhood of C is contactomorphic to a neighbourhood of the Legendrian circle $S^1 \times \{0\}$ in $(S^1 \times \mathbb{R}^2, \alpha_\beta)$. Here we identify S^1 with $\mathbb{R}/2\pi\mathbb{Z}$ (with the coordinate θ) and the contact form α_β is given by (3.5.6). □

Existence and classification of contact structures

As already indicated, in contact geometry there is a fundamental distinction between tight and overtwisted structures. Overtwisted structures are classified up to diffeomorphism by underlying topological data, while tight structures have a much finer classification that is still not well understood. For example, Eliashberg [177] proved the following theorem.

The isotopy classification of overtwisted contact structures on closed 3-manifolds coincides with their homotopy classification as tangent plane fields.

A remarkable paper by Borman–Eliashberg–Murphy [75] extends this result to all dimensions, in the process defining a new concept of overtwisted contact structures for higher-dimensional contact manifolds. Moreover, this paper solves the existence problem, showing that every almost contact manifold M^{2n+1} has a corresponding contact structure. Here an almost contact structure is a decomposition $TM = \mathbb{R} \oplus V$ for some complex bundle $V \to M$. Since the contact structures constructed in this way are overtwisted, there remains the very interesting question of the existence and classification of tight structures, where 'tight' is defined to mean 'not overtwisted' as in Definition 3.5.4.

An important simplifying feature of dimension three is that codimension one submanifolds are surfaces, and it turns out that contact structures can be understood by looking at the structures they induce on these surfaces. For example, Eliashberg proved in [182] that there is a unique tight contact structure on S^3 by thinking of S^3 as the union of a family S_r of disjoint 2-spheres, and analyzing the singular foliations $\xi \cap TS_r$ induced on these surfaces. For another exposition of Eliashberg's theorem, see Geiges [261]. There is much other work devoted to understanding tight contact 3-manifolds, that builds on the foundational work by Giroux [270] and Honda [329, 330].

Also in dimension three there is an operation called the **Lutz twist** that cuts out a small neighbourhood $\mathcal{N}_0$ of a closed transverse curve C, and glues back in a neighbourhood $\mathcal{N}_1$ that contains an overtwisted disc. Thus the function β_0 that determines the structure on $\mathcal{N}_0$ as in Exercise 3.5.22 (ii) has no zeros for $s = (x^2 + y^2)/2 > 0$ in $\mathcal{N}_0$, while the new function β_1 has one such zero. After such a twist the contact structure becomes overtwisted and so, if it was tight before, it is changed by this operation.

In higher dimensions there are many other interesting surgery operations in contact topology. For example, using surgery Geiges [258] constructed a contact structure on every simply connected 5-dimensional closed manifold M whose tangent bundle has the appropriate structure. Here TM must split as the sum of a real line bundle together with a rank two complex bundle, which happens exactly when its second Steifel–Whitney class w_2 has an integral lift. (Equivalently, the third integral Stiefel–Whitney class w_3 must vanish; cf. [261].) Contact surgery techniques also play a central role in the construction by Cieliebak–Eliashberg [111] of complex structures on Weinstein manifolds (Definition 7.4.5).

Relations with symplectic geometry

As we saw in Example 3.5.11 above, the circle bundle associated to an integral symplectic manifold carries a natural contact structure, which provides one way to go from the symplectic to the contact world. The resulting contact manifolds are rather special. In contact geometry itself there are other more important connections between the two worlds, one via **symplectizations** and another via **symplectic fillings**. We now discuss these notions, both of which have been used to great effect to throw light on the structure of the original contact manifold. We end with a brief discussion of open book decompositions.

Symplectization of contact manifolds

The symplectization of a transversally oriented contact manifold (M, ξ) is an exact symplectic manifold of one dimension higher, defined as follows.[17]

Proposition 3.5.23 (Symplectization) *Let (M, ξ) be a transversally oriented contact manifold.*

(i) *The set*

$$W := \left\{ (q, v^*) \;\middle|\; \begin{array}{l} q \in M,\ v^* \in T_q^*M,\ \ker v^* = \xi_q,\ \textit{and} \\ v \in T_qM,\ \alpha_q(v) > 0 \implies v^*(v) > 0 \end{array} \right\} \tag{3.5.13}$$

is a symplectic submanifold of $(T^*M, \omega_{\mathrm{can}})$.

(ii) *Let $\alpha \in \Omega^1(M)$ be a contact form and define W_α and $\omega_\alpha \in \Omega^2(W_\alpha)$ by*

$$W_\alpha := \mathbb{R} \times M, \qquad \omega_\alpha := -d(e^s\alpha) = -e^s(ds \wedge \alpha + d\alpha). \tag{3.5.14}$$

Here s denotes the coordinate on $\mathbb{R}$.[18] Then the map

$$\iota_\alpha : \mathbb{R} \times M \to W, \qquad \iota_\alpha(s, q) := e^s\alpha_q \tag{3.5.15}$$

is a symplectomorphism from $(W_\alpha, \omega_\alpha)$ to $(W, \omega_{\mathrm{can}}|_W)$.

Proof: It follows directly from the definitions that W is a submanifold of T^*M and that the map $\iota_\alpha : \mathbb{R} \times M \to W$ is a diffeomorphism. It follows also from the definitions, with $\dim M = 2n + 1$, that

$$\begin{aligned} \omega_\alpha^{n+1} &= e^{(n+1)s}(-ds \wedge \alpha - d\alpha)^{n+1} \\ &= (-1)^{n+1}(n+1)e^{(n+1)s} ds \wedge \alpha \wedge (d\alpha)^n. \end{aligned}$$

Thus

$$\frac{1}{(n+1)!}\,\omega_\alpha^{n+1} = \frac{(-1)^{n+1}}{n!}\; e^{(n+1)s} ds \wedge \alpha \wedge (d\alpha)^n. \tag{3.5.16}$$

So ω_α is nondegenerate and exact and hence is a symplectic form on $\mathbb{R} \times M$. Moreover, it follows from the definition of the canonical 1-form $\lambda_{\mathrm{can}} \in \Omega^1(T^*M)$ in (3.1.12) that $\iota_\alpha^*\lambda_{\mathrm{can}} = e^s\alpha$. (See Proposition 3.1.18.) Hence

$$\iota_\alpha^*\omega_{\mathrm{can}} = -\iota_\alpha^* d\lambda_{\mathrm{can}} = -d(e^s\alpha) = \omega_\alpha.$$

This proves Proposition 3.5.23. □

[17] The choice of signs below is consistent with our previous choices, but as we explain in more detail in Remark 3.5.35 below does not agree with choices common in the literature.

[18] We abuse notation and identify $\alpha \in \Omega^1(M)$ with its pullback to a 1-form on $\mathbb{R} \times M$ and identify the letter s with the function $\mathbb{R} \times M \to \mathbb{R} : (s, q) \mapsto s$.

Definition 3.5.24 *The symplectic manifold* (W, ω), *defined by* (3.5.13) *and*

$$\omega := \omega_{\mathrm{can}}|_W \in \Omega^2(W),$$

is called the **(intrinsic) symplectization of** (M, ξ). *Thus* $W \subset T^*M$ *is the real line bundle of contact elements on* (M, ξ) *whose sections are the contact forms. For every contact form* α *the symplectic manifold* $(W_\alpha, \omega_\alpha) = (\mathbb{R} \times M, -d(e^s\alpha))$ *is called the* **(extrinsic) symplectization** *of* (M, α).

By Proposition 3.5.23 the intrinsic and extrinsic symplectizations are symplectomorphic.

Example 3.5.25 **(i)** The symplectization of the unit sphere $M = S^{2n-1}$ with the standard contact form $\alpha = \frac{1}{2}\sum_{j=1}^n (y_j dx_j - x_j dy_j)$ (see Example 3.5.9) is symplectomorphic to $\mathbb{C}^n \setminus \{0\}$ with the standard symplectic form ω_0. An explicit symplectomorphism is given by

$$\mathbb{R} \times S^{2n-1} \to \mathbb{C}^n \setminus \{0\} : (s, x, y) \mapsto e^{s/2}(x + iy).$$

(ii) The symplectization of the unit cotangent bundle of the torus $M = \mathbb{T}^n \times S^{n-1}$ with the standard contact form $\alpha = \sum_{j=1}^n y_j dx_j$ (see Example 3.5.8) is symplectomorphic to the complement $\mathbb{T}^n \times (\mathbb{R}^n \setminus \{0\})$ of the zero section in the cotangent bundle with the standard symplectic form ω_{can}. An explicit symplectomorphism is given by

$$\mathbb{R} \times \mathbb{T}^n \times S^{n-1} \to \mathbb{T}^n \times (\mathbb{R}^n \setminus \{0\}) : (s, x, y) \mapsto (x, e^s y).$$

This extends to the unit cotangent bundle of any Riemannian manifold (see Example 3.5.7).

(iii) The symplectization of the Euclidean space $M = \mathbb{R}^{2n+1}$ with the standard contact form $\alpha = \sum_{j=1}^n y_j dx_j - dz$ (see equation (3.5.3)) is symplectomorphic to $(\mathbb{R}^{2n+2}, \omega_0)$. An explicit symplectomorphism is given by

$$\mathbb{R} \times \mathbb{R}^{2n+1} \to \mathbb{R}^{2n+2} : (s, x, y, z) \mapsto (s, e^s z, x_1, e^s y_1, \dots, x_n, e^s y_n).$$

This carries over to general 1-jet bundles (see Example 3.5.6).

(iv) Let $(V, \omega = -d\lambda)$ be an exact symplectic manifold and consider the circle bundle $M := V \times S^1$ with the contact form $\alpha := \lambda - dt$ (Example 3.5.13). Denote by $z = x + iy$ the coordinate on $\mathbb{C}^* := \mathbb{C} \setminus \{0\}$. Then the symplectization of $V \times S^1$ is symplectomorphic to $V \times \mathbb{C}^*$ with the symplectic form

$$\omega := \tfrac{1}{\pi} dx \wedge dy - d\left((x^2 + y^2)\lambda\right).$$

The map $\mathbb{R} \times V \times S^1 \to V \times \mathbb{C}^* : (s, q, e^{i\theta}) \mapsto (q, e^{s/2+i\theta})$ is an explicit symplectomorphism.

(v) The construction in part (iv) carries over to any circle bundle P over an integral symplectic manifold (V, ω) with Euler class $-[\omega]$ (see Example 3.5.11).

In that case the symplectization of (P, α) is symplectomorphic to the complement of the zero section in the associated complex line bundle

$$L := P \times_{S^1} \mathbb{C},$$

where the equivalence relation is given by $[p, z] \equiv [pe^{-i\theta}, e^{i\theta}z]$. The 1-form

$$\widetilde{\lambda} := (x^2 + y^2)\alpha + \tfrac{1}{2\pi}(ydx - xdy)$$

on $P \times \mathbb{C}$ descends to a 1-form on L, still denoted by $\widetilde{\lambda}$, and minus its differential

$$\widetilde{\omega} := -d\widetilde{\lambda} = \tfrac{1}{\pi}dx \wedge dy - d\left((x^2 + y^2)\alpha\right).$$

descends to a symplectic form on L, still denoted by $\widetilde{\omega}$ (see Theorem 6.3.3 below). The embedding $\iota : \mathbb{R} \times P \to L$ is given by

$$\iota(s, p) := [p, e^{s/2}].$$

It satisfies $\iota^*\widetilde{\lambda} = e^s\alpha$ and hence $\iota^*\widetilde{\omega} = -d(e^s\alpha)$. □

The fact that every contact manifold has a symplectization allows one to use the methods of symplectic geometry in order to investigate contact structures. In particular, many contact invariants such as the Hutchings–Taubes [340] embedded contact homology (ECH) for 3-dimensional contact manifolds have been defined this way, although this particular invariant now has an intrinsic description via Seiberg–Witten Floer homology by Taubes [613, 614, 615, 616, 617].

Exercise 3.5.26 Let (M, ξ) be a contact manifold. Choose a contact form α with Reeb field Y, and let $(W_\alpha, \omega_\alpha) := (\mathbb{R} \times M, -d(e^s\alpha))$ be the extrinsic symplectization.

(i) Let $f : M \to \mathbb{R}$ be a positive function so that $f\alpha$ is another contact form. Prove that $(W_\alpha, \omega_\alpha)$ and $(W_{f\alpha}, \omega_{f\alpha})$ are symplectomorphic via $(s, q) \mapsto (s - \log(f(q)), q)$.

(ii) Prove that $L \subset M$ is a Legendrian submanifold of (M, ξ) if and only if $\mathbb{R} \times L$ is a Lagrangian submanifold of the symplectization $(W_\alpha, \omega_\alpha)$.

(iii) Prove that $\psi : M \to M$ is a contactomorphism of (M, ξ) with $\psi^*\alpha = e^h\alpha$ if and only if the map $\widetilde{\psi}(s, q) := (s - h(q), \psi(q))$ is a symplectomorphism of $(W_\alpha, \omega_\alpha)$.

(iv) Let $X = X_H \in \mathcal{X}(M)$ be the contact vector field generated by $H := \iota(X)\alpha$ as in Lemma 3.5.14. Prove that the Hamiltonian function $\widetilde{H}(s, q) := e^s H(q)$ on W_α generates the Hamiltonian vector field $\widetilde{X}(s, q) = (-dH(q)(Y(q)), X(q))$.

(v) Prove that the Poisson bracket of $\widetilde{F} = e^s F$ and $\widetilde{G} = e^s G$ is $\{\widetilde{F}, \widetilde{G}\} = e^s\{F, G\}$. □

Liouville vector fields

Above we have shown how to build a symplectic manifold from a contact manifold. In the reverse process, that goes from the symplectic to the contact world, the key element is a Liouville vector field.

Definition 3.5.27 *A vector field X on a symplectic manifold (W, ω) is called a* **Liouville vector field** *if*

$$\mathcal{L}_X \omega = d\iota(X)\omega = \omega.$$

The flow ψ_t of a Liouville vector field satisfies $\psi_t^* \omega = e^t \omega$ wherever it is defined.

Since the flow of a Liouville vector field increases the volume, a closed symplectic manifold does not admit a global Liouville vector field. However, such vector fields do exist locally and they exist globally on certain noncompact manifolds. Here are three examples of global complete Liouville vector fields.

Example 3.5.28 **(i)** The radial vector field

$$X_0 := \tfrac{1}{2} \sum_{j=1}^{n} \left(x_i \frac{\partial}{\partial x_i} + y_i \frac{\partial}{\partial y_i} \right)$$

on $\mathbb{R}^{2n}$ is a Liouville vector field for $\omega_0 = \sum_j dx_j \wedge dy_j$. It is transverse to S^{2n-1} and the 1-form $\lambda_0 := -\iota(X_0)\omega_0 = \frac{1}{2}\sum_j (y_j dx_j - x_j dy_j)$ agrees with the contact form of Example 3.5.9 on S^{2n-1}.

(ii) The vector field

$$X_{\mathrm{can}} := \sum_{j=1}^{n} y_j \frac{\partial}{\partial y_i}$$

on $\mathbb{R}^n \times \mathbb{R}^n = T^*\mathbb{R}^n$ satisfies $\iota(X_{\mathrm{can}})\omega_{\mathrm{can}} = -\sum_{j=1}^n y_j dx_j = -\lambda_{\mathrm{can}}$ and hence is a Liouville vector field for $\omega_{\mathrm{can}} = \sum_j dx_j \wedge dy_j$. This example extends to any cotangent bundle T^*M and the flow of X_{can} is then given by $\psi_t(q, v^*) = (q, e^t v^*)$.

(iii) Let (M, ξ) be a contact manifold with contact form α and consider the extrinsic symplectization $(W_\alpha, \omega_\alpha) := (\mathbb{R} \times M, -d(e^s \alpha))$. The vector field

$$X_\alpha := \frac{\partial}{\partial s}$$

on W_α satisfies $\iota(X_\alpha)\omega_\alpha = -e^s \alpha$ and hence is a Liouville vector field. Its flow is given by $\psi_t(s, q) = (s + t, q)$ and its pushforward under the symplectomorphism $W_\alpha \to W : (s, q) \mapsto e^s \alpha_q$ is the restriction of the vector field $X_{\mathrm{can}} \in \mathcal{X}(T^*M)$ in part (ii) above to the intrinsic symplectization $W \subset T^*M$. □

We now explain in a series of exercises how to construct a Liouville vector field on any compact connected symplectic 2-manifold with nonempty boundary. As a first step, consider the following example.

Example 3.5.29 Consider the annulus $W := \mathbb{R}/\mathbb{Z} \times [-1,1]$ with coordinates (x,y) and symplectic form $\omega = dx \wedge dy$. Fix a real number r. The vector field

$$X_r := (r+y)\frac{\partial}{\partial y}$$

satisfies $\iota(X_r)\omega = -(y+r)dx$ and so is a Liouville vector field for all $r \in \mathbb{R}$. For $|r| < 1$ it points out on both boundary components, while for $|r| > 1$ it points in on one and points out on the other. □

Exercise 3.5.30 (i) A Liouville vector field on $(\mathbb{R}^2, \omega_0)$ with a hyperbolic critical point at the origin is given by

$$X := 2y\frac{\partial}{\partial y} - x\frac{\partial}{\partial x}.$$

Let $N \subset \mathbb{R}^2$ be a neighbourhood of the origin, diffeomorphic to an octagon, such that the four diagonal boundary arcs are tangent to the flow and separated by two incoming arcs on the left and right and two outgoing arcs at the top and bottom. (This is a manifold with corners; see Appendix A.1, and for an illustration see Figure 11.1.) Join the tangential arcs pairwise by rectangles with appropriate symplectic forms and Liouville vector fields, to obtain a Liouville vector field on a symplectic pair-of-pants that points out on two boundary components and points in on one, or vice versa.

(ii) Construct a Liouville vector field on a symplectic pair-of-pants that points out on the boundary. **Hint:** Glue a pair-of-pants with one incoming end as in part (i) to the annulus of Example 3.5.29 with two outgoing ends.

(iii) Use Moser isotopy to show that any two symplectic pairs-of-pants with the same volume are symplectomorphic. (See Exercise 3.2.6.)

(iv) Prove that every compact connected symplectic 2-manifold (Σ, ω) with nonempty boundary admits a global Liouville vector field that points out on the boundary. **Hint 1:** Choose a pair-of-pants decomposition. Then use parts (i), (ii), and (iii). **Hint 2:** Here is another argument explained to us by Janko Latschev. Choose a 1-form $\alpha \in \Omega^1(M)$ such that $d\alpha = \omega$ near $\partial\Sigma$ and $\int_{\partial\Sigma}\alpha = \int_\Sigma \omega$. Find a 1-form $\beta \in \Omega^1(\Sigma)$ with support in the interior of Σ such that $d\beta = \omega - d\alpha$. Define the vector field X by $\iota(X)\omega = \alpha + \beta$. With the appropriate choice of the 1-form α this vector field points out on the boundary. □

Contact type hypersurfaces

The following result explains when a hypersurface in a symplectic manifold has an induced contact structure. If it does, the induced structure is well-defined by Exercise 3.5.36 (i).

Proposition 3.5.31 *Let (W, ω) be a symplectic manifold and $M \subset W$ be a compact hypersurface. Then the following are equivalent.*

(i) *There exists a contact form α on M such that $-d\alpha = \omega|_M$.*

(ii) *There exists a Liouville vector field $X : U \to TW$, defined in a neighbourhood $U \subset W$ of M, which is transverse to M.*

Proof: First assume (ii) and define $\alpha := -\iota(X)\omega$. Then

$$-d\alpha = d\iota(X)\omega + \iota(X)d\omega = \mathcal{L}_X\omega = \omega.$$

Since T_qM is odd-dimensional there exists a nonzero vector $Y \in T_qM$ such that $\omega_q(Y, v) = 0$ for every $v \in T_qM$. Since ω is nondegenerate we must therefore have $\alpha_q(Y) = -\omega_q(X(q), Y) \neq 0$. Hence the subspaces

$$\xi_q := \{v \in T_qM \,|\, \omega_q(X(q), v) = 0\}, \qquad q \in M,$$

form a hyperplane field on M (the kernel of $\alpha|_M$) and Y is transversal to ξ. In fact ξ_q is the symplectic complement of $\mathrm{span}\{X(q), Y(q)\}$ for each $q \in M$. This implies that $\omega = -d\alpha$ is nondegenerate on ξ and hence α restricts to a contact form on M.

Conversely, suppose that $\alpha \in \Omega^1(M)$ is a contact form such that $-d\alpha = \omega|_M$. Let $Y \in \mathcal{X}(M)$ be the Reeb field of α, i.e. $\iota(Y)d\alpha = 0$ and $\alpha(Y) = 1$. Choose a section X of the vector bundle $T_MW \to M$ such that

$$\omega(X, Y) = -1, \qquad \omega(X, \xi) = 0 \tag{3.5.17}$$

on M. (First choose any section X_0 of T_MW such that $\omega(X_0, Y) = -1$. This section is necessarily transverse to M. Then, for every $q \in M$, there exists a unique vector $X_1(q) \in \xi_q$ such that $\omega(X_0(q) + X_1(q), v) = 0$ for every $v \in \xi_q$. Define $X := X_0 + X_1$.) It follows from (3.5.17) that

$$\omega(X(q), v) = -\alpha(v) \qquad \text{for } q \in M \text{ and } v \in T_qM. \tag{3.5.18}$$

Define $\phi : \mathbb{R} \times M \to W$ by $\phi(s, q) := \exp_q(sX(q))$. Then

$$d\phi(0, q)(\widehat{s}, \widehat{q}) = \widehat{s}X(q) + \widehat{q}, \qquad \text{for } \widehat{s} \in \mathbb{R} \text{ and } \widehat{q} \in T_qM.$$

Hence it follows from (3.5.18) that $\phi^*\omega|_{\{0\}\times M} = -ds \wedge \alpha - d\alpha$. By Moser isotopy (Lemma 3.2.1) there exists an embedding $\psi : (-\varepsilon, \varepsilon) \times M \to W$ such that

$$\psi(q, 0) = q, \qquad \psi^*\omega = e^s(-ds \wedge \alpha - d\alpha).$$

The required Liouville vector field is $\psi_*\partial/\partial s$. This proves Proposition 3.5.31. □

Definition 3.5.32 *A hypersurface M in a symplectic manifold (W, ω) is said to be of* **contact type** *if it satisfies the equivalent conditions in Proposition 3.5.31. It is said to be of* **restricted contact type** *if the Liouville vector field in (ii) can be defined on all of W. If (W, ω) is a compact symplectic manifold, its boundary $M := \partial W$ has contact type, and the Liouville vector field X in (ii) can be chosen to point out on the boundary, then (W, ω) is called a* **symplectic manifold with convex boundary**. *If, in addition, the Liouville vector field X is defined on all of W then the triple (W, ω, X) is called a* **Liouville domain**.

Proposition 3.5.33 *Let (W, ω) be a symplectic manifold of dimension $2n \geq 4$ and let $M \subset W$ be a compact connected hypersurface of contact type. Then M has a preferred positive side into which any transverse Liouville vector field points. In particular, there is no orientation-reversing diffeomorphism $\phi : M \to M$ that preserves the restriction $\omega|_M$.*

Proof: The proof was explained to us by Janko Latschev. Let X and X' be Liouville vector fields near M that are transverse to M. Consider the contact forms

$$\alpha := -\iota(X)\omega|_M, \qquad \alpha' := -\iota(X')\omega|_M,$$

and denote their Reeb vector fields by $Y, Y' \in \mathcal{X}(M)$. They both span the kernel of the 2-form $d\alpha = d\alpha' = \omega|_M$. Hence there is a nonvanishing function $\lambda : M \to \mathbb{R}$ such that $Y' = \lambda Y$. Moreover, for all $p \in M$ and $v \in T_pW$, we have $v \in T_pM$ if and only if $\omega(Y(p), v) = 0$. Since $\omega(Y, X) = \alpha(Y) = 1$ and $\omega(Y, \lambda X') = \alpha'(Y') = 1$, it follows that $X(p) - \lambda(p)X'(p) \in T_pM$ for all $p \in M$. It remains to prove that $\lambda > 0$. To see this, choose an orientation of M, observe that $\alpha - \alpha'$ is closed, and use Stokes' theorem and the fact that $n \geq 2$ to obtain

$$\int_M (\alpha - \alpha') \wedge (d\alpha)^{n-1} = \int_M (d\alpha - d\alpha') \wedge \alpha \wedge (d\alpha)^{n-2} = 0. \qquad (3.5.19)$$

If $\lambda < 0$ then $\iota(Y)(\alpha - \alpha') = 1 - \lambda^{-1} > 0$ and so $(\alpha - \alpha') \wedge (d\alpha)^{n-1}$ is a volume form, contradicting the fact that its integral is zero. Hence $\lambda > 0$ as claimed. □

Example 3.5.29 shows that the hypothesis dim $W \geq 4$ cannot be removed in Proposition 3.5.33. Another consequence is that no local symplectomorphism of W fixes M while interchanging its two sides, which means that in dimensions four and above the symplectic category has no connected sum operation. However there is such an operation in dimension two, that one can use in parametrized form to define the fibre connected sum discussed in Section 7.2.

Remark 3.5.34 (Orientations) We have thus far avoided discussing the orientations of contact manifolds. In particular, the integral in equation (3.5.19) vanishes and so the choice of orientation is immaterial. When (M, ξ) is a cooriented contact manifold of dimension $2n - 1$, a natural intrinsic choice of the orientation of M within our system of sign conventions is as follows. Choose a contact form $\alpha \in \Omega^1(M)$ with $\xi = \ker \alpha$ that is compatible with the coorientation of ξ and declare the $(2n-1)$-form

$$\mathrm{dvol}_\alpha := (-\alpha) \wedge \frac{(-d\alpha)^{n-1}}{(n-1)!} \in \Omega^{2n-1}(M) \qquad (3.5.20)$$

to be a positive volume form on M. When $M \subset W$ is a contact hypersurface of a symplectic manifold, X is a local Liouville vector field near M that is transverse to M, $\alpha := -\iota(X)\omega|_M$ is the associated contact form, and $\xi := \ker\alpha$ is the associated contact structure, then

$$\mathrm{dvol}_\alpha = -\alpha \wedge \left.\frac{\omega^{n-1}}{(n-1)!}\right|_M = \left.\iota(X)\frac{\omega^n}{n!}\right|_M.$$

Thus our intrinsic orientation convention asserts that a basis $e_2, \dots, e_{2n}$ of T_pM is positive if and only if $X(p), e_2, \dots, e_{2n}$ is a positive basis of T_pW. In particular, this is the boundary orientation when (W, ω) is a symplectic manifold with convex boundary $M = \partial W$ (see Definition 3.5.32). In other words, the intrinsic and extrinsic orientations of M agree. Proposition 3.5.33 asserts that when $n \geq 2$, the extrinsic orientation of M is independent of the choice of the Liouville vector field X used to define it. Moreover, equation (3.5.19) in the proof of Proposition 3.5.33 asserts that the total volume

$$\mathrm{Vol}(M) := \int_M \mathrm{dvol}_\alpha = \int_M (\iota(X)\omega) \wedge \frac{\omega^{n-1}}{(n-1)!} \tag{3.5.21}$$

is also independent of the choice of X. More precisely, equation (3.5.19) shows that when M is oriented with respect to α, and α' is the contact form associated to another Liouville vector field X', then $\int_M \mathrm{dvol}_{\alpha'} = \int_M \mathrm{dvol}_\alpha > 0$ so that X and X' induce the same orientation of M. □

Remark 3.5.35 (Signs) In this section our choice of signs has been in large part determined by our choice of the symplectic form $\omega_{\mathrm{can}} := -d\lambda_{\mathrm{can}}$ on cotangent bundles: see equation (3.1.11). Correspondingly, we have defined the standard contact form on $\mathbb{R}^{2n+1}$ by $\sum_j y_j dx_j - dz$ in (3.5.3), on the 1-jet space J^1L by $\lambda_{\mathrm{can}} - dz$ in Example 3.5.6, and on the symplectization $\mathbb{R} \times M$ of (M, ξ, α) by $-d(e^s\alpha)$ in (3.5.14). For consistency, we must then define the symplectic form near a contact hypersurface by $-d\alpha$ (in Proposition 3.5.31) and the orientation on (M, ξ, α) via the form $-\alpha$ as in (3.5.20). This minus sign may seem awkward, however, it can only be eliminated at the expense of new minus signs elsewhere.

The one universal sign convention in symplectic topology is that the Hamiltonian differential equations have the form $\dot{q} = \partial H/\partial p$, $\dot{p} = -\partial H/\partial q$. If the symplectic form on a cotangent bundle is chosen to be $\omega = d\lambda_{\mathrm{can}} = dp \wedge dq$, this implies the formula $\iota(X_H)\omega = -dH$ for Hamiltonian vector fields, which is used by many authors. With this convention the symplectic form on the intrinsic symplectization $W \subset T^*M$ is $d\lambda_{\mathrm{can}}|_W$ and the resulting symplectic form on the extrinsic symplectization $\mathbb{R} \times M$ is $d\iota_\alpha^*\lambda_{\mathrm{can}} = d(e^s\alpha)$ (see Proposition 3.5.23).

The rule of thumb for translating between these sign conventions is that the symplectic form changes sign while the Hamiltonian vector field X_H associated to H and the contact 1-forms remain unchanged. Here it is important to observe that the symplectic forms ω and $-\omega$ have the same Liouville vector fields so that the notion of convexity in Definition 3.5.32 remains unchanged as well. This change of sign conventions makes no essential difference to the basic geometric relations between symplectic and contact structures as given in Exercise 3.5.26, though some signs in these formulas then need to be changed.

Exercise 3.5.36 **(i)** Show that if a compact hypersurface $M \subset W$ has contact type, different choices of contact forms α such that $-d\alpha = \omega|_M$ give rise to isotopic contact structures on M.
(ii) Show that every simply connected hypersurface of contact type in fact has restricted contact type (see Definition 3.5.32) whenever ω is exact.
(iii) Consider a compact Lagrangian submanifold L of Euclidean space $(\mathbb{R}^{2n}, -d\lambda_0)$. By Theorem 3.4.13, a neighbourhood $\mathcal{N}$ of the zero section in T^*L embeds symplectically into $\mathbb{R}^{2n}$. For small r, the sphere bundle $S_r(T^*L)$ of radius r is contained in $\mathcal{N}$ and so also embeds into $\mathbb{R}^{2n}$. Show that these hypersurfaces have contact type. Using Gromov's results on the nonexactness of Lagrangian submanifolds of $\mathbb{R}^{2n}$ (see Theorem 13.2.3 below), Sean Bates [57] showed that they never have restricted contact type.
(iv) Show that if M is the boundary of the Liouville domain (W, ω, X) then its volume as defined in (3.5.21) equals the volume of (W, ω). □

The following exercise describes a useful generalization of the notion of a hypersurface of contact type that first appeared in [325] and subsequently in papers such as [186] on symplectic field theory.

Exercise 3.5.37 A **stable Hamiltonian structure** on a closed manifold M^{2n-1} is a pair (Ω, λ), consisting of a closed 2-form Ω and a 1-form λ, such that

$$\lambda \wedge \Omega^{n-1} \neq 0, \qquad \ker \Omega \subset \ker d\lambda, \tag{3.5.22}$$

where $\ker \Omega := \{v \in TM \mid \iota(v)\Omega = 0\}$. The corresponding **Reeb vector field** is the unique vector field Y on M such that

$$\iota(Y)\Omega = 0, \qquad \lambda(Y) = 1. \tag{3.5.23}$$

The first condition (3.5.22) asserts that such a vector field exists and is unique, and the second condition in (3.5.22) asserts that $\iota(Y)d\lambda = 0$. The existence of a stable Hamiltonian structure implies that M is orientable. Every contact form α on M determines a stable Hamiltonian structure (Ω, λ) with $\lambda := \alpha$ and $\Omega := -d\alpha$ and the orientation in Remark 3.5.34 is then determined by the volume form $\mathrm{dvol} := -\alpha \wedge \Omega^{n-1}/(n-1)!$.

(i) Given a stable Hamiltonian structure (Ω, λ) on M, show that the Reeb flow (i.e. the flow of Y) preserves the forms Ω, λ, and that the 2-form $\omega := \Omega - d(r\lambda)$ on $(-\varepsilon, \varepsilon) \times M$ is symplectic for sufficiently small $\varepsilon > 0$. Here r denotes the coordinate on $(-\varepsilon, \varepsilon)$. Show that in the contact case this corresponds to the symplectization via $s := \log(1 + r)$. Show that the characteristic line field $\ker(\omega|_{\{r\}\times M})$ is independent of r.

(ii) Conversely, let ω be a symplectic form on $(-\varepsilon, \varepsilon) \times M$ such that characteristic foliation on $\{r\} \times M$ is stable (i.e. independent of r). Show that M has a stable Hamiltonian structure (Ω_0, λ_0) such that Ω_0 is the pullback of ω under the inclusion $x \mapsto (0, x)$. Show that there is a $\delta > 0$ and an embedding $\phi : (-\delta, \delta) \times M \to (-\varepsilon, \varepsilon) \times M$ such that $\phi|_{\{0\}\times M} = \mathrm{id}$ and $\phi^*\omega = \Omega_0 - d(r\lambda_0)$. **Hint:** Write $\omega = \Omega_r - dr \wedge \lambda_r$ where $\Omega_r \in \Omega^2(M)$ and $\lambda_r \in \Omega^1(M)$. Show that Ω_r is closed and $\partial_r \Omega_r + d\lambda_r = 0$. Deduce that (Ω_r, λ_r) is a stable Hamiltonian structure for each r. Use Exercise 3.4.17 to find ϕ.

(iii) Every hypersurface $M \subset W$ of contact type in a symplectic manifold (W, ω) has a stable Hamiltonian structure with $\Omega = -d\lambda$ (see Definition 3.5.32). If (V, ω) is a closed symplectic manifold, then $M := \mathbb{R}/\mathbb{Z} \times V$ has a stable Hamiltonian structure (Ω, λ)

with $\lambda = dt$ and $\Omega = \omega - dt \wedge dH$, where $H : \mathbb{R}/\mathbb{Z} \times M$ is any smooth function. In each case, find the Reeb vector field Y. □

Symplectic filling

The following notions have proved very helpful in understanding contact structures.

Definition 3.5.38 *A closed contact manifold* (M, ξ) *is called* **strongly fillable** *if it is the boundary of a compact symplectic manifold* (W, ω) *that admits a Liouville vector field* X *near* $M = \partial W$ *which points out on the boundary such that* $\alpha := -\iota(X)\omega|_M$ *is a contact form for* ξ. *It is called* **weakly fillable** *if it is the boundary of a compact symplectic manifold* (W, ω) *that admits an almost complex structure* J *satisfying the following three conditions.*
(a) J *is* ω*-tame, i.e.* $\omega(v, Jv) > 0$ *for every nonzero tangent vector* $v \in TW$.
(b) $\xi = TM \cap JTM$.
(c) $M = \partial W$ *is* J**-convex**, *i.e. there is a neighbourhood* U *of* ∂W *and a smooth function* $f : U \to (-1, 0]$ *such that* 0 *is a regular value of* f, $M = f^{-1}(0)$, *and* $\omega_f := -d(df \circ J)$ *satisfies* $\omega_f(v, Jv) > 0$ *for all* $q \in U$ *and all* $v \in T_qW \setminus \{0\}$.

This definition of *weak fillability* is due to Massot–Niederkrüger–Wendl [440] and every strongly fillable contact manifold is weakly fillable. There is a stronger notion of *weak fillability* which asserts that ω **dominates** ξ in the sense that there exists a contact form α on (M, ξ) such that $\omega|_\xi = -d\alpha|_\xi$ and the boundary orientation of $M = \partial W$ agrees with the orientation induced by α as in Remark 3.5.34. This is equivalent to the above notion in dimension 3 and, by a theorem of McDuff [451], is equivalent to strong fillability in dimensions greater than or equal to 5. Examples by Eliashberg [184] show that weak and strong fillability differ in dimension 3 and Massot–Niederkrüger–Wendl [440] proved that they differ in all higher dimensions. By a result of Eliashberg [178] and Gromov [287] overtwisted contact 3-manifolds are never weakly fillable, and a similar result holds in higher dimensions by Borman–Eliashberg–Murphy [75]. In contrast, Etnyre–Honda [209] and Eliashberg [185] showed that every contact 3-manifold is the **concave** boundary of some compact symplectic 4-manifold (i.e. the Liouville vector field points inwards along the boundary). See Eliashberg–Murphy [192] for extensions of this result to higher dimensions.

An elementary 2-dimensional argument shows that a Liouville domain can have a disconnected (convex) boundary (Example 3.5.29 and Definition 3.5.32). That this phenomenon also occurs in higher dimensions was shown by McDuff [451] and Eliashberg–Thurston [197]. Many more examples are contained in the book of Geiges [261]. Thus fillable contact manifolds do not have to be connected. Using these results, Etnyre [208] constructed examples of nonseparating contact hypersurfaces in certain closed 4-dimensional symplectic manifolds. However not all contact manifolds support such an embedding, and also there are restrictions on the ambient symplectic manifold (see Albers–Bramham–Wendl [19] and Wendl [687]).

Now assume that (M, ξ) is a strongly fillable closed contact manifold and let (W, ω) and X be as in Definition 3.5.38, so that (W, ω) is a compact symplectic manifold, X is a Liouville vector field near $M = \partial W$ that points out on the boundary, and ξ is the kernel of the contact form

$$\alpha := -\iota(X)\omega|_M.$$

Then W can be extended (or *completed*) to a manifold

$$\widehat{W} := W \cup_M ([0, \infty) \times M).$$

The completed manifold $\widehat{W}$ carries a symplectic form $\widehat{\omega}$ that agrees with ω on W and with $-d(e^s\alpha)$ on the cylindrical end $[0, \infty) \times M$. In other words, the manifold $\widehat{W}$ has a cylindrical end on which the symplectic form expands, whereas the negative half of $\mathbb{R} \times M$ where the symplectic form contracts has been replaced by the compact manifold W. The symplectic manifold $(\widehat{W}, \widehat{\omega})$ is convex at infinity as defined by Eliashberg–Gromov [187].

Definition 3.5.39 (Convex at infinity) *A symplectic manifold* (W, ω) *without boundary is called* **convex at infinity** *if there exists an exhausting sequence of compact subsets* $W_i \subset W$ *with smooth convex boundary such that*

$$W_i \subset \mathrm{int}(W_{i+1}) \quad \forall\, i \in \mathbb{N} \qquad \text{and} \qquad W = \bigcup_{i=1}^{\infty} W_i.$$

This definition can be strengthened by requiring, in addition, the existence of a Liouville vector field X on the complement of a compact set such that X points out of W_i on the boundary ∂W_i for each i. Examples of noncompact symplectic manifolds that are convex at infinity even in the stronger sense include $(\mathbb{R}^{2n}, \omega_0)$, cotangent bundles of closed manifolds, and (more generally) all completions of compact symplectic manifolds with convex boundaries. Noncompact symplectic manifolds that are convex at infinity play a central role in symplectic topology and their study has much in common with that of closed symplectic manifolds (see Eliashberg–Gromov [187]). There is a very important subclass called **Weinstein manifolds** that also carry an adapted generalized Morse function. For further information, see Definition 7.4.5 and the discussion that follows.

Studying symplectic fillings has proved to be a very fruitful way of understanding tight contact structures. For example, the standard contact structure on the unit sphere $S^{2n-1} \subset \mathbb{R}^{2n}$ is fillable. Moreover, by parts (viii) and (ix) of Remark 4.5.2 in Chapter 4, any filling of the standard contact sphere by a manifold with vanishing π_2 must be diffeomorphic to a ball. Using this, Eliashberg [181] constructed contact structures on spheres of dimension $4k+1$, which may be filled by manifolds other than the ball and so must be nonstandard. For further results and references on the symplectic filling problem see the book by Cieliebak–Eliashberg [111] on Weinstein manifolds and the papers by Albers–Bramham–Wendl [19], Latschev–Wendl [401], and Wendl [683,685,686,687].

Open book decompositions

Symplectic and contact geometry are also very closely connected via the notion of an **open book**, which is a way of constructing a closed contact manifold of dimension $2n+1$ from a compact exact symplectic $2n$-manifold (Σ, ω) with convex boundary of restricted contact type together with a symplectomorphism ϕ of (Σ, ω) with compact support. We only discuss the basic definition. For further results and references see Geiges [261, Chapter 7.3]. Throughout we denote by $\mathbb{D} \subset \mathbb{C}$ the closed unit disc with coordinates $x + iy$.

Let Σ be a compact oriented $2n$-manifold with boundary

$$C := \partial\Sigma$$

and let $\phi : \Sigma \to \Sigma$ be a diffeomorphism with compact support so that ϕ is equal to the identity near the boundary. The **mapping torus** of (Σ, ϕ) is the $(2n+1)$-dimensional quotient manifold

$$M_\phi := \frac{\Sigma \times \mathbb{R}}{(z, t+1) \sim (\phi(z), t)}.$$

The elements of M_ϕ will be denoted by $[z,t]_\phi$ for $z \in \Sigma$ and $t \in \mathbb{R}$. Since $\phi = \mathrm{id}$ near the boundary, the boundary of M_ϕ is given by

$$\partial M_\phi = C \times \mathbb{R}/\mathbb{Z} \cong C \times S^1.$$

Thus one can construct a closed oriented manifold by forming the disjoint union $M_\phi \sqcup (C \times \mathbb{D})$ and identifying the equivalence class $[z,t]_\phi = [z,t+1]_\phi \in \partial M_\phi$ with the pair $(z, e^{2\pi i t}) \in C \times \mathbb{D}$. Denote this manifold by

$$M := M(\Sigma, \phi) := \frac{M_\phi \sqcup (C \times \mathbb{D})}{[z,t]_\phi \cong (z, e^{2\pi i t}),\ z \in C,\ t \in \mathbb{R}}. \tag{3.5.24}$$

(Since the dimension of C is odd, the orientation of $C \times S^1$ as the boundary of $C \times \mathbb{D}$ is opposite to the product orientation, whereas its orientation as the boundary of $\Sigma \times S^1$ coincides with the product orientation.)

Definition 3.5.40 *The manifold M in* (3.5.24) *is called a* **smooth open book**, *the submanifolds*

$$\Sigma_t := (\Sigma \times \{t\}) \cup (C \times \{re^{2\pi i t} \,|\, 0 < r \leq 1\})$$

are called the **pages**, *and the submanifold*

$$B := C \times \{0\} \subset C \times \mathbb{D}$$

is called the **binding**. *The pages have a natural coorientation. An* **open book decomposition** *of a closed oriented odd-dimensional manifold is an orientation-preserving diffeomorphism to a manifold of the form* (3.5.24).

Now assume that (Σ, ω) is a compact symplectic manifold with boundary $C = \partial\Sigma$ and that $\phi : \Sigma \to \Sigma$ is a symplectomorphism, equal to the identity near the boundary. Assume also that there exists a global Liouville vector field $X \in \mathcal{X}(\Sigma)$ which points out on the boundary so that

$$\lambda := -\iota(X)\omega \in \Omega^1(\Sigma)$$

restricts to a contact form on C. (For surfaces see Exercise 3.5.30 (iv).) In this situation we wish to construct a contact form on the smooth open book M defined by (3.5.24).

Choose a smooth function $\beta : \mathbb{R} \to \mathbb{R}$ such that $\beta(t+1) = \beta(t) + 1$ for all t and β vanishes near the origin, so $\beta(t) = k$ near every integer k. Then there is a unique smooth one parameter family of 1-forms $\lambda_t \in \Omega^1(\Sigma)$ for $t \in \mathbb{R}$, such that

$$\lambda_t = (1 - \beta(t))\lambda + \beta(t)\phi^*\lambda \qquad \text{for } 0 \le t \le 1,$$

and $\lambda_{t+1} = \phi^*\lambda_t$ for all $t \in \mathbb{R}$. These 1-forms also satisfy

$$\omega = -d\lambda_t, \qquad \lambda_k = (\phi^k)^*\lambda$$

for $t \in \mathbb{R}$ and $k \in \mathbb{Z}$, where $\phi^0 := \mathrm{id}$, $\phi^k := \phi \circ \cdots \circ \phi$ (k times), and $\phi^{-k} := (\phi^k)^{-1}$ for $k \in \mathbb{N}$. Now define

$$\alpha_\phi := \lambda_t - dt \in \Omega^1(\Sigma \times \mathbb{R}). \tag{3.5.25}$$

Since $\lambda_{t+1} = \phi^*\lambda_t$ for all t, this 1-form descends to the mapping torus M_ϕ, where it is still denoted by α_ϕ. Its differential is

$$d^{\Sigma\times\mathbb{R}}\alpha_\phi = d^\Sigma\lambda_t - \partial_t\lambda_t \wedge dt = -\omega - \partial_t\lambda_t \wedge dt.$$

Hence α_ϕ is a contact form on M_ϕ whose Reeb vector field is transverse to the fibres $\Sigma \times \{t\}$. Moreover it agrees with $\lambda - dt$ in a neighbourhood of the boundary. A matching contact form on $C \times \mathbb{D}$ is given by

$$\alpha|_{C\times\mathbb{D}} = \lambda + \tfrac{1}{2\pi}(ydx - xdy). \tag{3.5.26}$$

These 1-forms agree on the common boundary

$$\partial M_\phi = C \times \mathbb{R}/\mathbb{Z} \cong C \times \partial\mathbb{D}$$

under the identification $[z,t]_\phi \cong (z, e^{2\pi it})$. The coordinate charts near the boundary can be chosen such that α_ϕ and $\alpha|_{C\times\mathbb{D}}$ define a smooth contact form on M. The key to doing this is to use the Liouville vector field X to identify a neighbourhood of $\partial\Sigma$ with $(-\varepsilon, 0] \times C$. For more details see Geiges [261, Chapter 4.4.2].

Definition 3.5.41 *A contact manifold (M, ξ) is said to be* **supported by an open book decomposition** *if M admits a smooth open book decomposition and a contact form α such that the Reeb vector field Y_α is positively transverse to the pages and is tangent to the binding.*

The contact structure determined by (3.5.25) and (3.5.26) on the smooth open book M given by (3.5.24) is an example of a contact structure that is supported by the given open book decomposition. In dimension three it follows from Exercise 3.5.30 (iv) and the above construction that the set of contact structures on a smooth open book that are supported by the given open book decomposition is nonempty. One can also show that it is connected and hence, by Gray's theorem, there is a unique such structure up to isotopy.

It is a deep theorem by Giroux [271] (see also [272]) that every closed contact manifold (M, ξ) is supported by an open book decomposition. The proof is based on an adaptation of Donaldson's almost complex techniques (Section 7.4) to the contact setting.

4

ALMOST COMPLEX STRUCTURES

Every symplectic manifold admits compatible almost complex structures, but none of these need be integrable. When one is, the manifold is said to be Kähler. In this chapter we discuss basic results about almost complex structures and Kähler manifolds. We may think of these as global analogues of the results in Chapter 2 which relate symplectic to complex vector spaces and vector bundles.

In the early days it was an open question whether every symplectic manifold was in fact Kähler. The first nonKähler symplectic manifold was found by Thurston (Example 3.1.17), the first simply connected examples were constructed by McDuff (Exercise 7.1.33) in dimensions 10 and above, and a large class of other examples was then constructed by Gompf (Section 7.2). Most of these examples have nontrivial fundamental groups that distinguish them from Kähler manifolds. With the help of Seiberg–Witten theory, simply connected nonKähler symplectic four-manifolds were then found by Gompf–Mrowka [279], Fintushel–Stern [219], Vidussi [641, 642], and many others. It turns out that for many results and methods in modern symplectic topology it is immaterial whether the almost complex structure J is integrable or not. For example Gromov's theory of pseudoholomorphic curves (Section 4.5) is valid for nonintegrable almost complex structures and naturally extends the study of complex curves in Kähler manifolds to the symplectic world. Likewise, Donaldson's theory of almost holomorphic sections and symplectic hypersurfaces (Section 7.4) extends the theory of holomorphic sections of line bundles from the complex to the almost complex setting. It can be viewed as the dual picture, where one studies almost holomorphic functions *on* a symplectic manifold instead of pseudoholomorphic functions *with values in* a symplectic manifold. Both viewpoints merge in Taubes–Seiberg–Witten theory (Section 13.3), a third area where techniques from Kähler geometry carry over to the symplectic setting, although at the expense of highly nontrivial analysis.

We begin in Section 4.1 with a general discussion of almost complex structures on symplectic manifolds and in Section 4.2 we shall address the problem of integrability. In Sections 4.3 and 4.4 we discuss a variety of examples of Kähler manifolds, in particular those of complex dimension two, and show how to compute the Chern classes and Betti numbers of hypersurfaces in $\mathbb{C}\mathrm{P}^n$. Section 4.5 is a brief introduction to the theory of J-holomorphic curves. Sections 4.1, 4.3, and 4.4 contain useful background material for Chapter 7 on the construction of symplectic manifolds, and the material in Section 4.5 will be used in Section 13.3. However, the results of this chapter are not needed for Chapter 5 on symplectic reduction and Chapters 8–12 on symplectomorphisms and symplectic invariants.

4.1 Almost complex structures

An **almost complex structure** on a smooth manifold M is an automorphism $J : TM \to TM$ of the tangent bundle such that $J^2 = -\mathbb{1}$. An **almost complex manifold** is a pair (M, J) consisting of a manifold M and an almost complex structure J on M. Every almost complex manifold is oriented and has even dimension. Denote the space of almost complex structures on a manifold M by

$$\mathcal{J}(M) := \left\{J \in C^\infty(M, \mathrm{End}(TM)) \,|\, J^2 = -\mathbb{1}\right\}.$$

Now let M be a smooth manifold and let $\omega \in \Omega^2(M)$ be a nondegenerate 2-form. An almost complex structure J on M is called **ω-tame** (or **tamed by** ω) if

$$\omega(v, Jv) > 0 \qquad \text{for every nonzero tangent vector } v \in TM. \tag{4.1.1}$$

It is called **ω-compatible** (or **compatible with** ω) if it satisfies (4.1.1) and

$$\omega(Jv, Jw) = \omega(v, w) \qquad \text{for all } q \in M \text{ and all } v, w \in T_qM. \tag{4.1.2}$$

Thus J is compatible with ω if and only if the bilinear form

$$\langle v, w\rangle := \omega(v, Jw) \tag{4.1.3}$$

on the tangent bundle defines a Riemannian metric on M. Denote the spaces of ω-compatible and of ω-tame almost complex structures on M by

$$\mathcal{J}(M, \omega) := \left\{J \in \mathcal{J}(M) \,\middle|\, J \text{ is } \omega\text{-compatible}\right\},$$
$$\mathcal{J}_\tau(M, \omega) := \left\{J \in \mathcal{J}(M) \,\middle|\, J \text{ is } \omega\text{-tame}\right\}.$$

A Riemannian metric $g(v, w) = \langle v, w\rangle$ is called **compatible** with J if it satisfies $\langle Jv, Jw\rangle = \langle v, w\rangle$ for $v, w \in T_qM$. In this case the 2-form $\omega(v, w) := \langle Jv, w\rangle$ is nondegenerate and compatible with J and the triple (ω, J, g) is called **compatible**. Thus there is a one-to-one correspondence of nondegenerate 2-forms compatible with J and Riemannian metrics compatible with J. The space of such Riemannian metrics is convex and hence contractible.

Proposition 4.1.1 *Let M be a smooth manifold.*

(i) *For every nondegenerate 2-form ω on M the spaces $\mathcal{J}(M, \omega)$ and $\mathcal{J}_\tau(M, \omega)$ are nonempty and contractible.*

(ii) *Let J be an almost complex structure on M. Then the space of nondegenerate 2-forms on M that are compatible with J (respectively that tame J) is nonempty and contractible.*

(iii) *Fix a Riemannian metric g on M. Then the map $\omega \mapsto J_{g,\omega}$ in Proposition 2.5.6 induces a homotopy equivalence from the space of nondegenerate 2-forms on M to the space of almost complex structures on M.*

Proof: Proposition 4.1.1 follows immediately from Proposition 2.6.4. □

Remark 4.1.2 The terminology *ω-tame* is due to Gromov [287]. It is motivated by the behaviour of J-holomorphic curves in M when ω is a symplectic form. The basic observation is that, while one has no control over the energy of J-holomorphic curves for arbitrary almost complex structures, the energy is a topological invariant when the almost complex structure is tamed by some symplectic form ω. This energy bound was used by Gromov to establish his celebrated compactness theorem. This observation and its consequences for symplectic topology will be discussed more fully in Section 4.5. □

Remark 4.1.3 Let (M, ω) be a closed symplectic four-manifold. In [156] Donaldson posed the question of whether every ω-tame almost complex structure J on M is compatible with some symplectic form (in the same cohomology class as ω when $b^+ = 1$). For $M = \mathbb{C}P^2$ this question was answered in the affirmative by Taubes [618] and for $M = S^2 \times S^2$ it was answered by Li–Zhang [425]. In general this question is still open.

Such a result cannot hold in higher dimensions. Heuristically, this can be seen as follows. The space of tame almost complex structures is an open subset of the space of sections of a bundle with fibre $\mathrm{GL}(2n, \mathbb{R})/\mathrm{GL}(n, \mathbb{C})$ and hence of real dimension $2n^2$, while, by Remark 2.5.9, the space of compatible pairs is a space of sections of a bundle with fibre $\mathrm{GL}(2n, \mathbb{R})/U(n)$ and hence of real dimension $3n^2$. Therefore, in some sense there are n^2 extra functions that we can play with in order to satisfy the conditions. On the other hand, the fact that ω is closed means that it is locally exact and hence depends only on $2n$ functions rather than on its $n(2n-1)$ components. Therefore this imposes $n(2n-3)$ constraints, a number that is strictly bigger than n^2 when $n > 2$. □

Not every oriented even-dimensional manifold admits an almost complex structure. For instance, S^2 and S^6 are the only spheres that do, as we see in the examples below. Proposition 4.1.1 shows that M admits an almost complex structure if and only if it carries a nondegenerate 2-form, and that the homotopy classes of almost complex structures on M are in one-to-one correspondence with the homotopy classes of nondegenerate 2-forms.

Definition 4.1.4 *Let M be a manifold and let $\omega \in \Omega^2(M)$ be a nondegenerate 2-form. By Proposition 4.1.1 the space $\mathcal{J}(M, \omega)$ of ω-compatible almost complex structures is nonempty and contractible. Hence the first Chern class of (TM, J) is independent of $J \in \mathcal{J}(M, \omega)$. It is called the* **first Chern class of** *ω and is denoted by $c_1(\omega) := c_1(TM, J) \in H^2(M; \mathbb{Z})$ for $J \in \mathcal{J}(M, \omega)$.*

Example 4.1.5 Every oriented hypersurface $\Sigma \subset \mathbb{R}^3$ carries an almost complex structure J which it inherits from the vector product

$$\mathbb{R}^3 \times \mathbb{R}^3 \to \mathbb{R}^3 : (u, v) \mapsto u \times v.$$

Let $\nu : \Sigma \to S^2$ be the Gauss map which associates to every point $x \in \Sigma$ the outward unit normal vector

$$\nu(x) \perp T_x\Sigma.$$

Then the almost complex structure is given by the formula

$$J_x u := \nu(x) \times u.$$

Such a hypersurface also carries a natural Riemannian metric $g(u,v) = \langle u, v\rangle$ induced by the metric on $\mathbb{R}^3$ and a natural 2-form $\omega = \iota(\nu(x))\Omega$, where $\Omega(u,v,w)$ is the determinant of the matrix whose columns are the vectors u, v, and w. In other words,

$$\omega_x(v,w) := \langle \nu(x), v \times w\rangle$$

and hence J is compatible with ω. □

Exercise 4.1.6 Calculate the local coordinate representation of the above almost complex structure on S^2 using stereographic projection (see Exercise 4.3.4). □

Example 4.1.7 Let V be a finite-dimensional real vector space equipped with an inner product $\langle \cdot, \cdot \rangle$. A **vector product** on V is a skew-symmetric bilinear form

$$V \times V \to V : (u,v) \mapsto u \times v$$

such that for all $u, v \in V$, the vector $u \times v$ is orthogonal to u and v and its norm is the area of the parallelogram spanned by u and v, i.e.

$$\langle u \times v, u\rangle = \langle u \times v, v\rangle = 0, \qquad |u \times v|^2 = |u|^2\,|v|^2 - \langle u, v\rangle^2. \tag{4.1.4}$$

Equivalently, the vector product and inner product satisfy the equations

$$\langle u \times v, w\rangle = \langle u, v \times w\rangle, \tag{4.1.5}$$

$$u \times (v \times w) + u \times (v \times w) = \langle u, w\rangle v + \langle v, w\rangle u - 2\langle u, v\rangle w \tag{4.1.6}$$

for all $u, v, w \in V$. Vector products exist only in dimensions 0, 1, 3, 7 (see Husemoller [335] and also Salamon–Walpuski [560]). They vanish in dimensions zero and one, and they determine the inner product and orientation uniquely in dimensions three and seven. Namely, the norm squared of a nonzero vector u is the unique nonzero eigenvalue of the linear operator $w \mapsto -u \times (u \times w)$ and its orthogonal complement is the corresponding eigenspace. For every unit vector $u \in V$ it follows from (4.1.5) and (4.1.6) that the linear map $v \mapsto u \times v$ is a complex structure on $u^\perp$.

Here is the multiplication table of Calabi [91] for a vector product on $\mathbb{R}^7$ in the standard basis.

	e_1	e_2	e_3	e_4	e_5	e_6	e_7
e_1	0	e_3	$-e_2$	e_5	$-e_4$	e_7	$-e_6$
e_2	$-e_3$	0	e_1	$-e_6$	e_7	e_4	$-e_5$
e_3	e_2	$-e_1$	0	e_7	e_6	$-e_5$	$-e_4$
e_4	$-e_5$	e_6	$-e_7$	0	e_1	$-e_2$	e_3
e_5	e_4	$-e_7$	$-e_6$	$-e_1$	0	e_3	e_2
e_6	$-e_7$	$-e_4$	e_5	e_2	$-e_3$	0	e_1
e_7	e_6	e_5	e_4	$-e_3$	$-e_2$	$-e_1$	0

Any other vector product on $\mathbb{R}^7$ is related to this example by an orthogonal automorphism. The vector product on $\mathbb{R}^7$ is related to the Cayley numbers. As in the case of the complex numbers and the quaternions, $\mathbb{R}^7$ can be identified with the imaginary Cayley numbers. The vector product on $\mathbb{R}^7$ is the imaginary part of the product of two purely imaginary vectors in $\mathbb{R}^8$. Note that as one progresses from the reals, via the complex numbers and quaternions to the Cayley numbers, the multiplicative structure loses the properties of ordering (from reals to complex numbers), commutativity (from complex numbers to quaternions), and associativity (from quaternions to Cayley numbers).

It follows as in Example 4.1.5 that every oriented hypersurface $M \subset \mathbb{R}^7$ carries an almost complex structure; namely, choose a Gauss map $\nu : M \to S^6$ and define the almost complex structure J on M by

$$J_x u := \nu(x) \times u,$$

for $x \in M$ and $u \in T_x M = \nu(x)^\perp$. As before, such a hypersurface also carries a Riemannian metric g, induced by the metric on $\mathbb{R}^7$ and a compatible non-degenerate 2-form ω defined by

$$\omega_x(u, v) = \langle J_x u, v\rangle = \langle \nu(x), u \times v\rangle.$$

This discussion shows that S^6 is an almost complex manifold and carries a canonical nondegenerate 2-form. Since $H^2(S^6) = 0$ this 2-form cannot be closed. Calabi showed that none of the almost complex structures on S^6 constructed by this method are integrable. It is still unknown whether S^6 admits the structure of a complex manifold. □

Exercise 4.1.8 Prove that the 2-form $\omega_x(u, v) = \langle x, u \times v\rangle$ is nondegenerate on the orthogonal complement of $x \in \mathbb{R}^7$. □

Exercise 4.1.9 Let (M, J) be an almost complex manifold of dimension $4k$. Find an identity connecting its top Pontryagin class p_k with the Chern class c_{2k} of the complex bundle (TM, J). Deduce that none of the spheres S^{4k} admits an almost complex structure. Obtain a similar result for the spheres S^{4k+2} for $k \geq 2$ by using Bott's integrality theorem, which asserts that for any complex vector bundle E over S^{2n} the class $c_n(E)/(n-1)! \in H^{2n}(S^{2n})$ is integral (cf. Husemoller [335, Chapter 18.9.8]). □

Remark 4.1.10 Let M be a closed oriented smooth 4-manifold, let J be an almost complex structure on M, and denote by $c := c_1(TM, J) \in H^2(M; \mathbb{Z})$ the first Chern class (see Remark 2.7.2). The **Hirzebruch signature theorem** asserts that the integer $c^2 := \langle c \cup c, [M]\rangle \in \mathbb{Z}$ satisfies the identity

$$c^2 = 2\chi + 3\sigma, \tag{4.1.7}$$

where χ is the Euler characteristic of M (i.e. the alternating sum of Betti numbers) and σ is its signature (i.e. the signature of the quadratic form on $H^2(M)$ given by the cup product). Moreover, for every homology class $A \in H_2(M; \mathbb{Z})$ the integer $\langle c, A\rangle - A \cdot A \in \mathbb{Z}$ is even. (To see this, represent the class A by an oriented embedded surface and consider the splitting of TM into a direct sum of the tangent bundle and the normal bundle.) If there is no 2-torsion in $H^2(M;\mathbb{Z})$ then this condition asserts that c is an integral lift of the second Stiefel–Whitney class $w_2(TM) \in H^2(M;\mathbb{Z}/2\mathbb{Z})$.

Conversely, it is a classical result of Ehresmann and Wu that a closed oriented smooth 4-manifold M has an almost complex structure J with first Chern class $c \in H^2(M;\mathbb{Z})$ if and only if c satisfies equation (4.1.7) and is an integral lift of the second Stiefel–Whitney class. □

Exercise 4.1.11 **(i)** Show that the n-fold connected sum $M := \#n\mathbb{C}\mathrm{P}^2$ has an almost complex structure if and only if n is odd. (Compare Lemma 7.2.2 below, and see Van de Ven [638], for example.)

(ii) Let M be a closed orientable smooth four-manifold with nonzero Euler characteristic and let J_1, J_2 be almost complex structures on M with the same first Chern class. Show that J_1 and J_2 induce the same orientation of M.

(iii) Let M be a 4-torus or a ruled surface over a base of genus one (see Example 4.4.2 below) or a product of the 2-torus with a closed orientable 2-manifold. Show that M admits a pair of almost complex structures with the same first Chern class that induce opposite orientations of M. □

Remark 4.1.12 Let M be a closed oriented smooth 4-manifold and let

$$c \in H^2(M;\mathbb{Z})$$

be an integral lift of the second Stiefel–Whitney class of M that satisfies equation (4.1.7). Denote by

$$\mathcal{J}(M,c) := \left\{ J \in \mathcal{J}(M) \;\middle|\; \begin{array}{l} c_1(TM,J) = c, \\ J \text{ is compatible with the orientation} \end{array} \right\}$$

the set of almost complex structures with first Chern class c that induce the given orientation on M. By the Ehresmann–Wu theorem this space is nonempty (see Remark 4.1.10). If M is simply connected then $\mathcal{J}(M,c)$ has precisely two connected components. In general, there is a bijection

$$\pi_0(\mathcal{J}(M,c)) \cong \mathrm{Tor}_2(H^2(M;\mathbb{Z})) \times \left(\mathbb{Z}/2\mathbb{Z} \oplus \frac{H^3(M;\mathbb{Z})}{H^1(M;\mathbb{Z}) \cup c} \right). \qquad (4.1.8)$$

The first factor is the set

$$\mathrm{Tor}_2(H^2(M;\mathbb{Z})) := \left\{ a \in H^2(M;\mathbb{Z}) \,|\, 2a = 0 \right\}.$$

It characterizes the isomorphism classes of spin^c structures on TM with first Chern class c (see Lawson–Michelsohn [403, Appendix D]). The second factor characterizes the set of homotopy classes of almost complex structures on M whose canonical spin^c structures are isomorphic to a given spin^c structure with first Chern class c. (This can be proved with a standard Pontryagin manifold type construction as in Milnor [484, §7].) In particular, the $\mathbb{Z}/2\mathbb{Z}$-factor arises from the fact that

$$\pi_4(S^2) = \mathbb{Z}/2\mathbb{Z}.$$

This is relevant because, by Proposition 2.5.2, the space of linear complex structures on $\mathbb{R}^4$ that are compatible with the standard orientation is homotopy equivalent to the 2-sphere. Using this observation, Donaldson showed in [142,143] that there is a free involution on $\pi_0(\mathcal{J}(M,c))$ that reverses the *homological orientation*, i.e. the orientation of $H^0(M) \oplus H^1(M) \oplus H^{2,+}(M)$, and it was shown by Conolly–Lê–Ono [126] that Donaldson's involution preserves the canonical spin^c structure of J. □

Example 4.1.13 By Proposition 2.5.2 the space $\mathcal{J}^+(\mathbb{R}^4)$ of all real 4×4-matrices J that satisfy $J^2 = -\mathbb{1}$ and induce the standard orientation of $\mathbb{R}^4$ is homotopy equivalent to S^2. Hence the space of all almost complex structures on $\mathbb{T}^4$ that are compatible with the standard orientation is homotopy equivalent to the space of maps $\mathbb{T}^4 \to S^2$ and so is not connected. (The induced map on H_2 is a nontrivial invariant; it is half the first Chern class, which is even and has square zero by Remark 4.1.10. For every such class c the connected components of the space $\mathcal{J}(\mathbb{T}^4, c)$ are in bijective correspondence to the set $\mathbb{Z}/2\mathbb{Z} \times H^3(\mathbb{T}^4;\mathbb{Z})/(H^1(\mathbb{T}^4;\mathbb{Z}) \cup c)$ by Remark 4.1.12.) By Proposition 4.1.1 the space of those almost complex structures which are compatible with a given nondegenerate 2-form ω is connected. Until the early 1990s it was an open question whether every nondegenerate 2-form ω on $\mathbb{T}^4$ is homotopic to a symplectic form (within the space of nondegenerate forms). By a result of Taubes, a necessary condition is that ω has first Chern class zero. (See Proposition 13.3.11.) However, at the time of writing it is still an open question whether any two symplectic forms on the 4-torus that induce the same orientation are homotopic as nondegenerate 2-forms. A classical conjecture asserts that the space of symplectic forms on $\mathbb{T}^4$ that represent a given cohomology class is connected. □

Lemma 4.1.14 *Let M be a smooth manifold, let ω be a nondegenerate 2-form on M, let $J \in \mathcal{J}(M,\omega)$ be an ω-compatible almost complex structure, denote by $\langle\cdot,\cdot\rangle := \omega(\cdot, J\cdot)$ the Riemannian metric determined by ω and J, and let ∇ be the Levi-Civita connection of this metric. Then*

$$(\nabla_v J)J + J(\nabla_v J) = 0, \tag{4.1.9}$$

$$\langle(\nabla_u J)v, w\rangle + \langle v, (\nabla_u J)w\rangle = 0 \tag{4.1.10}$$

for $q \in M$ and $u, v, w \in T_qM$, and the 3-form $d\omega \in \Omega^3(M)$ is given by

$$d\omega(u,v,w) = \langle(\nabla_u J)v, w\rangle + \langle(\nabla_v J)w, u\rangle + \langle(\nabla_w J)u, v\rangle. \tag{4.1.11}$$

If ω is closed, then

$$(\nabla_{Jv} J) = -J(\nabla_v J) \tag{4.1.12}$$

for every $q \in M$ and every $v \in T_qM$.

Proof 1: Equation (4.1.9) follows by differentiating the identity $J^2 = -\mathbb{1}$. To prove equation (4.1.10), choose vector fields X, Y, Z such that

$$X(q) = u, \qquad Y(q) = v, \qquad Z(q) = w,$$

and differentiate the identity $\langle JY, Z\rangle + \langle Y, JZ\rangle = 0$ in the direction of the vector field X. To prove equation (4.1.11), recall the identity

$$\begin{aligned} d\omega(X,Y,Z) &= \mathcal{L}_X(\omega(Y,Z)) + \mathcal{L}_Y(\omega(Z,X)) + \mathcal{L}_Z(\omega(X,Y)) \\ &\quad + \omega([X,Y],Z) + \omega([Y,Z],X) + \omega([Z,X],Y). \end{aligned} \tag{4.1.13}$$

Inserting $\omega = \langle J\cdot,\cdot\rangle$ and using the formula $[X,Y] = \nabla_Y X - \nabla_X Y$ and the Leibniz rule, one finds that all the covariant derivatives of the vector fields cancel and this proves equation (4.1.11).

To prove equation (4.1.12), define the linear map $\mathcal{X}(M) \to \Omega^2(M) : X \mapsto \tau_X$ by

$$\tau_X(Y,Z) := \langle (\nabla_X J)Y, Z\rangle$$

for $X, Y, Z \in \mathcal{X}(M)$. That τ_X is a 2-form for every vector field X follows from equation (4.1.10). Moreover, it follows from (4.1.9) and (4.1.11) with $d\omega = 0$ that

$$\tau_X(Y,Z) + \tau_X(JY,JZ) = 0$$

and

$$\tau_X(Y,Z) + \tau_Y(Z,X) + \tau_Z(X,Y) = 0$$

for all $X, Y, Z \in \mathcal{X}(M)$. Hence

$$\begin{aligned} 2\tau_X(Y,Z) &= \tau_X(Y,Z) - \tau_X(JY,JZ) \\ &= -\tau_Y(Z,X) - \tau_Z(X,Y) + \tau_{JY}(JZ,X) + \tau_{JZ}(X,JY). \end{aligned}$$

This implies

$$\begin{aligned} 2\tau_{JX}(JY,Z) &= -\tau_{JY}(Z,JX) - \tau_Z(JX,JY) - \tau_Y(JZ,JX) - \tau_{JZ}(JX,Y) \\ &= -\tau_{JY}(JZ,X) + \tau_Z(X,Y) + \tau_Y(Z,X) - \tau_{JZ}(X,JY) \\ &= -2\tau_X(Y,Z). \end{aligned}$$

The formula $\tau_X(Y,Z) + \tau_{JX}(JY,Z) = 0$ for all $X, Y, Z \in \mathcal{X}(M)$ is equivalent to (4.1.12). This proves Lemma 4.1.14. □

Proof 2: Here is alternative proof of equations (4.1.9), (4.1.10), and (4.1.11) by explicit calculations. In local coordinates $(x_1, \ldots, x_{2n})$ the metric g is represented by the matrix $g_{ij}(x) = g_{ji}(x)$ and the matrix J has entries J^i_j with

$$\sum_\nu J^i_\nu J^\nu_j = -\delta^i_j.$$

The 2-form ω is given by

$$\omega = \sum_{i<j} a_{ij}(x)dx_i \wedge dx_j,$$

with $a_{ij}(x) + a_{ji}(x) = 0$ and

$$d\omega = \sum_{i<j<k} \left(\frac{\partial a_{ij}}{\partial x_k} + \frac{\partial a_{jk}}{\partial x_i} + \frac{\partial a_{ki}}{\partial x_j} \right) dx_i \wedge dx_j \wedge dx_k.$$

The matrices g, J and a are related by

$$\sum_\nu g_{i\nu} J^\nu_j = -a_{ij}.$$

The covariant derivative of J can in local coordinates be expressed in the form

$$\langle (\nabla_v J)w, u \rangle = \sum_{i,j,k} b_{ijk} u_i v_j w_k,$$

where

$$b_{ijk} = \sum_\nu g_{i\nu} \left(\frac{\partial J^\nu_k}{\partial x_j} + \sum_\mu \left(\Gamma^\nu_{j\mu} J^\mu_k - J^\nu_\mu \Gamma^\mu_{jk} \right) \right).$$

Here the Γ^k_{ij} are the Christoffel symbols

$$\Gamma^k_{ij} = \sum_\nu g^{k\nu} \Gamma_{\nu ij}, \qquad \Gamma_{kij} = \frac{1}{2} \left(\frac{\partial g_{ki}}{\partial x_j} + \frac{\partial g_{kj}}{\partial x_i} - \frac{\partial g_{ij}}{\partial x_k} \right).$$

Using this, it is easy to prove equation (4.1.9). Furthermore,

$$\begin{aligned}
b_{ijk} &= \sum_\nu g_{i\nu} \frac{\partial J^\nu_k}{\partial x_j} + \sum_\nu \Gamma_{ij\nu} J^\nu_k + \sum_\nu \Gamma_{\nu jk} J^\nu_i \\
&= \sum_\nu g_{i\nu} \frac{\partial J^\nu_k}{\partial x_j} + \frac{1}{2} \sum_\nu \left(\frac{\partial g_{ij}}{\partial x_\nu} + \frac{\partial g_{i\nu}}{\partial x_j} - \frac{\partial g_{j\nu}}{\partial x_i} \right) J^\nu_k \\
&\quad + \frac{1}{2} \sum_\nu \left(\frac{\partial g_{\nu j}}{\partial x_k} + \frac{\partial g_{\nu k}}{\partial x_j} - \frac{\partial g_{jk}}{\partial x_\nu} \right) J^\nu_i \\
&= -\frac{\partial a_{ik}}{\partial x_j} + \frac{1}{2} \sum_\nu \left(\frac{\partial g_{ij}}{\partial x_\nu} J^\nu_k - \frac{\partial g_{kj}}{\partial x_\nu} J^\nu_i \right) \\
&\quad + \frac{1}{2} \sum_\nu \left(\frac{\partial g_{j\nu}}{\partial x_k} J^\nu_i - \frac{\partial g_{j\nu}}{\partial x_i} J^\nu_k + \frac{\partial g_{k\nu}}{\partial x_j} J^\nu_i - \frac{\partial g_{i\nu}}{\partial x_j} J^\nu_k \right).
\end{aligned}$$

This expression is anti-symmetric in i and k, which implies equation (4.1.10). Moreover,

$$b_{ijk} + b_{jki} + b_{kij} = \frac{\partial a_{ij}}{\partial x_k} + \frac{\partial a_{jk}}{\partial x_i} + \frac{\partial a_{ki}}{\partial x_j}.$$

This proves (4.1.11). The proof of (4.1.12) in explicit coordinates is omitted. □

Exercise 4.1.15 Find an example of a nondegenerate 2-form ω which is not closed and does not satisfy equation (4.1.12). □

Exercise 4.1.16 A submanifold $L \subset M$ is called **totally real** if it is of half the dimension of M and $T_qL \cap J_qT_qL = \{0\}$ for $q \in L$.
(i) Let (ω, J, g) be a compatible triple. Show that any Lagrangian submanifold L is totally real, but not conversely. In fact, L is Lagrangian if and only if JTL is the g-orthogonal complement of TL.
(ii) Prove that if L is a totally real submanifold of (M, J) then there exists a Riemannian metric g on M such that g is compatible with J, JTL is the orthogonal complement of TL, and L is totally geodesic (see Exercise 1.1.27 and Proposition 2.5.4). Note that we cannot necessarily arrange that the corresponding 2-form ω be closed.
(iii) Show that if L is a Lagrangian submanifold of (M, ω) then there is an ω-compatible J such that L is totally geodesic with respect to the corresponding metric g_J. □

4.2 Integrability

A **complex manifold** of complex dimension n is a smooth manifold equipped with an atlas whose coordinate charts take values in $\mathbb{C}^n$ and have holomorphic transition maps. A **complex curve** (also called a **Riemann surface**) is a complex manifold of complex dimension one, and a **complex surface** is a complex manifold of complex dimension two. Every complex manifold is equipped with a canonical almost complex structure, given by multiplication with $i = \sqrt{-1}$ in local coordinates, or with the matrix J_0 in (1.1.10) if we identify $\mathbb{C}^n$ with $\mathbb{R}^{2n}$.

Conversely, let M be a $2n$-manifold. An almost complex structure J on M is called **integrable** if M can be covered by coordinate charts $\phi : U \to \phi(U) \subset \mathbb{R}^{2n}$ such that the almost complex structure is represented by the matrix J_0 in local coordinates, i.e.

$$d\phi(q) \circ J_q = J_0 \circ d\phi(q) : T_qM \to \mathbb{R}^{2n}$$

for all $q \in U$. Theorem 4.2.2 below asserts that the integrability condition can be characterized in terms of the vanishing of the Nijenhuis tensor, which is defined as follows. The **Nijenhuis tensor** of an almost complex structure $J \in \mathcal{J}(M)$ is the vector valued 2-form $N_J \in \Omega^2(M, TM)$ defined by

$$N_J(X, Y) := [X, Y] + J[JX, Y] + J[X, JY] - [JX, JY] \tag{4.2.1}$$

for two vector fields $X, Y : M \to TM$. The next proposition shows that N_J is indeed a family of bilinear maps $T_qM \times T_qM \to T_qM$ for $q \in M$.

Proposition 4.2.1 **(i)** *N_J is a tensor, i.e. $N_J(fX, gY) = fgN_J(X, Y)$ for all $X, Y \in \mathcal{X}(M)$ and all $f, g \in C^\infty(M)$.*
(ii) *$N_J(X, JX) = 0$ for all $X \in \mathcal{X}(M)$.*
(iii) *The Nijenhuis tensor is natural in the sense that*

$$N_{\phi^*J}(\phi^*X, \phi^*Y) = \phi^*N_J(X, Y)$$

for $X, Y \in \mathcal{X}(M)$ and $\phi \in \mathrm{Diff}(M)$.
(iv) *$N_{J_0} = 0$.*

Proof: According to the sign conventions of Remark 3.1.6 we can express the Lie bracket in local coordinates as $[X,Y] = \partial_Y X - \partial_X Y$, where

$$\partial_X Y = \sum_j \xi_j \frac{\partial Y}{\partial x_j} \quad \text{when} \quad X = \sum_j \xi_j \frac{\partial}{\partial x_j}.$$

Hence the Nijenhuis tensor of an almost complex structure $J : U \to \mathbb{R}^{2n\times 2n}$ on an open set $U \subset \mathbb{R}^{2n}$ is given by

$$N_J(X,Y) = (\partial_{JX}J)Y - (\partial_{JY}J)X + (\partial_X J)JY - (\partial_Y J)JX.$$

This proves statement (i). Statement (iii) follows from the identities

$$(\phi^*J)(\phi^*X) = \phi^*(JX), \qquad \phi^*[X,Y] = [\phi^*X, \phi^*Y].$$

Statements (ii) and (iv) are obvious. □

Let $U \subset \mathbb{R}^{2n}$ be an open set and let $J : U \to \mathbb{R}^{2n\times 2n}$ be the local model of an almost complex structure. Then integrability translates into the existence of a local diffeomorphism ϕ such that

$$\phi^*J = d\phi(z)^{-1}J(\phi(z))d\phi(z) = J_0.$$

(Here and subsequently, ϕ^* denotes the natural pullback operation of ϕ on almost complex structures.) Hence Proposition 4.2.1 shows that the Nijenhuis tensor vanishes whenever J is integrable. It is a nontrivial result of complex analysis that the converse holds.

Theorem 4.2.2 (Integrability theorem) *An almost complex structure J is integrable if and only if $N_J = 0$.*

Proof: See [371, Appendix 8] and [160, Chapter 2]. □

The proof is based on a version of the Frobenius integrability theorem whenever J is real analytic. (See Appendix 8 in [371].) In general the analysis is much more delicate. A proof may be found in Donaldson and Kronheimer [160, Chapter 2]. In (real) dimension two Proposition 4.2.1 implies that every almost complex structure is integrable. This is discussed further at the end of this section.

To make the integrability theorem more plausible we rephrase the condition $N_J = 0$ in more geometric terms. First observe that on $\mathbb{C}^n$ the vector fields

$$\frac{\partial}{\partial z_j} = \frac{1}{2}\left(\frac{\partial}{\partial x_j} - i\frac{\partial}{\partial y_j}\right), \qquad \frac{\partial}{\partial \bar{z}_j} = \frac{1}{2}\left(\frac{\partial}{\partial x_j} + i\frac{\partial}{\partial y_j}\right) \tag{4.2.2}$$

may be written in terms of J_0 as

$$\frac{\partial}{\partial z_j} = \frac{1}{2}(1 - iJ_0)\frac{\partial}{\partial x_j}, \qquad \frac{\partial}{\partial \bar{z}_j} = \frac{1}{2}(1 + iJ_0)\frac{\partial}{\partial x_j}.$$

More generally, a complex-valued vector field Z on an almost complex manifold (M,J) is a section of the complexified tangent bundle $TM \otimes \mathbb{C}$. Such a vector field Z is said to have **type** $(1,0)$ if $JZ = iZ$ and **type** $(0,1)$ if $JZ = -iZ$. Thus, in the integrable case $\partial/\partial z_j$ has type $(1,0)$ and $\partial/\partial \bar{z}_j$ has type $(0,1)$.

Exercise 4.2.3 Check that the type $(1,0)$ vector fields on (M,J) are precisely those of the form $(\mathbb{1}-iJ)X$, where X is a real vector field on M. Deduce that in the integrable case they have the form $\sum_j a^j \partial/\partial z_j$, where the a^j are complex-valued functions. □

The above exercise shows that in the integrable case the set of vector fields of type $(1,0)$ is closed under the Lie bracket. The following lemma shows that this is sufficient for integrability. In other words, the vanishing of the Nijenhuis tensor can be viewed as a Frobenius integrability condition.

Lemma 4.2.4 *The set of vector fields of type* $(1,0)$ *on* (M,J) *is closed under the Lie bracket if and only if* $N_J = 0$.

Proof: If X and Y are real-valued vector fields on M, then

$$[X - iJX, Y - iJY] = [X,Y] - [JX,JY] - i\left([JX,Y] + [X,JY]\right).$$

This vector field is of the form $Z - iJZ$ if and only if

$$J\left([X,Y] - [JX,JY]\right) = [JX,Y] + [X,JY].$$

This is equivalent to $N_J(X,Y) = 0$. □

One important geometric difference between almost complex and complex manifolds is that an almost complex manifold may have no local almost complex diffeomorphisms, while a complex manifold has many. In fact, a 1-parameter group ϕ_t of diffeomorphisms preserves the almost complex structure J if and only if it is the integral of a vector field X such that

$$\mathcal{L}_X J = 0. \tag{4.2.3}$$

The formula

$$J[X,Y] - [X,JY] = \mathcal{L}_X(JY) - J\mathcal{L}_X Y = (\mathcal{L}_X J)Y$$

shows that $\mathcal{L}_X J = 0$ if and only if the operator $\mathcal{L}_X$ commutes with J or, equivalently, $[X,JY] = J[X,Y]$ for all Y. But for an arbitrary J there may well be no vector field X with this property. The formula for the Nijenhuis tensor also shows that if $\mathcal{L}_X$ commutes with J then $\mathcal{L}_{JX}$ will not commute with J unless $N_J(X,Y) = 0$ for all Y. This implies, for example, that any holomorphic circle action on a complex manifold extends to a holomorphic action of $\mathbb{C}^*$. But this does not hold in the nonintegrable case.

A similar phenomenon occurs for symplectic forms. The condition that corresponds to integrability is that ω be closed. If ω is merely nondegenerate then the manifold (M,ω) usually has no local diffeomorphisms which preserve ω.

We now show that for symplectic manifolds (M,ω) a compatible complex structure J is integrable if and only if it is covariant constant with respect to the Levi-Civita connection of the associated metric g_J. In fact if we assume only that ω is nondegenerate, the condition that J is covariant constant with respect to g_J implies that both J and ω are 'integrable'.

Lemma 4.2.5 *Let ω be a nondegenerate 2-form on M and let $J \in \mathcal{J}(M,\omega)$. Let ∇ denote the Levi-Civita connection associated to the Riemannian metric $g_J := \omega(\cdot, J\cdot)$. Then the following are equivalent.*

(i) $\nabla J = 0$.

(ii) *J is integrable and ω is closed.*

Proof: We prove that (i) implies (ii). If $\nabla J = 0$ then it follows from equation (4.1.11) in Lemma 4.1.14 that $d\omega = 0$. Since $[X, Y] = \nabla_Y X - \nabla_X Y$ and $\nabla_X(JY) = (\nabla_X J)Y + J\nabla_X Y$ the Nijenhuis tensor can be expressed in the form

$$N_J(X,Y) = (\nabla_{JX} J)Y - (\nabla_{JY} J)X + (\nabla_X J)JY - (\nabla_Y J)JX. \tag{4.2.4}$$

Hence $\nabla J = 0$ implies $N_J = 0$.

We prove that (ii) implies (i). Using Lemma 4.1.14 and (4.2.4) we obtain

$$\begin{aligned}\langle N_J(X,Y), Z\rangle &= \langle (\nabla_{JX} J)Y - (\nabla_Y J)JX + (\nabla_X J)JY - (\nabla_{JY} J)X, Z\rangle \\ &= d\omega(JX, Y, Z) + d\omega(X, JY, Z) + 2\langle J(\nabla_Z J)X, Y\rangle.\end{aligned}$$

Here the last equation follows from (4.1.9), (4.1.10), and (4.1.11). Hence $\nabla J = 0$ whenever $d\omega = 0$ and $N_J = 0$. □

The two-dimensional case

We conclude this section with a discussion of the integrability theorem in the case where j is an almost complex structure on a manifold Σ of real dimension two. In this case the Nijenhuis tensor is obviously zero.

Theorem 4.2.6 *Every almost complex structure on a two-dimensional manifold is integrable.*

Proof: See page 165 and also [470, Appendix C.5] for another proof. □

We begin with a brief discussion of complex structures on the plane with coordinates $(x, y) \in \mathbb{R}^2$. A Riemannian metric

$$g = Edx^2 + 2Fdx \otimes dy + Gdy^2$$

on $\mathbb{R}^2$ is compatible with the standard symplectic structure $\omega_0 = dx \wedge dy$ if and only if

$$EG - F^2 = 1.$$

The associated complex structure on $\mathbb{R}^2$ is then given by

$$j = \begin{pmatrix} -F & -G \\ E & F \end{pmatrix}. \tag{4.2.5}$$

(Compare Exercise 2.5.19.) The space $\mathcal{J}^+(\mathbb{R}^2) = \mathcal{J}(\mathbb{R}^2, \omega_0)$ of such complex structures can be conveniently parametrized by the complex number

$$\tau = \frac{F + i}{E}.$$

Since g is positive definite we have $E, G > 0$ and so $\operatorname{Im}\tau > 0$. Hence $\mathcal{J}^+(\mathbb{R}^2)$ is diffeomorphic to the upper half plane $\mathbb{H} = \{z \in \mathbb{C} \,|\, \operatorname{Im} z > 0\}$.

Now an almost complex structure $j : \mathbb{R}^2 \to \mathbb{R}^{2\times 2}$ is integrable if, for every point q in $\mathbb{R}^2$, there is a diffeomorphism w of an open neighbourhood U of q onto some open set $V \subset \mathbb{C}$ such that $w^*i = j$ or, more explicitly,

$$dw(z)j(z) = idw(z) \tag{4.2.6}$$

for $z = (x, y) \in U$. Of course, the solutions of this equation are not unique. If $w : U \to V$ is any solution of (4.2.6) and $\phi : V \to \mathbb{C}$ is holomorphic, then $\phi \circ w$ also satisfies (4.2.6). Write $w(z) = u(z) + iv(z)$ and denote the components of $j(z)$ by $E(z)$, $F(z)$, and $G(z)$ as above. Then (4.2.6) is equivalent to

$$\frac{\partial u}{\partial x} = -F\frac{\partial v}{\partial x} + E\frac{\partial v}{\partial y}, \qquad -\frac{\partial u}{\partial y} = G\frac{\partial v}{\partial x} - F\frac{\partial v}{\partial y}.$$

Eliminating u from this equation, we see that v must satisfy the **Beltrami equation**

$$\frac{\partial}{\partial x}\left(G\frac{\partial v}{\partial x} - F\frac{\partial v}{\partial y}\right) + \frac{\partial}{\partial y}\left(-F\frac{\partial v}{\partial x} + E\frac{\partial v}{\partial y}\right) = 0. \tag{4.2.7}$$

This is an elliptic partial differential equation. The function u can be recovered from (4.2.7) by Poincaré's lemma. In order for (u, v) to be a local diffeomorphism we must find a nontrivial solution v of (4.2.7) which also satisfies the condition

$$G\left(\frac{\partial v}{\partial x}\right)^2 - 2F\frac{\partial v}{\partial x}\frac{\partial v}{\partial y} + E\left(\frac{\partial v}{\partial x}\right)^2 \neq 0. \tag{4.2.8}$$

In other words, once v is found it follows from (4.2.7) that there exists a function u which satisfies (4.2.6). Condition (4.2.8) guarantees that the transformation $(x, y) \mapsto (u, v)$ is a local diffeomorphism. The existence of local solutions of (4.2.7) and (4.2.8) is proved in Bers–John–Schechter [62] and Courant–Hilbert [128]. We shall only sketch the main idea.

Proof of Theorem 4.2.6: First assume that the almost complex structure $j : U \to \mathbb{R}^{2\times 2}$ with entries $E(z)$, $F(z)$, and $G(z)$ as in equation (4.2.5) is real analytic, and define $\tau(z) := (F(z) + i)/E(z)$. Then (4.2.6) can be rewritten as

$$\tau\frac{\partial w}{\partial x} - \frac{\partial w}{\partial y} = 0. \tag{4.2.9}$$

Since τ is a real analytic function on U, it extends to a holomorphic function on an open neighbourhood $\widehat{U} \subset \mathbb{C}^2$ of U. Assume without loss of generality that $0 \in U$. Shrinking $\widehat{U}$ if necessary, we shall prove that there is a unique holomorphic function $w : \widehat{U} \to \mathbb{C}$ which satisfies (4.2.9) and

$$w(0, y) = y. \tag{4.2.10}$$

If w satisfies (4.2.9) and (4.2.10), then y can be expressed as a function of x on the level set $w(x, y) = w_0$. On this level set the identity

$$\frac{\partial w}{\partial x} + \frac{\partial w}{\partial y}\frac{dy}{dx} = 0$$

shows that $x \mapsto y(x)$ is a solution of the ordinary differential equation

$$\frac{dy}{dx} + \frac{1}{\tau(x,y)} = 0, \qquad y(0) = w_0. \tag{4.2.11}$$

The solutions of (4.2.11) are the **characteristics** of (4.2.9) and they determine w uniquely.

To prove existence let $x \mapsto \phi(x, w_0)$ denote the unique solution of (4.2.11). Since $\phi(0, w_0) = w_0$ the equation $\phi(x, w_0) = y$ can be solved for w_0 locally near $x = y = 0$. In other words, there exists a holomorphic function $w = w(x, y)$ on a neighbourhood of $(0, 0)$ such that

$$w_0 = w(x,y) \qquad \iff \qquad y = \phi(x, w_0).$$

This is the required solution of (4.2.9) and (4.2.10).

If j is not real analytic, then the function τ cannot be extended to a holomorphic function on an open domain in $\mathbb{C}^2$. In this case define $\mu : U \to \mathbb{C}$ by $\mu(z) = (i + \tau(z))^{-1}(i - \tau(z))$. Then Equation (4.2.9) is equivalent to

$$\frac{\partial w}{\partial \bar{z}} = \mu(z)\frac{\partial w}{\partial z}. \tag{4.2.12}$$

This equation admits local solutions which are also local diffeomorphisms whenever $|\mu(z)| \le 1 - \delta$ for some positive number $\delta > 0$. To find a local solution, extend μ to the entire complex plane and write

$$w(z) = z - \frac{1}{\pi}\int_{\mathbb{C}} \frac{\mu(z)v(z)}{\zeta - z}\, d\zeta.$$

Then

$$\frac{\partial w}{\partial \bar{z}} = \mu v, \qquad \frac{\partial w}{\partial z} = 1 - \frac{1}{\pi}\int_{\mathbb{C}} \frac{\mu(\zeta)v(\zeta) - \mu(z)v(z)}{(\zeta - z)^2} d\zeta.$$

Hence (4.2.12) reduces to the integral equation

$$v(z) = 1 - \frac{1}{\pi}\int_{\mathbb{C}} \frac{\mu(\zeta)v(\zeta) - \mu(z)v(z)}{(\zeta - z)^2} d\zeta. \tag{4.2.13}$$

This equation has a unique smooth solution $v : \mathbb{C} \to \mathbb{C}$ whenever $|\mu| \le 1 - \delta$ and the corresponding function w is a diffeomorphism. This completes the outline of the proof of Theorem 4.2.6. For details see [62] and [128, pp. 350–357]. □

Exercise 4.2.7 Given $\tau \in \mathbb{H}$, denote by $j_\tau \in \mathbb{R}^{2\times 2}$ the complex structure associated to τ as above, and define $\Psi_\tau : \mathbb{R}^2 \to \mathbb{C}$ by $\Psi_\tau(x, y) = x + \tau y$. Prove that $\Psi_\tau \circ j_\tau = i \circ \Psi_\tau$ or, in other words, $\Psi_\tau{}^* i = j_\tau$. Prove that every linear isomorphism $\Psi : \mathbb{R}^2 \to \mathbb{C}$ factors uniquely as $\Psi = \lambda\Psi_\tau$, where $\lambda \in \mathbb{C}^*$ and $\tau \in \mathbb{H}$. Deduce that the space $\mathbb{H} \cong \mathcal{J}^+(\mathbb{R}^2)$ is diffeomorphic to the homogeneous space $\mathrm{GL}^+(2, \mathbb{R})/\mathbb{C}^* \cong \mathrm{SL}(2, \mathbb{R})/S^1$. (Compare Proposition 2.5.2.) □

Exercise 4.2.8 Two Riemannian metrics g_1 and g_2 on M are called **conformally equivalent** if there exists a positive function $\lambda : M \to \mathbb{R}$ such that $g_2 = \lambda g_1$. An orientation-preserving diffeomorphism $f : (M_1, g_1) \to (M_2, g_2)$ of Riemannian manifolds is called **conformal** if f^*g_2 is conformally equivalent to g_1. This means that f preserves angles but perhaps not lengths. A metric g is called **compatible** with an almost complex structure J if $g(Jv, Jw) = g(v, w)$. In the case $\dim M = 2$ prove that any two metrics g_1 and g_2 which are compatible with J are conformally equivalent.

Let (Σ_1, j_1) and (Σ_2, j_2) be Riemann surfaces with compatible Riemannian metrics g_1 and g_2, respectively. Prove that a diffeomorphism $\phi : \Sigma_1 \to \Sigma_2$ is holomorphic if and only if it is conformal. □

4.3 Kähler manifolds

A **Kähler manifold** is a triple (M, J, ω), where (M, ω) is a symplectic manifold and J is an integrable almost complex structure on M that is compatible with ω. If (M, ω, J) is a triple, where ω is a nondegenerate 2-form on M and J is an almost complex structure on M that is compatible with ω, then M is equipped with a Riemannian metric $g(\cdot, \cdot) := \omega(\cdot, J\cdot)$ and, by Lemma 4.2.5, the closedness of ω and the integrability of J together are equivalent to the condition that J is parallel under the Levi-Civita connection of g.

The simplest example of a Kähler manifold is Euclidean space $(\mathbb{R}^{2n}, J_0, \omega_0)$ (see Example 4.3.1 below). A class of examples is given by Riemann surfaces; they all carry Kähler forms. The most important example is complex projective space with its standard Kähler form ω_{FS} (see Example 4.3.3 below).

Example 4.3.1 (Euclidean space) The Kähler manifold $(\mathbb{R}^{2n}, J_0, \omega_0)$ can be identified with $\mathbb{C}^n$ in such a way that J_0 corresponds to multiplication by i. In complex geometry on $\mathbb{C}^n$ it is convenient to deal with the complex-valued functions $z_1, \ldots, z_n, \bar{z}_1, \ldots, \bar{z}_n$ as if they were independent variables. Thus we introduce the differential forms

$$dz_j = dx_j + i dy_j, \qquad d\bar{z}_j = dx_j - i dy_j.$$

Note that these are complex-valued 1-forms on $\mathbb{R}^{2n} = \mathbb{C}^n$. For complex-valued differential forms on $\mathbb{C}^n$ the differential $d : \Omega^k \to \Omega^{k+1}$ can be conveniently expressed in the form $d = \partial + \bar{\partial}$, where

$$\partial = \sum_{j=1}^n \frac{\partial}{\partial z_j} dz_j, \qquad \bar{\partial} = \sum_{j=1}^n \frac{\partial}{\partial \bar{z}_j} d\bar{z}_j,$$

and $\partial/\partial z_j$ and $\partial/\partial \bar{z}_j$ are defined by (4.2.2). The condition $d^2 = 0$ implies that

$$\partial^2 = 0, \qquad \bar{\partial}^2 = 0, \qquad \partial\bar{\partial} + \bar{\partial}\partial = 0.$$

The standard symplectic form can now be written as

$$\omega_0 = \frac{i}{2}\partial\bar{\partial} f = \frac{i}{2}\sum_{j=1}^n dz_j \wedge d\bar{z}_j, \qquad f(z) = \sum_{j=1}^n \bar{z}_j z_j.$$

□

Exercise 4.3.2 Express the chain rule in terms of the operators $\partial/\partial z_j$ and $\partial/\partial \bar{z}_j$. Prove that $\phi : \mathbb{C}^n \to \mathbb{C}$ is holomorphic if and only if $\bar{\partial}\phi = 0$. Prove that if $\phi : \mathbb{C}^n \to \mathbb{C}^n$ is holomorphic then

$$\phi^*\partial\omega = \partial\phi^*\omega, \qquad \phi^*\bar{\partial}\omega = \bar{\partial}\phi^*\omega$$

for every complex-valued differential form ω on $\mathbb{C}^n$. □

Example 4.3.3 (Complex projective space) This is the space $\mathbb{C}\mathrm{P}^n$ of complex lines in $\mathbb{C}^{n+1}$. Thus a point in $\mathbb{C}\mathrm{P}^n$ is the equivalence class of a nonzero complex $(n+1)$-vector $[z] = [z_0 : \cdots : z_n]$ under the equivalence relation

$$[z_0 : \cdots : z_n] \equiv [\lambda z_0 : \cdots : \lambda z_n]$$

for $\lambda \neq 0$. The tangent space of $\mathbb{C}\mathrm{P}^n$ at $[z]$ can be naturally identified with the space of complex linear maps from $\mathbb{C}z$ to the orthogonal complement of z, i.e.

$$T_{[z]}\mathbb{C}\mathrm{P}^n = \mathrm{Hom}^{\mathbb{C}}(\mathbb{C}z, z^{\perp}).$$

The complex structure is multiplication by i under this identification. Let $U_j \subset \mathbb{C}\mathrm{P}^n$ denote the coordinate patch where $z_j \neq 0$ and define

$$\phi_j : U_j \to \mathbb{C}^n : [z_0 : \cdots : z_n] \mapsto \left(\frac{z_0}{z_j}, \ldots, \frac{z_{j-1}}{z_j}, \frac{z_{j+1}}{z_j}, \ldots, \frac{z_n}{z_j}\right).$$

The transition maps $\phi_k \circ {\phi_j}^{-1}$ are holomorphic and hence the almost complex structure J is integrable. Now consider the 2-form

$$\rho_{\mathrm{FS}} := \frac{i}{2\left(\sum_{\nu=0}^n \bar{z}_\nu z_\nu\right)^2} \sum_{k=0}^n \sum_{j\neq k} \left(\bar{z}_j z_j \, dz_k \wedge d\bar{z}_k - \bar{z}_j z_k \, dz_j \wedge d\bar{z}_k\right). \tag{4.3.1}$$

This is a real valued 2-form on $\mathbb{C}^{n+1} \setminus \{0\}$ and it descends to a symplectic form on $\mathbb{C}\mathrm{P}^n$, which is denoted by ω_{FS} and called the **Fubini–Study form**. In other words, $\rho_{\mathrm{FS}} = \mathrm{pr}^*\omega_{\mathrm{FS}}$, where $\mathrm{pr} : \mathbb{C}^{n+1} \setminus \{0\} \to \mathbb{C}\mathrm{P}^n$ is the natural projection; cf. Exercise 5.1.3. On the open set where $z_j \neq 0$ the 2-form ω_{FS} can be written as

$$\omega_{\mathrm{FS}} = \frac{i}{2}\partial\bar{\partial}f_j, \qquad f_j(z) := \log\left(\frac{\sum_{\nu=0}^n \bar{z}_\nu z_\nu}{\bar{z}_j z_j}\right).$$

Hence ω_{FS} is closed. Moreover, by Exercise 4.3.4 (ii) below, the restriction of ρ_{FS} to the unit sphere S^{2n+1} agrees with the standard symplectic form

$$\omega_0 = \sum_k dx_k \wedge dy_k.$$

Hence ω_{FS} is nondegenerate. It is compatible with the complex structure, and the induced metric on $\mathbb{C}\mathrm{P}^n$ is called the **Fubini–Study metric**. □

The following exercise fills in some of the necessary details.

Exercise 4.3.4 **(i)** Abbreviate $dz\wedge d\bar{z} := \sum_j dz_j \wedge d\bar{z}_j$ and $\bar{z}\cdot dz := \sum_j \bar{z}_j \cdot dz_j$. Define the 1-form α_{FS} and the 2-form ρ_{FS} on $\mathbb{C}^{n+1}\setminus\{0\}$ by

$$\alpha_{\mathrm{FS}} := \frac{i}{4|z|^2}\left(z\cdot d\bar{z} - \bar{z}\cdot dz\right), \qquad \rho_{\mathrm{FS}} := \frac{i}{2}\left(\frac{dz\wedge d\bar{z}}{|z|^2} - \frac{\bar{z}\cdot dz\wedge z\cdot d\bar{z}}{|z|^4}\right).$$

(See equation (4.3.1).) Prove that $d\alpha_{FS} = \rho_{FS} = \frac{i}{2}\partial\bar{\partial}\log|z|^2$.

(ii) Define the map $\phi:\mathbb{C}^{n+1}\setminus\{0\}\to S^{2n+1}$ by $\phi(z) := |z|^{-1}z$. Prove that $\rho_{\mathrm{FS}} = \phi^*\omega_0$, where $\omega_0 := \sum_{k=0}^n dx_k\wedge dy_k$.

(iii) In the case $n=1$ prove that the symplectic form ω_{FS} on $\mathbb{CP}^1\cong\mathbb{C}\cup\{\infty\}$ in the usual coordinates $z = x+iy$ on $\mathbb{C}$ is given by

$$\omega_{\mathrm{FS}} = \frac{dx\wedge dy}{(1+x^2+y^2)^2}.$$

Let $S^2 := \left\{(x_1,x_2,x_3)\in\mathbb{R}^3 \,|\, x_1^2+x_2^2+x_3^2 = 1\right\}$ be the unit sphere with the area form

$$\mathrm{dvol}_{S^2} := x_1 dx_2\wedge dx_3 + x_2 dx_3\wedge dx_1 + x_3 dx_1\wedge dx_2$$

and total area 4π. (See Exercise 3.1.4.) Let $\phi : S^2\to\mathbb{CP}^1$ be the **stereographic projection** from the south pole given by $\phi(x_1,x_2,x_3) := [1+x_3 : x_1+ix_2]$. Prove that $\phi^*\omega_{\mathrm{FS}} = \frac{1}{4}\mathrm{dvol}_{S^2}$ and deduce that $\int_{\mathbb{CP}^1}\omega_{\mathrm{FS}} = \pi$.

(iv) Let $B\subset\mathbb{C}^n$ be the open unit ball and define the coordinate charts $\psi_j : U_j\to B$, $j=0,\dots,n$, on $\mathbb{CP}^n$ by

$$\psi_j([z_0:\cdots:z_n]) := \frac{|z_j|}{|z|}\left(\frac{z_0}{z_j},\dots,\frac{z_{j-1}}{z_j},\frac{z_{j+1}}{z_j},\dots,\frac{z_n}{z_j}\right). \tag{4.3.2}$$

Show that

$$\psi_j^{-1}(\zeta) = \left[\zeta_1:\cdots:\zeta_j:\sqrt{1-|\zeta|^2}:\zeta_{j+1}:\cdots:\zeta_n\right]$$

for $\zeta = (\zeta_1,\dots,\zeta_n)\in B$. Show that the ψ_j are Darboux charts on $\mathbb{CP}^n$. This shows again that $\mathbb{CP}^1\subset\mathbb{CP}^n$ has area π.

(v) Let $f:\mathbb{C}^n = \mathbb{R}^{2n}\to\mathbb{R}$ be any smooth function. Prove that

$$\omega := \frac{i}{2}\partial\bar{\partial}f = -\frac{1}{4}d(df\circ J_0).$$

Thus ω is nondegenerate if and only if f is plurisubharmonic. (See Definition 7.4.7 below.) In that case ω is a symplectic form compatible with J_0. □

Exercise 4.3.5 **(i)** Show that the **real projective space**

$$\mathbb{RP}^n := \{[z_0:\cdots:z_n]\in\mathbb{CP}^n \,|\, z_0,\dots,z_n\in\mathbb{R}\}$$

and the **Clifford torus**

$$\mathbb{T}^n := \{[z_0:\cdots:z_n]\in\mathbb{CP}^n \,|\, |z_0| = |z_1| = \cdots = |z_n|\}$$

are Lagrangian submanifolds of $(\mathbb{CP}^n,\omega_{\mathrm{FS}})$.

(ii) Prove that a complex submanifold of a Kähler manifold is a Kähler manifold. □

Example 4.3.6 Let (V,ω) be a $2n$-dimensional symplectic vector space. Then the space $\mathcal{J}(V,\omega)$ of ω-compatible linear complex structures on V is a smooth manifold whose tangent space at $J \in \mathcal{J}(V,\omega)$ is given by

$$T_J\mathcal{J}(V,\omega) = \left\{ \widehat{J} \in \mathrm{End}(V) \,\middle|\, \widehat{J}J + J\widehat{J} = 0,\ \omega(\widehat{J}\cdot,\cdot) + \omega(\cdot,\widehat{J}\cdot) = 0 \right\}.$$

The manifold $\mathcal{J}(V,\omega)$ carries a natural Kähler structure. The almost complex structure is given by

$$T_J\mathcal{J}(V,\omega) \to T_J\mathcal{J}(V,\omega) : \widehat{J} \mapsto -J\widehat{J}, \tag{4.3.3}$$

the symplectic form $\Omega \in \Omega^2(\mathcal{J}(V,\omega))$ is given by

$$\Omega_J(\widehat{J}_1,\widehat{J}_2) := \tfrac{1}{2}\mathrm{trace}\left(\widehat{J}_1 J \widehat{J}_2\right) \tag{4.3.4}$$

and the corresponding Riemannian metric on $\mathcal{J}(V,\omega)$ is given by

$$\langle \widehat{J}_1,\widehat{J}_2 \rangle := \tfrac{1}{2}\mathrm{trace}\left(\widehat{J}_1\widehat{J}_2\right). \tag{4.3.5}$$

That this is an inner product follows by taking $(V,\omega,J) = (\mathbb{R}^{2n},\omega_0,J_0)$ (see Proposition 2.5.4). The Levi-Civita connection of this metric is given by

$$\nabla_t\widehat{J}(t) = \partial_t\widehat{J}(t) - \tfrac{1}{2}J(t)\left(\widehat{J}(t)\dot{J}(t) + \dot{J}(t)\widehat{J}(t)\right). \tag{4.3.6}$$

Here $\mathbb{R} \to \mathcal{J}(V,\omega) : t \mapsto J(t)$ is a smooth curve and $\widehat{J}(t) \in T_{J(t)}\mathcal{J}(V,\omega)$ is a smooth vector field along this curve. That (4.3.6) defines a torsion-free Riemannian connection on $\mathcal{J}(V,\omega)$ follows by direct calculation. Moreover, it follows from (4.3.6) that $\nabla_t(J\widehat{J}) = J\nabla_t\widehat{J}$ and so the Levi-Civita connection preserves the almost complex structure. Hence, by Lemma 4.2.5, the almost complex structure (4.3.3) is integrable and the 2-form (4.3.4) is closed. □

Exercise 4.3.7 **(i)** Verify that (4.3.6) is the Levi-Civita connection of the Riemannian metric (4.3.5). Deduce that the geodesics are the solutions $\mathbb{R} \to \mathcal{J}(V,\omega) : t \mapsto J(t)$ of the differential equation

$$\ddot{J} - J\dot{J}^2 = 0$$

and have the form $J(t) = J\exp(-J\widehat{J}t)$. Thus $\mathcal{J}(V,\omega)$ is geodesically complete. Prove that the Riemann curvature tensor of the metric (4.3.5) is given by

$$R(\widehat{J}_1,\widehat{J}_2)\widehat{J}_3 = -\frac{1}{4}\big[[\widehat{J}_1,\widehat{J}_2],\widehat{J}_3\big]. \tag{4.3.7}$$

Prove that $\mathcal{J}(V,\omega)$ has nonpositive sectional curvature.

(ii) Prove that the map $\mathcal{S}_n \to \mathcal{J}(\mathbb{R}^{2n},\omega_0) : Z \mapsto J(Z)$ in Lemma 2.5.12 is a holomorphic diffeomorphism and that the pullback Kähler metric on $\mathcal{S}_n$ is given by the formula $|\widehat{Z}|_Z^2 = \mathrm{trace}((Y^{-1}\widehat{X})^2 + (Y^{-1}\widehat{Y})^2)$. □

Kähler geometry is a subject where complex geometry and symplectic topology intersect and can be studied from either viewpoint. From the complex viewpoint a fundamental question is whether a given complex manifold (M, J) admits a Kähler form ω. If it does, one can study the Kähler cone of cohomology classes that can be represented by symplectic forms that are compatible with J, and ask, for example, whether a given Kähler class contains a Kähler form that is the pullback of the Fubini–Study form $\omega_{\rm FS}$ in Example 4.3.3 under a suitable embedding of (M, J) into complex projective space, or whether the Kähler form in a given Kähler class can be chosen such that the resulting Riemannian metric $\omega(\cdot, J\cdot)$ has constant scalar curvature or is Kähler–Einstein. There are many deep theorems and open problems around this circle of questions. Here are some sample results.

- Voisin [650, 651] found Kähler manifolds such that the complex structure cannot be deformed to one that admits a holomorphic embedding into a projective space;
- a precise description of the Kähler cone is unknown for the n-point blowup of $\mathbb{C}{\rm P}^2$ with $n \geq 9$ (the Nagata conjecture), while as we explain in Example 13.4.4, the corresponding symplectic cone is known;
- a celebrated theorem of Yau asserts that every Kähler class on a closed complex manifold with $c_1 = 0$ contains a Ricci flat Kähler form; and
- the existence problem for constant scalar curvature Kähler metrics was solved for Fano manifolds by Chen–Donaldson–Sun [106, 107, 108] in 2013, but is open in general.

One special feature of Kähler geometry is the **Hodge decomposition**

$$H^k(M; \mathbb{C}) = \bigoplus_{p+q=k} H^{p,q}(M) \tag{4.3.8}$$

of the de Rham cohomology of a Kähler manifold (M, J, ω). This arises from a splitting $\Omega^k(M, \mathbb{C}) = \bigoplus_{p+q=k} \Omega^{p,q}(M)$ of the space of complex valued differential forms into those of type (p, q), which are complex linear in p variables and complex anti-linear in q variables. In the Kähler case this splitting descends to the space of harmonic forms because the Hodge Laplace operator preserves the splitting (see [285]). The **Hodge numbers** $h^{p,q} := \dim_{\mathbb{C}} H^{p,q}(M)$ satisfy the symmetry conditions $h^{p,q} = h^{q,p} = h^{n-q,n-p}$ and sum to the Betti numbers $b_k := \dim H^k(M; \mathbb{R}) = \sum_{p+q=k} h^{p,q}$. In particular, the first Betti number $b_1 = h^{1,0} + h^{0,1}$ is even, which explains why Thurston's manifold in Example 3.1.17 does not admit any Kähler structure. A Kähler form always has type $(1, 1)$ and so represents a class in $H^{1,1}(M; \mathbb{R})$. A deep theorem in Kähler geometry asserts that the Hodge numbers are invariant under deformation. However, they are not topological invariants; first examples were found in 1959 by Borel and Hirzebruch in complex dimension five. In complex dimension two, the **geometric genus** of a Kähler surface (M, J, ω) is the Hodge number $p_g := h^{2,0} = \frac{1}{2}(b^+ - 1)$ and so is determined by the topology and orientation of M.

A symplectic structure is much more malleable than a Kähler structure; it has more automorphisms and is supported by a much wider class of manifolds. Besides manifolds such as the Kodaira–Thurston manifold (see Example 3.1.17) that fail obvious homological or homotopy-theoretic tests, we describe in Example 13.4.6 a construction that yields many symplectic four-manifolds that are homeomorphic to Kähler surfaces but yet have no Kähler structure. Nevertheless, many features that are central to Kähler geometry have analogues in symplectic topology, such as holomorphic curves (Section 4.5), ample divisors, Lefschetz fibrations, and Kodaira embeddings (Section 7.4), the canonical divisor (Section 13.3), and the blowup construction (Section 7.1). Further, sometimes features in complex geometry that have no direct equivalent in symplectic geometry can be seen more indirectly. For example, the inequivalent complex structures on $S^2 \times S^2$ discovered by Hirzebruch are seen in the symplectic world via their relation to the changing homotopy type of the family of symplectomorphism groups $\mathrm{Symp}(S^2 \times S^2, \lambda \mathrm{pr}_1^*\sigma \oplus \mathrm{pr}_2^*\sigma)$, for $\lambda \geq 1$; see Example 13.4.2.

4.4 Kähler surfaces

Complex manifolds of complex dimension one are Riemann surfaces and they all admit Kähler forms. Moreover, closed Riemann surfaces are classified up to diffeomorphism by their genus g. Our knowledge of the classification of complex surfaces X is not so complete, though it is known that a closed complex surface admits a Kähler metric if and only if its first Betti number $b_1 = \dim H^1(X;\mathbb{Q})$ is even. Apart from the fundamental group $\pi_1(X)$, the most important invariants associated to X are the intersection form Q_X, the canonical class K and the Kodaira dimension $\mathrm{Kod}(X)$. We precede our discussion of these topics with some preliminary remarks about Poincaré duality.

Poincaré duality on a closed oriented n-dimensional manifold M sets up an isomorphism between the groups $H_{n-k}(M;\mathbb{Z})$ and $H^k(M;\mathbb{Z})$. The general definition of this correspondence is quite subtle. However, if we are content to neglect torsion, for example by working over $\mathbb{R}$, we can identify $H^k(M;\mathbb{R})$ with the group of homomorphisms $H_k(M;\mathbb{R}) \to \mathbb{R}$ and can then give the following geometric interpretation of this isomorphism. Recall that every compact oriented codimension-k submanifold N of a compact oriented n-dimensional manifold M carries a fundamental cycle because it can be triangulated. It therefore represents an $(n-k)$-dimensional homology class $[N] \in H_{n-k}(M;\mathbb{Z})$. Via Poincaré duality, the submanifold N also represents a k-dimensional cohomology class. One can think of this in two ways. On the one hand, the intersection number of N with a k-cycle in M determines a homomorphism $\phi_N : C_k(M;\mathbb{Z}) \to \mathbb{Z}$ (defined on singular chains) which is a cocycle and determines a cohomology class $a_N = [\phi_N]$ in the image of $H^k(M;\mathbb{Z})$ in $H^k(M;\mathbb{R})$. On the other hand, one can construct a closed k-form $\alpha_N \in \Omega^k(M)$ which is supported near N and is a volume form, multiplied by a cutoff function, on the fibres of a normal bundle with integral 1 on each transverse slice. It is then easy to see that the de Rham cohomology class of α_N agrees with the class a_N. More explicitly, this means that

$$\int_\Sigma \alpha_N = N \cdot \Sigma = \int_M \alpha_N \wedge \alpha_\Sigma$$

for any k-dimensional submanifold (or cycle) Σ in M.[19] It follows easily from de Rham's theorem and Hodge theory that the correspondence $[N] \mapsto a_N$ determines a well-defined surjective vector space homomorphism

$$\mathrm{PD} : H_{n-k}(M; \mathbb{R}) \to H^k(M; \mathbb{R}).$$

Namely, if N is homologous to zero then $N \cdot \Sigma = 0$ for every k-cycle Σ, and so de Rham's theorem asserts that α_N is exact. If PD is not surjective then there is a nonzero harmonic k-form $\alpha \in \Omega^k(M)$ that is L^2-orthogonal to its image. Hence $\int_N *\alpha = 0$ for every closed oriented $(n-k)$-submanifold $N \subset M$ and so, by de Rham's theorem, $*\alpha$ is exact. Since α is harmonic this implies $\alpha = 0$. Thus PD is surjective. Since $H^k(M; \mathbb{R}) \cong \mathrm{Hom}(H_k(M; \mathbb{Z}), \mathbb{R})$, it now follows for dimensional reasons that PD is an isomorphism. The cohomology class $a_N \in H^k(M; \mathbb{R})$ is called the **Poincaré dual** of $[N] \in H_{n-k}(M; \mathbb{R})$ and is denoted by $a_N = \mathrm{PD}([N])$. The inverse map will also be denoted by PD and so we write $[N] = \mathrm{PD}(a_N)$. Poincaré duality lifts to an isomorphism

$$\mathrm{PD} : H_{n-k}(M; \mathbb{Z}) \to H^k(M; \mathbb{Z})$$

and in particular, in the 4-dimensional case, the groups $H_2(M; \mathbb{Z})$ and $H^2(M; \mathbb{Z})$ are isomorphic. For more details see Bott–Tu [78].

Example 4.4.1 Let $L \to M$ be a complex line bundle with section $s : M \to L$, that meets the zero section transversally. Then its zero set Z_s is a codimension two submanifold of M, which is Poincaré dual to the first Chern class $c_1(L)$ by Theorem 2.7.5. □

The above description of Poincaré duality shows that it takes the intersection product to the cup product. In particular, if N and N' are closed 2-dimensional submanifolds of an oriented closed 4-manifold X, we define

$$Q_X([N], [N']) = [N] \cdot [N'] = \int_X \mathrm{PD}(N) \cup \mathrm{PD}(N').$$

It follows from Poincaré duality that Q_X is a unimodular quadratic form on $H_2(M; \mathbb{Z}) \cong H^2(M; \mathbb{Z})$, whose rank is the second Betti number

$$b_2 = b_2(X) = \dim H^2(X; \mathbb{R}),$$

and whose signature is $b^+(X) - b^-(X)$, where $b^+(X)$ and $b^-(X)$ are the numbers of positive and negative entries, respectively, in a diagonalization of Q_X over the reals. Thus

[19] We can in fact restrict to the case when N is a submanifold, since it is known that there is a basis of $H_{n-k}(M; \mathbb{R})$ that can be represented by submanifolds. The case $n - k = 2$ will concern us most, and here it is easy to show using standard general position arguments that every class in $H_2(M; \mathbb{Z})$ can be represented by a submanifold.

$$b_2(X) = b^+(X) + b^-(X).$$

The intersection form Q_X contains a great deal of topological information. For example, it determines the homotopy type of X if X is simply connected.

If X has an almost complex structure J its tangent bundle TX and its cotangent bundle T^*X are complex rank 2 bundles. The **canonical class** $K = K(X)$ is defined to be the first Chern class of the cotangent bundle

$$K = c_1(T^*X, J) = -c_1(TX, J) \in H^2(X; \mathbb{Z}).$$

We shall often write $Q_X(K, K) = K \cdot K = K^2$. The cohomology class K also appears as the first Chern class of the complex line bundle

$$\mathcal{K} := \mathcal{K}_X := \Lambda^{2,0}T^*X,$$

called the **canonical bundle** of (X, J), whose sections are the $(2, 0)$-forms on X. In the integrable case the canonical bundle carries a natural holomorphic structure which it inherits from the holomorphic structure of the tangent bundle.

When trying to understand the structure of complex surfaces it is useful to restrict to the **minimal** case. A complex surface X is said to be minimal if it cannot be blown down.[20] By the well-known Castelnuovo–Enriques criterion, this is equivalent to saying that X contains no holomorphically embedded 2-spheres with self-intersection number -1. (Such spheres are often called -1 rational curves.) It follows easily from Exercise 7.1.4 that if $\widetilde{X}$ denotes the blowup of X at a single point, then

$$K(\widetilde{X})^2 = K(X)^2 - 1, \qquad b^+(\widetilde{X}) = b^+(X), \qquad b^-(\widetilde{X}) = b^-(X) + 1.$$

The classification of Kähler surfaces was established by Enriques and Kodaira in the mid 20th century. A key ingredient is the **Kodaira dimension**

$$\mathrm{Kod}(X) = \limsup_{m \to \infty} \frac{\log \dim\, H^0(X, \mathcal{O}_X(m\mathcal{K}_X))}{\log\, m},$$

where $\mathcal{O}_X(m\mathcal{K}_X)$ denotes the sheaf of germs of holomorphic sections of the mth power $m\mathcal{K}_X$ of the canonical bundle. The classification theory asserts that this number takes one of the values $-\infty$, 0, 1, or 2 and that it is a diffeomorphism invariant (although it is not a homeomorphism invariant). Minimal surfaces with fixed Kodaira dimension have the following properties.

The case $\mathrm{Kod}(X) = -\infty$**:** This corresponds to the case where the canonical bundle $\mathcal{K}$ (and any power of it) has no holomorphic sections. Thus the manifold supports no holomorphic 2-forms. Minimal Kähler surfaces with this property are either rational (that is, equal to $\mathbb{C}P^2$ or $S^2 \times S^2$) or ruled: see Example 4.4.2 below. Note that there are three cases: $K^2 > 0$ (the rational case), $K^2 = 0$ (ruled

[20] The process of blowing up and down in the complex category is described in Section 7.1.

surfaces over $\mathbb{T}^2$), and $K^2 < 0$ (ruled surfaces over Riemann surfaces of higher genus).

The case $\mathrm{Kod}(X) = 0$**:** In this case the space $H^0(X, \mathcal{O}_X(m\mathcal{K}_X))$ of holomorphic sections of the mth tensor power of the canonical bundle has dimension either 0 or 1 for every m. This is the case when $K(X)$ is a torsion class. The only Kähler surfaces with this property are finite quotients of either the 4-torus or the K3-surface. The latter surface is discussed further in Example 4.4.3 below.

The case $\mathrm{Kod}(X) = 1$**:** In this case K is a nontorsion cohomology class with

$$K \cdot K = 0, \qquad K \cdot \omega > 0,$$

and X is an elliptic surface, i.e. there exists a holomorphic map $f : X \to \mathbb{C}\mathrm{P}^1$ whose generic fibre is a 2-torus and with finitely many exceptional fibres. One can get examples here by adding multiple fibres to the rational elliptic surface $\mathbb{C}\mathrm{P}^2\#9\overline{\mathbb{C}\mathrm{P}}^2$. The latter manifolds, when simply connected, are known as Dolgachev surfaces. For more details see Example 7.1.7.

The case $\mathrm{Kod}(X) = 2$**:** Minimal Kähler surfaces with Kodaira dimension 2 satisfy

$$K \cdot K > 0, \qquad K \cdot \omega > 0,$$

and they are said to be **of general type**. There is much current research on these manifolds.

Examples

We end this section with some examples. The discussion uses the **Lefschetz theorem on hyperplane sections** which states that, if V is a complex k-dimensional submanifold of $\mathbb{C}\mathrm{P}^n$ and if $H \cong \mathbb{C}\mathrm{P}^{n-1}$ is a hyperplane in $\mathbb{C}\mathrm{P}^n$, then the inclusion

$$\iota : V \cap H \to V$$

induces an isomorphism on homotopy groups π_i for $i < k - 1$ and a surjection on π_{k-1}. A proof may be found in Milnor [483]. See also Remark 7.4.4 below.

Example 4.4.2 (Ruled surfaces) The ruled surfaces form an interesting but fairly simple family of Kähler surfaces. These are complex surfaces X that fibre holomorphically over a Riemann surface Σ with fibre equal to $\mathbb{C}\mathrm{P}^1$ or, equivalently, S^2. We now describe an easy way to construct families of ruled surfaces, which gives all of them when the genus $g(\Sigma)$ of the base is 0 but not otherwise. Start with a holomorphic line bundle $L \to \Sigma$, take the Whitney sum with a trivial bundle to get the rank 2-bundle $L \oplus \mathbb{C} \to \Sigma$, and then projectivize the fibres. This gives a holomorphic bundle

$$\mathbb{P}(L \oplus \mathbb{C}) \to \Sigma$$

whose fibre at $z \in \Sigma$ is the space $\mathbb{P}(L_z \oplus \mathbb{C}) \cong \mathbb{C}\mathrm{P}^1$ of all lines through the origin in $L_z \oplus \mathbb{C}$.

The classification of ruled surfaces up to diffeomorphism is discussed in Lemma 6.2.3, where it is shown that for every base manifold Σ there are exactly two diffeomorphism classes of bundles $X \to \Sigma$ with fibre S^2, namely the trivial bundle $X = \Sigma \times S^2 \to S^2$ and one other, which we denote by $X_\Sigma \to \Sigma$. For example, when the base Σ is S^2, it follows from Example 7.1.5 that X_Σ is the one point blowup of $\mathbb{C}P^2$. Observe also that by Exercise 6.2.4, line bundles L with even first Chern number give rise to the trivial bundle and those with odd first Chern number to the nontrivial bundle.

A ruled surface is said to be **rational** if its base is the rational curve S^2 and **irrational** otherwise. It is a much more delicate matter to classify ruled surfaces up to complex biholomorphism. Even in the rational case the complex structure is not unique. Indeed, a celebrated theorem of Hirzebruch asserts that the Chern number of L is a complete holomorphic invariant. Thus there is a countable infinity of different complex structures on both $S^2 \times S^2$ and the one point blowup of $\mathbb{C}P^2$. The classification problem for ruled surfaces is fully understood. The interested reader can consult Griffiths–Harris [285] or Barth–Peters–Van de Ven [53] for more information. □

Example 4.4.3 Consider the hypersurface of degree d in $\mathbb{C}P^3$

$$X_d := \left\{[z_0 : \cdots : z_3] \in \mathbb{C}P^3 \,\middle|\, z_0^d + z_1^d + z_2^d + z_3^d = 0\right\}.$$

This is a smooth manifold. The Lefschetz hyperplane theorem implies that the inclusion $X_d \hookrightarrow \mathbb{C}P^3$ induces isomorphisms on π_0 and π_1 and hence X_d is connected and simply connected. To see this, let $L \to \mathbb{C}P^3$ be a holomorphic line bundle of degree d and choose a basis $s_0, \dots, s_n$ of the space of holomorphic sections of L. Then for every $z \in \mathbb{C}P^3$ there exists a j such that $s_j(z) \neq 0$. Hence the sections determine an embedding

$$\mathbb{C}P^3 \to \mathbb{C}P^n : z \mapsto [s_0(z) : \cdots : s_n(z)].$$

The image of this map is a complex submanifold $V \subset \mathbb{C}P^n$ and the image of X_d is the intersection of V with some hyperplane $H \subset \mathbb{C}P^n$. Hence

$$X_d \cong V \cap H$$

is connected and simply connected as claimed. This implies that the second homology is generated by $\pi_2(X_d)$. The dimension of $H_2(X_d; \mathbb{Z})$ is

$$b_2 = d^3 - 4d^2 + 6d - 2. \tag{4.4.1}$$

Moreover, the signature of Q_X is given by

$$\operatorname{sign}(Q_X) = \frac{1}{3}(4 - d^2)d. \tag{4.4.2}$$

The proof of these formulas in Milnor [482] relies on the following observations.

(i) The canonical generator of $H^2(\mathbb{C}\mathrm{P}^3;\mathbb{Z})$ is the first Chern class

$$h = c_1(H) \in H^2(\mathbb{C}\mathrm{P}^3;\mathbb{Z})$$

of the canonical line bundle H whose fibre over $\ell \in \mathbb{C}\mathrm{P}^3$ is the space $H_\ell = \ell^*$ of complex linear functionals on ℓ. To see this, fix a nonzero vector $w \in \mathbb{C}^4$ and consider the section $s : \mathbb{C}\mathrm{P}^3 \to H$ which assigns to $\ell \in \mathbb{C}\mathrm{P}^3$ the restriction of the linear functional $v \mapsto \langle w, v\rangle$ to ℓ. This section is transverse to the zero section and its zero set is a copy of $\mathbb{C}\mathrm{P}^2$ in $\mathbb{C}\mathrm{P}^3$. Hence h is the Poincaré dual of the hyperplane class

$$[\mathbb{C}\mathrm{P}^2] = \mathrm{PD}(h) \in H_4(\mathbb{C}\mathrm{P}^3;\mathbb{Z}).$$

Note also that h is the cohomology class of the standard symplectic structure τ_0 on $\mathbb{C}\mathrm{P}^3$ up to a **positive** factor.

(ii) The tangent bundle of $\mathbb{C}\mathrm{P}^3$ satisfies

$$T\mathbb{C}\mathrm{P}^3 \oplus \mathbb{C} \cong H^4 = H \oplus H \oplus H \oplus H.$$

This holds because $T_\ell\mathbb{C}\mathrm{P}^3 \cong \mathrm{Hom}(\ell, \ell^\perp)$ and $\mathbb{C} \cong \mathrm{Hom}(\ell, \ell)$.

(iii) The normal bundle ν_X is a complex line bundle over X and its first Chern class satisfies

$$c_1(\nu_X) = d\iota^* h,$$

where $\iota : X_d \to \mathbb{C}\mathrm{P}^3$ denotes the natural embedding. To prove this, let $\Sigma \subset X$ be a submanifold such that $\iota(\Sigma) \subset \mathbb{C}\mathrm{P}^3$ represents a 2-dimensional homology class of degree k. Then the intersection number of X and Σ is $X \cdot \Sigma = dk$. Now the intersection number is also the oriented number of zeros of a generic section $s : X \to \nu_X$ when restricted to Σ. In view of Theorem 2.7.5, this means that $\langle c_1(\nu_X), [\Sigma]\rangle = dk = d\langle \iota^* h, [\Sigma]\rangle$.

(iv) The cohomology class $\iota^* h^2 \in H^4(X;\mathbb{Z})$ is given by

$$\langle \iota^* h^2, [X]\rangle = d.$$

This holds because h^2 is the generator of $H^4(\mathbb{C}\mathrm{P}^3;\mathbb{Z})$ and its Poincaré dual is a line. Any such line intersects X in d points, counted with multiplicity.

(v) The Hirzebruch signature theorem asserts that

$$\mathrm{sign}(Q_X) = \frac{1}{3}\langle c_1(TX)^2 - 2c_2(TX), [X]\rangle \tag{4.4.3}$$

for any almost complex 4-manifold X. (See Remark 4.1.10.) This is a deeper formula which cannot be proved by an elementary argument.

We have $T_X\mathbb{C}\mathrm{P}^3 = TX \oplus \nu_X$, and hence

$$c(T_X\mathbb{C}\mathrm{P}^3) = c(TX)c(\nu_X),$$

where $c = 1 + c_1 + c_2$ denotes the total Chern class. By (ii) we have $c(T_X\mathbb{C}\mathrm{P}^3) = (1 + \iota^*h)^4$ and, by (iii), $c(\nu_X) = 1 + d\iota^*h$. Hence

$$(1 + \iota^*h)^4 = (1 + c_1(TX) + c_2(TX))(1 + d\iota^*h).$$

Solving this equation, first for c_1 and then for c_2, we obtain

$$c_1(TX) = (4 - d)\iota^*h, \qquad c_2(TX) = (6 - 4d + d^2)\iota^*h^2. \tag{4.4.4}$$

Now use (iv) and the fact that $\langle c_2(TX), [X]\rangle = 2+b_2$ is the Euler characteristic to obtain (4.4.1). Finally, (4.4.2) follows from the Hirzebruch signature theorem (v).

The equations (4.4.1) and (4.4.2) together show that the positive and negative parts of Q_X are of dimension

$$b^+(X) = \frac{1}{3}(d^3 - 6d^2 + 11d - 3), \qquad b^-(X) = \frac{1}{3}(2d^3 - 6d^2 + 7d - 3).$$

Since both numbers are positive (unless $d = 1$) the intersection form is indefinite. Now the first Chern class of any Kähler surface X satisfies

$$\langle c_1(TX), \alpha\rangle \equiv Q_X(\alpha, \alpha) \pmod 2$$

for $\alpha \in H_2(X; \mathbb{Z})$. In view of (4.4.4) this means that Q_X is **even** (i.e. $Q_X(\alpha, \alpha)$ is even for all $\alpha \in H_2(X)$) if and only if d is even. Hence the classification theorem of indefinite quadratic forms (cf. [487]) shows that the intersection form of X_d, in the odd case, is diagonalizable and so agrees with that of $X_d' = \ell\mathbb{C}\mathrm{P}^2\#m\overline{\mathbb{C}\mathrm{P}}^2$, where $\ell = b^+(X)$ and $m = b^-(X)$. As a side remark we point out that Donaldson's polynomial invariants of these manifolds are different, and so X_d and X_d' are not diffeomorphic for $d \geq 5$ (cf. [160]), although they are homotopy equivalent because they are simply connected and have isomorphic intersection forms. The case $d = 3$ is an exception. One can prove that X_3 is in fact diffeomorphic to $\mathbb{C}\mathrm{P}^2\#6\overline{\mathbb{C}\mathrm{P}}^2$. (This is $\mathbb{C}\mathrm{P}^2$ with six points blown up; see Chapter 7.)

It is interesting to distinguish the cases $d \leq 3$, $d = 4$, and $d \geq 5$. The manifolds $X_1 \cong \mathbb{C}\mathrm{P}^2$, $X_2 \cong S^2 \times S^2$, and $X_3 \cong \mathbb{C}\mathrm{P}^2\#6\overline{\mathbb{C}\mathrm{P}}^2$ are **positive** in the sense that the cohomology class $[\omega]$ of the restriction of the Fubini–Study form is a positive multiple of the first Chern class. (They are **monotone symplectic manifolds.**) The manifold X_4 is a compact, connected, simply connected 4-dimensional Kähler manifold whose first Chern class vanishes. All 4-manifolds with these properties are diffeomorphic: they have second Betti number $b_2 = 22$ and are called **K3-surfaces**. These surfaces have played an important role in 4-dimensional topology [160]. See Example 13.4.7 for more details about their properties as symplectic manifolds. The manifolds X_d for $d \geq 5$ are surfaces of **general type**. In this case, the first Chern class of TX_d is a negative multiple of the cohomology class $\iota^*h = [\omega]$ of the standard symplectic structure. □

Example 4.4.4 A similar example in three complex dimensions is given by the following family of hypersurfaces in $\mathbb{C}\mathrm{P}^4$:

$$Z_d := \left\{[z_0 : \cdots : z_4] \in \mathbb{C}\mathrm{P}^4 \,\middle|\, z_0^d + z_1^d + z_2^d + z_3^d + z_4^d = 0\right\}.$$

These manifolds are simply connected and have Betti numbers

$$b_2 = b_4 = 1, \qquad b_3 = d^4 - 5d^3 + 10d^2 - 10d + 4.$$

The identities $b_2 = b_4 = 1$ follow as before from the Lefschetz theorem on hyperplane sections. Hence $\pi_2(Z_d) = \mathbb{Z}$. The first Chern class of Z_d is given by

$$c_1 = (5 - d)\iota^* h,$$

where $h \in H^2(\mathbb{C}\mathrm{P}^4; \mathbb{Z})$ is the standard generator of the cohomology of $\mathbb{C}\mathrm{P}^4$ and $\iota : Z_d \to \mathbb{C}\mathrm{P}^4$ is the natural embedding of Z_d as a hypersurface in $\mathbb{C}\mathrm{P}^4$.

Now let $A \in \pi_2(Z_d)$ be the generator of the homotopy group with $\omega(A) > 0$. An explicit representative of A is given, for example, by the holomorphic curve $[z_0 : z_1] \mapsto [z_0 : z_1 : e^{i\pi/d} z_0 : e^{i\pi/d} z_1 : 0]$. Evaluating the first Chern class on this generator gives $c_1(A) = 5 - d$. So for $d \leq 4$ the manifold Z_d is positive (see previous example).

The manifold Z_5 has Betti numbers $b_2 = 1$ and $b_3 = 204$ and is an example of a **Calabi–Yau manifold** of three complex dimensions. These manifolds can be characterized by the existence of a nonzero holomorphic 3-form or the vanishing of the first Chern class of TZ_5. They have interesting algebraic geometric properties. In around 1990, Calabi–Yau manifolds gained the attention of theoretical physicists (see the work of Candelas and Ossa [95]) and this gave birth to the subject of **mirror symmetry.** □

Example 4.4.5 (Adjunction formula) Let X be a Kähler surface (of 4 real dimensions) and $\Sigma \subset X$ an embedded connected complex curve. Then, by Theorem 2.7.5, the first Chern class of its normal bundle ν_Σ agrees with the self-intersection number of Σ. Since $T_\Sigma X = T\Sigma \oplus \nu_\Sigma$ the genus $g = g(\Sigma)$ is given by the **adjunction formula**

$$2g(\Sigma) - 2 = \Sigma \cdot \Sigma - \langle c_1(TX), [\Sigma] \rangle. \tag{4.4.5}$$

This continues to hold in the almost complex case. As an example consider the hypersurface

$$\Sigma_d = \left\{[z_0 : z_1 : z_2] \in \mathbb{C}\mathrm{P}^2 \,|\, z_0^d + z_1^d + z_2^d = 0\right\}$$

of degree d in $\mathbb{C}\mathrm{P}^2$. This is an embedded surface whose genus is given by

$$2g(\Sigma_d) - 2 = d^2 - 3d. \tag{4.4.6}$$

In this case the Lefschetz hyperplane theorem asserts that Σ_d is connected. Moreover, as in Example 4.4.3, we have $T\mathbb{C}\mathrm{P}^2 \oplus \mathbb{C} \cong H^3$ and hence $c_1(T\mathbb{C}\mathrm{P}^2) = 3h$. This shows that $\langle c_1(T\mathbb{C}\mathrm{P}^2), [\Sigma_d] \rangle = 3d$ and moreover $\Sigma_d \cdot \Sigma_d = d^2$. Hence (4.4.6) follows from the adjunction formula. □

Exercise 4.4.6 Compute the Chern classes and Betti numbers of a complex hypersurface $M \subset \mathbb{C}\mathrm{P}^{n+1}$ of degree d. **Hint:** Use the techniques of Example 4.4.3 to compute the Chern classes of TM in terms of $\iota^* h$, where $\iota : M \to \mathbb{C}\mathrm{P}^{n+1}$ denotes the inclusion of M and $h \in H^2(\mathbb{C}\mathrm{P}^{n+1}; \mathbb{Z})$ is the canonical generator. Moreover, use the fact that by the Lefschetz theorem on hyperplane sections, the Betti numbers b_j of M agree with those of $\mathbb{C}\mathrm{P}^{n+1}$ for $j < n$. □

As we saw in Example 3.1.17, there are symplectic manifolds with no Kähler structure. In fact, there are restrictions on the topological type of smooth manifolds which admit Kähler structures, such as those imposed on its cohomology ring by the Hard Lefschetz theorem, and certain restrictions on the fundamental group. In particular, every Kähler manifold M has an even first Betti number and hence the abelianization $\pi_1(M)/[\pi_1(M), \pi_1(M)]$ of the fundamental group must have even rank. There is no such restriction on π_1 for symplectic 4-manifolds, by a theorem of Gompf (see Section 7.2). On the other hand, a result of Taubes and Liu asserts that every minimal symplectic 4-manifold with $K^2 < 0$ must be a ruled surface over a Riemann surface of genus greater than one, and hence can be considered as a Kähler manifold, although usually the underlying complex structure is not unique. Further, a result of Donaldson (cf. Section 7.4) shows that symplectic manifolds are like Kähler surfaces in that they support a symplectic version of the Lefschetz pencils discussed in Example 7.1.8. This brief discussion already shows that in dimension four symplectic manifolds form a rather interesting category between Kähler surfaces and general smooth 4-manifolds. For further information see Section 13.4.

4.5 J-holomorphic curves

The subject of symplectic topology was revolutionized in 1985 by Gromov's foundational paper [287] on pseudoholomorphic curves in symplectic manifolds. This notion has also led to many new developments in other fields like enumerative geometry, four-manifold topology, and string theory. The origin of the theory of pseudoholomorphic or J-holomorphic curves goes back to algebraic geometry, where holomorphic curves play a central role in the study of Kähler surfaces, Fano manifolds, Calabi–Yau manifolds, and algebraic varieties in general. For an overview of the subject from the algebro geometric viewpoint, see Clemens–Kollar–Mori [119]. The discussion in the present section is partly inspired by Donaldson's beautiful introductory article [155].

One can think of a holomorphic curve in a complex manifold as a subvariety of complex dimension one which can be described as the space of solutions of a system of equations that is locally defined on the manifold. This means that locally it is the common zero set of finitely many holomorphic functions. From another viewpoint, such a subvariety can be parametrized as the image of a holomorphic map from a Riemann surface to the complex manifold. More precisely, let (M, J) be a complex manifold of real dimension $2n$ and let (Σ, j) be a closed Riemann surface. A holomorphic curve, in the parametrized sense, is then a smooth map $u : \Sigma \to M$ whose differential $du(z) : T_z\Sigma \to T_{u(z)}M$ intertwines the complex structure $j(z)$ on the tangent space $T_z\Sigma$ with the complex structure $J(u(z))$ on the tangent space $T_{u(z)}M$ for every $z \in \Sigma$, i.e. $du \circ j = J \circ du$. Equivalently, u is a solution of the nonlinear Cauchy–Riemann equation

$$\bar{\partial}_J(u) := \tfrac{1}{2}(du + J \circ du \circ j) = 0. \tag{4.5.1}$$

Because of the integrability of almost complex structures in two dimensions, this notion carries over naturally to the almost complex world, in sharp contrast to the description of a holomorphic curve as a common zero set of holomorphic functions. An almost complex manifold of real dimension $2n > 2$ will in general not admit any nonconstant local holomorphic functions. In fact, the integrability condition is equivalent to the existence of sufficiently many local holomorphic functions near each point (see Section 4.2). In contrast, equation (4.5.1) makes perfect sense for almost complex target manifolds (M, J), and the local theory of its solutions has much in common with the local theory of holomorphic curves in complex manifolds. Here the word *local* has two meanings. On one hand it refers to the study of the properties of the solutions of (4.5.1) in a local coordinate chart in M (see [470, Chapter 2]) and on the other hand it refers to the study of the space of solutions of (4.5.1) near some fixed solution (see [470, Chapter 3]). From the first viewpoint, J-holomorphic curves in the almost complex setting share many properties with holomorphic curves in the integrable case, such as unique continuation and positivity of intersections in dimension four. From the second viewpoint, equation (4.5.1) is a nonlinear elliptic equation, the linearized operators are Fredholm, and the space of all equivalence classes of solutions that represent a fixed homology class (modulo automorphisms of (Σ, j)) can be parametrized by finite-dimensional moduli spaces.

Here is a somewhat more precise outline of this picture in the case $\Sigma = \mathbb{C}\mathrm{P}^1$. One can think of the space of all maps $u : \mathbb{C}\mathrm{P}^1 \to M$ representing a fixed homology class $A \in H_2(M;\mathbb{Z})$ as an *infinite-dimensional manifold*

$$\mathcal{B} := \left\{u : \Sigma \to M \,\middle|\, u \text{ is smooth and } u_*[\mathbb{C}\mathrm{P}^1] = A\right\},$$

whose tangent spaces

$$T_u\mathcal{B} = \Omega^0(\mathbb{C}\mathrm{P}^1, u^*TM)$$

consist of smooth sections of the pullback bundles u^*TM. The standard complex structure on $\mathbb{C}\mathrm{P}^1$ and the almost complex structure J on M determine an infinite-dimensional vector bundle $\mathcal{E} \to \mathcal{B}$ whose fibre

$$\mathcal{E}_u := \Omega^{0,1}_J(\mathbb{C}\mathrm{P}^1, u^*TM)$$

over $u \in \mathcal{B}$ is the space of all complex anti-linear vector bundle homomorphisms $T\mathbb{C}\mathrm{P}^1 \to u^*TM$. The left-hand side of equation (4.5.1) defines a section $\bar{\partial}_J : \mathcal{B} \to \mathcal{E}$ of this bundle, with zero set given by the space

$$\mathcal{M}(A, J) := \left\{u : \mathbb{C}\mathrm{P}^1 \to M \,\middle|\, u \text{ is smooth, } u_*[\mathbb{C}\mathrm{P}^1] = A,\ \bar{\partial}_J(u) = 0\right\}$$

of solutions of equation (4.5.1) in the homology class A. The intrinsic differential of the section $\bar{\partial}_J$ at a J-holomorphic curve $u : \mathbb{C}\mathrm{P}^1 \to M$ in the class A is a real linear Cauchy–Riemann operator

$$D_u = D\bar{\partial}_J(u) : T_u\mathcal{B} \to \mathcal{E}_u.$$

The Riemann–Roch theorem implies that it is a Fredholm operator of index

$$\mathrm{index}(D_u) = 2n + 2c_1(A), \tag{4.5.2}$$

where $c_1(A) := \langle c_1(TM, J), A\rangle$ and $2n$ is the real dimension of M. In favourable cases one can prove that for a generic almost complex structure J, the linear operator D_u is surjective for all $u \in \mathcal{M}(A, J)$, and it then follows from the infinite-dimensional implicit function theorem that $\mathcal{M}(A, J)$ is a submanifold of $\mathcal{B}$ of dimension $2n + 2c_1(A)$ whose tangent space at u is the kernel of D_u. The 6-dimensional reparametrization group

$$\mathrm{G} := \mathrm{PSL}(2, \mathbb{C})$$

of Möbius transformations of $\mathbb{C}\mathrm{P}^1$ acts on $\mathcal{M}(A, J)$ by composition on the right, so that the moduli space $\mathcal{M}(A, J)/\mathrm{G}$ is a manifold (or orbifold) of dimension

$$\dim\, \mathcal{M}(A, J)/\mathrm{G} = 2n + 2c_1(A) - 6. \tag{4.5.3}$$

This *local theory* is largely parallel to that of holomorphic curves in complex manifolds, except that dropping the integrability condition on the almost complex structure J gives much greater flexibility. Note that by varying the almost complex structure one can obtain a cobordism between moduli spaces associated to different almost complex structures J_0 and J_1 in the same homotopy class.

Remark 4.5.1 A dimension formula similar to that in equation (4.5.3) holds when the domain is a Riemann surface (Σ, j) of genus g. If the complex structure on the domain is fixed, one gets a moduli space of dimension $n(2-2g)+2c_1(A)$. When $g > 1$ and the complex structure on Σ varies over the $(6g-6)$-dimensional Teichmüller space, the dimension count is

$$n(2 - 2g) + 2c_1(A) + 6g - 6 = (n - 3)(2 - 2g) + 2c_1(A).$$

Since the group of automorphisms of a Riemann surface of genus greater than one is finite, this number $(n - 3)(2 - 2g) + 2c_1(A)$ is also the dimension of the moduli space of J-holomorphic curves of genus g representing the class A modulo the automophisms of (Σ, j), which will in favourable cases be an orbifold. This last formula also works for the cases $g = 0, 1$. Indeed, when $g = 0$ one gets $2n + 2c_1(A) - 6$, which is the dimension of the quotient space $\mathcal{M}(A, J)/\mathrm{G}$, while when $g = 1$ the formula holds because both Teichmüller space and the reparametrization group have dimension two. □

The theory discussed so far is *local* in that for a given J-holomorphic curve the local structure of the moduli space can be used, in good cases, to describe the nearby J-holomorphic curves and to show that the J-holomorphic curves persist under small deformations of the almost complex structure. However, this theory does not show how the J-holomorphic curves behave under arbitrary perturbations of the almost complex structure. Even curves that persist under small deformations of J might simply disappear or have very complicated limits under large perturbations. It was Gromov's great insight that if the almost complex structure J is compatible with a symplectic form ω then the moduli spaces have tractable compactifications. The key reason for this is the **energy identity**

$$E(u) := \tfrac{1}{2}\int_\Sigma |du|^2 \,\mathrm{dvol}_\Sigma = \int_\Sigma u^*\omega \tag{4.5.4}$$

for J-holomorphic curves $u : (\Sigma, j) \to (M, J)$. Here the pointwise norm of the derivative of du is determined by the Riemannian metric $g_J = \omega(\cdot, J\cdot)$ on M and the choice of an area form $\mathrm{dvol}_\Sigma \in \Omega^2(\Sigma)$, with associated Riemannian metric $\mathrm{dvol}_\Sigma(\cdot, j\cdot)$ on Σ. (However, the energy of u is independent of the choice of the volume form on Σ.) The second equality in (4.5.4) follows from the nondegeneracy condition and the compatibility of ω and J. Since ω is closed the energy is in fact a topological invariant $\omega(A) := \langle[\omega], A\rangle$, depending only on the homology class A represented by u. It is this energy control that allowed Gromov to compactify the moduli spaces of curves and thereby obtain invariants of symplectic manifolds. As a result he proved various remarkable theorems about symplectic manifolds. The following extended remark, which takes up the next three pages, describes some of Gromov's applications of the theory of J-holomorphic curves to the study of symplectic manifolds, as well as some related results. Some of these results refer to **minimal** symplectic 4-manifolds, i.e. those that do not contain symplectically embedded 2-spheres with self-intersection number minus one; cf. Remark 7.1.32.

Remark 4.5.2 (i) Nonsqueezing. In [287] Gromov used holomorphic curves to prove his celebrated nonsqueezing theorem (see Theorem 1.2.3). He showed that for every almost complex structure J on the cylinder $Z^{2n}(R)$ and any interior point $z_0 \in Z^{2n}(R)$ there is a J-holomorphic curve in $Z^{2n}(R)$ passing through z_0, whose boundary lies in the boundary of the cylinder and that has area at most πR^2. In particular, this holds when z_0 is the image of the origin under a symplectic embedding $\iota : B^{2n}(r) \to Z^{2n}(R)$ and J pulls back to the standard complex structure. Then the curve pulls back to a properly embedded holomorphic curve through the centre of the unit ball. Since such holomorphic curves are minimal surfaces with respect to the Euclidean metric, it follows from the monotonicity property of minimal surfaces that its area is at least πr^2. This implies $r \le R$. See Chapter 12 for a proof based on generating functions and the Hofer–Zehnder capacity, as well as other related results.

(ii) Obstructions to Lagrangian embeddings. *The standard symplectic vector space* $(\mathbb{R}^{2n}, \omega_0)$ *does not contain an exact closed Lagrangian submanifold.*

For exact Lagrangian submanifolds see Definition 9.3.5. Gromov proved that if $L \subset \mathbb{R}^{2n}$ is any compact Lagrangian submanifold then there exists a nonconstant J_0-holomorphic map $u : \mathbb{D} \to \mathbb{R}^{2n}$ on the closed unit disc $\mathbb{D} \subset \mathbb{C}$ that sends the boundary of the disc to the Lagrangian submanifold. It then follows from the energy identity that the restriction of a 1-form $\lambda \in \Omega^1(\mathbb{R}^{2n})$ with $d\lambda = -\omega_0$ to the Lagrangian submanifold cannot be exact. An exposition of Gromov's proof and a discussion of related developments are contained in [470, Section 9.2]. We return to applications of this theorem in Section 13.2 (see Theorem 13.2.3).

(iii) Lagrangian intersections. *Let* (M, ω) *be a symplectic manifold that is convex at infinity (Definition 3.5.39). Let* $L \subset M$ *be a compact Lagrangian submanifold such that* $[\omega]$ *vanishes on* $\pi_2(M, L)$. *Let* $\psi : M \to M$ *be a Hamiltonian symplectomorphism. Then*

$$\psi(L) \cap L \neq \emptyset.$$

Gromov reduced this result to the theorem in part (ii) via his '*figure eight trick*' in [288, 2.3.B_3']. Another proof based directly on moduli spaces of perturbed J-holomorphic curves with Lagrangian boundary conditions is contained in [470, Theorem 9.2.16]. This result has many consequences. For example it shows that the image of the zero section of the cotangent bundle of a closed manifold under a Hamiltonian symplectomorphism must intersect the zero section. This is a partial result towards the Arnold conjecture for cotangent bundles (Theorem 11.3.8). We give a proof on page 439 which is based on generating functions. Another consequence is that if (M, ω) is a closed symplectic manifold with $\pi_2(M) = 0$, then every Hamiltonian symplectomorphism of (M, ω) has a fixed point. This is a partial result towards another version of the Arnold conjecture (see Conjecture 1.2.4) which was later proved for many symplectic manifolds by Floer using his Floer homology theory. Chapter 11 contains a discussion of the Arnold conjecture and various related results, a generating function proof in some special cases, and and an outline of Floer theory.

(iv) Symplectic structures on $\mathbb{C}\mathrm{P}^2$. *If* ω *is a symplectic form on* $\mathbb{C}\mathrm{P}^2$ *and* C *is an embedded symplectic sphere with area* π *and self-intersection number one, then there exists a symplectomorphism*

$$\phi : (\mathbb{C}\mathrm{P}^2, C, \omega) \to (\mathbb{C}\mathrm{P}^2, \mathbb{C}\mathrm{P}^1, \omega_{\mathrm{FS}}).$$

Gromov proved this by choosing an ω-compatible almost complex structure J on $\mathbb{C}\mathrm{P}^2$ such that C is a J-holomorphic curve. The proof is then based on the fact that the J-holomorphic curves in the same homology class as C sweep out $\mathbb{C}\mathrm{P}^2$ and hence can be used to construct the required symplectomorphism. For an exposition see [470, Section 9.4]. The existence of a symplectically embedded sphere C with self-intersection number one was later established by Taubes, using Seiberg–Witten theory (see Section 13.3). Thus *any two symplectic forms on* $\mathbb{C}\mathrm{P}^2$ *with the same total volume are diffeomorphic.*

(v) Symplectic structures on $S^2 \times S^2$. *If (M, ω) is a closed connected minimal symplectic 4-manifold that contains two symplectically embedded spheres C_1, C_2 with self-intersection number $C_i \cdot C_i = 0$ and intersection number*

$$C_1 \cdot C_2 = 1,$$

then there is a diffeomorphism

$$\psi : M \to S^2 \times S^2$$

such that

$$\psi^*(\lambda_1 \mathrm{pr}_1^* \sigma + \lambda_2 \mathrm{pr}_2^* \sigma) = \omega, \qquad \psi(C_1) = S^2 \times \{\mathrm{pt}\}, \qquad \psi(C_2) = \{\mathrm{pt}\} \times S^2.$$

Here σ is an area form on S^2 with total area 1 and $\lambda_i := \int_{C_i} \omega > 0$.

This theorem is due to Gromov and McDuff. An exposition is contained in [470, Theorem 9.4.7]. The proof uses the moduli spaces of J-holomorphic curves in the homology classes of C_1 and C_2, which are compact for a generic almost complex structure and determine fibrations of M over S^2 with fibres S^2. By positivity of intersections they give rise to a diffeomorphism to $S^2 \times S^2$ and then Moser isotopy furnishes a symplectomorphism.

(vi) Rational and ruled surfaces. Here is a related theorem by McDuff.

Every closed connected minimal symplectic four-manifold (M, ω) that contains an embedded symplectic sphere $C \subset M$ with nonnegative self-intersection number is symplectomorphic to either $\mathbb{C}\mathrm{P}^2$ or a ruled surface, i.e. a 2-sphere bundle over a Riemann surface with symplectic fibres.

The first step of the proof is to establish, under the assumptions of this theorem, the existence of a symplectically embedded sphere $C \subset M$ with self-intersection number zero or one. If the self-intersection number is one, then (M, C, ω) is symplectomorphic to $(\mathbb{C}\mathrm{P}^2, \mathbb{C}\mathrm{P}^1, \lambda\omega_{\mathrm{FS}})$ for some constant $\lambda > 0$, as in Gromov's theorem in part (iv). If the self-intersection number is zero then the moduli space

$$\Sigma := \mathcal{M}(A, J)/\mathrm{G}$$

of J-holomorphic spheres in the homology class $A = [C]$ is a closed oriented 2-manifold for a generic ω-compatible almost complex structure J. Here compactness follows from the fact that (M, ω) is minimal. Using positivity of intersections one can then show that the evaluation map

$$\mathrm{ev} : \mathcal{M}(A, J) \times_{\mathrm{G}} \mathbb{C}\mathrm{P}^1 \to M$$

is a diffeomorphism, giving M the structure of a symplectic fibration over Σ. For details of the proof see McDuff [449], Lalonde–McDuff [390], or McDuff–Salamon [470, Theorem 9.4.1]. For a more detailed discussion of rational and ruled surfaces see Section 13.4. For an in depth discussion of symplectic fibrations see Chapter 6.

(vii) Symplectomorphisms of $S^2 \times S^2$ and $\mathbb{C}\mathrm{P}^2$. *The symplectomorphism group of* $(\mathbb{C}\mathrm{P}^2, \omega_{\mathrm{FS}})$ *retracts onto the isometry group* $\mathrm{PSU}(3)$. *The group of symplectomorphisms of* $(S^2 \times S^2, \sigma \oplus \sigma)$ *that induce the identity on homology retracts onto the subgroup* $\mathrm{SO}(3) \times \mathrm{SO}(3)$. *In particular, every symplectomorphism of* $(\mathbb{C}\mathrm{P}^2, \omega_{\mathrm{FS}})$ *is Hamiltonian, and a symplectomorphism of* $(S^2 \times S^2, \sigma \oplus \sigma)$ *is Hamiltonian if and only if it induces the identity on homology.*

Gromov's proof is based on the moduli spaces of embedded holomorphic spheres that appear in parts (iv) and (v) above, the fact that these moduli spaces are compact, and the fact that the space $\mathcal{J}(M, \omega)$ is contractible. For an exposition see [470, Section 9.5]. In particular, Gromov's theorem asserts that the square of the generalized Dehn twist about the anti-diagonal in $S^2 \times S^2$ (a Lagrangian sphere) is symplectically isotopic to the identity, in contrast to analogous symplectomorphisms of many other symplectic four-manifolds which by a theorem of Seidel are often not symplectically isotopic to the identity (see Example 10.4.3). Section 10.4 contains a more wide ranging discussion of what is known about the topology of various symplectomorphism groups.

(viii) Symplectic structures on $\mathbb{R}^4$. *If* (M, ω) *is a connected symplectic* 4-*manifold with* $\pi_2(M) = 0$ *and which, outside of a compact subset, is symplectomorphic to a neighbourhood of infinity in Euclidean space* $(\mathbb{R}^4, \omega_0)$ *then* (M, ω) *is symplectomorphic to* $(\mathbb{R}^4, \omega_0)$ *by a symplectomorphism which agrees with the given symplectomorphism outside of a compact set.*

Gromov proved this result by constructing two families of embedded holomorphic curves that can be used as coordinates for a diffeomorphism to $\mathbb{R}^4$. He then used Moser isotopy to obtain a symplectomorphism. An exposition of Gromov's argument is contained in [470, Section 9.4]. In [449] McDuff removed the hypothesis $\pi_2(M) = 0$ and proved that (M, ω) is symplectomorphic to a symplectic blowup of $(\mathbb{R}^4, \omega_0)$ at finitely many points (see Section 7.1). In particular, if M is minimal then (M, ω) is symplectomorphic to $(\mathbb{R}^4, \omega_0)$.

(ix) Symplectic structures on $\mathbb{R}^{2n}$. In higher dimensions there is a somewhat weaker result due to Eliashberg, Floer, and McDuff [449].

If (M, ω) *is a connected symplectic manifold with* $\langle [\omega], \pi_2(M) \rangle = 0$ *and which, outside of a compact subset, is symplectomorphic to a neighbourhood of infinity in Euclidean space* $(\mathbb{R}^{2n}, \omega_0)$ *then* M *is diffeomorphic to* $\mathbb{R}^{2n}$.

In dimensions $2n \geq 6$ it is an open question if the diffeomorphism from M to $\mathbb{R}^{2n}$ can be chosen to agree with the given symplectomorphism outside of some compact set. In dimensions $2n \geq 6$ it is also an open question whether the symplectic structure on $\mathbb{R}^{2n}$ induced by this diffeomorphism is diffeomorphic to the standard symplectic structure. □

Since Gromov's groundbreaking paper the theory of J-holomorphic curves has been developed in two main directions. On the one hand it has been used by many authors as a geometric tool to study the topology of symplectic manifolds, as well as their Lagrangian submanifolds and their symplectomorphism groups, in the spirit of the results discussed in Remark 4.5.2.

On the other hand this theory has led to numerical invariants of symplectic manifolds, called the *Gromov–Witten invariants*, and to a study of their intricate algebraic properties. In special cases these invariants take the form of homology classes

$$c_{A,k,\omega} \in H_{2n+2c_1(A)+2k-6}(M^k; \mathbb{Q})$$

represented by the evaluation maps

$$\mathrm{ev}_{A,k,J} : \mathcal{M}_k(A, J) := \mathcal{M}(A, J) \times_{\mathrm{G}} (S^2)^k \to M^k. \tag{4.5.5}$$

Here the group $\mathrm{G} = \mathrm{PSL}(2, \mathbb{C})$ of Möbius transformations acts contravariantly on $\mathcal{M}(A, J) \times (S^2)^k$ by $\phi^*(u, z_1, \ldots, z_k) := (u \circ \phi, \phi^{-1}(z_1), \ldots, \phi^{-1}(z_k))$. The general theory of Gromov–Witten invariants involves suitable compactifications of the moduli spaces $\mathcal{M}_k(A, J)$ (via Kontsevich's notion of *stable maps*) and the construction of so-called *virtual fundamental classes.* All of this goes much beyond the scope of the present book, even in the simplest cases, and the general theory has still to be fully understood. For the easier cases the interested reader is referred to McDuff–Salamon [470] and the references cited therein. The Gromov–Witten invariants can be assembled in generating functions, such as the *Gromov–Witten potential*, that satisfy remarkable partial differential equations, for instance the WDVV-equation. This equation is proved via gluing theorems for J-holomorphic curves (see [470, Chapters 10 and 11] and Ruan–Tian [549].) It corresponds to interesting relations between the Gromov–Witten invariants that can sometimes be formulated as recursion formulas and allow for highly nontrivial computations in enumerative geometry (see Kontsevich–Manin [373]).

The theory developed by Floer combines these two directions. Its central ingredients are J-holomorphic curves on cylinders, with suitable Hamilton perturbations, or J-holomorphic curves on strips with Lagrangian boundary conditions, that give rise to symplectic Floer homology groups associated to symplectomorphisms or pairs of Lagrangian submanifolds. This led to the notion of the *Donaldson* and *Fukaya categories* with their remarkable algebraic structures. These are related to *quantum cohomology* as defined via the Gromov–Witten invariants, and are a main player in Kontsevich's formulation of *homological mirror symmetry.* (See Chapter 11 for an outline of Floer theory, some of its applications to symplectic topology, and copious references to the literature.)

Returning to the realm of applications of J-holomorphic curves in symplectic topology, an important variant of Gromov–Witten theory is the *Gromov-invariant* of a symplectic four-manifold introduced by Taubes [610] in 1994–1995. This invariant counts embedded J-holomorphic curves (not necessarily connected and of any genus) in a fixed homology class in a symplectic four-manifold and is equal to the Seiberg–Witten invariant [608, 609, 611]. This theory has many applications in symplectic topology, especially because it provides a method for establishing the existence of J-holomorphic curves in dimension four. A more detailed discussion of these invariants and their applications in symplectic topology is contained in Section 13.3.

PART II

SYMPLECTIC MANIFOLDS

5

SYMPLECTIC GROUP ACTIONS

In this chapter we discuss symplectic manifolds with group actions. There is a basic process called symplectic reduction, which is a formalization of the well-known classical fact that if there is a symmetry group of dimension d acting on a system then the number of degrees of freedom of the system may be reduced by d. This means that the dimension of the phase space is reduced by $2d$ as both the position and the momentum coordinates are reduced by d. The more abstract underlying principle is that every coisotropic submanifold of a symplectic manifold is foliated by isotropic leaves and the quotient (if it is a manifold) inherits a symplectic structure. There are many significant examples of this process of symplectic reduction, because we can allow the group to be either a finite-dimensional Lie group or (provided one sets up a correct framework) an infinite-dimensional group such as the group of symplectomorphisms of a manifold or a gauge group. In some situations (such as when constructing generating functions for Lagrangian submanifolds of cotangent bundles) we even lose sight of the group. Thus the theory can apply to arbitrary symplectic manifolds, not just those with extra symmetry.

Before delving into the general theory we begin in Section 5.1 by discussing circle actions. Because the group is abelian, everything in this case is very explicit and easy to understand. As a foretaste of the general theory presented in Chapter 6, we discuss the relationship between S^1-reduction and the construction of S^2-bundles in some detail, showing how it can be treated at various levels of universality. This chapter contains two other sections that deal with abelian group actions. First, in Section 5.5 we give a proof of the Atiyah–Guillemin–Sternberg convexity theorem about the image of the moment map in the case of torus actions, and second in Section 5.6 we use equivariant cohomology to prove the Duistermaat–Heckman localization formula for S^1-actions. In both cases, versions of these results are valid for nonabelian groups, but we restrict to the abelian case for simplicity.

The other sections concern the action of more general groups. In Section 5.2 we explain the construction of moment maps for actions of Lie groups on a symplectic manifold, while in Section 5.3 we work out some explicit examples in both finite and infinite dimensions. Section 5.4 discusses the construction of the reduced space, known in many situations as the Weinstein–Marsden quotient. Finally, Section 5.7 provides an overview of geometric invariant theory (GIT) which grows out of the interplay between the actions of a real Lie group and its complexification.

Because of its connection with problems in mechanics, the subject of symplectic group actions has a long history, and, as pointed out by Weinstein [677], many of the basic concepts were developed by Sophus Lie. We will not touch on its many applications in mechanics and refer interested readers to Guillemin–Sternberg [297] and Marsden [438].

5.1 Circle actions

A **Hamiltonian action** of $S^1 = \mathbb{R}/\mathbb{Z}$ on (M, ω) is just a 1-parameter subgroup

$$\mathbb{R} \to \mathrm{Symp}(M) : t \mapsto \psi_t$$

of symplectomorphisms of M which is 1-periodic, i.e. $\psi_1 = \mathrm{id}$, and which is the integral of a Hamiltonian vector field X_H. The Hamiltonian function $H : M \to \mathbb{R}$ in this case is called the **moment map** (or sometimes momentum map) of the action. Note that H is well-defined only up to addition by a constant. If M is compact, it can be normalized so that $\int_M H\omega^n = 0$. The simplest example is the action of S^1 on S^2 by rotations about a vertical axis. As we saw in Example 3.1.7, the associated Hamiltonian is the height function. To make the motion 1-periodic, this should be multiplied by the constant 2π. Thus the associated moment map μ is $2\pi x_3$. (See Fig. 5.1.)

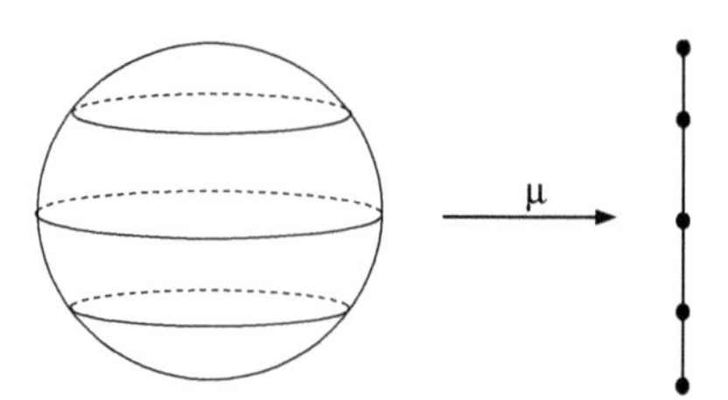

FIG. 5.1. Action of S^1 on S^2.

Remark 5.1.1 Different authors have different sign conventions. Some use the same identity $\iota(X_H)\omega = dH$ as we do, and some (e.g. Guillemin and Lerman) use the identity $\iota(X_H)\omega = -dH$. This means that their moment map is the negative of ours. Note that the signs here are also affected by the fact that we define the canonical symplectic form on a cotangent bundle to be $\omega_{\mathrm{can}} = -d\lambda_{\mathrm{can}}$. □

Lemma 5.1.2 *Suppose that S^1 acts freely on a regular level set $H^{-1}(\lambda)$ of H. Then the quotient manifold*

$$B_\lambda := H^{-1}(\lambda)/S^1$$

carries a symplectic form τ_λ which is characterized by the property that its pullback to $H^{-1}(\lambda)$ is the restriction $\omega|_{H^{-1}(\lambda)}$. The manifold $(B_\lambda, \tau_\lambda)$ is called the **symplectic quotient** *of (M, ω) at $\lambda \in \mathbb{R}$.*

Proof: By Example 3.1.12, the hypersurface $Q := H^{-1}(\lambda)$ is coisotropic and the orbits of S^1 are tangent to the null line field TQ^ω. Therefore, at each point $p \in Q$, ω_p descends to a nondegenerate 2-form $\overline{\omega}_p$ on the quotient T_pQ/T_pQ^ω. The invariance of ω implies that $\psi_t^*(\overline{\omega}_{\psi_t(p)}) = \overline{\omega}_p$ for all t. It follows that the bilinear maps

$$\overline{\omega}_p : \frac{T_pQ}{T_pQ^\omega} \times \frac{T_pQ}{T_pQ^\omega} \to \mathbb{R}$$

for $p \in Q$ determine a well-defined nondegenerate 2-form τ_λ on the quotient manifold $B_\lambda = H^{-1}(\lambda)/S^1$. Its pullback to $H^{-1}(\lambda)$ is closed, and hence so is τ_λ. This proves Lemma 5.1.2. □

The manifold $(B_\lambda, \tau_\lambda)$ is called the **symplectic quotient** or **reduced space** of (M, ω) at $\lambda \in \mathbb{R}$. Thus the reduced spaces in the case of the S^1 action on S^2 considered above are just single points. Here is a more interesting example.

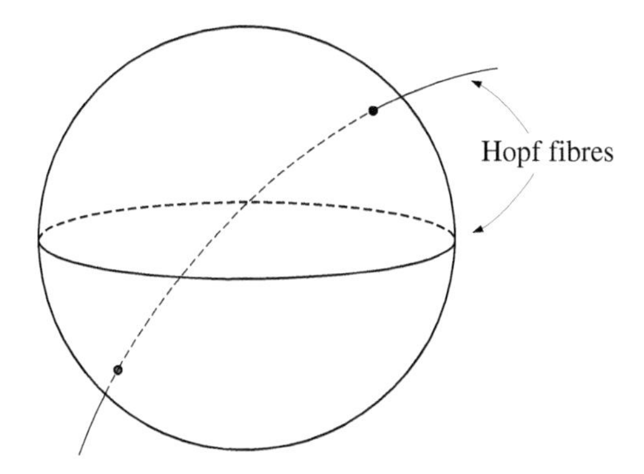

FIG. 5.2. The Hopf fibration. The 2-sphere $S^2 = S^3 \cap \{y_2 = 0\}$ contains one whole Hopf fibre, the equator. The other fibres intersect S^2 in pairs of antipodal points.

Exercise 5.1.3 Consider the action of S^1 on $(\mathbb{R}^{2n+2}, \omega_0)$ which, under the usual identification of $\mathbb{R}^{2n+2}$ with $\mathbb{C}^{n+1}$, corresponds to multiplication by $e^{2\pi i t}$. By Exercise 1.1.22 this action is generated by the function

$$H(z) = -\pi |z|^2.$$

Prove that the symplectic quotient at $\lambda = -\pi$ is $\mathbb{C}\mathrm{P}^n$ with the Fubini–Study form ω_{FS} defined in Example 4.3.3. This construction shows that ω_{FS} is $\mathrm{U}(n+1)$-invariant. **Hint:** Let $\mathrm{pr} : \mathbb{C}^{n+1} \setminus \{0\} \to \mathbb{C}\mathrm{P}^n$ denote the obvious projection and prove that

$$\mathrm{pr}^*\omega_{\mathrm{FS}} = \frac{i}{2}\partial\bar{\partial} f, \qquad f(z) := \log\left(\sum_{\nu=0}^{n} z_\nu \bar{z}_\nu\right),$$

and the restriction of $\mathrm{pr}^*\omega_{\mathrm{FS}}$ to S^{2n+1} agrees with ω_0 (Exercise 4.3.4). □

The restriction of pr to the unit sphere is the **Hopf fibration**

$$\pi_H : S^{2n+1} \to \mathbb{C}\mathrm{P}^n.$$

In the case $n = 1$ one gets the celebrated **Hopf map** $S^3 \to S^2$ which is not homotopic to a constant but induces the zero homomorphism on homology (except in dimension zero).

Exercise 5.1.4 We saw in Exercise 4.3.4 (iii) that

$$\int_{\mathbb{C}\mathrm{P}^1} \omega_{\mathrm{FS}} = \pi.$$

Find another proof by interpreting this integral in terms of the Hopf fibration

$$\pi_H : S^3 \to S^2 = \mathbb{C}\mathrm{P}^1.$$

Show that it equals the integral of ω_0 over a disc in $\mathbb{R}^4$ whose boundary lies along one of the fibres of π_H. (See Fig. 5.2.) □

Exercise 5.1.5 Let (M, ω) be a closed connected symplectic manifold and

$$\mathbb{R}/\mathbb{Z} \times M \to M : (t, p) \mapsto \psi_t(p)$$

be a Hamiltonian circle action, generated by a nonconstant Hamiltonian $H : M \to \mathbb{R}$.
(i) Prove that there exists an almost complex structure $J \in \mathcal{J}(M, \omega)$ that is compatible with ω and is invariant under the circle action, i.e.

$$\psi_t^* J = J$$

for every $t \in \mathbb{R}$. Prove that the space of S^1-invariant ω-compatible almost complex structures on M is contractible. **Hint:** Prove that M admits an S^1-invariant Riemannian metric and use Proposition 2.5.6.
(ii) Let $J \in \mathcal{J}(M, \omega)$ be S^1-invariant and denote by

$$\nabla H := JX_H$$

the gradient of H with respect to the Riemannian metric $g_J := \omega(\cdot, J\cdot)$. Prove that the flows of X_H and ∇H commute. Deduce that the circle action on M extends to a unique J-holomorphic $\mathbb{C}^*$-action

$$\mathbb{C}^* \times M \to M : (e^{2\pi(s+it)}, p) \mapsto \psi_{s,t}(p),$$

such that

$$\partial_s \psi_{s,t} = -\nabla H \circ \psi_{s,t}, \qquad \partial_t \psi_{s,t} = X_H \circ \psi_{s,t}, \qquad \psi_{0,t} = \psi_t. \tag{5.1.1}$$

Hint: Use the condition $\mathcal{L}_{X_H} J = 0$ to deduce that the Lie bracket of the vector fields X_H and ∇H vanishes.

(iii) Let J and $\psi_{s,t}$ be as in (ii), choose $p \in M$ such that $X_H(p) \neq 0$, and define $u : \mathbb{C}^* \to M$ by

$$u(e^{2\pi(s+it)}) := \psi_{s,t}(p).$$

Prove that u extends to a nonconstant J-holomorphic sphere. **Hint:** Show that there are fixed points $x_0, x_\infty \in M$ of the circle action such that

$$\lim_{s\to-\infty} \psi_{s,t}(p) = x_0, \qquad \lim_{s\to\infty} \psi_{s,t}(p) = x_\infty.$$

Deduce that the extension of u to $\mathbb{C} \cup \{\infty\}$, defined by $u(0) := x_0$ and $u(\infty) := x_\infty$, is continuous. Show that the energy of u is equal to

$$\int_{\mathbb{C}\setminus\{0\}} u^*\omega = H(x_0) - H(x_\infty).$$

Use removal of singularities for J-holomorphic curves in a closed symplectic manifold. (See McDuff–Salamon [470, Theorem 4.1.2].) □

We shall come across further examples of symplectic circle actions as we proceed.

Conditions for a circle action to be Hamiltonian

The question obviously arises as to which circle actions are Hamiltonian. An obvious necessary condition is that a Hamiltonian action on a compact symplectic manifold must have fixed points which correspond to the critical points of H. For Kähler manifolds, or more generally for manifolds for which the map $\wedge\omega^{n-1}$ induces an isomorphism from $H^1(M;\mathbb{R})$ to $H^{2n-1}(M;\mathbb{R})$,[21] this condition is necessary and sufficient. This is the assertion of the following theorem, which is due to Ono [513]. An earlier version was proved by Frankel [238].

Theorem 5.1.6 *Let (M, ω) be a closed connected symplectic manifold such that the map*

$$\wedge\omega^{n-1} : H^1(M;\mathbb{R}) \to H^{2n-1}(M;\mathbb{R})$$

is an isomorphism. Then a symplectic circle action on (M, ω) is Hamiltonian if and only if it has fixed points.

Proof: Any Hamiltonian function H on a compact manifold M has critical points. These give rise to zeros of the vector field X_H and hence to fixed points of the circle action ψ_t.

Now let X be a symplectic vector field which generates an S^1-action $\psi_t = \psi_{t+1} \in \mathrm{Symp}(M, \omega)$. Denote by $\alpha_X \in H_1(M,\mathbb{Z})$ the homology class which is represented by the orbits of the S^1 action. All these orbits are homologous because M is connected. Suppose that the action is not Hamiltonian. This means that the 1-form $\iota(X)\omega$ is not exact. Our hypothesis implies that the closed $(2n-1)$-form

[21] Kähler manifolds have this property by the hard Lefschetz theorem: see [285]. Symplectic manifolds with this property are sometimes said to be 'of Lefschetz type'.

$\iota(X)\omega^n$ is also not exact. We shall prove that the cohomology class of $\iota(X)\omega^n$ agrees up to a positive multiple with the Poincaré dual of α_X. In fact we shall assume without loss of generality that $\int_M \omega^n = 1$ and prove that

$$[\iota(X)\omega^n] = \mathrm{PD}(\alpha_X) \in H^{2n-1}(M, \mathbb{Z}).$$

It then follows that α_X is a nonzero homology class and so the action cannot have any fixed points.

To understand the Poincaré dual of α_X let $\gamma(t) = \psi_t(q)$ be a nonconstant orbit of the action and choose a $2n$-form $\sigma \in \Omega^{2n}(M)$ which is supported in a small neighbourhood of γ and which has integral 1. By averaging over S^1 we may assume that σ is S^1-invariant, i.e. $\psi_t^*\sigma = \sigma$ for all t. Since the volume of M with respect to both σ and ω^n is 1 it follows that $\sigma - \omega^n$ is exact, i.e. $\sigma - \omega^n = d\beta$ for some $\beta \in \Omega^{2n-1}(M)$. Since σ and ω^n are invariant we have $\sigma - \omega^n = d(\psi_t^*\beta)$ for all t and hence, averaging again over S^1, we may assume that β is S^1-invariant, i.e.

$$\sigma = \omega^n + d\beta, \qquad \beta = \psi_t^*\beta \quad \forall t.$$

This implies $\mathcal{L}_X\beta = 0$ and hence

$$\iota(X)\sigma - \iota(X)\omega^n = \iota(X)d\beta = -d\iota(X)\beta.$$

In other words, the cohomology class of $\iota(X)\sigma$ agrees with that of $\iota(X)\omega^n$. But it is easy to see that the $(2n-1)$-form $\iota(X)\sigma$ represents the Poincaré dual of α_X. (See Bott and Tu [78, Chapter 1] or page 172.) This proves Theorem 5.1.6. □

If M is a compact 4-manifold then the reduced spaces are 2-dimensional and the situation becomes quite easy to understand. The following result was proved by McDuff [447] by looking at what happens to the reduced spaces as one passes a critical level.

Proposition 5.1.7 *A symplectic S^1 action on a closed connected 4-manifold is Hamiltonian if and only if it has fixed points.*

The same paper contains an example of a symplectic but nonHamiltonian action of S^1 on a closed connected 6-manifold (M, ω) with nonempty fixed point set given by a union of two-dimension tori. Thus one needs *some* condition either on M or on the action to extend the result in the preceding proposition to higher dimensions. Tolman–Weitsman [630] have shown that in any dimension a semi-free S^1 action with isolated fixed points is always Hamiltonian. (An action is called semi-free if it is free outside the fixed point set.) However, the attractive conjecture that a symplectic circle action with isolated fixed points must be Hamiltonian has now been disproved by Tolman, who constructs in [629] a six-dimensional example with 32 fixed points in which some of the reduced spaces are K3 surfaces. (These are defined in Example 13.4.7.) Further information on this question may be found in Cho–Hwang–Suh [109]. Hamiltonian circle actions on 4-manifolds have been classified up to equivariant symplectomorphism by Karshon [354, 355].

Another simple situation in which a circle action is Hamiltonian was discovered by Lupton and Oprea [434]. In this paper they investigate what can be said about Hamiltonian actions from a purely topological point of view. Thus all they retain of the symplectic form is its cohomology class a, which of course has the property that $a^n > 0$. Now a symplectic manifold is called **monotone** if this cohomology class $a = [\omega]$ is a positive multiple of the first Chern class $c_1(M)$.[22] This is again a purely topological property and, by exploiting the universal properties of Chern classes, they prove the following.

Proposition 5.1.8 [434] *Any symplectic circle action on a monotone symplectic manifold is Hamiltonian.*

Another question concerns the homotopy class of the orbits of a nonHamiltonian circle action on a compact manifold. It is easy to see that the Kodaira–Thurston manifold $\mathbb{R}^4/\Gamma$ in Example 3.1.17 admits a free circle action with orbits which are null-homologous. No example is known of a free symplectic circle action with contractible orbits, but no one has yet proved that one cannot exist. (Note that a free action cannot be Hamiltonian.)

S^1-reduction and S^2-fibrations

By Lemma 5.1.2, a Hamiltonian action of S^1 on (M, ω) with Hamiltonian H gives rise to a family of (possibly singular) symplectic manifolds $(B_\lambda, \tau_\lambda)$. Let us now look at how the B_λ change as λ moves in an interval of regular values of H. For simplicity we will assume that the action is **semi-free**. This means that it is free away from the fixed point set or, equivalently, that there are no finite isotropy groups. In this case the reduced spaces B_λ have no singularities. Furthermore, because the diffeomorphism type of B_λ can change only when λ goes through a critical value, we may identify all the B_λ with a single manifold B. This is spelled out in the next proposition. Part (i) is Sternberg's *minimal coupling* [597].

Proposition 5.1.9 (i) *Let $I \subset \mathbb{R}$ be an interval and $\{\tau_\lambda\}_{\lambda \in I}$ be a family of symplectic forms on B such that*

$$[\tau_\lambda] = [\tau_\mu] + (\lambda - \mu)c_I, \tag{5.1.2}$$

for all $\lambda, \mu \in I$, where $c_I \in H^2(B;\mathbb{Z})$. Moreover, let $\pi : P \to B$ be a principal circle bundle with first Chern class c_I. Then there is an S^1-invariant symplectic form ω on the manifold $P \times I$ with Hamiltonian function H equal to the projection $P \times I \to I$ and with reduced spaces (B, τ_λ).

(ii) *Conversely, any compact connected symplectic manifold (M, ω) with a free Hamiltonian action of S^1 is equivariantly symplectomorphic to a manifold of the above form. Moreover, (M, ω) is determined up to an equivariant symplectomorphism by the manifold B and the family of forms τ_λ.*

[22] Note that with this definition the torus $\mathbb{T}^{2n}$ is not monotone. Also, some authors require only that a be positively proportional to c_1 on $\pi_2(M)$.

Proof: Let $\phi_t(p) = p \cdot e^{2\pi it}$ denote the principal circle action on P and let $X : P \to TP$ denote the vector field which generates this action, namely

$$X(p) := p \cdot (2\pi i) \in T_pP.$$

Now let $\tau_\lambda \in \Omega^2(B)$ be a family of symplectic forms which satisfies (5.1.2). Then the closed form

$$\dot{\tau}_\lambda = \frac{d}{d\lambda}\tau_\lambda$$

represents the cohomology class c_I and hence there exists a connection 1-form on P whose curvature (multiplied by $i/2\pi$) is the pullback $\pi^*\dot{\tau}_\lambda$. More explicitly, this means that there exists an invariant 1-form $\alpha_\lambda \in \Omega^1(P)$ (a connection form multiplied by $1/2\pi i$) such that

$$\alpha_\lambda(X) = 1, \qquad d\alpha_\lambda = -\pi^*(\dot{\tau}_\lambda). \tag{5.1.3}$$

(See Theorem 2.7.4.) Now the formula

$$\omega = \pi^*\tau_\lambda + \alpha_\lambda \wedge d\lambda \tag{5.1.4}$$

defines an S^1-invariant symplectic form on $P \times I$. A moment's thought shows that the moment map of the circle action on $P \times I$ is the projection $P \times I \to I$. This proves (i).

To prove (ii), we assume that ω is a symplectic form on a compact connected manifold M with a free Hamiltonian circle action. The image of the corresponding Hamiltonian $H : M \to \mathbb{R}$ is a compact interval $I \subset \mathbb{R}$ and for $\lambda \in I$ let $(B_\lambda, \tau_\lambda)$ denote the reduced spaces. Then the level set

$$P_\lambda := H^{-1}(\lambda)$$

is the total space of a circle bundle $\pi_\lambda : P_\lambda \to B_\lambda$ for $\lambda \in I$. Since M is connected and H has no critical points, the manifolds P_λ are all equivariantly diffeomorphic. To see this, choose a Riemannian metric on M which is invariant under the circle action and use the corresponding gradient flow of H to identify the level sets of H. This gives an equivariant diffeomorphism

$$\psi : P \times I \to M$$

such that $H \circ \psi : P \times I \to I$ is the projection onto the second factor. Here $P = P_{\lambda_0}$ can be chosen to be a fixed level set of H. We examine the pullback of the symplectic form

$$\psi^*\omega = \beta_\lambda + \alpha_\lambda \wedge d\lambda$$

with $\beta_\lambda \in \Omega^2(P)$ and $\alpha_\lambda \in \Omega^1(P)$. Since the circle action on $P \times I$ is Hamiltonian with respect to this symplectic form and the moment map is the projection $P \times I \to I$, it follows that $\alpha_\lambda(X) = 1$ and β_λ descends to a 2-form on $B = P/S^1$. Hence $\psi^*\omega$ is of the form (5.1.4) and the formula (5.1.3) follows from the fact that ω is closed. Finally, since the 1-forms $2\pi i\alpha_\lambda \in \Omega^1(P, i\mathbb{R})$ are connection forms on P, it follows from Chern–Weil theory and (5.1.3) that the cohomology class of $\dot{\tau}_\lambda$ is the first Chern class of the bundle $P \to B$ (see Theorem 2.7.4).

The proof of uniqueness in (ii) is based on an equivariant version of Moser isotopy. Consider two different symplectic forms

$$\omega = \pi^*\tau_\lambda + \alpha_\lambda \wedge d\lambda, \qquad \omega' = \pi^*\tau_\lambda + \alpha'_\lambda \wedge d\lambda$$

on $P \times I$ which are both invariant under the S^1-action. Then the 1-forms $(1-t)\alpha_\lambda + t\alpha'_\lambda$ all satisfy (5.1.3). Hence the 2-forms

$$\omega_t = \pi^*\tau_\lambda + ((1-t)\alpha_\lambda + t\alpha'_\lambda) \wedge d\lambda$$

are all closed and nondegenerate. Moreover, the 1-form $\alpha'_\lambda - \alpha_\lambda$ is invariant and horizontal and therefore descends to B. In other words, there exists a smooth family of 1-forms $\beta_\lambda \in \Omega^1(B)$ such that $\alpha'_\lambda - \alpha_\lambda = \pi^*\beta_\lambda$ for all $\lambda \in I$.

Since $\alpha'_\lambda - \alpha_\lambda$ is closed so is β_λ. Hence we have

$$\omega_t = \pi^*\tau_\lambda + (\alpha_\lambda + t\pi^*\beta_\lambda) \wedge d\lambda, \qquad d\beta_\lambda = 0,$$

and so

$$\frac{d}{dt}\omega_t = \pi^*\beta_\lambda \wedge d\lambda = -d\sigma, \qquad \sigma = \pi^*\rho_\lambda, \qquad \rho_\lambda = \int_{\lambda_0}^{\lambda} \beta_s ds.$$

Here each form ρ_λ is a closed 1-form on B, and the form $\sigma = \pi^*\rho_\lambda$ is to be understood as the 1-form on $P \times I$ that is given by $\pi^*\rho_\lambda$ on each slice $P \times \{\lambda\}$ and whose $d\lambda$-component is zero. Now choose a family of vector fields $\xi_\lambda : B \to TB$ such that

$$\iota(\xi_\lambda)\tau_\lambda = \rho_\lambda$$

and let $Y_{\lambda,t} : P \to TP$ be the unique lift of ξ_λ such that

$$\iota(Y_{\lambda,t})(\alpha_\lambda + t\pi^*\beta_\lambda) = 0.$$

Then $Y_{\lambda,t}$ is a family of S^1-equivariant vector fields on P which for each t fit together to form an equivariant vector field Y_t on $P \times I$. By construction, this vector field satisfies $\iota(Y_t)\omega_t = \sigma$, and so

$$\mathcal{L}_{Y_t}\omega_t + \frac{d}{dt}\omega_t = d(\iota(Y_t)\omega_t - \sigma) = 0.$$

Hence the flow $\psi_t : P \times I \to P \times I$ of Y_t preserves the S^1-action and satisfies $\psi_t^*\omega_t = \omega_0$ as required. This proves Proposition 5.1.9. □

The formula (5.1.2) can be interpreted as a special case of the localization formula by Duistermaat–Heckman, which we shall discuss in detail in Section 5.6. There we shall give a proof which is based on equivariant cohomology. Alternatively, one can prove this formula by using (5.1.2) and examining the change in B_λ as λ passes through a critical level (see Guillemin and Sternberg [298]). Here we will just look at what happens near the minimum (or maximum) of H.

Lemma 5.1.10 *Let $\mathbb{R}/\mathbb{Z} \to \mathrm{Symp}(M, \omega) : t \mapsto \psi_t$ be a Hamiltonian circle action generated by a proper function $H : M \to \mathbb{R}$ that is bounded below. Let $\lambda_0 := \min H$. Then the level set $Z := H^{-1}(\lambda_0)$ is a compact symplectic submanifold consisting entirely of fixed points. Moreover, the action ψ_t of S^1 rotates vectors v normal to Z in the clockwise direction, i.e. the derivative*

$$\left.\frac{d}{dt}\right|_{t=0} \omega(v, d\psi_t(p)v) < 0.$$

Proof: Because λ_0 is the minimum of H, the differential dH vanishes on the set $Z = H^{-1}(\lambda_0)$. Hence $X_H = 0$ on Z and all points of Z are fixed by S^1. The rest of the proof is an easy exercise which is left to the reader. (See Lemma 5.5.7 and Lemma 5.5.8 below.) □

We now explain the converse in the case of a codimension-two symplectic submanifold of M. Thus, given a complex line bundle $L \to Z$ over a closed symplectic manifold (Z, τ) with a clockwise action of S^1 in the fibres of L, we aim to construct an invariant closed 2-form Ω on L that extends the pullback of τ to the zero section Z_0, is nondegenerate near Z_0, and is such that the corresponding moment map has a minimum along Z_0. To this end, let $P \to Z$ be the principal S^1-bundle associated to L, choose a closed 2-form ρ on Z that represents its first Chern class, and let $\alpha \in \Omega^1(P)$ be the associated 1-form as in (5.1.3). Thus

$$\iota(X)\alpha = 1, \qquad d\alpha = -\pi^*\rho,$$

where $X : P \to TP$ is the generator of the S^1-action. (Thus $2\pi i\alpha$ is a connection form on P with curvature $-2\pi i\rho$.) If we now identify L with the line bundle $P \times_{S^1} \mathbb{C} \to Z$ with radial coordinate $r : L \to [0, \infty)$, the 2-form $\pi^*\tau + d(r^2\alpha)$ on L is smooth, S^1-invariant and symplectic near Z_0. This is the **local normal form** for symplectic structures near Z. Moreover, the moment map for the clockwise S^1-action is r^2. (Exercise: check these assertions.)

Example 5.1.11 With this information, we can now build S^1-invariant symplectic forms on S^2-bundles $\pi : M \to B$ which are associated to principal circle bundles $P \to B$, provided that B supports a suitable family of symplectic forms. We assume that S^1 acts on S^2 in the obvious way, by rotation about the north–south axis, and consider the associated S^1-action on $P \times S^2$ given by

$$\theta \cdot (p, z) = (p \cdot \theta^{-1}, \theta \cdot z),$$

for $\theta \in S^1$, $p \in P$ and $z \in S^2 = \mathbb{C} \cup \{\infty\}$. Then M is the quotient manifold $P \times_{S^1} S^2$. Let $h : S^2 \to [0, 1]$ be an S^1-invariant height function, that takes the south pole to 0 and the north pole to 1. There is an induced map $H : M \to [0, 1]$ such that the level sets $H^{-1}(\lambda)$ are all diffeomorphic to P, except for the two critical levels

$$Z_0 = H^{-1}(0), \qquad Z_1 = H^{-1}(1).$$

These critical levels are sections of the fibration $\pi : M \to B$.

Now suppose that $\tau_\lambda \in \Omega^2(B)$, $0 \le \lambda \le 1$, is a family of symplectic forms on B that satisfies equation (5.1.2), where c_I is the first Chern class of the circle bundle $P \to B$. By slightly perturbing this family, we may assume in addition that $\tau_\lambda = \tau_0 + \lambda\rho$ for λ near 0, resp. $\tau_\lambda = \tau_1 + (\lambda - 1)\rho$ for λ near 1, where ρ represents the first Chern class of $P \to B$. By Proposition 5.1.9, there is an S^1-invariant symplectic form ω on $M \setminus (Z_0 \cup Z_1) = H^{-1}(0, 1)$ that reduces to the family τ_λ. It remains to check that ω restricts near Z_0 (and Z_1) to the local normal form constructed above, so that it can be smoothly extended over the rest of M by setting it equal to the pullback of τ_i on Z_i for $i = 0, 1$. Note that the function H is just the moment map of the S^1-action on (M, ω). Moreover, the fibres S^2 are equipped with an S^1-invariant symplectic form σ with corresponding moment map h.

When B is a Riemann surface Σ we have already met these bundles in Example 4.4.2. There they were written as the projectivized bundles $\mathbb{P}(L \oplus \mathbb{C})$. Note that with the present conventions the section called Z_0 above has self-intersection number $-k$ (where k is the Chern number of L) and should be identified with the section $\mathbb{P}(L \oplus \{0\})$, while the section Z_1 has self-intersection number k and should be identified with $\mathbb{P}(\{0\} \oplus \mathbb{C})$. □

The above example can be viewed a little differently. For simplicity, let us suppose that the family τ_λ has the special form $\tau_\lambda = \tau_0 + \lambda\rho$, where ρ is a closed 2-form on B that represents the first Chern class of the bundle $P \to B$. As above, choose a 1-form $\alpha \in \Omega^1(P)$ such that $\iota(X)\alpha = 1$ and $d\alpha = -\pi^*\rho$, where $X : P \to TP$ is the generator of the S^1-action. Now consider the 2-form

$$\Omega := \pi_B^*\tau_0 - d(H\alpha) + \pi_S^*\sigma \in \Omega^2(P \times S^2),$$

where $H(p, z) = h(z)$ is the height function on S^2, π_B and π_S denote the projections of $P \times S^2$ onto B and S^2, and σ is the S^1-invariant volume form on S^2.

Exercise 5.1.12 Show that Ω has maximal rank on the odd-dimensional manifold $P \times S^2$, and that its kernel consists of all vectors tangent to the S^1-orbits. Deduce as in Lemma 5.1.2 that there is an induced symplectic form on the quotient $M = P \times_{S^1} S^2$. Adapt the above construction to deal with arbitrary families of forms τ_λ that satisfy (5.1.2), and then identify the resulting form on M with the one constructed in Example 5.1.11. □

We now move up a level, and claim that the symplectic form on M that was constructed in Exercise 5.1.12 is the reduction of an S^1-invariant symplectic form $\widetilde{\Omega}$ which lives on a symplectic manifold W containing $P \times S^2$ as a hypersurface. This manifold W is made by a very elegant universal construction, which is due to Sternberg [597] and Weinstein [672]. Here we only consider a special case. The full construction is carried out in Theorem 6.3.3.

Because P fibres over B, its tangent bundle TP contains a vertical subbundle $V \subset TP$ whose fibre at p is the subspace $V_p \subset T_pP$ of all vectors which are tangent to the fibre. Let V^*P denote the dual of this vertical subbundle. The connection form α picks out a horizontal space

$$H_p := \{v \in T_pP \,|\, \alpha_p(v) = 0\}$$

at each point $p \in P$ such that $T_pP = H_p \oplus V_p$. This horizontal subbundle determines an injection $\iota_\alpha : V^*P \to T^*P$ of the 'vertical cotangent bundle' into the full cotangent bundle. Namely, a vertical cotangent vector $\lambda \in V_p^*P$ is a linear functional $V_p \to \mathbb{R}$ and extends naturally to a linear functional on T_pP which vanishes on the horizontal subspace H_p. Now the manifold $W := V^*P \times S^2$ carries a natural symplectic form

$$\widetilde{\Omega} = \pi_B^*\tau_0 + \iota_\alpha^*\omega_{\mathrm{can}} + \pi_S^*\sigma,$$

where $\pi_B : W \to B$ and $\pi_S : W \to S^2$ are the obvious projections.

Exercise 5.1.13 Prove that $\widetilde{\Omega}$ is symplectic and is invariant under the diagonal action of S^1. Show that V^*P is equivariantly diffeomorphic to $P \times \mathbb{R}$ and that the moment map $\mu : W := P \times \mathbb{R} \times S^2 \to \mathbb{R}$ is given by $\mu(p, \eta, z) = h(z) - \eta$, where $h : S^2 \to \mathbb{R}$ is the height function used above. Show further that 0 is a regular value of μ and that the level set $\mu^{-1}(0)$ can be identified with the manifold $P \times S^2$ by a map which takes $\widetilde{\Omega}$ to Ω. Thus (M, ω) is the symplectic reduction of $(W, \widetilde{\Omega})$ at 0. □

5.2 Moment maps

Hamiltonian group actions

Let G be a Lie group with Lie algebra $\mathfrak{g} := \mathrm{Lie}(\mathrm{G})$ which acts covariantly on a symplectic manifold (M, ω) by symplectomorphisms. This means that there is a smooth group homomorphism

$$\mathrm{G} \to \mathrm{Symp}(M, \omega) : g \mapsto \psi_g.$$

In other words, the map $\mathrm{G} \times M \to M : (g, p) \mapsto \psi_g(p)$ is smooth, $\psi_g : M \to M$ is a symplectomorphism for every $g \in \mathrm{G}$, and

$$\psi_{gh} = \psi_g \circ \psi_h, \qquad \psi_{\mathbb{1}} = \mathrm{id},$$

for all $g, h \in \mathrm{G}$. The infinitesimal action determines a Lie algebra homomorphism

$$\mathfrak{g} \to \mathcal{X}(M, \omega) : \xi \mapsto X_\xi$$

defined by

$$X_\xi := \left.\frac{d}{dt}\right|_{t=0} \psi_{\exp(t\xi)}.$$

Since ψ_g is a symplectomorphism for every $g \in \mathrm{G}$ it follows that each X_ξ is a symplectic vector field. This means that the 1-form $\iota(X_\xi)\omega$ is closed for every ξ. Moreover, it follows by straightforward calculation that

$$X_{g^{-1}\xi g} = \psi_g^* X_\xi, \qquad X_{[\xi,\eta]} = [X_\xi, X_\eta] \tag{5.2.1}$$

for $\xi, \eta \in \mathfrak{g}$ and $g \in \mathrm{G}$. Here we use the notation

$$g^{-1}\xi g := \mathrm{Ad}(g^{-1})\xi := \left.\frac{d}{dt}\right|_{t=0} g^{-1}\exp(t\xi)g,$$

and the sign conventions of Remark 3.1.6.

The action of G on M is called **weakly Hamiltonian** if each vector field X_ξ is Hamiltonian. This means that the 1-form $\iota(X_\xi)\omega$ is exact for every $\xi \in \mathfrak{g}$. In other words, for every ξ there is a corresponding Hamiltonian function H_ξ. However, this function H_ξ is only determined up to a constant. It is easy to see that the constants can be chosen such that the map $\xi \mapsto H_\xi$ is linear. The action is called **Hamiltonian** if the map

$$\mathfrak{g} \to C^\infty(M) : \xi \mapsto H_\xi$$

can be chosen to be G-equivariant with respect to the adjoint action of G on its Lie algebra $\mathfrak{g}$. The next lemma characterizes Hamiltonian group actions.

Lemma 5.2.1 *Let* $\mathrm{G} \to \mathrm{Symp}(M,\omega) : g \mapsto \psi_g$ *be a symplectic action by a compact Lie group, let* $\mathfrak{g} \to \mathcal{X}(M,\omega) : \xi \mapsto X_\xi$ *be the infinitesimal action, and let* $\mathfrak{g} \to C^\infty(M) : \xi \mapsto H_\xi$ *be a linear map such that* $\iota(X_\xi)\omega = dH_\xi$ *for every* $\xi \in \mathfrak{g}$. *Consider the following assertions.*

(i) *The map* $\xi \mapsto H_\xi$ *is* G*-equivariant, i.e.*

$$H_\xi \circ \psi_g = H_{g^{-1}\xi g}$$

for all $\xi \in \mathfrak{g}$ *and* $g \in \mathrm{G}$.

(ii) *The map* $\xi \mapsto H_\xi$ *is a Lie algebra homomorphism with respect to the Poisson structure on* $C^\infty(M,\mathbb{R})$, *i.e.*

$$\{H_\xi, H_\eta\} = H_{[\xi,\eta]}$$

for all $\xi, \eta \in \mathfrak{g}$.

Then (i) implies (ii). If G *is connected then (i) and (ii) are equivalent.*

Proof: We assume first that G is connected and prove that (ii) implies (i). Recall that $X_{g^{-1}\xi g} = \psi_g^* X_\xi$ and hence the functions $H_{g^{-1}\xi g}$ and $H_\xi \circ \psi_g$ generate the same Hamiltonian vector field. So their difference is constant and, by (ii), this implies that

$$\begin{aligned} H_{g^{-1}[\xi,\eta]g} &= \{H_{g^{-1}\xi g}, H_{g^{-1}\eta g}\} \\ &= \{H_\xi \circ \psi_g, H_\eta \circ \psi_g\} \\ &= \{H_\xi, H_\eta\} \circ \psi_g \\ &= H_{[\xi,\eta]} \circ \psi_g \end{aligned}$$

for $\xi, \eta \in \mathfrak{g}$ and $g \in \mathrm{G}$. Now let $g_1 \in \mathrm{G}$ and $\xi \in \mathfrak{g}$ be given. Since G is connected there exists a smooth path $g : [0,1] \to \mathrm{G}$ connecting $g(0) = \mathbb{1}$ to $g(1) = g_1$. Define $\eta(t) := \dot{g}(t)g(t)^{-1} \in \mathfrak{g}$. Then

$$\frac{d}{dt}\psi_g = X_\eta \circ \psi_g, \qquad \frac{d}{dt}g^{-1}\xi g = g^{-1}[\xi,\eta]g,$$

and hence

$$\begin{aligned}\frac{d}{dt}(H_\xi\circ\psi_g - H_{g^{-1}\xi g}) &= d(H_\xi\circ\psi_g)\psi_g^*X_\eta - H_{g^{-1}[\xi,\eta]g} \\ &= \omega(X_{H_\xi\circ\psi_g},\psi_g^*X_\eta) - H_{[\xi,\eta]}\circ\psi_g \\ &= \omega(X_{H_\xi\circ\psi_g},X_{H_\eta\circ\psi_g}) - \{H_\xi,H_\eta\}\circ\psi_g \\ &= \{H_\xi\circ\psi_g,H_\eta\circ\psi_g\} - \{H_\xi,H_\eta\}\circ\psi_g \\ &= 0.\end{aligned}$$

This implies $H_\xi\circ\psi_{g_1} = H_{g_1^{-1}\xi g_1}$. Thus we have proved that (ii) implies (i) when G is connected. That (i) implies (ii) follows by differentiating the identity

$$H_\xi\circ\psi_{\exp(t\eta)} = H_{\exp(-t\eta)\xi\exp(t\eta)}$$

at $t=0$. This proves Lemma 5.2.1. □

Exercise 5.2.2 Assume that the symplectic form ω is exact (and so M is not compact). Choose a 1-form λ such that $\omega=-d\lambda$. A symplectic action of a Lie group G on M is called **exact** if $\psi_g^*\lambda=\lambda$ for every $g\in\mathrm{G}$. Prove that every exact group action is Hamiltonian with

$$H_\xi := \iota(X_\xi)\lambda$$

for $\xi\in\mathfrak{g}$. □

In general, a weakly Hamiltonian group action (see page 203) need not be Hamiltonian. The following lemma shows that there is an obstruction in the form of a Lie algebra cocycle (cf. [298]).

Lemma 5.2.3 *Assume that the action is weakly Hamiltonian and choose a linear map $\mathfrak{g}\to C^\infty(M):\xi\mapsto H_\xi$ such that $X_\xi=X_{H_\xi}$ for all $\xi\in\mathfrak{g}$. Then there is a (unique) bilinear map $\tau:\mathfrak{g}\times\mathfrak{g}\to\mathbb{R}$ such that for all $\xi,\eta\in\mathfrak{g}$,*

$$\{H_\xi,H_\eta\} - H_{[\xi,\eta]} = \tau(\xi,\eta).$$

This function τ satisfies, for all $\xi,\eta,\zeta\in\mathfrak{g}$,

$$\tau([\xi,\eta],\zeta)+\tau([\eta,\zeta],\xi)+\tau([\zeta,\xi],\eta)=0. \tag{5.2.2}$$

Proof: The identity

$$X_{H_{[\xi,\eta]}} = X_{[\xi,\eta]} = [X_\xi,X_\eta] = [X_{H_\xi},X_{H_\eta}] = X_{\{H_\xi,H_\eta\}}$$

shows that the function $H_{[\xi,\eta]}-\{H_\xi,H_\eta\}$ is always constant. This implies

$$\{H_{[\xi,\eta]},H_\zeta\} = \{\{H_\xi,H_\eta\},H_\zeta\}.$$

Hence (5.2.2) follows from the Jacobi identity for the Poisson bracket. □

Exercise 5.2.4 Show that when G is abelian and M is compact the orbits of a weakly Hamiltonian action of G are isotropic, i.e. $\omega(X_\xi,X_\eta)=0$ for all $\xi,\eta\in\mathfrak{g}$. Give an example to show that this is not always true for symplectic actions of abelian groups. **Hint:** See the proof of Lemma 5.2.3 and note that the Poisson brackets $\{H_\xi,H_\eta\}$ must vanish somewhere when M is compact because they have zero mean (cf. Exercise 3.1.9). □

Condition (5.2.2) asserts that the bilinear map $\tau : \mathfrak{g} \times \mathfrak{g} \to \mathbb{R}$ is a Lie algebra **cocycle**. It is determined by the action of G up to a **coboundary**. Such a coboundary is a map of the form

$$\tau(\xi, \eta) = \sigma([\xi, \eta]),$$

where $\sigma : \mathfrak{g} \to \mathbb{R}$ is linear. Thus every weakly Hamiltonian action of G on M determines a class $[\tau] \in H^2(\mathfrak{g}; \mathbb{R})$ in the **Lie algebra cohomology**. This class vanishes whenever τ itself is a coboundary and, when G is connected, this is precisely the condition for the action of G to be Hamiltonian. In fact, if H_ξ and τ are as in Lemma 5.2.3, and $\sigma : \mathfrak{g} \to \mathbb{R}$ is a linear map such that $\tau(\xi, \eta) = \sigma([\xi, \eta])$ for all $\xi, \eta \in \mathfrak{g}$, then the map

$$\mathfrak{g} \to C^\infty(M) : \xi \mapsto H_\xi + \sigma(\xi)$$

is the required Lie algebra homomorphism. This homomorphism is uniquely determined up to an additive linear functional $\sigma \in \mathfrak{g}^*$ such that $\sigma([\xi, \eta]) = 0$ for all ξ, η. Such a linear functional determines a class in $H^1(\mathfrak{g}; \mathbb{R})$. To sum up, when G is connected and the Lie algebra cohomology class of τ vanishes, the action is Hamiltonian by Lemma 5.2.1.

The moment map

Now assume that the action of G on M is Hamiltonian. Then a **moment map** for the action is a G-equivariant smooth map

$$\mu : M \to \mathfrak{g}^*$$

such that the Hamiltonian vector fields associated to the Hamiltonian functions $H_\xi : M \to \mathbb{R}$, defined by

$$H_\xi(p) := \langle \mu(p), \xi \rangle,$$

generate the action. Here $\langle \cdot, \cdot \rangle$ denotes the pairing between $\mathfrak{g}^*$ and $\mathfrak{g}$. The moment map condition can be expressed in the form

$$\langle d\mu(p)v, \xi \rangle = \omega(X_\xi(p), v) \tag{5.2.3}$$

for all $\xi \in \mathfrak{g}$, $p \in M$, $v \in T_pM$. The equivariance condition asserts that

$$\mu(\psi_g(p)) = \mathrm{Ad}(g^{-1})^* \mu(p) =: g\mu(p)g^{-1} \tag{5.2.4}$$

for $p \in M$ and $g \in$ G. Equivalently, the map $\xi \mapsto H_\xi$ is G-equivariant. By Lemma 5.2.1, this implies that it is a Lie algebra homomorphism with respect to the Poisson bracket, i.e.

$$\langle \mu(p), [\xi, \eta] \rangle = \omega(X_\xi(p), X_\eta(p)) \tag{5.2.5}$$

for all $\xi, \eta \in \mathfrak{g}$ and $p \in M$.

For later reference it is convenient to rewrite the above identities in a slightly different form. For $p \in M$ let

$$I_p : T_pM \to T_p^*M$$

be the isomorphism induced by ω and let

$$L_p : \mathfrak{g} \to T_pM$$

be the infinitesimal action. Thus

$$I_p(v) := \omega_p(v, \cdot), \qquad L_p\xi := X_\xi(p) \tag{5.2.6}$$

for $v \in T_pM$ and $\xi \in \mathfrak{g}$.

Lemma 5.2.5 *Let* $\mathrm{G} \times M \to M : (g, p) \mapsto \psi_g(p)$ *be a Hamiltonian group action, generated by an equivariant moment map* $\mu : M \to \mathfrak{g}^*$, *and let* $p \in M$. *Then the following holds.*

(i) *The dual of the linear map* $d\mu(p) : T_pM \to \mathfrak{g}^*$ *is given by*

$$d\mu(p)^* = I_p \circ L_p : \mathfrak{g} \to T_p^*M. \tag{5.2.7}$$

(ii) *The symplectic complement of the kernel of* $d\mu(p)$ *is the tangent space of the group orbit* $\mathcal{O}(p) := \{\psi_g(p) \,|\, g \in \mathrm{G}\}$, *i.e.*

$$(\ker\, d\mu(p))^\omega = \mathrm{im} L_p. \tag{5.2.8}$$

(iii) *Let* $\xi \in \mathfrak{g}$ *and denote by* $\mathrm{ad}(\xi)^* : \mathfrak{g}^* \to \mathfrak{g}^*$ *the dual of the linear map* $\mathrm{ad}(\xi) : \mathfrak{g} \to \mathfrak{g}$ *given by* $\mathrm{ad}(\xi) := [\xi, \cdot]$. *Then*

$$d\mu(p)L_p\xi = -\mathrm{ad}(\xi)^*\mu(p). \tag{5.2.9}$$

Proof: Equation (5.2.7) follows directly from (5.2.3). Moreover, by (5.2.3),

$$\mathrm{im}\, L_p \subset (\ker\, d\mu(p))^\omega$$

and, by (5.2.7),

$$\dim \mathrm{im} L_p = \dim \mathrm{im} d\mu(p) = \mathrm{codim} \ker d\mu(p) = \dim\, (\ker\, d\mu(p))^\omega .$$

Hence $\mathrm{im}\, L_p = (\ker\, d\mu(p))^\omega$. By (5.2.3) and (5.2.5),

$$\langle d\mu(p)L_p\xi, \eta\rangle = \omega(L_p\eta, L_p\xi) = -\langle \mu(p), [\xi, \eta]\rangle = -\langle \mu(p), \mathrm{ad}(\xi)\eta\rangle$$

for all $p \in M$ and $\xi, \eta \in \mathfrak{g}$. This proves (5.2.9) and Lemma 5.2.5. □

In most of our examples G is a connected subgroup of the unitary group U(n), in which case its Lie algebra $\mathfrak{g}$ carries an invariant inner product defined by the trace via $\langle \xi, \eta\rangle = \mathrm{trace}(\xi^*\eta)$, where ξ^* denotes the conjugate transpose of ξ. Here invariance means that the adjoint action of G on $\mathfrak{g}$ is by isometries. In this case, we may identify $\mathfrak{g}^*$ with $\mathfrak{g}$ and $\mathrm{Ad}(g)^*$ with $\mathrm{Ad}(g^{-1})$, and the last expression in (5.2.4) can be interpreted literally as matrix multiplication.

Exercise 5.2.6 Prove that these definitions are consistent with those in Section 5.1 where G is the circle group S^1. □

From a dynamical systems point of view the significance of the moment map is as follows. If $H : M \to \mathbb{R}$ is a Hamiltonian function which is invariant under the action of the group G, then H is an integral for the Hamiltonian flow of H_ξ for every $\xi \in \mathfrak{g}$ and hence $\{H, H_\xi\} = 0$. This implies that the moment map is constant along the integral curves of the Hamiltonian flow generated by H. The angular momentum in $\mathbb{R}^3$ is an example of this, which is the origin of the term *moment map*.

5.3 Examples

Example 5.3.1 (Angular momentum) Consider the diagonal action of

$$\mathrm{G} = \mathrm{SO}(3)$$

on the phase space $\mathbb{R}^6 = \mathbb{R}^3 \times \mathbb{R}^3$ (with the standard symplectic structure) by

$$\psi_\Phi(x, y) := (\Phi x, \Phi y)$$

for $\Phi \in \mathrm{SO}(3)$. A simple calculation shows that this action is exact, and so is Hamiltonian by Exercise 5.2.2, and is generated by the functions

$$H_A(x, y) := \langle y, Ax \rangle$$

for $A = -A^{\mathrm{T}} \in \mathfrak{so}(3)$. To examine the moment map note that the Lie algebra $\mathfrak{so}(3)$ can be identified with $\mathbb{R}^3$ via the map

$$\mathbb{R}^3 \to \mathfrak{so}(3) : \xi \mapsto A_\xi := \begin{pmatrix} 0 & -\xi_3 & \xi_2 \\ \xi_3 & 0 & -\xi_1 \\ -\xi_2 & \xi_1 & 0 \end{pmatrix}. \tag{5.3.1}$$

Thus $A_\xi x = \xi \times x$ for $x, \xi \in \mathbb{R}^3$, and

$$[A_\xi, A_\eta] = A_{\xi \times \eta}, \qquad A_{\Phi\xi} = \Phi A_\xi \Phi^{-1}, \qquad \mathrm{trace}(A_\xi^{\mathrm{T}} A_\eta) = 2\langle \xi, \eta \rangle.$$

The last identity implies that the standard inner product on $\mathbb{R}^3$ induces an invariant inner product on $\mathfrak{so}(3)$ and the dual $\mathfrak{so}(3)^*$ can be identified with $\mathfrak{so}(3)$ via this inner product. With this notation the Hamiltonian function H_{A_ξ} can be written in the form

$$H_{A_\xi}(x, y) = \langle y, \xi \times x \rangle = \langle x \times y, \xi \rangle,$$

and hence the moment map $\mu : \mathbb{R}^3 \times \mathbb{R}^3 \to \mathfrak{so}(3)$ is given by

$$\mu(x, y) = A_{x \times y}.$$

If we think of x as the position and y as the momentum coordinate, then $x \times y$ is the **angular momentum**. Now if $H : \mathbb{R}^6 \to \mathbb{R}$ is any Hamiltonian function which depends only on $|x|$, $|y|$, and $\langle x, y \rangle$ (e.g. in the case of motion in a central force field), then the function H is invariant under the diagonal action of SO(3), and hence in this case the angular momentum determines three independent integrals of the motion. (See Example 1.1.7.) □

Exercise 5.3.2 There is a natural double cover $\mathrm{SU}(2) \to \mathrm{SO}(3)$. To see this, identify $\mathrm{SU}(2)$ with the unit quaternions $S^3 \subset \mathbb{R}^4 \cong \mathbb{H}$ via the map $S^3 \to \mathrm{SU}(2) : x \mapsto U_x$ defined by

$$U_x := \begin{pmatrix} x_0 + ix_1 & x_2 + ix_3 \\ -x_2 + ix_3 & x_0 - ix_1 \end{pmatrix}. \tag{5.3.2}$$

Now the unit quaternions act on the imaginary quaternions by conjugation. Define $q(x) := x_0 + ix_1 + jx_2 + kx_3$ and $q(\xi) := i\xi_1 + j\xi_2 + k\xi_3$ for $x \in S^3$ and $\xi \in \mathbb{R}^3$. Then the map $S^3 \to \mathrm{SO}(3) : x \mapsto \Phi_x$ is defined by

$$q(\Phi_x \xi) := q(x)q(\xi)\overline{q(x)}.$$

(i) Prove that the matrix $\Phi_x \in \mathrm{SO}(3)$ is given by the formula

$$\Phi_x = \begin{pmatrix} x_0^2 + x_1^2 - x_2^2 - x_3^2 & 2(x_1x_2 - x_0x_3) & 2(x_1x_3 + x_0x_2) \\ 2(x_2x_1 + x_0x_3) & x_0^2 - x_1^2 + x_2^2 - x_3^2 & 2(x_2x_3 - x_0x_1) \\ 2(x_3x_1 - x_0x_2) & 2(x_3x_2 + x_0x_1) & x_0^2 - x_1^2 - x_2^2 + x_3^2 \end{pmatrix} \tag{5.3.3}$$

for $x \in S^3$. Deduce that the map $\mathrm{SU}(2) \to \mathrm{SO}(3) : U_x \mapsto \Phi_x$ is a group homomorphism and a double cover.

(ii) Prove that the differential of the group homomorphism $U_x \mapsto \Phi_x$ is the map $\mathfrak{su}(2) \to \mathfrak{so}(3) : u_\xi \mapsto A_\xi$, where

$$u_\xi = \frac{1}{2}\begin{pmatrix} i\xi_1 & \xi_2 + i\xi_3 \\ -\xi_2 + i\xi_3 & i\xi_1 \end{pmatrix}$$

for $\xi \in \mathbb{R}^3$. Prove directly that the map $u_\xi \mapsto A_\xi$ is a Lie algebra homomorphism and identifies the two invariant inner products. **Hint:** Prove that

$$[u_\xi, u_{\xi'}] = u_{\xi \times \xi'}, \qquad \mathrm{trace}(u_\xi^* u_{\xi'}) = \tfrac{1}{2}\langle \xi, \xi' \rangle.$$

Moreover, prove that the map $\xi \mapsto A_\xi$ in Example 5.3.1 is half the differential of the map $x \mapsto \Phi_x$ at $x = (1, 0, 0, 0)$. (Note that the Lie bracket on the imaginary quaternions is twice the cross product on $\mathbb{R}^3$.) □

Example 5.3.3 Consider the unit sphere $S^2 \subset \mathbb{R}^3$ with the standard volume form $\omega = \mathrm{dvol}_{S^2} \in \Omega^2(S^2)$, given by

$$\omega_x(\widehat{x}_1, \widehat{x}_2) := \langle x \times \widehat{x}_1, \widehat{x}_2 \rangle.$$

The group $\mathrm{SO}(3)$ acts on S^2 by isometries and the infinitesimal action is given by the vector fields $X_\xi(x) = \xi \times x$ for $\xi \in \mathbb{R}^3 \cong \mathfrak{so}(s)$. Hence

$$\omega_x(X_\xi(x), \widehat{x}) = \omega_x(\xi \times x, \widehat{x}) = \omega_x(\xi, x \times \widehat{x}) = \langle \xi, \widehat{x} \rangle$$

for $\xi \in \mathbb{R}^3$ and $\widehat{x} \in T_x S^2$. This shows that X_ξ is the Hamiltonian vector field of the function $H_\xi : S^2 \to \mathbb{R}$ given by $H_\xi(x) := \langle \xi, x \rangle$. Hence a moment map for the action is the inclusion $\mu : S^2 \to \mathbb{R}^3$, i.e. $\mu(x) = x$. Here we identify $\mathbb{R}^3$ with the Lie algebra of $\mathrm{SO}(3)$ and its dual via the isomorphism (5.3.1). □

Exercise 5.3.4 Consider the diagonal action of SO(3) on the n-fold product $(S^2)^n$ via the isometries

$$\psi_\Phi(x_1,\dots,x_n) := (\Phi x_1,\dots,\Phi x_n)$$

for $\Phi \in \mathrm{SO}(3)$ and $x = (x_1,\dots,x_n) \in (S^2)^n$. Prove that this action is generated by the moment map $\mu : (S^2)^n \to \mathbb{R}^3$ given by

$$\mu(x_1,\dots,x_n) = x_1 + \cdots + x_n.$$

(See equation (5.3.1) for the isomorphism $\mathbb{R}^3 \cong \mathfrak{so}(3)$.) Prove that zero is a regular value of μ if and only if n is odd. **Hint:** An n-tuple $x = (x_1,\dots,x_n) \in (S^2)^n$ is a critical point of μ if and only if the vectors $x_1,\dots,x_n$ are collinear. □

Example 5.3.5 Let (V,ω) be a symplectic vector space. Then the symplectic linear group $\mathrm{G} := \mathrm{Sp}(V,\omega)$ acts on the space $\mathcal{J}(V,\omega)$ of ω-compatible linear complex structures by Kähler isometries (see Example 4.3.6). The Lie algebra

$$\mathfrak{g} := \mathfrak{sp}(V,\omega) := \mathrm{Lie}(\mathrm{Sp}(V,\omega))$$

is the space of endomorphisms $A : V \to V$ that satisfy $\omega(A\cdot,\cdot)+\omega(\cdot,A\cdot) = 0$. The infinitesimal action of $\mathfrak{g}$ on $\mathcal{J}(V,\omega)$ is given by the vector fields X_A on $\mathcal{J}(V,\omega)$ defined by

$$X_A(J) := [A,J] \in T_J\mathcal{J}(V,\omega)$$

for $A \in \mathfrak{sp}(V,\omega)$ and $J \in \mathcal{J}(V,\omega)$. They satisfy

$$\Omega(X_A(J),\widehat{J}) = \tfrac{1}{2}\mathrm{trace}(X_A(J)J\widehat{J}) = \tfrac{1}{2}\mathrm{trace}([A,J]J\widehat{J}) = -\mathrm{trace}(A\widehat{J}).$$

Thus X_A is the Hamiltonian vector field of the function $H_A : \mathcal{J}(V,\omega) \to \mathbb{R}$ given by $H_A(J) := -\mathrm{trace}(AJ)$. Hence a moment map for the action of $\mathrm{Sp}(V,\omega)$ on $\mathcal{J}(V,\omega)$ is the function $\mu : \mathcal{J}(V,\omega) \to \mathfrak{sp}(V,\omega)^*$ given by

$$\langle \mu(J), A\rangle := -\mathrm{trace}(JA)$$

for $J \in \mathcal{J}(V,\omega)$ and $A \in \mathfrak{sp}(V,\omega)$. □

Exercise 5.3.6 Let (V,ω) be a symplectic vector space and $\mathrm{G} := \mathrm{Sp}(V,\omega)$ as in Example 5.3.5. Prove that the obvious action of $\mathrm{Sp}(V,\omega)$ on V is generated by the moment map $\mu : V \to \mathfrak{sp}(V,\omega)^*$, given by

$$\langle \mu(v), A\rangle = \tfrac{1}{2}\omega(Av,v)$$

for $v \in V$ and $A \in \mathfrak{sp}(V,\omega)$. □

In Example 5.3.5 and Exercise 5.3.6 there does not exist an invariant inner product[23] on the Lie algebra $\mathfrak{g} = \mathfrak{sp}(V,\omega)$. Hence the moment map in these examples cannot be written as a map with values in the Lie algebra, but only as a map with values in the dual of the Lie algebra. In Exercise 5.3.6 it is interesting to fix an almost complex structure $J \in \mathcal{J}(V,\omega)$ and restrict the action to the unitary subgroup $\mathrm{U}(V,\omega,J) = \mathrm{Sp}(V,\omega)\cap\mathrm{GL}(V,J)$. For the standard triple $(V,\omega,J) = (\mathbb{R}^{2n},\omega_0,J_0)$ this is the content of the next example.

[23] Note that by definition inner products are assumed to be positive definite.

Example 5.3.7 Consider the obvious action of $\mathrm{G} = \mathrm{U}(n)$ on $\mathbb{C}^n = \mathbb{R}^{2n}$ with the standard symplectic structure ω_0. Denote the elements of G by $U = X + iY$. Then U acts on $\mathbb{R}^{2n}$ by the linear symplectomorphism

$$\Psi_U = \begin{pmatrix} X & -Y \\ Y & X \end{pmatrix}.$$

The Lie algebra $\mathfrak{u}(n)$ consists of skew Hermitian matrices $\zeta = \xi + i\eta$, where $\xi = -\xi^{\mathrm{T}} \in \mathbb{R}^{n\times n}$ and $\eta = \eta^{\mathrm{T}} \in \mathbb{R}^{n\times n}$. The infinitesimal action is generated by the Hamiltonian functions

$$H_\zeta(z) = -\tfrac{1}{2}\langle x, \eta x\rangle + \langle y, \xi x\rangle - \tfrac{1}{2}\langle y, \eta y\rangle = \tfrac{i}{2}\, z^*\zeta z.$$

This defines a Lie algebra homomorphism $\mathfrak{u}(n) \to C^\infty(\mathbb{C}^n) : \zeta \mapsto H_\zeta$. The moment map $\mu : \mathbb{C}^n \to \mathfrak{u}(n)$ is related to the Hamiltonian functions H_ζ for $\zeta \in \mathfrak{u}(n)$ by the formula $H_\zeta(z) = \mathrm{tr}\,(\mu(z)^*\zeta)$. Hence

$$\mu(z) = -\tfrac{i}{2} zz^* \tag{5.3.4}$$

for $z \in \mathbb{C}^n$. Here we identify the Lie algebra $\mathfrak{u}(n)$ with its dual via the inner product $\langle \zeta_1, \zeta_2\rangle = \mathrm{trace}(\zeta_1^*\zeta_2)$. The explicit formula (5.3.4) shows that

$$\mu(Uz) = U\mu(z)U^{-1}$$

for $z \in \mathbb{C}^n$ and $U \in \mathrm{U}(n)$. □

Exercise 5.3.8 Show that the obvious action of $\mathrm{U}(n)$ on $(\mathbb{C}\mathrm{P}^{n-1}, \omega_{\mathrm{FS}})$ is Hamiltonian, and that the formula

$$\langle \mu([z]), \xi\rangle := \frac{\langle z, i\xi z\rangle}{2|z|^2}$$

for $z \in \mathbb{C}^n \setminus \{0\}$ and $\xi \in \mathfrak{u}(n)$ defines a moment map for the action. Here $\langle \cdot, \cdot\rangle$ denotes the standard real inner product on $\mathbb{C}^n$. **Hint:** Use Exercise 5.1.3 and Example 5.3.7. □

Before discussing the next example (the action of G on its cotangent bundle $T^*\mathrm{G}$ by right translation), we introduce some notation. For every $g \in \mathrm{G}$ right multiplication by g determines a diffeomorphism $R_g : \mathrm{G} \to \mathrm{G}$ defined by

$$R_g(h) := hg.$$

By a slight abuse of notation, the linearized map at the identity will also be denoted by $R_g = dR_g(\mathbb{1}) : \mathfrak{g} \to T_g\mathrm{G}$. It is defined by

$$R_g\eta := \eta g = \left.\frac{d}{dt}\right|_{t=0} \exp(t\eta)g$$

for $\eta \in \mathfrak{g}$. In the case where $\mathrm{G} \subset \mathrm{U}(n)$ is a matrix group the notation ηg can be understood as matrix multiplication. The diffeomorphism $L_h : \mathrm{G} \to \mathrm{G}$ and its linearization $L_h : \mathfrak{g} \to T_h\mathrm{G}$ are defined analogously by left multiplication with h.

Example 5.3.9 Consider the action of a Lie group G on its cotangent bundle $T^*\mathrm{G}$ which is induced by the right translations $R_{g^{-1}}$. In view of Exercise 3.1.21 the corresponding symplectomorphism $\psi_g : T^*\mathrm{G} \to T^*\mathrm{G}$ is given by

$$\psi_g(h, v^*) = (hg^{-1}, dR_g(hg^{-1})^* v^*)$$

for $v^* \in T_h^*\mathrm{G}$. Now for every $\xi \in \mathfrak{g}$ the 1-parameter group $t \mapsto R_{\exp(-t\xi)}$ of diffeomorphisms of G is generated by the vector field

$$\mathrm{G} \to T\mathrm{G} : h \mapsto -L_h\xi = -h\xi,$$

and it follows again from Exercise 3.1.21 that the flow $t \mapsto \psi_{\exp(t\xi)}$ is generated by the Hamiltonian function

$$H_\xi(h, v^*) = -\langle v^*, L_h\xi\rangle = -\langle L_h^* v^*, \xi\rangle.$$

In the second expression $\langle\cdot,\cdot\rangle$ denotes the pairing between $T_h^*\mathrm{G}$ and $T_h\mathrm{G}$, while in the last expression it denotes the pairing between $\mathfrak{g}^*$ and $\mathfrak{g}$. The reader may check that $H_\xi = \iota(X_\xi)\lambda_{\mathrm{can}}$ and, by Exercise 3.1.21,

$$\psi_g^*\lambda_{\mathrm{can}} = \lambda_{\mathrm{can}}.$$

Hence it follows from Exercise 5.2.2 that the action is Hamiltonian. The moment map $\mu : T^*\mathrm{G} \to \mathfrak{g}^*$ is given by

$$\mu(h, v^*) = -L_h^* v^*.$$

The reader may check directly that this map satisfies condition (5.2.4).

This example is easier to understand if we trivialize the bundle $T^*\mathrm{G}$. The above formulas suggest that we should identify the cotangent space $T_h^*\mathrm{G}$ with $\mathfrak{g}^*$ by means of the left invariant forms on G. In other words, we shall use the diffeomorphism

$$T^*\mathrm{G} \to \mathrm{G} \times \mathfrak{g}^* : (h, v^*) \mapsto (h, L_h^* v^*). \tag{5.3.5}$$

Then the action of G on $\mathrm{G} \times \mathfrak{g}^*$ is given by

$$\psi_g(h, \eta) = (hg^{-1}, Ad(g^{-1})^*\eta).$$

This action is generated by the Hamiltonian functions $H_\xi : \mathrm{G} \times \mathfrak{g}^* \to \mathbb{R}$, where

$$H_\xi(h, \eta) = -\langle \eta, \xi\rangle.$$

Hence the moment map $\mu : \mathrm{G} \times \mathfrak{g}^* \to \mathfrak{g}^*$ is just minus the projection onto the second component. This map obviously satisfies (5.2.4). □

Exercise 5.3.10 Identify the tangent space $T_h\mathrm{G}$ with the Lie algebra $\mathfrak{g}$ by means of left translations $T_h\mathrm{G} \to \mathfrak{g} : \widehat{h} \mapsto L_h^{-1}\widehat{h} = h^{-1}\widehat{h}$. Prove that the canonical 1-form λ_{can} on $T^*\mathrm{G}$ is the pullback, under the diffeomorphism $T^*\mathrm{G} \to \mathrm{G} \times \mathfrak{g}^*$ in (5.3.5), of the 1-form λ on $\mathrm{G} \times \mathfrak{g}^*$ given by

$$\lambda_{(h,\eta)}(\widehat{h}, \widehat{\eta}) = \langle \eta, h^{-1}\widehat{h}\rangle$$

for $h \in \mathrm{G}$, $\widehat{h} \in T_h\mathrm{G}$, and $\eta, \widehat{\eta} \in \mathfrak{g}^*$. Prove the identity $H_\xi = \iota(X_\xi)\lambda$ in Example 5.3.9. Check that the moment map satisfies (5.2.4). □

Example 5.3.11 (Coadjoint orbits) Let G be a connected Lie group and let $\mathcal{O} \subset \mathfrak{g}^*$ be an orbit under the coadjoint action of G. Then $\mathcal{O}$ carries a natural symplectic structure. To see this, note that the tangent space to $\mathcal{O}$ at η is given by

$$T_\eta\mathcal{O} = \{\mathrm{ad}(\xi)^*\eta \,|\, \xi \in \mathfrak{g}\},$$

where $\mathrm{ad}(\xi) : \mathfrak{g} \to \mathfrak{g}$ denotes the linear map $\mathrm{ad}(\xi)\xi' = [\xi, \xi']$. The symplectic structure on $\mathcal{O}$ is given by

$$\omega_\eta(\mathrm{ad}(\xi)^*\eta, \mathrm{ad}(\xi')^*\eta) = \langle \eta, [\xi, \xi'] \rangle \tag{5.3.6}$$

for $\xi, \xi' \in \mathfrak{g}$. Thus the dual Lie algebra $\mathfrak{g}^*$ is a union of symplectic manifolds. It is not itself a symplectic manifold, but is a Poisson manifold. This means that its function ring $C^\infty(\mathfrak{g}^*, \mathbb{R})$ carries a Lie algebra structure defined by a Poisson bracket. The Lie algebra $\mathfrak{g}$ embeds into $C^\infty(\mathfrak{g}^*, \mathbb{R})$ as the subspace of linear functions, and the restriction of this Poisson bracket to $\mathfrak{g}$ is just the standard Lie bracket. (For more details on Poisson manifolds see [676], for example.)

Now the group G acts on $\mathcal{O}$ by the symplectomorphisms

$$\psi_g(\eta) = \mathrm{Ad}(g^{-1})^*\eta,$$

and the infinitesimal action is given by the symplectic vector fields

$$X_\xi(\eta) = -\mathrm{ad}(\xi)^*\eta$$

with corresponding Hamiltonian functions

$$H_\xi(\eta) = \langle \eta, \xi \rangle.$$

Hence the moment map $\mu : \mathcal{O} \to \mathfrak{g}^*$ is the natural inclusion. □

Exercise 5.3.12 Show that Example 5.3.3 is a special case of Example 5.3.11. □

Exercise 5.3.13 Prove that the 2-form ω on $\mathcal{O}$ defined by (5.3.6) is closed. Prove that $X_\xi(\eta) = -\mathrm{ad}(\xi)^*\eta$ is the Hamiltonian vector field generated by $H_\xi(\eta) = \langle \eta, \xi \rangle$. Prove that the action of G on $\mathcal{O}$ is Hamiltonian. □

Exercise 5.3.14 For every $\eta \in \mathcal{O}$ there is a natural diffeomorphism

$$\mathrm{G}/\mathrm{G}_\eta \cong \mathcal{O}, \qquad \mathrm{G}_\eta := \{g \in \mathrm{G} \,|\, \mathrm{Ad}(g)^*\eta = \eta\},$$

induced by the map $g \mapsto \mathrm{Ad}(g^{-1})^*\eta$. The Lie algebra of G_η is given by

$$\mathfrak{g}_\eta := \{\xi \in \mathfrak{g} \,|\, \mathrm{ad}(\xi)^*\eta = 0\}.$$

Prove that $\mathfrak{g}_\eta$ is the kernel of the skew form

$$\mathfrak{g} \times \mathfrak{g} \to \mathbb{R} : (\xi, \xi') \mapsto \langle \eta, [\xi, \xi'] \rangle.$$

Give a direct proof that this 2-form determines a symplectic structure on $\mathrm{G}/\mathrm{G}_\eta$. □

Exercise 5.3.15 Show that a symplectic action of a connected semi-simple group is always Hamiltonian. **Hint:** Use the fact that in this case the Lie algebra cohomology groups $H^1(\mathfrak{g};\mathbb{R})$ and $H^2(\mathfrak{g};\mathbb{R})$ both vanish. See [590] for more details. □

Exercise 5.3.16 Suppose that G acts in a Hamiltonian way on the symplectic manifolds (M_j, ω_j) for $j = 1, 2$ with moment maps $\mu_j : M_j \to \mathfrak{g}^*$. Prove that the diagonal action $\mathrm{G} \to \mathrm{Symp}(M_1 \times M_2)$ is Hamiltonian, with moment map $\mu : M_1 \times M_2 \to \mathfrak{g}^*$ given by

$$\mu(p_1, p_2) = \mu_1(p_1) + \mu_2(p_2)$$

for $p_j \in M_j$. □

Exercise 5.3.17 Find a moment map $\mu_n : \mathbb{C}^n \to \mathbb{R}^n$ of the action of the n-torus $\mathbb{T}^n = \mathbb{R}^n/\mathbb{Z}^n$ on $\mathbb{C}^n$ given by

$$(\theta_1, \ldots, \theta_n) \cdot (z_1, \ldots, z_n) = (e^{2\pi i\theta_1} z_1, \ldots, e^{2\pi i\theta_n} z_n).$$

If $\iota : \mathbb{T}^k \to \mathbb{T}^n$ is a linear embedding and $\pi : \mathbb{R}^n \to \mathbb{R}^k$ is the dual projection, show that

$$\mu_k := \pi \circ \mu_n : \mathbb{C}^n \to \mathbb{R}^k$$

is a moment map for the induced action of $\mathbb{T}^k$. □

Infinite-dimensional examples

There are many infinite-dimensional analogues of Hamiltonian group actions that arise naturally in various geometric settings. Three model examples are discussed below. The first example concerns the action of the group of gauge transformations on the space of connections over a closed oriented 2-manifold and is related to flat connections. The second example concerns the action of the symplectomorphism group on the space of compatible complex structures and is related to Kähler–Einstein metrics. The third example concerns the action of the group of volume preserving diffeomorphisms on the space of embeddings and is related to Lagrangian submanifolds.

Example 5.3.18 (Flat connections over Riemann surfaces) This example is due to Atiyah and Bott [27]. Let Σ be a compact Riemann surface and G be a compact Lie group with Lie algebra $\mathfrak{g} = \mathrm{Lie}(\mathrm{G})$. Consider the space

$$\mathcal{A} := \Omega^1(\Sigma, \mathfrak{g})$$

of Lie algebra valued 1-forms on Σ. Think of such 1-forms $A \in \mathcal{A}$ as connections on the trivial bundle $\Sigma \times \mathrm{G}$. Then $\mathcal{A}$ is an infinite-dimensional Kähler manifold with symplectic form and complex structure given by

$$\Omega(\alpha, \beta) := \int_\Sigma \langle \alpha \wedge \beta \rangle, \qquad \alpha \mapsto *\alpha$$

for two infinitesimal connections $\alpha, \beta \in \Omega^1(\Sigma, \mathfrak{g}) = T_A\mathcal{A}$. Here $\langle \cdot, \cdot \rangle$ denotes an invariant inner product on the Lie algebra $\mathfrak{g}$ and $*$ denotes the Hodge $*$-operator on Σ.

The group

$$\mathcal{G} := \mathrm{Map}(\Sigma, \mathrm{G})$$

of gauge transformations acts contravariantly on the space $\mathcal{A}$ by

$$g^*A := g^{-1}dg + g^{-1}Ag$$

for $A \in \mathcal{A}$ and $g \in \mathcal{G}$. This action preserves the Kähler structure on $\mathcal{A}$ and is Hamiltonian. For every $\xi \in \Omega^0(\Sigma, \mathfrak{g}) = \mathrm{Lie}(\mathcal{G})$ the infinitesimal (covariant) action is given by the vector field $\mathcal{A} \to T\mathcal{A} : A \mapsto -d_A\xi$, where

$$d_A\xi := d\xi + [A, \xi]$$

denotes the covariant derivative of ξ. One checks easily that for each fixed ξ this is a Hamiltonian vector field with Hamiltonian function

$$\mathcal{A} \to \mathbb{R} : A \mapsto \int_\Sigma \langle F_A, \xi \rangle, \qquad F_A := dA + \tfrac{1}{2}[A \wedge A].$$

Here $F_A \in \Omega^2(\Sigma, \mathfrak{g})$ denotes the curvature of A. One can formally think of $\Omega^2(\Sigma, \mathfrak{g})$ as a kind of dual space of the Lie algebra $\Omega^0(\Sigma, \mathfrak{g}) = \mathrm{Lie}(\mathcal{G})$ via the pairing $\langle \xi, F \rangle := \int_\Sigma \langle \xi \wedge F \rangle$ for $\xi \in \Omega^0(\Sigma, \mathfrak{g})$ and $F \in \Omega^2(\Sigma, \mathfrak{g})$. It then follows that a moment map is given by the curvature

$$\mathcal{A} \to \Omega^2(\Sigma, \mathfrak{g}) : A \mapsto F_A.$$

Thus the zero set of the moment map is the space of flat connections on the trivial bundle $\Sigma \times \mathrm{G}$. This example generalizes naturally to nontrivial bundles as well as higher-dimensional Kähler manifolds (in place of Riemann surfaces). It also generalizes to product spaces $\mathcal{A} \times C^\infty(\Sigma, E)$ consisting of pairs (A, s), where A is a connection and s is a section of an associated vector bundle. If one considers general fibre bundles associated to Hamiltonian group actions on symplectic manifolds, one obtains the *symplectic vortex equations* of Cieliebak–Gaio–Salamon [114, 115] and Mundet-i-Riera [498]. □

Example 5.3.19 (Scalar curvature) Suppose that (M, ω) is a closed symplectic $2n$-manifold. Then the space $\mathcal{J}(M, \omega)$ of all almost complex structures on M that are compatible with ω is itself an infinite-dimensional analogue of a Kähler manifold. Its tangent space at J is the space of all smooth sections $\widehat{J} \in \Omega^0(M, \mathrm{End}(TM))$ of the endomorphism bundle that satisfy

$$\widehat{J}J + J\widehat{J} = 0, \qquad \omega(\cdot, \widehat{J}\cdot) + \omega(\widehat{J}\cdot, \cdot) = 0.$$

The metric and symplectic form on $T_J\mathcal{J}(M, \omega)$ are given by

$$\|\widehat{J}\|^2 := \tfrac{1}{2}\int_M \mathrm{trace}(\widehat{J}^2)\frac{\omega^n}{n!}, \qquad \Omega(\widehat{J}_1, \widehat{J}_2) := \tfrac{1}{2}\int_M \mathrm{trace}(\widehat{J}_1 J \widehat{J}_2)\frac{\omega^n}{n!},$$

and the complex structure is $\widehat{J} \mapsto -J\widehat{J}$ (see Example 4.3.6). The group

$$\mathcal{G} := \mathrm{Symp}(M, \omega)$$

of symplectomorphisms acts on $\mathcal{J}(M, \omega)$ by pullback $J \mapsto \phi^*J$ and this action preserves the Kähler structure.

Let us now assume that the space $\mathcal{J}_{\mathrm{int}}(M,\omega)$ of integrable complex structures compatible with ω is nonempty. Thus (M,ω,J) is a Kähler manifold for every $J \in \mathcal{J}_{\mathrm{int}}(M,\omega)$. Denote by

$$\mathcal{G}^{\mathrm{ex}} := \mathrm{Ham}(M,\omega) \subset \mathcal{G}$$

the normal subgroup of Hamiltonian symplectomorphisms and identify the Lie algebra of $\mathcal{G}^{\mathrm{ex}}$ with the space $C_0^\infty(M)$ of smooth functions with mean value zero. The L^2 inner product on this space is $\mathcal{G}$-invariant and can be used to identify its dual space with the quotient $C^\infty(M)/\mathbb{R}$. It turns out that the action of $\mathcal{G}^{\mathrm{ex}}$ on $\mathcal{J}_0(M,\omega)$ is Hamiltonian, and that a moment map for this action is the function

$$\mathcal{J}_{\mathrm{int}}(M,\omega) \to C^\infty(M)/\mathbb{R} : J \mapsto s_J$$

that assigns to each complex structure $J \in \mathcal{J}_{\mathrm{int}}(M,\omega)$ the scalar curvature s_J of the corresponding Kähler metric $g_J := \omega(\cdot, J\cdot)$. Explicitly this means that

$$\int_M H\dot{s}\omega^n = \tfrac{1}{2}\int_M \mathrm{trace}(J(\mathcal{L}_{X_H}J)\dot{J})\omega^n \tag{5.3.7}$$

for every Hamiltonian function $H : M \to \mathbb{R}$ and every smooth curve of complex structures $\mathbb{R} \to \mathcal{J}_{\mathrm{int}}(M,\omega) : t \mapsto J(t)$, where $\dot{J} := \partial_t J$ and $\dot{s} := \partial_t s_J$. The moment map is $\mathcal{G}$-equivariant and its zero set consists of all complex structures $J \in \mathcal{J}_{\mathrm{int}}(M,\omega)$ for which the metric g_J has constant scalar curvature.

That the scalar curvature can be interpreted as an infinite-dimensional analogue of the moment map was discovered, independently, by Fujiki [244] and Donaldson [147]. Donaldson's paper contains many interesting observations in which this interpretation of the scalar curvature as the moment map forms the starting point for a new *symplectic approach* to Kähler geometry. In [151, 152], Donaldson formulated a conjecture relating Tian's notion of K-stability to the existence of a constant scalar curvature Kähler metric, refining earlier conjectures by Yau [692] and Tian [626]. Yau's conjecture applies to Fano manifolds and relates K-stability to the existence of Kähler–Einstein metrics. It was confirmed by Chen–Donaldson–Sun [105, 106, 107, 108]. The Yau–Tian–Donaldson conjecture in the nonFano case, i.e. where the first Chern class is not a multiple of the Kähler class, is still open.

For toric manifolds, the Donaldson–Fujiki moment map picture outlined above is closely related to Abreu's equation [9], which has since been widely studied. This moment map picture was also used by Abreu–Granja–Kitchloo [10] to study the topology of the symplectomorphism groups of rational ruled surfaces. For general symplectic manifolds of finite volume, the Donaldson–Fujiki moment map was used in the almost complex setting by Shelukhin [580] in his construction of nontrivial quasimorphisms (see page 411) on the universal cover of the group of Hamiltonian symplectomorphisms. □

Example 5.3.20 (Lagrangian embeddings) This example is due to Donaldson [149, 153]. Let (M,ω) be a symplectic $2n$-manifold and S be a closed n-manifold equipped with a volume form $\sigma \in \Omega^n(S)$ with $\int_S \sigma = 1$. Define

$$\mathcal{F} := \{f : S \to M \,|\, f \text{ is an embedding and } f^*\omega \text{ is exact}\}.$$

This space is a Fréchet manifold with tangent spaces $T_f\mathcal{F} = \Omega^0(S, f^*TM)$ at $f \in \mathcal{F}$. It carries a natural symplectic structure given by

$$\Omega_f(\widehat{f}_1, \widehat{f}_2) := \int_S \omega(\widehat{f}_1, \widehat{f}_2)\sigma, \qquad \widehat{f}_1, \widehat{f}_2 \in T_f\mathcal{F}.$$

The group $\mathcal{G} := \mathrm{Diff}(S,\sigma)$ of volume preserving diffeomorphisms of S acts contravariantly on $\mathcal{F}$ via $\mathcal{G} \times \mathcal{F} \to \mathcal{F} : (f,\phi) \mapsto f \circ \phi$ and the action preserves the symplectic form Ω.

The Lie algebra of $\mathcal{G}$ is the space

$$\mathcal{X}(S,\sigma) := \{\xi \in \mathcal{X}(S) \,|\, \mathcal{L}_\xi\sigma = d\iota(\xi)\sigma = 0\}$$

of divergence-free vector fields on S, and the covariant infinitesimal action of $\mathcal{X}(S,\sigma)$ on $\mathcal{F}$ is given by the vector fields $f \mapsto -df \circ \xi$ on $\mathcal{F}$. Contracting such a vector field with the symplectic form Ω on $\mathcal{F}$, we obtain the 1-form

$$T_f\mathcal{F} \to \mathbb{R} : \widehat{f} \mapsto \Omega_f(-df \circ \xi, \widehat{f}) = \int_S \omega(\widehat{f}, df\cdot) \wedge \iota(\xi)\sigma. \tag{5.3.8}$$

This 1-form is exact whenever ω is exact or $H^{n-1}(S) = 0$. When these conditions are not satisfied one still obtains an exact 1-form, if one replaces the group $\mathcal{G}$ of volume preserving diffeomorphisms by the subgroup

$$\mathcal{G}^{\mathrm{ex}} \subset \mathcal{G}$$

of **exact volume preserving diffeomorphisms** (see Donaldson [153]). Its Lie algebra is the subspace

$$\mathcal{X}^{\mathrm{ex}}(S,\sigma) := \{\xi \in \mathcal{X}(S) \,|\, \iota(\xi)\sigma \in \mathrm{im}\, d\}$$

of all **exact divergence-free vector fields** ξ on S, i.e. those where the $(n-1)$-form $\iota(\xi)\sigma$ is exact. The subgroup $\mathcal{G}^{\mathrm{ex}}$ is defined as the group of all diffeomorphisms $\phi : S \to S$ that are generated by a smooth 1-parameter family of vector fields $\xi_t \in \mathcal{X}^{\mathrm{ex}}(S,\sigma)$, $0 \le t \le 1$, via the differential equation

$$\partial_t\phi_t = \xi_t \circ \phi_t, \qquad \phi_0 = \mathrm{id}, \qquad \phi_1 = \phi, \qquad \iota(\xi_t)\sigma = d\tau_t.$$

The group $\mathcal{G}^{\mathrm{ex}}$ is a connected normal subgroup of the identity component $\mathcal{G}_0$ of $\mathcal{G}$, just as $\mathrm{Ham}(M,\omega)$ is a normal subgroup of $\mathrm{Symp}_0(M,\omega)$ (see Exercise 3.1.14). In fact, if $n = 2$ then $\mathcal{G}^{\mathrm{ex}} = \mathrm{Ham}(S,\sigma)$. If $H^{n-1}(S;\mathbb{R}) = 0$ then $\mathcal{G}^{\mathrm{ex}} = \mathcal{G}_0$.

The dual space of the Lie algebra $\mathcal{X}^{\mathrm{ex}}(S,\sigma)$ of exact divergence-free vector fields can formally (i.e. heuristically) be identified with the space of exact 2-forms on S as follows. Every exact 2-form $\beta \in \Omega^2(S)$ determines a linear functional

$$\mathcal{X}^{\mathrm{ex}}(S,\sigma) \to \mathbb{R} : \xi \mapsto \int_S \beta \wedge \tau_\xi, \qquad d\tau_\xi = \iota(\xi)\sigma. \tag{5.3.9}$$

Here τ_ξ is an $(n-2)$-form chosen such that its differential agrees with the exact $(n-1)$-form $\iota(\xi)\sigma$. It is not uniquely determined by ξ. However, any two choices of τ_ξ differ by a closed $(n-2)$-form, and the integral in (5.3.9) is independent of the choice, because β is exact (in fact if and only if β is exact). In other words, every exact 2-form determines a linear functional on the space $\Omega^{n-2}(S)/\ker d$. As pointed out by Donaldson [153], one should more precisely consider the space of all currents on $\Omega^{n-2}(S)$ that vanish on the closed forms. Fortunately, however, the moment map will take values in the space of exact 2-forms, so currents are not needed.

It follows from Cartan's formula that for every $\xi \in \mathcal{X}^{\mathrm{ex}}(S,\sigma)$, the 1-form in (5.3.8) is the differential of the function $H_\xi : \mathcal{F} \to \mathbb{R}$ given by

$$H_\xi(f) := \int_S f^*\omega \wedge \tau_\xi, \qquad d\tau_\xi = \iota(\xi)\sigma. \tag{5.3.10}$$

Hence the action of $\mathcal{G}^{\mathrm{ex}}$ on $\mathcal{F}$ is a Hamiltonian group action and is generated by the moment map $\mu^{\mathrm{ex}} : \mathcal{F} \to \mathcal{X}^{\mathrm{ex}}(S,\sigma)^* \cong \left\{\beta \in \Omega^2(S) \,|\, \beta \in \operatorname{im} d\right\}$, given by

$$\mu^{\mathrm{ex}}(f) := f^*\omega.$$

The moment map is equivariant under the action of the full group $\mathcal{G}$ in that $\mu^{\mathrm{ex}}(f \circ \phi) = \phi^* f^*\omega = \phi^*\mu^{\mathrm{ex}}(f)$ for all $f \in \mathcal{F}$ and all $\phi \in \mathcal{G}$. The upshot is that the moment map vanishes if and only if $f^*\omega = 0$, i.e. f is a Lagrangian embedding. Dividing the zero set of the moment map by the group action of $\mathcal{G}^{\mathrm{ex}}$, we obtain the symplectic quotient (see Section 5.4 below)

$$\mathcal{F}/\!\!/\mathcal{G}^{\mathrm{ex}} = \frac{\{f : S \to M \,|\, f \text{ is a Lagrangian embedding}\}}{\mathcal{G}^{\mathrm{ex}}}.$$

If one divides the zero set of the moment map instead by the full group $\mathcal{G}$, then one obtains the space $\mathscr{L}$ of pairs (L,ρ), where $L = f(S) \subset M$ is a Lagrangian submanifold diffeomorphic to S and $\rho = (f^{-1})^*\sigma$ is a volume form on L with volume 1. In the case $H^{n-1}(S;\mathbb{R}) = 0$, where $\mathcal{G}^{\mathrm{ex}} = \mathcal{G}_0$, the quotient $\mathcal{F}/\!\!/\mathcal{G}_0$ is a covering of $\mathscr{L}$ equipped with an action of the mapping class group $\mathrm{MCG}(S) := \mathcal{G}/\mathcal{G}_0$, so that the quotient of $\mathcal{F}/\!\!/\mathcal{G}_0$ by $\mathrm{MCG}(S)$ is isomorphic to $\mathscr{L}$. In general the symplectic quotient $\mathcal{F}/\!\!/\mathcal{G}^{\mathrm{ex}}$ is a fibre bundle over the aforementioned covering space, whose fibre is isomorphic to the group $\mathcal{G}_0/\mathcal{G}^{\mathrm{ex}} \cong H^{n-1}(S;\mathbb{R})/\Gamma_\sigma$, where Γ_σ is the image of the homomorphism $\pi_1(\mathcal{G}) \to H^{n-1}(S;\mathbb{R}) : \{\phi_t\} \mapsto \mathrm{PD}(\{\phi_t(\mathrm{pt}\})$. (See Exercise 10.2.23.) For interesting variants of this setup and its relation to the Weil–Peterson metric on the moduli space of Riemann surfaces, as well as hyperKähler extensions thereof, we refer to Donaldson's beautiful paper [153].

Now assume that $\omega = -d\lambda$ is exact, and think of the formal dual space of $\mathcal{X}(S,\sigma)$ (isomorphic to the space of closed $(n-1)$-forms via $\xi \mapsto \iota(\xi)\sigma$) as the quotient space $\Omega^1(S)/\mathrm{im}\, d$. It follows from Cartan's formula that for every $\xi \in \mathcal{X}(S,\sigma)$, the 1-form in (5.3.8) is the differential of the function $H_\xi : \mathcal{F} \to \mathbb{R}$ given by

$$H_\xi(f) := -\int_S f^*\lambda \wedge \iota(\xi)\sigma. \tag{5.3.11}$$

Thus the $\mathcal{G}$-action on $\mathcal{F}$ is Hamiltonian and the moment map $\mu : \mathcal{F} \to \Omega^1(S)/\mathrm{im}\, d$ assigns to an embedding f the equivalence class $\mu(f) := [-f^*\lambda] \in \Omega^1(S)/\mathrm{im}\, d$. This moment map is again $\mathcal{G}$-equivariant. An element $f \in \mathcal{F}$ belongs to the zero set of this moment map if and only if $f^*\lambda$ is exact. Thus the zero set of the moment map consists of the exact Lagrangian embeddings $f : S \to M$ (see Definition 9.3.5). In this case the symplectic quotient $\mathcal{F}/\!/\mathcal{G}$ is isomorphic to the space $\mathscr{L}^{\mathrm{ex}}$ of all pairs (L,ρ), where $L = f(S)$ is an exact Lagrangian submanifold of $(M,-d\lambda)$ diffeomorphic to S and $\rho = (f^{-1})^*\sigma$ is a volume form on L with total volume one. □

5.4 Symplectic quotients

The general construction of symplectic quotients is based on the observation that every coisotropic submanifold of a symplectic manifold M is foliated by isotropic leaves and that the quotient, if it is smooth, is again a symplectic manifold. Its dimension differs from the dimension of M by twice the codimension of the original coisotropic submanifold. The linear version of this was discussed in Section 2.1. One important example of a coisotropic submanifold is the zero set of a moment map which arises from a Hamiltonian action of a compact Lie group. In this case the isotropic leaves are the orbits of the group action and this leads to the **Marsden–Weinstein quotient** which we shall discuss below.

Coisotropic submanifolds

Let (M,ω) be a symplectic manifold and $Q \subset M$ be a closed coisotropic submanifold.[24] Then the subspaces $T_pQ^\omega \subset T_pQ$ are all isotropic and thus determine an isotropic distribution in TQ. The next lemma asserts that this distribution is integrable. (Compare Exercise 3.4.9.)

Lemma 5.4.1 *For every coisotropic submanifold $Q \subset M$ the distribution TQ^ω is integrable.*

Proof: Let X and Y be vector fields on Q with values in TQ^ω and fix a point $p \in Q$. Given a tangent vector $v \in T_pQ$ choose any vector field $Z : Q \to TQ$ such that $Z(p) = v$. Since ω is closed, we have

[24] Recall from the beginning of Section 3.4 that we only consider properly embedded submanifolds that inherit their topology from the ambient manifold.

$$
\begin{aligned}
0 &= d\omega(X,Y,Z) \\
&= \mathcal{L}_X(\omega(Z,Y)) + \mathcal{L}_Y(\omega(X,Z)) + \mathcal{L}_Z(\omega(Y,X)) \\
&\quad + \omega([Y,Z],X) + \omega([Z,X],Y) + \omega([X,Y],Z) \\
&= \omega([X,Y],Z).
\end{aligned}
$$

In the last equality we have used the fact that $\omega(X,Z) = \omega(Y,Z) = 0$ for every vector field Z on Q. It follows that $\omega_p([X,Y](p),v) = 0$ for every $v \in T_pQ$. Hence $[X,Y](p) \in T_pQ^\omega$. □

Remark 5.4.2 More generally, every submanifold $Q \subset M$, coisotropic or not, is foliated by isotropic leaves which are tangent to the distribution $TQ \cap TQ^\omega$, provided that it is of constant dimension. If Q is symplectic these leaves are just points, while if Q is Lagrangian they are connected components of Q. □

It follows from Lemma 5.4.1 and Frobenius' theorem that every coisotropic submanifold $Q \subset M$ is foliated by isotropic leaves. We define an equivalence relation $\sim$ on Q by $p_0 \sim p_1$ iff p_0 and p_1 lie on the same leaf, i.e. iff there exists a smooth path $\gamma : [0,1] \to Q$ such that $\gamma(0) = p_0$, $\gamma(1) = p_1$, and $\dot\gamma(t) \in T_{\gamma(t)}Q^\omega$ for every t. The equivalence class of a point $p \in Q$ will be denoted by $[p]$. A coisotropic submanifold Q is called **regular** if it satisfies the following condition.

(R) *For every $p_0 \in Q$ there exists a submanifold $S \subset Q$ containing p_0 (called a* **local slice through** *p_0) that intersects every isotropic leaf of Q at most once and is such that $T_pQ = T_pS \oplus T_pQ^\omega$ for every $p \in S$. Moreover, the quotient space $Q/{\sim}$ is Hausdorff.*

If (R) holds then every leaf of Q is a submanifold. The converse is false. It is a general fact about foliations (surprisingly nontrivial to prove) that closed leaves are submanifolds (e.g. [544, Section 1.9.7]). However, even if all the isotropic leaves are compact, the coisotropic submanifold Q need not be regular. If it is, then the quotient $\overline{Q} := Q/{\sim}$ has a unique structure of a smooth manifold such that the projection $Q \to \overline{Q}$ is a submersion. The coordinate charts are determined by the local slices S in (R).

Example 5.4.3 (The linear case) Let (V,ω) be a symplectic vector space and $W \subset V$ be a coisotropic linear subspace. Then W is a regular coisotropic submanifold (with noncompact leaves) and the symplectic structure on $\overline{W} = W/W^\omega$ is the bilinear symplectic form $\overline{\omega} : \overline{W} \times \overline{W} \to \mathbb{R}$ of Lemma 2.1.7. □

Exercise 5.4.4 **(i)** Consider the standard Euclidean space $(\mathbb{R}^4,\omega_0)$ and the coisotropic submanifold $Q_a := \{(x_1,x_2,y_1,y_2) \in \mathbb{R}^4 \mid x_1^2 + y_1^2 + a(x_2^2 + y_2^2) = 1\}$ with $a > 0$. Show that the isotropic leaves are $L_{x,y} := \{(\cos(t)x_1, \sin(t)y_1, \cos(at)x_2, \sin(at)y_2) \mid t \in \mathbb{R}\}$. Thus the isotropic leaves are compact if and only if a is rational. Show that the isotropic foliation is regular if and only if $a = 1$.

(ii) Consider the cotangent bundle $T^*\mathbb{T}^2 = \mathbb{T}^2 \times \mathbb{R}^2$ of the standard 2-torus $\mathbb{T}^2 = \mathbb{R}^2/\mathbb{Z}^2$ and the coisotropic submanifold $Q_a := \{([x_1,x_2],(y_1,y_2)) \in \mathbb{T}^2 \times \mathbb{R}^2 \mid y_1 + ay_2 = 0\}$ with $a \in \mathbb{R}$. Show that the isotropic leaves are $L_{x,y} := \{([x_1+t, x_2+at],(y_1,y_2)) \mid t \in \mathbb{R}\}$. Show that the isotropic foliation is regular if and only if a is rational.

(iii) Consider the standard Euclidean space $(\mathbb{R}^4, \omega_0)$ and the coisotropic submanifold $Q := \{(x_1, x_2, y_1, y_2) \in \mathbb{R}^4 \setminus \{0\} \,|\, x_1y_1 + x_2y_2 = 0\}$. Show that the isotropic leaves are $L_{x,y} := \{(\lambda x, \lambda^{-1}y) \,|\, \lambda > 0\}$, that the isotropic foliation admits local slices as in (R), and that the quotient $\overline{Q} = Q/\sim$ is not Hausdorff.

(iv) Interpret the examples in (i), (ii), (iii) in terms of Hamiltonian group actions. □

Proposition 5.4.5 *If $Q \subset M$ is a regular coisotropic submanifold then the quotient $\overline{Q} = Q/\sim$ has a unique symplectic structure $\overline{\omega}$ whose pullback under the projection $Q \to \overline{Q}$ is the restriction of ω to Q.*

Proof: Assume $\dim M = 2n$ and $\mathrm{codim} Q = k \in \{0, \dots, n\}$. Fix a point $p_0 \in Q$ and a local slice $S_0 \subset Q$ through p_0. By the theorem of Frobenius (e.g. [544, Section 1.8]), there is a Q-open neighbourhood $U \subset Q$ of p_0 and a coordinate chart $\phi : U \to \mathbb{R}^{2n-2k} \times \mathbb{R}^k$ such that $\phi(p_0) = 0$, the intersection $\phi(U) \cap (\{x\} \times \mathbb{R}^k)$ is connected for every $x \in \mathbb{R}^{2n-2k}$, and

$$v \in T_pQ^\omega \qquad \Longleftrightarrow \qquad d\phi(p)v \in \{0\} \times \mathbb{R}^k$$

for all $p \in U$ and $v \in T_pQ$. Shrinking U, if necessary, we may assume that there exist constants $\varepsilon > 0$ and $\delta > 0$ such that

$$\Omega := \phi(U) = B_\delta^{2n-2k} \times B_\varepsilon^k$$

(where $B_r^\ell \subset \mathbb{R}^\ell$ denotes the ball of radius r centred at the origin) and $\phi(U \cap S_0)$ is the graph of a smooth map $f_0 : B_\delta^{2n-2k} \to B_\varepsilon^k$ with $f_0(0) = 0$. Then it follows from the regularity condition (R) that the intersection of each isotropic leaf with U is connected. Moreover, the pushforward of ω under ϕ is a closed 2-form

$$\tau := (\phi^{-1})^*\omega \in \Omega^2(\Omega)$$

such that, for all $z \in \Omega$ and $\zeta \in \mathbb{R}^{2n-2k} \times \mathbb{R}^k$,

$$\tau_z(\zeta, \cdot) = 0 \qquad \Longleftrightarrow \qquad \zeta \in \{0\} \times \mathbb{R}^k. \tag{5.4.1}$$

Thus the restriction of τ to $\Omega \cap (\{x\} \times \mathbb{R}^k)$ vanishes for every $x \in \mathbb{R}^{2n-2k}$ and the restriction of τ to $\phi(U \cap S_0)$ is a symplectic form.

Now let S_1 be any other local slice through a point $p_1 \in U \cap [p_0]$ such that $f(U \cap S_1)$ is the graph of a smooth map $f_1 : B_\delta^{2n-2k} \to B_\varepsilon^k$. Let

$$\psi : S_0 \cap U \to S_1 \cap U$$

be the unique diffeomorphism which preserves the leaves. Then

$$\phi \circ \psi \circ \phi^{-1}(x, f_0(x)) = (x, f_1(x)), \qquad x \in B_\delta^{2n-2k}.$$

Hence it follows from (5.4.1) that

$$(\phi \circ \psi \circ \phi^{-1})^*\tau|_{\phi(S_1 \cap U)} = \tau|_{\phi(S_1 \cap U)},$$

and so

$$\psi^* \,\omega|_{S_1 \cap U} = \omega|_{S_0 \cap U}\,.$$

This shows that the transition maps between nearby local slices are symplectomorphisms. Hence, by a standard open and closed argument, the transition maps between any two local slices through the same leaf are symplectomorphisms. This shows that there is a unique symplectic form $\overline{\omega}$ on $\overline{Q}$ such that $\omega|_S = (\pi|_S)^*\overline{\omega}|_{\overline{S}}$ for every local slice $S \subset Q$, where $\overline{S} := \{[p] \,|\, p \in S\} \subset \overline{Q}$ is the open set of all leaves passing through S and $\pi : Q \to \overline{Q}$ denotes the obvious projection. In other words, the tangent space of $\overline{Q}$ at a point $[p]$, for $p \in Q$, can be naturally identified with the quotient space

$$T_{[p]}\overline{Q} = \frac{T_pQ}{T_pQ^\omega},$$

and the symplectic form ω_p induces a nondegenerate 2-form $\overline{\omega}_{[p]}$ on this space (by Lemma 2.1.7). The collection of these local 2-forms defines a symplectic form $\overline{\omega} \in \Omega^2(\overline{Q})$, uniquely determined by the condition $\pi^*\overline{\omega} = \omega|_Q$. This proves Proposition 5.4.5. □

Now let $L \subset M$ be a Lagrangian submanifold of M. The submanifolds L and Q are said to **intersect cleanly** if their intersection is a submanifold of Q and

$$T_p(L \cap Q) = T_pL \cap T_pQ$$

for all $p \in L \cap Q$. Let $\widetilde{L}$ be the union of all isotropic leaves of Q which pass through L, and denote the image of $\widetilde{L}$ in $\overline{Q}$ by

$$\overline{L} := \widetilde{L}/\!\sim .$$

Proposition 5.4.7 below asserts that $\overline{L}$ is an immersed Lagrangian submanifold of $\overline{Q}$ whenever Q is a regular coisotropic submanifold of M.

Remark 5.4.6 Let M be of dimension $2n$ and Q be of dimension $2n - k$. Then the isotropic leaves of Q are of dimension k and $\overline{Q}$ has dimension $2n - 2k$. In the extreme case of transverse intersection, $L \cap Q$ has dimension $n - k$ and is transverse to the isotropic leaves. In this case the proof of Proposition 5.4.7 shows that the map $L \cap Q \to \overline{Q}$ is a Lagrangian immersion whose image is $\overline{L}$. The other extreme is the case $L \subset Q$, and then $L = \widetilde{L}$ is itself foliated by the leaves of Q and the induced map $L \to \overline{L}$ is a submersion. □

Proposition 5.4.7 *Let $Q \subset M$ be a regular coisotropic submanifold and $L \subset M$ be a Lagrangian submanifold which intersects Q cleanly. Then the following holds.*

(i) *$\overline{L}$ is the image of a Lagrangian immersion.*

(ii) *If L and Q intersect transversely, then the map $L \cap Q \to \overline{Q}$ is a Lagrangian immersion with image $\overline{L}$.*

(iii) *If the intersection of L with each isotropic leaf of Q is connected then $\overline{L}$ is the image of an injective Lagrangian immersion.*

Proof: Let $\dim M = 2n$ and $\dim Q = 2n - k$. Assume first that $L \cap Q$ is connected and hence is a submanifold of M of constant dimension

$$\dim(L \cap Q) = n - k + \ell, \qquad 0 \le \ell \le k.$$

The subspaces $T_pL + T_pQ = T_pL \oplus (T_pQ \cap T_p(L \cap Q)^\perp)$ form a smooth subbundle of $TM|_{L\cap Q}$, by clean intersection, and $\dim(T_pL + T_pQ) = 2n - \ell$ for all $p \in L \cap Q$. Its symplectic complement is a smooth subbundle $E \subset T(L \cap Q)$ with fibres

$$E_p := (T_pL + T_pQ)^\omega = T_pL \cap T_pQ^\omega \subset T_p(L \cap Q)$$

over $p \in L \cap Q$ of dimension $\dim E_p = \ell$. By Lemma 5.4.1 these subspaces form an integrable distribution on $L \cap Q$. Each leaf has dimension ℓ and is contained in an isotropic leaf of Q. Define an equivalence relation $\sim$ on $L \cap Q$ by $p_0 \sim p_1$ iff there exists a smooth path $\gamma : [0, 1] \to L \cap Q$ such that $\gamma(0) = p_0$, $\gamma(1) = p_1$, and $\dot\gamma(t) \in E_{\gamma(t)}$ for every t. Since the isotropic foliation of Q is regular, it follows that the foliation induced by E on $L \cap Q$ admits local slices as in condition (R) on page 219. (Given any element $p_0 \in L \cap Q$, choose a submanifold $\Sigma \subset L \cap Q$ through p_0 such that $T_{p_0}(L \cap Q) = T_{p_0}\Sigma \oplus (T_{p_0}L \cap T_{p_0}Q^\omega)$; then extend Σ to a submanifold $S \subset Q$ such that $T_{p_0}Q = T_{p_0}S \oplus T_{p_0}Q^\omega$; now shrink S, if necessary, to obtain a local slice $S_0 \subset S$ that intersects each isotropic leaf of Q at most once; then $\Sigma_0 := \Sigma \cap S_0$ is the required local slice for $L \cap Q$.) This shows that the quotient space $\Lambda := (L \cap Q)/\sim$ is a manifold (possibly nonHausdorff). Since $L \cap Q$ has dimension $n - k + \ell$ and the leaves have dimension ℓ, the quotient manifold has dimension $\dim \Lambda = n - k$. The resulting map $\iota : \Lambda \to \overline{Q}$ is smooth and its differential at a point $[p] \in \Lambda$ is the inclusion

$$(T_pL \cap T_pQ)/(T_pL \cap T_pQ^\omega) \to T_pQ/T_pQ^\omega.$$

The image of this map is the subspace $((T_pL \cap T_pQ) + T_pQ^\omega)/T_pQ^\omega$ of T_pQ/T_pQ^ω. By part (ii) of Lemma 2.1.7 this is a Lagrangian subspace of T_pQ/T_pQ^ω. Hence the map $\iota : \Lambda \to \overline{Q}$ is a Lagrangian immersion. Since Λ can be covered by countably many coordinate charts, it follows that $\overline{L}$ is the image of a Lagrangian immersion defined on a second countable Hausdorff manifold (even if Λ itself should not be Hausdorff). This proves part (i) in the case where $L \cap Q$ is connected. To obtain the result in general, apply the same argument to each connected component of $L \cap Q$. If L and Q intersect transversally, then $\ell = 0$ and $\Lambda = L \cap Q$ in the above discussion and this proves part (ii). Moreover, it follows directly from the definitions that the immersion $\iota : \Lambda \to \overline{Q}$ is injective if and only if the intersection of L with each isotropic leaf of Q is connected. In this case it also follows that Λ is Hausdorff. This proves part (iii) and Proposition 5.4.7. □

Exercise 5.4.8 Consider the standard Euclidean space $(\mathbb{R}^6, \omega_0)$, the coisotropic submanifold $Q := \{(x, y) \in \mathbb{R}^6 \setminus \{0\} \,|\, y_2 = y_3 = 0\}$, and the Lagrangian submanifold $L := \{(x, y) \in \mathbb{R}^6 \,|\, x_1 = x_2 = y_3 = 0\}$. Show that Q satisfies (R) and intersects L cleanly. Show that in this example the manifold Λ in the proof of Proposition 5.4.7 is not Hausdorff. □

Example 5.4.9 Consider the unit sphere $Q := S^3$ in $(\mathbb{C}^2, \omega_0)$. The isotropic leaves are the Hopf circles $\{(e^{it}z_0, e^{it}z_1) \,|\, t \in \mathbb{R}\}$ and $\overline{Q} = \mathbb{C}\mathrm{P}^1$. For $s \in \mathbb{R}$ the affine subspace $L_s := \{(x_0 + is, x_1) \,|\, x_0, x_1 \in \mathbb{R}\}$ is a Lagrangian submanifold of $\mathbb{C}^2$. It does not intersect Q for $|s| > 1$ and is transverse to Q for $|s| < 1$. For $s = 0$ it intersects each isotropic leaf through L_0 in precisely two points and hence the natural map $L_0 \cap Q \to \overline{L} = \mathbb{R}\mathrm{P}^1$ is a double covering of a great circle. For $0 < |s| < 1$ the map $L_s \cap Q \to \mathbb{C}\mathrm{P}^1$ is an immersion with a single double point at $[1:0]$. Moreover, if $0 < |s| < 1$ the map $(L_s \setminus \{(\sqrt{1-s^2}+is, 0)\}) \cap Q \to \mathbb{C}\mathrm{P}^1$ is an injective immersion, but not an embedding because it is not a homeomorphism to its image with the subspace topology. □

Example 5.4.10 (Generating functions) Consider the cotangent bundle of a closed manifold Y. In [402], Laudenbach and Sikorav show that given any Lagrangian submanifold $L \subset T^*Y$ that is Hamiltonian isotopic to the zero section there exists a fibre bundle $E \to Y$ and a smooth function $S : E \to \mathbb{R}$ such that L can be realized as the symplectic reduction of the Lagrangian submanifold $\mathrm{graph}(dS)$ of T^*E with respect to the coisotropic manifold consisting of all elements of T^*E that vanish on the fibres of $E \to Y$. The function S is then called a **generating function** of the Lagrangian submanifold L. This example is discussed in detail in Section 9.4; see in particular Proposition 9.4.4. □

Example 5.4.11 Let (M_0, ω_0), (M_1, ω_1) be symplectic manifolds, let $L_0 \subset M_0$ be a Lagrangian submanifold, and let $\psi : M_0 \to M_1$ be a symplectomorphism. Consider the symplectic manifold

$$M := M_0 \times M_0 \times M_1, \qquad \omega := \omega_0 \oplus (-\omega_0) \oplus \omega_1$$

and let $\Delta_0 \subset M_0 \times M_0$ denote the diagonal. Then

$$Q := \Delta_0 \times M_1 \subset M$$

is a coisotropic submanifold with isotropic leaves $\Delta_0 \times \{\mathrm{pt}\}$ and quotient $\overline{Q} = M_1$. The set $L := L_0 \times \mathrm{graph}(\psi) \subset M$ is a Lagrangian submanifold intersecting Q transversally and the reduced Lagrangian submanifold is $\overline{L} = L_1 := \psi(L_0) \subset M_1$. (Compare with Example 2.1.8.) □

Exercise 5.4.12 Use a construction similar to that in Example 5.4.11 to interpret the composition of symplectomorphisms in terms of symplectic quotients. □

These constructions are generalized to the notion of **Lagrangian correspondence** in Weinstein [680, 681] and Wehrheim–Woodward [664, 665]. A Lagrangian correspondence between (M_0, ω_0) and (M_1, ω_1) is simply a Lagrangian submanifold $L_{01} \subset (M_0 \times M_1, (-\omega_0) \oplus \omega_1)$. The composite of two Lagrangian correspondences L_{01} and $L_{12} \subset (M_1 \times M_2, (-\omega_1) \oplus \omega_2)$ is the image of $L_{01} \times L_{12}$ under the projection $Q := M_0 \times \Delta_1 \times M_2 \to M_0 \times M_2$, where Δ_1 is the diagonal in $(M_1 \times M_1, \omega_1 \oplus (-\omega_1))$ as above. It is well-defined whenever $L_{01} \times L_{12}$ and Q intersect cleanly; however, it will in general be immersed, rather than embedded.

The Marsden–Weinstein quotient

Let $\mathrm{G} \to \mathrm{Symp}(M) : g \mapsto \psi_g$ be a Hamiltonian group action on (M, ω) generated by a G-equivariant moment map $\mu : M \to \mathfrak{g}^*$. Since the zero element $0 \in \mathfrak{g}^*$ is a fixed point of the coadjoint action, its preimage $\mu^{-1}(0)$ is invariant under G. If zero is a regular value of the moment map then $\mu^{-1}(0)$ is a smooth submanifold of M. It turns out that this submanifold is always coisotropic and that the isotropic leaves are the G-orbits. If the group action is proper, in particular if G is compact, it follows that the isotropic leaves are smooth submanifolds of $\mu^{-1}(0)$. However, even if zero is a regular value of μ, it does not follow that $\mu^{-1}(0)$ is a *regular* coisotropic submanifold. This requires the following stronger hypothesis.

(RG) *The group* G *acts freely and properly on* $\mu^{-1}(0)$.

Proposition 5.4.13 *Assume (RG). Then zero is a regular value of the moment map, its preimage* $\mu^{-1}(0)$ *is a regular coisotropic submanifold of* M, *the isotropic leaves are the orbits of* G, *and the quotient space*

$$M/\!\!/\mathrm{G} := \mu^{-1}(0)/\mathrm{G}$$

is a symplectic manifold. It is called the **Marsden–Weinstein quotient** *and has dimension*

$$\dim M/\!\!/\mathrm{G} = \dim M - 2\dim \mathrm{G}.$$

Proof: The proof has three steps.

Step 1. *If zero is a regular value of* μ, *then* $\mu^{-1}(0)$ *is a coisotropic submanifold of* M *and the isotropic leaves are the* G*-orbits in* $\mu^{-1}(0)$.

Denote the G-orbit of $p \in M$ by $\mathcal{O}(p) := \{\psi_g(p) \,|\, g \in \mathrm{G}\}$. Then

$$\mathrm{im}\, L_p = \{X_\xi(p) \,|\, \xi \in \mathfrak{g}\} = T_p\mathcal{O}(p)$$

(see equation (5.2.6)). If $\mu(p) = 0$ then $T_p\mu^{-1}(0) = \ker\, d\mu(p)$ and it follows from equations (5.2.8) and (5.2.9) in Lemma 5.2.5 that

$$(\ker\, d\mu(p))^\omega = \mathrm{im} L_p \subset \ker d\mu(p).$$

When zero is a regular value of μ, it follows that $\mu^{-1}(0)$ is a coisotropic submanifold and the isotropic subbundle of the tangent bundle of $\mu^{-1}(0)$ is given by the tangent spaces of the G-orbits. This proves Step 1.

Step 2. *Zero is a regular value of* μ *if and only if the stabilizer subgroup*

$$\mathrm{G}_p := \{g \in \mathrm{G} \,|\, \psi_g(p) = p\}$$

is discrete for every $p \in \mu^{-1}(0)$.

By equation (5.2.7) in Lemma 5.2.5, the linear map $d\mu(p) : T_pM \to \mathfrak{g}^*$ is surjective if and only if the infinitesimal action $L_p : \mathfrak{g} \to T_pM$ is injective. Since $\ker L_p = \mathrm{Lie}(\mathrm{G}_p)$ is the Lie algebra of G_p the linear map L_p is injective if and only if G_p is a discrete subgroup of G. This proves Step 2.

Step 3. *We prove Proposition 5.4.13.*

Assume (RG). Then $G_p = \{1\}$ for every $p \in \mu^{-1}(0)$ and hence zero is a regular value of the moment map by Step 2. By Step 1 this implies that $\mu^{-1}(0)$ is a coisotropic submanifold and the isotropic leaves are the G-orbits. The local slice theorem for free and proper Lie group actions asserts that $\mu^{-1}(0)$ satisfies the slice condition in (R). Moreover, $\mu^{-1}(0)/G$ is Hausdorff. (If $p, q \in \mu^{-1}(0)$ belong to different G-orbits, and there is a sequence $p_i \in \mu^{-1}(0)$ converging to p and a sequence $g_i \in G$ such that $g_i p_i$ converges to q, then g_i cannot have a convergent subsequence in contradiction to (RG).) This shows that $\mu^{-1}(0)$ is a regular coisotropic submanifold and so the assertion follows from Proposition 5.4.5. □

Remark 5.4.14 For every $p \in \mu^{-1}(0)$ there is a chain complex

$$0 \longrightarrow \mathfrak{g} \xrightarrow{L_p} T_pM \xrightarrow{d\mu(p)} \mathfrak{g}^* \longrightarrow 0. \tag{5.4.2}$$

If zero is a regular value of the moment map this complex is exact at $\mathfrak{g}$ and $\mathfrak{g}^*$ and its homology at T_pM is isomorphic to the tangent space of $M/\!\!/G$:

$$T_{[p]}M/\!\!/G = T_{[p]}\frac{\mu^{-1}(0)}{G} \cong \frac{T_p\mu^{-1}(0)}{T_p\mathcal{O}(p)} = \frac{\ker d\mu(p)}{\operatorname{im} L_p}. \tag{5.4.3}$$

The skew-symmetric bilinear form $\omega_p : T_pM \times T_pM \to \mathbb{R}$ descends to a nondegenerate skew-symmetric bilinear form $\overline{\omega}_p$ on this quotient space by Lemma 2.1.7. If G acts freely and properly on $\mu^{-1}(0)$ then, by Proposition 5.4.5, these skew forms fit together to a unique symplectic form $\overline{\omega}$ on $M/\!\!/G$ whose pullback under the obvious projection $\mu^{-1}(0) \to M/\!\!/G$ is the restriction of ω to $\mu^{-1}(0)$. □

There is a more general construction of symplectic quotients of the form

$$M_{\mathcal{O}} := \mu^{-1}(\mathcal{O})/G,$$

where $\mathcal{O} \subset \mathfrak{g}^*$ is a coadjoint orbit under the action of G. In fact it follows from (5.2.4) that $\mu^{-1}(\mathcal{O})$ is invariant under G. Moreover, the map μ is transverse to $\mathcal{O}$ if and only if every point in $\mathcal{O}$ is a regular value of μ and in this case $\mu^{-1}(\mathcal{O})$ is a submanifold of M. If in addition G acts freely and properly on $\mu^{-1}(\mathcal{O})$ then the next proposition asserts that the quotient space $M_{\mathcal{O}}$ is a symplectic manifold.

Proposition 5.4.15 *Assume* G *acts freely and properly on* $\mu^{-1}(\mathcal{O})$. *Then* μ *is transverse to* $\mathcal{O}$, $\mu^{-1}(\mathcal{O})$ *is a smooth submanifold of* M, *and the quotient space*

$$M_{\mathcal{O}} := \mu^{-1}(\mathcal{O})/G$$

is a symplectic manifold of dimension $\dim M_{\mathcal{O}} = \dim M + \dim \mathcal{O} - 2\dim G$. *The symplectic form is given by*

$$\begin{aligned} \overline{\omega}_{[p]}([v_1],[v_2]) := &\ \omega(v_1, v_2) - \langle \mu(p), [\xi_1, \xi_2] \rangle \\ = &\ \omega(v_1 - L_p\xi_1, v_2 - L_p\xi_2) \end{aligned} \tag{5.4.4}$$

for $v_1, v_2 \in T_p\mu^{-1}(\mathcal{O})$, *where* $\xi_1, \xi_2 \in \mathfrak{g}$ *are chosen such that*

$$\operatorname{ad}(\xi_i)^*\mu(p) + d\mu(p)v_i = d\mu(p)(v_i - L_p\xi_i) = 0, \qquad i = 1, 2.$$

Proof: Consider the diagonal action of G on the product symplectic manifold $(\widetilde{M}, \widetilde{\omega})$ defined by

$$\widetilde{M} := M \times \mathcal{O}, \qquad \widetilde{\omega} := \omega \oplus (-\omega_{\mathcal{O}}).$$

Here $\omega_{\mathcal{O}}$ is the symplectic form on $\mathcal{O}$ defined in Example 5.3.11. By Example 5.3.11, this action is Hamiltonian and, by Exercise 5.3.16, it is generated by the moment map $\widetilde{\mu} : \widetilde{M} \to \mathfrak{g}^*$ given by

$$\widetilde{\mu}(p, \eta) := \mu(p) - \eta.$$

There is a G-equivariant diffeomorphism

$$\mu^{-1}(\mathcal{O}) \to \widetilde{\mu}^{-1}(0) : p \mapsto (p, \mu(p)).$$

By assumption, the Lie group G acts freely and properly on $\mu^{-1}(\mathcal{O})$ and therefore also on $\widetilde{\mu}^{-1}(0)$. Hence it follows from Proposition 5.4.13 that $\widetilde{\mu}^{-1}(0)$ is a regular coisotropic submanifold of $M \times \mathcal{O}$ whose isotropic leaves are the G-orbits, and the quotient manifold $(M \times \mathcal{O})/\!\!/\mathrm{G} = \widetilde{\mu}^{-1}(0)/\mathrm{G}$ inherits a symplectic form from the restriction of $\widetilde{\omega}$ to the submanifold

$$\widetilde{\mu}^{-1}(0) = \{(p, \mu(p)) \,|\, p \in M,\ \mu(p) \in \mathcal{O}\} \subset M \times \mathcal{O}.$$

For $\widetilde{p} := (p, \mu(p)) \in \widetilde{\mu}^{-1}(0)$ the tangent space of $\widetilde{\mu}^{-1}(0)$ at $\widetilde{p}$ is given by

$$T_{\widetilde{p}}\widetilde{\mu}^{-1}(0) = \left\{ (v, d\mu(p)v) \,\middle|\, \begin{array}{l} v \in T_pM \text{ and } \exists \xi \in \mathfrak{g} \text{ s.t.} \\ v - L_p\xi \in \ker\, d\mu(p) \end{array} \right\}. \tag{5.4.5}$$

(Note that $v - L_p\xi \in \ker\, d\mu(p)$ if and only if $\mathrm{ad}(\xi)^*\mu(p) + d\mu(p)v = 0$.) The restriction of the symplectic form $\widetilde{\omega}$ to $\widetilde{\mu}^{-1}(0)$ is given by the formula (5.4.4) (see Example 5.3.11). This proves Proposition 5.4.15. □

Exercise 5.4.16 Let $\mu : M \to \mathfrak{g}^*$ be an equivariant moment map of a Hamiltonian group action and let $\mathcal{O} \subset \mathfrak{g}^*$ be a coadjoint orbit.

(i) Let $p \in \mu^{-1}(\mathcal{O})$. Verify directly that the skew-form (5.4.4) on $T_p\mu^{-1}(\mathcal{O})$ descends to a nondegenerate skew-form on the quotient $T_p\mu^{-1}(\mathcal{O})/\mathrm{im}L_p$.

(ii) Prove that μ is transverse to $\mathcal{O}$ if and only if $\mathcal{O}$ contains a regular value of μ if and only if every element of $\mathcal{O}$ is a regular value of μ.

(iii) Assume that G acts freely and properly on $\mu^{-1}(\mathcal{O})$. Let $\eta \in \mathcal{O}$ and denote the stabilizer subgroup of η by $\mathrm{G}_\eta := \{g \in \mathrm{G} \,|\, \mathrm{Ad}(g)^*\eta = \eta\}$.

(a) Prove that G_η acts freely and properly on $\mu^{-1}(\eta)$.

(b) Prove that there is a unique symplectic form ω_η on the quotient manifold

$$M_\eta := \mu^{-1}(\eta)/\mathrm{G}_\eta$$

whose pullback under the projection $\mu^{-1}(\eta) \to M_\eta$ is the restriction of ω to $\mu^{-1}(\eta)$.

(c) Prove that the inclusion $\mu^{-1}(\eta) \to \mu^{-1}(\mathcal{O})$ descends to a symplectomorphism of the quotient manifolds $\mu^{-1}(\eta)/\mathrm{G}_\eta \cong \mu^{-1}(\mathcal{O})/\mathrm{G} = M_{\mathcal{O}}$, where the symplectic structure on $M_{\mathcal{O}}$ is as in Proposition 5.4.15. □

Exercise 5.4.17 (Grassmannians) Consider the obvious action of $\mathrm{U}(k)$ on the space $\mathbb{C}^{n\times k}$ of complex $n \times k$-matrices with the standard symplectic structure. Identify the Lie algebra $\mathfrak{u}(k)$ with its dual as above and prove that the moment map of the action is given by

$$\mu(B) = \frac{1}{2i} B^* B$$

for $B \in \mathbb{C}^{n\times k}$. Deduce that $\mu^{-1}(\mathbb{1}/2i)$ is the space of unitary k-frames $B \in \mathbb{C}^{n\times k}$ with $B^*B = \mathbb{1}$ and the quotient

$$\mu^{-1}(\mathbb{1}/2i)/\mathrm{U}(k) = \mathrm{G}(k, n)$$

is the Grassmannian. □

Exercise 5.4.18 (Toric manifolds) A closed connected symplectic $2m$-manifold (M, ω) is called **toric** if it admits a Hamiltonian group action of the torus $\mathbb{T}^m = \mathbb{R}^m/\mathbb{Z}^m$ such that the infinitesimal action $L_p : \mathbb{R}^m \to T_pM$ in equation (5.2.6) is injective for some $p \in M$. This notion can be extended to noncompact manifolds, but such extensions will not be considered here. Toric manifolds have many fascinating properties: see Audin [32], Givental [274], and also Example 5.5.2 and the discussion thereafter.

(i) Consider an action of the k-torus $\mathbb{T}^k$ on $\mathbb{C}^n$ that is induced by some inclusion $\iota : \mathbb{T}^k \to \mathbb{T}^n$ as in Exercise 5.3.17. Assume that the moment map $\mu_k : \mathbb{C}^n \to \mathbb{R}^k$ is proper. Let $\tau \in \mathbb{R}^k$ be a regular value of μ_k and suppose that $\mathbb{T}^k$ acts freely on $\mu^{-1}(\tau)$. Prove that $M/\!\!/\mathbb{T}^k(\tau) := \mu^{-1}(\tau)/\mathbb{T}^k$ is a toric manifold, and compute its moment map.

(ii) Show that any product of the form

$$(S^2 \times \cdots \times S^2, \lambda_1\sigma \oplus \cdots \oplus \lambda_m\sigma)$$

is toric, where σ is an area form on S^2 and $\lambda_i > 0$.

(iii) Show that any product of projective spaces is toric. □

Exercise 5.4.19 Examine the manifold $M_{\mathcal{O}} = \mu^{-1}(\mathcal{O})/\mathrm{G}$ in the case where

$$M = T^*\mathrm{G} \cong \mathrm{G} \times \mathfrak{g}^*$$

with the G-action of Exercise 5.3.9. □

Exercise 5.4.20 Suppose that G acts in a Hamiltonian way on a symplectic manifold (M, ω) with moment map $\mu : M \to \mathfrak{g}^*$. Consider the diagonal action of G on the product manifold (M', ω') defined by

$$M' := M \times T^*\mathrm{G}, \qquad \omega' := \omega \oplus \omega_{\mathrm{can}}.$$

By Exercise 5.3.16 this action is Hamiltonian. If we identify $T^*\mathrm{G}$ with the product $\mathrm{G} \times \mathfrak{g}^*$ as in Example 5.3.9, then the function

$$\mu' : M \times \mathrm{G} \times \mathfrak{g}^* \to \mathfrak{g}^*,$$

defined by

$$\mu'(p, h, \eta) := \mu(p) - \eta$$

for $p \in M$, $h \in \mathrm{G}$, and $\eta \in \mathfrak{g}^*$, is a moment map for the action. Prove that the Marsden–Weinstein quotient can be identified with (M, ω). □

Exercise 5.4.21 Let (M, ω) be a symplectic manifold equipped with a Hamiltonian G-action that is generated by an equivariant moment map $\mu : M \to \mathfrak{g}^*$. Suppose that G acts freely and properly on $\mu^{-1}(0)$ and denote by $\pi_\mu : \mu^{-1}(0) \to M/\!\!/\mathrm{G} := \mu^{-1}(0)/\mathrm{G}$ the projection onto the Marsden–Weinstein quotient. Prove that the map

$$\iota_\mu := \mathrm{id} \times \pi_\mu : \mu^{-1}(0) \to M \times M/\!\!/\mathrm{G} =: \widetilde{M}$$

is a Lagrangian embedding with respect to the symplectic form

$$\widetilde{\omega} := \omega \oplus (-\overline{\omega}) = \mathrm{pr}_1^*\omega - \mathrm{pr}_2^*\overline{\omega}.$$

Here $\overline{\omega} \in \Omega^2(M/\!\!/\mathrm{G})$ denotes the symplectic form of Proposition 5.4.13. (See Audin–Lalonde–Polterovich [35].) □

Exercise 5.4.22 Let (M, ω) be a $2n$-dimensional symplectic manifold and define

$$\widetilde{M} := M \times \mathbb{C}\mathrm{P}^{n-1}, \qquad \widetilde{\omega}_r := \mathrm{pr}_1^*\omega - r^2\mathrm{pr}_2^*\omega_{\mathrm{FS}} \in \Omega^2(\widetilde{M}).$$

Prove that $(\widetilde{M}, \widetilde{\omega}_r)$ contains a Lagrangian sphere for $r > 0$ sufficiently small. **Hint:** Prove that the set

$$L_r := \left\{ (z_0, z_1, \ldots, z_n, [z_0 : z_1 : \cdots : z_n]) \,\middle|\, z_i \in \mathbb{C}, \sum_{i=0}^{n} |z_i|^2 = r^2 \right\}$$

is a Lagrangian submanifold of $(\mathbb{C}^n \times \mathbb{C}\mathrm{P}^{n-1}, \widetilde{\omega}_r)$, where

$$\widetilde{\omega}_r := \mathrm{pr}_1^*\omega_0 - r^2\mathrm{pr}_2^*\omega_{\mathrm{FS}} \in \Omega^2(\mathbb{C}^n \times \mathbb{C}\mathrm{P}^{n-1}).$$

Relate this to Exercise 5.4.21. □

Exercise 5.4.23 Examine the Marsden–Weinstein quotient in Example 5.3.18. It is the moduli space

$$\mathcal{M}^{\mathrm{flat}} := \mathcal{A}^{\mathrm{flat}}/\mathcal{G}$$

of gauge equivalence classes of flat G-connections on Σ. Zero is not a regular value of the moment map, the moduli space $\mathcal{M}^{\mathrm{flat}}$ has singularities, and the twisted de Rham complex

$$0 \longrightarrow \Omega^0(\Sigma, \mathfrak{g}) \xrightarrow{d_A} \Omega^1(\Sigma, \mathfrak{g}) \xrightarrow{d_A} \Omega^2(\Sigma, \mathfrak{g}) \longrightarrow 0$$

is the analogue of (5.4.2). Even though the setup is infinite-dimensional, the quotient is finite-dimensional. If G is simply connected the quotient can be identified with the space of conjugacy classes of representations $\rho : \pi_1(\Sigma) \to \mathrm{G}$. □

Comments on the literature

The conditions which we have imposed on symplectic quotients are rather stringent, and there are many other interesting cases. Our hypothesis that the Lie group G acts freely on $\mu^{-1}(0)$ guarantees that the Marsden–Weinstein quotient is a smooth manifold. If instead one only assumes that zero is a regular value of the moment map, the quotient is a symplectic orbifold, and if even that condition is removed the quotient can be a highly singular space. The corresponding study of singular quotients in the algebraic geometric setting is the subject of geometric invariant theory and will be briefly discussed in Section 5.7; for a study of singular quotients from a symplectic viewpoint see e.g. Lerman–Sjamaar [411] and Lerman–Montgomery–Sjamaar [412]. Further interesting references on the subject include the papers by Goldman [275], Weinstein [679], and Jeffrey–Kirwan [347]. We also point out that the reduction of Proposition 5.4.15 at coadjoint orbits plays an important role in representation theory (see Guillemin–Sternberg [296] and Guillemin–Lerman–Sternberg [292]).

5.5 Convexity

In this section we consider the situation of a Hamiltonian torus action

$$\mathbb{T}^m = \mathbb{R}^m/\mathbb{Z}^m \to \mathrm{Symp}(M,\omega) \;:\; \theta \mapsto \psi_\theta$$

on a compact connected symplectic manifold (M,ω) of dimension $2n$. This means that there exists a moment map $\mu : M \to \mathbb{R}^m$ such that for every $\theta = (\theta_1,\dots,\theta_m) \in \mathbb{R}^m$ the Hamiltonian function $H_\theta := \langle\theta,\mu\rangle$ generates the Hamiltonian flow $t \mapsto \psi_{t\theta}$. In other words,

$$\frac{d}{dt}\psi_{t\theta} = X_{H_\theta} \circ \psi_{t\theta}.$$

Here we identify the Lie algebra $\mathfrak{g} = \mathbb{R}^m$ of $\mathbb{T}^m$ with its dual $\mathfrak{g}^* = \mathbb{R}^m$ via the standard inner product. Our goal in this section is to prove the Atiyah–Guillemin–Sternberg convexity theorem (cf. [26], [295]). This result was generalized by Kirwan [369, 370] to the case of nonabelian compact Lie groups. See also the paper by Lerman, Meinrenken, Tolman, and Woodward [410] that extends the result to actions of compact Lie groups on symplectic orbifolds, and the monograph [294] and Karshon–Bjorndahl [356] for a general approach to convexity theorems.

Theorem 5.5.1 (Atiyah–Guillemin–Sternberg) *Let (M,ω) be a closed connected symplectic manifold and let $\mathbb{T}^m \to \mathrm{Symp}(M,\omega) : \theta \mapsto \psi_\theta$ be a Hamiltonian torus action generated by a moment map $\mu : M \to \mathbb{R}^m$. Then the image $\mu(M) \subset \mathbb{R}^m$ of the moment map is a convex polytope. More precisely, the fixed point set of the action is a finite union of pairwise disjoint connected symplectic submanifolds $C_1,\dots,C_N$, the moment map is constant on each connected component C_j of the fixed point set, and the image of μ is the convex hull of the points $\eta_j := \mu(C_j) \in \mathbb{R}^m$, i.e.*

$$\mu(M) = K(\eta_1,\dots,\eta_N) := \left\{ \sum_{j=1}^N \lambda_j\eta_j \;\middle|\; \sum_{j=1}^N \lambda_j = 1,\ \lambda_j \geq 0 \right\}.$$

Proof: See page 237. □

Examples

Before embarking on the proof we illustrate this theorem by some examples.

Example 5.5.2 Consider the action of the torus $\mathbb{T}^n$ on $\mathbb{C}\mathrm{P}^n$ given by

$$(\theta_1, \dots, \theta_n) \cdot [z_0 : z_1 : \cdots : z_n] = [z_0 : e^{-2\pi i\theta_1} z_1 : \cdots : e^{-2\pi i\theta_n} z_n].$$

By Exercise 5.3.16, the moment map $\mu : \mathbb{C}\mathrm{P}^n \to \mathbb{R}^n$ has the form

$$\mu([z_0 : z_1 : \cdots : z_n]) = \pi \left(\frac{|z_1|^2}{\|z\|^2}, \dots, \frac{|z_n|^2}{\|z\|^2} \right),$$

where $\|z\|^2 := \sum_{i=0}^n |z_i|^2$. Thus the image of μ is the simplex

$$\Delta = \left\{ (x_1, \dots, x_n) \in \mathbb{R}^n \,\middle|\, 0 \le \sum_i x_i \le \pi \right\}.$$

This action has $n+1$ isolated fixed points $p_i = [0 : \cdots : 0 : 1 : 0 : \cdots : 0]$ for $i = 0, \dots, n$ which are mapped by μ to the vertices of Δ. Thus the convex hull of the points $\mu(p_0), \dots, \mu(p_n)$ is precisely the image of μ. □

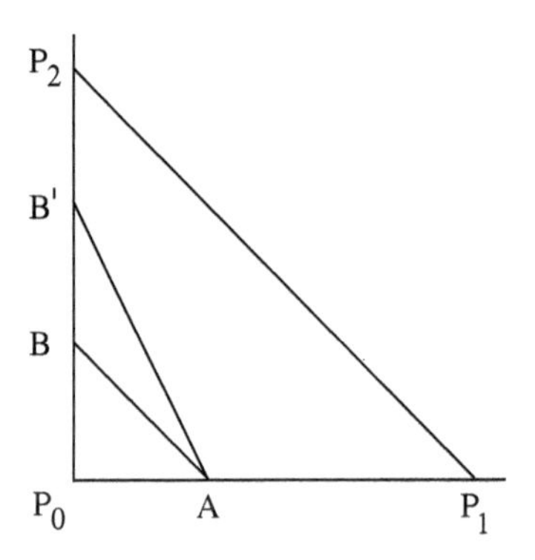

FIG. 5.3. The image of the moment map.

Exercise 5.5.3 Consider the case $n = 2$ in Example 5.5.2. Show that the inverse image of any vertex P_i of Δ is a single point, of any point on an edge is S^1, and of any point in the interior is $\mathbb{T}^2$. What is the inverse image of an edge, of line segments such as AB, AB' in Fig. 5.3, of the triangle ABP_0? □

Similarly, any $2m$-dimensional toric symplectic manifold (M, ω) (cf. Exercise 5.4.18) admits an action of $\mathbb{T}^m$ and hence maps to a convex polytope in $\mathbb{R}^m$. Delzant [132] showed that this polytope Δ_M completely determines the diffeomorphism type of M and its symplectic structure. Moreover, there are rather stringent requirements on the polytopes that arise. Some very nice explicit examples are worked out in Karshon's paper [355] on the classification of symplectic 4-manifolds with Hamiltonian circle action.

Symplectic toric manifolds form an accessible class of symplectic manifolds whose properties have been much studied, both from topological and from symplectic points of view. See McDuff–Salamon [470, Section 11.3] for a more detailed description of their construction and structure, and McDuff [466] for some general results about their topology. A generic fibre $L_a := \mu^{-1}(a)$ of the moment map $\mu : M \to \Delta_M \subset \mathbb{R}^n$ of a closed toric manifold is a $\mathbb{T}^n$-orbit and hence is a Lagrangian torus. In particular, the moment map is a singular Lagrangian fibration, which means that toric manifolds are good examples in which to study mirror symmetry. Much work has also been devoted to understanding the properties of the tori L_a. A first question is: which of them are **displaceable** by a Hamiltonian isotopy? In other words, for which $a \in \mathrm{int}(\Delta_M)$ is there a Hamitonian isotopy ϕ_t such that $\phi_1(L_a) \cap L_a = \emptyset$? Pioneering work by Entov–Polterovich [200] using Calabi quasimorphisms (cf. the discussion after Theorem 10.3.7) shows that when $M = \mathbb{C}\mathrm{P}^n$ the fibre over the centre of gravity of Δ_M is nondisplaceable. One can also address this problem using methods such as Lagrangian Floer theory, as in Fukaya–Oh–Ohta–Ono [250]. However, it has also been approached by more traditional methods. For example, Abreu–Macarini [11] exploit the fact that a toric manifold $\mathbb{C}^n /\!\!/ \mathbb{T}^k$ can be constructed from $\mathbb{C}^n$ by successive reduction by subtori of $\mathbb{T}^k$ to deduce general nondisplaceability results from those known for simpler manifolds such as products of projective spaces. Gonzalez and Woodward use a somewhat similar method in [281] to relate the toric minimal model program for a given manifold M to the positions of the nondisplaceable Lagrangian tori L_a in Δ_M.

Another classical example is provided by the results of Schur and Horn on the eigenvalues of Hermitian matrices.

Example 5.5.4 Schur [566] proved that if $\lambda_1, \dots, \lambda_n$ are the eigenvalues of a Hermitian matrix $A = A^* \in \mathbb{C}^{n\times n}$, then the vector $a = (a_1, \dots, a_n) \in \mathbb{R}^n$ of diagonal entries of A lies in the convex hull of the points

$$\sigma_*\lambda = (\lambda_{\sigma(1)}, \dots, \lambda_{\sigma(n)})$$

over all permutations $\sigma \in S_n$. Conversely, it was proved by Horn [332] that for any vector a in this convex hull there exists a Hermitian matrix A with diagonal entries a_j and eigenvalues λ_j. As noted by Atiyah [26], these results can be obtained as a special case of Theorem 5.5.1.

Let $\mathrm{G} = \mathrm{U}(n)$ be the unitary group, let $\mathbb{T} \subset \mathrm{U}(n)$ be the subgroup of diagonal matrices, and denote its Lie algebra by $\mathfrak{t} := \mathrm{Lie}(\mathbb{T})$. Thus $\mathfrak{t}$ consists of all diagonal matrices in the Lie algebra $\mathfrak{g} := \mathrm{Lie}(\mathrm{G}) = \mathfrak{u}(n)$. Denote the standard invariant inner product on $\mathfrak{g}$ by $\langle \xi, \eta \rangle = \mathrm{trace}(\xi^* \eta)$ and use it to identify $\mathfrak{g}$ with its dual. Then the orbit $\mathcal{O}_\lambda := \{ g\lambda g^{-1} \,|\, g \in \mathrm{G} \}$ of an element $\lambda \in \mathfrak{t}$ corresponds to a coadjoint orbit (Example 5.3.11) and so carries a symplectic form given by

$$\omega_\eta([\xi_1, \eta], [\xi_2, \eta]) = \langle \eta, [\xi_1, \xi_2] \rangle$$

for $\eta \in \mathcal{O}_\lambda$ and $\xi_1, \xi_2 \in \mathfrak{g}$. Example 5.3.11 shows that the adjoint action of G on $\mathcal{O}_\lambda$ is Hamiltonian and that a moment map for this action (under the above identification of $\mathfrak{g}$ with its dual $\mathfrak{g}^*$) is the obvious inclusion $\mu_{\mathrm{G}} : \mathcal{O}_\lambda \to \mathfrak{g}$. The moment map of the corresponding action of the n-torus $\mathbb{T} \subset \mathrm{G}$ is given by composing μ_{G} with the orthogonal projection $\mathfrak{g} \to \mathfrak{t}$ onto the subspace $\mathfrak{t} \subset \mathfrak{g}$ of diagonal matrices. It assigns to a skew-Hermitian matrix $A = g\lambda g^{-1} \in \mathcal{O}_\lambda$ its vector of diagonal entries, i.e.

$$\mu_{\mathbb{T}}(A) = \mathrm{diag}(A)$$

for $A \in \mathcal{O}_\lambda$. The fixed points of the torus action on $\mathcal{O}_\lambda$ are the diagonal matrices in $\mathcal{O}_\lambda$, and these are just the matrices obtained by permuting the diagonal entries of λ. Theorem 5.5.1 asserts that the set $\mu_{\mathbb{T}}(\mathcal{O}_\lambda)$ is the convex hull of these vectors and this is precisely the assertion of the theorems of Schur and Horn. (Replace $A = g\lambda g^{-1}$ by the Hermitian matrix iA.) □

The reasoning in the previous example can of course be generalized to any maximal torus in any compact Lie group. This was done by Kostant before Atiyah and Guillemin–Sternberg proved their convexity theorem (cf. [374]).

Morse–Bott functions

The proof of Theorem 5.5.1 is based on the connectedness of the level sets of the functions $H_\theta = \langle \mu, \theta \rangle$ which generate the action of the torus. These are not Morse functions in the traditional sense since their critical points need not be isolated. However, they are Morse–Bott functions and we shall briefly digress into a discussion of their properties. More background information can be found in Bott [77].

For the purpose of this discussion let M be any closed connected Riemannian manifold. A smooth function $f : M \to \mathbb{R}$ is called a **Morse–Bott function** if its critical set $\mathrm{Crit}(f) := \{x \in M \,|\, df(x) = 0\}$ is a submanifold of M and its tangent space at every point $x \in \mathrm{Crit}(f)$ is the kernel of the Hessian

$$d^2 f(x) : T_xM \times T_xM \to \mathbb{R}.$$

Thus $T_x\mathrm{Crit}(f) = \ker \nabla^2 f(x)$ for all $x \in \mathrm{Crit}(f)$, where $\nabla^2 f(x) : T_xM \to T_xM$ denotes the linear operator obtained from the Hessian via the Riemannian metric. More generally, for every $x \in M$, the operator

$$\nabla^2 f(x) : T_xM \to T_xM$$

is defined as the covariant derivative of the gradient vector field ∇f with respect to the Levi-Civita connection. It is always self-adjoint with respect to the inner product on T_xM. If f is a Morse–Bott function then its critical set decomposes into finitely many connected **critical manifolds**. Each critical manifold C is **normally hyperbolic** with respect to the negative gradient flow $\phi_t : M \to M$ of f which is defined by

$$\frac{d}{dt}\phi_t = -\nabla f \circ \phi_t, \qquad \phi_0 = \mathrm{id}.$$

More precisely, the tangent space T_xM at a point $x \in C$ decomposes as a direct sum

$$T_xM = T_xC \oplus E_x^+ \oplus E_x^-,$$

where E_x^+ and E_x^- are spanned by the positive and negative eigenspaces of $\nabla^2 f(x)$, respectively. It follows that the points $y \in M$ whose trajectories $\phi_t(y)$ converge to C as $t \to \infty$ form a manifold called the **stable manifold** $W^s(C)$, which is tangent to the bundle $TC \oplus E^+$ along C. Similarly, the points whose trajectories converge to C as $t \to -\infty$ form the **unstable manifold** $W^u(C)$, which is tangent to $TC \oplus E^-$ along C. (See Fig. 5.4.)

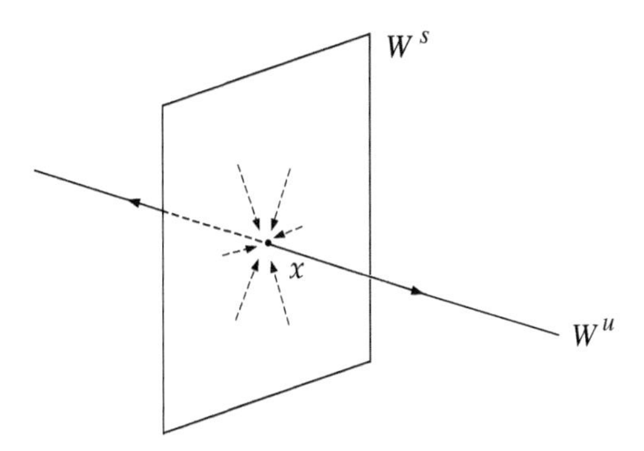

FIG. 5.4. Stable and unstable manifolds.

The **index** of a connected critical submanifold C is defined by

$$n^-(C) = \dim W^u(C) - \dim C = \operatorname{codim} W^s(C),$$

and agrees with the dimension of the negative eigenspace of the Hessian of f on the normal bundle of C. Similarly, the **coindex** of C is defined by

$$n^+(C) = \dim W^s(C) - \dim C = \operatorname{codim} W^u(C)$$

and agrees with the dimension of the positive eigenspace of the Hessian on the normal bundle of C. Since f decreases along the orbits of ϕ_t, every trajectory $\phi_t(p)$ must converge to some critical manifold as $t \to \infty$. In fact, more is true: one can show that each gradient flow line converges exponentially to a critical point in forward and backward time. (For a proof see for example the excellent textbook on dynamical systems by Zehnder [693] .) It follows that

$$M = \bigcup_C W^s(C).$$

Similarly, M is the union of all the unstable manifolds.

Lemma 5.5.5 *Let $f : M \to \mathbb{R}$ be a Morse–Bott function on a closed connected manifold M whose critical manifolds all have index and coindex $n^\pm(C) \neq 1$. Then the level set $f^{-1}(c)$ is connected for every $c \in \mathbb{R}$.*

Proof: We prove first that there is precisely one connected critical manifold of index zero. To see this, denote by C_0 the union of all the critical manifolds of index zero. Then the complement of $W^s(C_0)$ is a union of submanifolds of codimension at least 2; namely the stable manifolds of all the other critical manifolds. Hence the open set $W^s(C_0) \subset M$ is connected and hence C_0 is also connected. Similarly, there is a unique connected critical manifold of coindex 0.

Now let $c_0 < c_1 < \cdots < c_N$ be the critical levels of f. Then the critical manifold of index zero is $C_0 := f^{-1}(c_0)$ and the critical manifold of coindex zero is $C_N := f^{-1}(c_N)$. They are both connected. We prove first that the level set $f^{-1}(c)$ is connected for $c_0 < c < c_1$. To see this, note that any two points $x_0, x_1 \in f^{-1}(c)$ can be connected by flowlines to points in C_0 and these can be connected by a path in C_0. Since $\operatorname{codim} C_0 = n^+(C_0) \geq 2$, the resulting path can be disjoined from C_0 by a standard transversality argument (as in Milnor [484] or Guillemin–Pollack [293]), and then moved up to the level c by the gradient flow.

Now we shall prove that $f^{-1}(c)$ is connected for every regular value c of f. Suppose, by induction, that the level set $f^{-1}(c)$ is connected for $c < c_j$, where $j < N$. Consider two points $x_0, x_1 \in f^{-1}(c_j + \varepsilon)$. Connect them by paths in $f^{-1}(c_j + \varepsilon)$ to points in $W^s(C_0)$ and move them to the level $c_j - \varepsilon$ using the flow. Now the resulting points x_0' and x_1' can, by the induction hypothesis, be connected by a path $\gamma' : [0,1] \to f^{-1}(c_j - \varepsilon)$. This path can be chosen transverse to all the unstable manifolds. Since these all have codimension at least 2, except for $W^u(C_N)$, the path must be contained in $W^u(C_N)$. Now use the flow to move the path up to the level $c_j + \varepsilon$ to obtain the required path $\gamma : [0,1] \to f^{-1}(c_j + \varepsilon)$ which connects x_0 and x_1. This proves that $f^{-1}(c_j + \varepsilon)$ is connected and completes the induction argument. Thus we have proved that the regular level sets of f are connected.

To prove that a singular level set $f^{-1}(c_j)$ is connected for $0 < j < N$, choose some regular value $c > c_j$ such that the interval $(c_j, c]$ consists entirely of regular values. Then there is a continuous surjection

$$\psi : f^{-1}(c) \to f^{-1}(c_j),$$

defined as follows. Denote by $\phi^t : M \to M$ the downward gradient flow of f. Let $x \in f^{-1}(c)$. If $f(\phi^t(x)) > c_j$ for all $t > 0$, define

$$\psi(x) := \lim_{t \to \infty} \phi^t(x).$$

Otherwise choose $t(x) > 0$ such that $f(\phi^{t(x)}(x)) = c_j$ and define $\psi(x) = \phi^{t(x)}(x)$. Then the Morse–Bott assumption shows that ψ is surjective. Moreover, it follows from standard considerations about the limits of gradient flow lines (every sequence of flow lines has a subsequence converging in a suitable sense to a broken flow line) that ψ is continuous. Hence $f^{-1}(c_j)$ is connected for every j. □

We remark that our proof of Atiyah–Guillemin–Sternberg convexity only uses the fact that the regular level sets in Lemma 5.5.5 are connected.

Proof of convexity

We begin the proof of Theorem 5.5.1 by constructing an almost complex structure which is invariant under the action. The argument actually works for any symplectic action of any compact group.

Lemma 5.5.6 *There exists an almost complex structure J on M that is compatible with ω and invariant under the action of the torus in the sense that*

$$\psi_\theta^* J = J$$

for every $\theta \in \mathbb{T}^m$.

Proof: Fix a Riemannian metric g_0 on M and average the metrics $\psi_\theta^* g_0$ over the torus to obtain an invariant metric g. The image of this metric under the map $\mathfrak{Met}(V) \to \mathcal{J}(V, \omega) : g \mapsto J_{g,\omega}$ of Proposition 2.5.6 is an invariant almost complex structure which is compatible with ω. Note that we cannot average the almost complex structures directly. $\square$

Lemma 5.5.7 *Let $\mathrm{G} \subset \mathbb{T}^m$ be a subgroup. Then the fixed point set*

$$\mathrm{Fix}(\mathrm{G}) = \bigcap_{\theta \in \mathrm{G}} \mathrm{Fix}(\psi_\theta)$$

is a symplectic submanifold of M.

Proof: Let $x \in \mathrm{Fix}(\mathrm{G})$ and, for $\theta \in \mathrm{G}$, denote the differential of the symplectomorphism ψ_θ at x by

$$\Psi_\theta := d\psi_\theta(x) : T_x M \to T_x M.$$

These maps determine a unitary action of G on the complex symplectic vector space $(T_x M, \omega_x, J_x)$. Now consider the exponential map $\exp_x : T_x M \to M$ with respect to the invariant metric $g(v, w) = \omega(v, Jw)$. This map is equivariant in the sense that

$$\exp_x(\Psi_\theta \xi) = \psi_\theta(\exp_x(\xi))$$

for $\theta \in \mathrm{G}$ and $\xi \in T_x M$. (If $\gamma : \mathbb{R} \to M$ is a geodesic then so is $\psi_\theta \circ \gamma$.) Hence the fixed points of ψ_θ near x correspond to the fixed points of Ψ_θ on the tangent space $T_x M$. In other words,

$$T_x \mathrm{Fix}(\mathrm{G}) = \bigcap_{\theta \in \mathrm{G}} \ker(\mathbb{1} - \Psi_\theta).$$

Since the linear maps Ψ_θ are unitary transformations of $T_x M$, it follows that the eigenspace with eigenvalue 1 is invariant under J_x and is therefore a symplectic subspace. This proves Lemma 5.5.7. $\square$

We now examine the Hamiltonian functions $H_\theta = \langle \mu, \theta \rangle$ which generate the action. The next lemma shows that these are Morse–Bott functions with even-dimensional critical manifolds of even index and so, by Lemma 5.5.5, all its level sets are connected.

Lemma 5.5.8 *For every $\theta \in \mathbb{R}^m$ the function $H = H_\theta = \langle \theta, \mu \rangle : M \to \mathbb{R}$ is a Morse–Bott function with even-dimensional critical manifolds of even index. Moreover, the critical set*

$$\mathrm{Crit}(H_\theta) = \bigcap_{\tau \in T_\theta} \mathrm{Fix}(\psi_\tau)$$

is a symplectic submanifold. Here

$$T_\theta := \mathrm{cl}(\{t\theta + k \,|\, t \in \mathbb{R},\, k \in \mathbb{Z}^m\}/\mathbb{Z}^m)$$

is the closed subtorus generated by θ.

Proof: Assume first that θ has rationally independent components so that the vectors $t\theta + k$ for $t \in \mathbb{R}$ and $k \in \mathbb{Z}^m$ form a dense set in $\mathbb{R}^m$. Then every critical point x of $H = \langle \theta, \mu \rangle$ is fixed under the action of the whole group $\mathbb{T}^m$, that is

$$\mathrm{Crit}(H) = \bigcap_{\tau \in \mathbb{T}^m} \mathrm{Fix}(\psi_\tau).$$

Hence it follows from Lemma 5.5.7 with $\mathrm{G} = \mathbb{T}^m$ that the critical set of H is a symplectic submanifold of M (with finitely many components). We must prove that it is normally hyperbolic with respect to the gradient flow of H, in other words that its normal bundle is the sum of the tangent bundles to the stable and unstable manifolds, and that these both have even index. To see this, consider the Hessian

$$S_x = \nabla^2 H(x) : T_x M \to T_x M$$

of H at x. The linear vector field $dX_H(x) = -J_x S_x$ on $T_x M$ generates the 1-parameter group $\Psi_{t\theta} := \exp(-tJ_x S_x)$ and hence the kernel of S_x corresponds to the fixed points of the matrices $\Psi_{t\theta}$. Since θ has rationally independent components these are the common fixed points of the matrices Ψ_τ for $\tau \in \mathbb{T}^m$ and hence

$$T_x \mathrm{Crit}(H) = \bigcap_{\tau \in \mathbb{T}^m} \ker(\mathbb{1} - \Psi_\tau) = \ker S_x.$$

Since the group generated by $-J_x S_x$ is unitary the linear map S_x commutes with J_x. Hence all its eigenspaces are invariant under J_x and are therefore even-dimensional. This proves that the manifold of critical points of H is normally hyperbolic and has even index with respect to the gradient flow of H. This also shows again that the tangent space $T_x \mathrm{Crit}(H) = \ker S_x$ is a complex and hence symplectic subspace of $T_x M$.

Thus we have proved the lemma under the assumption that θ has rationally independent components. The general case follows by restricting to the action of the closed subtorus $T_\theta \subset \mathbb{T}^m$. □

Denote the components of the moment map $\mu : M \to \mathbb{R}^m$ by

$$\mu = (\mu_1, \dots, \mu_m).$$

Call μ **irreducible** if the 1-forms $d\mu_1, \dots, d\mu_m$ are linearly independent and **reducible** otherwise. If μ is reducible then the function

$$H_\theta = \sum_{j=1}^m \theta_j \mu_j \tag{5.5.1}$$

is constant for some nonzero vector $\theta \in \mathbb{R}^m$. In this case the action reduces to that of an $(m-1)$-torus. To see this, simply neglect a component with $\theta_j \neq 0$. More precisely, there exists an action

$$\mathbb{T}^{m-1} \to \mathrm{Symp}(M, \omega) : \tau \mapsto \psi'_\tau$$

with moment map $\mu' : M \to \mathbb{R}^{m-1}$ and an integer matrix $A \in \mathbb{Z}^{(m-1)\times m}$ such that

$$\psi_\theta = \psi'_{A\theta}, \qquad \mu(x) = A^{\mathrm{T}} \mu'(x)$$

for $\theta \in \mathbb{T}^m$ and $x \in M$. Hence in the reducible case the convexity of $\mu(M)$ follows from the convexity of $\mu'(M)$.

Proof of Theorem 5.5.1: Our goal is to prove that the image $\mu(M)$ is convex by induction over the dimension m of the torus. Following Atiyah [26], we shall prove first, again by induction over m, that the level set $\mu^{-1}(\eta)$ is connected for every regular value $\eta \in \mathbb{R}^m$. In the case $m = 1$ this is precisely the assertion of Lemma 5.5.5 for the function $f = \mu : M \to \mathbb{R}$ which, by Lemma 5.5.8, satisfies the assumptions of Lemma 5.5.5. Assume, by induction, that the assertion has been proved for every Hamiltonian action of an $(m-1)$-torus on M. If our action of $\mathbb{T}^m$ is reducible then there is nothing to prove because $\mu^{-1}(\eta) = \emptyset$ for every regular value of μ. Hence assume that μ is irreducible. Then the function $H_\theta : M \to \mathbb{R}$ in (5.5.1) is nonconstant for every nonzero vector $\theta \in \mathbb{R}^m$. In this case the set

$$Z = \bigcup_{\theta \neq 0} \mathrm{Crit}(H_\theta)$$

is a countable union of proper submanifolds of M. (We shall see later that it is actually a finite union.) To see this, recall from Lemma 5.5.8 that the critical points of H_θ are the fixed points of the action of the closed subtorus $T_\theta \subset \mathbb{T}^m$ and form even-dimensional proper submanifolds. Since the fixed point set decreases as the torus increases it suffices to consider 1-dimensional subtori or, equivalently, integer vectors θ. This proves the assertion about Z. Hence it follows from Baire's category theorem that the complement $M \setminus Z$ is dense in M. Moreover, $x \in M \setminus Z$ if and only if the linear functionals $d\mu_1(x), \dots, d\mu_m(x)$ are linearly independent and hence $M \setminus Z$ is open.

We now claim that the set of regular values of μ intersects the image of μ in a dense subset of $\mu(M)$. To see this, let $\eta := \mu(x) \in \mu(M)$ and approximate x by a sequence $x_j \in M \setminus Z$. Then the image $\mu(M)$ contains a neighbourhood of $\mu(x_j)$. By Sard's theorem, there exists a regular value $\eta_j \in \mathbb{R}^m$ of μ which is arbitrarily close to $\mu(x_j)$ and so $\mu^{-1}(\eta_j) \neq \emptyset$. Hence $\eta_j \in \mu(M)$ is a sequence of regular values of μ that converges to η, which proves the claim. A similar argument shows that the set of all points $\eta \in \mu(M)$ such that $(\eta_1, \dots, \eta_{m-1})$ is a regular value of the reduced moment map $(\mu_1, \dots, \mu_{m-1})$ is dense in $\mu(M)$.

Now let $\eta = (\eta_1, \dots, \eta_m) \in \mathbb{R}^m$ be a regular value of μ and assume, in addition, that $(\eta_1, \dots, \eta_{m-1})$ is a regular value of $(\mu_1, \dots, \mu_{m-1})$. Under these assumptions we will prove that the submanifold $\mu^{-1}(\eta)$ is connected. By the induction hypothesis, the manifold

$$Q := \bigcap_{j=1}^{m-1} \mu_j^{-1}(\eta_j)$$

is connected. Consider the function $\mu_m|_Q : Q \to \mathbb{R}$. A point $x \in Q$ is critical for $\mu_m|_Q$ if and only if there exist real numbers $\theta_1, \dots, \theta_{m-1}$ such that

$$\sum_{j=1}^{m-1} \theta_j d\mu_j(x) + d\mu_m(x) = 0.$$

This means that x is a critical point of the function $H_\theta = \langle \mu, \theta \rangle : M \to \mathbb{R}$, where $\theta = (\theta_1, \dots, \theta_{m-1}, 1) \in \mathbb{R}^m$. By Lemma 5.5.8, all the critical sets of H_θ are even-dimensional manifolds and have even index. Let $C \subset M$ be the critical manifold of H_θ which contains x. We now prove that C intersects Q transversally at x, i.e. $T_xM = T_xC + T_xQ$, which holds precisely if the linear functionals $d\mu_j(x) : T_xM \to \mathbb{R}$, $j = 1, \dots, m-1$, remain linearly independent when restricted to the subspace T_xC. To this end, observe that by Lemma 5.5.8, C is a symplectic submanifold of M. Moreover, the Hamiltonian flow of μ_j commutes with that of H_θ and so must preserve the critical manifold C. Thus the Hamiltonian vector fields

$$X_j := X_{\mu_j} : M \to TM$$

are all tangent to C, and so $X_1(x), \dots, X_{m-1}(x) \in T_xC$. By assumption, these vectors are linearly independent. Because T_xC is a symplectic subspace of T_xM there exists, for every nonzero vector $\lambda \in \mathbb{R}^{m-1}$, a nonzero vector $\xi \in T_xC$ such that

$$0 \neq \omega\left(\sum_{j=1}^{m-1} \lambda_j X_j(x), \xi \right) = \sum_{j=1}^{m-1} \lambda_j d\mu_j(x)\xi.$$

This shows that the linear functionals $d\mu_j(x) : T_xC \to \mathbb{R}$, $j = 1, \dots, m-1$, are linearly independent, so that C is transverse to Q as claimed.

This implies that the subspace $T_xQ \cap T_xC^{\perp}$ is a complement of T_xC in T_xM. Hence the Hessian of H_θ is nondegenerate on this space with even index and coindex. In other words, $C \cap Q$ is a critical manifold for $H_\theta|_Q$ of even index and coindex. The same holds for $\mu_m|_Q$ since it only differs from H_θ by the constant $\sum_{j=1}^{m-1} \eta_j \theta_j$. Thus we have proved that the function $\mu_m : Q \to \mathbb{R}$ has only critical manifolds of even index and coindex. Moreover, η_m is a regular value of $\mu_m : Q \to \mathbb{R}$ because η is a regular value of μ. Hence it follows from Lemma 5.5.5 that the level set

$$\mu^{-1}(\eta) = Q \cap \mu_m^{-1}(\eta_m)$$

is connected. Thus we have proved that the set $\mu^{-1}(\eta)$ is connected whenever η is a regular value of μ and $(\eta_1, \ldots, \eta_{m-1})$ is a regular value of the reduced moment map $(\mu_1, \ldots, \mu_{m-1})$. But the set V of such points is dense in $\mathbb{R}^m$. Now let $U \subset \mathbb{R}^m$ be the (open and dense) set of regular values of μ and denote by $\nu(\eta) := \#\pi_0(\mu^{-1}(\eta))$ the number of connected components of $\mu^{-1}(\eta)$ for $\eta \in U$. Then $\nu : U \to \mathbb{N}_0$ is a locally constant function on U and $\nu(\eta) \in \{0, 1\}$ for $\eta \in V$. Since V is a dense subset of U by Sard's theorem, it follows that $\nu(\eta) \in \{0, 1\}$ for every $\eta \in U$. Hence $\mu^{-1}(\eta)$ is connected for every regular value η of μ.

Now we prove, by induction over m, that the set $\mu(M)$ is convex. In the case $m = 1$ convexity follows from the fact that $\mu(M)$ is connected. Now assume that convexity has been proved for any Hamiltonian action of $\mathbb{T}^{m-1}$. Then, for every reducible action of $\mathbb{T}^m$, convexity follows directly from the induction hypothesis. Hence assume that our action of $\mathbb{T}^m$ is irreducible. Under this assumption we have seen that the set of regular values of μ is dense in $\mu(M)$. Choose an injective integer matrix $A \in \mathbb{Z}^{m \times (m-1)}$ and consider the torus action

$$\mathbb{T}^{m-1} \to \mathrm{Symp}(M, \omega) : \theta \mapsto \psi_{A\theta}$$

with moment map

$$\mu_A = A^{\mathrm{T}} \mu : M \to \mathbb{R}^{m-1}.$$

Our assumption implies that this action is also irreducible and hence the set of regular values of μ_A is dense in $\mu_A(M)$. Moreover, by the first part of the proof, the set $\mu_A^{-1}(\eta)$ is connected for every regular value $\eta \in \mathbb{R}^{m-1}$ of μ_A. But for any given point x_0 with $A^{\mathrm{T}} \mu(x_0) = \eta$ this set can be written in the form

$$\mu_A^{-1}(\eta) = \left\{ x \in M \,|\, \mu(x) - \mu(x_0) \in \ker A^{\mathrm{T}} \right\}.$$

Since $\ker A^{\mathrm{T}}$ is 1-dimensional, this implies that if $\mu(x_1) - \mu(x_0) \in \ker A^{\mathrm{T}}$ and $\eta = A^{\mathrm{T}} \mu(x_0)$ is a regular value of μ_A, then every convex combination of $\mu(x_0)$ and $\mu(x_1)$ belongs to the image of μ, i.e. $(1 - t)\mu(x_0) + t\mu(x_1) \in \mu(M)$ for $0 \le t \le 1$. To see this, just connect x_0 and x_1 by a path in $\mu_A^{-1}(\eta)$. Now any two points $x_0, x_1 \in M$ can be approximated arbitrarily closely by points x_0', x_1' such that $\mu(x_1') - \mu(x_0') \in \ker A^{\mathrm{T}}$ for some injective integer matrix $A \in \mathbb{Z}^{m \times (m-1)}$. With a further perturbation we may assume that $\eta = A^{\mathrm{T}} \mu(x_0')$ is a regular value of μ_A, and hence every convex combination of $\mu(x_0')$ and $\mu(x_1')$ lies in the image of μ. Taking the limits $x_0' \to x_0$ and $x_1' \to x_1$, we obtain that $\mu(M)$ is convex.

Now, by Lemma 5.5.7, the fixed point set C of the action decomposes into finitely many connected symplectic submanifolds $C_1, \dots, C_N$ of M. The moment map μ is constant on each of these sets because $C_j \subset \mathrm{Crit}(H_\theta)$ for every j and every $\theta \in \mathbb{R}^m$. Hence $\eta_j = \mu(C_j) \in \mathbb{R}^m$ is a single point for each j. By what we have proved so far the convex hull of the points η_j is contained in the image of μ. Conversely, let $\eta \in \mathbb{R}^m$ be a point which is not in the convex hull of the η_j and choose a vector $\theta \in \mathbb{R}^m$ with rationally independent components such that $\langle \eta_j, \theta\rangle < \langle \eta, \theta\rangle$ for all j. Then the subtorus T_θ generated by θ as in Lemma 5.5.8 is the whole of $\mathbb{T}^n$, so that any point fixed by T_θ is in fact fixed by $\mathbb{T}^n$. Hence Lemma 5.5.8 implies that the function $H_\theta = \langle \mu, \theta\rangle$ attains its maximum on one of the sets C_j. Thus

$$\sup_{p\in M} \langle \mu(p), \theta\rangle < \langle \eta, \theta\rangle,$$

so that $\eta \notin \mu(M)$. This shows that $\mu(M)$ is the convex hull of the points η_j, which completes the proof of Theorem 5.5.1. □

5.6 Localization

We shall now return to circle actions on a closed symplectic $2n$-manifold (M, ω). Let $H : M \to \mathbb{R}$ be a smooth function that generates a circle action

$$\mathbb{R}/\mathbb{Z} \to \mathrm{Ham}(M) : t \mapsto \psi_t$$

via

$$\partial_t \psi_t = X \circ \psi_t, \qquad \psi_0 = \mathrm{id}, \qquad \iota(X)\omega = dH.$$

Our goal in this section is to prove the following beautiful localization theorem by Duistermaat and Heckman [168]. It expresses an integral over M that can be interpreted as the *Fourier transform* of H in terms of a sum over contributions from the fixed points of the action. This formula has attracted considerable interest. It can be viewed as a special case of a *stationary phase formula* (see Guillemin–Sternberg [299]) and, among many other applications, can be used to compute the ring structure of the cohomology of symplectic quotients (see, for example, Jeffrey–Weitsman [348] and Martin [439]). There are many other localization formulas that compute other characteristic classes in terms of the fixed points set of a group action; for a typical application to the study of S^1-actions on almost complex manifolds see Li–Liu [426].

Theorem 5.6.1 (Duistermaat–Heckman)
Consider a circle action on a closed manifold (M, ω) that is generated by a Morse function $H : M \to \mathbb{R}$. Then

$$\int_M e^{-\hbar H} \frac{\omega^n}{n!} = \sum_p \frac{e^{-\hbar H(p)}}{\hbar^n e(p)} \tag{5.6.1}$$

for every $\hbar \in \mathbb{C}$, where the sum runs over all critical points of H and $e(p) \in \mathbb{Z}$ is the product of the **weights** *at p.*

Proof: See page 245. □

Remark 5.6.2 **(i)** The action of S^1 on T_pM is conjugate to an n-fold product of circle actions on $\mathbb{C}$ by $z \mapsto e^{-2\pi i k_j t} z$ for $t \in S^1$ (see the proof of Lemma 5.5.8) and we define $e(p) = k_1 \cdots k_n$. Here we use the conjugate action to make the formula (5.6.1) consistent with that in Duistermaat and Heckman [168] and Atiyah and Bott [28]. In particular, a minimum has positive weights while a maximum has negative weights.

(ii) Both sides of (5.6.1) can be interpreted as power series in an indeterminate $\hbar$ and then the equation decomposes into a sequence of equations by comparing coefficients. For example,

$$\int_M \omega^n = (-1)^n \sum_p \frac{H(p)^n}{e(p)} \tag{5.6.2}$$

and

$$\sum_p \frac{H(p)^k}{e(p)} = 0$$

for $k = 0, \ldots, n-1$.

(iii) By Lemma 5.5.8, the Hamiltonian function $H : M \to \mathbb{R}$ of a circle action is always a Morse–Bott function with even-dimensional critical manifolds of even index and coindex. The normal bundle of every such manifold C splits into complex line bundles $\nu_C = L_1 \oplus \cdots \oplus L_m$ on which the circle acts with weights $k_1, \ldots, k_m$. In this case the contribution of C to the integral is

$$e^{-\hbar H(C)} \int_C \prod_{j=1}^m \frac{1}{c_1(L_j) + k_j \hbar}.$$

Here we take the formal inverse

$$\frac{1}{c_1(L) + k\hbar} = \frac{1}{k\hbar} \sum_{i=0}^{\dim C/2} \left(-\frac{c_1(L)}{k\hbar} \right)^i .$$

(iv) An interesting test case for the Duistermaat–Heckman formula is the standard circle action on S^2. If we take S^2 to be the standard 2-sphere in $\mathbb{R}^3$ with the standard area form ω, then it has area

$$\int_{S^2} \omega = 4\pi.$$

Moreover, the Hamiltonian function $H(x) = 2\pi x_3$ generates a circle action of period 1. (It rotates the equator with speed 2π.) The critical points are the north and south poles $p_N = (0,0,1)$ and $p_S = (0,0,-1)$, with

$$H(p_N) = 2\pi, \qquad e(p_N) = -1, \qquad H(p_S) = -2\pi, \qquad e(p_S) = 1.$$

(v) More generally, consider the circle action on $\mathbb{C}\mathrm{P}^n$ given by

$$[z_0 : \cdots : z_n] \mapsto [z_0 : e^{-2\pi it} z_1 : \cdots : e^{-2\pi int} z_n].$$

This is generated by the Hamiltonian

$$H([z_0 : \cdots : z_n]) = \frac{\sum_{k=1}^n \pi k |z_k|^2}{\sum_{k=0}^n |z_k|^2}$$

and has fixed points $p_k = [0 : \cdots : 1 : \cdots : 0]$, for $0 \le k \le n$. Thus $H(p_k) = \pi k$, and it is easy to check that $e(p_k) = (-1)^k k!(n-k)!$. Therefore formula (5.6.2) becomes

$$\int_{\mathbb{C}\mathrm{P}^n} \omega^n = (-1)^n \sum_{k=1}^n \frac{(-1)^k (\pi k)^n}{k!(n-k)!}.$$

Since

$$\int_{\mathbb{C}\mathrm{P}^n} \omega^n = n! \mathrm{Vol}\,(\mathbb{C}\mathrm{P}^n) = \pi^n,$$

this is equivalent to the identity

$$(-1)^n n! = \sum_{k=1}^n \binom{n}{k} (-1)^k k^n. \qquad \Box$$

In our proof of Theorem 5.6.1 we follow Berline and Vergne [61] and Atiyah and Bott [28]. The proof is based on a generalization of Proposition 5.1.9 to the case of finite isotropy subgroups. The most elegant formulation of this generalization is in terms of equivariant cohomology. We now explain the de Rham version of this. Denote by $\Omega^k_{S^1}(M)$ the space of smooth k-forms α on M which are invariant under the action of S^1. This means that

$$\mathcal{L}_X \alpha = d\iota(X)\alpha + \iota(X)d\alpha = 0.$$

Also denote by $\Omega^*_{S^1}(M)[\hbar]$ the ring of polynomials in $\hbar$ with coefficients in $\Omega^*_{S^1}(M)$. Here $\hbar$ is to be understood as a generator of degree 2 and so an element of $\Omega^k_{S^1}(M)[\hbar]$ can be written in the form

$$\alpha = \alpha_k + \hbar\alpha_{k-2} + \hbar^2\alpha_{k-4} + \cdots,$$

where $\alpha_j \in \Omega^j_{S^1}(M)$. Note that this is a finite sum. Consider the differential

$$d_\hbar = d + \iota(X)\hbar$$

on $\Omega^*_{S^1}(M)[\hbar]$. This is an operator of degree 1 and it is easy to see that $d_\hbar \circ d_\hbar = 0$. This construction works for any S^1 action on any manifold and the cohomology of the resulting equivariant de Rham complex can be naturally identified with the equivariant cohomology $H^*_{S^1}(M)$. Here we shall only use the fact that this construction is natural with respect to smooth equivariant maps. For example, the operator $d_\hbar$ commutes with pullback by an equivariant map f and, if f is equivariantly homotopic to the identity, then

$$d_\hbar\alpha = 0 \qquad \Longrightarrow \qquad \alpha - f^*\alpha \in \operatorname{im} d_\hbar$$

for every $\alpha \in \Omega^*_{S^1}(M)[\hbar]$. The crucial point is the following localization lemma which implies that every top-dimensional equivariant differential form is equivalent to a form which is concentrated near the fixed points.

Lemma 5.6.3 (Localization) *Assume that the action has isolated fixed points and*

$$\tau = \tau_{2n} + \hbar\tau_{2n-2} + \cdots + \hbar^n\tau_0$$

is a $d_\hbar$-closed $2n$-form such that τ_0 vanishes on the fixed points of the action. Then τ is $d_\hbar$-exact. In particular,

$$\int_M \tau_{2n} = 0.$$

Proof: Assume first that the action has no fixed points. Then the vector field X has no zeros and so there exists a 1-form $\alpha \in \Omega^1(M)$ such that

$$\iota(X)\alpha = 1, \qquad \iota(X)d\alpha = 0.$$

To see this just choose any 1-form with $\alpha(X) = 1$ and average over the group action to obtain an invariant such form. Now a $2n$-form $\tau \in \Omega^{2n}_{S^1}(M)[\hbar]$ is in the kernel of $d_\hbar$ if and only if

$$d\tau_{2k-2} + \iota(X)\tau_{2k} = 0$$

for every k. Define $\sigma = \sigma_{2n-1} + \hbar\sigma_{2n-3} + \cdots + \hbar^{n-1}\sigma_1$ inductively by

$$\sigma_1 = \tau_0\alpha, \qquad \sigma_{2k+1} = \tau_{2k} \wedge \alpha - \sigma_{2k-1} \wedge d\alpha.$$

Then a simple inductive calculation shows that

$$\tau_{2k} = d\sigma_{2k-1} + \iota(X)\sigma_{2k+1}$$

as claimed.

Now assume that the action has fixed points and τ_0 vanishes on these. In this case choose an equivariant smooth map $\phi : M \to M$ which is isotopic to the identity and maps a neighbourhood of the fixed point set onto the fixed point set. (For example, use the exponential map with respect to an invariant metric.) Then $\phi^*\tau - \tau$ is $d_\hbar$-exact and $\phi^*\tau$ vanishes on a neighbourhood of the fixed point set. Hence the first part of the proof shows that $\phi^*\tau$ is $d_\hbar$-exact and so is τ. □

The previous lemma holds for every S^1-action with isolated fixed points and does not make use of the symplectic structure or the Hamiltonian function.

However, in the case of a Hamiltonian action there is an interesting equivariantly closed 2-form

$$\omega - \hbar H \in \ker d_\hbar$$

and Theorem 5.6.1 is a localization formula for the exponential of this form. Note that in the equivariant case this exponential is an infinite sum. To prove the localization theorem we need the following lemma, which asserts the existence of a *pushforward*. To understand this geometrically, consider the space $M_{S^1} := M \times_{S^1} ES^1$ as a fibre bundle over $BS^1 = \mathbb{C}\mathrm{P}^\infty$ with fibres symplectomorphic to M. For each isolated fixed point $p \in \mathrm{Fix}$ there is an obvious section $f_p : \mathbb{C}\mathrm{P}^\infty \to M_{S^1}$ and the submanifold $N_p := f_p(\mathbb{C}\mathrm{P}^\infty) \subset M_{S^1}$ can be represented by a $2n$-form $\tau_p \in \Omega^{2n}(M_{S^1})$. This form can be interpreted as the *Poincaré dual* of the homology class $[N_p]$ or as the *pushforward* of the class $\mathbb{1} \in H^0(N_p)$ under the map f_p. Explicitly, this form is supported near N_p, and restricts to a top degree form of total volume 1 on each fibre of the normal bundle. The important fact is that the pullback

$$f_p{}^* f_{p*} \mathbb{1} = e(p)\hbar^n \in H^{2n}(\mathbb{C}\mathrm{P}^\infty)$$

represents the Euler class of the normal bundle. Here $\hbar \in H^2(\mathbb{C}\mathrm{P}^\infty)$ is to be understood as the standard generator and $e(p)$ is the product of the weights as above. The following lemma rephrases this idea in terms of equivariant de Rham cohomology. The proof was explained to us by Shaun Martin.

Lemma 5.6.4 *Assume that the circle action is Hamiltonian and the critical points of H are all nondegenerate. Then for every fixed point p there exists an equivariant differential form*

$$\tau_p = \tau_{p,2n} + \hbar\tau_{p,2n-2} + \cdots + \hbar^n \tau_{p,0} \quad \in \quad \Omega^{2n}_{S^1}(M)[\hbar]$$

which is supported in an arbitrarily small neighbourhood of p and satisfies

$$\int_M \tau_{p,2n} = 1, \qquad \tau_{p,0}(p) = e(p), \qquad d_\hbar \tau_p = 0.$$

Proof: Consider the splitting

$$T_pM = L_1 \oplus \cdots \oplus L_n$$

into complex subspaces (with respect to an invariant complex structure) such that S^1 acts on L_j by multiplication with θ^{-k_j}. Now let

$$\phi_j : L_j \to S^2$$

be a smooth map which maps 0 to the south pole p_S, maps the complement of the unit ball to the north pole p_N, and is equivariant with respect to the k_j-fold rotational action of S^1 on S^2. Let ω_0 be the standard area form on S^2 with total area one, and let $H_0 : S^2 \to [-1, 0]$ be the Hamiltonian function with

$$H_0(p_S) = -1, \qquad H_0(p_N) = 0,$$

which generates the standard rotation. Then the equivariant 2-form

$$\omega_0 - k_j \hbar H_0$$

is equivariantly closed with respect to the k_j-fold cover of the standard action. Now the equivariant differential form

$$\sigma_p = \prod_{j=1}^{n} {\phi_j}^*(\omega_0 - k_j \hbar H_0)$$

on T_pM has integral 1 (the degree-$2n$ term), and evaluates to $\hbar^n e(p)$ at 0 (the degree-0 term). Since a neighbourhood of 0 in T_pM can be identified with a neighbourhood of p in M this proves Lemma 5.6.4. □

Proof of Theorem 5.6.1: For $k \geq n$ consider the equivariant differential form

$$\sigma = \hbar^{n-k}(\omega - \hbar H)^k - \sum_p \frac{(-H(p))^k}{e(p)} \tau_p$$

in $\Omega^{2n}_{S^1}(M)[\hbar]$, where $\tau_p \in \Omega^{2n}_{S^1}(M)[\hbar]$ is the differential form constructed in Lemma 5.6.4. Then σ is equivariantly closed and its degree-0 term vanishes on the fixed points. By Lemma 5.6.3, this implies that the integral of its degree-$2n$ term must also vanish. Thus

$$\binom{k}{n} \int_M (-H)^{k-n} \omega^n = \sum_p \frac{(-H(p))^k}{e(p)}$$

for $k \geq n$. Since for $k = n$ the integral on the left is independent of the choice of the Hamiltonian function H which generates the action, it follows that the sum on the right for $k = n$ remains unchanged if a constant is added to H. This implies that the sum vanishes for $0 \leq k \leq n - 1$. Now these identities can be conveniently expressed in the form

$$\int_M (\omega - \hbar H)^k = \sum_p \frac{(-\hbar H(p))^k}{\hbar^n e(p)},$$

where the expression on the left is to be understood as the integral of the degree-$2n$ term. This formula holds for every $k \geq 0$. Now divide by $k!$ and take the sum over all k to obtain (5.6.1). □

5.7 Remarks on GIT

Let G be a compact Lie group (with Lie algebra $\mathfrak{g} := \mathrm{Lie}(\mathrm{G})$) that acts on a closed Kähler manifold (M, ω, J) by Kähler isometries. In this situation it is interesting to extend the group action to an action of the complexified Lie group G^c. When G is a Lie subgroup of $\mathrm{U}(n)$ its complexification is the subgroup

$$\mathrm{G}^c := \{\exp(i\eta)u \,|\, \eta \in \mathfrak{g},\, u \in \mathrm{G}\} \tag{5.7.1}$$

of $\mathrm{GL}(n, \mathbb{C})$. In general the complexified group is characterized as the unique (up to isomorphism) complex Lie group G^c that contains G as a maximal compact subgroup, has a connected quotient G^c/G, and whose Lie algebra

$$\mathfrak{g}^c := \mathrm{Lie}(\mathrm{G}^c) = \mathfrak{g} \oplus i\mathfrak{g}$$

is the complexification of $\mathfrak{g} = \mathrm{Lie}(\mathrm{G})$. Equivalently, every Lie group homomorphism from G to a complex Lie group extends uniquely to a Lie group homomorphism from G^c to that complex Lie group. Taking the group of holomorphic diffeomorphisms of (M, J) as the target group, one obtains by the discussion after equation (4.2.3) a holomorphic action of G^c on M. A complex Lie group that is the complexification of a compact subgroup is called **reductive**. The action of such groups on Kähler manifolds or algebraic varieties and the study of their quotients is the subject of geometric invariant theory (GIT).[25] There is a vast literature on this subject; see for example Mumford–Fogarty–Kirwan [497], Kempf [365], Kempf–Ness [366], Ness [503, 504] and the references therein. A key observation is that the quotient M/G^c need not even be Hausdorff and it is necessary to restrict to a suitable subset of *stable* or *semi-stable* points to obtain a *good* quotient which is again an algebraic variety. In the Kähler setting, where the G-action is Hamiltonian and generated by a moment-map $\mu : M \to \mathfrak{g}^*$, the stability conditions can be defined as follows. An element $p \in M$ is called

μ-**unstable** iff $\overline{\mathrm{G}^c(p)} \cap \mu^{-1}(0) = \emptyset$,
μ-**semistable** iff $\overline{\mathrm{G}^c(p)} \cap \mu^{-1}(0) \neq \emptyset$,
μ-**polystable** iff $\mathrm{G}^c(p) \cap \mu^{-1}(0) \neq \emptyset$,
μ-**stable** iff $\mathrm{G}^c(p) \cap \mu^{-1}(0) \neq \emptyset$ and $\mathrm{G}^c_p := \{g \in \mathrm{G}^c \,|\, gp = p\}$ is discrete.

These conditions depend on the choice of the moment map. The sets M^{ss} of semistable points and M^{s} of stable points are open, while the set M^{ps} of polystable points need not be open.

Example 5.7.1 The most basic example is given by the action of $\mathrm{G} = S^1$ on $S^2 = \mathbb{C} \cup \{\infty\}$ by rotation about the origin. In this case $\mathrm{G}^c = \mathbb{C}^*$ has precisely two fixed points at $0, \infty$, all other points lying on a single G^c-orbit. Thus the naive quotient S^2/G^c consists of three points with a nonHausdorff

[25] In GIT the reductive group is usually denoted by G (instead of G^c) and the compact subgroup by K (instead of G).

topology. If we take a moment map $\mu : S^2 \to \mathbb{R}$ in which $\mu^{-1}(0)$ is a circle, then $M^{\mathrm{s}} = M^{\mathrm{ss}} = M^{\mathrm{ps}} = \mathbb{C} \setminus \{0\}$, while M^{us} consists of the two fixed points. However, if μ is normalized so that $\mu(0) = 0$, then there are no stable points, and we have $M^{\mathrm{ps}} = \{0\} \subset M^{\mathrm{ss}} = \mathbb{C}$ while $M^{\mathrm{us}} = \{\infty\}$. See Exercise 5.7.6 below for a more thorough investigation of this example. □

In geometric invariant theory one studies the complex quotient $M^{\mathrm{ss}}/\mathrm{G}^c$, also called the **GIT quotient**. The subset $M^{\mathrm{ps}}/\mathrm{G}^c$ is naturally isomorphic to the symplectic quotient $M/\!\!/\mathrm{G}$. Its *smooth part*

$$M^{\mathrm{s}}/\mathrm{G}^c \cong M^{\mathrm{s}}/\!\!/\mathrm{G} := (\mu^{-1}(0) \cap M^{\mathrm{s}})/\mathrm{G}$$

is an orbifold corresponding to the open subset $M^{\mathrm{s}}/\!\!/\mathrm{G}$ of $M/\!\!/\mathrm{G}$. Geometric invariant theory was originally developed to study complex quotients and in the original treatment of Mumford the symplectic form and moment map were not used. However, as mentioned by Lerman [409], Mumford became aware of the existence of a relation between the complex quotient and the Marsden–Weinstein quotient, a connection that was subsequently explored in much greater depth by several authors (see Kirwan [369], Guillemin–Sternberg [296], and Ness [504]). This relation is a source of many deep connections between symplectic and complex geometry. One can approach the entire subject of GIT from either an algebro geometric or a symplectic viewpoint (see [262] for an exposition of the latter). The first emphasizes the complex structure of the quotient while the latter emphasizes the symplectic structure. A value of the symplectic approach is that it may sometimes be easier to understand quotients by a compact group than by a noncompact group. Moreover, one can study the stable and semistable points via the gradient flow of the norm squared of the moment map, an approach that often carries over to suitable infinite-dimensional settings, as in Atiyah–Bott [27].

An important part of GIT is the study of the stability conditions. They can be characterized in terms of the **Mumford numerical invariants**

$$w_\mu(p, \xi) := \lim_{t \to \infty} \langle \mu(\exp(it\xi)p), \xi \rangle \tag{5.7.2}$$

associated to a point $p \in M$ and an element $\xi \in \Lambda$, where

$$\Lambda := (\xi \in \mathfrak{g} \setminus \{0\} \mid \exp(\xi) = 1\} .$$

When one chooses an invariant inner product on the Lie algebra $\mathfrak{g}$ of the compact Lie group G, and uses it to identify $\mathfrak{g}$ with its dual $\mathfrak{g}^*$, the Mumford numerical invariants are related to the norm of the moment map on the complexified group orbit by the **moment-weight inequality**

$$\sup_{\xi \in \Lambda} \frac{-w_\mu(p, \xi)}{|\xi|} \le \inf_{g \in \mathrm{G}^c} |\mu(gp)|. \tag{5.7.3}$$

Equality holds whenever the right hand side is positive. Moreover, the weights $w_\mu(p, \xi)$ are well-defined for all $\xi \in \mathfrak{g} \setminus \{0\}$, and the supremum in (5.7.3) agrees with the supremum over all $\xi \in \mathfrak{g} \setminus \{0\}$. (See [262]; the proof uses the fact that the function $\xi \mapsto w_\mu(p,\xi)$ is continuous for torus actions, although it is not continuous in general.) It follows from (5.7.3) that every critical point of the norm squared of the moment map minimizes the norm of the moment map on its complexified group orbit (see Ness [504] for the projective case). It also follows from (5.7.3) that a semistable element $p \in M^{\mathrm{ss}}$ cannot have any negative weight. In fact the **Hilbert–Mumford criterion** asserts the following.

Theorem 5.7.2 *For every $p \in M$ the following holds.*

(i) *p is μ-semistable if and only if $w_\mu(p,\xi) \geq 0$ for all $\xi \in \Lambda$.*

(ii) *p is μ-polystable if and only if $w_\mu(p,\xi) \geq 0$ for all $\xi \in \Lambda$ and*

$$w_\mu(p,\xi) = 0 \qquad \Longrightarrow \qquad \lim_{t\to\infty} \exp(it\xi)p \in \mathrm{G}^c(p).$$

(iii) *p is μ-stable if and only if $w_\mu(p,\xi) > 0$ for all $\xi \in \Lambda$.*

Proof: See Mumford–Fogarty–Kirwan [497] in the algebraic geometric setting and [262] for Hamiltonian group actions on general closed Kähler manifolds. □

The proof of Theorem 5.7.2 uses the gradient flow of the norm squared of the moment map and of the **Kempf–Ness function** $\Phi_p : \mathrm{G}^c/\mathrm{G} \to \mathbb{R}$ associated to a point $p \in M$. The invariant inner product on $\mathfrak{g}$ determines a complete Riemannian metric with nonpositive sectional curvature on the quotient manifold G^c/G, and the geodesics have the form $\gamma(t) = [g\exp(it\eta)]$ for $g \in \mathrm{G}^c$ and $\eta \in \mathfrak{g}$ (see e.g. [262, Appendix B]). The lift of the Kempf–Ness function to G^c (denoted by the same letter Φ_p) is the unique real valued function on G^c that satisfies

$$d\Phi_p(g)\widehat{g} = -\langle \mu(g^{-1}p), \mathrm{Im}(g^{-1}\widehat{g})\rangle, \qquad \Phi_p|_{\mathrm{G}} \equiv 0, \tag{5.7.4}$$

for $\widehat{g} \in T_g\mathrm{G}^c$. It descends to the quotient G^c/G, is convex along geodesics, has critical points $[g]$ where $\mu(g^{-1}p) = 0$, and the map $\mathrm{G}^c \to \mathrm{G}^c(p) : g \mapsto g^{-1}p$ intertwines the negative gradient flow of the lifted Kempf–Ness function Φ_p (given by the differential equation $g^{-1}\dot{g} = i\mu(g^{-1}p)$ on G^c) with the negative gradient flow of the function

$$f := \tfrac{1}{2}|\mu|^2 : M \to \mathbb{R}$$

(given by the differential equation $\dot{p} = -JL_p\mu(p)$; see equation (5.2.6)). The function Φ_p is asymptotically linear along each geodesic and $w_\mu(p,\xi)$ is its asymptotic slope along the geodesic ray $[0,\infty) \to \mathrm{G}^c/\mathrm{G} : t \mapsto [\exp(-it\xi)]$. The generalized Kempf–Ness theorem asserts that a point $p \in M$ is μ-semistable if and only if Φ_p is bounded below, and that it is μ-stable if and only if Φ_p is bounded below and proper (see [262, Theorem 8.1]).

The stability conditions can also be defined in purely algebraic geometric terms in the case when the moment map determines a lift of the G^c action to a holomorphic line bundle $L \to M$ with first Chern class $c_1(L) = -[\omega]$. (For the relation between such a lift and the moment map, see [262, Section 10].) The

Mumford numerical invariant $w_\mu(p, \xi)$ is then the weight of the action of the one-parameter subgroup determined by ξ on the *central fibre* of L over the limit point $p^+ := \lim_{t\to\infty} \exp(it\xi)p$ of the $\mathbb{C}^*$ orbit of p. Moreover, in this situation the Kempf–Ness theorem asserts that p is μ-polystable if and only if the corresponding orbit in L is closed. For a full exposition of this story see the original work of Mumford–Fogarty–Kirwan [497], Kempf–Ness [366], and Ness [503, 504], as well as Thomas [622], Woodward [690], and Georgoulas–Robbin–Salamon [262].

Example 5.7.3 The archetypal example is a linear group action on projective space. More precisely, let V be a finite-dimensional complex vector space and $\mathrm{G} \subset \mathrm{GL}(V)$ be a compact Lie group with complexification $\mathrm{G}^c \subset \mathrm{GL}(V)$. Call a nonzero vector $v \in V$

unstable iff $0 \in \overline{\mathrm{G}^c(v)}$,

semistable iff $0 \notin \overline{\mathrm{G}^c(v)}$,

polystable iff $\mathrm{G}^c(v) = \overline{\mathrm{G}^c(v)}$,

stable iff $\mathrm{G}^c(v) = \overline{\mathrm{G}^c(v)}$ and $\mathrm{G}^c_v := \{g \in \mathrm{G}^c \,|\, gv = v\}$ is discrete.

This fits into our discussion as follows. The linear action of G^c on V induces an action on the projectivization

$$M := \mathbb{P}(V).$$

Fix a G-invariant Hermitian structure on V. As in [262] we choose a scaling factor [26] $\hbar > 0$ and restrict the symplectic structure to the sphere of radius $r := \sqrt{2\hbar}$. It is S^1-invariant and descends to $\mathbb{P}(V)$. (This is $2\hbar$ times the Fubini–Study form. Its integral over the positive generator of $\pi_2(\mathbb{P}(V))$ is $2\pi\hbar$.) The standard moment map for this action is given by

$$\langle \mu(p), \xi \rangle = \hbar \frac{\langle v, i\xi v \rangle}{|v|^2}$$

for $\xi \in \mathfrak{g}$, $v \in V \setminus \{0\}$ and $p = [v] \in \mathbb{P}(V)$ (see Exercise 5.3.8). The Kempf–Ness function associated to $p = [v] \in \mathbb{P}(V)$ is given by

$$\Phi_p(g) = \hbar\Big(\log|g^{-1}v| - \log|v|\Big).$$

The Kempf–Ness theorem in its original form asserts that $p = [v]$ is μ-semistable (respectively μ-polystable or μ-stable) if and only if v is semistable (respectively

[26] Think of $\hbar$ as Planck's constant. If the symplectic form ω is replaced by a positive real multiple $c\omega$ and $\hbar$ is replaced by $c\hbar$, all the formulas which follow remain correct. Our choice of $\hbar$ is consistent with the physics notation in that $\omega, \mu, \hbar$ have the units of action and the Hamiltonian $H_\xi = \langle \mu, \xi \rangle$ has the units of energy.

polystable or stable); see Kempf–Ness [366] and also [262, Theorem 9.5]. Moreover, the Mumford numerical invariants of $p = [v]$ are given by

$$w_\mu(p, \xi) = \hbar \max_{v_j \neq 0} \lambda_j,$$

where $\lambda_1 < \cdots < \lambda_k$ are the eigenvalues of $i\xi$, the $V_j \subset V$ are the corresponding eigenspaces, and

$$v = \sum_{j=1}^{k} v_j, \qquad v_j \in V_j.$$

This shows that $w_\mu(p, \xi) < 0$ if and only if v is contained in the direct sum of the negative eigenspaces of $i\xi$. Thus the Hilbert–Mumford criterion for semistability asserts in this case that $0 \in \overline{\mathrm{G}^c(v)}$ (i.e. v is not semistable) if and only if there exists an element $\xi \in \Lambda$ such that $\lim_{t\to\infty} \exp(it\xi)v = 0$ (i.e. $w_\mu(p, \xi) < 0$). This is the Hilbert–Mumford criterion in its original form. □

Remark 5.7.4 In the appendix of [504] Mumford used GIT to prove Atiyah–Guillemin–Sternberg convexity for linear torus actions on algebraic varieties. □

Exercise 5.7.5 Verify the moment-weight inequality and the Hilbert–Mumford criterion for linear torus actions on $\mathbb{C}\mathrm{P}^n$. □

Exercise 5.7.6 Find a formula for the Kempf–Ness function associated to the standard circle action on the 2-sphere, where the moment map is the height function and the quotient space G^c/G is the real axis. Examine how the Kempf–Ness function changes as one adds a constant to the height function and how it depends on the point in the 2-sphere. Prove the generalized Kempf–Ness theorem (which characterizes the μ-stability conditions in terms of the properties of the Kempf–Ness function) in this example. □

Exercise 5.7.7 Show that $\mathrm{PSL}(2, \mathbb{C})$ is the complexification of $\mathrm{SO}(3)$ and that the complexification of the standard action of $\mathrm{SO}(3)$ on S^2 is the action of $\mathrm{PSL}(2, \mathbb{C})$ by Möbius transformations. **Hint:** Stereographic projection. □

Exercise 5.7.8. (The Mumford Quotient) Let $n \in \mathbb{N}$ and consider the action of $\mathrm{G} := \mathrm{SO}(3)$ on $M := (S^2)^n$ with the moment map

$$\mu(x) = \sum_{i=1}^{n} x_i.$$

(See Example 5.3.4.) Prove that the Mumford numerical invariants are given by

$$w_\mu(x, \xi) = \#\{i \,|\, x_i \neq -\xi\} - \#\{i \,|\, x_i = -\xi\}$$

for $x = (x_1, \ldots, x_n) \in (S^2)^n$ and $\xi \in S^2 \subset \mathbb{R}^3 \cong \mathfrak{so}(3)$. Verify that the function $\mathfrak{g}\backslash\{0\} \to \mathbb{R} : \xi \mapsto w_\mu(x, \xi)$ is not continuous. If n is odd prove that an element $x \in (S^2)^n$ is semistable if and only if it is polystable if and only if it is stable. Prove that the three notions of stability do not agree when n is even. Prove the moment-weight inequality and the Hilbert–Mumford criterion in this example. Describe the Marsden–Weinstein quotient (called the Mumford quotient in this example) geometrically. □

Infinite-dimensional versions of GIT

There are many remarkable infinite-dimensional analogues of GIT in various areas of geometry which at first glance may have little connection to symplectic geometry. The discovery of this connection is in many cases due to Donaldson. One example is the Donaldson–Uhlenbeck–Yau correspondence between stable holomorphic vector bundles and Hermitian Yang–Mills connections over Kähler manifolds. (This is a special case of the Kobayashi–Hitchin correspondence.) The moment map picture in this setting goes back to Atiyah–Bott [27] for vector bundles over Riemann surfaces (see Example 5.3.18 and Exercise 5.4.23). In this case the analogue of the Hilbert–Mumford criterion is the Narsimhan–Seshadri theorem [502] relating stable bundles to flat connections over Riemann surfaces (see Donaldson [139] for another proof). In [140] Donaldson established a correspondence between stable bundles over Kähler surfaces and anti-self-dual instantons and used it to prove nontriviality of the Donaldson invariants for Kähler surfaces. Donaldson's result was extended to higher dimensions by Uhlenbeck–Yau [632].

Another infinite-dimensional analogue of geometric invariant theory is Donaldson's program [147,151,152] for the study of constant scalar curvature Kähler (cscK) metrics (see Example 5.3.19). In this example the analogue of the moment map is the scalar curvature, the analogue of the Kempf–Ness function is the Mabuchi functional [436], the analogues of the weights are the Futaki invariants [255], and the analogue of the Hilbert–Mumford criterion is the Donaldson–Tian–Yau conjecture relating K-stability to the existence of a cscK metric. In the Fano case (where a cscK metric is a Kähler–Einstein metric) this conjecture has been confirmed by Chen–Donaldson–Sun [105, 106, 107, 108].

In symplectic geometry, Abreu–Granja–Kitchloo [10] use Donaldson's ideas in Example 5.3.19 to study the action of the symplectomorphism group of $S^2 \times S^2$ on the space of compatible integrable complex structures from a GIT point of view, with the aim of understanding the change of the homotopy type of the symplectomorphism group $\mathrm{Symp}(S^2 \times S^2, \omega_\lambda)$ as the form $\omega_\lambda := \lambda \mathrm{pr}_1^*\omega_{\mathrm{FS}} + \mathrm{pr}_2^*\omega_{\mathrm{FS}}$ on $S^2 \times S^2$ varies.

A fourth infinite-dimensional example goes back to a construction of Richard Thomas [621] and is related to special Lagrangian submanifolds of Calabi–Yau manifolds. In this setting, Jake Solomon [595] defined an analogue of the Kempf–Ness function which here takes the form of a real valued function on (the universal cover of) a Hamiltonian isotopy class of Lagrangian submanifolds; its critical points are the special Lagrangian submanifolds in this isotopy class.

For further infinite-dimensional examples and related topics, the reader is referred to the papers by Donaldson [149, 153, 156, 157, 158], Fine et al. [159, 213, 214, 215], and Thomas [621, 622]. All these geometric interconnections and results illustrate the importance of ideas from symplectic geometry in other fields of mathematics.

6

SYMPLECTIC FIBRATIONS

This chapter begins with a general discussion of symplectic fibrations. Its main feature is a simple construction of a symplectic form on the total space of a symplectic fibration that is due to Thurston. The next section describes in detail the basic example of symplectic ruled surfaces. These are symplectic 2-sphere bundles over Riemann surfaces. Section 6.3 develops the notions of symplectic connection and holonomy, which were already implicit in Thurston's proof. We explain the Sternberg–Weinstein universal construction for fibre bundles that was mentioned in Chapter 5, and discuss Seidel's construction of generalized Dehn twists. In Section 6.4 we show that symplectic connections with Hamiltonian holonomy play a special role. They give rise to the notion of a coupling form, which is due to Guillemin, Lerman, and Sternberg and is related to the reduction techniques developed in Chapter 5. Finally, in Section 6.5 we study Hamiltonian fibrations, that is, fibrations whose structural group reduces to the Hamiltonian group, and show that they are precisely the symplectic bundles that are trivial over the 1-skeleton of the base and that have a closed connection 2-form.

6.1 Symplectic fibrations

Let F be a smooth manifold. A **locally trivial fibration with fibre** F is a smooth map $\pi : M \to B$ between smooth manifolds equipped with an open cover $\{U_\alpha\}_\alpha$ of B and a collection of diffeomorphisms $\phi_\alpha : \pi^{-1}(U_\alpha) \to U_\alpha \times F$ such that the following diagram commutes for every α:

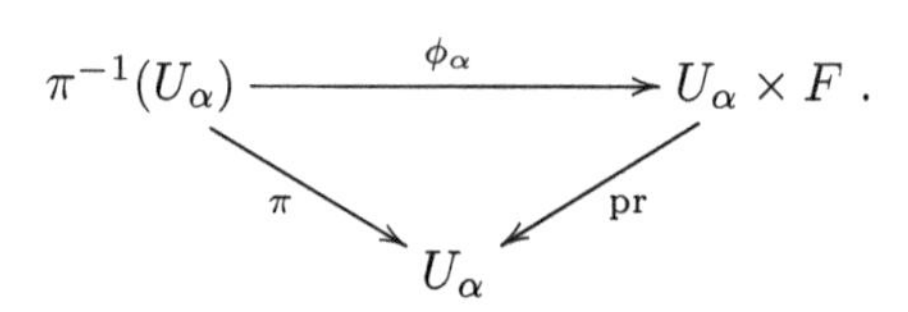

The maps ϕ_α are called **local trivializations**. We denote by $F_b = \pi^{-1}(b)$ the fibre over $b \in B$ and by $\phi_\alpha(b) : F_b \to F$ the restriction of ϕ_α to F_b followed by the projection onto F. The maps $\phi_{\beta\alpha} : U_\alpha \cap U_\beta \to \mathrm{Diff}(F)$ defined by

$$\phi_{\beta\alpha}(b) = \phi_\beta(b) \circ \phi_\alpha(b)^{-1}$$

are called the **transition functions**. For each $b \in B$ the inclusion of the fibre is denoted by $\iota_b : F_b \to M$. Throughout we shall assume that M is connected and compact without boundary unless explicit mention is made to the contrary.

Introduction to Symplectic Topology. Dusa McDuff, Dietmar Salamon. ©Oxford University Press 2017. Published 2017 by Oxford University Press.

Exercise 6.1.1 Let M be a closed connected smooth manifold. Prove that a smooth map $\pi : M \to B$ is a locally trivial fibration if and only if it is a surjective submersion, i.e. it is surjective and the differential $d\pi(p) : T_pM \to T_{\pi(p)}B$ is surjective for every element $p \in M$. □

In many interesting examples the transition functions preserve an additional structure on the fibre F. A fibration is said to have **structure group**

$$\mathrm{G} \subset \mathrm{Diff}(F)$$

if the transition functions all take values in G. For example, if the fibre $F = V$ is a vector space then a locally trivial fibration with structure group $\mathrm{G} = \mathrm{Aut}(V)$ is a vector bundle. Another example is where F is an oriented manifold and the transition maps all preserve the orientation. In this case $\pi : M \to F$ is called an **oriented fibration.** In this section we are interested in the case where the fibre is a compact symplectic manifold (F, σ) and the structure group G is contained in the group of symplectomorphisms $\mathrm{Symp}(F, \sigma)$, i.e.

$$\phi_{\beta\alpha}(b) \in \mathrm{Symp}(F, \sigma)$$

for all α, β and all $b \in U_\alpha \cap U_\beta$. In this case $\pi : M \to B$ is called a **symplectic fibration.** If $\pi : M \to B$ is a symplectic fibration then each fibre F_b carries a natural symplectic structure $\sigma_b \in \Omega^2(F_b)$ defined by

$$\sigma_b = \phi_\alpha(b)^*\sigma$$

for $b \in U_\alpha$. It follows from the definition that this form is independent of the choice of α.

Definition 6.1.2 *Let $\pi : M \to B$ be a symplectic fibration. A symplectic form $\omega \in \Omega^2(M)$ is called* **compatible with the symplectic fibration** *if, for every $b \in B$, the restriction of ω to the fibre F_b agrees with σ_b. In terms of the local trivializations $\phi_\alpha : \pi^{-1}(U_\alpha) \to U_\alpha \times F$ this means that the maps $\phi_\alpha(b)^{-1} : (F, \sigma) \to (M, \omega)$ are symplectic embeddings for all α and all $b \in U_\alpha$. In this situation we also say that the* **fibration $\pi : M \to B$ is compatible with ω.**

Lemma 6.1.3 *Let $\pi : M \to B$ be a locally trivial fibration with connected base and $\omega \in \Omega^2(M)$ be a symplectic form such that the fibres are all symplectic submanifolds of M. Then $\pi : M \to B$ admits the structure of a symplectic fibration which is compatible with ω.*

Proof: This is an exercise with hints. First use Stokes' theorem to prove that the symplectic forms

$$\sigma_b = \iota_b^*\omega \in \Omega^2(F_b)$$

all represent the same cohomology class in $H^2(F)$ under the local trivializations $\phi_\alpha(b) : F_b \to F$. Then use Moser's stability theorem to show that the fibres (F_b, σ_b) are all symplectomorphic to a standard fibre (F, σ). Deduce that the structure group of the fibration can be reduced to $\mathrm{Symp}(F, \sigma)$. Note that this argument works only because M is assumed closed. □

Our goal in this section is to understand the following problem. Given a fibration $\pi : M \to B$ with symplectic fibre (F, σ) and structure group

$$\mathrm{G} \subset \mathrm{Symp}(F, \sigma),$$

find conditions under which M admits a compatible symplectic form ω. We would also like to know which cohomology classes on M can be represented by compatible symplectic forms, and the relation between two compatible symplectic forms in a given cohomology class. There is an obvious necessary condition for existence. Namely, there must be a cohomology class $a \in H^2(M)$ which pulls back to the cohomology classes

$$[\sigma_b] \in H^2(F_b)$$

under the injections

$$\iota_b : F_b \to M.$$

The next result shows that if M is compact and the base B is a symplectic manifold as well, then this condition is also sufficient.

Theorem 6.1.4 (Thurston) *Let $\pi : M \to B$ be a compact symplectic fibration with symplectic fibre (F, σ) and connected symplectic base (B, β). Denote by $\sigma_b \in \Omega^2(F_b)$ the canonical symplectic form on the fibre F_b and suppose that there is a cohomology class $a \in H^2(M)$ such that*

$$\iota_b^* a = [\sigma_b]$$

for some (and hence every) $b \in B$. Then, for every sufficiently large real number $K > 0$, there exists a symplectic form $\omega_K \in \Omega^2(M)$ which is compatible with the fibration π and represents the class $a + K[\pi^*\beta]$.

Proof: Let $\tau_0 \in \Omega^2(M)$ be any closed 2-form which represents the class $a \in H^2(M)$. For any α, denote by $\sigma_\alpha \in \Omega^2(U_\alpha \times F)$ the 2-form obtained from σ via pullback under the projection $U_\alpha \times F \to F$. If the open sets $U_\alpha \subset B$ in the cover are chosen to be contractible then it follows from our assumption on the class a that the 2-forms

$$\phi_\alpha^* \sigma_\alpha - \tau_0 \in \Omega^2(\pi^{-1}(U_\alpha))$$

are exact. Choose a collection of 1-forms $\lambda_\alpha \in \Omega^1(\pi^{-1}(U_\alpha))$ such that

$$\phi_\alpha^* \sigma_\alpha - \tau_0 = d\lambda_\alpha.$$

Now choose a partition of unity $\rho_\alpha : B \to [0, 1]$ which is subordinate to the cover $\{U_\alpha\}_\alpha$ and define $\tau \in \Omega^2(M)$ by

$$\tau := \tau_0 + \sum_\alpha d((\rho_\alpha \circ \pi)\lambda_\alpha).$$

This 2-form is obviously closed and represents the class a. Moreover, the 1-form $d(\rho_\alpha \circ \pi)$ vanishes on vectors tangent to the fibre and hence

$$\begin{aligned}\iota_b^*\tau &= \iota_b^*\tau_0 + \sum_\alpha (\rho_\alpha \circ \pi)\iota_b^* d\lambda_\alpha \\ &= \sum_\alpha (\rho_\alpha \circ \pi)\iota_b^*(\tau_0 + d\lambda_\alpha) \\ &= \sum_\alpha (\rho_\alpha \circ \pi)\iota_b^*{\phi_\alpha}^*\sigma_\alpha \\ &= \sum_\alpha (\rho_\alpha \circ \pi)\sigma_b \\ &= \sigma_b.\end{aligned}$$

The form τ need not be nondegenerate. However, it is nondegenerate on the (vertical) subspaces

$$\mathrm{Vert}_x = \ker d\pi(x) \subset T_xM.$$

Hence τ determines a field of horizontal subspaces

$$\mathrm{Hor}_x = (\mathrm{Vert}_x)^\tau = \{\xi \in T_xM \mid \tau(\xi, \eta) = 0 \;\forall\, \eta \in V_x\}.$$

The subspace $\mathrm{Hor}_x \subset T_xM$ is a horizontal complement of Vert_x and the projection $d\pi(x) : \mathrm{Hor}_x \to T_{\pi(x)}B$ is a bijection. Now the form

$$\omega_K = \tau + K\pi^*\beta$$

agrees with τ on Vert_x, is nondegenerate on Hor_x for $K > 0$ sufficiently large, and satisfies $\omega_K(\xi, \eta) = 0$ for $\xi \in \mathrm{Hor}_x$ and $\eta \in \mathrm{Vert}_x$. This shows that ω_K is nondegenerate for K sufficiently large. □

Thurston's construction is very simple and appealing. However, as the following example shows, not every fibration $\pi : M \to B$ which has a compatible symplectic form has a symplectic manifold as its base.

Exercise 6.1.5 (Weinstein [674]) Identify $\mathbb{C}^4$ with the quaternionic plane $\mathbb{H}^2$ and consider the obvious fibration $\pi : \mathbb{C}\mathrm{P}^3 \to \mathbb{H}\mathrm{P}^1$. What are the fibres? Show that the Fubini–Study form ω_{FS} on $\mathbb{C}\mathrm{P}^3$ is nondegenerate on these fibres, so that π is a symplectic fibration. Show also that $\mathbb{H}\mathrm{P}^1 \cong S^4$ and hence does not carry a symplectic form. □

The proof of Thurston's Theorem 6.1.4 shows that there are two steps in the construction of the symplectic form ω on the total space M of a symplectic fibration. The first step is to find a closed 2-form $\tau \in \Omega^2(M)$ which restricts to the canonical symplectic forms σ_b on the fibres, i.e.

$$\iota_b^*\tau = \sigma_b, \qquad d\tau = 0. \tag{6.1.1}$$

The second step is to add the pullback of a form on the base B in order to get a nondegenerate form on M. This second step is obvious whenever the base B

is a symplectic manifold. Hence we shall concentrate now on the first step. The proof of Theorem 6.1.4 shows that the required form τ exists whenever there is a corresponding cohomology class $a \in H^2(M;\mathbb{R})$ with $\iota_b^* a = [\sigma_b]$ for $b \in B$. The following example shows that such a class a need not always exist.

Example 6.1.6 Consider the Hopf surface

$$M = S^3 \times S^1 = (\mathbb{C}^2 \setminus \{0\})/\mathbb{Z},$$

where the generator of $\mathbb{Z}$ acts on $\mathbb{C}^2 \setminus \{0\}$ by multiplication by 2. This manifold does not admit a symplectic structure because $H^2(M;\mathbb{R}) = 0$. However, it is easy to see that the projection

$$\pi : M \to \mathbb{C}\mathrm{P}^1,$$

defined by $\pi(z) = [z_0 : z_1]$ for $z = (z_0, z_1) \in \mathbb{C}^2 \setminus \{0\}$, is a symplectic fibration with fibre $\mathbb{T}^2 = (\mathbb{C} \setminus \{0\})/\mathbb{Z}$. Here we use the fact that every locally trivial fibration with a 2-dimensional fibre admits the structure of a symplectic fibration (see below). □

The following lemma gives a simple criterion for the existence of the required cohomology class a.

Lemma 6.1.7 *Let $\pi : M \to B$ be a symplectic fibration with fibres symplectomorphic to (F, σ). Assume that the first Chern class $c_1(TF) \in H^2(F;\mathbb{R})$ is a nonzero multiple of the class $[\sigma]$. Then there exists a cohomology class $a \in H^2(M;\mathbb{R})$ such that $\iota_b^* a = [\sigma_b]$ for every $b \in B$.*

Proof: By assumption, $c_1(TF) = \lambda[\sigma]$ for some constant $\lambda \neq 0$ and hence $c_1(TF_b) = \lambda[\sigma_b]$ for every $b \in B$. Now consider the vector bundle

$$\mathrm{Vert} \subset TM$$

whose fibre at $x \in M$ is the vertical subspace $\mathrm{Vert}_x = \ker d\pi(x) \subset T_x M$. This is a symplectic vector bundle whose pullback bundle under the inclusion $\iota_b : F_b \to M$ agrees with the tangent bundle $TF_b = \iota_b^* \mathrm{Vert}$. Hence its first Chern class $c = c_1(\mathrm{Vert}) \in H^2(M;\mathbb{Z})$ satisfies

$$\iota_b^* c = c_1(TF_b) = \lambda[\sigma_b]$$

for every $b \in B$. Thus the class $a = \lambda^{-1} c \in H^2(M;\mathbb{R})$ is as required. □

Corollary 6.1.8 *Let (F, σ) be a compact symplectic manifold whose first Chern class $c_1(TF) \in H^2(F;\mathbb{R})$ is a nonzero multiple of the class $[\sigma]$. Then the total space M of any symplectic fibration $\pi : M \to B$ with fibre (F, σ) and compact symplectic base B carries a compatible symplectic structure.*

Proof: Theorem 6.1.4 and Lemma 6.1.7. □

This result applies when F is any compact oriented surface except for $\mathbb{T}^2$. However, in higher dimensions the assumption on the fibres is very restrictive. Gotay et al. [282] investigated conditions which imply the existence of an extension $a \in H^2(M)$ of the classes $[\sigma_b] \in H^2(F_b)$. They showed that if the base B and fibre F of a fibre bundle are both simply connected and if the restriction maps $\iota_b^* : H^2(M;\mathbb{R}) \to H^2(F_b;\mathbb{R})$ are not surjective, then the groups $H^{2k}(F;\mathbb{R})$ are nonzero for all $k \geq 0$. In particular, this means that the fibre F must be infinite-dimensional. This shows that the total space M of a symplectic fibration carries a closed 2-form τ which satisfies (6.1.1) whenever F and B are compact and simply connected. We shall prove this in Section 6.4 assuming only that the fibre is simply connected. This result has deep interconnections with *symplectic gauge theory.*

6.2 Symplectic 2-sphere bundles

The above results give a fairly complete answer to the question of which fibrations over a symplectic base support a compatible symplectic structure. However, they do not throw any light on the question of which cohomology classes a on M can be represented by π-compatible symplectic forms on M. They also say nothing about the uniqueness question, the subtlety of which was demonstrated by Kasper in his thesis [360] where he gave an example of a symplectic form on a product space $M_1 \times M_2$ which is compatible with both projections, but is not isotopic to a product form. In this subsection we restrict attention to smooth 4-manifolds M that fibre over closed oriented 2-manifolds with fibres diffeomorphic to S^2. Such 4-manifolds are called **ruled surfaces**. We address the question of which cohomology classes on ruled surfaces can be represented by symplectic forms. (The uniqueness problem is discussed further in Section 13.4.) The reader is cautioned that the fibration is no longer *symplectic* but only *oriented* and thus compatibility has a new meaning. Note also that the total space of an oriented fibration over an oriented base is naturally oriented.

Definition 6.2.1 *Let $\pi : M \to B$ be an oriented fibration over an oriented base. A symplectic form $\omega \in \Omega^2(M)$ is called* **compatible with the oriented fibration** *if ω is compatible with the orientation of M and, for every $b \in B$, the restriction of ω to the fibre F_b is a symplectic form compatible with the orientation of F_b.*

Theorem 6.2.2 *Let F be a compact oriented 2-manifold of genus $g(F) \neq 1$. Then the total space of any oriented fibration $\pi : M \to B$ with fibre F and compact symplectic base B admits a symplectic form that is compatible with the oriented fibration.*

Proof: We first claim that π admits the structure of a symplectic fibration. The proof is based on the observation that the structure group of a principal G-bundle $P \to B$ can be reduced to the subgroup $\mathrm{H} \subset \mathrm{G}$ if and only if the associated G/H-bundle $Q = P \times_{\mathrm{G}} \mathrm{G/H}$ admits a section. Apply this to the case where $\mathrm{G} = \mathrm{Diff}^+(F)$, $\mathrm{H} = \mathrm{Symp}(F,\sigma)$, and $\mathcal{P} \to B$ is a $\mathrm{Diff}^+(F)$-bundle constructed as in Remark 6.4.11. Then the quotient $\mathrm{Diff}^+(F)/\mathrm{Symp}(F,\sigma)$ can be identified with the space of symplectic forms on F. In the 2-dimensional case this space is contractible, and so the structure group of the fibration reduces to $\mathrm{Symp}(F,\sigma)$. The result now follows from Corollary 6.1.8. □

In general, it is an open question as to which classes $a \in H^2(M)$ can be represented by symplectic forms (that are compatible with the oriented fibration). However, there is one case in which this problem has been completely solved, namely that of S^2-bundles over orientable 2-manifolds. We begin by describing the diffeomorphism types of these bundles.

Lemma 6.2.3 *For every closed oriented 2-manifold B there are exactly two orientable S^2-bundles with base B, namely the product $B \times S^2$ and the nontrivial bundle X_B.*

Proof: As in Section 2.6, one can prove that every S^2-bundle over a compact connected Riemann surface with boundary admits a trivialization. Hence, split $B = B_1 \cup_C B_2$ and trivialize the bundle over each half. Thus

$$X = (B_1 \times S^2) \cup (B_2 \times S^2)/\sim,$$

where $(b,z) \sim (b, \phi_b(z))$ for $b \in C = B_1 \cap B_2$ and $z \in S^2$. Here

$$C \cong S^1 \to \mathrm{Diff}^+(S^2) : b \mapsto \phi_b$$

is a loop in the structure group $\mathrm{Diff}^+(S^2)$ of orientation-preserving diffeomorphisms of the fibre. It is easy to see that diffeomorphism classes of bundles correspond to homotopy classes of loops. Thus the oriented bundles with fibre S^2 are classified by the group

$$\pi_1(\mathrm{Diff}^+(S^2)) \cong \pi_1(\mathrm{SO}(3)) \cong \mathbb{Z}/2\mathbb{Z}.$$

This proves the lemma.

Here is an alternative argument. S^2-bundles over any base B are classified by homotopy classes of maps $B \to B\mathrm{Diff}^+(S^2) = B\mathrm{SO}(3)$. Here, $B\mathrm{Diff}^+(S^2)$ denotes the classifying space of the group $\mathrm{Diff}^+(S^2)$. For any group G this space is defined as the quotient E/G, where E is a contractible space on which the group acts freely. The identity $B\mathrm{Diff}^+(S^2) = B\mathrm{SO}(3)$ follows from the fact that $\mathrm{Diff}^+(S^2)$ deformation retracts onto its linear part SO(3) (see Smale [591]). The space $E\mathrm{SO}(3)$ can be realized as the space of orthonormal 3-frames in $\mathbb{R}^\infty = \bigcup_n \mathbb{R}^n$. Since $E\mathrm{SO}(3)$ is contractible it follows from the homotopy exact sequence of the fibration $E\mathrm{SO}(3) \to B\mathrm{SO}(3)$ with fibre SO(3) that

$$\pi_1(B\mathrm{SO}(3)) \cong \pi_0(\mathrm{SO}(3)) = 0, \qquad \pi_2(B\mathrm{SO}(3)) \cong \pi_1(\mathrm{SO}(3)) = \mathbb{Z}/2\mathbb{Z}.$$

Hence it follows from elementary obstruction theory that for every compact oriented Riemann surface B, there are precisely two homotopy classes of maps from $B \to B\mathrm{SO}(3)$ giving rise to two diffeomorphism classes of S^2-bundles over B. For more details see Milnor and Stasheff [488]. □

An explicit construction for the bundles X_B is sketched in the following exercise. The space X_{S^2} is discussed from a different point of view in Example 7.1.5 below.

Exercise 6.2.4 Let $P \to B$ be an S^1-bundle over a closed oriented 2-manifold with Euler number $e = e(P) \in \mathbb{Z}$. (The Euler number agrees with the Chern number of the associated complex line bundle over B.) Define $M = P \times_{S^1} S^2$ as in Example 5.1.11. Prove that the S^2 bundle $M \to B$ is trivial if and only if the Euler number $e(P)$ is even, and is nontrivial if and only if it is odd. See Example 4.4.2. □

Exercise 6.2.4 shows that the product bundle admits sections $B_{2k} \subset B \times S^2$ with even self-intersection number $2k$, and X_B admits sections $B_{2k+1} \subset X_B$ with odd self-intersection number $2k+1$. In either case the corresponding homology classes satisfy $[B_{k+2}] = [B_k] + [F]$, where $[F] \in H_2(X;\mathbb{Z})$ denotes the homology class of the fibre. In the case of the trivial bundle $B \times S^2$, we choose a basis

$$b = \mathrm{PD}([B \times \mathrm{pt}]), \qquad f = \mathrm{PD}([\mathrm{pt} \times S^2])$$

of $H^2(B \times S^2;\mathbb{R})$, where PD denotes the Poincaré dual. In the case of the nontrivial bundle, we will work with the basis

$$b_+ = \mathrm{PD}([B_1]), \qquad b_- = -\mathrm{PD}([B_{-1}]) \tag{6.2.1}$$

of $H^2(X_B)$. Thus

$$\langle b_+, [B_1]\rangle = \langle b_-, [B_{-1}]\rangle = 1, \qquad \langle b_\pm, [B_{\mp 1}]\rangle = 0.$$

Moreover, $[B_1] \cdot [B_{-1}] = 0$ and hence we have the following relations:

$$\langle b_+ \cup b_+, [X_B]\rangle = 1, \qquad \langle b_- \cup b_-, [X_B]\rangle = -1, \qquad b_+ \cup b_- = 0.$$

A cohomology class $\mu_1 b + \mu_2 f$ on $B \times S^2$ can be represented by a symplectic form which is compatible with both fibrations if and only if $\mu_1 > 0$ and $\mu_2 > 0$. Now consider the nontrivial bundle X_B and suppose that the class

$$a := \mu_+ b_+ + \mu_- b_- \in H^2(X_B;\mathbb{R})$$

can be represented by a symplectic form which is compatible with the fibration. Then we have

$$0 < \langle a^2, [X_B]\rangle = \mu_+^2 - \mu_-^2,$$

and so $|\mu_+| > |\mu_-|$. Moreover, compatibility implies that a is positive on the class $[F] = [B_1] - [B_{-1}]$ of the fibre, hence $\mu_+ - \mu_- > 0$, and hence

$$\mu_+ > |\mu_-|.$$

However, note that μ_- need not be positive here. It turns out that in the case $B = S^2$ these classes a with $a^2 > 0$ do not all have compatible symplectic representatives, though they do for all other B. The following theorem is proved in McDuff [449, 454]. Recall the compatibility condition of Definition 6.2.1.

Theorem 6.2.5 *Let $a := \mu_+ b_+ + \mu_- b_- \in H^2(X_B;\mathbb{R})$, where $b_\pm \in H^2(X_B;\mathbb{R})$ are defined by* (6.2.1) *and $\mu_\pm \in \mathbb{R}$.*

(i) *If $B = S^2$ then there is a symplectic form ω on X_B in the class a that is compatible with the oriented fibration if and only if $\mu_+ > \mu_- > 0$.*

(ii) *If B has positive genus then there is a symplectic form ω on X_B in the class a that is compatible with the oriented fibration if and only if $\mu_+ > |\mu_-|$.*

Proof: See [449, 454] and the exercises below. □

The following exercises show how to construct symplectic forms on X_B in each of the classes a mentioned above. These symplectic forms are all homotopic. The other statement in (i) — that we must have $\mu_- > 0$ in order for a to be a symplectic class — is much harder to prove, since it uses the theory of J-holomorphic curves. The idea is to show that the Gromov invariant of the class B_{-1} is 1, so that this class always has a symplectic representative. If one adds the full force of Taubes–Seiberg–Witten theory, one can in fact prove that all cohomologous symplectic forms on manifolds such as $B \times S^2$ and X_B are diffeomorphic. These questions are discussed further in Section 13.3.

Exercise 6.2.6 Show that the methods of Example 5.1.11 give rise to an S^1-invariant symplectic form on X_B in each class a such that $\mu_+ > \mu_- > 0$. **Hint:** Consider the S^2-bundle $\pi : X_B \to B$ associated to the circle bundle with Euler class 1. Observe that the zero section of the associated complex line bundle has self-intersection 1 and so represents the class $[B_1]$. In Example 5.1.11, the zero section is the maximal critical level of the moment map and is denoted by Z_1. □

Exercise 6.2.7 Let $\pi : M \to B$ be an oriented fibration with closed oriented two-dimensional fibre F and closed oriented two-dimensional base B. Denote by $\mathcal{S}(M,\pi)$ the space of all symplectic forms on M that are compatible with the oriented fibration (see Definition 6.2.1). Prove that $\mathcal{S}(M,\pi)$ is connected. **Hint:** Let τ be a positive area form on B. If $\omega \in \mathcal{S}(M,\pi)$, show that $\omega + K\pi^*\tau \in \mathcal{S}(M,\pi)$ for all $K \geq 0$. If $\omega_0, \omega_1 \in \mathcal{S}(M,\pi)$, show that $(1-t)\omega_0 + t\omega_1 + K\pi^*\tau \in \mathcal{S}(M,\pi)$ for all $t \in [0,1]$, provided that $K > 0$ is chosen sufficiently large. □

Exercise 6.2.8 Prove that up to conjugacy, there exists a unique pair of matrices $U, V \in \mathrm{SU}(2)$ such that $UVU^{-1}V^{-1} = -\mathbb{1}$. Under the double cover $\mathrm{SU}(2) \to \mathrm{SO}(3)$ of Exercise 5.3.2 these correspond to orthogonal matrices $A, B \in \mathrm{SO}(3)$ such that $AB = BA$. Prove that A and B represent half-turns around two orthogonal axes. Prove that two such matrices in SO(3) cannot lift to commuting matrices in SU(2). □

Exercise 6.2.9 Check the details of the following construction of S^2-bundles over surfaces of higher genus. First consider the case when the base is the torus $\mathbb{T}^2$. Identify $\mathbb{T}^2$ with the quotient $\mathbb{R}^2/\mathbb{Z}^2$ and consider the representation

$$\mathbb{Z}^2 \to \mathrm{SO}(3) : (j,k) \mapsto A^j B^k,$$

where $A, B \in \mathrm{SO}(3)$ are as in Exercise 6.2.8, i.e. half-turns around two orthogonal axes. This determines an action of $\mathbb{Z}^2$ on S^2 and the resulting quotient

$$X_{\mathbb{T}} = \mathbb{R}^2 \times_{\mathbb{Z}^2} S^2$$

is the nontrivial S^2-bundle over $\mathbb{T}^2$. This holds because, in view of Exercise 6.2.8, the above homomorphism $\mathbb{Z}^2 \to \mathrm{SO}(3)$ does not lift to $\mathrm{SU}(2)$. Since $\mathbb{Z}^2$ acts by elements of order 2, the square

$$I_z = [0,2] \times [0,2] \times \{z\} \subset \mathbb{R}^2 \times S^2$$

(for $z \in S^2$) descends to a closed submanifold in X_T which represents the homology class

$$[I_z] = 4[B_+] - 2[F] = 2[B_+] + 2[B_-].$$

Now let σ be an area form on S^2 with total area 1. Then the symplectic form $\omega = \lambda_1 ds \wedge dt + \lambda_2 \sigma$ on $\mathbb{R}^2 \times S^2$ descends to a symplectic form on X_T in the class $a = \mu_+ b_+ + \mu_- b_-$, where

$$\mu_+ + \mu_- = 2\lambda_1, \qquad \mu_+ - \mu_- = \lambda_2.$$

Thus by a correct choice of the parameters λ_1 and λ_2 one can get a symplectic form in any cohomology class a such that $\mu_+ \neq \pm\mu_-$. The condition $\mu_+ > |\mu_-|$ enters if we require ω to be positive on the fibres (i.e. $\lambda_2 > 0$) and compatible with the usual orientation of X_T (i.e. $\lambda_1 \lambda_2 > 0$).

This construction can be adapted to the case when the base is a Riemann surface Σ of genus $g = g(\Sigma) > 1$. Here we can think of Σ as the quotient $\mathbb{H}/\Gamma$, where $\mathbb{H}$ is the upper half-plane

$$\mathbb{H} := \{z \in \mathbb{C} \,|\, \mathrm{Im}\, z > 0\},$$

and Γ is a suitable discrete subgroup of $\mathrm{PSL}(2,\mathbb{R})$ acting as in Exercise 2.5.11 (but in the case $n = 1$). With this identification $\mathbb{H}$ is the universal cover of the Riemann surface Σ and Γ is the corresponding group of deck transformations. Hence Γ is isomorphic to the fundamental group $\pi_1(\Sigma)$ and so has generators $\alpha_1, \dots, \alpha_g, \beta_1, \dots, \beta_g$ with the single relation

$$\prod_{j=1}^{g} [\alpha_j, \beta_j] = 1,$$

where

$$[\alpha, \beta] := \alpha\beta\alpha^{-1}\beta$$

denotes the commutator. A representation $\rho : \Gamma \to \mathrm{SO}(3)$ is given by the matrices $A_j = \rho(\alpha_j)$ and $B_j = \rho(\alpha_j)$ which satisfy the same relation as the α_j and β_j. Any such action determines a 4-manifold

$$X_\rho = \Sigma \times_\rho S^2.$$

The diffeomorphism type of X_ρ is determined by whether or not ρ lifts to a homomorphism $\Gamma \to \mathrm{SU}(2)$. If such a lift exists then

$$X_\rho \cong \Sigma \times S^2$$

and otherwise X_ρ is the nontrivial bundle. A homomorphism which does not lift to $\mathrm{SU}(2)$ can be obtained, for example, by choosing A_1 and B_1 as in Exercise 6.2.8, i.e. as half-turns around two orthogonal axes, and $A_j = B_j = \mathbb{1}$ for $j \geq 2$. The bundle so obtained is the pullback by a map of degree 1 of the above bundle on $\mathbb{T}^2$, and all the above remarks apply. □

6.3 Symplectic connections

For the sake of completeness, this section begins with a brief review of principal bundles, connections, and curvature. For more background information we refer the reader to Kobayashi and Nomizu [371] and Donaldson and Kronheimer [160]. We then discuss the Sternberg–Weinstein universal construction for fibre bundles whose structure group G acts on the fibre F in a Hamiltonian way. Next we show, following Guillemin, Lerman, and Sternberg, how symplectic connections are generated by so-called connection 2-forms, that is 2-forms that extend the given symplectic form on the fibre and are 'vertically closed'. The section ends by considering the generalized Dehn twists constructed by Seidel.

Principal bundles and connections

Let G be a Lie group. A **principal G-bundle** is a smooth manifold P with a smooth right G-action $P \times \mathrm{G} \to P : (p, g) \mapsto pg$, such that the action is free, the quotient space $B = P/\mathrm{G}$ is a manifold, and the natural projection map $\pi : P \to B$ is a locally trivial fibration. If G is compact then the first condition implies the second and third. We shall use the notation $T_pP \mapsto T_{pg}P : v \mapsto vg$ for the induced action on the tangent bundle, and

$$p\xi = \left.\frac{d}{dt}\right|_{t=0} p\exp(t\xi) \in T_pP, \qquad p \in P, \quad \xi \in \mathfrak{g},$$

for the infinitesimal action of the Lie algebra $\mathfrak{g} = \mathrm{Lie}(\mathrm{G})$. Thus the **vertical tangent bundle** $V \subset TP$ has fibres

$$V_p := \{p\xi \,|\, \xi \in \mathfrak{g}\} \subset T_pP.$$

(Compare this with Exercise 5.1.13.)

Exercise 6.3.1 Prove that the action of G on the cotangent bundle T^*P is Hamiltonian with moment map

$$\mu_P(p, v^*) = -L_p^* v^*,$$

where $L_p^* : T_p^*P \to \mathfrak{g}^*$ is the dual of the linear map $L_p : \mathfrak{g} \to T_pP$ defined by

$$L_p\xi := p \cdot \xi.$$

Hint: Use Exercise 3.1.21 to prove that the action of G on T^*P is generated by Hamiltonian functions $H_\xi(p, v^*) = -\langle v^*, p \cdot \xi\rangle$. See also Exercise 5.3.9 with $P = \mathrm{G}$. □

A **connection** on P is a field of horizontal subspaces $H_p \subset T_pP$ such that

$$T_pP = H_p \oplus V_p, \qquad H_{pg} = H_p \cdot g$$

for $p \in P$ and $g \in \mathrm{G}$. Every such horizontal subspace $H_p \subset T_pP$ determines a projection $T_pP \to T_pP/H_p \simeq V_p \simeq \mathfrak{g}$ and the collection of these defines a Lie algebra valued 1-form $A \in \Omega^1(P, \mathfrak{g})$. It is a simple matter to check that this 1-form satisfies

$$A_p(p\xi) = \xi, \qquad A_{pg}(vg) = g^{-1}A_p(v)g$$

for $v \in T_pP$, $g \in \mathrm{G}$, and $\xi \in \mathfrak{g}$. Any 1-form $A \in \Omega^1(P, \mathfrak{g})$ which satisfies these conditions is called a **connection 1-form** and we denote the space of such 1-forms by $\mathcal{A}(P)$. For any $A \in \mathcal{A}(P)$ the kernels $H_p = \ker A_p$ define a connection in the sense of a horizontal distribution. Note that connections on P form an affine space whose associated vector space $\Omega^1_{\mathrm{Ad}}(P, \mathfrak{g})$ consists of equivariant and horizontal Lie algebra valued 1-forms on P. This space can be identified with the space $\Omega^1(B, \mathfrak{g}_P)$ of 1-forms on B with values in the bundle $\mathfrak{g}_P = \mathrm{Ad}(P)$ that is associated to P via the adjoint action of G on its Lie algebra $\mathfrak{g}$.

Given a connection $A \in \mathcal{A}(P)$ every path $\gamma : [0, 1] \to B$ determines an equivariant isomorphism

$$\Phi_\gamma : P_{\gamma(0)} \to P_{\gamma(1)}$$

defined in terms of the horizontal lifts of γ. These isomorphisms are independent of the parametrization of γ and satisfy the obvious composition rule. They are called the **parallel transport** maps or the **holonomy** associated to A. These maps depend only on the homotopy classes of the paths γ (with fixed endpoints) if and only if the field of horizontal subspaces $H \subset TP$ is integrable. The **curvature** of a connection A is a 2-form $F_A \in \Omega^2(B, \mathfrak{g}_P)$ with values in the adjoint bundle which can be interpreted as an obstruction to the integrability of the field of horizontal subspaces. It is defined by

$$F_A := dA + \tfrac{1}{2}[A \wedge A] \in \Omega^2_{\mathrm{Ad}}(P, \mathfrak{g}),$$

and can be characterized as the unique Lie algebra valued 2-form on P which satisfies $(F_A)_p(p\xi, v) = 0$, for all $v \in T_pP$ and $\xi \in \mathfrak{g}$, and has the property that

$$F_A(X, Y) = A([X, Y])$$

for any two horizontal vector fields $X, Y : P \to H$. The reader may check that these two definitions of the curvature are equivalent.

An **automorphism** of P is a diffeomorphism $\phi : P \to P$ which is equivariant under the action of G and preserves the fibres. Any such automorphism has the form $\phi(p) = pu(p)$, where $u : P \to \mathrm{G}$ is a smooth map such that $u(pg) = g^{-1}u(p)g$ for $p \in P$ and $g \in \mathrm{G}$. Such maps $u : P \to \mathrm{G}$ are called **gauge transformations** and we denote by $\mathcal{G}(P)$ the group of gauge transformations. It is an easy matter to prove that the pullback ϕ^*A under the bundle automorphism ϕ can be expressed in terms of the gauge transformation u as

$$u^*A := u^{-1}du + u^{-1}Au.$$

This defines a contravariant action of the group $\mathcal{G}(P)$ on the space $\mathcal{A}(P)$. The induced action on the curvature is given by $F_{u^*A} = u^{-1}F_Au$.

Exercise 6.3.2 Associated to any homomorphism $\mathrm{G} \to \mathrm{Diff}(F)$ is a locally trivial fibre bundle $M = P \times_{\mathrm{G}} F$. Prove that every connection on P determines a connection (i.e. a field of horizontal subspaces) on M. □

We shall now investigate the properties of connections that preserve the fibrewise symplectic structure on a symplectic fibration. We begin with the following theorem which is due to Weinstein [672]. We give two proofs; the first is a direct computation, while the second follows Weinstein's original paper in using the properties of symplectic reduction.

Theorem 6.3.3 (Weinstein) *Let* $\mathrm{G} \to \mathrm{Symp}(F, \sigma) : g \mapsto \psi_g$ *be a Hamiltonian group action. Then every connection on a principal* G*-bundle* $P \to B$ *gives rise to a closed* 2*-form* τ *on the associated fibration* $P \times_{\mathrm{G}} F \to B$ *which restricts to the forms* σ_b *on the fibres.*

The first proof uses the fact that the connection defines a splitting of the tangent bundle of $M := P \times_{\mathrm{G}} F$ into a vertical and a horizontal subbundle. The 2-form τ can then be defined as the difference of two terms, where the first term is the 2-form σ on the vertical subbundle and the second term is a 2-form on the horizontal subbundle obtained by evaluating the moment map on the curvature of the connection. That the resulting 2-form is closed follows from the Bianchi identity and the fact that $\sigma - \mu$ is a closed 2-form in equivariant de Rham cohomology. This approach is motivated by symplectic vortex equations [114].

Proof 1: Denote by $\mathfrak{g} \to \mathrm{Vect}(F) : \xi \mapsto X_\xi$ the infinitesimal action of the Lie algebra $\mathfrak{g} := \mathrm{Lie}(\mathrm{G})$ and let $\mu : F \to \mathfrak{g}^*$ be a moment map for the action. Then

$$\langle \mathrm{d}\mu(x)\widehat{x}, \xi \rangle = \sigma(X_\xi(x), \widehat{x}), \qquad \langle \mu(x), [\xi, \eta] \rangle = \sigma(X_\xi(x), X_\eta(x)) \tag{6.3.1}$$

for $x \in F$, $\widehat{x} \in T_x F$, and $\xi, \eta \in \mathfrak{g}$. Choose a connection 1-form $A \in \mathcal{A}(P)$ and denote by $F_A \in \Omega^2_{\mathrm{Ad}}(P, \mathfrak{g})$ its curvature. Then

$$A_p(p\xi) = \xi, \qquad (F_A)_p(v_1, v_2) = (dA)_p(v_1, v_2) + [A_p(v_1), A_p(v_2)] \tag{6.3.2}$$

for $p \in P$, $\xi \in \mathfrak{g}$, and $v_1, v_2 \in T_p P$. Define $\pi_A : TP \times TF \to TF$ by

$$\pi_A(v, \widehat{x}) := \widehat{x} + X_{A_p(v)}(x)$$

for $v \in T_p P$ and $\widehat{x} \in T_x F$. Then the 2-form $\tau_A \in \Omega^2(P \times F)$, defined by

$$\tau_A := \sigma - d\langle \mu, A \rangle = \pi_A^* \sigma - \langle \mu, F_A \rangle, \tag{6.3.3}$$

is closed, G-invariant, and horizontal. (Here we slightly abuse notation and identify objects on P and F with their pullbacks to the product $P \times F$.) The second equation in (6.3.3) follows from (6.3.1) and (6.3.2). It shows that τ_A descends to a closed 2-form on $M = P \times_{\mathrm{G}} F$. The explicit formula

$$\begin{aligned} \tau_A\left((v_1, \widehat{x}_1), (v_2, \widehat{x}_2)\right) &= \sigma\left(\widehat{x}_1 + X_{A_p(v_1)}(x), \widehat{x}_2 + X_{A_p(v_2)}(x)\right) \\ &\quad - \langle \mu(x), (F_A)_p(v_1, v_2) \rangle \end{aligned}$$

for $v_1, v_2 \in T_p P$ and $\widehat{x}_1, \widehat{x}_2 \in T_x F$ shows that it restricts to the symplectic form on each fibre. This proves Theorem 6.3.3. □

Proof 2: First consider the case where F is the cotangent bundle $T^*\mathrm{G}$. By Exercise 5.3.9 the action of G on its cotangent bundle is Hamiltonian. Moreover, in this case the associated fibre bundle

$$V^* := P \times_{\mathrm{G}} T^*\mathrm{G}$$

can be naturally identified with the vertical cotangent bundle of P. The fibre of V^* over $b \in B$ is the cotangent bundle T^*P_b of the fibre.

Now a connection 1-form $A \in \mathcal{A}(P)$ determines a projection $TP \to V$ along the horizontal subbundle $H = \ker A$ and hence an injection

$$\iota_A : V^* \to T^*P.$$

By definition of a connection 1-form, this injection is equivariant under the action of G and hence the 2-form $\omega_A := \iota_A^*\omega_{\mathrm{can}} \in \Omega^2(V^*)$ is invariant under the action of G. It need not be nondegenerate but it is easy to see that ω_A restricts to the canonical symplectic form on the fibres of V^*. In fact, there is a canonical injection $\iota_p : T^*\mathrm{G} \to V^*$ and the composition $\iota_A \circ \iota_p : T^*\mathrm{G} \to T^*P$ is the canonical identification of $T^*\mathrm{G}$ with the cotangent bundle T^*P_b. Hence it preserves the canonical symplectic structures.

Now consider the product space

$$W := V^* \times F,$$

where (F, σ) is any symplectic manifold with a Hamiltonian G-action and moment map $\mu_F : F \to \mathfrak{g}^*$. Consider the closed form

$$\tau_A := \omega_A \oplus \sigma \in \Omega^2(W).$$

This form restricts to the standard 2-forms on each fibre $W_b = T^*P_b \times F$ of the bundle $W \to B$. The action of G on these fibres is Hamiltonian with moment map

$$\mu_W := (\mu_P \circ \iota_A) \oplus \mu_F : W \to \mathfrak{g}^*.$$

Because 0 is a regular value of μ_P, it follows as in Exercise 5.4.20 (iii) that 0 is a regular value of μ_W. The group G acts freely on $\mu_W^{-1}(0)$ and the quotient

$$M := \mu_W^{-1}(0)/\mathrm{G}$$

can be naturally identified with the space $P \times_{\mathrm{G}} F$. Since the 2-form τ_A is invariant under the action of G it descends to a 2-form on M (which we still denote by τ_A). This is proved as in Proposition 5.4.5. Moreover, symplectic reduction for the fibres $T^*P_b \times F \cong T^*\mathrm{G} \times F$ shows that the induced form on M restricts to the canonical symplectic forms in the fibres (see Exercise 5.4.20). Note that we do not need to assume here that F is compact. This completes the second proof of Theorem 6.3.3. □

Exercise 6.3.4 Check that when $\mathrm{G} = S^1$ this construction reduces to the situation studied in Exercise 5.1.13. □

Guillemin–Lerman–Sternberg [292] developed a beautiful generalization of this construction. In their book they apply these ideas to the important problem of studying group representations. The goal here is to investigate 2-forms τ on the total space of a symplectic fibration which restrict to the canonical 2-forms on the fibres. These 2-forms τ need not be either closed or nondegenerate.

Symplectic connections

Let $\pi : M \to B$ be a symplectic fibration with fibre (F, σ). For $x \in M$, denote by

$$\mathrm{Vert}_x := \ker d\pi(x) = T_x F_{\pi(x)}$$

the vertical tangent space to the fibre. A **connection** Γ on $\pi : M \to B$ is a field of horizontal subspaces $\mathrm{Hor}_x \subset T_x M$ such that

$$TM = \mathrm{Vert} \oplus \mathrm{Hor}.$$

Given such a splitting, every path $\gamma : [0,1] \to B$ determines a diffeomorphism

$$\Phi_\gamma : F_{\gamma(0)} \to F_{\gamma(1)}$$

which assigns to a point $x_0 \in F_{\gamma(0)}$ the endpoint $x_1 = \gamma^\sharp(1) \in F_{\gamma(1)}$ of the unique horizontal lift $\gamma^\sharp : [0,1] \to M$ of the path γ which starts at $\gamma^\sharp(0) = x_0$. The diffeomorphism Φ_γ is called the **holonomy** of the path γ. It is independent of the choice of the parametrization of γ and there is an obvious composition rule under juxtaposition of paths. Moreover, there is a one-to-one correspondence between such a collection of diffeomorphisms $\{\Phi_\gamma\}_\gamma$ (also called a **parallel transport structure**) and connections in the sense of horizontal distributions. The connection Γ is called **symplectic** if the associated diffeomorphisms Φ_γ preserve the symplectic structures in the fibres, i.e. $\Phi_\gamma^* \sigma_{\gamma(1)} = \sigma_{\gamma(0)}$ for every path $\gamma : [0,1] \to B$.

Now let $\tau \in \Omega^2(M)$ be a 2-form which is **compatible** with the fibration in the sense that

$$\iota_b^* \tau = \sigma_b$$

for every $b \in B$. Every such 2-form gives rise to a connection Γ_τ with horizontal distribution

$$\mathrm{Hor}_x = \mathrm{Vert}_x^\tau = \{\xi \in T_x M \,|\, \tau(\xi, \eta) = 0 \;\; \forall\, \eta \in \mathrm{Vert}_x\}. \tag{6.3.4}$$

That this subspace is a complement of Vert_x is equivalent to the condition that τ restricts to a nondegenerate form on each fibre. Conversely, every horizontal distribution $\mathrm{Hor} \subset TM$ can be represented in this way by a suitable 2-form $\tau \in \Omega^2(M)$. However, the 2-form τ is not unique. Two compatible 2-forms τ_0 and τ_1 determine the same horizontal distribution if and only if the vertical subbundle Vert is contained in the kernel of the difference $\tau_1 - \tau_0$. The following result from [282] characterizes symplectic connections in terms of the corresponding 2-form τ. We give two proofs, one that is conceptual and one that is computational.

Lemma 6.3.5 *Assume that $\tau \in \Omega^2(M)$ satisfies $\iota_b^*\tau = \sigma_b$ for every $b \in B$. Then the connection Γ_τ is symplectic if and only if τ is* **vertically closed** *in the sense that $d\tau(\eta_1, \eta_2, \cdot) = 0$ for all $x \in M$ and all vertical tangent vectors $\eta_1, \eta_2 \in \mathrm{Vert}_x$.*

Proof 1: Given a vector field $v : B \to TB$ on the base B denote by

$$v^\sharp : M \to TM$$

its horizontal lift. Then the flow of $v^\sharp$ preserves the fibres of π. The connection Γ_τ is symplectic if and only if the flow of every such horizontal vector field $v^\sharp$ preserves the restriction of τ to the fibres, or, equivalently, if and only if

$$\mathcal{L}_{v^\sharp}\tau(\eta_1, \eta_2) = 0$$

for every vector field $v : B \to TB$ and any two vertical tangent vectors $\eta_1, \eta_2 \in \mathrm{Vert}_x = \ker\, d\pi(x)$. This means that

$$d(\iota(v^\sharp)\tau)(\eta_1, \eta_2) + (\iota(v^\sharp)d\tau)(\eta_1, \eta_2) = 0.$$

But, by definition of Hor in (6.3.4), the 1-form $\alpha = \iota(v^\sharp)\tau$ vanishes on vertical vectors. This means that $\iota_b{}^*d\alpha = d\iota_b{}^*\alpha = 0$, where $\iota_b : F_b \to M$ denotes the inclusion of the fibre, and so $d\alpha$ vanishes on vertical vectors. Thus the first term in the above equation is zero. Now the vanishing of the second term is equivalent to $d\tau(\eta_1, \eta_2, \xi) = 0$ whenever $\eta_1, \eta_2 \in \mathrm{Vert}_x$ and $\xi \in \mathrm{Hor}_x$. But since the restriction of τ to each fibre is closed, this is equivalent to the condition $d\tau(\eta_1, \eta_2, \cdot) = 0$. □

Proof 2: It suffices to consider the trivial bundle $M = B \times F$ where $B \subset \mathbb{R}^m$ is an open set. Let $x = (x_1, \dots, x_m)$ denote the coordinate on B. Then a 2-form $\tau \in \Omega^2(B \times F)$ can be written in the form

$$\tau = \sigma + \sum_i \alpha_i \wedge dx_i + \sum_{i<j} f_{ij}\, dx_i \wedge dx_j, \tag{6.3.5}$$

where $\sigma(x) \in \Omega^2(F)$, $\alpha_i(x) \in \Omega^1(F)$, and $f_{ij}(x) = -f_{ji}(x) \in \Omega^0(F)$ for $x \in B$ and $i, j \in \{1, \dots, m\}$. The differential of τ is given by

$$\begin{aligned} d^{B\times F}\tau = d\sigma &+ \sum_i (\partial_i\sigma + d\alpha_i) \wedge dx_i \\ &+ \sum_{i<j} (\partial_j\alpha_i - \partial_i\alpha_j + df_{ij}) \wedge dx_i \wedge dx_j \\ &+ \sum_{i<j<k} (\partial_i f_{jk} + \partial_j f_{ki} + \partial_k f_{ij}) dx_i \wedge dx_j \wedge dx_k. \end{aligned} \tag{6.3.6}$$

On the right we abbreviate $\partial_i = \partial/\partial x_i$ and $d = d^F$ denotes the exterior differential on F. This shows that the form τ is vertically closed iff

$$d\sigma(x) = 0, \qquad \partial_i\sigma(x) + d\alpha_i(x) = 0. \tag{6.3.7}$$

This condition is independent of the functions f_{ij}.

Now suppose that $\sigma(x)$ is a symplectic form on F for every $x \in B$ and (6.3.7) holds. Then the horizontal lift of a tangent vector $\xi \in \mathbb{R}^m = T_x B$ at a point $(x,q) \in B \times F$ is the vector $\xi^\sharp = (\xi, X) \in \mathbb{R}^m \times T_q F = T_{x,q}(B \times F)$ given by

$$\iota(X)\sigma = \sum_{i=1}^m \xi_i \alpha_i.$$

Hence the holonomy along a path $t \mapsto x(t)$ in B is given by the diffeomorphisms $\Phi_t : F \to F$ defined by

$$\frac{d}{dt}\Phi_t = X_t \circ \Phi_t, \qquad \iota(X_t)\sigma_t = \alpha_t,$$

where $\alpha_t = \sum_i \alpha_i(x(t))\dot{x}_i(t)$. That parallel transport preserves the symplectic forms $\sigma_t = \sigma(x(t))$ follows from Moser's isotopy method. Namely, since τ is vertically closed, the α_i and σ satisfy (6.3.7), and hence

$$\partial_t \sigma_t + d\alpha_t = 0.$$

This implies that ${\Phi_t}^*\sigma_t = \sigma_0$ for all t, and hence parallel transport preserves the symplectic structures on the fibres. Conversely, if parallel transport preserves the symplectic structures, then the same argument shows that the 2-form τ satisfies (6.3.7). This proves Lemma 6.3.5. □

Any 2-form τ which satisfies the conditions of Lemma 6.3.5 is called a **connection 2-form** on M. The space of such forms will be denoted by $\mathcal{T}(M,\sigma)$. This space is clearly contractible. Moreover, the arguments that apply in the general theory of connections on principal bundles show that it is always nonempty. The following exercise gives an explicit construction, which closely follows the proof of Thurston's Theorem 6.1.4.

Exercise 6.3.6 Let $\pi : M \to B$ be a symplectic fibration with fibre (F, σ) and local trivializations

$$\phi_\alpha : \pi^{-1}(U_\alpha) \to U_\alpha \times F.$$

As in the proof of Theorem 6.1.4, denote by

$$\sigma_\alpha \in \Omega^2(U_\alpha \times F)$$

the 2-form obtained from σ via pullback under the projection $U_\alpha \times F \to F$. Let $\rho_\alpha : B \to \mathbb{R}$ be a partition of unity subordinate to the cover $\{U_\alpha\}_\alpha$ and define

$$\tau := \sum_\alpha (\rho_\alpha \circ \pi){\phi_\alpha}^*\sigma_\alpha \in \Omega^2(M).$$

Prove that the 2-form τ satisfies the requirements of Lemma 6.3.5 and hence determines a symplectic connection on M via (6.3.4). □

The condition in Lemma 6.3.5 that τ be vertically closed can be restated in geometric terms as the requirement that for every loop $\gamma : S^1 \to B$ the restriction of τ to $\pi^{-1}(\gamma)$ is closed. The proof of Lemma 6.3.5 can then be interpreted as showing that the holonomy of the characteristic foliation on $\pi^{-1}(\gamma)$ is symplectic. We shall see in Theorem 6.4.1 below that if τ is closed (and not just vertically closed), then the holonomy around every contractible loop in the base is a Hamiltonian symplectomorphism.

Generalized Dehn twists

The following example is due to Seidel [570] and plays a crucial role in his construction of singular symplectic fibrations. As we explain in Example 10.4.3, Seidel used this construction and computations of Floer homology to prove that large classes of symplectic 4-manifolds carry symplectomorphisms which are isotopic to the identity, but not symplectically so.

Exercise 6.3.7 Consider the singular symplectic fibration

$$\mathbb{C}^n \to \mathbb{C} : z = (z_1, \ldots, z_n) \mapsto z_1^2 + \cdots + z_n^2$$

with connection given by the standard symplectic form on $\mathbb{C}^n$. The fibre over $w = 0$ is singular and it is interesting to compute the holonomy around the loop

$$t \mapsto e^{2\pi i t}$$

in the base.

(i) Show that the fibre

$$F = \pi^{-1}(1) = \left\{ x + iy \in \mathbb{C}^n \,|\, |x|^2 - |y|^2 = 1,\ \langle x, y \rangle = 0 \right\}$$

can be naturally identified with the cotangent bundle of S^{n-1}. **Hint:** Think of T^*S^{n-1} explicitly as the symplectic submanifold of $\mathbb{R}^{2n}$ given by

$$T^*S^{n-1} = \left\{ \xi + i\eta \in \mathbb{C}^n \,|\, |\xi|^2 = 1,\ \langle \xi, \eta \rangle = 0 \right\}.$$

Prove that the map

$$F \to T^*S^{n-1} : (x, y) \mapsto (|x|^{-1}x, |x|y) \tag{6.3.8}$$

is a symplectomorphism.

(ii) Prove that with this identification of the fibre with T^*S^{n-1}, the holonomy of the fibration $\mathbb{C}^n \to \mathbb{C}$ around the loop $[0, 1] \to \mathbb{C} : t \mapsto e^{2\pi i t}$ is the symplectomorphism $\psi : T^*S^{n-1} \to T^*S^{n-1}$ given by $\psi(\xi_0, \eta_0) = (\xi_1, \eta_1)$, where $|\eta_0| = |\eta_1|$ and

$$\xi_1 + i\frac{\eta_1}{|\eta_1|} = -\exp\left(-\frac{2\pi i\, |\eta_0|}{\sqrt{1 + 4|\eta_0|^2}} \right) \left(\xi_0 + i\frac{\eta_0}{|\eta_0|} \right). \tag{6.3.9}$$

This map ψ is a **generalized Dehn twist**. Note that ψ is close to the identity when $|\eta|$ is very large and is equal to minus the identity on the zero section, i.e. $\psi(\xi, 0) = (-\xi, 0)$.

Hint: Prove that the horizontal subspace at $z \in \mathbb{C}^n$ is $\mathrm{Hor}_z = \{\lambda \bar{z} \,|\, \lambda \in \mathbb{C}\}$. Deduce that the horizontal lifts of the loop $t \mapsto e^{2\pi i t}$ satisfy the differential equation

$$\dot{z} = \frac{\pi i e^{2\pi i t}}{|z|^2} \bar{z}. \tag{6.3.10}$$

Show that the norm $|z(t)|$ is constant for every solution of (6.3.10), and that the function $f(t) = e^{-\pi i t} z(t)$ satisfies the differential equation $f'(t) = i\pi(-f + \lambda \overline{f})$, where $\lambda := |z(0)|^2$ is constant. Hence show that the solutions of (6.3.10) are

$$\begin{aligned} &\sqrt{1 - 1/|z|^2}\, x(t) + i\sqrt{1 + 1/|z|^2}\, y(t) \\ &= \exp\left(\pi i t (1 - \sqrt{1 - 1/|z|^4})\right) \left(\sqrt{1 - 1/|z|^2}\, x(0) + i\sqrt{1 + 1/|z|^2}\, y(0)\right), \end{aligned}$$

where $x(t) := \mathrm{Re}\, z(t)$ and $y(t) := \mathrm{Im}\, z(t)$. Show that a point $z := (x, y)$ lies in the fibre $F = \pi^{-1}(1)$ exactly if

$$|x| = \sqrt{\frac{|z|^2 + 1}{2}}, \qquad |y| = \sqrt{\frac{|z|^2 - 1}{2}}, \qquad \langle x, y \rangle = 0.$$

Using this, show that the time-1-map $F \to F : z_0 \mapsto z_1$ is determined by the identity

$$|y_1|\, x_1 + i|x_1|\, y_1 = -\exp\left(-\frac{2\pi i\, |x_0|\, |y_0|}{|z_0|^2}\right) (|y_0|\, x_0 + i|x_0|\, y_0), \tag{6.3.11}$$

where $|x_0| = |x_1|$ and $|y_0| = |y_1|$. (Check in particular that this map does have image in F.) Show finally that, under the diffeomorphism $F \to T^*S^{n-1}$ in (6.3.8), the map (6.3.11) is conjugate to (6.3.9). □

Generalized Dehn twists have many interesting applications. Suppose, for example, that a symplectic manifold M contains an embedded Lagrangian n-sphere $L \subset M$. Then there is a symplectic embedding of a neighbourhood of the zero section in T^*L into M. Modifying the generalized Dehn twist of Exercise 6.3.7 one obtains a symplectomorphism of T^*L which is equal to the antipodal map on the zero section and to the identity outside an arbitrarily small neighbourhood of the zero section. Hence there is a symplectomorphism

$$\psi_L : M \to M$$

which is equal to this modified Dehn twist in some neighbourhood of L and to the identity outside this neighbourhood. Seidel [570] studied such generalized Dehn twists and computed their Floer homology groups. By examining the ring structure of the Floer homology he was able to show that many symplectic 4-manifolds admit Dehn twists ψ_L whose squares are not symplectically isotopic to the identity. In contrast, if M is a symplectic 4-manifold, then the square

$$\psi_L^2 := \psi_L \circ \psi_L$$

is smoothly isotopic to the identity. Any such example gives rise to a symplectic fibration over S^1 which can be trivialized smoothly, but not symplectically.

It is interesting to note that the extended symplectomorphisms ψ_L described above also appear as the holonomy of a singular fibration over the 2-disc with a singular fibre over the origin. This fibration can be obtained as the union of the product $\mathbb{D} \times (M \setminus U)$, where U is a tubular neighbourhood of L, and the singular fibration of Example 6.3.7 with a suitable identification on the overlap. Such singular fibrations were used by Seidel [570] to compute the Floer homology groups of ψ_L.

In algebraic geometry generalized Dehn twists appear as the monodromy of Lefschetz fibrations. Recently Donaldson has generalized this circle of ideas to symplectic 4-manifolds [146, 148]. In this case the fibres are Riemann surfaces and the monodromy maps are, of course, ordinary Dehn twists. Donaldson proved that every symplectic 4-manifold X admits, after a suitable number of blowups (see Section 7.1 below), the structure of a topological Lefschetz fibration $X\#m\overline{\mathbb{C}\mathrm{P}}^2 \to S^2$, where the monodromy maps around the singular fibres are Dehn twists. This structure gives rise to finitely many Dehn twists whose composition is the identity. Conversely, any such collection of Dehn twists can be used to construct a symplectic Lefschetz fibration (see Section 7.4).

Remark 6.3.8 For $n \in \mathbb{N}$, let $\tau_n : T^*S^n \to T^*S^n$ be the modified generalized Dehn twist with compact support. For $n = 2$, arguments of Kronheimer and Seidel [570] show that its square is smoothly isotopic to the identity via a compactly supported isotopy. For $n = 6$ the same holds by an unpublished argument of Sevennec. By another unpublished argument of Biran and Giroux, one can use smooth open books (see Definition 3.5.40) to prove that τ_n^2 is not smoothly isotopic to the identity by a compactly supported isotopy whenever $n \neq 2, 6$. Namely, let $M_1 := M(T^*S^n, \mathrm{id})$ and $M_2 := M(T^*S^n, \tau_n^2)$ be the closed smooth $2n$-manifolds determined by the smooth open book construction in (3.5.24). A computation of their homology groups shows that M_1 and M_2 cannot be diffeomorphic unless $n = 2$ or $n = 6$. **Exercise:** Carry out the details. □

6.4 Hamiltonian holonomy and the coupling form

Lemma 6.3.5 shows that every symplectic connection Γ on a symplectic fibration $\pi : M \to B$ has the form Γ_τ for some vertically closed 2-form τ on M. However, it is not true that τ can always be taken to be closed. For example, the symplectic fibration $\pi : S^3 \times S^1 \to S^2$ of Example 6.1.6 with the fibre equal to a torus clearly does not admit a closed 2-form on the total space that is compatible with the fibration, since all closed 2-forms on $S^3 \times S^1$ are exact.

In this section we characterize fibrations for which τ may be taken to be closed. Our result is a mild generalization of a remarkable theorem due to Guillemin, Lerman, and Sternberg [292], who discovered an elegant way of constructing closed connection 2-forms τ using the holonomy of a symplectic connection. Their construction works whenever this holonomy is Hamiltonian around every contractible loop in the base. This is automatic in the case considered by Guillemin, Lerman, and Sternberg, since they consider only the case where the fibre F is simply connected.

We give two proofs of this theorem: one by direct calculation, and another more conceptual proof that follows very closely the original argument of Guillemin, Lerman, and Sternberg, and depends on general ideas about symplectic gauge transformations and the curvature of symplectic connections. Where necessary, we shall assume that the reader is familiar with the elementary properties of Hamiltonian symplectomorphisms and the flux homomorphism, as described in Chapter 10. Recall further that integration over a fibre of dimension $2k$ induces a map $H^j(M) \to H^{j-2k}(B)$ (cf. Bott and Tu [78]).

Theorem 6.4.1 *Let $\pi : M \to B$ be a symplectic fibration and Γ be a symplectic connection on M. Then the following are equivalent.*

(i) *There exists a closed connection 2-form $\tau \in \Omega^2(M)$ such that $\Gamma = \Gamma_\tau$.*

(ii) *The holonomy of Γ around any contractible loop in B is Hamiltonian.*

Moreover, if (ii) holds, then there exists a unique closed connection 2-form $\tau_\Gamma \in \Omega^2(M)$ which generates Γ via (6.3.4) and satisfies the normalization condition

$$\int_F \tau_\Gamma^{k+1} = 0 \in \Omega^2(B), \tag{6.4.1}$$

where $\int_F$ denotes integration over the fibre. Any other closed connection 2-form $\tau \in \Omega^2(M)$ which generates Γ is related to τ_Γ by

$$\tau - \tau_\Gamma = \pi^*\beta$$

for some closed form $\beta \in \Omega^2(B)$ that is uniquely determined by τ.

Proof: See pages 273 and 280. □

The 2-form τ_Γ in Theorem 6.4.1 is called the **coupling form** of Γ.

Remark 6.4.2 Condition (ii) in Theorem 6.4.1 asserts that the symplectomorphism $\Phi_\gamma : F_{\gamma(0)} \to F_{\gamma(0)}$, determined by a contractible loop $\gamma : S^1 \to B$, lies in the group $\mathrm{Ham}(F, \sigma)$ and so can be connected to the identity by a Hamiltonian isotopy. However, γ does not determine such an isotopy. Rather, it determines a family of symplectomorphisms

$$\Phi_t : F_{\gamma(0)} \to F_{\gamma(t)},$$

and to obtain an isotopy we must choose a trivialization of the bundle over γ. The proof given below shows that a Hamiltonian isotopy can be obtained by trivializing the fibration over any disc in B with boundary γ. Other trivializations of the bundle over γ may lead to paths with nonzero flux connecting the identity to Φ_γ. More precisely, if we change the trivialization by a loop of symplectomorphisms via $S^1 \times F \to S^1 \times F : (t, x) \mapsto (t, \psi_t(x))$, then the flux is changed by the element $\mathrm{Flux}(\{\psi_t\})$ of the flux group $\Gamma_\sigma \subset H^1(F; \mathbb{R})$. □

Corollary 6.4.3 *Let (F, σ) be a compact simply connected symplectic manifold. Then the total space M of any symplectic fibration $\pi : M \to B$ with fibre (F, σ) and compact symplectic base B carries a compatible symplectic structure.*

Proof: Condition (ii) in Theorem 6.4.1 is automatically satisfied whenever the fibre F is simply connected. Hence Theorem 6.4.1 shows that there exists a closed 2-form $\tau \in \Omega^2(M)$ such that $\iota_b^*\tau = \sigma_b$ for every $b \in B$. Let $\beta \in \Omega^2(B)$ be a symplectic form. Then, by Theorem 6.1.4, the 2-form $\omega := \tau + k\pi^*\beta \in \Omega^2(M)$ is symplectic for k sufficiently large. □

Proof of Theorem 6.4.1: '(i) implies (ii)' We first prove this by direct calculation, and then give a more geometric proof.

Assume that $\Gamma = \Gamma_\tau$ for some closed 2-form $\tau \in \Omega^2(M)$ such that $\iota_b^*\tau = \sigma_b$ for every $b \in B$. Let $z = x + iy$ denote the coordinate on the closed unit disc $\mathbb{D} \subset \mathbb{C}$ and let $u : \mathbb{D} \to B$ be any smooth map. Then the pullback bundle $u^*M \to \mathbb{D}$ admits a symplectic trivialization. This means that u lifts to a bundle map

$$\phi : \mathbb{D} \times F \to M$$

such that for each $z \in \mathbb{D}$, the function

$$(F, \sigma) \to (F_{u(z)}, \sigma_{u(z)}) : q \mapsto \phi(z, q)$$

is a symplectomorphism. Hence the pullback of the connection form τ under the trivialization $\phi : \mathbb{D} \times F \to M$ is given by

$$\phi^*\tau = \sigma + \alpha \wedge dx + \beta \wedge dy + f\, dx \wedge dy,$$

where $\alpha(z), \beta(z) \in \Omega^1(F)$ and $f(z) \in \Omega^0(F)$ for $z \in \mathbb{D}$. Since τ is closed and $\sigma(z) = \sigma$ for all $z \in \mathbb{D}$, we have

$$d\alpha = d\beta = 0, \qquad df = \partial_x\beta - \partial_y\alpha$$

for all $z \in \mathbb{D}$. Now the holonomy of the connection form $\phi^*\tau$ around the loop

$$z(t) := e^{2\pi i t} = x(t) + iy(t)$$

is the path of symplectomorphisms $\Phi_t : F \to F$ given by

$$\frac{d}{dt}\Phi_t = X_t \circ \Phi_t, \qquad \iota(X_t)\sigma = \alpha_t = \alpha(z)\dot{x} + \beta(z)\dot{y}.$$

The formula $df = \partial_x\beta - \partial_y\alpha$ shows that the differential of the 1-form $\alpha dx + \beta dy \in \Omega^1(\mathbb{D}, \Omega^1(F))$ is given by $df\, dx \wedge dy \in \Omega^2(\mathbb{D}, \Omega^1(F))$. Hence the 1-form

$$\int_0^1 \alpha_t\, dt = d\int_{\mathbb{D}} f(x, y)\, dxdy$$

is exact, and this implies that the flux of the path $\{\Phi_t\}_{0\le t\le 1}$ is zero:

$$\mathrm{Flux}(\{\Phi_t\}) = \int_0^1 [\alpha_t]\, dt = 0.$$

By Theorem 10.2.5 in Chapter 10, the map

$$\Phi_1 : F \to F$$

is a Hamiltonian symplectomorphism.

Alternatively, by Lemma 10.2.8, the flux of the path $\{\Phi_t\}$ vanishes if and only if $\sigma \in \Omega^2(F)$ integrates to zero over any 2-chain of the form

$$[0,1] \times S^1 \to F : (t,\theta) \mapsto \Phi_t(\gamma(\theta)),$$

where $\gamma : S^1 \to F$ is any loop. This can be expressed, equivalently, as the vanishing of the integral of $\phi^*\tau \in \Omega^2(\mathbb{D} \times F)$ over the 2-chain

$$[0,1] \times S^1 \to \mathbb{D} \times F : \quad (t,\theta) \mapsto \big(1, \Phi_t(\gamma(\theta))\big).$$

But this chain is homotopic, with fixed boundary, to the chain

$$v : [0,1] \times S^1 \to \mathbb{D} \times F : \quad (t,\theta) \mapsto \big(e^{2\pi i t}, \Phi_t(\gamma(\theta))\big).$$

Now observe that by the definition of Φ_t the pullback form $v^*\phi^*\tau$ vanishes identically, since the flow lines of Φ_t are the characteristic (or null) directions of the restriction of $\phi^*\tau$ to $\partial\mathbb{D} \times F$. Thus we again find that $\mathrm{Flux}(\{\Phi_t\}) = 0$. This proves (ii). □

'(ii) implies (i)' We also give two proofs of this implication. The first will be a rather brief sketch of a computational proof. The second follows the lines of argument of Guillemin, Lerman, and Sternberg. It is very geometric but will require considerable preparation.

In a local symplectic trivialization of the bundle $\pi : M \to B$ the coupling form $\tau = \tau_\Gamma$ can be expressed in the form (6.3.5), where σ is given and the forms $\alpha_i(x) \in \Omega^1(F)$ are determined by the connection Γ. By Exercise 6.4.9 below, the holonomy condition (ii) translates into the requirement that the 1-forms $\partial_i\alpha_j - \partial_j\alpha_i \in \Omega^1(F)$ are exact for all i and j and all x. Hence for all x and all i, j there exists a unique function $f_{ij}(x) : F \to \mathbb{R}$ which has mean value zero and satisfies $df_{ij} = \partial_i\alpha_j - \partial_j\alpha_i$. Using (6.3.6) one checks easily that τ is closed and satisfies (6.4.1) if and only if the f_{ij} in (6.3.5) agree with the functions just defined. This argument proves the local existence and uniqueness of the coupling form. Global existence follows immediately from uniqueness, since any two local coupling forms must agree on the intersection of their domains.

It remains to prove that any other closed connection 2-form $\tau \in \Omega^2(M)$ with $\Gamma_\tau = \Gamma$ has the form

$$\tau = \tau_\Gamma + \pi^*\beta,$$

where β is a closed2-form on B. To see this note that, in local coordinates on B and a local trivialization of $\pi : M \to B$, the 2-form $\tau - \tau_\Gamma$ has the form (6.3.5) with $\sigma = 0$ and $\alpha_i = 0$. Since $\tau - \tau_\Gamma$ is closed it follows that $df_{ij} = 0$ and hence f_{ij} is constant on each fibre. This implies that $\tau - \tau_\Gamma$ can locally be expressed as the pullback of a 2-form on B. Since this 2-form on B is uniquely determined by τ and τ_Γ it exists globally. Hence there is a unique 2-form $\beta \in \Omega^2(B)$ such that $\tau - \tau_\Gamma = \pi^*\beta$. Since $\tau - \tau_\Gamma$ is closed so is β. This completes the first proof of Theorem 6.4.1. □

Guillemin, Lerman, and Sternberg's construction of the coupling form τ_Γ is based on the following simple geometric idea. Because τ_Γ generates Γ, its values are determined except on pairs of Γ-horizontal vectors. Thus to define τ_Γ we just have to specify the values $\tau_\Gamma({v_1}^\sharp, {v_2}^\sharp)$ for any two vector fields v_1 and v_2 on B. We shall see that under the holonomy assumption (ii), the vertical part $[{v_1}^\sharp, {v_2}^\sharp]^{\mathrm{Vert}}$ of the commutator is a Hamiltonian vector field on each fibre F_b. Hence there exists a unique Hamiltonian function $H_b : F_b \to \mathbb{R}$ of mean value zero such that

$$dH_b = [{v_1}^\sharp, {v_2}^\sharp]^{\mathrm{Vert}}.$$

We then define

$$\tau_\Gamma({v_1}^\sharp(x), {v_2}^\sharp(x)) := H_{\pi(x)}(x).$$

The proof that τ_Γ is well-defined and closed reduces to some basic facts about connections, gauge transformations, and curvature on symplectic fibrations. We shall discuss these in a series of exercises.

Symplectic gauge transformations

Exercise 6.4.4 Let $\pi : M \to B$ be a locally trivial fibration. A diffeomorphism

$$\phi : M \to M$$

is said to **descend to** B if there exists a diffeomorphism $f : B \to B$ such that

$$\pi \circ \phi = f \circ \pi.$$

Such diffeomorphisms form a group whose Lie algebra consists of all those vector fields

$$X : M \to TM$$

which **descend to** B in the sense that there exists a vector field $v : B \to TB$ such that

$$d\pi \circ X = v \circ \pi.$$

Prove that if $Y : M \to TM$ is another vector field which descends to $w : B \to TB$ then

$$d\pi \circ [X, Y] = [v, w] \circ \pi.$$

Deduce that the vector fields which descend to B form a Lie algebra. □

Exercise 6.4.5 Let $\pi : M \to B$ be a symplectic fibration and $\tau \in \Omega^2(M)$ be a connection 2-form. A **fibrewise symplectomorphism** of M is a diffeomorphism $\phi : M \to M$ which descends to B and preserves the symplectic structures on the fibres. This means that

$$\phi^*\tau \overset{\text{fibre}}{=} \tau,$$

where we write $\overset{\text{fibre}}{=}$ to mean that the restrictions of the two sides to any fibre agree. This condition is independent of the choice of the connection 2-form τ. It depends only on the symplectic structures in the fibres.

The Lie algebra of the group of fibrewise symplectomorphisms is the space of vector fields $X : M \to TM$ which descend to B and satisfy

$$d(\iota(X)\tau) \overset{\text{fibre}}{=} 0.$$

Vector fields with this property are called **fibrewise symplectic**. Prove that these vector fields form a Lie algebra. **Hint:** By definition of a connection 2-form, the 1-form $\iota(X)d\tau \in \Omega^2(M)$ vanishes on vertical tangent vectors. Hence the condition on X to be a fibrewise symplectic vector field can be expressed in the form

$$\mathcal{L}_X\tau \overset{\text{fibre}}{=} 0.$$

Now use the identity[27]

$$\mathcal{L}_{[X,Y]}\tau = \mathcal{L}_Y\mathcal{L}_X\tau - \mathcal{L}_X\mathcal{L}_Y\tau$$

and prove that both terms on the right vanish on pairs of vertical tangent vectors. (Use the fact that the flows of X and Y preserve the vertical tangent bundle.) □

The group of fibrewise symplectomorphisms is the natural group of automorphisms of the symplectic vector bundle $\pi : M \to B$. An important subgroup consists of those fibrewise symplectomorphisms $\psi : M \to M$ which descend to the identity on B, i.e.

$$\pi \circ \psi = \pi.$$

These are called **symplectic gauge transformations** and they form a group $\mathrm{Diff}(M, \sigma)$. If $\Gamma = \Gamma_\tau$ is a symplectic connection generated by the 2-form $\tau \in \mathcal{T}(M, \sigma)$ via (6.3.4) then so is

$$\psi^*\Gamma_\tau = \Gamma_{\psi^*\tau}$$

for $\psi \in \mathrm{Diff}(M, \sigma)$. The corresponding horizontal distributions are also related by ψ in the obvious way. In other words, the group $\mathrm{Diff}(M, \sigma)$ of symplectic gauge transformations acts on the space $\mathcal{T}(M, \sigma) \subset \Omega^2(M)$ of connection 2-forms. The Lie algebra of the group $\mathrm{Diff}(M, \sigma)$ is the space of **vertical symplectic vector fields** $Y : M \to TM$. These are fibrewise symplectic vector fields which in addition are tangent to the fibres, i.e.

$$d\pi \circ Y = 0, \qquad d(\iota(Y)\tau) \overset{\text{fibre}}{=} 0.$$

The space of such vector fields will be denoted by $\mathcal{X}(M, \sigma)$.

Remark 6.4.6 The group $\mathrm{Diff}(M, \sigma)$ of symplectic gauge transformations is a normal subgroup of the group of fibrewise symplectomorphisms. In other words, if $\phi : M \to M$ is a fibrewise symplectomorphism and $\psi \in \mathrm{Diff}(M, \sigma)$ then

[27] See Remark 3.1.6 on sign conventions.

$\phi^{-1} \circ \psi \circ \phi \in \text{Diff}(M, \sigma)$. Differentiating with respect to $\psi \in \text{Diff}(M, \sigma)$ we obtain

$$Y \in \mathcal{X}(M, \sigma) \qquad \Longrightarrow \qquad \phi^* Y \in \mathcal{X}(M, \sigma)$$

for every fibrewise symplectomorphism ϕ. Differentiating again with respect to ϕ we obtain the infinitesimal statement

$$Y \in \mathcal{X}(M, \sigma) \qquad \Longrightarrow \qquad [X, Y] \in \mathcal{X}(M, \sigma)$$

for every fibrewise symplectic vector field X. Thus, the Lie bracket of a vertical symplectic vector field Y with a fibrewise symplectic vector field X is again a vertical symplectic vector field. This is especially important when $X = v^\sharp$ is the horizontal lift of a vector field v on B. □

Exercise 6.4.7 Let $Y : M \to TM$ be a **vertical Hamiltonian vector field** with Hamiltonian function $H : M \to \mathbb{R}$ in the sense that the 1-forms $\iota(Y)\tau$ and dH agree on the fibres

$$\iota(Y)\tau \overset{\text{fibre}}{=} dH.$$

Prove that for every fibrewise symplectic vector field $X : M \to TM$ the Lie bracket $[X, Y]$ is a vertical Hamiltonian vector field with Hamiltonian function $\mathcal{L}_X H$:

$$\iota([X, Y])\tau \overset{\text{fibre}}{=} d(\mathcal{L}_X H).$$

Hint: Prove first that if $\phi : M \to M$ is a fibrewise symplectomorphism then $\phi^* Y$ is a vertical Hamiltonian vector field with Hamiltonian function $H \circ \phi$. □

Symplectic curvature

Let $\Gamma = \Gamma_\tau$ be a symplectic connection on the symplectic fibration $\pi : M \to B$. If $X : M \to TM$ is a fibrewise symplectic vector field then it descends to a vector field $v : B \to TB$ and hence its horizontal part with respect to the decomposition $TM = \text{Hor} \oplus \text{Vert}$ agrees with the horizontal lift $X^{\text{Hor}} = v^\sharp$ of v. Hence the horizontal and vertical parts of a fibrewise symplectic vector field are again fibrewise symplectic vector fields. Apply this to the Lie bracket of two horizontal lifts $v_1^\sharp, v_2^\sharp : M \to TM$ of vector fields $v_1, v_2 : B \to TB$ to obtain a vertical symplectic vector field

$$\Omega_\Gamma(v_1, v_2) := [v_1^\sharp, v_2^\sharp]^{\text{Vert}} = [v_1^\sharp, v_2^\sharp] - [v_1, v_2]^\sharp.$$

The last identity follows from the fact that, in view of Exercise 6.4.4, the vector field $[v_1^\sharp, v_2^\sharp]$ descends to $[v_1, v_2]$, and so its horizontal part agrees with the horizontal lift $[v_1, v_2]^\sharp$. It is easy to see that the bilinear map

$$\Omega_\Gamma : \mathcal{X}(B) \times \mathcal{X}(B) \to \mathcal{X}(M, \sigma)$$

is bilinear over the functions on B and hence the restriction of $\Omega_\Gamma(v_1, v_2)$ to the fibre F_b depends only on $v_1(b)$ and $v_2(b)$. Hence Ω_Γ is a 2-form

$$\Omega_\Gamma : T_b B \times T_b B \to \mathcal{X}(F_b, \sigma_b)$$

called the **curvature** of the connection Γ. It vanishes if and only if the horizontal distribution $\text{Hor} \subset TM$ is integrable.

Lemma 6.4.8 (Curvature identity [292]) *Let $\tau \in \mathcal{T}(M,\sigma)$ be a connection 2-form. Then the curvature of the connection $\Gamma = \Gamma_\tau$ satisfies*

$$\iota([v_1^\sharp, v_2^\sharp]^{\mathrm{Vert}})\tau \overset{\mathrm{fibre}}{=} \iota(v_2^\sharp)\iota(v_1^\sharp)d\tau - d\iota(v_2^\sharp)\iota(v_1^\sharp)\tau$$

for $v_1, v_2 : B \to TB$.

Proof: The connection 2-form τ satisfies

$$\begin{aligned} d\tau(v_1^\sharp, v_2^\sharp, Y) &= \tau([v_1^\sharp, v_2^\sharp], Y) + \tau([v_2^\sharp, Y], v_1^\sharp) + \tau([Y, v_1^\sharp], v_2^\sharp) \\ &\quad + \mathcal{L}_Y(\tau(v_1^\sharp, v_2^\sharp)) + \mathcal{L}_{v_1^\sharp}(\tau(v_2^\sharp, Y)) + \mathcal{L}_{v_2^\sharp}(\tau(Y, v_1^\sharp)). \end{aligned}$$

Since $\tau(v^\sharp, Y) = 0$ for every vertical vector field Y and, by Remark 6.4.6, the Lie bracket $[v^\sharp, Y]$ of a horizontal and a vertical vector field is again vertical, the only nonvanishing terms on the right hand side are

$$\tau([v_1^\sharp, v_2^\sharp], Y) + \mathcal{L}_Y(\tau(v_1^\sharp, v_2^\sharp)) = \tau([v_1^\sharp, v_2^\sharp]^{\mathrm{Vert}}, Y) + \iota(Y)d\,\iota(v_2^\sharp)\iota(v_1^\sharp)\tau.$$

This proves the lemma. □

Exercise 6.4.9 Let $\tau \in \Omega^2(M)$ be a symplectic connection. Prove that the holonomy of Γ_τ around all contractible loops in B is Hamiltonian if and only if the curvature form

$$\Omega_\Gamma : T_bB \times T_bB \to \mathcal{X}(F_b, \sigma_b)$$

takes values in the space of Hamiltonian vector fields on F_b. **Hint:** In a local trivialization of the bundle $\pi : M \to B$, the 2-form τ can be written in the form (6.3.5). Prove that under the resulting identification of (F_b, σ_b) with (F, σ), the horizontal lift of the vector field $\partial/\partial x_i$ on B is given by $(\partial/\partial x_i, v_i)$, where $v_i \in \mathcal{X}(F, \sigma)$ satisfies

$$\iota(v_i)\sigma = \alpha_i.$$

Deduce that the curvature 2-form is given by

$$\Omega = v_{ij} dx_i \wedge dx_j, \qquad \iota(v_{ij})\sigma = \partial_j \alpha_i - \partial_i \alpha_j + d(\sigma(v_i, v_j)).$$

Prove that the holonomy around contractible loops in B is Hamiltonian if and only if the 1-forms $\partial_i \alpha_j - \partial_j \alpha_i$ are exact for all i and j. □

Exercise 6.4.10 Consider the connection form $\tau \in \Omega^2(\mathbb{R}^m \times F)$ given by

$$\tau = \sigma + \sum_i dH_i \wedge dx_i + \sum_{i<j} (\partial_i H_j - \partial_j H_i)\, dx_i \wedge dx_j.$$

Prove that τ is closed and that the curvature of Γ_τ is the 2-form $\Omega_\tau \in \Omega^2(\mathbb{R}^m, \mathcal{X}(F, \sigma))$ given by

$$\Omega_\tau = \sum_{i<j} X_{F_{ij}} dx_i \wedge dx_j, \qquad F_{ij} = \partial_j H_i - \partial_i H_j + \{H_i, H_j\}.$$

Prove that the parallel transport in the x_i-direction is given by the integral curves of the Hamiltonian vector field X_{H_i}. □

Remark 6.4.11 We now show that the above concepts of symplectic connections and curvature can be interpreted as a special case of connections and curvature on a principal bundle with structure group $\mathrm{G} = \mathrm{Symp}(F, \sigma)$. Thus, we can think of Theorem 6.4.1 as a generalization of Weinstein's Theorem 6.3.3 to the case of an infinite-dimensional Lie group.

The bundle in question is the **symplectic frame bundle**

$$\mathcal{P} \to B$$

whose fibre over $b \in B$ consists of all symplectomorphisms $f : F \to F_b$ with

$$f^*\sigma_b = \sigma.$$

The tangent space $T_f\mathcal{P} \subset C^\infty(f^*TM)$ is the space of all vector fields

$$\widehat{f}(x) \in T_{f(x)}M$$

along f such that $d\pi(f)\widehat{f} \equiv \mathrm{const}$ and the 1-form $\tau(\widehat{f}, df\cdot) \in \Omega^1(F)$ is closed. The reader may check that this space is independent of the choice of the connection 2-form τ with $\Gamma_\tau = \Gamma$. The Lie algebra of the structure group is the space $\mathfrak{g} = \mathcal{X}(F, \sigma)$ of symplectic vector fields and it acts on $\mathcal{P}$ by $f \mapsto df \circ X$. Thus the vertical tangent space

$$\mathcal{V}_f \subset T_f\mathcal{P}$$

consists of all vector fields of the form $\widehat{f} = df \circ X$ for $X \in \mathcal{X}(F, \sigma)$. Now any symplectic connection $TM = \mathrm{Hor} \oplus \mathrm{Vert}$ on M gives rise to a horizontal subspace

$$\mathcal{H}_f \subset T_f\mathcal{P}$$

which consists of the vector fields $\widehat{f} = v^\sharp \circ f$, where $v^\sharp : F_b \to T_{F_b}M$ is the horizontal lift of v defined by

$$d\pi(x)v^\sharp(x) = v, \qquad v^\sharp(x) \in \mathrm{Hor}_x,$$

for $x \in F_b$. Thus a symplectic connection on $\pi : M \to B$ gives rise to a splitting

$$T\mathcal{P} = \mathcal{H} \oplus \mathcal{V}$$

and hence a connection on the principal bundle $\mathcal{P}$.

The curvature of a connection on a principal bundle P is a 2-form with values in the bundle $\mathfrak{g}_P = \mathrm{Ad}(P)$ which is associated to P via the adjoint action of P on its Lie algebra. In the case of the symplectic frame bundle $\mathcal{P}$, this is the bundle

$$\mathrm{Ad}(\mathcal{P}) \to B$$

whose fibre over b is the space $\mathcal{X}(F_b, \sigma_b)$ of symplectic vector fields on the fibre. Hence the curvature 2-form of a symplectic connection $\Gamma = \Gamma_\tau$ is a 2-form

$$\Omega_\Gamma : T_bB \times T_bB \to \mathcal{X}(F_b, \sigma_b).$$

The reader may check that the above definition $\Omega_\Gamma(v_1, v_2) = [v_1^\sharp, v_2^\sharp]^{\mathrm{Vert}}$ is consistent with the usual definition of curvature for principal bundles. □

We are now ready to give our second proof of the second part of Theorem 6.4.1.

Proof of Theorem 6.4.1: '(ii) implies (i)' Let Γ be a symplectic connection on M such that the holonomy around contractible loops in B is Hamiltonian. We shall construct the coupling form τ_Γ in terms of the curvature Ω_Γ. Note that we only have to specify the values of τ_Γ on pairs of horizontal vectors since all its other values are determined by the requirement that $\iota_b{}^*\tau_\Gamma = \sigma_b$ and Hor is the complement of Vert with respect to τ_Γ. Hence it suffices to specify the functions $\tau_\Gamma(v_1^\sharp, v_2^\sharp) : M \to \mathbb{R}$ for any two vector fields $v_1, v_2 : B \to TB$. By assumption, the holonomy of Γ around every contractible loop in B is Hamiltonian. By Exercise 6.4.9 this implies that

$$\Omega_\Gamma(v_1, v_2) = [v_1^\sharp, v_2^\sharp]^{\mathrm{Vert}} \in \mathcal{X}(F_b, \sigma_b)$$

is a Hamiltonian vector field for all $v_1, v_2 \in T_bB$. Hence, for any two vector fields $v_1, v_2 : B \to TB$ there exists a unique Hamiltonian function

$$H_{v_1,v_2} : M \to \mathbb{R}$$

such that

$$\iota([v_1^\sharp, v_2^\sharp]^{\mathrm{Vert}})\sigma_b = d\,H_{v_1,v_2}|_{F_b}, \qquad \int_{F_b} H_{v_1,v_2}\sigma_b{}^k = 0$$

for every $b \in B$. Since the map $T_bB \times T_bB \to \mathbb{R} : (v_1, v_2) \mapsto H_{v_1,v_2}(x)$ is bilinear and skew-symmetric for all $x \in M$ with $\pi(x) = b$, there is a unique connection 2-form $\tau_\Gamma \in \mathcal{T}(M, \sigma)$ which generates the connection Γ and satisfies

$$\tau_\Gamma(v_1^\sharp, v_2^\sharp) = H_{v_1,v_2} \tag{6.4.2}$$

for any two vector fields $v_1, v_2 : B \to TB$. We prove that this form is closed. By definition, the restriction $\iota_b{}^*\tau_\Gamma = \sigma_b$ to the fibre is closed. Hence $d\tau_\Gamma$ vanishes on all triples of vertical vectors. Moreover, it follows from the definition of τ_Γ and the curvature identity in Lemma 6.4.8 that

$$\iota(v_2^\sharp)\iota(v_1^\sharp)d\tau_\Gamma \overset{\text{fibre}}{=} 0,$$

and hence $d\tau_\Gamma$ vanishes on triples of the form $(v_1^\sharp, v_2^\sharp, Y)$ where Y is vertical. By Lemma 6.3.5, $d\tau_\Gamma$ also vanishes on triples $(v^\sharp, Y_1, Y_2)$ where Y_1 and Y_2 are vertical. Hence it remains to prove that $d\tau_\Gamma$ vanishes on triples of horizontal vectors. To see this consider the vertical Hamiltonian vector field

$$Y = [[v_1^\sharp, v_2^\sharp], v_3^\sharp]^{\mathrm{Vert}} = \Omega_\Gamma([v_1, v_2], v_3) - [v_3^\sharp, \Omega_\Gamma(v_1, v_2)].$$

The last equality follows by a simple calculation involving the formula

$$[v_1^\sharp, v_2^\sharp] = [v_1, v_2]^\sharp + \Omega_\Gamma(v_1, v_2).$$

It follows from the definition of τ_Γ and Exercise 6.4.7 that Y is generated (fibrewise) by the Hamiltonian function

$$H = \tau_\Gamma([v_1^\sharp, v_2^\sharp], v_3^\sharp) - \mathcal{L}_{v_3^\sharp}\tau(v_1^\sharp, v_2^\sharp).$$

Here we have also used the formula $\tau_\Gamma([v_1^\sharp, v_2^\sharp]^{\mathrm{Vert}}, v_3^\sharp) = 0$. Now consider the corresponding terms after symmetric permutation of v_1, v_2, v_3, take their sum, and use the formula for $d\tau_\Gamma$ in the proof of Lemma 6.4.8 to obtain $d\tau_\Gamma(v_1^\sharp, v_2^\sharp, v_3^\sharp) = 0$. Thus we have proved that τ_Γ is closed and, in particular, that (ii) implies (i).

We prove that τ_Γ satisfies the normalization condition $\int_F {\tau_\Gamma}^{k+1} = 0$. This condition can be expressed in the form

$$\int_{F_b} \iota(v_2^\sharp)\iota(v_1^\sharp){\tau_\Gamma}^{k+1} = 0 \tag{6.4.3}$$

for all $b \in B$ and all $v_1, v_2 \in T_bB$. To prove this formula, note that

$$\iota(X)(\alpha \wedge \beta) = (\iota(X)\alpha) \wedge \beta + (-1)^{\deg(\alpha)}\alpha \wedge \iota(X)\beta$$

and hence, by induction,

$$\iota(Y)\iota(X){\tau_\Gamma}^{k+1} = (k+1)\left(\tau_\Gamma(X,Y)\tau^k - k\iota(X)\tau_\Gamma \wedge \iota(Y)\tau_\Gamma \wedge {\tau_\Gamma}^{k-1}\right).$$

Since $\tau_\Gamma(v^\sharp, Y) = 0$ for every vertical vector Y, (6.4.3) is equivalent to

$$\int_{F_b} \tau_\Gamma(v_1^\sharp, v_2^\sharp){\tau_\Gamma}^k = 0. \tag{6.4.4}$$

But this follows from the fact that H_{v_1,v_2} has mean value zero over each fibre. Hence τ_Γ satisfies (6.4.1). This proves the existence of the coupling form.

To prove uniqueness, suppose that $\tau \in \mathcal{T}(M,\sigma)$ is a closed connection 2-form which generates $\Gamma = \Gamma_\tau$ and satisfies (6.4.1). Then it follows from the curvature identity of Lemma 6.4.8 that

$$\iota([v_1^\sharp, v_2^\sharp]^{\mathrm{Vert}})\tau \overset{\text{fibre}}{=} d(\tau(v_1^\sharp, v_2^\sharp)) \tag{6.4.5}$$

for $v_1, v_2 : B \to TB$. This shows that the vector field $[v_1^\sharp, v_2^\sharp]^{\mathrm{Vert}}$ is Hamiltonian on each fibre F_b with generating Hamiltonian function $\tau(v_1^\sharp, v_2^\sharp)$. Moreover, we have seen that the normalization condition (6.4.1) translates into (6.4.4), and hence the Hamiltonian function $\tau(v_1^\sharp, v_2^\sharp)$ has mean value zero. Hence

$$\tau(v_1^\sharp, v_2^\sharp) = H_{v_1,v_2},$$

and, by (6.4.2), $\tau_\Gamma = \tau$.

The proof that any other closed connection 2-form $\tau \in \Omega^2(M)$ with $\Gamma_\tau = \Gamma$ has the form $\tau = \tau_\Gamma + \pi^*\beta$, where β is a closed 2-form on B, is as before. This completes the second proof of Theorem 6.4.1. □

Exercise 6.4.12 Let τ be the form on $M = P \times_{\mathrm{G}} F$ which was constructed in Theorem 6.3.3. Show that the connection Γ_τ agrees with the connection on M which is induced by $A \in \mathcal{A}(P)$ (see Exercise 6.3.2). Show also that τ is the coupling form of this connection. **Hint:** Use the uniqueness part of Theorem 6.4.1. □

Exercise 6.4.13 Show that in the fibration $\mathbb{C}\mathrm{P}^3 \to \mathbb{H}\mathrm{P}^1$ considered in Exercise 6.1.5, the Fubini–Study form ω_{FS} on $\mathbb{C}\mathrm{P}^3$ is the coupling form of a symplectic connection. □

A symplectic connection Γ on a symplectic fibration $\pi : M \to B$ with compact simply connected fibre (F, σ) is called **fat** if its coupling form τ_Γ is nondegenerate (cf. [674] and [407]). Intuitively, this means that the connection has *a lot of curvature*. Many of the fibrations which arise naturally in the study of the orbits of the coadjoint action of G admit fat connections, and Guillemin, Lerman, and Sternberg have found many important applications of the ideas presented above, which they describe in their book [292].

Finally we note that the theory of the coupling form has found a remarkable application in the work by Polterovich [528], in which he gives a geometric interpretation of the so-called coarse Hofer norm of a symplectomorphism in terms of the K-area of an associated symplectic fibration.

6.5 Hamiltonian fibrations

Let $\pi : M \to B$ be a symplectic fibration and let

$$\Gamma = \Gamma_\tau$$

be a symplectic connection on M. Γ is called a **Hamiltonian connection** if the holonomy around every loop in B is a Hamiltonian symplectomorphism. $\pi : M \to B$ is called a **Hamiltonian fibration** if it admits a Hamiltonian connection. Our aim in this section is to find necessary and sufficient conditions for a symplectic fibration to be Hamiltonian.

Since the Hamiltonian group $\mathrm{Ham}(F, \sigma)$ is connected, an obvious necessary condition for the existence of a Hamiltonian connection is that the restriction of the bundle to every loop in B admits a symplectic trivialization. The next two exercises clarify the meaning of this condition. Theorem 6.5.3 below characterizes Hamiltonian fibrations.

Exercise 6.5.1 Let $\pi : M \to B$ be a symplectic fibration and suppose that the base manifold B has been decomposed as a CW complex. (Those unfamiliar with the language of CW complexes can simply assume that B has been smoothly triangulated and that B_1 is the union of the 1-simplices in this triangulation.) Show that $\pi : M \to B$ admits a symplectic trivialization over the 1-skeleton B_1 of B if and only if, for every symplectic connection on M, the holonomy around every loop in B is symplectically isotopic to the identity. Show further that there exists a nontrivial symplectic fibration $M \to S^1$ with fibre (F, σ) if and only if the symplectomorphism group of (F, σ) is disconnected. **Hint:** By the connectedness of the space of connections, it is enough to find one connection whose holonomy is symplectically isotopic to the identity. □

Exercise 6.5.2 Consider the fibration $\pi : M \to \mathbb{T}^2$ with fibre $F = \mathbb{T}^2$, with total space M equal to the quotient of $\mathbb{R}^4$ by the equivalence relation

$$[s,t,x,y] \sim [s+i, t+j, x+it+k, y+\ell] \tag{6.5.1}$$

for $(s,t,x,y) \in \mathbb{R}^4$ and $i,j,k,\ell \in \mathbb{Z}$, with projection given by

$$\pi : M \to \mathbb{T}^2, \qquad \pi([s,t,x,y]) := [s,t],$$

and where the symplectic form on the fibre is $dx \wedge dy$. Prove the following.

(i) This fibration admits a symplectic trivialization over the 1-skeleton of $\mathbb{T}^2$.

(ii) The 2-form $\tau := dx \wedge dy - s\, dt \wedge dy$ on $\mathbb{R}^4$ descends to a connection 2-form on M.

(iii) M does not admit a closed connection 2-form.

(iv) M carries a symplectic form $\omega := ds \wedge dy + dx \wedge dt$ with respect to which the above fibres are Lagrangian.

(v) (M,ω) is symplectomorphic to Thurston's manifold $\mathbb{R}^4/\Gamma$ in Example 3.1.17.

(vi) The pullback of the standard complex structure on $\mathbb{C}^2$ under the diffeomorphism

$$\mathbb{R}^4 \to \mathbb{C}^2 : (s,t,x,y) \mapsto \left(s + \sqrt{-1}\,t,\, x + \sqrt{-1}\,y - \tfrac{st}{2} - \sqrt{-1}\,\tfrac{s^2+t^2}{4}\right)$$

is given by

$$J(s,t,x,y) = \begin{pmatrix} 0 & -1 & 0 & 0 \\ 1 & 0 & 0 & 0 \\ s & 0 & 0 & -1 \\ 0 & -s & 1 & 0 \end{pmatrix}.$$

This is an integrable complex structure on $\mathbb{R}^4$. It is preserved by the diffeomorphism $(s,t,x,y) \mapsto (s+1, t, x+t, y)$ and by any translation in the variables t,x,y. Hence it is invariant under the equivalence relation (6.5.1). It descends to a complex structure on M such that the projection $\pi : M \to \mathbb{T}^2$ is holomorphic with respect to the standard complex structure on $\mathbb{T}^2$. □

The next theorem characterizes the Hamiltonian fibrations among those that satisfy the equivalent conditions in Exercise 6.5.1 above. A necessary condition is that there is a closed connection 2-form τ on the total space M. Exercise 6.5.2 shows that this condition is nontrivial, while Exercise 6.5.5 shows that it is not sufficient. In order to explain why, we recall from Remark 6.4.2 that the flux of the holonomy round a loop γ in B is well-defined as an element of $H^1(F;\mathbb{R})$ once one has chosen a trivialization of the bundle over γ. If no trivialization is chosen, then the flux of the holonomy lies in the quotient group $H^1(F;\mathbb{R})/\Gamma_\sigma$, where Γ_σ denotes the flux subgroup of (F,σ) defined in Proposition 10.2.13. Hence the closed form τ defines a homomorphism $\pi_1(B) \to H^1(F;\mathbb{R})/\Gamma_\sigma$ given by the flux of the holonomy of the connection Γ_τ around the loops in B. This gives rise to a homomorphism $f_\tau : H_1(B;\mathbb{Z}) \to H^1(F;\mathbb{R})/\Gamma_\sigma$ that lifts to a homomorphism

$$\widetilde{f}_\tau : H_1(B;\mathbb{Z}) \to H^1(F;\mathbb{R})$$

if and only if it vanishes on all torsion elements in $H_1(B;\mathbb{Z})$.

Theorem 6.5.3 *Let $\pi : M \to B$ be a symplectic fibration whose restriction to every loop in B admits a symplectic trivialization. The following are equivalent.*

(i) *$\pi : M \to B$ is a Hamiltonian fibration.*

(ii) *The structure group reduces to* $\mathrm{Ham}(F, \sigma)$.

(iii) *There exists a closed 2-form $\tau \in \Omega^2(M)$ such that $\iota_b^*\tau = \sigma_b$ for every $b \in B$, and f_τ lifts to a homomorphism $\widetilde{f}_\tau : H_1(B; \mathbb{Z}) \to H^1(F; \mathbb{R})$.*

Proof: See page 285. □

Remark 6.5.4 The existence of the lift $\widetilde{f}_\tau$ does not depend on the choice of the closed connection form τ. To see this, fix a symplectic trivialization

$$\Phi : B_1 \times F \to M$$

over the 1-skeleton B_1 of B and a homology class $a \in H^2(M; \mathbb{R})$ such that $\iota_b^* a = [\sigma_b]$ for some (and hence every) $b \in B$. Each such pair (a, Φ) defines a homomorphism $\mu_{\Phi^*a} : H_1(B_1; \mathbb{Z}) \to H^1(F; \mathbb{R})$ via

$$\langle \mu_{\Phi^*a}([\beta]), [\gamma] \rangle := \int_{S^1 \times S^1} (\beta \times \gamma)^* \Phi^* a$$

for every pair of based loops $\beta : S^1 \to B_1$ and $\gamma : S^1 \to F$. If τ is a closed connection form in class a, then $\Phi^*\tau$-parallel translation around the loop β in $B_1 \times F$ gives a path $\{g_t\}_{0 \le t \le 1}$ in $\mathrm{Symp}(F)$ that starts at the identity and is such that $\mu_{\Phi^*a}([\beta]) = \mathrm{Flux}(\{g_t\})$. Hence μ_{Φ^*a} does lift the pullback

$$H_1(B_1; \mathbb{Z}) \longrightarrow H_1(B; \mathbb{Z}) \xrightarrow{f_\tau} H^1(F; \mathbb{R})/\Gamma_\sigma$$

of f_τ to the free group $H_1(B_1; \mathbb{Z})$. But we need a lift $\widetilde{f}_\tau$ that is defined on $H_1(B; \mathbb{Z})$, and this exists if and only if $\mu_{\Phi^*a}([\beta]) \in \Gamma_\sigma$ for every loop $\beta : S^1 \to B_1$ that represents a torsion class in $H_1(B; \mathbb{Z})$.

It remains to prove that the latter condition is independent of a. Let

$$\beta : S^1 = \mathbb{R}/\mathbb{Z} \to B_1$$

represent a torsion class in $H_1(B; \mathbb{Z})$. Then there is a positive integer k such that the loop $s \mapsto \beta(ks)$ is homologous to zero in B. This implies that the k-fold multiple of the cycle $\Phi \circ (\beta \times \gamma) : \mathbb{T}^2 \to M$ is homologous to a cycle $v : \mathbb{T}^2 \to F_{b_0}$ in the fibre. To be explicit, note that there is an oriented Riemann surface Σ with two boundary components $\partial_0\Sigma$ and $\partial_1\Sigma$, diffeomorphisms $\iota_0 : S^1 \to \partial_0\Sigma$ (orientation-reversing) and $\iota_1 : S^1 \to \partial_1\Sigma$ (orientation-preserving), and a smooth map $u : \Sigma \to B$ such that $u \circ \iota_1(s) = \beta(ks)$ and $u \circ \iota_0(s) = b_0$ for $s \in S^1$. Since the fibration M is trivial over B_1, one proves as in Proposition 2.6.7 that the given trivialization of M over β extends to a symplectic trivialization $\Psi : \Sigma \times F \to M$ over u such that $\Psi(\iota_1(s), q) = \Phi(ks, q)$ for $s \in S^1$ and $q \in F$. Hence the cycle

$$\mathbb{T}^2 \to M : (s,t) \mapsto \Phi(\beta(ks), \gamma(t)) = \Psi(\iota_1(s), \gamma(t))$$

is homologous to the cycle $\mathbb{T}^2 \to F_{b_0} : (s,t) \mapsto v(s,t) := \Psi(\iota_0(s), \gamma(t))$. It follows that

$$\langle \mu_{\Phi^* a}([\beta]), [\gamma] \rangle = \frac{1}{k} \int_{\mathbb{T}^2} v^* \sigma_{b_0}.$$

Since the map u and the trivialization Ψ along u are independent of a, so is the cycle v and the cohomology class $\mu_{\Phi^* a}([\beta]) \in H^1(F; \mathbb{Z})$.

Since the existence of the extension τ is equivalent to the existence of a cohomological extension a of $[\sigma_b]$ by Thurston's Theorem 6.1.4, it follows that condition (iii) in Theorem 6.5.3 is equivalent to the following purely cohomological statement.

(iv) *There exists a cohomology class $a \in H^2(M; \mathbb{R})$ such that $\iota_b^* a = [\sigma_b]$ for every $b \in B$, and the homomorphism $\mu_{\Phi^* a} : H_1(B_1; \mathbb{Z}) \to H^1(F; \mathbb{R})$ takes the elements of $H_1(B_1; \mathbb{Z})$ that map to torsion classes in $H_1(B; \mathbb{Z})$ to the subgroup $\Gamma_\sigma \subset H^1(F; \mathbb{R})$, for some (and hence every) symplectic trivialization Φ over the 1-skeleton.* □

Proof of Theorem 6.5.3. To prove that (i) implies (ii), use the Hamiltonian connection to construct the local trivializations. The converse follows from a similar construction as in Exercise 6.3.6, the details of which are left to the reader. That (i) implies (iii) follows immediately from Theorem 6.4.1. It remains to prove that (iii) implies (i). Assume that $\tau \in \Omega^2(M)$ is a closed 2-form such that $\iota_b^* \tau = \sigma_b$ for every $b \in B$. Then Γ_τ is a symplectic connection with Hamiltonian holonomy around contractible loops in B.

We shall prove that if the lift $\widetilde{f}_\tau$ exists there is a 2-form $\rho \in \Omega^2(M)$ such that

(a) $d\rho = 0$ and $\iota_b{}^* \rho = 0$ for every $b \in B$.

(b) The connection $\Gamma_{\tau - \rho}$ is Hamiltonian.

We now give a very explicit construction for ρ.[28] We suppose that B has been smoothly triangulated and denote by B_k the k-skeleton. Let U_1 be an open neighbourhood of B_1 which retracts onto B_1. By assumption there is a lift

$$\widetilde{f} : \pi_1(B_1) \to H^1(F; \mathbb{R})$$

of the homomorphism $\pi_1(B_1) \to H^1(F; \mathbb{R})/\Gamma_\sigma$ defined by the flux of the holonomy of Γ_τ, such that $\widetilde{f}(\gamma) = 0$ for every loop $\gamma \in \pi_1(B_1)$ that is homologous to zero in B. Our first aim is to construct a symplectic trivialization

$$\phi_1 : \pi^{-1}(U_1) \to U_1 \times F$$

such that $\widetilde{f}$ is given by taking the holonomy of Γ_τ with respect to this trivialization ϕ_1. To this end, choose a maximal tree T in B_1 (that is, a maximal connected

[28] One can also prove this theorem by much less explicit arguments: see McDuff [461].

contractible subgraph) and denote the remaining edges in $B_1 \setminus T$ by $e_1, \ldots, e_m$. Choose a symplectomorphism $F_{b_*} \to F$ for some arbitrary vertex $b_* \in B_0$ and then trivialize π over T by parallel translation along the edges of T with respect to the given connection Γ_τ. Each of the remaining edges e_j, $1 \le j \le m$, can be completed to a loop ℓ_j by adding edges from T, and our aim is to extend the trivialization over the e_j in such a way that the holonomy with respect to ϕ_1 along each loop ℓ_j has flux equal to $\tilde{f}(\ell_j)$. To do this, let $\gamma : [0,1] \to B_1$ parametrize e_j with endpoints $b_0 = \gamma(0)$ and $b_1 = \gamma(1)$. Denote by

$$\Phi_t : F_{\gamma(0)} \to F_{\gamma(t)}$$

the parallel transport of Γ_τ along γ, and let $\psi_0 : F_{b_0} \to F$ and $\psi_1 : F_{b_1} \to F$ denote the symplectomorphisms already constructed. Then, by Exercise 6.5.1, the symplectomorphism $\psi_1 \circ \Phi_1 \circ \psi_0^{-1} : (F, \sigma) \to (F, \sigma)$ is symplectically isotopic to the identity. Hence one can choose a symplectic isotopy $\chi_t : F \to F$ such that

$$\chi_0 = \mathrm{id}, \quad \chi_1 = \psi_1 \circ \Phi_1 \circ \psi_0{}^{-1}, \quad \mathrm{Flux}(\{\chi_t\}) = \tilde{f}(\ell_j).$$

It follows that $\psi_t = \chi_t \circ \psi_0 \circ \Phi_t^{-1} : F_{\gamma(t)} \to F$ is a symplectic trivialization over γ which connects ψ_0 and ψ_1 and satisfies the required conditions. This defines ϕ_1 over the 1-skeleton; it clearly extends over U_1.

Denote by $\sigma_1 \in \Omega^2(U_1 \times F)$ the pullback of the 2-form $\sigma \in \Omega^2(F)$ under the projection $U_1 \times F \to F$. Then the 2-form

$$\rho_1 := \tau - \phi_1^* \sigma_1 \in \Omega^2(\pi^{-1}(U_1))$$

satisfies $\iota_b{}^* \rho_1 = 0$ for every $b \in U_1$. Moreover, the holonomy of the connection $\Gamma_{\tau - \rho_1} = \Gamma_{\phi_1^* \sigma_1}$ around any loop in U_1 is the identity. We shall prove that there exists a closed 2-form $\rho \in \Omega^2(M)$ which satisfies $\iota_b^* \rho = 0$ for all $b \in B$ and agrees with ρ_1 in some neighbourhood of $\pi^{-1}(B_1)$. The construction of this extension is by induction over the k-skeleton $B_k \subset B$ for $k \ge 2$. To carry out the induction step let us choose a collection of neighbourhoods U_k of B_k in B such that

$$U_1 \subset U_2 \subset \cdots \subset U_m = B, \qquad \mathrm{cl}(U_{k+1} \setminus U_k) \cap B_k = \emptyset.$$

Here $m := \dim B$. For each k consider the open set $V_k := W_k \cap U_k$, where W_k is a small neighbourhood of the attaching set $\mathrm{cl}(U_k) \cap \mathrm{cl}(U_{k+1} \setminus U_k)$:

$$V_k = W_k \cap U_k, \qquad \mathrm{cl}(U_k) \cap \mathrm{cl}(U_{k+1} \setminus U_k) \subset W_k, \qquad \mathrm{cl}(V_k) \cap B_k = \emptyset.$$

Suppose that these open sets have been chosen sensibly, so that V_k is a disjoint union of connected open sets each of which retracts onto a contractible embedded k-sphere in B. Then, for $k = 2, \ldots, m = \dim B$, we shall construct 2-forms $\rho_k \in \Omega^2(\pi^{-1}(U_k))$ such that

(A) $d\rho_k = 0$ and $\iota_b^* \rho_k = 0$ for every $b \in U_k$.
(B) $\rho_k|_{\pi^{-1}(U_{k-1} \setminus V_{k-1})} = \rho_{k-1}$.
(C) The restriction $\rho_k|_{\pi^{-1}(V_k)}$ is exact.

The first and hardest step is the extension to the 2-skeleton. Note first that V_1 is homotopy equivalent to a collection of embedded circles in B which are homotopic to the boundaries of the 2-simplices in B and so are contractible in B. Hence it follows from the construction of ϕ_1 that the holonomy of τ around every loop $\beta : S^1 \to V_1$ has zero flux with respect to the trivialization ϕ_1. This implies that the integral of τ vanishes over all cycles of the form ${\phi_1}^{-1}(\beta \times \gamma)$, where $\beta : S^1 \to V_1$ and $\gamma : S^1 \to F$. (Prove this!) Hence

$$\int_{{\phi_1}^{-1}(\beta\times\gamma)} \rho_1 = \int_{{\phi_1}^{-1}(\beta\times\gamma)} \tau - \int_{\beta\times\gamma} \sigma_1 = 0.$$

Since every 2-cycle in $\pi^{-1}(V_1)$ is homologous to a sum of 2-cycles of the form $\phi_1^{-1}(\beta \times \gamma)$ and a vertical cycle in some fibre, it follows that the restriction of ρ_1 to $\pi^{-1}(V_1)$ is exact. Hence choose $\mu_1 \in \Omega^1(\pi^{-1}(V_1))$ such that $\rho_1|_{\pi^{-1}(V_1)} = d\mu_1$. Next choose a smooth cutoff function $\theta_1 : V_1 \to [0,1]$ such that $\theta_1 = 1$ near $\mathrm{cl}(V_1) \cap \mathrm{cl}(U_1 \setminus V_1)$ and $\theta_1 = 0$ near $\mathrm{cl}(V_1) \cap \mathrm{cl}(U_2 \setminus U_1)$. Then define

$$\rho_2 := \begin{cases} \rho_1, & \text{in } \pi^{-1}(U_1 \setminus V_1), \\ d((\theta_1 \circ \pi)\mu_1), & \text{in } \pi^{-1}(V_1), \\ 0, & \text{in } \pi^{-1}(U_2 \setminus U_1). \end{cases}$$

This 2-form obviously satisfies (A) and (B) but it will not, in general, satisfy (C). We shall prove, however, that there exists a 2-form $\beta_2 \in \Omega^2(U_2)$ such that the 2-form $\tilde{\rho}_2 := \rho_2 - \pi^*\beta_2$ satisfies (A), (B), and (C).

To construct β_2, fix a point $q_0 \in F$ and consider the section $s_1 : B_1 \to M$ defined by $s_1(b) := \phi_1^{-1}(b, q_0)$ for $b \in B_1$. Use the Hamiltonian connection $\Gamma_{\tau-\rho_2}$ to trivialize the bundle $\pi : M \to B$ over every 2-simplex $\Delta \subset B_2$. Denote this trivialization by

$$\phi_\Delta : \pi^{-1}(\Delta) \to \Delta \times F.$$

This gives rise to a loop $\partial\Delta \to \mathrm{Ham}(F, \sigma) : b \mapsto \psi_b$ of Hamiltonian symplectomorphisms given by

$$(b, \psi_b(q)) := \phi_\Delta \circ {\phi_1}^{-1}(b, q)$$

for $b \in \partial\Delta$ and $q \in F$. But, by Proposition 10.2.19, for every such Hamiltonian loop the path $\partial\Delta \to F : b \mapsto \psi_b(q_0)$ is homologous to zero in $H_1(F;\mathbb{R})$. This means that there is a function $u_\Delta : \Delta \setminus \{b_\Delta\} \to F$ such that $u(b) = \psi_b(q_0)$ for $b \in \partial\Delta$, and that as b approaches the barycentre b_Δ, the points $u(b)$ converge to a nullhomologous loop in the fibre F_{b_Δ}. Define the section

$$s_\Delta : \Delta \setminus \{b_\Delta\} \to \pi^{-1}(\Delta), \qquad s_\Delta(b) := {\phi_\Delta}^{-1}(b, u_\Delta(b)).$$

It satisfies $s_\Delta|_{\partial\Delta} = s_1|_{\partial\Delta}$. Now choose a closed 2-form $\beta_2 \in \Omega^2(U_2)$ which vanishes over U_1 and in a neighbourhood of each barycentre b_Δ, and satisfies

$$\int_\Delta \beta_2 = \int_\Delta {s_\Delta}^*\rho_2$$

for every 2-simplex Δ in B_2. Then $\int_\Delta {s_\Delta}^*(\rho_2 - \pi^*\beta_2) = 0$ for every 2-simplex Δ. Since $s_\Delta(b) = s_1(b)$ for every $b \in \partial\Delta$ and every 2-simplex Δ, and $\rho_2 - \pi^*\beta_2$

vanishes on the fibres, it follows that the 2-form $\tilde{\rho}_2 := \rho_2 - \pi^*\beta_2$ satisfies (A), (B), and (C). In particular, the integral of $\tilde{\rho}_2$ vanishes over every 2-cycle in $\pi^{-1}(V_2)$, because every such 2-cycle is homologous to a sum of *'vertical chains'* in the fibres and *'horizontal chains'* composed of the sections $s_\Delta : \Delta \setminus \{b_\Delta\} \to M$. (It is here that we use the fact that the loops $\partial\Delta \to F : b \mapsto \psi_b(q_0)$ are nullhomologous.) The integral of $\tilde{\rho}_2$ vanishes over both, and hence $\tilde{\rho}_2$ satisfies (C).

The rest of the proof is easy. In each induction step, choose $\mu_k \in \Omega^1(\pi^{-1}(V_k))$ such that $\rho_k|_{\pi^{-1}(V_k)} = d\mu_k$. Next choose a cutoff function $\theta_k : V_k \to [0,1]$ such that $\theta_k = 1$ near $\mathrm{cl}(V_k) \cap \mathrm{cl}(U_k \setminus V_k)$ and $\theta_k = 0$ near $\mathrm{cl}(V_k) \cap \mathrm{cl}(U_{k+1} \setminus U_k)$. Then define

$$\rho_{k+1} := \begin{cases} \rho_k, & \text{in } \pi^{-1}(U_k \setminus V_k), \\ d((\theta_k \circ \pi)\mu_k), & \text{in } \pi^{-1}(V_k), \\ 0, & \text{in } \pi^{-1}(U_{k+1} \setminus U_k). \end{cases}$$

For $k \geq 2$ this 2-form obviously satisfies (A), (B), and (C). With $k = m$, the 2-form $\rho := \rho_m$ satisfies (a) and (b), and this proves Theorem 6.5.3. □

Example 6.5.5 Consider the quotient

$$M := \frac{S^2 \times \mathbb{T}^2}{\mathbb{Z}_2},$$

where we think of $S^2 \subset \mathbb{R}^3$ as the unit sphere and $\mathbb{T}^2 = \mathbb{R}^2/\mathbb{Z}^2$ as the standard torus. The nontrivial element of $\mathbb{Z}_2$ acts by the involution

$$(x, y) \mapsto (-x, y + (1/2, 0)),$$

where $x \in S^2$ and $y \in \mathbb{T}^2$. The projection $M \to S^2/\mathbb{Z}_2 = \mathbb{R}\mathrm{P}^2$ is a symplectic fibration with fibre $\mathbb{T}^2$. The closed 2-form

$$\tau := dy_1 \wedge dy_2 \in \Omega^2(S^2 \times \mathbb{T}^2)$$

descends to a closed connection 2-form on M; its holonomy around each contractible loop in $\mathbb{R}\mathrm{P}^2$ is the identity and around each noncontractible loop is the symplectomorphism $(y_1, y_2) \mapsto (y_1 + 1/2, y_2)$. Thus M has a closed connection 2-form and the restriction of M to every loop in $\mathbb{R}\mathrm{P}^2$ admits a symplectic trivialization. However, this fibration is not Hamiltonian: one can show that the holonomy of every closed connection 2-form on M over every noncontractible loop in $\mathbb{R}\mathrm{P}^2$ differs from the half rotation $(y_1, y_2) \mapsto (y_1 + 1/2, y_2)$ by a Hamiltonian symplectomorphism and so cannot itself be Hamiltonian. □

Remark 6.5.6 Hamiltonian fibrations have interesting topological properties that can be studied using Gromov–Witten invariants: see Seidel [569], Lalonde–McDuff–Polterovich [398], and Lalonde–McDuff [396]. These questions are related to homotopy theoretic questions about symplectomorphism groups and their actions that are discussed further in Chapter 10. □

7

CONSTRUCTING SYMPLECTIC MANIFOLDS

In this chapter we look at various ways to construct symplectic manifolds and submanifolds. As Gromov points out in Section 3.4.4 of his book *Partial Differential Relations*, many of the constructions which work well for Kähler manifolds (e.g. blowing up and down, branched coverings, fibrations) have symplectic analogues. It is also possible to do a few kinds of surgery, though in this respect contact manifolds are more flexible.

We begin in Section 7.1 by studying blowing up and down in both the complex and the symplectic contexts. This has turned out to have many interesting applications. For example, it was used to give the first example of a simply connected nonKähler closed symplectic manifold (cf. Example 7.1.33) and is the key to understanding symplectic ball embeddings, especially in dimension 4. Section 7.2 is devoted to a discussion of a very special kind of surgery called fibre connected sum. We use it to describe Gompf's construction of symplectic 4-manifolds with arbitrary fundamental group. In Section 7.3 we explain Gromov's telescope construction. It shows that for open manifolds the h-principle rules and the inclusion of the space of symplectic forms into the space of nondegenerate 2-forms is a homotopy equivalence. In particular, every almost complex open manifold has a symplectic structure. Section 7.4 outlines Donaldson's construction of codimension two symplectic submanifolds and explains the associated decompositions of the ambient manifold. Its contact analogue is the open book decomposition of a contact manifold that we explained in Section 3.5.

7.1 Blowing up and down

We begin our discussion with a recollection of the blowup construction for complex manifolds. We then discuss in detail the construction of Kähler forms on a complex blowup as preparation for the symplectic blowup. This can be given a purely symplectic description in terms of ball embeddings; cf. equation (7.1.14). However, with this naive description it is hard to understand either the relation between the blowups given by different ball embeddings or the connection between the complex and symplectic blowup operations. For this we need to develop the rather intricate language of Theorem 7.1.21 (Existence) and Theorem 7.1.23 (Uniqueness). The section ends with applications to ball embedding problems.

Blowing up and down in the complex category

Intuitively, the complex blowup of a point replaces that point by the set of all lines through that point. This operation gives a complex manifold because the complement of a point in $\mathbb{C}^n$ can be identified with the complement of the zero section in the tautological line bundle over projective space $\mathbb{C}\mathrm{P}^{n-1}$. Here is the formal definition. The **(complex) blowup of $\mathbb{C}^n$ at the origin** is the space $\widetilde{\mathbb{C}}^n \subset \mathbb{C}\mathrm{P}^{n-1} \times \mathbb{C}^n$ of all pairs (ℓ, z) consisting of a one-dimensional complex linear subspace $\ell \subset \mathbb{C}^n$ and an element $z \in \ell$, i.e.

$$\widetilde{\mathbb{C}}^n := \left\{ (\ell, z) \,\middle|\, \begin{array}{l} \ell = [w_1 : \cdots : w_n] \in \mathbb{C}\mathrm{P}^{n-1}, \\ z = (z_1, \ldots, z_n) \in \mathbb{C}^n, \\ w_j z_k = w_k z_j \ \forall\, j, k \end{array} \right\}. \tag{7.1.1}$$

Here we identify the equivalence class $[w_1 : \cdots : w_n] \in \mathbb{C}\mathrm{P}^{n-1}$ of $w \in \mathbb{C}^n \setminus \{0\}$ with the line

$$\ell := \{(\lambda w_1, \ldots, \lambda w_n) \,|\, \lambda \in \mathbb{C}\} \subset \mathbb{C}^n.$$

Thus $\widetilde{\mathbb{C}}^n$ is a complex submanifold of $\mathbb{C}\mathrm{P}^{n-1} \times \mathbb{C}^n$ and there are two projections

$$\pi : \widetilde{\mathbb{C}}^n \to \mathbb{C}^n, \qquad \mathrm{pr} : \widetilde{\mathbb{C}}^n \to \mathbb{C}\mathrm{P}^{n-1}.$$

The projection $\mathrm{pr} : \widetilde{\mathbb{C}}^n \to \mathbb{C}\mathrm{P}^{n-1}$ exhibits $L := \widetilde{\mathbb{C}}^n$ as the tautological line bundle over $\mathbb{C}\mathrm{P}^{n-1}$ whose fibre over a point $\ell \in \mathbb{C}\mathrm{P}^{n-1}$ is the line ℓ itself. To construct local trivializations write the lines in $\mathbb{C}\mathrm{P}^{n-1}$ near ℓ as graphs of complex linear maps from ℓ to $\ell^\perp$. The tautological line bundle was examined in Exercise 2.7.8 and its adjoint is the canonical line bundle $H \to \mathbb{C}\mathrm{P}^{n-1}$ which played an important role in Example 4.4.3. The restriction of $\pi : \widetilde{\mathbb{C}}^n \to \mathbb{C}^n$ to the complement of the set

$$Z := \pi^{-1}(0) = \mathbb{C}\mathrm{P}^{n-1} \times \{0\} \subset \widetilde{\mathbb{C}}^n$$

is a holomorphic diffeomorphism from $\widetilde{\mathbb{C}}^n \setminus Z$ to $\mathbb{C}^n \setminus \{0\}$. Thus $\widetilde{\mathbb{C}}^n$ is obtained from $\mathbb{C}^n$ by replacing the origin in $\mathbb{C}^n$ by the set $Z \cong \mathbb{C}\mathrm{P}^{n-1}$ of all lines in through the origin (see Fig. 7.1). The set Z is called the **exceptional divisor**. It is a complex codimension-one submanifold of $\widetilde{\mathbb{C}}^n$ and agrees with the zero section of the tautological line bundle $L \to \mathbb{C}\mathrm{P}^{n-1}$. Note that the normal bundle $\nu_Z \to Z$ of the exceptional divisor in $\widetilde{\mathbb{C}}^n$ is isomorphic to the tautological line bundle.

Lemma 7.1.1 *Let $h \in H^2(\mathbb{C}\mathrm{P}^{n-1}; \mathbb{Z})$ be the positive generator. Then the first Chern class of the tautological line bundle $L = \widetilde{\mathbb{C}}^n \to \mathbb{C}\mathrm{P}^{n-1}$ is $c_1(L) = -h$. In particular, if $n = 2$ then the exceptional divisor $Z \subset \widetilde{\mathbb{C}}^2$ is diffeomorphic to $\mathbb{C}\mathrm{P}^1$ and has self-intersection number $Z \cdot Z = \langle c_1(\nu_Z), [Z] \rangle = -1$.*

Proof: This is Exercise 2.7.8. Here is another approach to this calculation, based on Theorem 2.7.5. Because the tautological line bundle $L \to \mathbb{C}\mathrm{P}^{n-1}$ restricts to

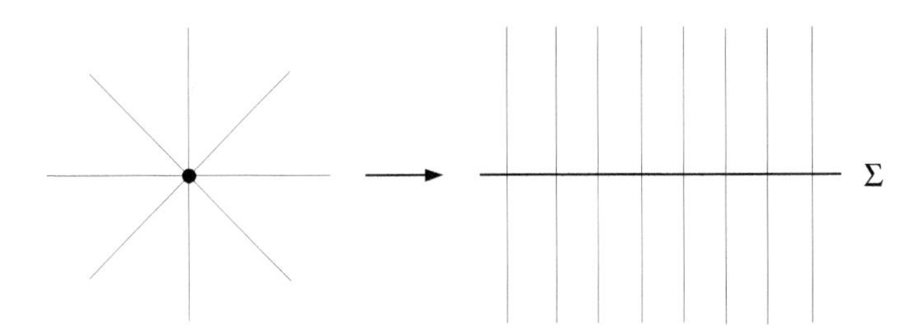

FIG. 7.1. The complex blowup.

the tautological bundle on $\mathbb{C}\mathrm{P}^1 \subset \mathbb{C}\mathrm{P}^{n-1}$, it suffices to prove that the first Chern number of the tautological bundle $L \to \mathbb{C}\mathrm{P}^1$ is minus one. To see this, consider the section $s : \mathbb{C}\mathrm{P}^1 \to L = \widetilde{\mathbb{C}}^2$ defined by

$$s([w_0 : w_1]) := \left([w_0 : w_1], \left(\frac{|w_0|^2}{|w_0|^2 + |w_1|^2}, \frac{\overline{w}_0 w_1}{|w_0|^2 + |w_1|^2}\right)\right)$$

for $[w_0 : w_1] \in \mathbb{C}\mathrm{P}^1$. This section has a unique zero at the point $[0 : 1]$ and the index of this zero is minus one. (Prove this!) Hence $\langle c_1(L), [\mathbb{C}\mathrm{P}^1]\rangle = -1$ as claimed. This proves Lemma 7.1.1. □

Part (ii) of the following exercise shows that the construction of $\widetilde{\mathbb{C}}^n$ is natural. In other words, any local biholomorphism $\psi : \mathbb{C}^n \to \mathbb{C}^n$ which preserves 0 lifts to a local biholomorphism of $\widetilde{\mathbb{C}}^n$.

Exercise 7.1.2 **(i)** Prove that $\widetilde{\mathbb{C}}^n$ is a complex submanifold of $\mathbb{C}\mathrm{P}^{n-1} \times \mathbb{C}^n$ and construct an atlas on $\widetilde{\mathbb{C}}^n$ with holomorphic transition maps. **Hint:** Define coordinate charts $\widetilde{\phi}_j : \widetilde{U}_j \to \mathbb{C}^n$ on $\widetilde{\mathbb{C}}^n$ by $\widetilde{U}_j := \{([w_1 : \cdots : w_n], z) \in \widetilde{\mathbb{C}}^n \mid w_j \neq 0\}$ and

$$\widetilde{\phi}_j([w_1 : \cdots : w_n], z) := \left(\frac{w_1}{w_j}, \dots, \frac{w_{j-1}}{w_j}, z_j, \frac{w_{j+1}}{w_j}, \dots, \frac{w_n}{w_j}\right).$$

(ii) Let $\Omega_0, \Omega_1 \subset \mathbb{C}^n$ be two open neighbourhoods of zero and let $\psi : (\Omega_0, 0) \to (\Omega_1, 0)$ be a biholomorphism (i.e. a holomorphic diffeomorphism satisfying $\psi(0) = 0$). Define

$$\widetilde{\Omega}_j := \widetilde{\mathbb{C}}^n \cap (\mathbb{C}\mathrm{P}^{n-1} \times \Omega_j), \qquad j = 0, 1.$$

Prove that there is a unique biholomorphism $\widetilde{\psi} : \widetilde{\Omega}_0 \to \widetilde{\Omega}_1$ such that

$$\pi \circ \widetilde{\psi} = \psi \circ \pi.$$

Prove that $\widetilde{\psi}(\ell, 0) = (d\psi(0)\ell, 0)$ for $\ell = [w_1 : \cdots : w_n] \in \mathbb{C}\mathrm{P}^{n-1}$.

(iii) Let $u = (u_1, \dots, u_n) : \Omega \to \mathbb{C}^n$ be a nonconstant holomorphic curve, defined on a connected open neighbourhood $\Omega \subset \mathbb{C}$ of zero, such that $u(0) = 0$. Prove that there is a unique holomorphic curve $\widetilde{u} : \Omega \to \widetilde{\mathbb{C}}^n$ such that $\pi \circ \widetilde{u} = u$. Prove that

$$\widetilde{u}(0) = ([u_1^{(k)}(0) : \cdots : u_n^{(k)}(0)], 0),$$

where $k \in \mathbb{N}$ is the smallest integer such that the kth derivative $u^{(k)}$ of u does not vanish at the origin. Cf. Example 7.1.9. □

Now let M be an n-dimensional complex manifold. Fix a point $p_0 \in M$ and denote the complex projectivization of $T_{p_0}M$ by

$$Z := \{\ell \subset T_{p_0}M \mid \ell \text{ is a one-dimensional complex subspace}\}.$$

The element of Z determined by a nonzero tangent vector $0 \neq v \in T_{p_0}M$ will be denoted by $[v] := \mathbb{C}v \in Z$.

Lemma 7.1.3 *Define*

$$\widetilde{M} := (M \setminus \{p_0\}) \cup Z, \qquad \pi_M : \widetilde{M} \to M$$

by $\pi_M|_{M\setminus\{p_0\}} := \mathrm{id}$ *and* $\pi_M(\widetilde{p}) := p_0$ *for* $\widetilde{p} \in Z$. *Then* $\widetilde{M}$ *admits the structure of a complex manifold such that the following holds.*
(i) π_M *is holomorphic and restricts to a biholomorphism from* $\widetilde{M}\setminus Z$ *to* $M\setminus\{p_0\}$.
(ii) *If* $\phi : U \to \mathbb{C}^n$ *is a holomorphic coordinate chart on a neighbourhood* $U \subset M$ *of* p_0 *such that* $\phi(p_0) = 0$, *then the map* $\widetilde{\phi} : \widetilde{U} \to \widetilde{\mathbb{C}}^n$, *defined by*

$$\widetilde{U} := \pi_M^{-1}(U), \qquad \widetilde{\phi}(\widetilde{p}) := \begin{cases} ([z], z), & \text{if } \widetilde{p} \in \widetilde{U} \setminus Z \text{ and } z = \phi \circ \pi_M(\widetilde{p}), \\ ([w], 0), & \text{if } \widetilde{p} = [v] \in Z \text{ and } w = d\phi(p_0)v, \end{cases} \tag{7.1.2}$$

is a biholomorphism onto its image $\widetilde{\phi}(\widetilde{U}) = \widetilde{\mathbb{C}}^n \cap (\mathbb{C}\mathrm{P}^{n-1} \times \phi(U))$.
(iii) *The inclusion* $\widetilde{\iota} : Z \hookrightarrow \widetilde{M}$ *is a holomorphic embedding.*
(iv) *Every nonconstant holomorphic curve* $u : \Sigma \to M$ *on a connected Riemann surface* Σ *has a unique holomorphic lift* $\widetilde{u} : \Sigma \to \widetilde{M}$ *such that* $\pi_M \circ \widetilde{u} = u$.
The complex manifold structure on $\widetilde{M}$ *is uniquely determined by (i) and (ii).*

The complex manifold $\widetilde{M}$ in Lemma 7.1.3 is called the **(complex) blowup of** M **at** p_0 and the complex submanifold Z is called the **exceptional divisor**.

Proof of Lemma 7.1.3: We construct an atlas on $\widetilde{M}$ with holomorphic transition maps. Associated to every holomorphic coordinate chart $\phi : U \to \mathbb{C}^n$ on M, defined on an open set $U \subset M \setminus \{p_0\}$, is a coordinate chart on $\widetilde{M}$ defined by

$$\widetilde{\phi} := \phi \circ \pi_M : \widetilde{U} := \pi_M^{-1}(U) \to \mathbb{C}^n.$$

Associated to every holomorphic coordinate chart $\phi = (\phi_1, \dots, \phi_n) : U \to \mathbb{C}^n$ on a neighbourhood $U \subset M$ of p_0 such that $\phi(p_0) = 0$, there are n coordinate charts $\widetilde{\phi}_j : \widetilde{U}_j \to \mathbb{C}^n$, $j = 1, \dots, n$, obtained by composition of the map $\widetilde{\phi} : \widetilde{U} \to \widetilde{\mathbb{C}}^n$ in (7.1.2) with the coordinate charts in Exercise 7.1.2 (i). They are defined by

$$\widetilde{U}_j := \left\{ \widetilde{p} \in \widetilde{M} \;\middle|\; \begin{array}{l} \pi_M(\widetilde{p}) \in U, \text{ and} \\ \text{if } \widetilde{p} \notin Z \text{ then } \phi_j(\pi_M(\widetilde{p})) \neq 0, \text{ and} \\ \text{if } \widetilde{p} = [v] \in Z,\ v \in T_{p_0}M \setminus \{0\}, \text{ then } d\phi_j(p_0)v \neq 0 \end{array} \right\},$$

$$\widetilde{\phi}_j(\widetilde{p}) := \begin{cases} \left(\frac{z_1}{z_j}, \dots, \frac{z_{j-1}}{z_j}, z_j, \frac{z_{j+1}}{z_j}, \dots, \frac{z_n}{z_j}\right), & \text{if } \widetilde{p} \in \widetilde{U}_j \setminus Z, z = \phi(f(\widetilde{p})), \\ \left(\frac{w_1}{w_j}, \dots, \frac{w_{j-1}}{w_j}, 0, \frac{w_{j+1}}{w_j}, \dots, \frac{w_n}{w_j}\right), & \text{if } \widetilde{p} = [v] \in Z, w = d\phi(p_0)v. \end{cases}$$

The inverse of this coordinate chart is the map $\widetilde{\psi}_j : \widetilde{\Omega}_j \to \widetilde{M}$, defined by

$$\widetilde{\Omega}_j := \{\zeta \in \mathbb{C}^n \,|\, \zeta_j(\zeta_1, \dots, \zeta_{j-1}, 1, \zeta_{j+1}, \dots, \zeta_n) \in \phi(U)\},$$

$$\widetilde{\psi}_j(\zeta) := \begin{cases} (\phi \circ \pi_M)^{-1}(\zeta_j w) \in \widetilde{M} \setminus Z, & \text{if } \zeta_j \neq 0, \\ [d\phi(p_0)^{-1} w] \in Z, & \text{if } \zeta_j = 0, \end{cases} \qquad w_i := \begin{cases} \zeta_i, & \text{for } i \neq j, \\ 1, & \text{for } i = j. \end{cases}$$

It follows from Exercise 7.1.2 (ii) that the transition maps for any pair of these coordinate charts are biholomorphisms. Hence they define a complex manifold structure on $\widetilde{M}$. That this complex manifold structure is uniquely determined by conditions (i) and (ii), and that it satisfies (iii), is obvious. That it satisfies (iv) follows from Exercise 7.1.2 (iii). This proves Lemma 7.1.3. □

The blowup construction in Lemma 7.1.3 extends to oriented smooth manifolds M, provided that one first specifies a complex structure, compatible with the orientation, on a neighbourhood of the point p_0. The resulting manifold $\widetilde{M}$ depends both on the choice of the point p_0 and on the choice of the complex structure in a neighbourhood of p_0. However, the next exercise shows that the diffeomorphism type of $\widetilde{M}$ is independent of these choices. Namely, $\widetilde{M}$ is diffeomorphic to the oriented connected sum of M with $\overline{\mathbb{CP}}^n$ (complex projective space with the orientation reversed).

Exercise 7.1.4 (Connected sums) The connected sum of two oriented n-manifolds is constructed by removing the interior of a closed embedded ball from each of the manifolds and identifying the resulting boundaries by an orientation-reversing diffeomorphism, which extends to an orientation-preserving diffeomorphism between neighbourhoods of these boundaries. Here is an explicit description, following Donaldson–Kronheimer [160, page 284]. Denote the closed unit ball in $\mathbb{R}^n$ by

$$B := B(1), \qquad B(r) := \{\xi \in \mathbb{R}^n \,|\, |\xi| \leq r\} \subset \mathbb{R}^n,$$

with interior $\operatorname{int} B$. Let M_1, M_2 be oriented n-manifolds and choose orientation-preserving embeddings $\iota_i : B \to M_i$ for $i = 1, 2$, an orientation-reversing orthogonal matrix $\Phi \in \mathrm{O}(n)$, and a constant $0 < \varepsilon < 1$. Consider the diffeomorphism $\phi_\varepsilon : \mathbb{R}^n \setminus \{0\} \to \mathbb{R}^n \setminus \{0\}$ defined by

$$\phi_\varepsilon(\xi) := \frac{\varepsilon^2}{|\xi|^2}\Phi\xi.$$

This diffeomorphism is orientation-preserving and carries the sphere of radius r to the sphere of radius ε^2/r. Hence ϕ_ε maps the open annulus $\operatorname{int} B(R\varepsilon) \setminus B(R^{-1}\varepsilon)$ to itself for every $R > 1$. Fix a constant R such that $1 < R < 1/\varepsilon$, denote $B_i := \iota_i(B(R^{-1}\varepsilon)) \subset M_i$ for $i = 1, 2$, and define the **connected sum** $M_1 \# M_2$ as the quotient space

$$M_1 \# M_2 := (M_1 \setminus B_1) \cup_{\phi_\varepsilon} (M_2 \setminus B_2),$$

where $\iota_2(\xi) \equiv \iota_1(\phi_\varepsilon(\xi))$ for $R^{-1}\varepsilon < |\xi| < R\varepsilon$. The connected sum $M_1 \# M_2$ inherits the orientations of M_1 and M_2. It depends on the choice of the embeddings $\iota_i : B \to M_i$, the matrix Φ, and the number ε (but not on the choice of R).

(i) Prove that the diffeomorphism type of $M_1 \# M_2$ is independent of the choices.

(ii) Prove that the complex blowup $\widetilde{M}$ of a complex manifold M at a point $p_0 \in M$ is diffeomorphic to the oriented connected sum $M \# \overline{\mathbb{C}\mathrm{P}}^n$, where $\overline{\mathbb{C}\mathrm{P}}^n$ denotes the manifold $\mathbb{C}\mathrm{P}^n$ with the opposite orientation. **Hint:** The normal bundle of $\mathbb{C}\mathrm{P}^{n-1}$ in $\mathbb{C}\mathrm{P}^n$ has first Chern class h and so is isomorphic to L^* (see Exercise 2.7.6 and Example 4.4.3).

(iii) Let $e \in H^2(\widetilde{M};\mathbb{Z})$ be the Poincaré dual of the exceptional divisor Z and let $\widetilde{\iota}: \mathbb{C}\mathrm{P}^{n-1} \to \widetilde{M}$ be a holomorphic embedding with image Z. Prove that $\langle e, \alpha \rangle = 0$ for every $\alpha \in H_2(\widetilde{M} \setminus Z; \mathbb{Z}$, and that $\widetilde{\iota}^* e = -h \in H^2(\mathbb{C}\mathrm{P}^{n-1};\mathbb{Z})$. □

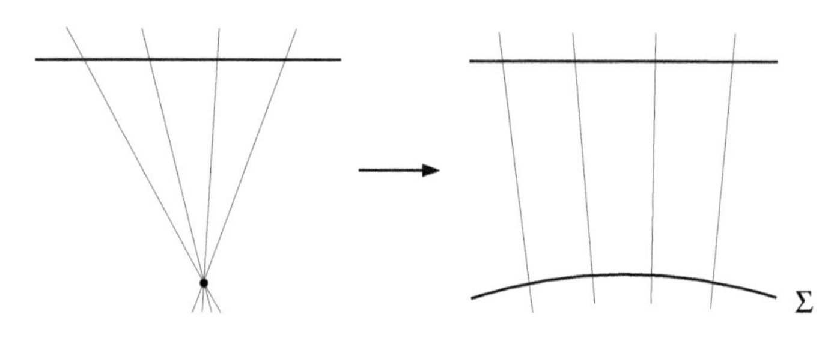

FIG. 7.2. Blowing up a point in $\mathbb{C}\mathrm{P}^2$.

Example 7.1.5 Consider the manifold

$$X := \mathbb{C}\mathrm{P}^2 \# \overline{\mathbb{C}\mathrm{P}}^2$$

obtained by blowing up $\mathbb{C}\mathrm{P}^2$ at the point $p_0 := [1:0:0]$. The family of lines in $\mathbb{C}\mathrm{P}^2$ through p_0 lifts to a family of disjoint lines, each of which meets the exceptional divisor Z at one point. (See Fig. 7.2.) Thus X fibres over $Z \cong S^2$ with fibres $F \cong \mathbb{C}\mathrm{P}^1 \cong S^2$. Since S^2-bundles over S^2 are classified up to diffeomorphism by elements of $\pi_1(\mathrm{Diff}(S^2)) = \pi_1(\mathrm{SO}(3)) = \mathbb{Z}/2\mathbb{Z}$, there are only two such bundles, the trivial one and a nontrivial one. To see which of these bundles X is, we calculate its homology as follows. Observe that X can be embedded in $\mathbb{C}\mathrm{P}^1 \times \mathbb{C}\mathrm{P}^2$ as the submanifold

$$X = \left\{([w_1 : w_2], [z_0 : z_1 : z_2]) \,|\, w_1 z_2 = w_2 z_1\right\}$$

with blowdown map given by the projection $\pi : X \to \mathbb{C}\mathrm{P}^2$ onto the second factor. With this identification, the exceptional divisor $Z := \pi^{-1}(p_0)$ is given by

$$Z = \mathbb{C}\mathrm{P}^1 \times \{p_0\}, \qquad p_0 = [1:0:0].$$

The fibres of the projection $f : X \to \mathbb{C}\mathrm{P}^1$ onto the first factor are given by

$$F_b = f^{-1}([1:b]) = \left\{([1:b],[1:z:bz]) \,|\, z \in \mathbb{C}\right\} \cup \left\{([1:b],[0:1:b])\right\}$$

for $b \in \mathbb{C}$. They are all diffeomorphic to $\mathbb{C}\mathrm{P}^1$. The homology $H_2(X;\mathbb{Z})$ is generated by the class $[Z]$ of the exceptional divisor and the class $[F]$ of the fibre. Their intersection numbers are

$$F \cdot F = 0, \qquad Z \cdot F = 1, \qquad Z \cdot Z = -1.$$

Namely, both Z and $F = F_b$ are complex submanifolds and intersect transversally at precisely one point, the class $[F]$ can be represented by two disjoint fibres, and the equation $Z \cdot Z = -1$ follows from Lemma 7.1.1. This implies that the intersection form of X is given by

$$Q_X = \begin{pmatrix} 1 & 0 \\ 0 & -1 \end{pmatrix}$$

with respect to the basis $[Z] + [F], [Z]$. Since

$$Q_{S^2 \times S^2} = \begin{pmatrix} 0 & 1 \\ 1 & 0 \end{pmatrix},$$

it follows that X is not diffeomorphic to $S^2 \times S^2$. □

Exercise 7.1.6 Prove that the fibre bundle $f : X \to \mathbb{C}\mathrm{P}^1$ is biholomorphic to the projectivization $\mathbb{P}(L \oplus \mathbb{C})$, where $\mathrm{pr} : L \to \mathbb{C}\mathrm{P}^1$ is the tautological line bundle. **Hint:** Use the above identification of X with a complex submanifold of $\mathbb{C}\mathrm{P}^1 \times \mathbb{C}\mathrm{P}^2$. □

Example 7.1.7 (Elliptic surfaces) A complex surface M is said to be **elliptic** if there is a holomorphic map $f : M \to \mathbb{C}\mathrm{P}^1$ whose generic fibre $f^{-1}(z)$ is an elliptic curve, i.e. is biholomorphic to the 2-torus equipped with a linear complex structure. In general, the map f will not be a fibration since a finite number of the fibres will be singular. In this example we show how to construct the simplest elliptic surface by blowing up points in $\mathbb{C}\mathrm{P}^2$.

Let V be the manifold obtained by blowing up $\mathbb{C}\mathrm{P}^2$ at nine distinct points. The complex structure on V varies depending on which set X of nine points one blows up. One case of particular interest is when the nine points are the points of intersection of two transverse nonsingular cubics Γ_1 and Γ_2. Then there is a pencil (i.e. a 1-parameter family) of cubics all passing through X. One way to see this is to express Γ_i as the zero set of a homogeneous cubic polynomial $P_i : \mathbb{C}^3 \to \mathbb{C}$ (such as $z_0^3 + z_1^3 + z_2^3$). Then, for each point $w = [\lambda : \mu] \in \mathbb{C}\mathrm{P}^1$, there is a corresponding cubic Γ_w which is the zero set of $\lambda P_1 + \mu P_2$. Since the intersection number of two cubic curves in $\mathbb{C}\mathrm{P}^2$ is 9, and since each pair of curves in the family intersects in the nine distinct points of X, these points of intersection must be transverse.[29] Hence the lifts $\widetilde{\Gamma}_w$ of these cubics to the blowup V are all *disjoint*. Now it is easy to check that exactly one cubic Γ_w goes through each point of $\mathbb{C}\mathrm{P}^2 \setminus X$.[30] It follows that each point of V lies on exactly one curve Γ_w, and hence there is a holomorphic map $f : V \to \mathbb{C}\mathrm{P}^1$ such that

[29] It is a general fact that points of intersection of two complex curves on a complex surface always contribute positively to the intersection number, and that nontransverse intersections such as tangencies always contribute *more* than 1. Similar results hold for J-holomorphic curves in symplectic 4-manifolds (see Gromov [287], McDuff [455], Micallef–White [479], Appendix E of [470], and the articles in [34], for example).

[30] This is a codimension-1 phenomenon. Generically, there is a 1-dimensional family of cubics passing through eight distinct points in $\mathbb{C}\mathrm{P}^2$. Hence the remaining point in our collection of nine points is determined by the first eight.

$\Gamma_w = f^{-1}(w)$. This map is not quite a fibration, since a finite number of the fibres will be singular. However, the generic fibre is a torus, since nondegenerate cubic curves in $\mathbb{C}P^2$ have genus one (see Example 4.4.5). For more details see Griffiths–Harris [285]. □

Example 7.1.8 (Lefschetz pencils) More generally all projective Kähler surfaces X support a Lefschetz pencil, which is a one-parameter family of complex curves, finitely many of them singular, that all go through the same finite number of points and are otherwise distinct. If X is blown up at these points, we obtain a **Lefschetz fibration.** This is a nontrivial holomorphic map $f : \widetilde{X} \to \mathbb{C}P^1$ that has connected fibres and nodal singularities. For all but a finite number of points, the fibres are Riemann surfaces of some genus g, and there are finitely many singular or exceptional fibres. One way of constructing such a map is to embed X as a complex submanifold into some projective space $\mathbb{C}P^n$, choose a generic plane P of complex codimension 2 that meets X transversally in a finite number of points and then consider the family of submanifolds of X formed by the intersection of X with the pencil of hyperplanes containing P. As above, these hyperplanes are indexed by a parameter $w \in \mathbb{C}P^1$. Moreover, there is exactly one curve $\Gamma_w = X \cap H_w$ through each point of $X \setminus P$. In good cases (which can be achieved for suitable embeddings into a projective space and suitable choices of P), any two of the curves Γ_w intersect transversally at the points of $X \cap P$. It then follows that the blowup $\widetilde{X}$ of X at the points of $X \cap P$ admits a holomorphic projection $f : \widetilde{X} \to \mathbb{C}P^1$ that takes the points in $\Gamma_w \setminus P$ to $w \in \mathbb{C}P^1$. (Cf. Theorem 7.4.12 and Example 7.4.13.) □

One of the main applications of blowing up in complex geometry is to resolve singularities.

Example 7.1.9 (Resolving singularities) (i) Consider the variety $V \subset \mathbb{C}^2$ defined by the equation $z_1 z_2 = 0$. This has a singularity at 0 where the two lines $z_1 = 0$ and $z_2 = 0$ cross. But, because these lines represent different points in the exceptional divisor Z, it follows that V has a nonsingular lift $\widetilde{V} \subset \widetilde{\mathbb{C}}^2$ consisting of two disjoint lines. In general, the lift (or **proper transform**) of a variety $V \subset \mathbb{C}$ to $\widetilde{\mathbb{C}}^2$ is defined as the closure of $\pi^{-1}(V \setminus \{0\})$, i.e.

$$\widetilde{V} := \overline{\pi^{-1}(V \setminus \{0\})} \subset \widetilde{\mathbb{C}}^2.$$

It consists of everything which lies over V except for the irrelevant parts of Z. If $u : \mathbb{C} \supset \Omega \to \mathbb{C}^2$ is a local holomorphic parametrization of V such that $u(0) = 0$, then $\widetilde{V}$ is parametrized by the holomorphic extension $\widetilde{u} : \Omega \to \widetilde{\mathbb{C}}^2$ of the lift $\pi^{-1} \circ u|_{\Omega \setminus \{0\}}$. We give some explicit examples below.

(ii) Suppose that V is defined by the equation ${z_1}^2 = {z_2}^3$. Then V has a cusp at 0, and can be parametrized by $u(t) = (t^3, t^2)$. Its lift has the parametrization

$$\widetilde{u}(t) = ([t^3 : t^2], t^3, t^2) = ([t : 1], t^3, t^2),$$

and so is nonsingular at $t = 0$. □

Exercise 7.1.10 **(i)** Consider the two curves $C_1 = \{z_1 = 0\}$ and $C_2 = \{z_1 = {z_2}^2\}$ in $\mathbb{C}^2$ which are tangent at 0. Show that their lifts to $\widetilde{\mathbb{C}}^2$ intersect transversally.

(ii) If $C_3 = \{z_1 = {z_2}^3\}$, show that you have to blow up twice before the lift of C_3 intersects C_1 transversally. What is the second point that you blow up? □

It is also possible to blow up along a complex submanifold S. Here one replaces the points of S by the space of lines in the normal bundle to S (see Griffiths–Harris [285, Chapter 4 §6]).

The opposite process to blowing up is blowing down, where one replaces the exceptional divisor Z by a single point. To do this, it is not enough just to find a copy Z of $\mathbb{C}\mathrm{P}^{n-1}$ whose normal bundle has the correct first Chern class. In order to blow down Z one needs a neighbourhood of Z to be biholomorphic to a neighbourhood of the zero section of the tautological line bundle L, since this is our model space. Because $H^1(\mathbb{C}\mathrm{P}^{n-1}, \mathcal{O}) = 0$, the complex structure on the normal bundle to Z is uniquely determined by its first Chern class, but this does not immediately imply that the complex structure on a neighbourhood of Z is unique. However, according to the Castelnuovo–Enriques criterion, this is the case when $n = 2$, and so one can blow down any holomorphic copy of $\mathbb{C}\mathrm{P}^1$ with self-intersection number minus one. Moreover, blowing up and down are algebraic operations, i.e. if the manifold M is algebraic, so is any manifold obtained from it by blowing points up and down.

Kähler forms on the blowup

We now study the special family of Kähler forms $\widetilde{\omega}_\lambda$ on the tautological line bundle $L = \widetilde{\mathbb{C}}^n$ over $\mathbb{C}\mathrm{P}^{n-1}$ defined by

$$\widetilde{\omega}_\lambda := \pi^*\omega_0 + \lambda^2 \mathrm{pr}^*\omega_{\mathrm{FS}} \in \Omega^2(L). \tag{7.1.3}$$

Here ω_{FS} is the standard Fubini–Study form on $\mathbb{C}\mathrm{P}^{n-1}$ for which $\mathbb{C}\mathrm{P}^1$ has area π (see Example 4.3.3). Thus $\widetilde{\omega}_\lambda$ is a combination of pullbacks of the standard forms over the two projections

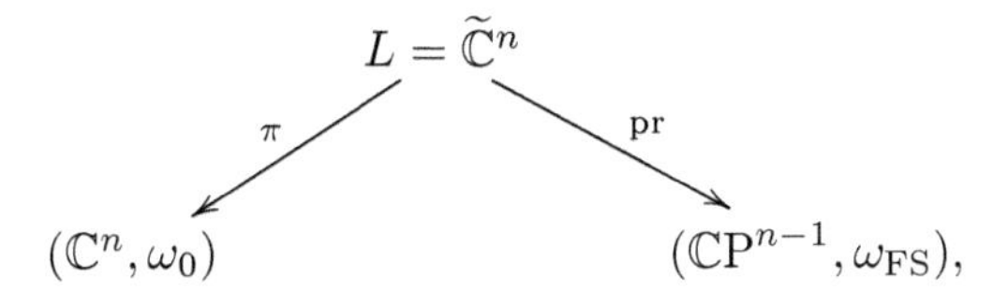

where the blowdown map π is a diffeomorphism over $\mathbb{C}^n \setminus \{0\}$ and pr is a vector bundle projection. Moreover, because $\int_{\mathbb{C}\mathrm{P}^1} \omega_{\mathrm{FS}} = \pi$ we find

$$\int_E \widetilde{\omega}_\lambda = \pi\lambda^2,$$

where E is a line in the exceptional divisor of $\widetilde{\mathbb{C}}^n$, or, equivalently, in the zero section of L.

Denote the closed ball of radius ε in $\mathbb{C}^n$ and its inverse image in L by

$$\begin{aligned} B(\varepsilon) &:= \{z \in \mathbb{C}^n \mid |z| \le \varepsilon\}, \\ L(\varepsilon) &:= \pi^{-1}(B(\varepsilon)) = \Big\{([w], z) \in L = \widetilde{\mathbb{C}}^n \,\big|\, |z| \le \varepsilon\Big\}. \end{aligned} \tag{7.1.4}$$

Thus $L(\varepsilon)$ is a closed neighbourhood of Z. We now show that the deleted neighbourhood

$$L(\varepsilon) \setminus Z = \pi^{-1}\big(B(\varepsilon) \setminus \{0\}\big) = \Big\{([w], z) \in L = \widetilde{\mathbb{C}}^n \,\big|\, 0 < |z| \le \varepsilon\Big\},$$

equipped with the symplectic form $\widetilde{\omega}_\lambda$ in (7.1.3), is symplectomorphic to the spherical shell $B(\sqrt{\lambda^2+\varepsilon^2}) \setminus B(\varepsilon)$ in $(\mathbb{C}^n, \omega_0)$.

Lemma 7.1.11 **(i)** *Let $f : (0,\infty) \to (0,\infty)$ be a smooth function with positive derivative and define $F : \mathbb{C}^n \setminus \{0\} \to \mathbb{C}^n$ by*

$$F(z) := f(|z|)\frac{z}{|z|}.$$

Then $F^*\omega_0$ is a Kähler form on $\mathbb{C}^n \setminus \{0\}$ for the standard complex structure i.
(ii) *Define the diffeomorphism $F_\lambda : \mathbb{C}^n \setminus \{0\} \to \mathbb{C}^n \setminus B(\lambda)$ by*

$$F_\lambda(z) = \sqrt{|z|^2+\lambda^2}\,\frac{z}{|z|} \tag{7.1.5}$$

for $z \in \mathbb{C}^n \setminus \{0\}$. Then $\pi^* F_\lambda^* \omega_0 = \widetilde{\omega}_\lambda$ on $L \setminus Z$. Thus $\big(L(\varepsilon) \setminus Z, \widetilde{\omega}_\lambda\big)$ is symplectomorphic to $\big(B(\sqrt{\lambda^2+\varepsilon^2}) \setminus B(\lambda), \omega_0\big)$.

Proof: We prove that $F^*\omega_0$ is of type $(1,1)$. To see this, note that

$$\begin{aligned} F^*dz_j &= \frac{f(|z|)}{|z|}dz_j + \frac{|z|f'(|z|) - f(|z|)}{2|z|^3} z_j \sum_{k=1}^n \big(\bar{z}_k dz_k + z_k d\bar{z}_k\big), \\ F^*d\bar{z}_j &= \frac{f(|z|)}{|z|}d\bar{z}_j + \frac{|z|f'(|z|) - f(|z|)}{2|z|^3} \bar{z}_j \sum_{k=1}^n \big(\bar{z}_k dz_k + z_k d\bar{z}_k\big). \end{aligned}$$

Since $\omega_0 = \frac{i}{2}\sum_{j=1}^n dz_j \wedge d\bar{z}_j$, we have

$$\begin{aligned} F^*\omega_0 &= \frac{i}{2}\frac{f(|z|)^2}{|z|^2}\sum_{j=1}^n dz_j \wedge d\bar{z}_j \\ &\quad + \frac{i}{2}\frac{|z|f(|z|)f'(|z|) - f(|z|)^2}{|z|^4} \sum_{j,k=1}^n \bar{z}_j z_k dz_j \wedge d\bar{z}_k. \end{aligned} \tag{7.1.6}$$

Hence $F^*\omega_0$ is of type $(1,1)$ as claimed. Moreover, F is a diffeomorphism onto its image whenever the derivative of f is positive. Hence the symmetric bilinear form

$F^*\omega_0(\cdot, i\cdot)$ is nondegenerate for every such f and so its signature is independent of f. Taking $f(|z|) = |z|^2$, we find that $F^*\omega_0(\cdot, i\cdot)$ is positive definite for some, and hence every, f with positive derivative. This proves (i).

To prove (ii), define the symplectic form ρ_λ on $\mathbb{C}^n \setminus \{0\}$ by

$$\begin{aligned}\rho_\lambda &:= \frac{i}{2}\partial\bar{\partial}\Big(|z|^2 + \lambda^2 \log(|z|^2)\Big) \\ &= \frac{i}{2}\left(dz \wedge d\bar{z} + \lambda^2 \frac{dz \wedge d\bar{z}}{|z|^2} - \lambda^2 \frac{\bar{z} \cdot dz \wedge z \cdot d\bar{z}}{|z|^4}\right) \\ &= \omega_0 + \lambda^2 \rho_{\mathrm{FS}}.\end{aligned} \tag{7.1.7}$$

Here $dz \wedge d\bar{z}$, $\bar{z} \cdot dz$, and ρ_{FS} are as in Exercise 4.3.4. We saw in Example 4.3.3 that $\pi^*\rho_{\mathrm{FS}} = \mathrm{pr}^*\omega_{FS}$ on $L \setminus Z$. Hence $\pi^*\rho_\lambda = \widetilde{\omega}_\lambda \in \Omega^2(L \setminus Z)$ by (7.1.3). Moreover, by (7.1.6) with $f(|z|) = \sqrt{|z|^2 + \lambda^2}$, the diffeomorphism F_λ in (7.1.5) satisfies $F_\lambda^*\omega_0 = \rho_\lambda$. Since $F_\lambda : B(\varepsilon) \setminus \{0\} \to B(\sqrt{\lambda^2 + \varepsilon^2}) \setminus B(\lambda)$ is a diffeomorphism, this proves Lemma 7.1.11. □

We next construct a smooth family of symplectic forms on $\widetilde{\mathbb{C}}^n$ that agree with the model form $\widetilde{\omega}_\lambda$ of Lemma 7.1.11 in a neighbourhood $L(\varepsilon)$ of the exceptional divisor and with $\pi^*\omega_0$ on $\widetilde{\mathbb{C}}^n \setminus L(\lambda + \varepsilon)$, which as we have just seen is symplectomorphic to the complement of the ball $B(\lambda + \varepsilon)$ in $\mathbb{C}^n$.

Definition 7.1.12 (Kähler forms on $L = \widetilde{\mathbb{C}}^n$) *Choose a smooth function*

$$(0, \infty)^2 \times (0, 1) \to \mathbb{R} : (r, \lambda, \varepsilon) \mapsto f_{\lambda,\varepsilon}(r)$$

such that

$$f_{\lambda,\varepsilon}(r) = \begin{cases} \sqrt{\lambda^2 + r^2}, & \text{for } r \le \varepsilon, \\ r, & \text{for } r \ge \lambda + \varepsilon, \end{cases} \qquad f'_{\lambda,\varepsilon}(r) > 0, \tag{7.1.8}$$

for all ε, λ, r, define the diffeomorphisms $F_{\lambda,\varepsilon} : \mathbb{C}^n \setminus \{0\} \to \mathbb{C}^n \setminus B(\lambda)$ by

$$F_{\lambda,\varepsilon}(z) := f_{\lambda,\varepsilon}(|z|)\frac{z}{|z|}, \tag{7.1.9}$$

and define the symplectic forms $\widetilde{\omega}_{\lambda,\varepsilon} \in \Omega^2(L)$ by

$$\widetilde{\omega}_{\lambda,\varepsilon}|_{L(\varepsilon)} := \widetilde{\omega}_\lambda|_{L(\varepsilon)}, \qquad \widetilde{\omega}_{\lambda,\varepsilon}|_{L\setminus Z} := \pi^* F_{\lambda,\varepsilon}^* \left(\omega_0|_{\mathbb{C}^n \setminus B(\lambda)}\right). \tag{7.1.10}$$

The smooth family of functions $f_{\lambda,\varepsilon}$ exists by Lemma A.3.1. The 2-forms $\widetilde{\omega}_{\lambda,\varepsilon}$ are well-defined by part (ii) of Lemma 7.1.11 and are Kähler forms by part (i) of Lemma 7.1.11.

The next proposition is the key result that will allow us to translate from the complex notion of blowup to the symplectic notion. We later use equation (7.1.12) below to determine the cohomology class of the blowup form $\widetilde{\omega}_\lambda$. This equation can also be understood in terms of the proper transforms of holomorphic curves through the blowup point $\{0\}$, explaining what happens to their (local) homology classes (see Example 7.1.9).

Proposition 7.1.13 *The Kähler forms $\widetilde{\omega}_{\lambda,\varepsilon} \in \Omega^2(\widetilde{\mathbb{C}}^n)$ in Definition 7.1.12 depend smoothly on $\lambda > 0$ and $0 < \varepsilon < 1$, and they satisfy the equations*

$$\widetilde{\omega}_{\lambda,\varepsilon}|_{L(\varepsilon)} = \pi^*\omega_0 + \lambda^2 \mathrm{pr}^*\omega_{\mathrm{FS}} = \widetilde{\omega}_\lambda, \qquad \widetilde{\omega}_{\lambda,\varepsilon}|_{\widetilde{\mathbb{C}}^n \setminus L(\lambda+\varepsilon)} = \pi^*\omega_0. \tag{7.1.11}$$

Moreover,

$$\int_\Omega u^*\omega_0 = \int_\Omega \widetilde{u}^*\widetilde{\omega}_{\lambda,\varepsilon} + m\pi\lambda^2 \tag{7.1.12}$$

for every holomorphic function $u : \Omega \to \mathbb{C}^n$ on a bounded open set $\Omega \subset \mathbb{C}$ which extends smoothly to the closure $\overline{\Omega}$, satisfies $u(\partial\Omega) \cap B(\lambda+\varepsilon) = \emptyset$, and has finitely many zeros of total multiplicity m. Here $\widetilde{u} : \Omega \to \widetilde{\mathbb{C}}^n$ denotes the unique holomorphic lift of u such that $\pi \circ \widetilde{u} = u$.

Proof: By definition the 2-forms $\widetilde{\omega}_{\lambda,\varepsilon}$ depend smoothly on λ and ε, and they satisfy equation (7.1.11) by definition of $F_{\lambda,\varepsilon}$. To prove (7.1.12), recall from Exercise 4.3.4 that $\rho_{\mathrm{FS}} = d\alpha_{\mathrm{FS}}$, where

$$\alpha_{\mathrm{FS}} := \frac{i}{4|z|^2}\left(z \cdot d\overline{z} - \overline{z} \cdot dz\right) \in \Omega^1(\mathbb{C}^n \setminus \{0\}).$$

Claim. *If $u : \Omega \to \mathbb{C}^n$ is a holomorphic function in a neighbourhood $\Omega \subset \mathbb{C}$ of the origin such that $u(z) = z^m v(z)$ for some integer $m > 0$ and some holomorphic function $v : \Omega \to \mathbb{C}^n$ that does not vanish at the origin, then*

$$\lim_{\delta \to 0} \int_{|z|=\delta} u^*\alpha_{\mathrm{FS}} = m\pi. \tag{7.1.13}$$

First observe that $u^*\alpha_{\mathrm{FS}} = |u|^{-2}u^*\alpha_0$ and hence $u^*\alpha_{\mathrm{FS}}|_{\partial B^2(\delta)} = \delta^{-2m}|v|^{-2}u^*\alpha_0$. If $v(z) \equiv v$ is constant then $|u'(z)| = m^2|z|^{2m-2}|v|^2$ and hence

$$\begin{aligned}
\int_{\partial B^2(\delta)} u^*\alpha_{\mathrm{FS}} &= \frac{1}{\delta^{2m}|v|^2}\int_{\partial B^2(\delta)} u^*\alpha_0 = \frac{1}{\delta^{2m}|v|^2}\int_{B^2(\delta)} u^*\omega_0 \\
&= \frac{1}{\delta^{2m}|v|^2}\int_{B^2(\delta)} |u'|^2 = \frac{2\pi m^2|v|^2}{\delta^{2m}|v|^2}\int_0^\delta r^{2m-1}\,dr \\
&= m\pi.
\end{aligned}$$

If v is nonconstant and $u_0(z) := z^m v(0)$, we have

$$\left|\int_{\partial B^2(\delta)} u^*\alpha_{\mathrm{FS}} - \int_{\partial B^2(\delta)} u_0^*\alpha_{\mathrm{FS}}\right| = O(\delta),$$

which proves the claim.

It remains to show that $\widetilde{\omega}_{\lambda,\varepsilon}$ satisfies (7.1.12). Let $\Omega \subset \mathbb{C}$ be a bounded open set and let $u : \Omega \to \mathbb{C}^n$ be a holomorphic function which extends smoothly to $\overline{\Omega}$, satisfies $u(\partial\Omega) \cap B(\lambda+\varepsilon) = \emptyset$, and vanishes at $z_1, \dots, z_\ell \in \Omega$. Let $m_1, \dots, m_\ell \in \mathbb{N}$ be the multiplicities of the zeros so that $u(z_j + z) = z^{m_j} v_j(z)$ where $v_j(0) \neq 0$. Assume without loss of generality that the boundary of Ω is smooth and let $\widetilde{u} : \overline{\Omega} \to \widetilde{\mathbb{C}}^n$ be the unique holomorphic lift of u.

For $\delta > 0$ sufficiently small denote $\Omega_\delta := \Omega \setminus \bigcup_{j=1}^{\ell} B_\delta(z_j)$, where $B_\delta(z_j) \subset \mathbb{C}$ denotes the closed ball of radius δ centred at z_j. Then

$$\begin{aligned}\int_\Omega \widetilde{u}^* \widetilde{\omega}_{\lambda\varepsilon} &= \lim_{\delta\to 0} \int_{\Omega_\delta} \widetilde{u}^* \pi^* F_{\lambda,\varepsilon}^* \omega_0 = \lim_{\delta\to 0} \int_{\partial\Omega_\delta} u^* F_{\lambda,\varepsilon}^* \alpha_0 \\ &= \int_{\partial\Omega} u^* \alpha_0 - \sum_{j=1}^{\ell} \lim_{\delta\to 0} \int_{\partial B_\delta(z_j)} u^*(\alpha_0 + \lambda^2 \alpha_{\mathrm{FS}}) \\ &= \int_\Omega u^* \omega_0 - \sum_{j=1}^{\ell} \lambda^2 m_j \pi.\end{aligned}$$

Here the penultimate equality follows from (7.1.11) and the last equality follows from the claim. This proves Proposition 7.1.13. □

Blowing up and down in the symplectic category

Next we discuss the blowup construction for symplectic manifolds. There are two ways to think about the symplectic blowup construction. On the one hand it is a method of constructing a symplectic form on the complex blowup (for a suitable choice of a complex structure in a neighbourhood of the blown up point). This was Gromov's original point of view in [288], finding applications in McDuff [444]. On the other hand there is an intrinsic construction in symplectic topology, where the role of a point is played by a symplectically embedded standard ball. This geometric point of view was first described in McDuff [450]. We will show that both constructions are equivalent. In particular, the diffeomorphism type of the blown up manifold is the same as in the complex category (provided that the manifold is Kähler), so that the manifold is just altered near one point. However the whole ball is needed in order to describe the symplectic form on the blowup. In fact, blowing up amounts to removing the interior of a symplectically embedded ball and collapsing the bounding sphere to the exceptional divisor by the Hopf map. Similarly, to blow down one removes the exceptional divisor and glues in a ball. The radius λ of the ball corresponds to the cohomology class of the restriction of the blowup form to the exceptional divisor, a large ball corresponding to a large divisor, and a small ball to a small divisor. Intuitively speaking, blowing down a submanifold Z simplifies the topology at the expense of increasing the volume by the area of Z. Conversely, blowing up increases the topology and decreases the volume.

We now give more precise descriptions of these operations.

Blowing down

Suppose that $Z \cong \mathbb{C}\mathrm{P}^{n-1}$ is symplectically embedded in (M, ω) with normal bundle isomorphic to the tautological bundle L. Then, by the Symplectic Neighbourhood Theorem 3.4.10, Z has a neighbourhood $\mathcal{N}(Z)$ that is symplectomorphic to the symplectic manifold $(L(\varepsilon), \widetilde{\omega}_\lambda)$ defined by (7.1.4) and (7.1.3). The **blow-down** of M along Z is obtained by cutting $\mathcal{N}(Z)$ out and sewing $B(\sqrt{\lambda^2+\varepsilon^2})$ back in. This construction is independent of the choice of ε and of the symplectomorphism between $\mathcal{N}(Z)$ and $L(\varepsilon)$, i.e. the blown down manifolds associated to different choices are symplectomorphic. The blowdown construction replaces the **exceptional divisor** Z by an embedded ball $B(\lambda)$ with the standard symplectic structure. Note that in the 4-dimensional case Z is just a symplectically embedded 2-sphere with self-intersection number minus one. Such a sphere is called an **exceptional sphere**. (One can also describe the blowdown in terms of the fibre connected sum; cf. the discussion just before Theorem 7.2.8.)

Blowing up

Conversely, a **symplectic blowup** $(\widetilde{M}_\psi, \widetilde{\omega}_\psi)$ of (M, ω) of **weight** λ is obtained from a symplectic embedding

$$\psi : B^{2n}(\lambda) \to M.$$

Extend ψ to a symplectic embedding of $B(\sqrt{\lambda^2+\varepsilon^2})$ for some $\varepsilon > 0$ (see Theorem 3.3.1). Then replace the image $\psi(B(\sqrt{\lambda^2+\varepsilon^2}))$ by the standard neighbourhood $L(\varepsilon)$. Finally define manifold $\widetilde{M}_\psi$ as the quotient space

$$\widetilde{M}_\psi := \Big(M \setminus \psi(B(\lambda))\Big) \sqcup L(\varepsilon)/_\sim, \tag{7.1.14}$$

where $([z], z) \in L(\varepsilon) \setminus Z$ is identified with $\psi(\sqrt{1+\lambda^2/|z|^2}z) \in M \setminus \psi(B(\lambda))$. The symplectic form $\widetilde{\omega}_\psi$ equals ω on $M \setminus \psi(B(\lambda))$ and equals $\widetilde{\omega}_\lambda$ on $L(\varepsilon)$. These symplectic forms match on the overlap $B(\sqrt{\lambda^2+\varepsilon^2}) \setminus B(\lambda) \cong L(\varepsilon) \setminus Z$ by part (ii) of Lemma 7.1.11. It turns out that the symplectomorphism class of $(\widetilde{M}_\psi, \widetilde{\omega}_\psi)$ is independent of the choice of the extension (see Theorem 7.1.23 below). As a topological space the manifold $\widetilde{M}_\psi$ can also be identified with the quotient $(M \setminus \psi(\mathrm{int}\, B(\lambda)))/\sim$, where the equivalence relation identifies two elements of the boundary $\psi(\partial B(\lambda))$ if and only if they belong to the same Hopf circle.

Notice that the terminology 'blowing up and down' is really not very suited to what happens in the symplectic category. Indeed, one can think of blowing down in the symplectic category as an expansion. The manifold M is slit open along Z, and then a large ball is sewn in. Conversely, symplectic blowing up is a collapsing process in which the interior of a large ball $B^{2n}(\lambda)$ is cut out and its boundary is collapsed to Z. (See Fig. 7.3.) This procedure has been formalized in Lerman's notion of symplectic cutting [408].

Exercise 7.1.14 The formula (7.1.14) shows that the volume of (M, ω) decreases when it is blown up to $(\widetilde{M}_\psi, \widetilde{\omega}_\psi)$. Verify this by a calculation in cohomology, using the diffeomorphism $\widetilde{M}_\psi \cong M \# \overline{\mathbb{C}\mathrm{P}}^n$ of Exercise 7.1.4. □

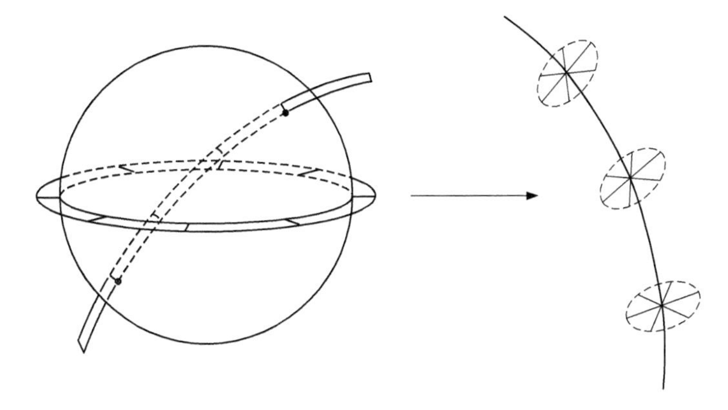

FIG. 7.3. The symplectic blowup.

Exercise 7.1.15 (Blowing up and the moment map) (i) The λ blowup of $\mathbb{R}^{2n}$ at the origin 0 is the total space L of the tautological line bundle over $\mathbb{C}\mathrm{P}^{n-1}$ with the symplectic form $\widetilde{\omega}_\lambda$ of (7.1.3). Prove that the standard action of $\mathrm{G} := \mathrm{U}(n)$ on $(L, \widetilde{\omega}_\lambda)$ is Hamiltonian and find a formula for the moment map $\mu = \mu_{\mathrm{G}}$. (Compare with Example 5.3.7.)

(ii) Find a formula for the moment map $\mu_{\mathbb{T}}$ of the action of the standard $\mathbb{T}^n \subset \mathrm{U}(n)$ on $(L, \widetilde{\omega}_\lambda)$. (Compare Example 5.5.2.) Show, in particular, that the image of $\mu_{\mathbb{T}}$ is a truncation of the first quadrant, i.e.

$$\operatorname{im} \mu_{\mathbb{T}} = \{(x_1, \ldots, x_n) \in \mathbb{R}^n \,|\, x_i \geq 0,\, x_1 + \cdots + x_n \geq \lambda'\},$$

and find a formula for λ' in terms of λ. Observe that with Delzant's technique [132] for constructing a toric manifold from the image of the moment map, one can construct the blowup from the truncated image of the moment map. □

The interplay described above between the processes of blowing up and symplectic reduction is an important component of the work of Guillemin, Lerman, and Sternberg [292]. Using a more elaborate version of this process, Fine and Panov [217] resolve certain orbifold singularities, thereby constructing new families of nonKähler symplectic 6-manifolds from the twistor space of a hyperbolic 4-orbifold.

Example 7.1.16 By Example 7.1.5, the manifold X obtained by blowing up one point of $(\mathbb{C}\mathrm{P}^2, \omega_{\mathrm{FS}})$ is the nontrivial S^2-bundle over S^2. Thus, we now have two ways of putting a symplectic form on X, one by symplectic blowup and the other by the methods of Section 6.1. (See, in particular, Exercise 6.2.6.) For any $\lambda < 1$ there is a symplectic embedding $\psi : B(\lambda) \to \mathbb{C}\mathrm{P}^2$ such that the corresponding blowup form $\widetilde{\omega}_\psi$ is nondegenerate on the fibres of this fibration. For example, one can take ψ to be the composite of the inclusion $B(\lambda) \hookrightarrow \operatorname{int} B(1)$ with the symplectomorphism $\iota : (\operatorname{int} B(1), \omega_0) \to (\mathbb{C}\mathrm{P}^2 \setminus \mathbb{C}\mathrm{P}^1, \omega_{\mathrm{FS}})$, given by

$$\iota(z_0, z_1) := [z_0 : z_1 : \sqrt{1 - |z_0|^2 - |z_1|^2}].$$

(See Exercises 4.3.4 and 5.1.3.) In the resulting blowup the parts of the complex lines through the origin which lie outside $\psi(B^{2n}(\lambda))$ correspond precisely to the fibres of the fibration $f: X \to S^2$. Thus it is possible to blow up in such a way that the resulting symplectic form is compatible with f. It then follows from Exercise 6.2.7 that all these symplectic forms are homotopic, and hence, by Theorem 13.3.32, that each cohomologous pair is isotopic. □

The relation between the symplectic and complex blowup

The symplectic blowup construction in equation (7.1.14) gives rise to different, albeit diffeomorphic, manifolds for different symplectic embeddings of balls. In contrast, it is sometimes useful to describe the symplectic blowup as the construction of different symplectic forms, associated to different symplectic embeddings of balls, on one and the same manifold. In particular, this will show that isotopic symplectic embeddings of a ball (of the same radius with fixed centre) give rise to isotopic symplectic forms on the blowup (i.e. the resulting symplectic forms are connected by a path of symplectic forms in the same cohomology class). Following the discussion in McDuff–Polterovich [469], we choose the complex blowup as our model on which different symplectic forms are constructed.

To describe the model, we begin by reviewing the complex blowup construction of Lemma 7.1.3 in the Kähler context. Thus we assume that there is an almost complex structure J which is integrable on all the relevant parts of M and which is compatible with a symplectic form ω. We also impose the condition that the Kähler metric near the blowup point is flat. More precisely, we assume that the following normalization condition holds.

Definition 7.1.17 *Let (M, ω) be a $2n$-dimensional connected symplectic manifold and J be an ω-compatible almost complex structure on M that is integrable near a point $p_0 \in M$. A* **normalization** *of (M, ω, J) at p_0 is the choice of a compact neighbourhood $U_0 \subset M$ of p_0, a constant $\delta > 0$, and a diffeomorphism $\psi_0 : B(\delta) \to U_0$, defined on the closed ball $B(\delta) := B^{2n}(\delta) \subset \mathbb{R}^{2n}$ of radius $\delta > 0$ centred at the origin, such that*

$$\psi_0(0) = p_0, \qquad \psi_0^*\omega = \omega_0, \qquad \psi_0^*J = J_0. \tag{7.1.15}$$

In particular, J is integrable in U_0 and the Kähler metric in U_0 is flat.

Remark 7.1.18 **(i)** By Theorem A.2.1, every ω-compatible almost complex structure on $U_0 = \psi_0(B(\delta))$ extends ω-compatibly to all of M.

(ii) The blowup of M defined below will depend (as an almost complex manifold) only on the manifold (M, J) and on the coordinate chart ψ_0. This definition only uses the fact that ψ_0 is holomorphic. The condition that ψ_0 is also a Darboux chart will be needed for the construction of symplectic forms on the blowup. If (M, ω) is equipped with N Darboux charts $\psi_i : B(\delta_i) \to U_i$ with disjoint images, one can use these to construct a model for the N-fold blowup of M.

(iii) If (M, ω, J) is a Kähler manifold it may be convenient to work with the complex blowup at p_0 and to choose ψ_0 to be a holomorphic coordinate chart.

In this case ψ_0 will not, in general, be a Darboux chart. It is then necessary to replace ω by a small exact perturbation $\omega' = \omega + d\alpha$ such that $\psi_0^*\omega' = \omega_0$. For example, when working on the complex blowup of $\mathbb{C}\mathrm{P}^n$ one must first modify the Fubini–Study form in a neighbourhood of the blown up point p_0 in order for there to be a normalization at p_0. □

Definition 7.1.19 *Assume that (M, ω, J) is normalized at p_0 by the choice of $\psi_0 : B(\delta) \to U_0$, and let*

$$Z := (T_{p_0}M \setminus \{0\})/\mathbb{C}^* \tag{7.1.16}$$

be the complex projectivization of $T_{p_0}M$ (where $z = x+iy \in \mathbb{C}^*$ acts on $v \in T_{p_0}M$ by $z \cdot v := xv + yJv$). The **almost complex blowup of** M **at** p_0 *is the almost complex manifold $(\widetilde{M}, \widetilde{J})$, defined by*

$$\widetilde{M} := (M \setminus \{p_0\}) \cup Z. \tag{7.1.17}$$

It is equipped with the projection $\pi_M : \widetilde{M} \to M$ defined by

$$\pi_M|_{M \setminus \{p_0\}} := \mathrm{id}, \qquad \pi_M(\widetilde{p}) := p_0 \quad \text{for} \quad \widetilde{p} \in Z. \tag{7.1.18}$$

The manifold structure on $\widetilde{M}$ and the almost complex structure $\widetilde{J}$ on $\widetilde{M}$ are determined by the condition that π_M restricts to a holomorphic diffeomorphism from $(\widetilde{M} \setminus Z, \widetilde{J})$ to $(M \setminus \{p_0\}, J)$ and that the lift

$$\widetilde{\psi}_0 : L(\delta) \to \widetilde{U}_0 := (U_0 \setminus \{p_0\}) \cup Z \tag{7.1.19}$$

of $\psi_0 : B(\delta) \to U_0$, defined by

$$\widetilde{\psi}_0([w], z) := \begin{cases} \pi_M^{-1}(\psi_0(z)) \in \widetilde{U}_0 \setminus Z, & \text{if } z \neq 0, \\ [d\psi_0(0)w] \in Z, & \text{if } z = 0, \end{cases} \tag{7.1.20}$$

for $([w], z) \in L(\delta)$ is a holomorphic diffeomorphism. (See equation (7.1.4) *for the definition of $L(\delta)$.) Thus there is a commutative diagram*

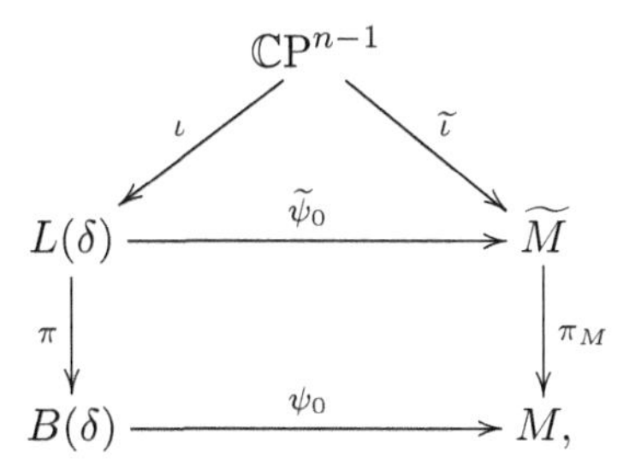

where $\iota : \mathbb{C}\mathrm{P}^{n-1} \to L(\delta)$ denotes the inclusion of the zero section and the map $\widetilde{\iota} : \mathbb{C}\mathrm{P}^{n-1} \to \widetilde{M}$ denotes the identification of $\mathbb{C}\mathrm{P}^{n-1}$ with the **exceptional divisor** $Z \subset \widetilde{M}$*, induced by the differential $d\psi_0(0) : \mathbb{C}^n \to T_{p_0}M$. All maps in the diagram are holomorphic.*

Definition 7.1.20 *Assume (M,ω,J) is normalized at p_0 with δ,ψ_0 as in Definition 7.1.17. A symplectic embedding $\psi:(B(\lambda),\omega_0)\to(M,\omega)$ is said to be* **based** *if $\psi(0)=p_0$, and is said to be* **normalized** *if it agrees with ψ_0 near the origin.*

The next theorem explains how a normalized symplectic embedding ψ of $B(\lambda)$ as in Definition 7.1.20 gives rise to a symplectic form $\widetilde{\omega}_\psi$ on $\widetilde{M}$ whose restriction to the exceptional divisor is isomorphic to $\lambda^2\omega_{\mathrm{FS}}$ under the diffeomorphism $\widetilde{\iota}:\mathbb{C}\mathrm{P}^{n-1}\to Z$. After showing how to normalize an arbitrary ball embedding in Proposition 7.1.22, we discuss the uniqueness of symplectic blowup in Theorem 7.1.23.

Theorem 7.1.21 (Symplectic blowup) *Assume (M,ω,J) is normalized at p_0 with δ,ψ_0 as in Definition 7.1.17 and let $(\widetilde{M},\widetilde{J})$ be the almost complex blowup of M at p_0 in Definition 7.1.19. For $\lambda>0$ and $0<\varepsilon<1$ let $\widetilde{\omega}_{\lambda,\varepsilon}\in\Omega^2(\widetilde{\mathbb{C}}^n)$ be as in Definition 7.1.12. Then, for every triple $(\psi,\lambda,\varepsilon)$ consisting of a normalized symplectic embedding $\psi:B(r)\to M$ and a pair of constants $\lambda>0$ and $0<\varepsilon<1$ satisfying*

$$\lambda+\varepsilon<r,\qquad \varepsilon<\delta,\qquad \psi|_{B(\varepsilon)}=\psi_0|_{B(\varepsilon)},\tag{7.1.21}$$

there exists a unique symplectic form $\widetilde{\omega}_{\psi,\lambda,\varepsilon}\in\Omega^2(\widetilde{M})$ such that

$$\begin{aligned}\widetilde{\omega}_{\psi,\lambda,\varepsilon}|_{\widetilde{\psi}(L(r))}&=(\widetilde{\psi}^{-1})^*\left(\widetilde{\omega}_{\lambda,\varepsilon}|_{L(r)}\right),\\ \widetilde{\omega}_{\psi,\lambda,\varepsilon}|_{\widetilde{M}\setminus\widetilde{\psi}(L(\lambda+\varepsilon))}&=\pi_M^*\left(\omega|_{M\setminus\psi(B(\lambda+\varepsilon))}\right).\end{aligned}\tag{7.1.22}$$

Here $\widetilde{\psi}:L(r)\to\widetilde{M}$ is the unique embedding that satisfies $\pi_M\circ\widetilde{\psi}=\psi\circ\pi$. The symplectic forms $\widetilde{\omega}_{\psi,\lambda,\varepsilon}$ have the following properties.

(i) *Let $\widetilde{\omega}_\lambda:=\pi^*\omega_0+\lambda^2\mathrm{pr}^*\omega_{\mathrm{FS}}\in\Omega^2(L)$ be the symplectic form in (7.1.3). Then*

$$\widetilde{\omega}_{\psi,\lambda,\varepsilon}|_{\widetilde{\psi}(L(\varepsilon))}=(\widetilde{\psi}^{-1})^*\left(\widetilde{\omega}_\lambda|_{L(\varepsilon)}\right).\tag{7.1.23}$$

(ii) *$(\widetilde{M},\widetilde{\omega}_{\psi,\lambda,\varepsilon})$ is symplectomorphic to the manifold $(\widetilde{M}_\psi,\widetilde{\omega}_\psi)$ in (7.1.14).*

(iii) *The map $(\psi,\lambda,\varepsilon)\mapsto\widetilde{\omega}_{\psi,\lambda,\varepsilon}$ is smooth.*

(iv) *If $\psi^*J=J_0$ on $B(\lambda+\varepsilon)$ then $\widetilde{\omega}_{\psi,\lambda,\varepsilon}$ is compatible with $\widetilde{J}$.*

(v) *The symplectic form $\widetilde{\omega}_{\psi,\lambda,\varepsilon}$ represents the cohomology class*

$$[\widetilde{\omega}_{\psi,\lambda,\varepsilon}]=\pi_M^*[\omega]-\pi\lambda^2\mathrm{PD}([\widetilde{Z}])\in H^2(\widetilde{M};\mathbb{R}).$$

Here $[Z]$ denotes the fundamental class of the exceptional divisor with its complex orientation and $\mathrm{PD}([Z])$ denotes its Poincaré dual in de Rham cohomology.

(vi) *The symplectic form $\widetilde{\omega}_{\psi,\lambda,\varepsilon}$ is homotopic, through a smooth family of (possibly noncohomologous) symplectic forms on $\widetilde{M}$, to one that is compatible with $\widetilde{J}$.*

(vii) *The isotopy class of the symplectic form $\widetilde{\omega}_{\psi,\lambda,\varepsilon}$ is independent of the choice of the function $f_{\lambda,\varepsilon}$ in Definition 7.1.12 (satisfying (7.1.8)), used to construct it.*

Proof: By equation (7.1.11) in Proposition 7.1.13,

$$\widetilde{\psi}_{\lambda,\varepsilon}|_{L(r)\setminus L(\lambda+\varepsilon)} = \pi^*(\omega_0|_{B(r)\setminus B(\lambda+\varepsilon)}) = \pi^*\psi^*(\omega|_{\psi(B(r)\setminus B(\lambda+\varepsilon))}).$$

Hence the two formulas for $\widetilde{\omega}_{\psi,\lambda,\varepsilon}$ in (7.1.22) agree on $\widetilde{\psi}(L(r)\setminus L(\lambda+\varepsilon))$. This proves the existence and uniqueness of the symplectic forms $\widetilde{\omega}_{\psi,\lambda,\varepsilon}$ in (7.1.22). They satisfy (i) by construction.

Now let $F_{\lambda,\varepsilon}:\mathbb{C}^n\setminus\{0\}\to\mathbb{C}^n\setminus B(\lambda)$ be the diffeomorphism in (7.1.9) and define the map $\widetilde{\phi}:\widetilde{M}\to\widetilde{M}_\psi$ by

$$\begin{aligned} \widetilde{\phi}([d\psi(0)w]) &:= [\psi(\lambda w)] && \text{for } w\in S^{2n-1},\\ \widetilde{\phi}(\pi_M(\psi(z))) &:= \psi(F_{\lambda,\varepsilon}(z)) && \text{for } z\in B^{2n}(r)\setminus\{0\},\\ \widetilde{\phi}(\pi_M(p)) &:= p && \text{for } p\in M\setminus\psi(B(\lambda+\varepsilon)). \end{aligned} \tag{7.1.24}$$

Then it follows from (7.1.23) that $\widetilde{\phi}:(\widetilde{M},\widetilde{\omega}_{\psi,\lambda,\varepsilon})\to(\widetilde{M}_\psi,\widetilde{\omega}_\psi)$ is a symplectomorphism. That the map $(\psi,\lambda,\varepsilon)\mapsto\widetilde{\omega}_{\psi,\lambda,\varepsilon}$ is smooth follows from Proposition 7.1.13. Hence the symplectic forms $\widetilde{\omega}_{\psi,\lambda,\varepsilon}$ satisfy (i), (ii), and (iii).

We prove (iv). By equation (7.1.23), $\widetilde{\omega}_{\psi,\lambda,\varepsilon}$ is compatible with $\widetilde{J}$ on the subsets $\widetilde{\psi}(L(\varepsilon))$ and $\widetilde{M}\setminus\widetilde{\psi}(L(\lambda+\varepsilon))$. Moreover, $\widetilde{\psi}^*\widetilde{\omega}_{\psi,\lambda,\varepsilon}=\pi^*F_{\lambda,\varepsilon}^*\omega_0$ on $L(\lambda+\varepsilon)\setminus Z$. If $\psi^*J=J_0$ then, by Lemma 7.1.11, this symplectic form is compatible with the complex structure

$$\pi^*J_0=\pi^*\psi^*J=(\psi\circ\pi)^*J=(\pi_M\circ\widetilde{\psi})^*J=\widetilde{\psi}^*\pi_M^*J=\widetilde{\psi}^*\widetilde{J}.$$

Hence $\widetilde{\omega}_{\psi,\lambda,\varepsilon}$ and $\widetilde{J}$ are also compatible on the set $\widetilde{\psi}(L(\lambda+\varepsilon))\setminus Z$ and hence on all of $\widetilde{M}$. This proves (iv).

Condition (v) asserts that the Poincaré dual of the exceptional divisor $Z\subset\widetilde{M}$ is the cohomology class of the closed 2-form $\widetilde{\omega}_{\psi,\lambda,\varepsilon}-\pi_M^*\omega$. This follows from equation (7.1.12) in Proposition 7.1.13, and the fact that m is the intersection number of the lift $\widetilde{u}$ with Z. Alternatively, one can argue as follows. By (7.1.23) the 2-form $\widetilde{\omega}_{\psi,\lambda,\varepsilon}-\pi_M^*\omega$ is supported in the tubular neighbourhood $\widetilde{\psi}(L(\lambda+\varepsilon))$ of Z and satisfies

$$\widetilde{\iota}^*(\widetilde{\omega}_{\psi,\lambda,\varepsilon}-\pi_M^*\omega)=\iota^*\widetilde{\omega}_{\lambda,\varepsilon}=\iota^*\widetilde{\omega}_\lambda=\lambda^2\omega_{\mathrm{FS}}.$$

Thus its integral over the 2-sphere $\widetilde{\iota}(\mathbb{C}\mathrm{P}^1)\subset Z$ is equal to $\pi\lambda^2$. Moreover, the first Chern class of the normal bundle of the exceptional divisor Z pulls back to the negative generator $-h\in H^2(\mathbb{C}\mathrm{P}^{n-1};\mathbb{Z})$ under the diffeomorphism $\widetilde{\iota}:\mathbb{C}\mathrm{P}^{n-1}\to Z$. Thus the Poincaré dual of Z is the class in $H^2(\widetilde{M};\mathbb{R})$ which vanishes on all 2-cycles in $M\setminus\{p_0\}\cong\widetilde{M}\setminus Z$ and which pulls back to $-h$ on $\mathbb{C}\mathrm{P}^{n-1}$ (see Exercise 7.1.4 and Bott and Tu [78]). Hence the closed 2-form $\widetilde{\omega}_{\psi,\lambda,\varepsilon}-\pi_M^*\omega$ represents the class $-\pi\lambda^2\mathrm{PD}([Z])$. This proves (v).

We prove (vi). Choose constants $0 < \lambda_0 < \lambda$ and $0 < \varepsilon_0 < \varepsilon$ such that $\lambda_0 + \varepsilon_0 < \varepsilon$, and define

$$\lambda_t := (1-t)\lambda_0 + t\lambda, \qquad \varepsilon_t := (1-t)\varepsilon_0 + t\varepsilon$$

for $0 \le t \le 1$. Then it follows from assertion (iv) that $\widetilde{\omega}_{\psi,\lambda_0,\varepsilon_0}$ is compatible with $\widetilde{J}$ and from (iii) that $t \mapsto \widetilde{\omega}_{\psi,\lambda_t,\varepsilon_t}$ is a smooth family of symplectic forms. This proves (vi).

Assertion (vii) follows from the fact that $\widetilde{\omega}_{\psi,\lambda,\varepsilon}$ depends smoothly on the function $f_{\lambda,\varepsilon}$ satisfying (7.1.8), that the set of such functions is convex, and that the cohomology class of $\widetilde{\omega}_{\psi,\lambda,\varepsilon}$ depends only on λ, by (v). This proves (vii) and Theorem 7.1.21. □

In order to define the blowup symplectic form associated to an arbitrary symplectic embedding $\psi : B(\lambda) \to M$ we must first modify ψ so that it is normalized. The next proposition explains how to do this. It is not clear whether this normalization is unique up to isotopy, though this does hold for simply connected M by Lemma 7.1.25. However part (ii) below shows that any path of based embeddings is homotopic to a normalized path.

Proposition 7.1.22 (Normalization of ball embeddings) *Let (M,ω) be a connected symplectic manifold without boundary.*

(i) *Let $r > 0$ and let $\psi_0, \psi_1 : B(r) \to M$ be symplectic embeddings. Then there exists a Hamiltonian symplectomorphism $\phi : M \to M$ such that $\phi \circ \psi_1$ agrees with ψ_0 near the origin. Moreover, if ψ_0, ψ_1 are based (i.e. $\psi_0(0) = \psi_1(0) = p_0$) the Hamiltonian isotopy to ϕ can be chosen based at p_0.*

(ii) *Let $r > 0$ and let $\psi_t : B(r) \to M$, $0 \le t \le 1$, be a smooth family of symplectic embeddings such that $\psi_t(0) = p_0$ for all t and ψ_1 agrees with ψ_0 near the origin. Then there exists a smooth family of symplectomorphisms*

$$\chi_t : B(r) \to B(r), \qquad 0 \le t \le 1,$$

such that $\chi_0 = \chi_1 = \mathrm{id}$ and $\psi_t \circ \chi_t$ agrees with ψ_0 near the origin for all t.

Proof: We prove (i). Choose a smooth path $\gamma : [0,1] \to M$ with endpoints $\gamma(0) = \psi_0(0)$ and $\gamma(1) = \psi_1(0)$ and let $\{\chi_t\}_{0 \le t \le 1}$ be a Hamiltonian isotopy such that $\chi_0 = \mathrm{id}$ and

$$\chi_t(\gamma(0)) = \gamma(t)$$

for all t. (The generating Hamiltonians must satisfy $dH_t(\gamma(t)) = \omega(\dot\gamma(t), \cdot)$.) Then $\chi_1^{-1} \circ \psi_1(0) = \psi_0(0)$. Thus ψ_0 and $\chi_1^{-1} \circ \psi_1$ are symplectic embeddings of $B(r)$ into M that agree at the origin. It follows from part (ii) of Theorem 3.3.1 that there exists a Hamiltonian isotopy ϕ_t such that $\phi_1 \circ \psi_0$ agrees with $\chi_1^{-1} \circ \psi_1$ near the origin. Hence $\phi := \phi_1^{-1} \circ \chi_1^{-1}$ is a Hamiltonian symplectomorphism such that $\phi \circ \psi_1$ agrees with ψ_0 near the origin. This proves (i). The proof of (ii) is based on the following.

Claim. *Let $0 < \varepsilon < r$ and suppose that $\iota_t : B(\varepsilon) \to \mathbb{R}^{2n}$, $0 \le t \le 1$ is a smooth family of symplectic embeddings such that*

$$\iota_0 = \iota_1 = \mathrm{id}, \qquad \iota_t(B(\varepsilon)) \subset \mathrm{int}(B(r)), \qquad \iota_t(0) = 0 \tag{7.1.25}$$

for all t. Then there exists a constant $\delta \in (0, \varepsilon)$ and a smooth family of symplectomorphisms $\chi_t : B(r) \to B(r)$, $0 \le t \le 1$, such that

$$\chi_0 = \chi_1 = \mathrm{id}, \qquad \chi_t|_{B(\delta)} = \iota_t|_{B(\delta)} \tag{7.1.26}$$

for all t.

To prove the claim, consider the loop $\Psi_t := d\iota_t(0) \in \mathrm{Sp}(2n)$ of symplectic matrices and define

$$P_t := \left(\Psi_t \Psi_t^T\right)^{1/2}, \qquad U_t := \left(\Psi_t \Psi_t^T\right)^{-1/2} \Psi_t.$$

Then $U_t \in \mathrm{U}(n)$, P_t is a symmetric positive definite symplectic matrix, and

$$\Psi_t = P_t U_t, \qquad P_0 = P_1 = U_0 = U_1 = \mathbb{1}. \tag{7.1.27}$$

Hence there is a smooth family of symplectomorphisms $\phi_t : B(r) \to B(r)$, supported in the interior of $B(r)$, and a constant $\delta > 0$ such that

$$\phi_t|_{B(\delta)} = P_t, \qquad \phi_0 = \phi_1 = \mathrm{id}. \tag{7.1.28}$$

Namely, for each t consider the Hamiltonian isotopy $s \mapsto P_{s,t} := \left(\Psi_t \Psi_t^T\right)^{s/2}$ and multiply the Hamiltonian functions by a smooth cutoff function to obtain a Hamiltonian isotopy $s \mapsto \phi_{s,t}$ that is supported in the interior of $B(r)$ and agrees with $P_{s,t}$ near the origin. Then the time-1 maps $\phi_t := \phi_{1,t}$ are Hamiltonian symplectomorphisms of $B(r)$ that satisfy (7.1.28). Use the same method, with $P_{s,t}$ replaced by $\iota_{s,t} := s^{-1}\iota_t(s\cdot)$, to construct a smooth family of Hamiltonian isotopies $s \mapsto \psi_{s,t}$, supported in the interior of $B(r)$, such that

$$\psi_{0,t}|_{B(\delta)} = \Psi_t, \qquad \psi_{1,t}|_{B(\delta)} = \iota_t, \qquad \psi_{s,0} = \psi_{s,1} = \mathrm{id}. \tag{7.1.29}$$

(Shrink δ if necessary.) By (7.1.27), (7.1.28), and (7.1.29) the symplectomorphisms $\chi_t := \psi_{1,t} \circ \psi_{0,t}^{-1} \circ \phi_t \circ U_t : B(r) \to B(r)$ satisfy (7.1.26). In particular, they agree with $\iota_t \circ \Psi_t^{-1} \circ P_t \circ U_t = \iota_t$ near the origin. This proves the claim.

We prove (ii). Let $\psi_t : B(r) \to M, 0 \le t \le 1$, be a smooth family of symplectic embeddings such that $\psi_t(0) = p_0$ for all t and ψ_0 and ψ_1 agree near the origin. Choose $\varepsilon > 0$ so small that

$$\psi_1|_{B(\varepsilon)} = \psi_0|_{B(\varepsilon)}, \qquad \psi_0(B(\varepsilon)) \subset \psi_t(\mathrm{int}(B(r)))$$

for all t. Then the symplectic embeddings $\iota_t := \psi_t^{-1} \circ \psi_0 : B(\varepsilon) \to \mathrm{int}(B(r))$ satisfy (7.1.25). By the claim there is a smooth family of symplectomorphisms $\chi_t : B(r) \to B(r)$ that satisfy (7.1.26). Hence

$$\psi_t \circ \chi_t|_{B(\delta)} = \psi_t \circ \iota_t|_{B(\delta)} = \psi_0|_{B(\delta)}$$

for all t and $\chi_0 = \chi_1 = \mathrm{id}$. Thus the symplectomorphisms χ_t satisfy the requirements of (ii). This completes the proof of Proposition 7.1.22. □

Theorem 7.1.23 (Uniqueness of blowup) *Assume (M, ω, J) is normalized at p_0 by $\psi_0 : B(\delta) \to U_0$, and let $\pi_M : \widetilde{M} \to M$ be the almost complex blowup in Definition 7.1.19. For $i = 1, 2$ let $\psi_i : B(r_i) \to M$ be normalized symplectic embeddings, fix positive constants λ_i, ε_i satisfying $\lambda_i + \varepsilon_i < r_i$, $\varepsilon_i < \delta$, and $\psi_i = \psi_0$ on $B(\varepsilon_i)$, and let $\widetilde{\omega}_{\psi_i,\lambda_i,\varepsilon_i}$ be the symplectic forms on $\widetilde{M}$ in Theorem 7.1.21. Then the following holds.*

(i) *The symplectic forms $\widetilde{\omega}_{\psi_i,\lambda_i,\varepsilon_i}$ can be joined by a path of symplectic forms.*

(ii) *The symplectic forms $\widetilde{\omega}_{\psi_i,\lambda_i,\varepsilon_i}$ are cohomologous if and only if $\lambda_1 = \lambda_2$.*

(iii) *If $\lambda_1 = \lambda_2 =: \lambda$ and the based embeddings $\psi_i|_{B(\lambda)} : (B(\lambda), 0) \to (M, p_0)$ are based symplectically isotopic then the symplectic forms $\widetilde{\omega}_{\psi_i,\lambda_i,\varepsilon_i}$ are isotopic.*

(iv) *If $\lambda_1 = \lambda_2 =: \lambda$ and the symplectic embeddings $\psi_i : B(\lambda) \to M$ are isotopic through symplectic embeddings $\psi_t : B(\lambda) \to M$, $1 \le t \le 2$, which do not fix the origin, then the symplectic forms $\widetilde{\omega}_{\psi_i,\lambda_i,\varepsilon_i}$ are diffeomorphic.*

Proof: The symplectic forms $\widetilde{\omega}_{\psi_1,\lambda_1,\varepsilon_1}$ and $\widetilde{\omega}_{\psi_2,\lambda_2,\varepsilon_2}$ agree whenever

$$\lambda_1 = \lambda_2 =: \lambda, \qquad \varepsilon_1 = \varepsilon_2 =: \varepsilon, \qquad \lambda + \varepsilon < \delta.$$

Hence assertion (i) follows from part (iii) of Theorem 7.1.21. Assertion (ii) follows from part (v) of Theorem 7.1.21.

We prove (iii). Assume the embeddings $\psi_i|_{B(\lambda)} : (B(\lambda), 0, \omega_0) \to (M, p_0, \omega)$ for $i = 1, 2$ are based symplectically isotopic. By part (ii) of Theorem 3.3.1 with $\lambda = 0$, the same holds for the embeddings $\psi_i|_{B(\lambda)} : (B(\lambda + \delta), 0, \omega_0) \to (M, p_0, \omega)$ for some $\delta > 0$. By part (ii) of Proposition 7.1.22 the isotopy can be chosen to agree with ψ_0 near the origin for each t. Hence, by Theorem 7.1.21 parts (iii) and (v), the symplectic forms $\widetilde{\omega}_{\psi_1,\lambda,\delta}$ and $\widetilde{\omega}_{\psi_2,\lambda,\delta}$ are isotopic. Moreover, it follows from assertion (ii) that the map

$$[\delta, \varepsilon_i] \to \Omega^2(\widetilde{M}) : \varepsilon \mapsto \widetilde{\omega}_{\psi_i,\lambda,\varepsilon}$$

is an isotopy for $i = 1, 2$. Hence, the symplectic forms $\widetilde{\omega}_{\psi_1,\lambda,\varepsilon_1}$ and $\widetilde{\omega}_{\psi_2,\lambda,\varepsilon_2}$ are isotopic.

We prove (iv). By part (ii) of Theorem 3.3.1 there exists a Hamiltonian symplectomorphism $\phi : M \to M$ and a constant $\delta > 0$ such that the composition $\phi \circ \psi_1$ agrees with ψ_2 on $B(\lambda + \delta)$. Hence ϕ induces a symplectomorphism

$$\widetilde{\phi} : (\widetilde{M}_{\psi_1}, \widetilde{\omega}_{\psi_1}) \to (\widetilde{M}_{\psi_2}, \widetilde{\omega}_{\psi_2})$$

of the symplectic blowups. By Theorem 7.1.21 (ii), $(\widetilde{M}, \widetilde{\omega}_{\psi_i,\lambda,\varepsilon_i})$ is symplectomorphic to $(\widetilde{M}_{\psi_i}, \widetilde{\omega}_{\psi_i})$. Thus $(\widetilde{M}, \widetilde{\omega}_{\psi_1,\lambda,\varepsilon_1})$ is symplectomorphic to $(\widetilde{M}, \widetilde{\omega}_{\psi_2,\lambda,\varepsilon_2})$. This proves Theorem 7.1.23 □

The process of blowing down is opposite to blowing up. The reader who is interested in a more detailed discussion of this is referred to [469, §5].

Blowing up and embeddings of balls

We have seen in the construction of equation (7.1.14) and in Theorems 7.1.21 and 7.1.23 that there is a close connection between symplectic embeddings of balls and symplectic forms on the blowup. To focus the discussion, assume that (M, ω) is a closed connected symplectic manifold, let

$$\pi_M : \widetilde{M} \to M$$

be the almost complex blowup of M associated to a normalization at a point $p_0 \in M$ and denote by $Z \subset \widetilde{M}$ the exceptional divisor (see Definitions 7.1.17 and 7.1.19). Thus $\widetilde{M}$ is diffeomorphic to the oriented connected sum $M \# \overline{\mathbb{C}\mathrm{P}}^n$. It is convenient to define the cohomology classes $a \in H^2(M; \mathbb{R})$, $c \in H^2(M; \mathbb{Z})$, $\widetilde{a} \in H^2(\widetilde{M}; \mathbb{Z})$, and $\widetilde{c}, e \in H^2(\widetilde{M}; \mathbb{Z})$ by

$$a := [\omega], \qquad c := c_1(\omega), \qquad \widetilde{a} := \pi_M^* a, \qquad \widetilde{c} := \pi_M^* c, \qquad e := \mathrm{PD}(Z).$$

Slightly abusing notation we sometimes write e as a de Rham cohomology class.

Existence

By Theorem 7.1.21 the symplectic form $\widetilde{\omega}_{\psi,\lambda,\varepsilon}$ associated to a normalized symplectic embedding $\psi : B(\lambda) \to M$ represents the cohomology class

$$[\widetilde{\omega}_{\psi,\lambda,\varepsilon}] = \widetilde{a} - \pi\lambda^2 e, \tag{7.1.30}$$

and it has the first Chern class

$$c_1(\widetilde{\omega}_{\psi,\lambda,\varepsilon}) = \widetilde{c} - (n-1)e. \tag{7.1.31}$$

Corollary 7.1.24 *Let $\lambda > 0$. If there exists a symplectic embedding $B(\lambda) \hookrightarrow M$, then the de Rham cohomology class $\widetilde{a} - \pi\lambda^2 e$ on $\widetilde{M}$ has a symplectic representative with first Chern class $\widetilde{c} - (n-1)e$.*

Proof: Every symplectic embedding $\psi : B(\lambda) \to M$ extends to $B(r)$ for some constant $r > \lambda$ by part (i) of Theorem 3.3.1, and can be normalized near the origin by part (i) of Proposition 7.1.22. Hence the assertion follows from part (v) of Theorem 7.1.21. □

In the language of Section 12.1, Corollary 7.1.24 asserts that the inequality

$$w_G(M, \omega) > \pi\lambda^2$$

for the Gromov width of M implies existence of a symplectic form on $\widetilde{M}$ in the cohomology class $\widetilde{a} - \pi\lambda^2 e$ with first Chern class $\widetilde{c} = \pi^* c - (n-1)e$. Conversely, one can ask whether every symplectic form on the blowup in the cohomology class $\widetilde{a} - \pi\lambda^2 e$ and with first Chern class $\widetilde{c} = \pi^* c - (n-1)e$ is isotopic to one obtained by blowup from (M, ω). This would imply that the Gromov width $w_G(M, \omega)$ (see equation (12.1.2)) depends only on the cohomology class of ω. It is unlikely to hold in general but does hold in some special cases (see Example 13.4.5).

Corollary 7.1.24 can sometimes be used to find obstructions to the existence of symplectic embeddings of balls from the nonexistence of symplectic forms on the blowup of M at one or several points. Conversely, when embeddings of balls can be constructed, Corollary 7.1.24 can be used to determine which cohomology classes on the blowup have symplectic representatives. For the one-point blowup of the four-torus this was carried out by Latschev–McDuff–Schlenk [400].

Uniqueness

By Theorem 7.1.23, any two normalized embeddings of $B(\lambda)$ into M induce cohomologous symplectic forms that can be joined by a path of (not necessarily cohomologous) symplectic forms; moreover, the forms are isotopic if the embeddings are isotopic as based symplectic embeddings. In order to understand when the latter condition holds, we need to investigate the space of ball embeddings.

We first show that when M is simply connected, two symplectic embeddings of $B(\lambda)$ into M are based symplectically isotopic if and only if they are symplectically isotopic. Denote the space of symplectic embeddings of a closed ball of radius λ into (M, ω) by

$$\mathrm{Emb}(B(\lambda), M) := \{\psi : B(\lambda) \hookrightarrow M \mid \psi \text{ is an embedding, } \psi^*\omega = \omega_0\}$$

and denote the space of based symplectic embeddings by

$$\mathrm{Emb}_{p_0}(B(\lambda), M) := \{\psi \in \mathrm{Emb}(B(\lambda), M) \mid \psi(0) = p_0\}.$$

Lemma 7.1.25 *If M is simply connected then the inclusion*

$$\mathrm{Emb}_{p_0}(B(\lambda), M) \hookrightarrow \mathrm{Emb}(B(\lambda), M)$$

induces an isomorphism on the space of path-components.

Proof 1: Let $\psi_0, \psi_1 \in \mathrm{Emb}_{p_0}(B(\lambda), M)$ and let $[0,1] \to \mathrm{Emb}(B(\lambda), M) : t \mapsto \psi_t$ be a smooth family of symplectic embeddings connecting ψ_0 to ψ_1. Since M is simply connected there exists a smooth map $\gamma : [0,1]^2 \to M$ such that

$$\gamma(0,t) = \gamma(s,0) = \gamma(s,1) = p_0, \qquad \gamma(1,t) = \psi_t(0).$$

Choose a smooth family of Hamiltonian isotopies $s \mapsto \phi_{s,t}$ such that

$$\phi_{0,t} = \phi_{s,0} = \phi_{s,1} = \mathrm{id}, \qquad \phi_{s,t}(p_0) = \gamma(s,t).$$

Then the path

$$[0,1] \to \mathrm{Emb}_{p_0}(B(\lambda), M) : t \mapsto \phi_{1,t}^{-1} \circ \psi_t$$

connects ψ_0 to ψ_1. □

Proof 2: Denote by ev : $\mathrm{Emb}(B(\lambda), M) \to M$ the evaluation map at the origin, i.e. $\mathrm{ev}(\psi) := \psi(0)$ for $\psi \in \mathrm{Emb}(B(\lambda), M)$. It follows from the Isotopy Extension Theorem 3.3.2 that this is a locally trivial fibration with fibre $\mathrm{Emb}_{p_0}(B(\lambda), M)$ over p_0.[31] Moreover, the group of Hamiltonian symplectomorphisms acts transitively on M and hence each fibre is isomorphic to $\mathrm{Emb}_{p_0}(B(\lambda), M)$ by a suitable Hamiltonian symplectmorphism of M. Hence there is an exact sequence

$$\begin{aligned} \ldots \to \pi_1\big(\mathrm{Emb}(B(\lambda), M)\big) \to \pi_1(M) \to \pi_0\big(\mathrm{Emb}_{p_0}(B(\lambda), M)\big) \\ \to \pi_0\big(\mathrm{Emb}(B(\lambda), M)\big) \to \pi_0(M), \end{aligned}$$

and the result follows. □

Corollary 7.1.26 *Assume M is simply connected and let $\psi_0, \psi_1 : B(\lambda) \to M$ be normalized symplectic embeddings. If ψ_0 and ψ_1 are isotopic through symplectic embeddings of $B(\lambda)$, then the symplectic forms $\widetilde{\omega}_{\psi_0,\lambda,\varepsilon}$ and $\widetilde{\omega}_{\psi_1,\lambda,\varepsilon}$ on $\widetilde{M}$, constructed in Theorem 7.1.21, are isotopic.*

Proof: By Lemma 7.1.25 the embeddings ψ_0, ψ_1 are based symplectically isotopic. Hence the assertion follows from part (iii) of Theorem 7.1.23. □

Exercise 7.1.27 **(i)** If $\psi \in \mathrm{Symp}(\mathbb{R}^{2n})$ fixes the origin, there is a canonical path joining it to a linear symplectomorphism given by $\psi_t(x) = t^{-1}\psi(tx)$, $0 < t \le 1$. Use this to show that the group $\mathrm{Symp}(\mathbb{R}^{2n})$ of symplectomorphisms of $\mathbb{R}^{2n}$ is path-connected.

(ii) Show that for any radius λ, the space of symplectic embeddings of the standard ball $B(\lambda)$ into $\mathbb{R}^{2n}$ is path-connected (see Banyaga [50]). □

Remark 7.1.28 (Isotopies on 4-manifolds with $b^+ = 1$) Let (M, ω) be a closed symplectic four-manifold with $b^+ = 1$. (Examples include all blowups of $\mathbb{C}\mathrm{P}^2$ and of ruled surfaces.) Under this assumption a theorem of McDuff [458] and Li–Liu [424, Proposition 4.11] asserts that any two cohomologous symplectic forms on M that can be joined by a path of symplectic forms are in fact isotopic. (For various related results see McDuff [449, 450, 453], Lalonde [389], Biran [65], McDuff–Opshtein [468].) The proof involves Taubes–Seiberg–Witten theory and the inflation technique developed by Lalonde and McDuff [389, 393, 394, 458].

It follows that any two symplectic forms $\widetilde{\omega}_{\psi_0,\lambda,\varepsilon_0}$ and $\widetilde{\omega}_{\psi_1,\lambda,\varepsilon_0}$ on the blowup, associated to two normalized symplectic embeddings ψ_0, ψ_1 and the same constant $\lambda > 0$, are isotopic. As is shown in [458, Corollary 1.5], this isotopy is constructed in such a way that one can conclude that the underlying symplectic embeddings ψ_0, ψ_1 are symplectically isotopic. Since any embedding is isotopic to a normalized embedding by Proposition 7.1.22, we conclude that when $b^+ = 1$ the symplectic embedding space $\mathrm{Emb}(B(\lambda), M)$ is path connected, whenever it is nonempty. This result extends to cohomologous symplectic forms on the k-point blowup of M associated to k normalized embeddings of balls with disjoint images. The analogous statement for $b^+ > 1$ and in higher dimensions is an open question. □

[31] A similar argument for diffeomorphisms may be found in Palais [522].

Remark 7.1.29 (Isotopy of symplectic embeddings) In dimension four, the space of symplectic embeddings of $B(\lambda)$ into $B(1)$ is connected for every $\lambda < 1$ by theorems of McDuff [450, 458] and Biran [65]. (This is closely related to the fact, mentioned in Remark 7.1.28, that $\mathrm{Emb}(B(\lambda), \mathbb{C}\mathrm{P}^2)$ is connected.) The analogous statement in higher dimensions is an open question, even for arbitrarily small (but fixed) values of λ. □

Remark 7.1.30 If (M, ω, J) is Kähler, we saw in Theorem 7.1.21 that the complex blowup $(\widetilde{M}, \widetilde{J})$ carries a symplectic form $\widetilde{\omega}$ which equals ω outside a small neighbourhood of p_0 and is compatible with $\widetilde{J}$. Thus blowups of Kähler manifolds are Kähler. This was first proved by Kodaira [372]. Note, however, that the Kähler form ω must first be modified so that the Kähler metric is flat near p_0. Hence in general this works only if one blows up by a sufficiently small amount, i.e. the radius of the corresponding symplectic ball will be small. However, by constructing appropriate ball embeddings, several cases are now known in which a one-point blowup of a Kähler manifold carries symplectic forms in a class with no Kähler representative. (See Example 13.4.9 and [400] and the references therein.) □

Remark 7.1.31 (Symplectic packing) Another geometric question to which these methods apply is the **symplectic packing problem**. For example, one can ask how large a part of the volume of a symplectic manifold (M, ω) can be filled by a symplectic embedding of k disjoint balls of the same radius. The answer is encoded in the number

$$v_k(M, \omega) := \sup \left\{ \frac{k\mathrm{Vol}(B^{2n}(r))}{\mathrm{Vol}(M, \omega)} \;\middle|\; \begin{array}{l} \text{there exist symplectic embeddings} \\ \iota_i : B^{2n}(r) \to M,\, i = 1, \dots, k, \\ \text{with pairwise disjoint images} \end{array} \right\},$$

and (M, ω) is said to have a **full filling by k balls** if $v_k(M, \omega) = 1$. The first results about this problem were found by Gromov [287] for $M = \mathbb{C}\mathrm{P}^2$, and his work was later extended by McDuff–Polterovich [469] and Biran [66]. In [67] Biran has shown that for every closed symplectic four-manifold (M, ω) with rational symplectic form, i.e. $[\omega] \in H^2(M; \mathbb{Q})$, there exists an integer N such that $v_k(M, \omega) = 1$ for all $k \geq N$. This property is known as **packing stability**. The rationality hypothesis was removed by Buse–Hind–Opshtein [90]. Although very little is known about symplectic embeddings in dimensions greater than four, Buse–Hind [89] extended the packing stability result to balls of any dimension. For B^4 Biran showed $N = 9$ (the same number as for $\mathbb{C}\mathrm{P}^2$), while for B^{2n} Buse–Hind obtain the upper bound $N \leq \lceil (8\frac{1}{36})^{\frac{n}{2}} \rceil$. In contrast, Entov–Verbitsky show in [205] that hyperKähler manifolds can always be fully packed by any number of balls. □

Remark 7.1.32 (Minimal reductions) A symplectic 4-manifold is called **minimal** if it contains no symplectically embedded 2-spheres of self-intersection number minus one. (These are called **exceptional spheres**. In the holomorphic case they are called **exceptional divisors**.) Just as in the complex case, one can always make a symplectic four-manifold minimal by blowing down a finite collection of disjoint exceptional spheres (see McDuff [449, §3]). Moreover, most manifolds have a unique minimal reduction. The archetypal exception is the two point blowup of the complex projective plane. It has three exceptional spheres representing the homology classes E_1, E_2, and $E_3 = L - E_1 - E_2$, with $E_1 \cdot E_2 = 0$ and $E_1 \cdot E_3 = E_2 \cdot E_3 = 1$. Blowing down E_1 and E_2 gives $\mathbb{C}\mathrm{P}^2$ and blowing down E_3 gives $\mathbb{C}\mathrm{P}^1 \times \mathbb{C}\mathrm{P}^1$. More generally, McDuff proved in [452] that the minimal reduction is unique up to symplectomorphism unless M is **rational** or **ruled**, i.e. it is a blowup of $\mathbb{C}\mathrm{P}^2$ or of an S^2-bundle over an oriented 2-manifold with a symplectic form that is compatible with the fibration. □

Symplectic blowing up along a submanifold

It is also possible to blow up a symplectic manifold along a symplectic submanifold S, by blowing up in the directions normal to S. Here we will briefly discuss the local theory; that is, how to blow up a little bit. The global theory has not yet been fully worked out. The whole question is very closely connected with the theory of symplectic fibrations, and, just as in the previous section, there are two approaches, a naive approach à la Thurston, and a more sophisticated approach using the Sternberg–Weinstein universal construction. The former method is worked out in McDuff [444]. Here we will use the latter method, because it is potentially more powerful.

The basic construction for blowing up is very simple. Let S be a compact symplectic submanifold in (M, ω) of codimension $2k$. Its normal bundle ν_S is a symplectic vector bundle, and so, by the results of Section 2.6, has a compatible complex structure. Therefore, we may consider it to be the bundle with fibre $(\mathbb{R}^{2k}, \omega_0)$ associated to a principal $\mathrm{U}(k)$-bundle $P \to S$. By Example 5.3.7 and Exercise 7.1.15, $\mathrm{U}(k)$ acts in a Hamiltonian way on $(\mathbb{R}^{2k}, \omega_0)$ and on its blowup $(\widetilde{\mathbb{R}}^{2k}_\epsilon, \widetilde{\omega}_\epsilon)$. Thus, we are in the situation of Theorem 6.3.3. Choose a connection β on P, and let α and $\widetilde{\alpha}_\epsilon$ be the corresponding 2-forms on the bundle $\nu_S \cong P \times_{\mathrm{U}(k)} \mathbb{R}^{2k}$ and its blowup

$$P \times_{\mathrm{U}(k)} \widetilde{\mathbb{R}}^{2k}.$$

Then, if π_S is the projection $\nu_S \to S$ and ω_S denotes the restriction of the form ω to S, the form

$$\alpha + \pi_S^* \omega_S$$

is nondegenerate on the fibre and equals ω_S on the zero section, and so is symplectic in some neighbourhood of the zero section. Therefore, by the Symplectic Neighbourhood Theorem 3.4.10, there is $\epsilon_0 > 0$ such that a neighbourhood $\mathcal{N}_{\epsilon_0}(S)$ of S in M is symplectomorphic to the disc bundle $\nu_S(\epsilon_0) = P \times_{\mathrm{U}(k)} B(\epsilon_0)$ with the form $\alpha + \pi^* \omega_S$. But, by construction, whenever $\epsilon < \epsilon_0$, the neighbourhood $\widetilde{B}(\epsilon_0)$ of the exceptional divisor in the ϵ-blowup $(\widetilde{\mathbb{R}}^{2k}, \widetilde{\omega}_\epsilon)$ has boundary symplectomorphic to $(\partial B(\epsilon_0), \omega_0)$. Thus the boundary of the blowup

$$(P \times_{\mathrm{U}(k)} \widetilde{B}(\epsilon_0), \widetilde{\alpha}_\epsilon)$$

is symplectomorphic to the boundary of the disc bundle $\nu_S(\epsilon_0)$, and we may construct the ϵ-blowup $(\widetilde{M}_S, \omega_{S,\epsilon})$ of M along S by cutting out $\mathcal{N}_{\epsilon_0}(S)$ and gluing back in $P \times_{\mathrm{U}(k)} \widetilde{B}(\epsilon_0)$.

This construction depends on a choice of connection β and of symplectomorphism $\phi : \mathcal{N}_{\epsilon_0}(S) \to \nu_S(\epsilon_0)$. It is easy to see that the diffeomorphism class of the manifold M_S is independent of these choices. Furthermore, given two different choices β, ϕ and β', ϕ', one can check that the corresponding forms $\widetilde{\omega}_\epsilon$ and $\widetilde{\omega}'_\epsilon$ on M_S are isotopic for sufficiently small ϵ. However, because there is a noncompact family of choices, it is not clear whether there is an $\epsilon > 0$ such that all ϵ-blowups of M along S are symplectomorphic. (In fact, as we saw above, this is not even known when S is a single point.)

Guillemin and Sternberg [298] use the blowup construction to explain what happens to the reduced manifolds B_λ described in Lemma 5.1.2 as λ passes through a critical value. This idea has been greatly expanded by Li–Ruan [427], who use it as the basis for a symplectic version of complex birational geometry. The next exercise is taken from [444] and sketches another application.

Exercise 7.1.33 This exercise shows how the blowing up techniques can be used to construct a **simply connected nonKähler** symplectic manifold. The manifold $M = \mathbb{R}^4/\Gamma$ in Example 3.1.17 has an integral symplectic form ω (i.e. $[\omega]$ lifts to an integral class in $H^2(M;\mathbb{Z})$). Hence it follows from Tischler [627] (see also Narasimham–Seshadri [501]) that (M, ω) symplectically embeds into $(\mathbb{C}\mathrm{P}^N, \tau_0)$ for some large N. (An improvement of this result due to Gromov [288] implies that one can take $N = 5$.) Let (X, τ_ϵ) be an ϵ-blowup of $\mathbb{C}\mathrm{P}^N$ along the image of M. Use van Kampen's theorem to show that X is simply connected. Show that $H^3(X)$ is isomorphic to $H^1(M)$ and so has rank 3. Thus X cannot be Kähler, since the odd Betti numbers of Kähler manifolds are even. □

7.2 Connected sums

The pointwise connected sum

The usual pointwise operation of the oriented connected sum is described in Exercise 7.1.4. In the symplectic category this connected sum construction works only in dimension 2. The technical reason is that only in dimension 2 is there a symplectic embedding $\mathcal{N}(S^{2n-1}) \to \mathbb{R}^{2n}$ of a neighbourhood of the unit sphere that takes the sphere onto itself while interchanging its two sides. This follows from Proposition 3.5.33. Alternatively, one can note that if there were such a symplectomorphism in higher dimensions it would give rise to a symplectic structure on the sphere S^{2n}, which obviously does not exist. However, in dimension 2 the symplectic connected sum of the manifolds (M_i, ω_i) can be defined as in Exercise 7.1.4, provided that we choose the embeddings $f_i : B \to M_i$ to be local Darboux charts and the diffeomorphism ϕ_ε to be area-preserving. Then the connected sum $M_1 \# M_2$ inherits a symplectic structure from ω_1 and ω_2.

Exercise 7.2.1 Fill in the details of this construction with the techniques of Exercises 3.2.8 and 7.1.4. Give an explicit formula for a symplectomorphism of the 2-dimensional annulus which interchanges the two boundary components. □

Note that this is a rather strict notion of the connected sum because we required the symplectic form ω on $M = M_1 \# M_2$ to restrict to the given forms ω_i on the two halves of M. However, matters do not become much better even if we weaken this requirement. The weakest condition which has some geometric meaning is to require that the form ω on $M_1 \# M_2$ restricts to a form τ_i on $M_i \setminus B_i$ which is isotopic to the given form ω_i. Because of Gromov's classification theorem for symplectic structures on open manifolds (see Section 7.3), this is equivalent to requiring that τ_i be isotopic to ω_i through nondegenerate but not necessarily closed 2-forms on $M_i \setminus B_i$. However, even this is impossible except in dimension 6: see Audin [31].

Lemma 7.2.2 *Let (M_1, ω_1) and (M_2, ω_2) be two compact symplectic manifolds of dimension $2n \neq 2, 6$. Then the connected sum $M := M_1 \# M_2$ does not carry a symplectic form which on each half $M_i \setminus B_i$ is isotopic to ω_i through nondegenerate forms.*

Proof: Assume by contradiction that such a form ω exists and let J be an ω-tame almost complex structure on M. Then its restriction J_i to $M_i \setminus B_i$ is homotopic to any ω_i-tame almost complex structure and hence to one which extends over B_i. It follows that J_i itself extends over B_i. To see this, think of J_i as a section of a bundle with contractible fibres as discussed in Chapter 2. The two extensions of J over the balls B_i fit together to give an almost complex structure on S^{2n}. But such a structure exists only for $n = 1$ and $n = 3$ (see Exercise 4.1.9). □

Example 7.2.3 Because there is an orientation-reversing diffeomorphism of the torus $\mathbb{T}^{2n}$, the connected sum $\mathbb{T}^{2n} \# \mathbb{C}\mathrm{P}^n$ is diffeomorphic to the one-point blowup $\mathbb{T}^{2n} \# \overline{\mathbb{C}\mathrm{P}}^n$ and hence carries a symplectic structure. □

Exercise 7.2.4 Use Example 4.1.7 to show that the connected sum of any two almost complex manifolds of dimension 6 has an almost complex structure. □

At the time of writing no examples are known where is it possible to put a symplectic form on the oriented connected sum $M_1 \# M_2$ of two 6-dimensional symplectic manifolds. (To avoid Example 7.2.3 above we assume that the connected sum is taken compatible with the symplectic orientations.) Neither are there any examples where it is known to be impossible. The only relevant result proved so far is described in Example 13.2.8.

Remark 7.2.5 In contrast, it is possible to put a contact structure on the connected sum of two contact manifolds. This was first proved by Meckert [477], and a simpler proof was found by Bennequin [60]. In fact, contact structures are more amenable to the usual kinds of surgery, which has meant that more progress has been made with constructing contact manifolds than symplectic ones. In fact, every orientable 3-manifold, every orientable simply connected 5-manifold with vanishing third integral Stiefel–Whitney class W_3, and indeed every closed odd-dimensional manifold with an almost contact structure supports a contact structure (see Eliashberg [179], Weinstein [678], Geiges [258, 261], and Borman, Eliashberg and Murphy's paper [75]). □

A few surgeries are well adapted to the symplectic context. Luttinger's work on surgery along Lagrangian tori is one example (see Example 3.4.16 and Luttinger [435], Eliashberg–Polterovich [194], Auroux–Donaldson–Katzarkov [44]). Another is the conifold transition, a six-dimensional surgery that replaces a Lagrangian 3-sphere by a symplectic 2-sphere, whose properties are explored by Smith–Thomas–Yau in [594]. Here we concentrate on describing the work of Gompf [277, 278] and McCarthy and Wolfson [442] on connected sums along codimension-2 symplectic submanifolds. This construction was mentioned by Gromov in his book [288], but its dramatic possibilities were first noticed by Gompf. Here we will only discuss the easiest case, that of a trivial normal bundle, and will illustrate its power by showing that a compact symplectic 4-manifold may have an arbitrary finitely presented fundamental group. This means that the class of symplectic 4-manifolds is very different from the class of Kähler surfaces, since there are significant, rather subtle restrictions on the fundamental groups of Kähler manifolds.

Fibre connected sums

Let (M_1, ω_1) and (M_2, ω_2) be symplectic manifolds of the same dimension $2n$ and (Q, τ) be a compact symplectic manifold of dimension $2n-2$. Suppose that

$$\iota_1 : Q \to M_1, \qquad \iota_2 : Q \to M_2$$

are symplectic embeddings such that their images $Q_j = \iota_j(Q) \subset M_j$ have trivial normal bundle. Then, by the Symplectic Neighbourhood Theorem 3.4.10 of Section 3.4, there exist symplectic embeddings

$$f_j : Q \times B^2(\varepsilon) \to M_j, \qquad {f_j}^*\omega_j = \tau \oplus dx \wedge dy,$$

such that $f_j(q, 0) = \iota_j(q)$ for $q \in Q$. Now consider the annulus

$$A(\delta, \varepsilon) := B^2(\varepsilon) \setminus \operatorname{int}(B^2(\delta))$$

for $0 < \delta < \varepsilon$ and choose an area- and orientation-preserving diffeomorphism

$$\phi : A(\delta, \varepsilon) \to A(\delta, \varepsilon)$$

that interchanges the two boundary components. Then the **fibre connected sum** is defined by

$$M_1 \#_Q M_2 = \Big(M_1 \setminus \mathcal{N}(Q_1)\Big) \cup_\phi \Big(M_2 \setminus \mathcal{N}(Q_2)\Big),$$

where $\mathcal{N}(Q_j) = f_j(Q \times B^2(\delta))$ and

$$f_2(q, z) \equiv f_1(q, \phi(z))$$

for $q \in Q$ and $\delta < |z| < \varepsilon$. Then the symplectic forms ω_1 and ω_2 agree on the overlap $Q \times A(\delta, \varepsilon)$ and hence induce a symplectic structure on $M_1 \#_Q M_2$.

This construction depends on the choice of the local symplectomorphisms f_j and on the constant δ. Different choices of δ obviously give rise to symplectic structures of different volume. More importantly, different (i.e. nonisotopic) choices of the framings f_j (see Exercise 3.4.12) will in general give rise to different manifolds $M_1 \#_Q M_2$.

Exercise 7.2.6 Let $V = E(1)$ be the elliptic surface described in Example 7.1.7, and let T be one of its fibres. Then T has trivial normal bundle, and so, given a symplectic 2-torus $T' \subset (X, \omega)$ with trivial normal bundle, we may form the connected sum

$$X' := X \#_{T'} V.$$

Here we must scale the form on V so that T and T' have the same area. Use van Kampen's theorem to show that $\pi_1(X') = \pi_1(X)/\langle \iota_*(\pi_1(T')) \rangle$, where $\langle \iota_*(\pi_1(T')) \rangle$ is the normal subgroup generated by the image of $\pi_1(T')$ by the inclusion $\iota : T' \to X$. □

Exercise 7.2.7 Let $V = E(1)$ be the elliptic surface in Exercise 7.2.6. As explained in Example 7.1.7, there is a holomorphic map

$$f : V \to \mathbb{C}\mathrm{P}^1$$

whose generic fibre is a torus $T = \mathbb{T}^2$. Since the normal bundle of this torus is trivial, we may form the fibre connected sum

$$W := V \#_T V.$$

(In contrast to the previous example, we take the obvious framings of the normal bundle, namely those which are pullbacks of a frame on the base. In fact Gompf [278] has shown that one gets nothing new in this case by considering twisted framings.)

(i) Use van Kampen's theorem and the Mayer–Vietoris sequence to prove that W is simply connected and that $H_2(W)$ has rank 22. Moreover, its first Chern class vanishes. (For help with this see [279].)

(ii) Consider the diagram

$$\begin{array}{ccc} & & V \\ & & \big\downarrow f \\ \mathbb{C}\mathrm{P}^1 & \xrightarrow{\phi} & \mathbb{C}\mathrm{P}^1 \end{array},$$

where $\phi : \mathbb{C}\mathrm{P}^1 \to \mathbb{C}\mathrm{P}^1$ is a 2-fold branched cover. Show that W is diffeomorphic to the pullback $\phi^* V = \{(z, x) \mid \phi(z) = f(x)\}$ of V by ϕ. Thus W has a complex structure, and hence also is an elliptic surface. By (i) and Example 4.4.3, it is a K3-surface.

Observe that the fibre connected sum is not a construction in the complex category. If Q_i are complex submanifolds of M_i with trivial normal bundle then one can sum them symplectically provided that the integral of the symplectic form is the same on both of them. However, one cannot always put a complex structure on $M_1 \#_Z M_2$ because the complex structures on neighbourhoods of Z_1 and Z_2 may not match up. See [278]. □

This construction has proved to be a very useful way to find new symplectic 4-manifolds. The blowdown can be considered as a special case. In fact, since the complement $\mathbb{C}P^n \setminus \mathbb{C}P^{n-1}$ of a hyperplane in $\mathbb{C}P^n$ can be identified with the ball (see Example 7.1.16), the blowdown of M along an exceptional divisor Σ is simply the fibre sum of (M, Σ) with the pair $(\mathbb{C}P^n, \mathbb{C}P^{n-1})$. (Here, of course, we are allowing a nontrivial normal bundle.) Similarly, the rational blowdown is best thought of as a fibre connected sum: see Symington [603].

We end this section by proving the following beautiful result of Gompf. Other applications of the fibre connected sum are discussed in Chapter 13; see Examples 13.2.11 and 13.4.6 for example.

Theorem 7.2.8 (Gompf [278]) *Every finitely presented group* G *is the fundamental group of some compact symplectic* 4*-manifold.*

Proof: Fix a presentation

$$\mathrm{G} = \langle g_1, \ldots, g_k | r_1, \ldots, r_\ell \rangle.$$

Let F be a compact oriented Riemann surface of genus k, and choose a standard collection of oriented simple closed curves $\alpha_1, \ldots, \alpha_k; \beta_1, \ldots, \beta_k$ which represent a homology basis for $H_1(F)$. Thus the curves α_i are disjoint, as are the β_j, and $\alpha_i \cdot \beta_j = \delta_{ij}$. Then the quotient $\pi_1(F)/\langle \beta_1, \ldots, \beta_k \rangle$ is the free group generated by the loops $\alpha_1, \ldots, \alpha_k$ if these are attached to some base point. For $i = 1, \ldots, \ell$ choose an immersed oriented closed curve γ_i on F which represents the relation $r_i(\alpha_1, \ldots, \alpha_k)$, and set $\gamma_{\ell+j} = \beta_j$, for $j = 1, \ldots, k$. Then

$$\mathrm{G} \cong \frac{\pi_1(F)}{\langle \gamma_1, \ldots \gamma_{\ell+k} \rangle}.$$

Thus it suffices to do symplectic surgeries which kill the loops γ_i of $\pi_1(F)$.

Let us suppose that there is a closed 1-form ρ on F which restricts to a volume form on each of the oriented loops γ_i. This means that it is strictly positive on the oriented tangent spaces to the γ_i. (We will sketch in Exercise 7.2.9 below how to construct such a form.) Consider the manifold

$$X = F \times \mathbb{T}^2$$

with a product symplectic form $\omega = \omega_1 \times \omega_2$. Choose an oriented simple closed curve α in $\mathbb{T}^2$ that is nontrivial in homology, and a 1-form θ on $\mathbb{T}^2$ that is positive on all the positively oriented tangent vectors to α. For $i = 1, \ldots, \ell + k$ let T_i be the immersed torus

$$T_i = \gamma_i \times \alpha \subset X.$$

Then T_i is Lagrangian with respect to ω, but is symplectic with respect to the form $\rho \wedge \theta$. Moreover, the 2-form

$$\omega' := \omega + \rho \wedge \theta$$

is symplectic on $X = F \times \mathbb{T}^2$ provided that the 1-forms ρ and θ have been chosen sufficiently small. Thus the symplectic manifold (X, ω') contains the symplectically immersed tori T_i as well as $\{z\} \times \mathbb{T}^2$. We now slightly perturb all these tori to make them disjoint and symplectically embedded. Since the T_i all lie in the 3-dimensional submanifold $F \times \alpha$ of X, and we may choose the point z to be disjoint from the γ_i, this is clearly possible. In fact, if we write $X = F \times S^1 \times \alpha$, it suffices to perturb the curves γ_i in $F \times S^1$ to make them disjointly embedded. Moreover, it is easy to see that all these tori have trivial normal bundle. Thus we may attach $\ell + k + 1$ copies of the elliptic surface V to X along these tori as in Exercise 7.2.6. This kills the homotopy classes represented by the loops γ_i and by $\pi_1(\mathbb{T}^2)$. Thus the resulting manifold has fundamental group $\pi_1 \cong \mathrm{G}$. □

Exercise 7.2.9 This exercise completes the proof of the above theorem. We may assume that the loops γ_i are in general position, that is that they intersect transversally in pairs. Thus we may consider their union to be an oriented, possibly disconnected, graph Γ on F. Let E be the collection of all oriented edges and isolated circles in this graph.

(i) Suppose that there is a closed 1-form ρ^* on F such that $\int_e \rho^* > 0$ for all $e \in E$. Show that there is a smooth function f on F which vanishes at every vertex of Γ and is such that $\rho^* + df$ restricts to a volume form on every γ_i.

(ii) Consider a 2-torus $T = S^1 \times S^1$ together with the curves

$$\alpha := S^1 \times \mathrm{pt}, \qquad \beta := \mathrm{pt} \times S^1.$$

Let γ be a curve parallel to β, and choose a little disc D disjoint from α and β meeting γ in an arc. Show that there is a closed 1-form ρ^* on $T \setminus D$ whose integral over α, β and $\gamma \setminus D$ is strictly positive, and which is identically zero near the boundary ∂D.

(iii) Not every graph Γ on a surface F admits a 1-form ρ^* which satisfies the conditions in (i) above. (See (iv) below.) Show that it is possible to modify F and Γ until it does by attaching a copy $T_e \setminus D_e$ of $T \setminus D$ to each edge e of Γ in such a way that $\gamma_e \setminus D_e$ is incorporated into the edge e, and the circles α_e and β_e are added to Γ. (See Fig. 7.4.) The 1-forms ρ_e^* may be extended to be zero over the rest of the surface.

(iv) There is a homological obstruction to the existence of such a closed 1-form ρ^*. Give an example where ρ^* does not exist. More generally, formulate the obstruction. □

It is possible that some of the manifolds constructed above are Kähler. However, Gompf also shows how to modify the construction so that all the manifolds obtained are definitely nonKähler because their Chern numbers cannot be realized by complex surfaces. The paper contains a wealth of other interesting examples, and is highly recommended reading for anyone interested in trying to understand the topology of symplectic manifolds. For later work that constructs a variety of symplectic manifolds, see Torres–Yazinski [631] and the references therein.

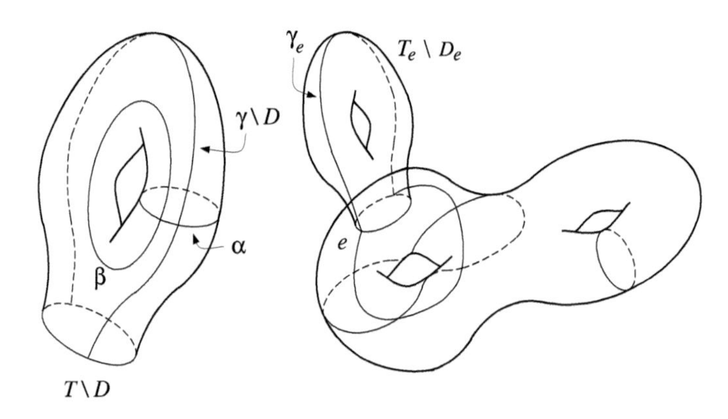

FIG. 7.4. Attaching the torus $T_e \setminus D_e$.

7.3 The telescope construction

In Proposition 4.1.1 we have seen that there is a one-to-one correspondence between homotopy classes of almost complex structures and homotopy classes of nondegenerate 2-forms. It is natural to ask whether every nondegenerate 2-form is homotopic (within the class of nondegenerate 2-forms) to a symplectic form. This question was completely resolved by Gromov for open manifolds in his thesis in 1969 [286]. He proved that in an open manifold[32] every homotopy class of nondegenerate 2-forms can be represented by a symplectic form and, moreover, that the symplectic form can be chosen exact. In fact he proved a more general theorem, namely that the inclusion of the space of symplectic forms on M, representing a given cohomology class a, into the space of all nondegenerate 2-forms on M is a homotopy equivalence. That this map induces an isomorphism on π_0 can be restated as follows.

Theorem 7.3.1 (Gromov) *Let M be an open $2n$-dimensional manifold. Let $\tau \in \Omega^2(M)$ be a nondegenerate 2-form and let $a \in H^2(M;\mathbb{R})$.*

(i) *There exists a smooth family of nondegenerate 2-forms τ_t on M such that $\tau_0 = \tau$ and τ_1 is a symplectic form that represents the class a.*

(ii) *If τ_t is a smooth family of nondegenerate 2-forms on M such that τ_0 and τ_1 are symplectic forms representing the cohomology class a, then there exists a smooth family of symplectic forms ω_t in the cohomology class a such that $\omega_0 = \tau_0$ and $\omega_1 = \tau_1$.*

Proof: We shall prove the theorem in the case $a = 0$. The strategy is to exhaust M by the sublevel sets

$$M^c := \{q \in M \mid f(q) \le c\}$$

[32] A manifold is called **open** if each connected component is either noncompact or has a nonempty boundary.

of a Morse function $f : M \to \mathbb{R}$ that is proper and bounded below, and does not have any critical points of index $2n$. For simplicity, we shall assume that there exists such a Morse function that in addition has only finitely many critical points. Without loss of generality we may also assume that there is only one critical point of index zero and that the critical points $q_0, q_1, \ldots, q_N$ all lie on different critical levels $c_j = f(q_j)$ with $c_0 < c_1 < \cdots < c_N$. Given a level $c > c_0$ we shall prove the following.

Claim: *There exists a smooth family of nondegenerate 2-forms $\tau_t \in \Omega^2(M)$ such that $\tau_0 = \tau$ and τ_1 is exact on M^c.*

Note first that if the claim has been proved for some regular level c with $c_j < c < c_{j+1}$, then it holds for every other level $c' \in (c_j, c_{j+1})$. To see this, use the gradient flow of f to construct a smooth isotopy $\phi_t : M \to M$ such that $\phi_0 = \mathrm{id}$ and $\phi_1(M^{c'}) = M^c$, and define $\tau'_t := \phi_t^* \tau_t$.

Secondly, we prove the claim for $c = c_0 + \varepsilon$, where $\varepsilon > 0$ is sufficiently small. By Theorem 2.1.3, there exists a local chart $\phi_0 : U_0 \to \mathbb{R}^{2n}$ on a neighbourhood of q_0 such that the 2-form $\phi_0^* \omega_0$ agrees with τ at the point q_0. If U_0 is sufficiently small then it follows that the 2-forms $\tau + t(\phi_0^* \omega_0 - \tau)$ are nondegenerate on U_0 for $0 \le t \le 1$. Now choose a cutoff function $\beta : M \to [0, 1]$ which is equal to 1 near q_0 and vanishes outside U_0. Then the 2-forms

$$\tau_t := \tau + t\beta({\phi_0}^* \omega_0 - \tau)$$

are nondegenerate for $0 \le t \le 1$ and τ_1 agrees with $\phi_0^* \omega_0$ in a neighbourhood of q_0. This proves the claim for $c_0 < c < c_1$.

The main point of the proof is to understand the construction of the isotopy as we pass a critical level c_j. Hence assume by induction that the claim has been proved for $c < c_j$. Then we may assume that τ is exact on $M^{c_j - \varepsilon}$, where $\varepsilon > 0$ is arbitrarily small. Let q_j be a critical point of index $m = m_j \le 2n - 1$. Choose local Morse coordinates $(x, y) \in I^m \times I^{2n-m}$ near q_j with $I = [-1, 1]$ in which

$$f(x, y) = |y|^2 - |x|^2 + c_j.$$

By assumption the 2-form τ is exact in a neighbourhood of $(\partial I^m) \times I^{2n-m}$. We must find an isotopy of nondegenerate 2-forms from τ to a 2-form which is exact on all of $I^m \times I^{2n-m}$. The crucial induction step is formulated in Lemma 7.3.2 below (in the case where the parameter space Λ is a single point). It implies that there exists an isotopy of nondegenerate 2-forms on I^{2n} which for $t = 0$ agrees with the given 2-form τ, for $t = 1$ is exact, and for all t agrees with τ near the *lower boundary* $(\partial I^m) \times I^{2n-m}$. Hence this isotopy extends to an isotopy $\tau_t \in \Omega^2(M^c)$ for some $c > c_j$ such that

$$\tau_0 = \tau|_{M^c}, \qquad \tau_1 \in \Omega^2(M^c) \text{ is exact.}$$

We now extend the restriction of τ_t to $M^{c-\varepsilon} \supset M^{c_j}$ to an isotopy on all of M which on $M \setminus M^c$ is given by $\tau_0 = \tau$. An explicit formula is

$$\widetilde{\tau}_t(x) := \begin{cases} \tau_0(x), & \text{if } c \le f(x), \\ \tau_{\varepsilon^{-1}t(c-f(x))}, & \text{if } c-\varepsilon \le f(x) \le c, \\ \tau_t(x), & \text{if } c-\varepsilon \ge f(x). \end{cases}$$

This is only continuous but can easily be approximated by a smooth isotopy. Thus we have constructed an isotopy of nondegenerate 2-forms from τ to a 2-form which is exact on M^c for some $c > c_j$. If this has been proved for $j = N$ we may use the fact that M is diffeomorphic to M^c for $c > c_N$ to prove (i).

Thus we have proved (i) modulo the proof of Lemma 7.3.2 below. To prove statement (ii) we must deform a given isotopy τ_λ, $0 \le \lambda \le 1$, of nondegenerate 2-forms where τ_0 and τ_1 are exact, into one where all the 2-forms are exact. A successive application of Lemma 7.3.2 gives an isotopy of paths $\tau_{\lambda,t}$ which for $t = 0$ agrees with τ_λ, and which for $t = 1$, $\lambda = 0$, and $\lambda = 1$ consists entirely of exact 2-forms. This gives the required isotopy of nondegenerate exact 2-forms from τ_0 to τ_1. The details of this argument are left to the reader.

The case where $a \in H^2(M;\mathbb{R})$ is a nontrivial cohomology class is proved by a simple modification of this argument. Just fix a closed 2-form $\alpha \in \Omega^2(M)$ which represents the class a and replace the condition on τ to be exact by the condition that $\tau - \alpha$ is exact. Wherever we use pullback it is by diffeomorphisms ϕ which are isotopic to the identity, and so $\phi^*\alpha - \alpha$ is always exact. Hence $\phi^*\tau - \alpha$ will be exact whenever $\tau - \alpha$ is. This proves the theorem. □

Lemma 7.3.2 *Let $\Lambda \subset \mathbb{R}^N$ be a compact set and let $1 \le m < 2n$. Assume that*

$$\Lambda \to \Omega^2(I^{2n}) : \lambda \mapsto \tau_\lambda$$

is a smooth family of nondegenerate 2-forms such that τ_λ is exact in a neighbourhood of $\partial I^m \times I^{2n-m}$ for every λ. Then there exists an isotopy

$$\Lambda \times [0,1] \to \Omega^2(I^{2n}) : (\lambda, t) \mapsto \tau_{\lambda,t}$$

of nondegenerate 2-forms such that

(i) $\tau_{\lambda,0} = \tau_\lambda$.
(ii) $\tau_{\lambda,1}$ *is exact on* I^{2n}.
(iii) $\tau_{\lambda,t} = \tau_\lambda$ *near* $(\partial I^m) \times I^{2n-m}$ *for every* t.
(iv) *If* τ_λ *is exact on* I^{2n} *then* $\tau_{\lambda,t}$ *is exact on* I^{2n} *for every* t.

Proof: The proof is by induction over m and involves the telescope construction. The first step is the case $m = 1$ and we shall first assume that Λ is a single point. Thus we are given a 2-form

$$\tau \in \Omega^2(I \times I^{2n-1})$$

whose restriction to a neighbourhood of $(\partial I) \times I^{2n-1} = \{-1, 1\} \times I^{2n-1}$ is exact. We must prove that τ is isotopic through nondegenerate 2-forms to an exact 2-form on $I \times I^{2n-1}$ where the isotopy is constant near $\partial I \times I^{2n-1}$.

To see this, choose $\varepsilon > 0$ so small that τ is exact on $[-1, -1+3\varepsilon] \times [-1, 1]^{2n-1}$ and on $[1 - 3\varepsilon, 1] \times [-1, 1]^{2n-1}$. Then, after shrinking ε if necessary, there exists a smooth family of nondegenerate exact 2-forms

$$\tau_\xi = d\sigma_\xi \in \Omega^2([\xi, \xi + 3\varepsilon] \times [-\varepsilon, \varepsilon]^{2n-1}), \qquad -1 \le \xi \le 1 - 3\varepsilon,$$

which at the point $(\xi, 0)$ agree with τ, and satisfy $\tau_{-1} = \tau|_{[-1,-1+3\varepsilon]\times[-\varepsilon,\varepsilon]^{2n-1}}$ and $\tau_{1-3\varepsilon} = \tau|_{[1-3\varepsilon,1]\times[-\varepsilon,\varepsilon]^{2n-1}}$. Choose $\varepsilon > 0$ so small that the 2-form τ_ξ is isotopic to τ through nondegenerate 2-forms on $[\xi, \xi + 3\varepsilon] \times [-\varepsilon, \varepsilon]^{2n-1}$. Given $\varepsilon > 0$ we choose $\delta > 0$ so small that whenever $\xi < \xi' < \xi + \delta$, we can homotop from σ_ξ to $\sigma_{\xi'}$ on an interval of length ε without losing nondegeneracy. We shall do this on the interval $[\xi', \xi' + \varepsilon]$. More precisely, choose a smooth cutoff function $\beta : \mathbb{R} \to [0, 1]$ such that

$$\beta(x) = \begin{cases} 0, & \text{if } x \le 0, \\ 1, & \text{if } x \ge \varepsilon, \end{cases}$$

define $\beta_{\xi'}(x, y) = \beta(x - \xi', y)$, and consider the 1-form

$$\sigma_{\xi,\xi'} := (1 - \beta_{\xi'})\sigma_\xi + \beta_{\xi'}\sigma_{\xi'}.$$

The differential of this 1-form is nondegenerate when $\delta > 0$ is sufficiently small, it agrees with τ_ξ on the domain on $[\xi, \xi'] \times [-\varepsilon, \varepsilon]^{2n-1}$, and it agrees with $\tau_{\xi'}$ on the domain $[\xi' + \varepsilon, \xi' + 3\varepsilon] \times [-\varepsilon, \varepsilon]^{2n-1}$.

Now we shall use the **telescope construction** to find an exact 2-form $d\sigma$ on $I \times [-\varepsilon, \varepsilon]^{2n-1}$, which is nondegenerate, agrees with τ on $\{\pm 1\} \times [-\varepsilon, \varepsilon]^{2n-1}$, and is isotopic to τ in the class of nondegenerate 2-forms where the isotopy is constant on $\{\pm 1\} \times [-\varepsilon, \varepsilon]^{2n-1}$. To construct the 1-form σ we choose a sequence

$$-1 = x_0 < x_1 < \cdots < x_{N-1} < x_N = 1 - 3\varepsilon$$

such that $x_{j+1} - x_j < \delta < \varepsilon$. It is convenient to introduce the notation

$$B_j := [x_j, x_{j+1} + \varepsilon] \times [-\varepsilon, \varepsilon]^{2n-1}, \qquad B := [-1, 1] \times [-\varepsilon, \varepsilon]^{2n-1},$$

and

$$A_j := [2j\varepsilon, (2j + 3)\varepsilon] \times [-\varepsilon, \varepsilon]^{2n-1}, \qquad A := [0, (2N + 3)\varepsilon] \times [-\varepsilon, \varepsilon]^{2n-1}.$$

For each j choose a zig-zag immersion $\phi_j : A_j \to B_j$ which shifts the left third $[2j\varepsilon, (2j + 1)\varepsilon] \times [-\varepsilon, \varepsilon]^{2n-1}$ onto $[x_j, x_j + \varepsilon] \times [-\varepsilon, \varepsilon]^{2n-1}$ and shifts the right third $[(2j+2)\varepsilon, (2j+3)\varepsilon] \times [-\varepsilon, \varepsilon]^{2n-1}$ onto $[x_{j+1}, x_{j+1} + \varepsilon] \times [-\varepsilon, \varepsilon]^{2n-1}$. Such a map can be constructed according to Fig. 7.5. Here the top line of the diagram represents the three parts of A_j, and the second line illustrates various stages of its image under the immersion. Observe that the maps ϕ_j and ϕ_{j+1} agree on their common domain $[(2j + 2)\varepsilon, (2j + 3)\varepsilon] \times [-\varepsilon, \varepsilon]^{2n-1}$. Hence the maps $\phi_0, \ldots, \phi_N$ fit together to form an immersion

$$\phi : A \to B.$$

This immersion is isotopic (through immersions) to the obvious affine diffeomorphism $\psi : A \to B$.

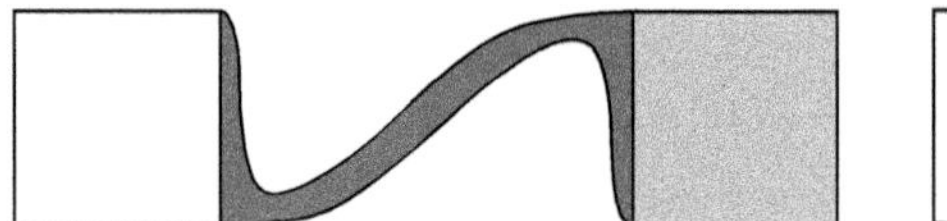

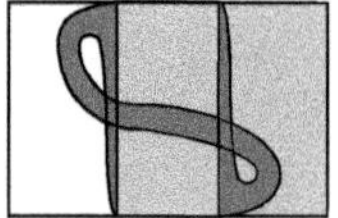

FIG. 7.5. The zig-zag immersion.

We identify A and B via the affine diffeomorphism ψ and hence consider the 2-form $\psi^*\tau \in \Omega^2(A)$. This 2-form is isotopic to $\phi^*\tau \in \Omega^2(A)$ through nondegenerate 2-forms and the isotopy is constant on $\{0, (2N+3)\varepsilon\} \times [-\varepsilon, \varepsilon]^{2n-1}$. We now claim that $\phi^*\tau$ is isotopic through nondegenerate 2-forms, modulo $\{0, (2N+3)\varepsilon\} \times [-\varepsilon, \varepsilon]^{2n-1}$, to an exact 2-form. To see this consider the 2-forms

$$\tau_j := \phi_j^*\tau_{x_j} \in \Omega^2(A_j).$$

By definition of τ_{x_j} these 2-forms are exact, namely

$$\tau_j = d\sigma_j, \qquad \sigma_j := \phi_j^*\sigma_{x_j} \in \Omega^1(A_j).$$

By our choice of $\varepsilon > 0$ each 2-form τ_j is isotopic to the restriction of $\phi^*\tau$ to A_j. Now the constant δ was chosen so small that σ_j and σ_{j+1} are arbitrarily close on their overlap domain $[(2j+2)\varepsilon, (2j+3)\varepsilon] \times [-\varepsilon, \varepsilon]^{2n-1}$ and hence can be glued without losing nondegeneracy. More precisely, define $\sigma \in \Omega^1(A)$ by

$$\sigma|_{A_j} := \begin{cases} \phi^*\sigma_{x_{j-1},x_j} & \text{on } [2j\varepsilon, (2j+1)\varepsilon] \times [-\varepsilon, \varepsilon]^{2n-1}, \\ \phi^*\sigma_{x_j} & \text{on } [(2j+1)\varepsilon, (2j+2)\varepsilon] \times [-\varepsilon, \varepsilon]^{2n-1}, \\ \phi^*\sigma_{x_j,x_{j+1}} & \text{on } [(2j+2)\varepsilon, (2j+3)\varepsilon] \times [-\varepsilon, \varepsilon]^{2n-1}. \end{cases}$$

Then σ is arbitrarily close to σ_j on A_j and hence $d\sigma$ is isotopic to $\phi^*\tau$ through nondegenerate 2-forms on A as claimed. Since A is diffeomorphic to B via ψ we deduce that τ is isotopic through nondegenerate 2-forms to an exact 2-form on $[-1, 1] \times [-\varepsilon, \varepsilon]^{2n-1}$ and the isotopy is constant on $\{-1, 1\} \times [-\varepsilon, \varepsilon]^{2n-1}$. One can now use a final isotopy to replace $[-\varepsilon, \varepsilon]^{2n-1}$ by $[-1, 1]^{2n-1}$.

Thus we have proved the lemma for $m = 1$ and in the unparametrized case. The extension of the proof to a nontrivial parameter space Λ is obvious. Moreover, the general case $m > 1$ can be reduced to the case $m - 1$ by the same argument which proves the case $m = 1$. The details are left to the reader. $\square$

The techniques of the above proof were developed by Gromov in his thesis [286]; a more elaborate version appeared in his book [288]. The version of the argument presented here was explained to us by Eliashberg. Other expositions can be found in Haefliger [303], Poenaru [526], Eliashberg–Mishachev [190]. The main ingredient of the proof can be expressed in terms of the sheaf $\mathcal{S}$, defined by

$$\mathcal{S}(U) := \{\alpha \in \Omega^1(U) \,|\, (d\alpha)^n \neq 0\}$$

for every open set $U \subset M$. Thus the set $\mathcal{S}(U)$ consists of all 1-forms on U such that $d\alpha$ is nondegenerate. We have used the following crucial facts.

(i) The condition is open, i.e. $\mathcal{S}(U)$ is an open subset of $\Omega^1(U)$.

(ii) The condition is natural, i.e. $\alpha \in \mathcal{S}(U)$ if and only if $\phi^*\alpha \in \mathcal{S}(\phi^{-1}(U))$ for every diffeomorphism $\phi : M \to M$.

Any other sheaf with these properties gives rise to a result similar to Theorem 7.3.1. This can be used to solve many interesting problems that can be expressed in terms of partial differential equations and concern the existence of geometric structures such as embeddings, foliations, contact structures, and divergence-free vector fields on open manifolds. The point is that an open manifold M has a handle decomposition with handles of index less than its dimension, and Gromov shows how to extend these structures over such handles. His constructions are geometric and, as in the proof of Lemma 7.3.2 above, require that there be an extra dimension to play with. Therefore, in many situations they fail for closed manifolds, since these must be completed by the addition of a final handle of top index. Closed manifolds form a borderline case, and interesting obstructions to existence often arise. If there are no such obstructions, one says that the h-**principle** holds, where the letter 'h' stands for homotopy; if there is existence for all k-dimensional families of objects then the parametric h-principle holds.

Such questions are discussed in many different geometric contexts in Gromov's book [288]. The book [190] by Eliashberg–Mishachev develops a simpler approach to many of Gromov's constructions, while Gromov's article [289] points out how the hard–soft dichotomy in symplectic geometry is a reflection of this phenomenon. Murphy's notion [500] of looseness in contact geometry led to a notion of **flexible Weinstein domain** (cf. Definition 7.4.5) that gives a new tool in trying to understand the demarcation between soft and hard in symplectic geometry. In particular, in [191] Eliashberg and Murphy develop an h-principle for embeddings of flexible Weinstein domains, that they use in [192] to construct symplectic structures on cobordisms with nonempty positive boundary and overtwisted negative boundary. This leads to the following sharpening of Gromov's existence result.

Example 7.3.3 Suppose that (M^{2n}, J, a) is a closed, connected, almost complex manifold with $a \in H^2(M;\mathbb{R})$ such that $a^n > 0$, and let p_0 be any point in M. In dimensions greater than four it is unknown whether such M must support a symplectic structure in class a. (This is false in dimension four: see

Example 13.3.14.) However, Gromov constructs a symplectic form ω on $M \setminus \{p_0\}$ in the class a that is tamed by some almost complex structure homotopic to J. One might hope to construct such a form that is controlled near p_0 in the sense that near p_0 it equals the negative symplectization of an overtwisted contact structure on S^{2n-1} in the standard homotopy class of plane fields (see Chapter 3.5 for relevant definitions). So far, such a claim is out of reach. However, in [192] Eliashberg–Murphy show that *if $n \geq 3$ and we remove a second point p_∞, choosing J on $M \setminus \{p_\infty\}$ to agree near p_∞ with the standard structure near infinity on $\mathbb{C}^n$, then we can choose the symplectic form to be standard near p_∞ and controlled as described above near p_0.*

Note that by part (ix) of Remark 4.5.2 the above statement would be false in general if one just removed the point p_∞. □

7.4 Donaldson submanifolds

In 1994 Donaldson proved the following remarkable existence theorem for symplectic submanifolds. Symplectic manifolds for which $[\omega]$ satisfies the condition below are called **integral symplectic manifolds.**

Theorem 7.4.1 (Donaldson) *Let (M, ω) be a closed symplectic $2n$-manifold and suppose that the cohomology class $[\omega] \in H^2(M; \mathbb{R})$ admits an integral lift. Then, for every sufficiently large integer k and every integral lift $a \in H^2(M; \mathbb{Z})$ of $[\omega]$, there exists a connected codimension-2 symplectic submanifold $Z_k \subset M$ that represents the Poincaré dual of the cohomology class ka. For large enough k this submanifold can be chosen such that the homomorphism $\pi_i(Z_k) \to \pi_i(M)$ is bijective for $0 \leq i \leq n-2$ and is surjective for $i = n-1$.*

Proof: See Donaldson [146] and also Auroux [36]. □

In complex geometry, when (M, ω, J) is a Kähler manifold, this result is well-known, and the submanifold Z_k can in fact be chosen to be complex.[33] In this case the proof goes along the following lines. Choose a holomorphic line bundle $L \to M$ with first Chern class $c_1(L) = a$. Then for large k the kth power $L^{\otimes k} := L \otimes \cdots \otimes L$ has 'many' holomorphic sections. If k is sufficiently large then one of these sections, say $s_k : M \to L^{\otimes k}$, is transverse to the zero section and the required complex submanifold is the intersection of s_k with the zero section

$$Z_k := \{q \in M \mid s_k(q) = 0\}.$$

Details of this argument can be found in [285]. (See also Remark 7.4.4 (ii) below.)

[33] One subtlety here is that the integrality condition means much more in the Kähler setting. Because nondegeneracy is an open condition, one can always slightly perturb a symplectic form to make its class rational and then rescale to make it integral. However, by Kodaira's embedding theorem, a (closed) Kähler manifold with an integral symplectic form is projective algebraic, i.e. a complex submanifold of a high-dimensional complex projective space, and therefore must contain many holomorphic curves. It follows that many Kähler manifolds are not projective: for example, any 4-torus that contains no compact holomorphic curves. In [650] Voisin proved the much stronger result that there are Kähler manifolds in any complex dimension ≥ 4 whose complex structure cannot be deformed to be projective; see also Voisin [651].

This reasoning must obviously fail in the almost complex case. Firstly, there is no such thing as a holomorphic line bundle over M, because an almost complex manifold will in general not support any nonconstant pseudoholomorphic functions, even on small open sets. So instead of a holomorphic vector bundle one might consider a complex line bundle with Chern class $c_1(L) = a$, and fix a Hermitian structure on L and a Hermitian connection $D : C^\infty(M, L) \to \Omega^1(M, L)$. Even in the almost complex case this operator splits into $D = \partial + \bar\partial$, where

$$\partial s := \tfrac{1}{2}(Ds - iDs \circ J), \qquad \bar\partial s := \tfrac{1}{2}(Ds + iDs \circ J)$$

denote the complex linear and complex anti-linear parts of Ds for $s \in C^\infty(M, L)$. But in general the bundle $L^{\otimes k}$ will have no pseudoholomorphic sections (i.e. sections $s : M \to L^{\otimes k}$ that satisfy $\bar\partial s = 0$) for any k. Now Donaldson's strategy in the proof of Theorem 7.4.2 is to find a section $s = s_k : M \to L^{\otimes k}$ such that

$$|\bar\partial s| < |\partial s| \tag{7.4.1}$$

pointwise for every $q \in M$ with $s(q) = 0$. Here ∂ and $\bar\partial$ are defined in terms of the induced connection on $L^{\otimes k}$. If such a section has been found then the first assertion in Theorem 7.4.1 follows from the next lemma.

Lemma 7.4.2 *Let $L \to M$ be a complex line bundle with Chern class $c_1(L) = a$. Assume L carries a Hermitian structure and let $D = \partial + \bar\partial$ be a Hermitian connection. If a section $s : M \to L$ satisfies (7.4.1) on its zero set then s is transverse to the zero section and the codimension-2 submanifold*

$$Z := \{q \in M \mid s(q) = 0\}$$

is symplectic and represents the cohomology class a.

Proof: Assume that L carries a Hermitian structure and let D be a Hermitian connection. Choose local coordinates $z = (x_1, \ldots, x_n, y_1, \ldots, y_n)$ on M such that the almost complex structure J is standard at $z = 0$ and the section $s : U \to \mathbb{C}$ (in a Hermitian trivialization) vanishes at $z = 0$. Then the complex linear and anti-linear 1-forms ∂s and $\bar\partial s$ at the point $z = 0$ are independent of the choice of the connection D and are given by the standard formulas

$$\partial s(0) = \sum_{j=1}^n a_j dz_j, \qquad \bar\partial s(0) = \sum_{j=1}^n b_j d\bar z_j,$$

where

$$a_j := \frac{\partial s}{\partial z_j}(0), \qquad b_j := \frac{\partial s}{\partial \bar z_j}(0).$$

Hence the tangent space of the submanifold N at zero is the subspace

$$T_0 Z = \left\{ \zeta \in \mathbb{C}^n \;\middle|\; \sum_j a_j \zeta_j + \sum_j b_j \bar\zeta_j = 0 \right\}.$$

The symplectic complement of this space is spanned, over the reals, by the vectors

$$u := (\bar{a}_1 - b_1, \ldots, \bar{a}_n - b_n), \qquad v := (i(\bar{a}_1 + b_1), \ldots, i(\bar{a}_n + b_n)).$$

Condition (7.4.1) shows that

$$\omega_0(u, v) = \sum_{j=1}^{n} \left(|a_j|^2 - |b_j|^2\right) > 0$$

and hence T_0Z is in fact a symplectic subspace of $\mathbb{C}^n$. Thus we have proved that Z is a symplectic submanifold. That Z represents the class $a = c_1(L)$ is a general fact about Chern classes. (See Remark 2.7.2.) □

The existence proof for a sequence of sections

$$s_k : M \to L^{\otimes k}$$

for large k that satisfy (7.4.1) on their zero sets is highly nontrivial and requires some subtle analysis. Donaldson first constructs exponentially decaying local sections which satisfy the required inequality on small balls. He then uses a patching argument and has to make sure that the cutoff functions do not destroy the inequality (7.4.1). This requires subtle estimates and a refinement of Sard's theorem. Geometrically, the effect of this construction is that the zero sets of s_k *are everywhere* or *fill out all of* M as k tends to infinity. More formally, the *currents*

$$k^{-1}Z_k, \qquad Z_k := s_k^{-1}(0),$$

converge to the 2-form ω. Details are given in [146]. The reader is also referred to [144] for a preliminary discussion of this subject, and to the paper [36] by Auroux in which Donaldson's construction is generalized to other bundles.

Remark 7.4.3 Donaldson proved his existence theorem for symplectic hypersurfaces in the spring of 1994, before the discovery of the Seiberg–Witten invariants. At the time the **generalized adjunction inequality**

$$2g(\Sigma) - 2 \geq \Sigma \cdot \Sigma + |c \cdot \Sigma| \tag{7.4.2}$$

for oriented embedded 2-dimensional submanifolds Σ with nonnegative self-intersection numbers in Kähler surfaces with

$$b^+ > 1$$

had been proved by Kronheimer–Mrowka [378, 379, 380] (with *Seiberg–Witten basic classes* replaced by *Donaldson basic classes* as defined by Kronheimer and Mrowka). It was also known that the canonical class of a Kähler surface is a basic class, and so Donaldson was able to prove the following.

If (X, ω, J) is a Kähler surface of general type (with $b^+ > 1$ and $[\omega] \cdot c < 0$, where $c := c_1(\omega)$ is minus the canonical class) then X does not admit a symplectic structure ω' with first Chern class c and $[\omega'] \cdot c > 0$.

If such a symplectic structure did exist we could assume, without loss of generality, that $[\omega']$ admits an integral lift. Theorem 7.4.1 would then, for large k, guarantee the existence of a connected symplectic submanifold

$$\Sigma_k \subset M$$

representing an integral lift of $k[\omega']$. This surface would have positive self-intersection and the genus would be given by the adjunction formula

$$2g(\Sigma_k) - 2 = \Sigma_k \cdot \Sigma_k - c \cdot \Sigma_k.$$

Since $c \cdot \Sigma_k = k\, c \cdot [\omega'] > 0$ this would violate the inequality (7.4.2).

This corollary was the first topological restriction on almost complex structures which are compatible with some symplectic form in a given cohomology class. In view of Proposition 4.1.1, this result can also be interpreted as a restriction on the cohomology classes of symplectic forms within a given connected component of the space of nondegenerate 2-forms. □

The application of Theorem 7.4.1 in Remark 7.4.3 was later superseded by the Seiberg–Witten invariants. It has been strengthened to the effect that the only symplectic 4-manifolds that satisfy

$$K \cdot [\omega] < 0$$

are blowups of rational or ruled surfaces (see page 531). However, Theorem 7.4.1 has many other applications, some of which are discussed below.

Remark 7.4.4 **(i)** In four dimensions, Theorem 7.4.1 can be interpreted as an existence statement for pseudoholomorphic curves (of high genus) and can be used in conjunction with Gromov's techniques of pseudoholomorphic curves in symplectic manifolds (cf. [287] and [470]). But, because one has no control over the parameter k, in the end it gives rather little useful direct information about these curves, especially when compared to Taubes–Seiberg–Witten theory. This is not surprising: one should think of Donaldson's result as an expression of symplectic flexibility, and hence not expect it would give much insight into such rigid phenomena as the behaviour of holomorphic curves.

(ii) As in algebraic geometry, one can interpret the submanifold Z_k in Theorem 7.4.1 geometrically as a hyperplane section. To see this, choose sections

$$s_0, \dots, s_N$$

of the line bundle $L^{\otimes k}$ and consider the associated map

$$M \to \mathbb{C}\mathrm{P}^N : x \mapsto [s_0(x) : \cdots : s_N(x)].$$

This will be an embedding if we choose sufficiently many of the s_i, and will be symplectic provided that the s_i satisfy a suitable refinement of (7.4.1). The zero set of any section

$$s = \sum \mu_i s_i$$

can be interpreted as the intersection of the image of M in $\mathbb{C}\mathrm{P}^N$ with the corresponding hyperplane. □

There have been many applications of Theorem 7.4.1. Here we concentrate on two of them, first discussing the decomposition of

$$M = (M \setminus Z) \cup Z$$

provided by a single hypersurface Z, and second describing an extension to families of hypersurfaces.

The hyperplane decomposition

The following definitions are due to Cieliebak–Eliashberg [111]. A smooth function $f : W \to \mathbb{R}$ on a manifold W is called a **generalized Morse function** if each critical point of f is either nondegenerate or embryonic. Here a critical point p is called **embryonic** if $f = c + x_1^3 - \sum_{i=2}^k x_i^2 + \sum_{i=k+1}^{2n} x_i^2$ in suitable coordinates near p. (Such a critical point is also called a **birth-death singularity**.)

Definition 7.4.5 *A* **Weinstein manifold** *is a quadruple* (W, ω, X, f)*, consisting of a noncompact connected symplectic manifold* (W, ω) *without boundary, a generalized Morse function* $f : W \to \mathbb{R}$ *that is bounded below and proper, and a complete Liouville vector field* X *on* W *that satisfies the following conditions.*

(a) *The inequality* $df \cdot X > 0$ *holds on* $W \setminus \mathrm{Crit}(f)$.

(b) *Choose a Riemannian metric on* W*. Then every critical point of* f *has a neighbourhood* U *in which the inequality*

$$df \cdot X \geq \delta(|X|^2 + |df|^2)$$

holds for some $\delta > 0$*. (This assertion is independent of the choice of the metric.)*

A **Weinstein domain** *consists of a compact connected symplectic manifold* (W, ω) *with boundary, a generalized Morse function* $f : W \to \mathbb{R}$ *that has no critical points on the boundary and satisfies* $f^{-1}(\max f) = \partial W$*, and a Liouville vector field* X *that satisfies (a) and (b).*

Remark 7.4.6 Every Weinstein domain (W, ω, f, X) is a symplectic manifold with convex contact type boundary ∂W with contact form $\alpha := -\iota(X)\omega|_{\partial W}$ (see Definition 3.5.32). If f is normalized so that $f(\partial W) = 0$ and $df \cdot X = 1$ near the boundary, it can be completed to a Weinstein manifold by attaching a cylindrical end $[0, \infty) \times \partial W$ with $\omega = -d(e^s\alpha)$, $X = \partial/\partial s$, and $f(s, q) = s$ (see page 148). Conversely, if (W, ω, f, X) is a Weinstein manifold and c is a regular value of f then $W^c := f^{-1}((-\infty, c])$ is a Weinstein domain. □

In Cieliebak–Eliashberg [111] it is shown that a Weinstein structure (ω, X, f) on W can be perturbed to one such that f is Morse. Assume f is Morse and denote by ϕ_X^t the flow of X. Then every critical point of f is nondegenerate, both as a critical point of f and as a zero of the vector field X. Hence the set

$$\Lambda_p := \left\{ q \in W \,\middle|\, I_q = \mathbb{R},\ \lim_{t\to\infty} \phi_X^t(q) = p \right\}$$

is a submanifold of W, called the **stable manifold of** p. Since X is a Liouville vector field, its flow enlarges the symplectic form. Since Λ_p is invariant under the flow of X and the flow converges to p, it follows that ω must vanish on Λ_p. Thus, in the Morse case, the stable manifolds of all the critical points are isotropic submanifolds of W, and therefore have dimensions at most n. It follows that the Morse indices of the critical points of f are less than or equal to the middle dimension of W. In particular, the boundary of each Weinstein domain of dimension $\dim W \geq 4$ is connected. This is in sharp contrast to Liouville domains (see Definition 3.5.32 and page 147).

Definition 7.4.7 *A function $f : W \to \mathbb{R}$ on a complex manifold (W, J) is called* **plurisubharmonic** *if the 2-form*

$$\omega_f := -d(df \circ J) \tag{7.4.3}$$

satisfies

$$\omega_f(v, Jv) > 0 \tag{7.4.4}$$

for every nonzero tangent vector $v \in TW$.

For any function $f : W \to \mathbb{R}$, the 2-form ω_f in (7.4.3) is of type $(1,1)$, i.e. it satisfies $\omega_f(J\cdot, J\cdot) = \omega_f$. Thus the taming condition (7.4.4) asserts that ω_f is a symplectic form compatible with J. The same argument as above shows that the critical points of a plurisubharmonic Morse function f can only have Morse indices up to the middle dimension and that their unstable manifolds are ω_f-isotropic. Another consequence of the plurisubharmonic condition is that the composition of a J-holomorphic curve with f is subharmonic (see [470, Lemma 9.2.10]) and hence cannot have any interior maximum. Thus (W, J) does not contain any nonconstant J-holomorphic curves defined on closed Riemann surfaces.

Definition 7.4.8 *A* **Stein manifold** *is a connected complex manifold (W, J) without boundary that admits a plurisubharmonic function $f : W \to \mathbb{R}$ that is bounded below and proper. A* **Stein domain** *is a complex manifold (W, J) with boundary that admits a plurisubharmonic function $f : W \to \mathbb{R}$ without critical points on the boundary satisfying $f^{-1}(\max f) = \partial W$.*

Stein manifolds can also be characterized as complex submanifolds of $\mathbb{C}^N$ that are closed as subsets of $\mathbb{C}^N$. A plurisubharmonic function is then given by $f(z) := \frac{1}{2}|z|^2$. This is the original notion of a Stein manifold. The equivalence of the two definitions was proved by Hans Grauert in 1958.

Remark 7.4.9 **(i)** A theorem of Eliashberg asserts that every Weinstein manifold admits a Stein structure and, conversely, that every Stein manifold (W, J) admits a symplectic form ω, a vector field X, and a Morse function $f : W \to \mathbb{R}$ such that the quadruple (W, ω, X, f) is a Weinstein manifold. In dimensions ≥ 6 (but not in dimension 4) the question of whether there is a Weinstein structure on a given Liouville domain (cf. Definition 3.5.32) is purely homotopy theoretic. There is also a notion of flexible Weinstein domain which, prompted by ideas in Murphy [500], has found many interesting applications. (See Cieliebak–Eliashberg [111] for proofs of these theorems and much more.)

(ii) An as yet unpublished theorem of Giroux asserts that for large enough k the Donaldson hypersurface $Z \subset M$ Poincaré dual to $k[\omega]$ can be chosen such that its complement $Z \subset M$ is a Weinstein domain (with a collar neighbourhood of the boundary attached).

(iii) In the Kähler case the decomposition of M into a complex hypersurface $Z \subset M$ and a Stein domain $W := M \setminus Z$ is a classical result in Kähler geometry. If s is a holomorphic section of a line bundle with zero set Z, then the function

$$f := -\log|s| : W \to \mathbb{R}$$

is plurisubharmonic. Hence, if f is a Morse function, its unstable manifolds are isotropic and they form the so-called **isotropic skeleton** or **Lagrangian skeleton** $\Delta \subset W$. In [68], Biran proved that the complement $M \setminus \Delta$ of the isotropic skeleton is a disc bundle over Z. This is the **Biran decomposition** of M. It was used by Biran in [65, 66, 67] to establish packing stability for all symplectic four-manifolds with integral symplectic forms, and in [68] to obtain intersection results for symplectically embedded balls and Lagrangian submanifolds (the *Lagrangian barrier* phenomenon). Another application is the Lagrangian circle bundle construction and its applications to questions of Lagrangian embeddings and the topology of Lagrangian submanifolds (see [69, 70, 72]). □

Symplectic Lefschetz fibrations

Definition 7.4.10 *Let (M, ω) be a closed symplectic 4-manifold. A* **symplectic Lefschetz fibration on** *M consists of a smooth function $\pi : M \to \mathbb{C}\mathrm{P}^1$ and an almost complex structure $J \in \mathcal{J}(M, \omega)$ satisfying the following conditions.*

(i) *The function π has finitely many critical values $t_1, \ldots, t_\ell \in \mathbb{C}\mathrm{P}^1$ and, for every $j \in \{1, \ldots, \ell\}$, there is a unique critical point $x_j \in \pi^{-1}(t_j)$. Moreover, each x_j is nondegenerate as a critical point of π.*

(ii) *The projection π is holomorphic and the almost complex structure J is integrable near each critical point x_j.*

(iii) *The symplectic form ω is nondegenerate on the vertical subspace* $\ker d\pi(x)$ *for every $x \in M \setminus \{x_1, \ldots, x_\ell\}$.*

Condition (iii) in Definition 7.4.10 implies that the restriction of the map π to $M \setminus \pi^{-1}(\{t_1, \ldots, t_\ell\})$ is a symplectic fibration whose fibres are Riemann surfaces. The symplectic form ω determines a symplectic connection. Moreover, conditions (i) and (ii) imply that in suitable local holomorphic coordinates the map π takes the form $(z_1, z_2) \mapsto z_1^2 + z_2^2$. Applying Exercise 6.3.7, we find that the holonomy around each singular fibre is a Dehn twist around some loop $\gamma \subset \Sigma$ in a regular fibre Σ. In fact it is a **positive Dehn twist**, where the positive condition means that a tubular neighbourhood of γ can be identified with the annulus $\mathbb{R}/2\pi\mathbb{Z} \times [-1, 1]$ by an orientation-preserving diffeomorphism in such a way that the Dehn twist is given by $(\theta, \eta) \mapsto (\theta + f(\eta), \eta)$, where $f'(\eta) \geq 0$ and $f(-1) = 0$, $f(0) = \pi$, and $f(1) = 2\pi$. This condition depends only on the orientation of the Riemann surface but not on the orientation of the circle.

Exercise 7.4.11 Prove that when $n = 2$ the holonomy in Exercise 6.3.7 is a positive Dehn twist in the sense described above. □

Now choose generators $\alpha_1, \ldots, \alpha_\ell$ of the fundamental group of the punctured 2-sphere $\mathbb{C}\mathrm{P}^1 \setminus \{t_1, \ldots, t_\ell\}$ with base point t_0 such that α_j encircles t_j precisely once. Then the holonomy around α_j determines a positive Dehn twist $\phi_j : \Sigma \to \Sigma$ of the regular fibre $\Sigma := \pi^{-1}(t_0)$. Since the composition of the loops α_j is contractible we obtain $\phi_1 \circ \phi_2 \circ \cdots \circ \phi_\ell \sim \mathrm{id}$, where $\sim$ means Hamiltonian isotopic. Thus each symplectic Lefschetz fibration gives rise to a relation of this form. (The precise relation depends on the choice of loops $\alpha_1, \ldots, \alpha_\ell$ and so is not unique; cf. Seidel [574].) Conversely, given any such relation one can build a corresponding Lefschetz fibration; see Gompf [280]. Donaldson [148] proved a version of the following theorem in arbitrary dimensions. However, for simplicity we state it only in dimension four.

Theorem 7.4.12 (Donaldson) *Let (M, ω) be a compact symplectic 4-manifold and suppose that the cohomology class $[\omega]$ admits an integral lift. Denote*

$$m := \int_M \omega \wedge \omega.$$

Then, for every sufficiently large integer k, the blowup $M \# mk^2 \overline{\mathbb{C}\mathrm{P}}^2$ admits the structure of a symplectic Lefschetz fibration

$$\pi : M \# mk^2 \overline{\mathbb{C}\mathrm{P}}^2 \to \mathbb{C}\mathrm{P}^1.$$

The general fibre is a Riemann surface Σ_k whose genus is given by

$$2g(\Sigma_k) - 2 = mk^2 + kK \cdot [\omega]$$

and the number of singular fibres is

$$\ell = 3mk^2 + 2kK \cdot [\omega] + \chi(M).$$

Proof: See Donaldson [148]. □

Theorem 7.4.12 is a refinement of Theorem 7.4.1. The main idea is to prove the existence of two sections s_0 and s_1 of the line bundle $L^{\otimes k} \to M$ that both satisfy (7.4.1). The sections must be chosen in such a way that the zero sets Σ_0 and Σ_1 intersect transversally with intersection number 1 at each intersection point. Then there are precisely $N = mk^2$ intersection points $x_1, \ldots, x_N$. The idea now is to blow up these points and define the projection

$$\pi : M\#N\overline{\mathbb{C}\mathrm{P}}^2 \to \mathbb{C}\mathrm{P}^1$$

by $\pi(x) = [s_0(x) : s_1(x)]$ for $x \in M \setminus \{x_1, \ldots, x_N\}$. The preimage of the point $t = [t_0 : t_1]$ under this map is the zero set of the section $s_t(x) = t_1 s_0(x) - t_0 s_1(x)$. The singular values of π are the points $t_j \in \mathbb{C}\mathrm{P}^1$ for which the curve Σ_{t_j} is not smooth. One of the difficulties in the proof is to show that s_0 and s_1 can be chosen such that each singular curve Σ_{t_j} has precisely one singular point $x_j \in \Sigma_{t_j}$ and that these singular points are nondegenerate critical points of π.

Example 7.4.13 Lefschetz fibrations are a well-known construct in algebraic geometry: see Example 7.1.8. To construct them for $M = \mathbb{C}\mathrm{P}^2$, choose two smooth degree d complex curves $\Sigma_0, \Sigma_1 \subset \mathbb{C}\mathrm{P}^2$ that intersect transversally. Then there is a complex 1-dimensional family of curves of degree d passing through the d^2 points $\Sigma_0 \cap \Sigma_1$. Explicitly, choose homogeneous polynomials P_0 and P_1 of degree d in the variables z_0, z_1, z_2 such that $\Sigma_j = \{P_j = 0\}$ for $j = 0, 1$. Then the 1-dimensional family of curves is given by $\Sigma_t := \{t_0 P_1 = t_1 P_0\}$ for $[t_0 : t_1] \in \mathbb{C}\mathrm{P}^1$. By the adjunction formula, the curves have genus $g(\Sigma_t) = \frac{(d-1)(d-2)}{2}$. In a generic such family, there will be $\ell = 3(d-1)^2$ values of t for which Σ_t is not smooth, and each such singular curve has a single double point. These curves form a Lefschetz fibration with total space $X = \mathbb{C}\mathrm{P}^2 \# d^2 \overline{\mathbb{C}\mathrm{P}}^2$. The case $d = 3$ gives the elliptic fibration of Exercise 7.1.7. Note also that if we do not blow up at the x_i then we obtain a **symplectic Lefschetz pencil**. □

Theorem 7.4.12 gives rise to a characterization of symplectic 4-manifolds in terms of finite collections of positive Dehn twists of a Riemann surface whose composition is the identity. By a gluing construction, every such collection of Dehn twists gives rise to a symplectic Lefschetz pencil (or fibration) of some symplectic 4-manifold, and the converse is the content of Theorem 7.4.12. The symplectic Lefschetz pencil on M is not unique but there is one for every sufficiently large integer k. To obtain invariants one has to introduce a suitable equivalence relation on these tuples of Dehn twists as explained in Donaldson's paper [148], with further calculation in, for example, Auroux–Katzarkov [43]. The upshot is that deformation classes of symplectic 4-manifolds correspond to certain equivalence classes of tuples of Dehn twists whose product is the identity. No one has yet succeeded in directly using this description to find new interesting invariants of symplectic 4-manifolds. Nevertheless, the fact that suitable blowups of symplectic 4-manifolds carry Lefschetz fibrations has been enormously influential. For example, it was used by Donaldson–Smith [162] to give a new proof of Taubes' theorem that for symplectic 4-manifolds with $b^+(M) > 1 + b_1(M)$ the canonical class $-c_1(M)$ can always be represented by a symplectically embedded submanifold. The versatility of the construction is illustrated in Auroux [41], which constructs many relatively uncomplicated but nonequivalent symplectic 4-manifolds with the same contact boundary. It also inspired the search for decompositions of 4-manifolds (including nonsymplectic manifolds and those with boundary) that relate more directly to the splittings of 3-manifolds used in Heegaard–Floer theory. One main example is the singular Lefschetz pencils or **broken fibrations** of Auroux–Donaldson–Katzarkov [45] and Lekili [406]. For a survey of symplectic Lefschetz fibrations and related open problems, see Donaldson [154].

Symplectic Lefschetz fibrations have also been useful in higher dimensions since Seidel [574] succeeded in understanding the effect of Dehn twists on invariants like the Fukaya category. Finally, we remark that Donaldson's almost complex techniques discussed in the present section were carried over to the contact setting by Giroux [271, 272], and used to prove that every closed contact manifold is supported by an open book decomposition (see Definition 3.5.41).

PART III

SYMPLECTOMORPHISMS

8

AREA-PRESERVING DIFFEOMORPHISMS

An essential feature of symplectic geometry is that there is a rich group of diffeomorphisms that preserve the symplectic structure. This part of the book develops the basic theory of such symplectomorphisms. We begin with the 2-dimensional case in which symplectomorphisms are just area- and orientation-preserving diffeomorphisms. These arise as Poincaré sections for Hamiltonian systems in a 4-dimensional phase space (in classical terminology with two degrees of freedom). A celebrated landmark in this subject is Poincaré's last geometric theorem, proved by Birkhoff in the 1920s, which asserts that an area-preserving twist map of the annulus must have at least two distinct fixed points. This result can be viewed as the first theorem in global symplectic geometry. We shall prove below the existence of one fixed point. We shall then discuss some application of these ideas to billiard problems. It turns out that the resulting twist maps satisfy a strong monotonicity property and therefore admit generating functions. Such twist maps are much easier to understand than general ones and should be compared to symplectomorphisms which are C^1-close to the identity.

The ideas presented in this chapter are only the beginnings of the subject of Hamiltonian dynamical systems. This is a rich and interesting field of mathematics which has been studied over the last three centuries by many great mathematicians, such as Newton, Weierstrass, Poincaré, Kolmogorov, and Siegel. For the interested reader we recommend the books by Abraham and Marsden [7], Siegel and Moser [584], Marsden [437], and Meyer and Hall [478] for further reading.

Much of the early work in Hamiltonian dynamics was motivated by the attempt to understand celestial mechanics and, for example, the dynamics of the planetary system. One important question to consider is whether such a system has periodic orbits, and this leads to problems such as the Weinstein conjecture discussed in Chapter 1. The dynamics near a periodic orbit can be described in terms of local symplectomorphisms called Poincaré section maps and we shall explain this in Section 8.1. Section 8.2 is devoted to the Poincaré–Birkhoff theorem and Section 8.3 to generating functions and the billiard problem.

8.1 Periodic orbits

Let $H : M \to \mathbb{R}$ be a Hamiltonian function on a $2n$-dimensional symplectic manifold (M, ω) and consider the Hamiltonian differential equation

$$\dot{p}(t) = X_H(p(t)). \tag{8.1.1}$$

Denote by $\phi_H^t : \Omega_t \to M$ the corresponding Hamiltonian flow and suppose that $p(t) = p(t+T) = \phi_H^t(p(0))$ is a nonconstant periodic solution of (8.1.1) on the energy surface $H = 0$. Choose any smooth function $G : M \to \mathbb{R}$ such that

$$G(p_0) = 0, \qquad \{G, H\}(p_0) \neq 0$$

for $p_0 := p(t_0)$ and consider the set

$$\Sigma := \{p \in M \,|\, H(p) = G(p) = 0\}.$$

If U is a sufficiently small neighbourhood of p_0 then $\Sigma \cap U$ is a smooth hypersurface of $H^{-1}(0)$. Moreover, there exists a unique smooth map $\tau : \Sigma \cap U \to \mathbb{R}$ such that

$$\phi_H^{\tau(p)}(p) \in \Sigma, \qquad \tau(p_0) = T.$$

To see this, we must solve the equation $G(\phi_H^t(p)) = 0$ for $t = \tau(p)$ in a neighbourhood of $p = p_0$. This is possible by the implicit function theorem if

$$\frac{d}{dt} G(\phi_H^t(p)) = \{G, H\}(p) \neq 0.$$

Now define the local diffeomorphism

$$\psi : \Sigma \cap U \to \Sigma : p \mapsto \phi_H^{\tau(p)}(p).$$

This is the **Poincaré section map** of the periodic solution $p(t)$. Note that fixed points of ψ correspond to periodic orbits of the Hamiltonian differential equation (8.1.1). In particular, $p_0 = \psi(p_0)$ is such a fixed point. The eigenvalues of the linearized map

$$d\psi(p_0) : T_{p_0}\Sigma \to T_{p_0}\Sigma$$

are called the **Floquet multipliers** of the periodic solution $p(t)$.

Lemma 8.1.1 *The Floquet multipliers of $p(t)$ are independent of the choice of the Poincaré section.*

Proof: Let $\Sigma' := \{p \in \Omega \,|\, G'(p) = H(p) = 0\}$ be another Poincaré section with $G'(p_0') = 0$, where $p_0' := p(t_0')$. Let U' be a sufficiently small neighbourhood of p_0' and let $\psi' : \Sigma' \cap U' \to \Sigma'$ be the corresponding Poincaré section map. For U' sufficiently small there is a natural diffeomorphism

$$\chi : \Sigma' \cap U' \to \Sigma$$

defined by $\chi(p) = \phi^{\tau(p)}(p)$ with $\tau(p_0') = t_0 - t_0'$. Then

$$\psi \circ \chi = \chi \circ \psi', \qquad \chi(p_0') = p_0.$$

Differentiating this identity at p_0' we obtain that the linearized maps $d\psi(p_0)$ and $d\psi'(p_0')$ are similar. □

Lemma 8.1.2 *The Poincaré section $\Sigma \cap U$ is a symplectic submanifold of M and the Poincaré section map $\psi : \Sigma \cap U \to \Sigma$ is a symplectomorphism.*

Proof: The hypersurface Σ is of dimension $2n-2$ and the tangent space at p is

$$T_p\Sigma = \{v \in T_pM \,|\, dG(p)v = dH(p)v = 0\}.$$

The condition $\{G, H\} = \omega(X_G, X_H) \neq 0$ shows that the 2-dimensional subspace spanned by $X_G(p)$ and $X_H(p)$ is a complement of $T_p\Sigma$. Now let $v \in T_p\Sigma$ and suppose that $\omega(v, w) = 0$ for all $w \in T_p\Sigma$. Then $\omega(X_H(p), v) = dH(p)v = 0$ and $\omega(X_G(p), v) = dG(p)v = 0$ and hence $v = 0$. Thus the 2-form ω is nondegenerate on the subspace $T_p\Sigma \subset \mathbb{R}^{2n}$.

To prove that ψ is a symplectomorphism we consider the 2-form

$$\omega_H := \omega + dH \wedge dt$$

on $\mathbb{R} \times M$. This is the **differential form of Cartan**. It has a 1-dimensional kernel consisting of those pairs $(\theta, v) \in \mathbb{R} \times T_pM$ which satisfy

$$v = \theta X_H(p).$$

Now let $\mathbb{D} \subset \mathbb{C}$ denote the closed unit disc in the complex plane and let $u : \mathbb{D} \to \Sigma$ be a 2-dimensional surface in Σ. We must prove that

$$\int_{\mathbb{D}} u^*\psi^*\omega = \int_{\mathbb{D}} u^*\omega.$$

To see this, consider the manifold with corners

$$\Omega := \{(t, z) \,|\, z \in \mathbb{D},\, 0 \leq t \leq \tau(u(z))\}$$

and define $v : \Omega \to \mathbb{R} \times M$ by

$$v(t, z) := \left(t, \phi^t(u(z))\right).$$

Denote $v_0(z) := v(0, z)$ and $v_1(z) := v(\tau(u(z)), z)$. Then $v_0^*\omega_H = u^*\omega$ and $v_1^*\omega_H = u^*\psi^*\omega$. Moreover, the tangent plane to the surface $v(\mathbb{R} \times \partial\mathbb{D})$ contains the kernel of ω_H. Hence the 2-form $v^*\omega_H$ vanishes on the surface $\mathbb{R} \times \partial\mathbb{D}$. Since ω_H is closed it follows from Stokes' theorem that

$$0 = \int_{\Omega} v^* d\omega_H = \int_{\partial\Omega} \omega_H = \int_{\mathbb{D}} u^*\psi^*\omega - \int_{\mathbb{D}} u^*\omega.$$

Hence ψ is a symplectomorphism. □

In particular, the previous lemma shows that $d\psi(p_0)$ is a symplectic linear transformation of the symplectic vector space $T_{p_0}\Sigma$. Hence λ is a Floquet multiplier of the periodic solution $p(t)$ if and only if λ^{-1} is a Floquet multiplier (see Lemma 2.2.2).[34] So the Floquet multipliers cannot all be of modulus less than 1 and this excludes asymptotic stability. The best possible notion of stability is that orbits starting sufficiently close to $p(t)$ remain near $p(t)$ indefinitely (in backward and forward time). A necessary condition for this is that all the Floquet multipliers lie on the unit circle. But this condition is by no means sufficient. The stability problem is extremely subtle and satisfactory results have so far only been found in the case of two degrees of freedom. These results require the existence of infinitely many 2-dimensional invariant tori 'surrounding' the periodic orbit. In the Poincaré section Σ these invariant tori appear as invariant circles. Existence results for such invariant circles are obtained from KAM theory and involve the hard implicit function theorem of Nash and Moser. We shall not pursue this direction here, but instead concentrate on the question of the existence of periodic orbits in the regions bounded by the invariant tori. These periodic orbits correspond to the fixed points of the Poincaré section map in the annulus region bounded by the corresponding invariant circles (see Fig. 8.1).

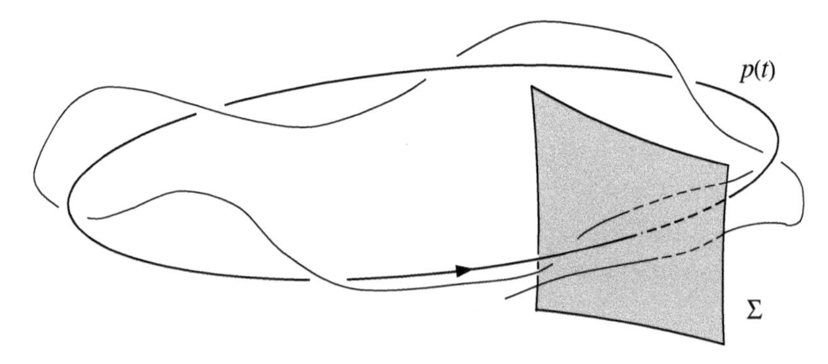

FIG. 8.1. A Poincaré section.

Exercise 8.1.3 Prove that two Poincaré section maps of the same periodic orbit are conjugate by a symplectomorphism. □

8.2 The Poincaré–Birkhoff theorem

In the case of two degrees of freedom the Poincaré section Σ is a 2-dimensional surface and we have seen that the Poincaré section map ψ preserves the area form ω. Now on any surface there exist, by Darboux's Theorem 3.2.2, local coordinates in which the area form agrees with the standard area form $\omega_0 = du \wedge dv$ (see also Exercise 3.2.8). Hence we can identify a neighbourhood of the fixed point p_0 in the surface Σ with an open neighbourhood of the origin in $\mathbb{R}^2$ in such a way that the area form ω on Σ corresponds to the standard area form $\omega_0 = du \wedge dv$ on $\mathbb{R}^2$.

[34] If $n = 2$ and hence the Poincaré section Σ is 2-dimensional then this assertion follows from the fact that $\det(d\psi(p_0)) = 1$.

In such coordinates the Poincaré section map is an area-preserving homeomorphism of an open neighbourhood of $0 \in \mathbb{R}^2$ with a fixed point at 0. If there exists an invariant circle near 0 then this invariant circle cuts out an invariant disc. Any two invariant circles cut out an annulus. Such an annulus can be identified with a standard annulus

$$A = \{(u,v) \in \mathbb{R}^2 \,|\, a \le u^2 + v^2 \le b\}$$

via an area-preserving transformation. **Poincaré's last geometric theorem** asserts that any area-preserving homeomorphism $\psi : A \to A$ which preserves the two boundary components and twists them in opposite directions must have at least two fixed points. This result was proved by Birkhoff in 1925, and so it is also known as the Poincaré–Birkhoff theorem.

It is convenient to introduce polar coordinates $x \in \mathbb{R}/\mathbb{Z}$ and $y > 0$ such that

$$u = \sqrt{y}\cos 2\pi x, \qquad v = \sqrt{y}\sin 2\pi x.$$

Then

$$du \wedge dv = -\pi dx \wedge dy,$$

and hence the homeomorphism ψ is still area-preserving in the coordinates x and y. In these coordinates the annulus is the strip $a \le y \le b$ and the homeomorphism ψ takes the form

$$\psi(x,y) = (f(x,y), g(x,y)),$$

where

$$f(x+1,y) = f(x,y) + 1, \qquad g(x+1,y) = g(x,y). \tag{8.2.1}$$

We assume that ψ preserves the two boundary components of A, i.e.

$$g(x,a) = a, \qquad g(x,b) = b, \tag{8.2.2}$$

and satisfies the **twist condition**

$$f(x,a) < x, \qquad f(x,b) > x. \tag{8.2.3}$$

Two points (x,y) and (x',y') are called **geometrically distinct** if either $y' \ne y$ or $x' - x \notin \mathbb{Z}$. We remark that any lift of a homeomorphism of the annulus to the universal cover $\mathbb{R} \times [a,b]$ satisfies (8.2.1) and that any lift that satisfies the twist condition (8.2.3) is orientation-preserving if and only if it preserves the two boundary components.

Theorem 8.2.1 (Poincaré–Birkhoff) *Let ψ be an area-preserving homeomorphism of the annulus $A := \{(x,y) \in \mathbb{R}^2 \,|\, a \le y \le b\}$ satisfying (8.2.1), (8.2.2), and (8.2.3). Then ψ has at least two geometrically distinct fixed points.*

Proof: See page 347 for the existence of one fixed point. □

Remark 8.2.2 The Poincaré–Birkhoff theorem is an archetypal example of a beautiful result in mathematics. It is easy to state, difficult to prove, none of the assumptions can be removed, and the conclusion is sharp. For example, the half rotation $\psi(x,y) := (x+1/2, y)$ is area- and orientation-preserving but violates the twist condition, the diffeomorphism $\psi(x,y) := (x+y-1/2, y/2+y^2/2)$ of the annulus $0 \le y \le 1$ is orientation-preserving, and satisfies the twist condition but is not area-preserving, and the diffeomorphism $\psi(x,y) := (x+y-1/3, 1-y)$ of the annulus $0 \le y \le 1$ is area-preserving and satisfies the twist condition but reverses the orientation. All three maps do not have any fixed points.

It is easy to construct an area-preserving diffeomorphism of the annulus which satisfies all the requirements of the Poincaré–Birkhoff theorem and has precisely two fixed points (see Fig. 8.2).

In Fig. 8.3 we illustrate the orbits of a typical twist map to give an idea of how complicated they can be. Notice the large chaotic regions. □

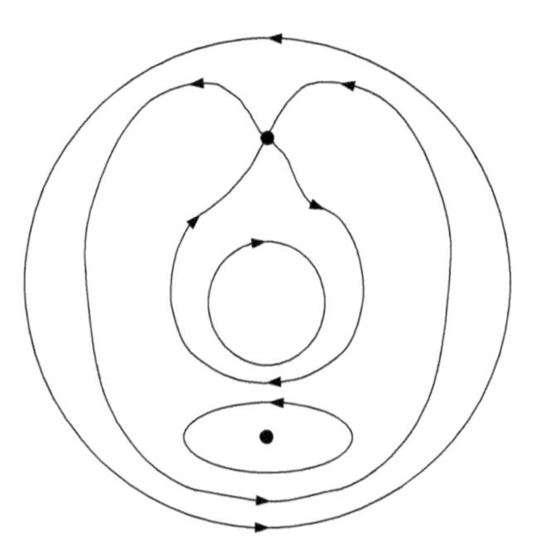

FIG. 8.2. A twist map with exactly two fixed points.

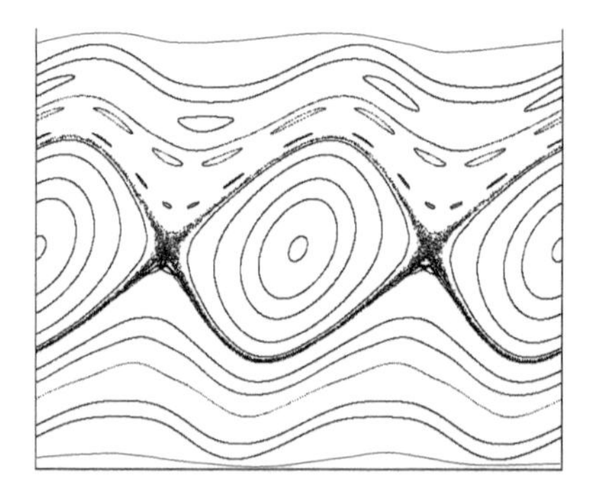

FIG. 8.3. Orbits of a twist map of the annulus, shown on the universal cover $\mathbb{R} \times [a,b]$.

The proof of Theorem 8.2.1 is elementary if ψ satisfies in addition the **monotone twist** condition

$$y < y' \qquad \Longrightarrow \qquad f(x,y) < f(x,y'). \tag{8.2.4}$$

If this holds then for every $x \in \mathbb{R}$ there exists a unique $y = w(x) \in (a,b)$ such that

$$f(x, w(x)) = x.$$

The map $w : \mathbb{R} \to \mathbb{R}$ is continuous and it follows from Equation (8.2.1) that $w(x+1) = w(x)$. By the area-preserving property the curve

$$\Gamma = \{(x, w(x)) \,|\, x \in \mathbb{R}\}$$

must intersect its image $\psi(\Gamma)$ at least twice (see Fig. 8.4). Each intersection point

$$(x, w(x)) \in \Gamma \cap \psi(\Gamma)$$

is a fixed point of ψ. Here we only need one of the conditions $g(x,a) = a$ or $g(x,b) = b$ but not both. This is related to the interpretation of ψ as a Poincaré section map with the fixed boundary component corresponding to the periodic solution $p(t)$.

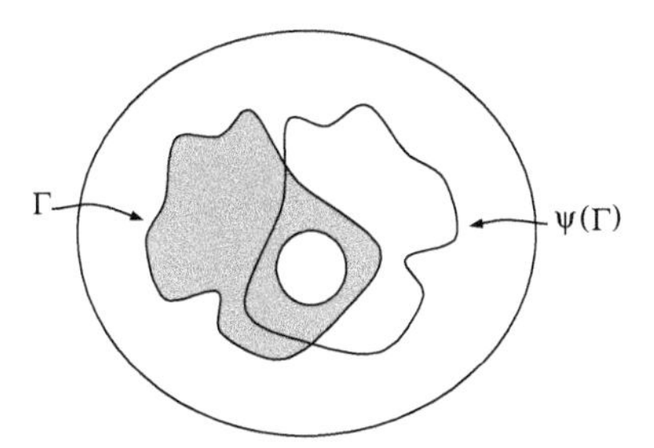

FIG. 8.4. The curves Γ and $\psi(\Gamma)$.

This beautiful argument fails completely if the monotone twist condition is not satisfied. It may be viewed as a *local* argument for homeomorphisms which are close to the standard map $\psi(x,y) = (x + y - c, y)$. In contrast, Poincaré's geometric theorem is a *global* result which asserts the existence of fixed points for every area-preserving twist homeomorphism regardless of whether or not it is close to any standard map.

Proof of Theorem 8.2.1: We shall only prove the existence of one fixed point. The second fixed point can be obtained by a simple modification of this argument, for which we refer to Birkhoff [74] and Brown and Neumann [80]. Franks [239] describes circumstances under which there are infinitely many fixed points.

The strategy of the proof is as follows. Assume that ψ has no fixed point. For any path $\gamma : [0,1] \to A$ with $\gamma(0) \in \mathbb{R} \times a$ and $\gamma(1) \in \mathbb{R} \times b$ we define the map

$$\rho_\gamma(t) := \frac{\psi(\gamma(t)) - \gamma(t)}{|\psi(\gamma(t)) - \gamma(t)|} \in S^1.$$

Then it follows from the twist condition that

$$\rho_\gamma(0) = -1, \qquad \rho_\gamma(1) = +1.$$

Choose any lift $\alpha_\gamma : [0,1] \to \mathbb{R}$ such that

$$\rho_\gamma(t) = e^{i\alpha_\gamma(t)}.$$

Then $\alpha_\gamma(1) - \alpha_\gamma(0) = (2k+1)\pi$ for some integer k. Now any two such paths γ_0 and γ_1 are homotopic by a homotopy γ_λ, and so are the corresponding lifts α_{γ_0} and α_{γ_1}. But the expression $\alpha_{\gamma_\lambda}(1) - \alpha_{\gamma_\lambda}(0)$ is independent of λ. Hence $\alpha_\gamma(1) - \alpha_\gamma(0)$ is either positive for every γ or negative for every γ. We define

$$\mu(\psi) = \operatorname{sign}(\alpha_\gamma(1) - \alpha_\gamma(0)) \in \{+1, -1\}.$$

This is well-defined whenever ψ is a continuous map which satisfies the twist condition (8.2.3) and has no fixed points. In the area-preserving case we shall prove that $\mu(\psi)$ must be both $+1$ and -1, giving the required contradiction.

We extend ψ to a homeomorphism of $\mathbb{R}^2$ by $f(x,y) := f(x,a)$ for $y \le a$, $f(x,y) := f(x,b)$ for $y \ge b$, and $g(x,y) := y$ in both cases. Note that ψ is area-preserving only on the annulus A. We assume that ψ has no fixed points in the annulus and hence no fixed points in $\mathbb{R}^2$.

Now define the translation

$$\tau_\varepsilon : \mathbb{R}^2 \to \mathbb{R}^2 : (x,y) \mapsto (x, y+\varepsilon).$$

Then the map $\psi_\varepsilon := \tau_\varepsilon \circ \psi$ is a homeomorphism of $\mathbb{R}^2$ which is area-preserving on the annulus A and has no fixed points for $\varepsilon > 0$ sufficiently small. Consider the strip

$$D_0 := \{(x,y) \in \mathbb{R}^2 \,|\, a \le y < a + \varepsilon\}.$$

Since the homeomorphism ψ_ε maps the upper half-plane $y \ge a$ into the smaller half-plane $y \ge a + \varepsilon$, we obtain that the strip D_0 does not intersect its images $D_j := {\psi_\varepsilon}^j(D_0)$ under the iterates of ψ_ε for any $j \ge 1$. Hence the strips D_j are pairwise disjoint. Also note that the closure of D_0 intersects D_1 in the upper boundary $y = a + \varepsilon$. Since ψ_ε is area-preserving, each strip D_j has area ε. Hence there must be an integer $k \ge 2$ such that $D_{k-1} \not\subset A$. Hence there exists a point $x_0 \in \mathbb{R}$ such that

$$y_k \ge b, \qquad (x_k, y_k) = {\psi_\varepsilon}^k(x_0, a).$$

We choose k to be the first integer with this property (see Fig. 8.5).

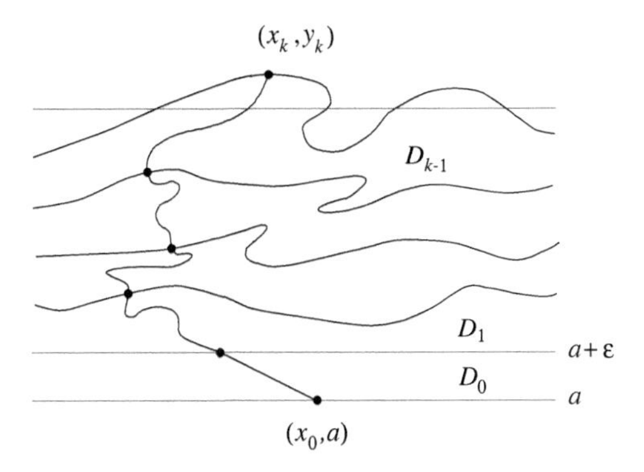

FIG. 8.5. The strips D_i and path γ.

Now define the path $\gamma : [0, k+1] \to \mathbb{R}^2$ by first connecting the point (x_0, a) with its image under ψ_ε by a straight line in D_0, i.e.

$$\gamma(t) := (1-t)(x_0, a) + t\psi_\varepsilon(x_0, a) \in D_0 \qquad \text{for } 0 \le t < 1,$$

and then iterating this path under ψ_ε, i.e.

$$\gamma(j+t) := {\psi_\varepsilon}^j(\gamma(t)) \in D_j \qquad \text{for } 0 \le t < 1.$$

Then $\gamma(t) \in A$ for $0 \le t \le k-1$ and $\gamma(k) = (x_k, y_k)$ with $y_k \ge b$. Let $T \in [k-1, k]$ be the first time such that $\gamma(t) \in \mathbb{R} \times b$ and define

$$\rho(t) := \frac{\psi_\varepsilon(\gamma(t)) - \gamma(t)}{|\psi_\varepsilon(\gamma(t)) - \gamma(t)|} = \frac{\gamma(t+1) - \gamma(t)}{|\gamma(t+1) - \gamma(t)|}, \qquad 0 \le t \le T.$$

Since the sets D_j are pairwise disjoint, it follows that $\gamma(s) \ne \gamma(t)$ for $s \ne t$. Hence ρ is homotopic to the path

$$\rho_0(t) := \begin{cases} \frac{\gamma(2t+1) - \gamma(0)}{|\gamma(2t+1) - \gamma(0)|}, & 0 \le t \le T/2, \\ \frac{\gamma(T+1) - \gamma(2t-T)}{|\gamma(T+1) - \gamma(2t-T)|}, & T/2 \le t \le T. \end{cases}$$

Now $\gamma(0) \in \mathbb{R} \times a$, $\gamma(T) \in \mathbb{R} \times b$, and $\gamma(t) \in A$ for $0 \le t \le T$ and hence $\rho_0(t)$ lies in the upper half-plane for every t. Choose a lift $\alpha_0(t) \in \mathbb{R}$ such that

$$\rho_0(t) = e^{i\alpha_0(t)}, \qquad \pi/2 < \alpha_0(0) < \pi.$$

Then $0 < \alpha_0(t) < \pi$ for $0 \le t \le T$ and, in particular,

$$\alpha_0(T) < \pi/2.$$

Note in fact that $\alpha_0(t)$ decreases from approximately π to approximately 0 as t runs from 0 to T. Taking the limit $\varepsilon \to 0$ we obtain $\mu(\psi) = -1$. A similar argument with τ_ε replaced by $\tau_{-\varepsilon}$ leads to a curve $\gamma(t)$, for which the vector $\rho_0(t)$ lies always in the lower half-plane and the corresponding lift $\alpha_0(t)$ increases from approximately $-\pi$ to approximately 0 as t runs from 0 to T. This results in the contradiction $\mu(\psi) = +1$ and hence ψ must have a fixed point. □

A **periodic point** of ψ is a fixed point of one of the iterates $\psi^q = \psi \circ \cdots \circ \psi$ of ψ.

Corollary 8.2.3 *Let ψ be an area-preserving homeomorphism of the annulus satisfying* (8.2.1) *and* (8.2.2), *and suppose that*

$$m = \max_x (f(x, a) - x) < \min_x (f(x, b) - x) = M.$$

Then ψ has infinitely many geometrically distinct periodic points.

Proof: Denote by $f^q(x, y)$ and $g^q(x, y)$ the first and second components of the qth iterate of ψ. Then

$$f^q(x, a) - x = \sum_{j=0}^{q-1} \left(f^{j+1}(x, a) - f^j(x, a) \right) \le qm,$$

$$f^q(x, b) - x = \sum_{j=0}^{q-1} \left(f^{j+1}(x, b) - f^j(x, b) \right) \ge qM.$$

If $q > 1/(M - m)$ then there exists an integer p such that

$$m < \frac{p}{q} < M.$$

This implies that

$$f^q(x, a) - p < f^q(x, a) - qm \le x \le f^q(x, b) - qM < f^q(x, b) - p.$$

Hence the qth iterate of ψ must have at least two fixed points for every integer $q > 1/(M - m)$. Now the periodic points corresponding to different rational numbers $p/q \ne p'/q'$ must be geometrically distinct. □

8.3 The billiard problem

The Poincaré–Birkhoff theorem has many important applications in Hamiltonian dynamical systems and in celestial mechanics (see Arnold [23], Meyer and Hall [478], Moser [493], and Siegel and Moser [584]). An interesting class of applications is to billiard problems. As we shall see below, these lead to area-preserving maps of the annulus which satisfy a strong monotone twist condition. Such maps admit a generating function and hence they are much easier to deal with than general area-preserving maps. We shall now explain this.

Generating functions

Let $\psi : (x_0, y_0) \mapsto (x_1, y_1)$ be an area-preserving diffeomorphism of the annulus $A := \mathbb{R}/\mathbb{Z} \times [a, b]$ given by

$$x_1 = f(x_0, y_0), \qquad y_1 = g(x_0, y_0).$$

In the smooth case ψ is area-preserving if and only if

$$\frac{\partial f}{\partial x_0}\frac{\partial g}{\partial y_0} - \frac{\partial f}{\partial y_0}\frac{\partial g}{\partial x_0} = 1. \tag{8.3.1}$$

We assume that ψ also satisfies (8.2.1), (8.2.2), (8.2.3), and the strong monotone twist condition

$$\frac{\partial f}{\partial y_0} > 0. \tag{8.3.2}$$

This implies that we can solve the equation $x_1 = f(x_0, y_0)$ for y_0. For convenience, we will consider ψ to be a diffeomorphism on the domain $\Omega := \mathbb{R} \times [a, b]$, which is the universal cover of A. Let U be the domain

$$U := \left\{(x_0, x_1) \in \mathbb{R}^2 \,|\, f(x_0, a) \le x_1 \le f(x_0, b)\right\}.$$

Lemma 8.3.1 *Under the above assumptions there exists a function $h : U \to \mathbb{R}$ such that, for all $(x_0, y_0) \in \Omega$ and $(x_1, y_1) \in \mathbb{R}^2$ with $(x_0, x_1) \in U$, we have*

$$(x_1, y_1) = \psi(x_0, y_0) \quad \Longleftrightarrow \quad y_0 = -\frac{\partial h}{\partial x_0}(x_0, x_1),\ y_1 = \frac{\partial h}{\partial x_1}(x_0, x_1).$$

Proof: By assumption the map

$$\Omega \to U : (x_0, y_0) \mapsto (x_0, f(x_0, y_0))$$

has a smooth inverse $U \to \Omega : (x_0, x_1) \mapsto (x_0, -u(x_0, x_1))$, where $u : U \to \mathbb{R}$. Consider the map $U \to \mathbb{R}^2 : (x_0, x_1) \mapsto (y_0, y_1)$ defined by

$$y_0 = -u(x_0, x_1), \qquad y_1 = v(x_0, x_1) := g(x_0, -u(x_0, x_1)).$$

Since ψ is a symplectomorphism, the 1-form

$$y_1 dx_1 - y_0 dx_0 = u dx_0 + v dx_1$$

on U is closed. Since U is simply connected this implies that there exists a function $h : U \to \mathbb{R}$ such that

$$\frac{\partial h}{\partial x_0} = u, \qquad \frac{\partial h}{\partial x_1} = v.$$

This proves the lemma. □

Lemma 8.3.1 shows that the area-preserving monotone twist diffeomorphism ψ is represented by the function $h : U \to \mathbb{R}$ via

$$(x_1, y_1) = \psi(x_0, y_0) \qquad \Longleftrightarrow \qquad y_0 = -\frac{\partial h}{\partial x_0}, \quad y_1 = \frac{\partial h}{\partial x_1}. \tag{8.3.3}$$

Differentiating the identity

$$\frac{\partial h}{\partial x_0}(x_0, f(x_0, y_0)) = -y_0$$

with respect to y_0 gives

$$\frac{\partial^2 h}{\partial x_0 \partial x_1} < 0. \tag{8.3.4}$$

In the following we shall use the notation

$$h_1 := \frac{\partial h}{\partial x_0}, \qquad h_2 := \frac{\partial h}{\partial x_1}, \qquad h_{12} := \frac{\partial^2 h}{\partial x_0 \partial x_1}.$$

Note that U is invariant under the translation $(x_0, x_1) \mapsto (x_0 + 1, x_1 + 1)$.

Lemma 8.3.2

$$h(x_0 + 1, x_1 + 1) = h(x_0, x_1). \tag{8.3.5}$$

Proof: First note that

$$u(x_0 + 1, x_1 + 1) = u(x_0, x_1), \qquad v(x_0 + 1, x_1 + 1) = v(x_0, x_1).$$

Hence

$$\frac{\partial}{\partial x_i}\left(h(x_0 + 1, x_1 + 1) - h(x_0, x_1)\right) = 0$$

for $i = 0, 1$, and so the function $h(x_0+1, x_1+1) - h(x_0, x_1)$ is constant. Integrating the 1-form dh along the curve $\gamma(t) = (x_0 + t, f(x_0 + t, a))$ we obtain

$$\begin{aligned} h(x_0 + 1, f(x_0, a) + 1) - h(x_0, f(x_0, a)) &= \int_\gamma dh \\ &= \int_\gamma \frac{\partial h}{\partial x_1} dx_1 + \int_\gamma \frac{\partial h}{\partial x_0} dx_0 \\ &= \int_\gamma y_1 dx_1 - \int_\gamma y_0 dx_0 \\ &= 0. \end{aligned}$$

The last identity holds because $y_0 = y_1 = a$ along the curve γ. □

A discrete variational problem

As we will see in more detail in Chapter 9, the generating function h determines a discrete-time variational problem as follows. Let

$$\mathbf{x} = (x_0, x_1, \ldots, x_\ell) \in \mathbb{R}^{\ell+1}$$

be a finite sequence of real numbers such that $(x_j, x_{j+1}) \in U$ for all j. Then there exists a sequence of real numbers $y_0, y_1, \ldots, y_\ell$ such that $a \leq y_j \leq b$ and $(x_j, y_j) = \psi^j(x_0, y_0)$ if and only if

$$h_2(x_{j-1}, x_j) + h_1(x_j, x_{j+1}) = 0. \tag{8.3.6}$$

The corresponding sequence of y_j's can be recovered from the x_j's via $y_j = h_2(x_{j-1}, x_j) = -h_1(x_j, x_{j+1})$. Now (8.3.6) is the Euler–Lagrange equation for the discrete variational problem

$$I_\ell(\mathbf{x}) := \sum_{j=0}^{\ell-1} h(x_j, x_{j+1})$$

subject to fixed boundary conditions at $j = 0$ and $j = \ell$. In other words, the solutions of equation (8.3.6) are the critical points of I_ℓ when it is considered as a function of $x_1, \ldots, x_{\ell-1}$. This variational formulation has led to many important results about periodic and quasi-periodic solutions known as Aubry–Mather theory: see, for example, Mather [441] and Golé [276]. In particular, it gives rise to yet another proof of Poincaré's geometric theorem in the monotone twist case: under the twist condition (8.2.3) the diagonal $x_0 = x_1$ is contained in the domain U of h. By (8.3.5) the function $x \mapsto h(x, x)$ is of period 1 and hence must have at least two critical points (a maximum and a minimum). Each critical point corresponds to a fixed point of ψ.

Exercise 8.3.3 Prove that every function $h : U \to \mathbb{R}$ which satisfies (8.3.4) and (8.3.5) determines an area-preserving monotone twist diffeomorphism ψ via Equation (8.3.3). Which condition on h corresponds to the invariance of the boundary circles? □

The billiard problem

Let $\gamma : \mathbb{R} \to \mathbb{R}^2$ be a smooth curve which is of period 1 and parametrized by arc-length:

$$\gamma(s+1) = \gamma(s), \qquad |\dot{\gamma}(s)| = 1.$$

We assume that the region B enclosed by γ is convex. This means that for each $s \in \mathbb{R}$ the tangent $\gamma(s) + t\dot{\gamma}(s)$ intersects the curve γ only at the point $\gamma(s)$. Now consider a ball rolling through B with constant velocity and bouncing off the boundary with the usual law of reflection (see Fig. 8.6). This determines a map on the space of pairs $(s, t) \in \mathbb{R}/\mathbb{Z} \times [0, \pi]$, where $\gamma(s)$ is the point at which the ball hits the boundary and $t \in [0, \pi]$ is the angle at which the ball leaves the point $\gamma(s)$.

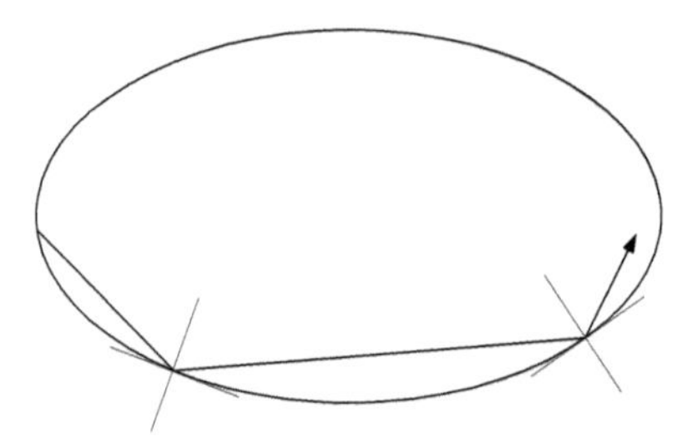

FIG. 8.6. A billiard trajectory.

Let the function $h : \mathbb{R}^2 \to \mathbb{R}$ be defined by

$$h(s_0, s_1) := -|\gamma(s_1) - \gamma(s_0)|$$

and let (s_0, t_0) and (s_1, t_1) be successive points on the orbit determined by the ball. Then

$$\frac{\partial h}{\partial s_0}(s_0, s_1) = -\left\langle \frac{\gamma(s_1) - \gamma(s_0)}{|\gamma(s_1) - \gamma(s_0)|}, \dot{\gamma}(s_1) \right\rangle = -\cos t_1,$$

and similarly

$$\frac{\partial h}{\partial s_1}(s_0, s_1) = \left\langle \frac{\gamma(s_1) - \gamma(s_0)}{|\gamma(s_1) - \gamma(s_0)|}, \dot{\gamma}(s_0) \right\rangle = \cos t_0.$$

If the curve γ is uniformly convex then we obtain

$$\frac{\partial^2 h}{\partial s_0 \partial s_1} = \sin t_1 \frac{\partial t_1}{\partial s_0} < 0,$$

since the angle t_1 is decreasing with s_0 (see Fig. 8.7). We conclude that h is the generating function of an area-preserving monotone twist diffeomorphism ψ on the annulus $\mathbb{R}/\mathbb{Z} \times [-1, 1]$ in the variables

$$x := s \in \mathbb{R}/\mathbb{Z}, \qquad y := -\cos t \in [-1, 1].$$

The correspondence between h and ψ is given by (8.3.3). So the functions f and g are defined by

$$f(s_0, -\cos t_0) := s_1, \qquad g(s_0, -\cos t_0) := -\cos t_1.$$

There is an ambiguity in the definition of f since s_1 is only determined up to an additive integer. We may choose this integer such that

$$f(x, -1) = x, \qquad f(x, 1) = x + 1.$$

Hence it follows from Corollary 8.2.3 that for every rational number $0 < p/q < 1$ there exists a periodic orbit $(x_j, y_j) = \psi^j(x_0, y_0)$ such that

$$x_{j+q} = x_j + p, \qquad y_{j+q} = y_j.$$

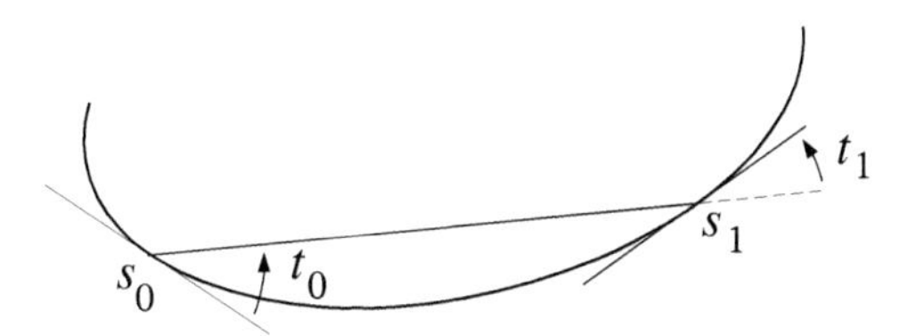

FIG. 8.7. The angles of reflection.

There has been much work on billiard problems from many different points of view. Besides containing many further references, the paper [20] by Alkoumi–Schlenk develops a method for finding the shortest closed billiard orbits on convex tables B as considered above, and shows that this invariant for B agrees with the Hofer–Zehnder capacity (see Section 12.4) of the domain $B \times D^2$ in $\mathbb{R}^4 = T^*\mathbb{R}^2$. Here we consider $B \subset \mathbb{R}^2$ as a subset of the base, and D^2 as the unit disc in the cotangent fibre.

Exercise 8.3.4 Give a geometric interpretation of these periodic orbits for different values of p and q. Use the corresponding variational problem to prove existence. □

Exercise 8.3.5 Consider the billiard problem in the domain enclosed by the ellipse

$$\Gamma := \left\{(x, y) \in \mathbb{R}^2 \,|\, \alpha x^2 + \beta y^2 = 1\right\}.$$

Find the associated monotone twist diffeomorphism. Prove that this billiard problem can be obtained as the limit of the geodesic flows on the ellipsoids

$$\Sigma_\varepsilon := \left\{(x, y, z) \in \mathbb{R}^3 \,|\, \alpha x^2 + \beta y^2 + \varepsilon z^2 = 1\right\}$$

as ε tends to zero. □

9

GENERATING FUNCTIONS

We have seen in Chapter 8 how generating functions can be used to prove existence theorems for periodic orbits of monotone twist maps. Some of the most exciting developments in symplectic geometry, such as the Arnold conjecture, the Weinstein conjecture, Gromov's nonsqueezing theorem, symplectic rigidity, and Hofer's metric, are centred around such existence statements. For example, we shall see in Chapter 11 that the Arnold conjecture is a natural generalization of the Poincaré–Birkhoff theorem, and that it provides an example of symplectic rigidity.

In this chapter we shall discuss generating functions in more detail. There are many interconnections between generating functions, symplectic action, Legendre transformation, the Hamilton–Jacobi equation, and variational principles. We shall discuss two classical generating functions for symplectomorphisms of $\mathbb{R}^{2n}$, which we shall call S and V. The function S is closely related to the symplectic action, but it exists only for symplectomorphisms which satisfy a certain nondegeneracy condition, such as the Legendre condition discussed in Section 1.1 or the monotone twist condition of Section 8.3. In contrast the function V exists for every symplectomorphism which is sufficiently close to the identity. Following Robbin and Salamon [540, 543] we show in Section 9.2 how generating functions of type V give rise to a discrete-time analogue of the symplectic action functional and hence lead to a discrete variational problem which does not require the aforementioned nondegeneracy condition. This will form the basis for our work in Chapters 11 and 12. In Section 9.3, these results are placed into the more general context of Hamiltonian symplectomorphisms on arbitrary symplectic manifolds. In the final section we examine generating functions for exact Lagrangian submanifolds in cotangent bundles.

9.1 Generating functions and symplectic action

Let $\psi : \Omega \to \Omega'$ be a symplectomorphism defined on a simply connected open set $\Omega \subset \mathbb{R}^{2n}$. Write the equation $(x_1, y_1) = \psi(x_0, y_0)$ in the form[35]

$$x_1 = u(x_0, y_0), \qquad y_1 = v(x_0, y_0).$$

Denote by $U \subset \mathbb{R}^{2n}$ the set of all vectors of the form $(x_0, u(x_0, y_0)) \in \mathbb{R}^{2n}$ for $(x_0, y_0) \in \Omega$ and assume that the map

[35] Warning: Here $x_0 = (x_{01}, \ldots, x_{0n}) \in \mathbb{R}^n$ is a vector and not a coordinate of a vector. Similarly for x_1, y_0, and y_1.

$$\Omega \to U : (x_0, y_0) \mapsto (x_0, u(x_0, y_0))$$

is a diffeomorphism. By the implicit function theorem, a symplectomorphism $\psi : \Omega \to \mathbb{R}^{2n}$ satisfies this condition in a neighbourhood of a point $z_0 = (x_0, y_0)$ if and only if

$$d\psi(z_0) = \begin{pmatrix} A & B \\ C & D \end{pmatrix}, \qquad \det(B) \neq 0. \tag{9.1.1}$$

This is the analogue of the Legendre condition (1.1.7). Geometrically, this condition means that the diffeomorphism ψ maps each fibre $\{x_0\} \times \mathbb{R}^n$ to a submanifold of $\mathbb{R}^{2n}$ which is transverse to the fibres $\{x\} \times \mathbb{R}^n$ (see Fig. 9.1). This is a generalization of the monotone twist condition (8.2.4).

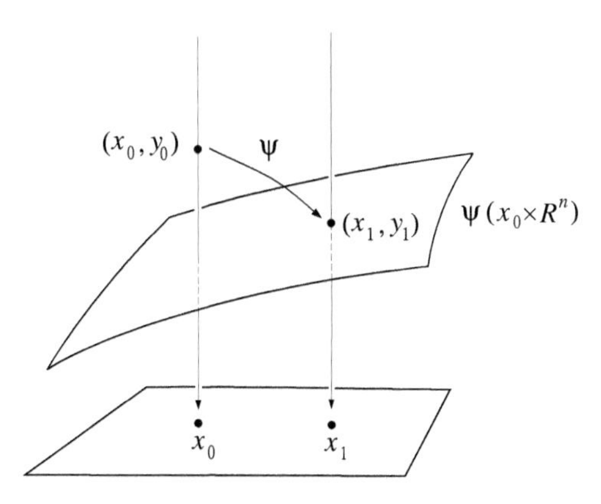

FIG. 9.1. For every pair x_0, x_1 there is a unique y_0 so that $\psi(x_0, y_0) \in x_1 \times \mathbb{R}^n$.

Lemma 9.1.1 *Under the above assumptions there exists a function $S : U \to \mathbb{R}$ such that, for all $(x_0, y_0) \in \Omega$ and $(x_1, y_1) \in \mathbb{R}^{2n}$ with $(x_0, x_1) \in U$, we have*

$$(x_1, y_1) = \psi(x_0, y_0) \quad \iff \quad y_0 = -\frac{\partial S}{\partial x_0}(x_0, x_1), \; y_1 = \frac{\partial S}{\partial x_1}(x_0, x_1).$$

Any function S with these properties is called a **generating function of type** S *for ψ.*

Proof: By assumption, the map $\Omega \to U : (x_0, y_0) \to (x_0, u(x_0, y_0))$ has a smooth inverse $U \to \Omega : (x_0, x_1) \mapsto (x_0, -f(x_0, x_1))$, where $f : U \to \mathbb{R}^n$. Consider the map $U \to \mathbb{R}^{2n} : (x_0, x_1) \mapsto (y_0, y_1)$ defined by

$$y_0 = -f(x_0, x_1), \qquad y_1 = g(x_0, x_1) := v(x_0, f(x_0, x_1)).$$

Since ψ is a symplectomorphism, the 1-form

$$\langle y_1, dx_1 \rangle - \langle y_0, dx_0 \rangle = \langle f, dx_0 \rangle + \langle g, dx_1 \rangle$$

on U is closed. Since U is simply connected this implies that there exists a function $S : U \to \mathbb{R}$ such that $f = \partial S / \partial x_0$ and $g = \partial S / \partial x_1$. This proves the lemma. □

Exercise 9.1.2 Prove the converse of Lemma 9.1.1. More precisely, let $U \subset \mathbb{R}^{2n}$ be an open set and $U \to \mathbb{R} : (x_0, x_1) \mapsto S(x_0, x_1)$ be a smooth function. Assume that the map

$$U \to \Omega : (x_0, x_1) \mapsto (x_0, -\partial_{x_0} S(x_0, x_1))$$

is a diffeomorphism onto an open set Ω. Prove that the map $\psi : \Omega \to \mathbb{R}^{2n}$ defined by

$$\psi(x_0, -\partial_{x_0} S(x_0, x_1)) := (x_1, \partial_{x_1} S(x_0, x_1))$$

is a symplectomorphism (with generating function S). □

Example 9.1.3 A linear symplectomorphism of the form

$$\Psi = \begin{pmatrix} A & B \\ C & D \end{pmatrix} \tag{9.1.2}$$

admits a generating function $S = S(x_0, x_1)$ if and only if $\det(B) \neq 0$. A generating function is given by

$$S(x_0, x_1) := \tfrac{1}{2}\langle x_0, B^{-1}Ax_0\rangle - \langle x_0, B^{-1}x_1\rangle + \tfrac{1}{2}\langle x_1, DB^{-1}x_1\rangle$$

for $x_0, x_1 \in \mathbb{R}^n$. □

We now show that if the symplectomorphism $\psi = \phi_H^{t_1,t_0}$, generated by a time-dependent family of Hamiltonian functions $H_t : \mathbb{R}^{2n} \to \mathbb{R}$, admits a generating function of type S then such a generating function is given by the action integral

$$\mathcal{A}_H(z) = \int_{t_0}^{t_1} \Big(\langle y, \dot{x}\rangle - H(t, x, y) \Big)\, dt \tag{9.1.3}$$

introduced in Chapter 1. Our assumption on ψ implies that for each pair (x_0, x_1) in a suitable domain $U = U_{t_0,t_1} \subset \mathbb{R}^n \times \mathbb{R}^n$ there is a unique pair of conjugate variables $(y_0, y_1) \in \mathbb{R}^n \times \mathbb{R}^n$ such that $\psi(x_0, y_0) = (x_1, y_1)$. In this case the curve

$$z(t) := (x(t), y(t)) := \phi_H^{t,t_0}(x_0, y_0)$$

is the unique solution of Hamilton's equations

$$\dot{x} = \frac{\partial H}{\partial y}, \qquad \dot{y} = -\frac{\partial H}{\partial x} \tag{9.1.4}$$

with boundary conditions

$$x(t_0) = x_0, \qquad x(t_1) = x_1. \tag{9.1.5}$$

This determines a smooth map $U \to C^\infty([t_0, t_1], \mathbb{R}^{2n}) : (x_0, x_1) \mapsto z$. Define

$$S_H(x_0, x_1) := \mathcal{A}_H(z) \tag{9.1.6}$$

for $(x_0, x_1) \in U$, where $z = (x, y) : [t_0, t_1] \to \mathbb{R}^{2n}$ is the unique solution of (9.1.4) and (9.1.5) and $\mathcal{A}_H(z)$ is given by (9.1.3).

Lemma 9.1.4 *The function S_H is a generating function for $\psi = \phi_H^{t_1,t_0}$.*

Proof: Choose $(x_0, x_1) \in U$ and $(\xi_0, \xi_1) \in \mathbb{R}^n \times \mathbb{R}^n$. For small $s \in \mathbb{R}$ denote by $z_s(t) = (x_s(t), y_s(t))$ the unique solution of Hamilton's equations (9.1.4) with boundary conditions $x_s(t_0) = x_0 + s\xi_0$ and $x_s(t_1) = x_1 + s\xi_1$. Differentiate the formula $S_H(x_0+s\xi_0, x_1+s\xi_1) = \mathcal{A}_H(z_s)$ with respect to s. From equation (1.1.9) we get

$$\frac{\partial S_H}{\partial x_0}\xi_0 + \frac{\partial S_H}{\partial x_1}\xi_1 = \langle y_1, \xi_1\rangle - \langle y_0, \xi_0\rangle,$$

where $y_j := y_s(t_j)$ for $s = 0$ and $j = 0, 1$. Since $\psi(x_0, y_0) = (x_1, y_1)$ this says that S_H is a generating function for ψ. □

A generating function for ψ is determined only up to an additive constant. To distinguish the generating function of Lemma 9.1.4 we may call it the **generating function determined by** H on the interval $[t_0, t_1]$ (cf. [540]).

Hamilton–Jacobi equation

Fix $(t_0, x_0) \in \mathbb{R} \times \mathbb{R}^n$ and let $S(t, x)$ denote the value at (x_0, x) of the generating function determined by H on the interval $[t_0, t]$ via (9.1.3), (9.1.4), (9.1.5), and (9.1.6). Then S satisfies the **Hamilton–Jacobi equation**

$$\partial_t S + H(t, x, \partial_x S) = 0. \tag{9.1.7}$$

To see this, let $I \to \mathbb{R}^{2n} : t \mapsto (x(t), y(t))$ be a solution of Hamilton's equation (9.1.4) with $x(t_0) = x_0$ and $(x_0, x(t)) \in U_{t_0,t}$ for all $t \in I$. Then

$$S(t, x(t)) = \int_{t_0}^{t} \Big(\langle y(s), \dot{x}(s)\rangle - H(s, x(s), y(s)) \Big)\, ds$$

for all $t \in I$. Differentiate this equation with respect to t and use $y = \partial S/\partial x$.

Solutions of the Hamilton–Jacobi equation play an important role in Hamiltonian dynamics. For example, the following exercise shows that in the time-independent case such solutions correspond to Lagrangian submanifolds which are invariant under the Hamiltonian flow. In the integrable case the invariant tori are examples of such invariant Lagrangian submanifolds. If, moreover, the Hamiltonian function arises from a Legendre transformation where the Lagrangian satisfies

$$\frac{\partial^2 L}{\partial v^2} > 0,$$

then the solutions of the Hamiltonian flow which lie on an invariant Lagrangian submanifold *minimize* the action $I(x) = \int L(x, \dot{x})\, dt$. We shall give a proof of this important and beautiful observation in Lemma 9.1.6 below. It can be used to prove the nonexistence of invariant tori. Moreover, since solutions of the Hamilton–Jacobi equation always exist locally (see Exercise 9.1.7 below), this fact can also be used to prove that geodesics locally minimize the energy.

Exercise 9.1.5 For every smooth function $S : \mathbb{R}^n \to \mathbb{R}$ define the set

$$L_S := \{(x, \partial_x S(x)) \,|\, x \in \mathbb{R}^n\} \subset \mathbb{R}^{2n}.$$

Prove that this set is a **Lagrangian submanifold** in the sense that the symplectic form ω_0 vanishes on the tangent bundle TL_S. Prove that L_S is invariant under the Hamiltonian flow of $H : \mathbb{R}^{2n} \to \mathbb{R}$ if and only if S satisfies the Hamilton–Jacobi equation

$$H(x, \partial_x S) = c \tag{9.1.8}$$

for some constant $c \in \mathbb{R}$. The analogous statement in the time-dependent case is that $S = S(t,x) = S_t(x)$ satisfies the Hamilton–Jacobi equation (9.1.7) (with 0 on the right hand side replaced by any constant) if and only if

$$\phi_H^{t_1,t_0}(L_{S_{t_0}}) = L_{S_{t_1}}$$

for all t_0 and t_1. □

Let $S : \Omega \to \mathbb{R}$ be a solution of the Hamilton–Jacobi equation (9.1.8), where $\Omega \subset \mathbb{R}^n$ and the Hamiltonian function $H : \Omega \times \mathbb{R}^n \to \mathbb{R}$ satisfies the Legendre condition

$$\frac{\partial^2 H}{\partial y^2} > 0. \tag{9.1.9}$$

Moreover, assume that for every $x \in \Omega$ the map $\mathbb{R}^n \to \mathbb{R}^n : y \mapsto \partial_y H(x,y)$ has a global inverse so that the inverse Legendre transformation gives rise to a Lagrangian $L : \Omega \times \mathbb{R}^n \to \mathbb{R}$. Define $f : \Omega \to \mathbb{R}^n$ by either of the following equivalent equations:

$$f(x) = \partial_y H(x, \partial_x S(x)), \qquad \partial_x S(x) = \partial_v L(x, f(x)). \tag{9.1.10}$$

Since S satisfies the Hamilton–Jacobi equation, it follows from Exercise 9.1.5 that if $\dot{x} = f(x)$ and $y(t) = \partial_x S(x(t))$ then $x(t)$ and $y(t)$ satisfy Hamilton's equations (9.1.4). Hence every solution of $\dot{x} = f(x)$ satisfies the Euler–Lagrange equations of L. In fact every such solution necessarily minimizes the action integral.

Lemma 9.1.6 *If $x : [t_0, t_1] \to \Omega$ satisfies $\dot{x} = f(x)$ and $\xi : [t_0, t_1] \to \Omega$ is any function such that $\xi(t_0) = x(t_0) = x_0$ and $\xi(t_1) = x(t_1) = x_1$, then*

$$\int_{t_0}^{t_1} L(x, \dot{x})\, dt \le \int_{t_0}^{t_1} L(\xi, \dot{\xi})\, dt.$$

Proof: Since the matrix $\partial^2 L/\partial v^2$ is positive definite there is an inequality

$$L(\xi, f(\xi)) + \langle \partial_v L(\xi, f(\xi)), v - f(\xi) \rangle \le L(\xi, v) \tag{9.1.11}$$

for all $\xi \in \Omega$ and $v \in \mathbb{R}^n$. Moreover, by definition of the Legendre transformation and the Hamilton–Jacobi equation, we have

$$L(x, f(x)) = \langle \partial_x S(x), f(x) \rangle - c.$$

Hence

$$\begin{aligned}\int_{t_0}^{t_1} L(x,\dot{x})\,dt &= \int_{t_0}^{t_1} \left(\langle \partial_x S(x), \dot{x}\rangle - c\right) dt \\ &= S(x_1) - S(x_0) - c(t_1 - t_0) \\ &= \int_{t_0}^{t_1} \Big(\langle \partial_x S(\xi), \dot{\xi} - f(\xi)\rangle + L(\xi, f(\xi))\Big)\,dt \\ &\le \int_{t_0}^{t_1} L(\xi,\dot{\xi})\,dt.\end{aligned}$$

The last step follows from (9.1.10) and (9.1.11). This proves Lemma 9.1.6. □

Exercise 9.1.7 Let $H : \Omega \times \mathbb{R}^n$ be a smooth Hamiltonian function. Let $\Gamma_0 \subset \Omega$ be a hypersurface and $S_0 : \Gamma_0 \to \mathbb{R}$ be smooth. A point $x^* \in \Gamma_0$ is called **regular** if there exists a vector $y^* \in \mathbb{R}^n$ such that

$$H(x^*, y^*) = c, \quad \partial_y H(x^*, y^*) \notin T_{x^*}\Gamma_0, \quad \langle y^*, \xi\rangle = dS_0(x^*)\xi \text{ for } \xi \in T_{x^*}\Gamma_0.$$

Prove that there exists a unique solution $S : U \to \mathbb{R}$ of the Hamilton–Jacobi equation (9.1.8) in a neighbourhood U of x^* such that $S|_{\Gamma_0 \cap U} = S_0$ and $\partial_x S(x^*) = y^*$. **Hint:** Prove first that there exists a unique function $f : \Gamma_0 \cap U \to \mathbb{R}^n$ such that $f(x^*) = y^*$ and the above holds for every $x_0 \in \Gamma_0 \cap U$ with y^* replaced by $y_0 = f(x_0)$. Define S by the formula

$$S(x(T)) = S_0(x_0) + \int_0^T \left(\langle y, \dot{x}\rangle - H(x,y)\right) dt,$$

where T is small and $x(t)$ and $y(t)$ satisfy Hamilton's equations with $x(0) = x_0 \in \Gamma_0 \cap U$ and $y(0) = f(x_0)$. Note in fact that $H(x(t), y(t)) = c$ for all t. Prove that with this notation $\partial_x S(x(T)) = y(T)$. □

Exercise 9.1.8 Generalize Lemma 9.1.6 to the time-dependent case. □

Exercise 9.1.9 Prove that geodesics locally minimize the energy. (See the discussion in Example 1.1.23.) □

Exercise 9.1.10 Find a Hamiltonian function of the form

$$H(x,y) = \frac{1}{2}|y|^2 + V(x)$$

for which there does not exist any global solution $S : \mathbb{R}^n \to \mathbb{R}$ of the Hamilton–Jacobi equation. Choose V such that $V(x+j) = V(x)$ for $j \in \mathbb{Z}^n$ and interpret the result in terms of the nonexistence of Lagrangian invariant tori. **Hint:** Choose V such that for some point $x_0 \in \mathbb{R}^n$ every solution $z(t) = (x(t), y(t))$ of (9.1.4) with $x(t_0) = x_0$ does not satisfy the minimality condition of Lemma 9.1.6, where

$$L(x,v) = \frac{1}{2}|v|^2 - V(x).$$

Note that any solution satisfies the minimality condition on a sufficiently small interval but it may not be minimal if the interval is chosen too large. □

Discrete-time variational problems

Just as one can think of a map as a discrete-time analogue of a flow, one can think of a symplectomorphism as a discrete-time analogue of a Hamiltonian flow. We introduced Hamiltonian flows as the solution to a variational problem. In an analogous way symplectomorphisms arise naturally from discrete-time variational problems. This is related to the generating function S of Lemma 9.1.1 as follows.

Assume that $\psi : \Omega \to \mathbb{R}^{2n}$ is a symplectomorphism with generating function $S : U \to \mathbb{R}$ and let

$$(x_j, y_j) = \psi^j(x_0, y_0), \qquad 1 \le j \le \ell,$$

be an orbit of length ℓ under ψ. It follows from Lemma 9.1.1 that this orbit is uniquely determined by the sequence

$$\mathbf{x} := (x_0, x_1, \ldots, x_\ell).$$

To see this, note that a given sequence $\mathbf{x}$ with $(x_j, x_{j+1}) \in U$ for all j corresponds to an orbit $(x_j, y_j) = \psi^j(x_0, y_0)$ of ψ which starts at $(x_0, y_0) \in \Omega$ and ends at (x_ℓ, y_ℓ) if and only if

$$\partial_2 S(x_{j-1}, x_j) + \partial_1 S(x_j, x_{j+1}) = 0 \tag{9.1.12}$$

for $j = 1, \ldots, \ell - 1$. Here, $\partial_1 S$ denotes the gradient vector of S with respect to the first n variables (previously called x_0) and $\partial_2 S$ denotes the gradient vector with respect to the last n variables (previously called x_1). If $\mathbf{x}$ satisfies (9.1.12) then the sequence $\mathbf{y} := (y_0, \ldots, y_\ell)$ can be recovered from $\mathbf{x}$ via

$$y_j = \partial_2 S(x_{j-1}, x_j) = -\partial_1 S(x_j, x_{j+1})$$

(see Fig. 9.2).

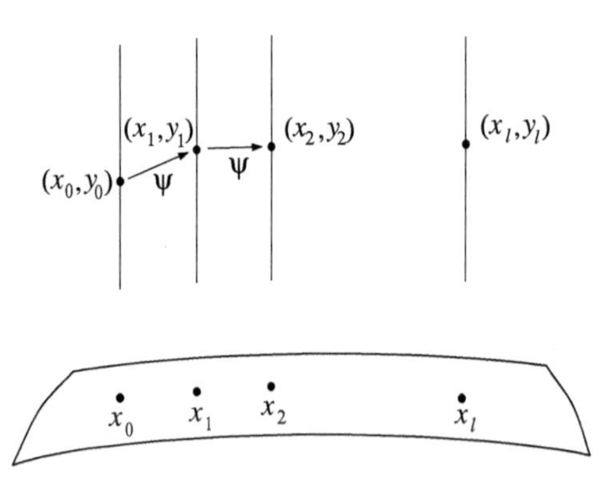

FIG. 9.2. The orbit is determined by $x_0, \ldots, x_\ell$.

Now (9.1.12) is the Euler–Lagrange equation for the discrete-time variational problem

$$I(\mathbf{x}) := \sum_{j=0}^{\ell-1} S(x_j, x_{j+1})$$

subject to variations $\mathbf{x} + \xi$, where $\xi = (\xi_0, \xi_1, \ldots, \xi_\ell)$ with $\xi_0 = \xi_\ell = 0$. In fact, this is the precise analogue of the Legendre transformation discussed at the beginning of Chapter 1. The sum $I(\mathbf{x})$ corresponds to the action integral $I(x) = \int L(x, \dot{x})\, dt$ in the continuous-time case and the discrete-time dynamical system $j \mapsto \psi^j$ corresponds to the Hamiltonian flow $t \mapsto \phi_H^t$ generated by (9.1.4) via the Legendre transformation. The Legendre condition (1.1.7) is the analogue of condition (9.1.1) which guarantees the existence of a generating function of type S. We have already encountered this variational formulation for the orbits of ψ in our discussion of monotone twist maps of the annulus in Chapter 8.

Exercise 9.1.11 Assume that the Hamiltonian function $H : \Omega \to \mathbb{R}$ satisfies (1.1.7). Prove that the Hamiltonian flow ϕ_H^t satisfies (9.1.1) for small nonzero t. □

9.2 Discrete Hamiltonian mechanics

We have seen above that the orbits under a symplectomorphism $\psi : \mathbb{R}^{2n} \to \mathbb{R}^{2n}$ can be expressed as the solutions of a discrete-time variational problem, provided that ψ admits a generating function of type S. Furthermore, the condition (9.1.1) for the existence of such a generating function is analogous to the Legendre condition (1.1.7) in the continuous-time case. Now, in the continuous-time case the symplectic action gives rise to an alternative variational formulation of Hamiltonian systems which does not require the Legendre condition (see Lemma 1.1.8). Similarly, in the discrete-time case, there is a variational formulation for the orbits of ψ even if ψ does not admit a generating function of type S. To make this work one has to assume that ψ is C^1-close to the identity. But every symplectic isotopy can be decomposed by sufficiently small time steps into a sequence of symplectomorphisms which satisfy this condition and so, in contrast to (9.1.1), it does not pose a serious restriction. As a result we obtain a discrete-time version of the symplectic action functional. This function was used by Chaperon [97] in his proof of the Conley–Zehnder theorem (see Chapter 11). It also appears in the work of Robbin and Salamon [540, 543], Givental [273], and Golé [276]. We shall see below that the discrete symplectic action functional can be a powerful tool in symplectic topology. Here is how this works.

Let $\psi : \mathbb{R}^{2n} \to \mathbb{R}^{2n}$ be a global symplectomorphism and, as before, write the equation $(x_1, y_1) = \psi(x_0, y_0)$ in the form

$$x_1 = u(x_0, y_0), \qquad y_1 = v(x_0, y_0).$$

If the first derivatives of ψ satisfy

$$\|d\psi(z) - \mathbb{1}\| \le \tfrac{1}{2} \tag{9.2.1}$$

for all $z \in \mathbb{R}^{2n}$, then condition (9.1.1) need not be satisfied. However, the matrix A in (9.1.1) will necessarily be nonsingular and it follows that there is another kind of generating function. The map $\mathbb{R}^{2n} \to \mathbb{R}^{2n} : (x_0, y_0) \mapsto (u(x_0, y_0), y_0)$ has a global inverse and hence we can use the coordinates $x_1 = u(x_0, y_0)$ and y_0 as independent variables replacing x_0 and y_0.

Lemma 9.2.1 *If ψ satisfies (9.2.1) then there is a smooth function $V : \mathbb{R}^{2n} \to \mathbb{R}$ such that $(x_1, y_1) = \psi(x_0, y_0)$ if and only if*

$$x_1 - x_0 = \frac{\partial V}{\partial y}(x_1, y_0), \qquad y_1 - y_0 = -\frac{\partial V}{\partial x}(x_1, y_0).$$

Conversely, if $V : \mathbb{R}^{2n} \to \mathbb{R}$ is a C^2-function with sufficiently small second derivatives, then these equations determine a symplectomorphism

$$\mathbb{R}^{2n} \to \mathbb{R}^{2n} : (x_0, y_0) \mapsto (x_1, y_1) = \psi(x_0, y_0).$$

Any function with these properties is called a **generating function of type** V.

Proof: By assumption there exists a smooth map

$$\mathbb{R}^{2n} \to \mathbb{R}^n : (x_1, y_0) \mapsto x_0 = f(x_1, y_0)$$

such that $x_0 = f(x_1, y_0) = f(u(x_0, y_0), y_0)$. for all $(x_1, y_0) \in \mathbb{R}^n \times \mathbb{R}^n$ (see Exercise 9.2.2 below). Define $g : \mathbb{R}^{2n} \to \mathbb{R}^n$ by

$$y_1 = g(x_1, y_0) := v(f(x_1, y_0), y_0).$$

Since ψ is a symplectomorphism the 1-form

$$\langle y_1, dx_1 \rangle + \langle x_0, dy_0 \rangle = \langle g, dx_1 \rangle + \langle f, dy_0 \rangle$$

is closed and hence there exists a function $W : \mathbb{R}^{2n} \to \mathbb{R}$ such that

$$x_0 = f(x_1, y_0) = \frac{\partial W}{\partial y}(x_1, y_0), \qquad y_1 = g(x_1, y_0) = \frac{\partial W}{\partial x}(x_1, y_0).$$

The function $V(x_1, y_0) := \langle x_1, y_0 \rangle - W(x_1, y_0)$ is as required. The proof of the converse statement is left to the reader. □

Exercise 9.2.2 Let $f : \mathbb{R}^n \to \mathbb{R}^n$ be a smooth function such that

$$\|df(x) - \mathbb{1}\| \le \tfrac{1}{2}$$

for all $x \in \mathbb{R}^n$. Prove that f has a global smooth inverse. **Hint:** Consider the fixed point problem $x = y + \phi(x)$, where $\phi = \mathrm{id} - f$. □

Example 9.2.3 A symplectic matrix $\Psi \in \mathrm{Sp}(2n)$ of the form (9.1.2) admits a generating function $V = V(x_1, y_0)$ if and only if $\det(A) \neq 0$. If this holds then a generating function is given by

$$V(x_1, y_0) = -\tfrac{1}{2}\langle x_1, CA^{-1}x_1 \rangle + \langle y_0, (\mathbb{1} - A^{-1})x_1 \rangle + \tfrac{1}{2}\langle y_0, A^{-1}By_0 \rangle$$

for $x_1, y_0 \in \mathbb{R}^n$. □

Note that the form of the difference equation in Lemma 9.2.1 is analogous to that of the Hamiltonian differential equation (9.1.4), with the derivatives $\dot{x}$ and $\dot{y}$ replaced by the differences $x_1 - x_0$ and $y_1 - y_0$, respectively. Thus the generating function V plays a role in discrete-time Hamiltonian mechanics similar to that of the Hamiltonian H in the continuous-time case. Now there are two ways to relate continuous-time Hamiltonian systems to discrete time. One is to begin with a symplectomorphism which is sufficiently close to the identity (for example by sampling a Hamiltonian flow) and then find the corresponding function V to construct a discrete-time Hamiltonian system. The second possibility is to begin with the function $V = \tau H$ in Lemma 9.2.1 for a sufficiently small number τ, iterate the resulting symplectomorphism ϕ_H^τ a large number of times (for example t/τ times), and let τ converge to zero. Then the symplectomorphisms $(\phi_H^\tau)^{t/\tau}$ will converge to the Hamiltonian flow ϕ_H^t. This direction was pursued by Robbin and Salamon [540] to give a rigorous interpretation of Feynman path integrals (with the symplectic action as the 'Lagrangian'), relating these to the Maslov index and the metaplectic representation. Here we shall discuss the first approach.

Let $H(t,x,y) = H_t(x,y)$ be a time-dependent Hamiltonian of class C^2 such that

$$\sup_{t\in\mathbb{R}} \sup_{z\in\mathbb{R}^{2n}} \left(\left\| d^2H_t(z) \right\|_{\mathbb{R}^{2n\times 2n}} + |dH_t(z)|_{\mathbb{R}^{2n}} \right) < \infty. \tag{9.2.2}$$

This implies that the solution operators $\phi_H^{t_1,t_0}$ of the Hamiltonian system (9.1.4) are globally defined for all t_0 and t_1. Fix $t_0 < t_1$ and a sufficiently large integer N and define

$$\psi = \phi_H^{t_1,t_0}, \qquad \psi_j = \phi_H^{\tau_{j+1},\tau_j}, \qquad \tau_j = t_0 + \frac{j}{N}(t_1 - t_0),$$

for $j = 0, \ldots, N-1$. Then

$$\psi = \psi_{N-1} \circ \psi_{N-2} \circ \cdots \circ \psi_0.$$

If N is sufficiently large, each symplectomorphism ψ_j satisfies the requirements of Lemma 9.2.1. Hence for every j there exists a smooth function $V_j : \mathbb{R}^{2n} \to \mathbb{R}$ that generates the symplectomorphism

$$(x_{j+1}, y_{j+1}) = \psi_j(x_j, y_j)$$

via the **Hamiltonian difference equations**

$$x_{j+1} - x_j = \frac{\partial V_j}{\partial y}(x_{j+1}, y_j), \qquad y_{j+1} - y_j = -\frac{\partial V_j}{\partial x}(x_{j+1}, y_j). \tag{9.2.3}$$

Conversely, it follows from Lemma 9.2.1 that, for any function $V_j : \mathbb{R}^{2n} \to \mathbb{R}$ with sufficiently small first and second derivatives, the equations (9.2.3) determine a symplectomorphism ψ_j.

Discrete symplectic action

It turns out that the solutions of (9.2.3) can be described by a variational problem just as in the continuous-time case. Denote by $\mathcal{P} = \mathbb{R}^{2nN+n}$ the set of *discrete paths* $\mathbf{z} = (x_0, \ldots, x_N, y_0, \ldots, y_{N-1})$, where x_j and y_j are real n-vectors. (See Fig. 9.3.) Define the **discrete symplectic action**

$$\Phi : \mathcal{P} \to \mathbb{R}$$

by the formula

$$\Phi(\mathbf{z}) := \sum_{j=0}^{N-1} \Big(\langle y_j, x_{j+1} - x_j \rangle - V_j(x_{j+1}, y_j) \Big) \tag{9.2.4}$$

for $\mathbf{z} \in \mathcal{P}$. Consider first the case of fixed boundary values x_0, x_N. Thus we think of Φ as a function of $x_1, \ldots, x_{N-1}, y_0, \ldots, y_{N-1}$. Denote the variations by $\zeta = (\xi_0, \ldots, \xi_N, \eta_0, \ldots, \eta_{N-1})$.

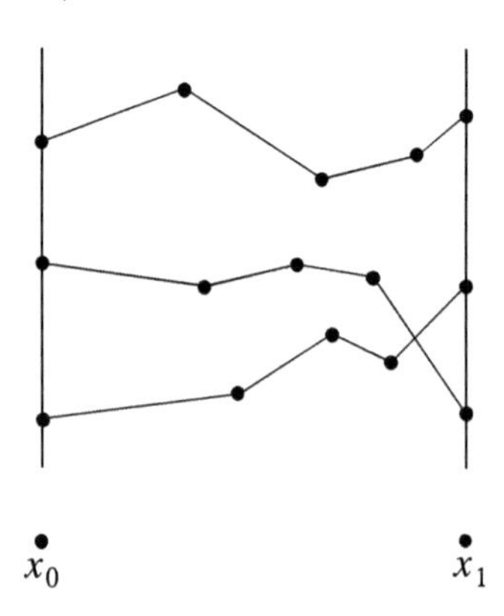

FIG. 9.3. Elements of $\mathcal{P}$.

Lemma 9.2.4 *A point $\mathbf{z} \in \mathcal{P}$ is critical for Φ with respect to variations of the form $\mathbf{z} + \zeta \in \mathcal{P}$ with $\xi_0 = \xi_N = 0$ if and only if $\mathbf{z}$ satisfies the Hamiltonian difference equations* (9.2.3) *(the first equation for $0 \leq j \leq N-1$ and the second for $0 \leq j \leq N-2$).*

Proof: The derivatives of Φ are given by

$$\frac{\partial \Phi}{\partial x_j} = y_{j-1} - y_j - \frac{\partial V_{j-1}}{\partial x}(x_j, y_{j-1}),$$

$$\frac{\partial \Phi}{\partial y_j} = x_{j+1} - x_j - \frac{\partial V_j}{\partial y}(x_{j+1}, y_j)$$

for $1 \leq j \leq N-1$. The second equation in (9.2.3) continues to hold for $j = 0$. □

Note that for any critical point $\mathbf{z} \in \mathcal{P}$ there is a unique vector $y_N \in \mathbb{R}^n$ determined by the second equation in (9.2.3) for $j = N-1$. Now assume that the symplectomorphism $\psi = \psi_{N-1} \circ \cdots \circ \psi_0$ admits a global generating function of type S. Then for every pair $(x_0, x_N) \in \mathbb{R}^{2n}$ there is a unique pair $(y_0, y_N) \in \mathbb{R}^{2n}$ such that $\psi(x_0, y_0) = (x_N, y_N)$. Hence the discrete path $(x_{j+1}, y_{j+1}) = \psi_j(x_j, y_j)$ is the unique solution of the Hamiltonian difference equation (9.2.3) with boundary values x_0 and x_N. This determines a smooth map

$$\mathbb{R}^{2n} \to \mathcal{P} : (x_0, x_N) \mapsto \mathbf{z}.$$

Consider the function $S : \mathbb{R}^{2n} \to \mathbb{R}$ defined by

$$S(x_0, x_N) := \Phi(\mathbf{z}), \tag{9.2.5}$$

where $\mathbf{z} \in \mathcal{P}$ is the unique solution of (9.2.3) with endpoints x_0, x_N and $\Phi(\mathbf{z})$ is defined by (9.2.4). As in the continuous-time case, this is a generating function for ψ of type S.

Exercise 9.2.5 Denote by $\mathcal{P}(x_0, x_N)$ the affine space of discrete paths with fixed endpoints and let $\mathbf{z} \in \mathcal{P}(x_0, x_N)$ be a critical point of

$$\Phi : \mathcal{P}(x_0, x_N) \to \mathbb{R}.$$

Prove that $\mathbf{z}$ is nondegenerate as a critical point of Φ if and only if

$$d\psi(z_0) = \begin{pmatrix} A & B \\ C & D \end{pmatrix}, \qquad \det(B) \neq 0,$$

i.e. ψ admits a generating function of type S in a neighbourhood of $z_0 = (x_0, y_0)$. □

Now consider the case of periodic boundary conditions $x_0 = x_N$. The subspace of those discrete paths $\mathbf{z} \in \mathcal{P}$ which satisfy this condition can obviously be identified with the set $\mathcal{P}_{\mathrm{per}}$ of all periodic sequences $\mathbf{z} = \{z_j\}_{j \in \mathbb{Z}}$ which satisfy $z_{j+N} = z_j$ for all $j \in \mathbb{Z}$. The discrete symplectic action

$$\Phi : \mathcal{P}_{\mathrm{per}} \to \mathbb{R}$$

is still given by (9.2.4). In this case it is convenient to extend V_j and ψ_j periodically to all $j \in \mathbb{Z}$ via

$$V_{j+N} = V_j, \qquad \psi_{j+N} = \psi_j,$$

for $j \in \mathbb{Z}$.

Lemma 9.2.6 *A periodic sequence $\mathbf{z}$ is a critical point of $\Phi : \mathcal{P}_{\mathrm{per}} \to \mathbb{R}$ if and only if $z_0 = (x_0, y_0)$ is a fixed point of ψ and $z_{j+1} = \psi_j(z_j)$ for all $j \in \mathbb{Z}$. Moreover, $\mathbf{z}$ is nondegenerate as a critical point of Φ if and only if z_0 is nondegenerate as a fixed point of ψ.*

Proof: To prove the first assertion, note that the identities in the proof of Lemma 9.2.4 continue to hold for all $j \in \mathbb{Z}$. Hence $\mathbf{z} \in \mathcal{P}_{\mathrm{per}}$ is a critical point of Φ if and only if $z_{j+1} = \psi_j(z_j)$ for all $j \in \mathbb{Z}$. To prove the second assertion, note that the Hessian of Φ at $\mathbf{z}$ is the symmetric linear operator $\zeta \mapsto \zeta'$ given by

$$\xi'_{j+1} = \eta_j - \eta_{j+1} - \frac{\partial^2 V_j}{\partial x^2}\xi_{j+1} - \frac{\partial^2 V_j}{\partial x \partial y}\eta_j,$$
$$\eta'_j = \xi_{j+1} - \xi_j - \frac{\partial^2 V_j}{\partial y \partial x}\xi_{j+1} - \frac{\partial^2 V_j}{\partial y^2}\eta_j.$$

Here all the partial derivatives of V_j are evaluated at the pair $(x_{j+1}, y_j) \in \mathbb{R}^{2n}$. Hence a sequence $\zeta = \{\zeta_j\}_j$ is in the kernel of the Hessian if and only if

$$\zeta_{j+1} = d\psi_j(z_j)\zeta_j$$

for every j. In view of the boundary condition $\zeta_N = \zeta_0$ this means that

$$\zeta_0 = d\psi(z_0)\zeta_0.$$

Hence the kernel of $d^2\Phi(\mathbf{z})$ is isomorphic to the kernel of $\mathbb{1} - d\psi(z_0)$. □

Remark 9.2.7 A nondegenerate fixed point $z_0 = \psi(z_0)$ corresponds to a nondegenerate periodic solution

$$z(t) = z(t+1) = \phi_H^{t,0}(z_0)$$

of the underlying continuous-time Hamiltonian system ϕ_H^{t,t_0} associated to a periodic Hamiltonian function $H : \mathbb{R}/\mathbb{Z} \times M \to \mathbb{R}$ and discretized at a partition $0 = t_0 < t_1 < \cdots < t_N = 1$. Every such periodic solution has a **Conley–Zehnder index** $\mu_{\mathrm{CZ}}(\Psi_z)$ (cf. [125], [563], [541]). The Conley–Zehnder index is an adaptation of the Maslov index to paths $\Psi : [0,1] \to \mathrm{Sp}(2n)$ of symplectic matrices that satisfy

$$\Psi(0) = \mathbb{1}, \qquad \det(\mathbb{1} - \Psi(1)) \neq 0.$$

The path Ψ_z is obtained from the periodic solution z by linearizing the Hamiltonian flow along $z(t)$, i.e.

$$\Psi_z(t) := d\phi_H^{t,0}(z_0), \qquad 0 \le t \le 1.$$

If N is sufficiently large then the signature of the Hessian of Φ at the corresponding critical point $\mathbf{z}$ agrees with the Conley–Zehnder index

$$\mathrm{sign}\, d^2\Phi(\mathbf{z}) = \mu_{\mathrm{CZ}}(\Psi_z).$$

This formula can be proved by methods similar to those in Robbin–Salamon [540] where an analogous identity is proved for the Hessian associated to the boundary condition $\xi_0 = \xi_N = 0$. □

9.3 Hamiltonian symplectomorphisms

There are several ways to extend the above results. Here we will describe the extension to symplectomorphisms of arbitrary exact manifolds. We shall discuss the extension to generating functions for Lagrangian submanifolds of cotangent bundles in the next section.

We begin with a general statement about isotopies of exact symplectic manifolds. Recall from Section 3.1 that a family of symplectomorphisms ϕ_t, $0 \le t \le 1$, which starts at $\phi_0 = \mathrm{id}$ is called a Hamiltonian isotopy if it is generated by Hamiltonian functions $H_t : M \to \mathbb{R}$ via

$$\frac{d}{dt}\phi_t = X_t \circ \phi_t, \qquad \iota(X_t)\omega = dH_t. \tag{9.3.1}$$

Proposition 9.3.1 *Suppose $\omega = -d\lambda$ and $\phi_t \in \mathrm{Diff}(M)$ is an isotopy starting at the identity $\phi_0 = \mathrm{id}$. Then ϕ_t is a symplectic isotopy if and only if the 1-form $\phi_t^*\lambda - \lambda$ is closed for every t. It is a Hamiltonian isotopy if and only if*

$$\phi_t^*\lambda - \lambda = dF_t$$

for a smooth family of functions $F_t : M \to \mathbb{R}$. In this case the functions F_t are related to the Hamiltonian functions H_t in (9.3.1) by

$$F_t = \int_0^t (\iota(X_s)\lambda - H_s) \circ \phi_s \, ds \tag{9.3.2}$$

(up to additive constants).

Proof: Let ϕ_t be a Hamiltonian isotopy generated by Hamiltonian vector fields $X_t : M \to TM$ as in equation (9.3.1). Then

$$\partial_t \phi_t^*\lambda = \phi_t^*(\iota(X_t)d\lambda + d\iota(X_t)\lambda) = \phi_t^* d(\iota(X_t)\lambda - H_t)$$

and the identity $\phi_t^*\lambda - \lambda = dF_t$ holds with F_t given by (9.3.2). Conversely, let ϕ_t be generated by X_t and assume that $\phi_t^*\lambda - \lambda = dF_t$. Define

$$H_t := \iota(X_t)\lambda - \left(\frac{d}{dt}F_t\right) \circ {\phi_t}^{-1}.$$

Then

$$0 = \partial_t(\phi_t^*\lambda - dF_t) = \phi_t^*(\iota(X_t)d\lambda + d\iota(X_t)\lambda) - d\partial_t F_t = \phi_t^*(\iota(X_t)d\lambda + dH_t),$$

and hence $\iota(X_t)\omega = dH_t$. This proves the assertion about Hamiltonian isotopies. The proof of the assertion about symplectic isotopies is obvious. □

Definition 9.3.2 *A symplectic manifold (M, ω) is called* **exact** *if ω is exact. A symplectomorphism ϕ of an exact symplectic manifold $(M, -d\lambda)$ is called* **exact (with respect to λ)** *if $\phi^*\lambda - \lambda$ is exact.*

In general, the notion of exact symplectomorphism may depend on the 1-form λ with $\omega = -d\lambda$. If, however, ϕ induces the identity on $H^1(M;\mathbb{R})$ (for example if ϕ is isotopic to the identity or if M is simply connected) then this definition is independent of choice of λ. Moreover, Proposition 9.3.1 asserts that every Hamiltonian symplectomorphism is exact with respect to any 1-form λ. Conversely, if there exists an isotopy ϕ_t from the identity $\phi_0 = \mathrm{id}$ to $\phi_1 = \phi$ such that every ϕ_t is λ-exact then the functions F_t with $\phi_t^*\lambda - \lambda = dF_t$ can be chosen to depend smoothly on t and hence, again by Proposition 9.3.1, ϕ_t is a Hamiltonian isotopy. Thus we have proved the following result which, in the next chapter, we will extend to the case where the symplectic form is not exact.

Corollary 9.3.3 *Let (M, ω) be an exact symplectic manifold and let ϕ_t be a symplectic isotopy. If ϕ_t is Hamiltonian for each t then ϕ_t is a Hamiltonian isotopy, i.e. it is generated by Hamiltonian vector fields.*

Observe that the value of the function F_t in (9.3.2) at a point $x \in M$ is the integral of the action form $\lambda - H_s\, ds$ along the path $[0, t] \to M : s \mapsto \phi_s(x)$. This action integral made its first appearance in Section 1.1 as the variational principle which prescribes the flow ϕ_t. It also played an important role in our discussion of generating functions of type S in Section 9.1.

Remark 9.3.4 The generating function S in Lemma 9.1.1 can easily be recovered from F in the special case

$$M = \mathbb{R}^{2n}, \qquad \lambda = y_0 dx_0 := \sum_j y_{0j} dx_{0j}.$$

With the notation $(x_1, y_1) = \phi(x_0, y_0)$ we have

$$\phi^*\lambda = y_1 dx_1,$$

and hence the condition $\phi^*\lambda - \lambda = dF$ becomes $y_1 dx_1 - y_0 dx_0 = dF$. Here we think of x_0 and y_0 as the independent variables on $\mathbb{R}^{2n}$. If ϕ satisfies the condition (9.1.1) then there exists a kind of Green's function $G : \mathbb{R}^n \times \mathbb{R}^n \to \mathbb{R}^n$ which assigns to each pair of boundary data (x_0, x_1) a vector $y_0 = G(x_0, x_1)$ such that the corresponding solution $[0, 1] \mapsto \mathbb{R}^{2n} : t \mapsto (x(t), y(t))$ of the Hamiltonian system (9.1.4) with $x(0) = x_0$ and $y(0) = y_0$ satisfies $x(1) = x_1$. The generating function S is then given by

$$S(x_0, x_1) := F(x_0, G(x_0, x_1)),$$

and the formula

$$y_1 dx_1 - y_0 dx_0 = dS,$$

now with x_0 and x_1 as independent coordinates, shows that $y_0 = -\partial S/\partial x_0$ and $y_1 = \partial S/\partial x_1$ as in Lemma 9.1.1 and the proof of Lemma 9.1.4. □

There are different versions of the symplectic action functional corresponding to different 1-forms λ with $\omega = -d\lambda$ and these give rise to different kinds of generating function. In some cases it may be convenient to choose two different such 1-forms λ_0 and λ_1 on the source and target of the symplectomorphism ϕ and consider the equation $\phi^*\lambda_1 - \lambda_0 = dF$. More generally, one can even consider 1-forms on the product manifold $M \times M$ with symplectic structure $(-\omega) \oplus \omega$ which do not respect the product structure. This will be important, for example, in order to recover the generating function of type V. In fact, these ideas have a long history (see Weinstein [670] and Abraham and Marsden [7]), and we follow Bates [56] in our exposition.

Definition 9.3.5 *Let $(M, -d\lambda)$ be an exact symplectic manifold. A Lagrangian embedding $\iota : L \to M$ is called* **exact (with respect to λ)** *if $\iota^*\lambda$ is exact. A Lagrangian submanifold of M is called* **exact (with respect to λ)** *if it is the image of an exact Lagrangian embedding or, equivalently, if the restriction of λ to the submanifold is exact.*

As before, this definition depends on the 1-form λ with $\omega = -d\lambda$ unless the homomorphism $\iota^* : H^1(M;\mathbb{R}) \to H^1(L;\mathbb{R})$ vanishes. With this definition a symplectomorphism $\phi : M \to M$ is exact with respect to the 1-form $\lambda \in \Omega^1(M)$ (see Definition 9.3.2) if and only if its graph

$$gr_\phi : M \to M \times M, \qquad gr_\phi(q) := (q, \phi(q)), \tag{9.3.3}$$

is an exact Lagrangian embedding with respect to the 1-form $\alpha := (-\lambda) \oplus \lambda$.

Corollary 9.3.6 *Let $(M, -d\lambda)$ be an exact symplectic manifold, let $\iota : L \to M$ be an exact Lagrangian embedding, and let $\phi \in \mathrm{Ham}(M)$. Then $\phi \circ \iota : L \to M$ is an exact Lagrangian embedding.*

Proof: $(\phi \circ \iota)^*\lambda = \iota^*(\phi^*\lambda - \lambda) + \iota^*\lambda$ is exact by Proposition 9.3.1. □

Corollary 9.3.7 *Let (M, ω) be an exact symplectic manifold, let $\Delta \subset M \times M$ denote the diagonal, and let $\alpha \in \Omega^1(M \times M)$ be a 1-form that satisfies*

$$-d\alpha = (-\omega) \oplus \omega, \qquad \alpha|_\Delta = 0. \tag{9.3.4}$$

If $\phi \in \mathrm{Ham}(M, \omega)$ then the 1-form $gr_\phi^*\alpha \in \Omega^1(M)$ is exact.

Proof: The inclusion of the diagonal $\iota : M \to M \times M$ is an exact Lagrangian embedding of M into $(M \times M, -d\alpha)$ because $\iota^*\alpha = 0$ and $\psi := \mathrm{id} \times \phi$ is a Hamiltonian symplectomorphism of $(M \times M, -d\alpha)$. Hence $gr_\phi = \psi \circ \iota$ is an exact Lagrangian embedding of M into $(M \times M, -d\alpha)$ by Corollary 9.3.6. □

Definition 9.3.8 *Let (M, ω) be a connected symplectic manifold without boundary, let $\alpha \in \Omega^1(M \times M)$ be a 1-form that satisfies (9.3.4), and let $\phi : M \to M$ be a symplectomorphism. ϕ is called* **α-exact** *if the 1-form $gr_\phi^*\alpha \in \Omega^1(M)$ is exact. A function $S : M \to \mathbb{R}$ is called an* **α-generating function for ϕ** *if $dS = gr_\phi^*\alpha$. When ϕ has compact support and is α-exact, the function S can be chosen with compact support, is then uniquely determined by α and ϕ, and is denoted by $S_{\alpha,\phi}$.*

Corollary 9.3.7 asserts that a Hamiltonian symplectomorphism of an exact manifold $(M, \omega = -d\lambda)$ is α-exact for every 1-form α that satisfies (9.3.4). The function $F : M \to \mathbb{R}$ of Proposition 9.3.1 which assigns to every point $x \in M$ the symplectic action of the path $[0,1] \to M : t \mapsto \phi_t(x)$ appears as the special case $\alpha = (-\lambda) \oplus \lambda$. In this case $gr_\phi^*\alpha = \phi^*\lambda - \lambda = dF$ and hence

$$S_{(-\lambda)\oplus\lambda,\phi}(x) = \int_0^1 \Big(\lambda(\dot\gamma(t)) - H_t(\gamma(t))\Big)\, dt, \qquad \gamma(t) = \phi_t(x). \qquad (9.3.5)$$

In the general case $S_{\alpha,\phi}$ can also be written in this form, but in the product space $M \times M$ and with the symplectic action defined in terms of α.

Remark 9.3.9 The generating functions of type V in Lemma 9.2.1 are made from 1-forms α which do not respect the product structure of $M \times M$. Recall from Section 3.4 that the diagonal Δ is a Lagrangian submanifold of the product manifold $(M \times M, (-\omega) \oplus \omega)$. By Theorem 3.4.13, there is a neighbourhood $\mathcal{N}(\Delta)$ in $M \times M$ which is symplectomorphic to a neighbourhood of the zero section M_0 in the cotangent bundle T^*M. When $M = \mathbb{R}^{2n}$ this local symplectomorphism extends to a global symplectomorphism $\Psi : \mathbb{R}^{2n} \times \mathbb{R}^{2n} \to T^*\mathbb{R}^{2n}$ as follows. On the cotangent bundle $T^*\mathbb{R}^{2n}$ we use the coordinates (q_1, q_2, p_1, p_2), where the q's are in the base and p's are in the fibre. The symplectic form is

$$\omega_{\mathrm{can}} = dq_1 \wedge dp_1 + dq_2 \wedge dp_2.$$

Then with coordinates (x_0, y_0, x_1, y_1) on $\mathbb{R}^{2n} \times \mathbb{R}^{2n}$, the map

$$\Psi(x_0, y_0, x_1, y_1) = (q_1, q_2, p_1, p_2),$$

defined by

$$q_1 = x_1, \qquad q_2 = y_0, \qquad p_1 = y_1 - y_0, \qquad p_2 = x_0 - x_1,$$

is a symplectomorphism which takes the diagonal to the zero section. Moreover, the canonical 1-form $\lambda_{\mathrm{can}} = p_1 dq_1 + p_2 dq_2$ pulls back to

$$\alpha_0 = \Psi^*\lambda_{\mathrm{can}} = (y_1 - y_0)dx_1 + (x_0 - x_1)dy_0.$$

Now let $\phi : \mathbb{R}^{2n} \to \mathbb{R}^{2n}$ be a symplectomorphism and write $\phi(x_0, y_0) = (x_1, y_1)$. Suppose that ϕ is sufficiently C^1-close to the identity for the variables x_1, y_0 to be independent. Then there is a Green's function $G : \mathbb{R}^n \times \mathbb{R}^n \to \mathbb{R}^n$ such that $x_0 = G(x_1, y_0)$ if and only if $\phi(x_0, y_0) = (x_1, y_1)$ for some y_1. Define the function $V : \mathbb{R}^n \times \mathbb{R}^n \to \mathbb{R}$ by

$$V(x_1, y_0) = S_{\alpha_0,\phi}(G(x_1, y_0), y_0).$$

Then the formula $dV = \alpha_0$ becomes

$$x_1 - x_0 = \frac{\partial V}{\partial y_0}(x_1, y_0), \qquad y_1 - y_0 = -\frac{\partial V}{\partial x_1}(x_1, y_0)$$

as in Lemma 9.2.1. □

The action spectrum

Although each ϕ has a whole family of possible generating functions, it turns out that these functions do share some important properties which can be used to define symplectic invariants for ϕ. The following lovely observation was made by Bates [56].

Proposition 9.3.10 *Let (M,ω) be an exact manifold with $H^1(M;\mathbb{R}) = 0$ and let $\phi \in \mathrm{Ham}_c(M)$ be a Hamiltonian symplectomorphism with compact support. Then all its generating functions $S_{\alpha,\phi}$ of compact support agree on the set* $\mathrm{Fix}(\phi)$ *of fixed points of ϕ. Here α is any 1-form on $M \times M$ which satisfies* (9.3.4).

Proof: First consider the case where $\alpha = (-\lambda) \oplus \lambda$ with $\omega = -d\lambda$ and suppose that $\alpha' = (-\lambda') \oplus \lambda'$ is another such form. Then $\beta := \lambda' - \lambda$ is closed and it follows from equation (9.3.5) that the corresponding generating functions $S, S' : M \to \mathbb{R}$ satisfy

$$S'(z) - S(z) = \int_\gamma \beta, \qquad \gamma(t) := \phi_t(z),$$

where ϕ_t is a compactly supported Hamiltonian isotopy from $\phi_0 = \mathrm{id}$ to $\phi_1 = \phi$. Since $H^1(M;\mathbb{R}) = 0$ the 1-form β is exact and, since γ is a loop, the integral is zero. Hence $S'(z) = S(z)$. The general case can be reduced to this one by replacing M with $M \times M$, ϕ with $\mathrm{id} \times \phi$, and λ with α. This proves Proposition 9.3.10.

Alternatively, the proof can be based on the observation that, for any two 1-forms α_1, α_2 on $M \times M$ which satisfy (9.3.4), there is a function $f : M \times M \to \mathbb{R}$ which vanishes on Δ and is such that $\alpha_1 - \alpha_2 = df$. This holds because the closed 1-form $\alpha_1 - \alpha_2$ vanishes on the diagonal and so represents a relative cohomology class in $H^1(M \times M, \Delta)$. By assumption this cohomology group is zero and so f exists by the relative Poincaré lemma (see Bott and Tu [78]). □

Example 9.3.11 This example shows that the condition $H^1(M;\mathbb{R}) = 0$ in Proposition 9.3.10 cannot be removed. Let $M := \mathbb{R}/\mathbb{Z} \times \mathbb{R} \cong T^*S^1$ be the cotangent bundle of the circle with coordinates $x \in \mathbb{R}/\mathbb{Z}$ and $y \in \mathbb{R}$. Let $H(x,y) = h(y)$ be a compactly supported Hamiltonian function on M that is independent of the x-variable. Then the Hamiltonian differential equation has the form

$$\dot{x} = h'(y), \qquad \dot{y} = 0$$

and the Hamiltonian symplectomorphism ϕ is given by $\phi(x,y) = (x + h'(y), y)$. Consider the 1-forms $\lambda_r := (y + r)dx$ with $-d\lambda_r = dx \wedge dy$. The generating function $S_r : M \to \mathbb{R}$ that satisfies $dS_r = \phi^*\lambda_r - \lambda_r$ is given by

$$S_r(x,y) = (y + r)h'(y) - h(y).$$

The fixed points of ϕ are the pairs $(x,y) \in M$ with $h'(y) \in \mathbb{Z}$. The corresponding periodic solution of Hamilton's equation is contractible if and only if $h'(y) = 0$. For such fixed points the value $S_r(x,y) = -h(y)$ is independent of r. However, in the noncontractible case the value $S_r(x,y)$ (which is still the symplectic action of the periodic solution with respect to λ_r) does depend on r. □

Definition 9.3.12 (Action spectrum) *Let (M,ω) be an exact symplectic manifold with $H^1(M;\mathbb{R})=0$ and suppose that $\phi: M\to M$ is a compactly supported Hamiltonian symplectomorphism. For every fixed point z denote the common value of all the compactly supported generating functions by*

$$\mathcal{A}_\phi(z) := S_{\alpha,\phi}(z), \qquad z=\phi(z). \tag{9.3.6}$$

This number is called the **action** *of the fixed point z. The set of these numbers as z ranges over the fixed point set is called the* **action spectrum** *of ϕ.*

Of course, Proposition 9.3.10 was well-known for special classes of generating functions and underlies the work of Viterbo [646] and Hofer–Zehnder [325]. In the case $\alpha=(-\lambda)\oplus\lambda$ the result of Proposition 9.3.10 can be rephrased in a different way. In view of (9.3.5) the number $\mathcal{A}(z)$ can be expressed as the symplectic action of the path $t\mapsto\phi_t(z)$ but the definition of the number is independent of the choice of this isotopy.

Corollary 9.3.13 *Let (M,ω) be an exact manifold with $H^1(M;\mathbb{R})=0$ and let $\phi: M\to M$ be a compactly supported Hamiltonian symplectomorphism. Choose a 1-form $\lambda\in\Omega^1(M)$ such that $\omega=-d\lambda$ and a Hamiltonian isotopy $\phi_t: M\to M$ with compactly supported Hamiltonian functions $H_t: M\to\mathbb{R}$ such that $\phi_0=\mathrm{id}$ and $\phi_1=\phi$. Then the corresponding action integral*

$$\mathcal{A}_H(z) := \int_0^1 \Big(\lambda(\dot\gamma(t)) - H_t(\gamma(t))\Big)\,dt, \qquad \gamma(t)=\phi_t(z),$$

for $z\in M$ is independent of the Hamiltonian isotopy used to define it. If $z=\phi(z)$ is a fixed point then it is also independent of the 1-form λ.

Proof: Proposition 9.3.1 and Proposition 9.3.10. □

We now give a geometric interpretation of the action of a fixed point which will be useful in our discussion of the Hofer metric in Section 12.3. This is a Finsler metric on the space $\mathrm{Ham}_c(M)$ of compactly supported Hamiltonian symplectomorphisms. It is induced by the norm on the space of Hamiltonian functions $C_0^\infty(M,\mathbb{R})$ given by

$$\|H\| = \sup_{x\in M} H(x) - \inf_{x\in M} H(x).$$

The length $\mathcal{L}(\{\phi_t\})$ of the path $[0,1]\to\mathrm{Ham}_c(M): t\mapsto\phi_t$ generated by the Hamiltonian $[0,1]\to C_0^\infty(M,\mathbb{R}): t\mapsto H_t$ is defined to be

$$\mathcal{L}(\{\phi_t\}) = \int_0^1 \|H_t\|\,dt.$$

The next corollary shows how this length, under some rather special hypotheses, is related to the action spectrum. We shall see in Chapter 12 that these hypotheses amount to the assumption that the Hamiltonian isotopy in question is a geodesic in the group $\mathrm{Ham}_c(M)$ with respect to the Hofer metric. These were investigated in depth by Bialy and Polterovich [64] and Lalonde and McDuff [391].

Corollary 9.3.14 *Let (M, ω) be a connected, simply connected, and exact symplectic manifold and let $\phi_t \in \mathrm{Ham}_c(M)$ be the compactly supported Hamiltonian isotopy generated by $H_t \in C_0^\infty(M, \mathbb{R})$. Suppose that there are points $p, P \in M$ such that*

$$H_t(p) = \inf_{x \in M} H_t(x), \qquad H_t(P) = \sup_{x \in M} H_t(x)$$

for every t. Then p and P are fixed points of ϕ_t for every t, and

$$\mathcal{L}(\{\phi_t\}) = \int_0^1 \left(H_t(P) - H_t(p)\right) dt = \mathcal{A}_{\phi_1}(p) - \mathcal{A}_{\phi_1}(P).$$

Moreover, for any other common fixed point z of all the symplectomorphisms ϕ_t we have

$$\mathcal{A}_{\phi_1}(P) \le \mathcal{A}_{\phi_1}(z) \le \mathcal{A}_{\phi_1}(p).$$

Proof: This is an obvious consequence of the fact that whenever $z = \phi_t(z)$ for all t then $X_t(\phi_t(z)) = 0$ for all t and hence the formula (9.3.5) shows that

$$\mathcal{A}_{\phi_1}(z) = -\int_0^1 H_t(z)\, dt.$$

Hence the first statement follows by integrating the identity

$$\|H_t\| = H_t(P) - H_t(p)$$

and the second statement by integrating the inequality

$$H_t(p) \le H_t(z) \le H_t(P)$$

for all t. □

Exercise 9.3.15 Let $\phi : \mathbb{R}^2 \to \mathbb{R}^2$ be the time-1 map of the Hamiltonian function H which is given in polar coordinates by

$$H(r, \theta) = h(r)$$

for some function h which equals 1 near $r = 0$ and equals 0 for $r \ge 1$. Calculate $\mathcal{A}_\phi(0)$ and verify the statement of Corollary 9.3.14 in this case. □

The next proposition shows that in the simply connected case, the action $\mathcal{A}_\phi(z)$ of a fixed point $z = \phi(z)$ agrees with the area enclosed by the loop made of the two paths $\gamma : [0, 1] \to M$ and $\phi \circ \gamma$, where $\gamma(0) = z_0$ is some point outside the support of ϕ and $\gamma(z) = z$.

Proposition 9.3.16 *Assume that (M,ω) is a connected, simply connected, and exact symplectic manifold. Let $\phi \in \mathrm{Ham}_c(M)$ be a compactly supported Hamiltonian symplectomorphism with fixed point z. Choose a compactly supported Hamiltonian isotopy $\phi_t \in \mathrm{Ham}_c(M)$ connecting the identity $\phi_0 = \mathrm{id}$ to $\phi_1 = \phi$ with generating Hamiltonian $H_t \in C_0^\infty(M)$, and choose $z_0 \in M$ such that $H_t(z_0) = 0$ and $dH_t(z_0) = 0$ for all t. Then every smooth map $u : [0,1] \times [0,1] \to M$ with*

$$u(s,1) = \phi(u(s,0)), \qquad u(0,t) = z_0, \qquad u(1,t) = z, \tag{9.3.7}$$

satisfies

$$\mathcal{A}_\phi(z) = \int u^*\omega.$$

Proof: We first prove that the integral is independent of the choice of u. To see this note that if u_0 and u_1 are any two maps which satisfy the required conditions then, since M is simply connected, we can fill in the loop consisting of $u_0(s,0)$ and $u_1(s,0)$ by a map $v : [0,1] \times [0,1] \to M$ such that

$$v(\theta,0) = z_0, \quad v(\theta,1) = z, \quad v(0,s) = u_0(s,0), \quad v(1,s) = u_1(s,0).$$

Then the four squares v, u_0, $\phi \circ v$, and u_1 form a sphere and so the integral of ω over it must vanish. Taking proper account of the orientations and using $\int v^*\phi^*\omega = \int v^*\omega$, we obtain $\int u_0^*\omega = \int u_1^*\omega$ as claimed.

Now choose a smooth path $\beta : [0,1] \to M$ such that $\beta(0) = z_0$ and $\beta(s) = z$ for $1/2 \le s \le 1$ and choose a smooth map $u_0 : [0,1] \times [0,1] \to M$ that satisfies (9.3.7) and in addition

$$u_0(s,t) = \phi_t(\beta(s)) \qquad \text{for } 0 \le s \le 1/2, \tag{9.3.8}$$

$$u_0(s,0) = u_0(s,1) = z \qquad \text{for } 1/2 \le s \le 1. \tag{9.3.9}$$

Then $u_0(0,t) = z_0$, $u_0(1/2,t) = \phi_t(z)$, and $u_0(1,t) = z$ for all t. Abbreviate $X_t := X_{H_t}$. Differentiate equation (9.3.8) to obtain

$$\frac{\partial u_0}{\partial s} = d\phi_t(\beta(s))\dot\beta(s), \qquad \frac{\partial u_0}{\partial t} = d\phi_t(\beta(s))\big(\phi_t^*X_t(\beta(s))\big),$$

for $0 \le s \le 1/2$. Since $\phi_t^*\omega = \omega$, this gives

$$\begin{aligned}
\int_{[0,1/2]\times[0,1]} u_0^*\omega &= \int_0^{1/2}\int_0^1 \omega\big(\dot\beta(s), \phi_t^*X_t(\beta(s))\big)\,dtds \\
&= -\int_0^{1/2}\int_0^1 d(H_t\circ\phi_t)(\beta(s))\dot\beta(s)\,dtds \\
&= -\int_0^1 H_t(\phi_t(z))\,dt.
\end{aligned}$$

In the second half of the integral, use Stokes' theorem with $\omega = -d\lambda$ to obtain

$$\int_{[1/2,1]\times[0,1]} u_0^*\omega = \int_0^1 \lambda(X_t(\phi_t(z)))\,dt.$$

Hence the assertion of Proposition 9.3.16 follows from (9.3.5) and (9.3.6). □

The assertion of Proposition 9.3.16 continues to hold when $\pi_1(M) \neq 0$ provided that ϕ is the time-1 map of an isotopy ϕ_t such that the fixed point z is **contractible**, in other words so that the loop $\gamma(t) = \phi_t(z)$ is contractible. In this case the first part of the proof (the independence of u) works only if we fix the homotopy class of the path $s \mapsto u(s, 0)$ from z_0 to z. But this does not affect the assertion because the second part of the proof shows that for every such homotopy class we can find a map u_0 whose area agrees with the action.

Local generating functions

If (M, ω) is not exact then there is no 1-form α whose differential agrees with $(-\omega) \oplus \omega$ on the whole of $M \times M$. However, such 1-forms do exist on suitable neighbourhoods $\mathcal{N}(\Delta)$ of the diagonal Δ and, as we now show, we can extend Corollary 9.3.7 to define α-generating functions for Hamiltonian symplectomorphisms that are connected to the identity by a sufficiently small Hamiltonian isotopy. One obvious choice for α is a 1-form given by $\alpha := \Psi^*\lambda_{\mathrm{can}}$, where

$$\Psi : \mathcal{N}(\Delta) \to \mathcal{N}(M_0)$$

is a symplectomorphism from a neighbourhood of the diagonal in $M \times M$ to a neighbourhood of the zero section in T^*M (see Theorem 3.4.13). Then, given a Hamiltonian isotopy of $\phi_t : M \to M$ that is sufficiently close to the identity, Proposition 9.3.1 implies that there is a family of functions $F_t : M \to \mathbb{R}$ such that $gr_{\phi_t}^* \Psi^* \lambda_{\mathrm{can}} = dF_t$. On the other hand, the composition $\Psi \circ gr_{\phi_t} : M \to T^*M$ is C^1-close to the zero section and its image is therefore the graph of a 1-form. Since ϕ_t is a symplectomorphism its graph is a Lagrangian submanifold and so this 1-form is closed. In other words, there exists a diffeomorphism $f_t : M \to M$ and a closed 1-form $\sigma_t : M \to T^*M$ such that $\Psi \circ gr_{\phi_t} = \sigma_t \circ f_t$. By Proposition 3.1.18, we have $\sigma_t^* \lambda_{\mathrm{can}} = \sigma_t$ and hence $f_t^* \sigma_t = dF_t$. Hence σ_t is exact and so the graph of ϕ_t is an exact Lagrangian submanifold of $(\mathcal{N}(\Delta), -\Psi^*\lambda_{\mathrm{can}})$. The next proposition extends this observation to a more general setting.

Proposition 9.3.17 *Let (M, ω) be a symplectic manifold without boundary, let $U \subset M \times M$ be an open neighbourhood of the diagonal, and let $\alpha \in \Omega^1(U)$ satisfy* (9.3.4) *on U. Let $\phi_t \in \mathrm{Ham}(M, \omega)$ be a Hamiltonian isotopy such that $\phi_0 = \mathrm{id}$ and $\mathrm{graph}(\phi_t) \subset U$ for all t. Then $gr_{\phi_t}^* \alpha \in \Omega^1(M)$ is exact for all t.*

Proof: Define $U_t := (\mathrm{id} \times \phi_t)^{-1}(U) \cap U$ and consider the Hamiltonian isotopy $\psi_t := \mathrm{id} \times \phi_t : U_t \to U$. Proposition 9.3.1 extends to this setting and so there is a smooth family of functions $F_t : U_t \to \mathbb{R}$ such that $\psi_t^* \alpha - \alpha = dF_t$ for all t. Since U_t is a neighbourhood of the diagonal for all t, it follows as in the proof of Corollary 9.3.7 that $gr_{\phi_t}^* \alpha$ is exact for all t. □

All our previous results about the action spectrum can be adapted to this context. In particular, Corollary 9.3.14 continues to hold on any manifold for Hamiltonian isotopies $\phi_t : M \to M$ that are sufficiently close to the identity. When $M = \mathbb{R}^{2n}$, this can also be proved by using the Hamilton–Jacobi equation (see Bialy and Polterovich [64]).

9.4 Lagrangian submanifolds

In this section we discuss the construction of generating functions for Lagrangian embeddings $\iota : L \to T^*L$ in cotangent bundles. We shall mostly be concerned with Hamiltonian perturbations of the zero section L_0, and explain a natural class of generating functions for such embeddings. The starting point is the observation that a Lagrangian submanifold that is sufficiently close to the zero section L_0 must be the graph of a closed 1-form (see Proposition 3.4.2). The graphs of exact 1-forms $\Lambda = \mathrm{graph}(dS)$ form an important special class. They are examples of *exact Lagrangian submanifolds* with *generating function* S. More generally, an exact Lagrangian embedding $\iota : L \to T^*L$ can be characterized by the condition that $\iota^*\lambda_{\mathrm{can}}$ is exact (see Definition 9.3.5). We shall see below that the image of the zero section under a Hamiltonian symplectomorphism always satisfies this condition.

If $\iota : L \to T^*L$ is an exact Lagrangian embedding then, according to Eliashberg, the attempt to find a generating function can be explained in terms of the following diagram:

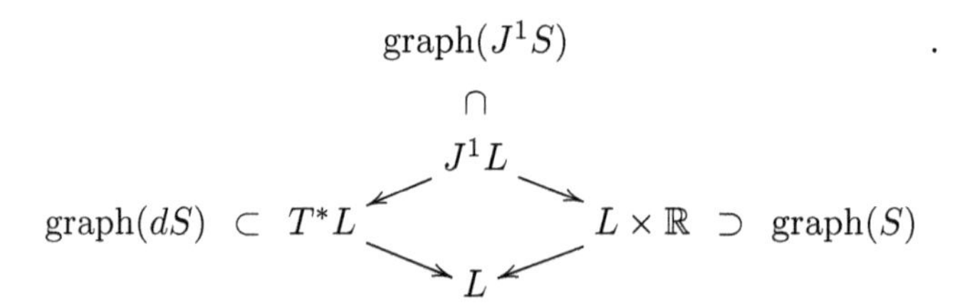

Begin with an exact Lagrangian submanifold $\Lambda \subset T^*L$ and lift it to a Legendrian submanifold of the 1-jet bundle $J^1L = T^*L \times \mathbb{R}$ and finally project to $L \times \mathbb{R}$. This works when $\Lambda = \mathrm{graph}(dS)$ is the graph of a differential dS. Then lifting means finding the value of the function S, given its first derivatives, and projecting means forgetting the derivative. But in general this process works only locally and the resulting generating function will be multivalued. Even in the 1-dimensional case where $L = S^1$, the problem of defining a suitable function S quickly becomes highly complicated and quite intractable. This technique is related to the analysis of wave fronts and Cerf diagrams [176].

In this section we will describe a more general type of generating function which circumvents these complications in a simple and elegant way. They are single valued, but in exchange are defined on large-dimensional fibre bundles over L. Eliashberg calls this 'trading complexity for dimension'. These generating functions are a powerful tool in symplectic topology and have been used to great effect, notably by Viterbo [646]. They are closely related to the discrete variational techniques described in Section 9.2. (See Robbin–Salamon [540].)

One important application of these global techniques concerns the question of Lagrangian intersections. If Λ is the graph of an exact 1-form dS then its intersection points with the zero section L_0 are precisely the critical points of S. Hence, by Morse theory, the number of such intersection points is at least the sum of the Betti numbers. That this estimate should continue to hold for general exact Lagrangian submanifolds is a version of the Arnold conjecture and will be discussed in more detail in Chapter 11. One way of proving this result is to use the generating function approach sketched here (see Viterbo [646]). In fact there is a more general assertion that the number of intersection points should be bounded below by the minimal number of critical points of a function on L. In the nonsimply connected case this number can be considerably larger than the sum of the Betti numbers. At present, this stronger version of the Arnold conjecture is still wide open, although there are some partial results by Rudyak [551] and Oprea–Rudyak [517].

We will begin by discussing exact Lagrangian sections $\Lambda \subset T^*L$ in more detail. In particular, we will derive a formula for the generating function S in terms of the symplectic action. From now on we use the notation

$$\iota : L \to T^*L, \qquad \pi : T^*L \to L$$

for the injection of the zero section and the canonical projection, and $L_0 = \iota(L)$ for the zero section itself. The next lemma is a special case of Corollary 9.3.6. Its proof contains an explicit formula that motivates the discussion below.

Lemma 9.4.1 *Let $\psi_t : T^*L \to T^*L$ be a Hamiltonian isotopy. Then the restriction of its time-1 map ψ to the zero section induces an exact Lagrangian embedding $\psi \circ \iota : L \to T^*L$.*

Proof: Recall from Proposition 9.3.1 that every Hamiltonian symplectomorphism $\psi = \psi_1$ of a cotangent bundle T^*L satisfies

$$\psi_t^*\lambda_{\mathrm{can}} - \lambda_{\mathrm{can}} = dF_t, \qquad F_t := \int_0^t (\iota(X_s)\lambda_{\mathrm{can}} - H_s) \circ \psi_s \, ds,$$

where $H_t : T^*L \to \mathbb{R}$ is the time-dependent Hamiltonian which generates ψ_t via the vector fields $X_t := X_{H_t}$. Hence

$$\iota^*\psi^*\lambda_{\mathrm{can}} = d(F \circ \iota),$$

where $F := F_1$. □

Observe that the function

$$F = \int_0^1 (\iota(X_t)\lambda_{\mathrm{can}} - H_t) \circ \psi_t \, dt$$

in the proof of Lemma 9.4.1 assigns to every point $z \in T^*L$ the symplectic action of the path $\gamma(t) = \psi_t(z)$, $0 \le t \le 1$, in T^*L. However, it is not in itself a generating function for the exact Lagrangian submanifold $\Lambda = \psi_1(L_0)$, since in general the submanifold Λ cannot be reconstructed from F. The next proposition shows that if $\Lambda = \psi_1(L_0)$ is a section of T^*L, then it is the graph of the 1-form dS, where S is a function on L which is very closely related to $F \circ \iota$.

Proposition 9.4.2 *Let ψ_t be a Hamiltonian isotopy of T^*L, generated by the Hamiltonian functions $H_t : T^*L \to \mathbb{R}$, such that $\Lambda := \psi_1(L_0) \subset T^*L$ is a section of the cotangent bundle. Define the function $S : L \to \mathbb{R}$ by*

$$S(q) := \int_0^1 \Big(\lambda_{\mathrm{can}}(\dot\gamma(t)) - H_t(\gamma(t))\Big)\, dt, \qquad (9.4.1)$$

where $\gamma : [0,1] \to T^*L$ is the unique solution of the boundary value problem

$$\dot\gamma(t) = X_{H_t}(\gamma(t)), \qquad \gamma(0) \in L_0, \qquad \gamma(1) \in T^*_qL. \qquad (9.4.2)$$

Then $\Lambda = \mathrm{graph}(dS)$.

Proof: Denote $\psi = \psi_1$ and note first that the map

$$f = \pi \circ \psi \circ \iota : L \to L$$

is a diffeomorphism (and hence $\Lambda = \psi(L_0)$ is a section of T^*L) if and only if the boundary value problem (9.4.2) has a unique solution γ. (See Fig. 9.4.) Secondly, this solution is given by $\gamma(t) = \psi_t(f^{-1}(q), 0)$. This means that the function $S : L \to \mathbb{R}$, defined by (9.4.1) and (9.4.2), is related to F by the identity

$$S \circ f = F \circ \iota.$$

By assumption, the submanifold $\Lambda = \psi(L_0)$ is the graph of the 1-form

$$\sigma := \psi \circ \iota \circ f^{-1},$$

and this satisfies $\sigma^*\lambda_{\mathrm{can}} = \sigma$. Since $\psi^*\lambda_{\mathrm{can}} - \lambda_{\mathrm{can}} = dF$ and $\iota^*\lambda_{\mathrm{can}} = 0$, we obtain $f^*dS = \iota^*dF = \iota^*\psi^*\lambda_{\mathrm{can}}$ and hence

$$dS = (\psi \circ \iota \circ f^{-1})^*\lambda_{\mathrm{can}} = \sigma^*\lambda_{\mathrm{can}} = \sigma$$

as required. □

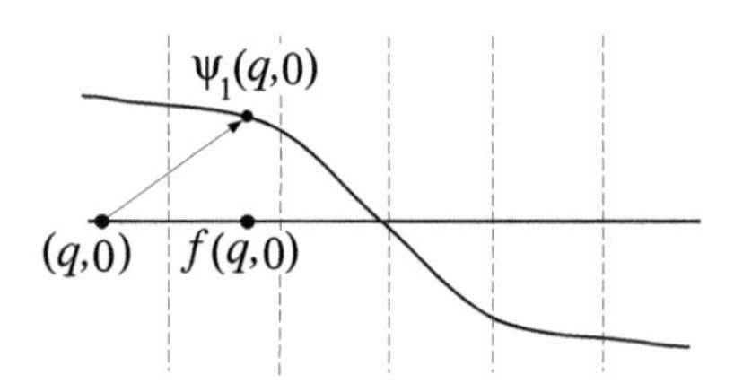

FIG. 9.4. The map f.

The assumption of Proposition 9.4.2 that Λ is a section of the cotangent bundle is very restrictive. In general, the image of an exact Lagrangian embedding will not be the graph of a 1-form but may be extremely complicated. However, there is a more general notion of a generating function which still applies in such cases. We now explain this.

A **variational family** is a pair

$$\pi : E \to L, \qquad \Phi : E \to \mathbb{R},$$

consisting of a fibre bundle E over L and a smooth function Φ on E. Denote the fibre of E over $q \in L$ by $E_q := \pi^{-1}(q)$ and consider the set

$$\mathcal{C} = \mathcal{C}(E, \Phi)$$

of **fibre critical points**. These are the points $c \in E$ which are critical points of the restriction of Φ to the fibre E_q with $q = \pi(c)$. In other words, the level set of Φ through c is tangent to E_q (see Fig. 9.5). At such a point the differential $d\Phi(c)$ vanishes on the vertical $T_cE_q = \ker\, d\pi(c)$. This means that there exists a unique cotangent vector $v^* \in T_q^*L$ such that

$$d\Phi(c) = v^* \circ d\pi(c). \tag{9.4.3}$$

Exercise 9.4.3 Prove that $c \in E_q$ is a critical point of the function $\Phi|_{E_q} : E_q \to \mathbb{R}$ if and only if there exists a cotangent vector $v^* \in T_q^*L$ that satisfies equation (9.4.3). The solution v^* of equation (9.4.3) is called the **Lagrange multiplier**. □

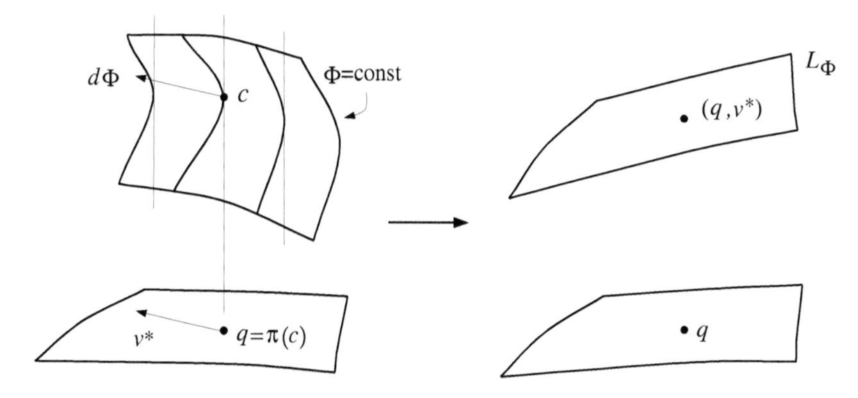

FIG. 9.5. A fibre critical point.

Consider the map $\iota_\Phi : \mathcal{C}(E, \Phi) \to T^*L$ which assigns to each fibre critical point $c \in \mathcal{C}(E, \Phi)$ the corresponding pair (q, v^*), where $q := \pi(c)$ and v^* is the Lagrange multiplier. Denote the image of this map by

$$L_\Phi := \{(q, v^*) \in T^*L \,|\, \exists\, c \in E_q \text{ such that } d\Phi(c) = v^* \circ d\pi(c)\}.$$

An extreme case is where $E = L$ and $\pi : E \to L$ is the identity. Then L_Φ is the graph of $d\Phi$ and is therefore a Lagrangian submanifold. More generally, the variational family (E, Φ) is called **transversal** if the graph in T^*E of $d\Phi$ intersects the fibre normal bundle

$$N_E := \{(c, \eta) \in T^*E \,|\, \eta \perp \ker d\pi(c)\}$$

transversally. The following proposition is due to Hörmander [331]. It asserts that *the manifold of Lagrange multipliers is a Lagrangian submanifold.*

Proposition 9.4.4 *If the variational family* (E, Φ) *is transversal then the set* $\mathcal{C}(E, \Phi)$ *of fibre critical points is a manifold of dimension* $n = \dim L$ *and the map* $\iota_\Phi : \mathcal{C}(E, \Phi) \to T^*L$ *is an exact Lagrangian immersion.*

Proof: The fibre normal bundle N_E is a coisotropic submanifold of T^*E of codimension $N := \dim E - \dim L$. The isotropic leaves are the fibres of E and hence the symplectic quotient (see Proposition 5.4.5) of N_E can be identified with T^*L. Now the manifold graph$(d\Phi) \subset T^*E$ is Lagrangian and, by Proposition 5.4.7, its intersection with N_E determines a Lagrangian submanifold of T^*L.

Here is a more explicit argument. Choose local coordinates x on L, y on T_x^*L, and ξ on the fibre of E over x. Then the fibre critical points are the pairs (x, ξ) with

$$\partial_\xi \Phi(x, \xi) = 0,$$

and the corresponding point in $(x, y) \in L_\Phi$ is defined by the equation

$$y = \partial_x \Phi(x, \xi).$$

Now denote the Hessian of Φ at (x, ξ) by

$$\mathrm{d}^2\Phi(x, \xi) = \begin{pmatrix} A & B \\ B^{\mathrm{T}} & C \end{pmatrix},$$

where $A = A^{\mathrm{T}} = \partial_x \partial_x \Phi \in \mathbb{R}^{n \times n}$, $B = \partial_x \partial_\xi \in \mathbb{R}^{n \times N}$, and $C = C^{\mathrm{T}} = \partial_\xi \partial_\xi \Phi \in \mathbb{R}^{N \times N}$. That the family (E, Φ) is transversal at (x, ξ) means that the matrix $(B^{\mathrm{T}}\ C) \in \mathbb{R}^{N \times (n+N)}$ has rank N. The tangent space $T_{(x,\xi)}\mathcal{C}$ is the kernel of this matrix and the differential of the map $(x, \xi) \mapsto y$ is given by the matrix $(A\ B)$. Hence the following lemma completes the proof. □

Lemma 9.4.5 *Let* $A = A^{\mathrm{T}} \in \mathbb{R}^{n \times n}$ *and* $C = C^{\mathrm{T}} \in \mathbb{R}^{N \times N}$ *be symmetric matrices and* $B \in \mathbb{R}^{n \times N}$. *Then*

$$\Lambda := \left\{(x, Ax + B\xi) \,|\, B^{\mathrm{T}}x + C\xi = 0\right\} \subset \mathbb{R}^{2n}$$

is a Lagrangian subspace with respect to the standard symplectic structure.

Proof: Note that $(x, y) \in \Lambda^\omega$ if and only if

$$B^{\mathrm{T}}x' + C\xi' = 0 \qquad \Longrightarrow \qquad \langle x', y\rangle - \langle Ax' + B\xi', x\rangle = 0.$$

This can be rephrased as

$$B^{\mathrm{T}}x' + C\xi' = 0 \qquad \Longrightarrow \qquad \langle x', y - Ax\rangle + \langle \xi', -B^{\mathrm{T}}x\rangle = 0.$$

But the vector $(y - Ax, -B^{\mathrm{T}}x)$ is perpendicular to the kernel of $(B^{\mathrm{T}}\ C)$ if and only if it is in the image of the transposed matrix $(B^{\mathrm{T}}\ C)^{\mathrm{T}}$. This, however, means that there exists a vector $\xi \in \mathbb{R}^N$ such that

$$B\xi = y - Ax, \qquad C\xi = -B^{\mathrm{T}}x.$$

This is equivalent to $(x, y) \in \Lambda$. □

Remark 9.4.6 Lemma 9.4.5 can be viewed as the special case of Proposition 9.4.4, where

$$L = \mathbb{R}^n, \qquad E = \mathbb{R}^n \times \mathbb{R}^N,$$

and

$$\Phi(x,\xi) = \tfrac{1}{2}\langle x, Ax\rangle + \langle x, B\xi\rangle + \tfrac{1}{2}\langle \xi, C\xi\rangle.$$

Note that in this case transversality is not required. In fact L_Φ is still an immersed Lagrangian submanifold when graph$(d\Phi)$ and N_E intersect cleanly in the sense that the tangent space of the intersection agrees with the intersection of the tangent spaces. □

Remark 9.4.7 Let (E, Φ) be a transversal variational family. Then the intersections of the Lagrangian submanifold $L_\Phi \subset T^*L$ with the zero section L_0 correspond to the critical points of the function $\Phi : E \to \mathbb{R}$ via

$$L_\Phi \cap L_0 = \{\iota_\Phi(c) \,|\, c \in E,\, d\Phi(c) = 0\}.$$

This can be used to prove the Arnold conjecture for cotangent bundles. In Chapter 11 we shall give a proof of the Arnold conjecture for the torus which is based on this observation. □

Example 9.4.8 Consider the case where E is the set of all paths $\gamma : [0,1] \to T^*L$ with $\gamma(0) \in L_0$, and the projection $\pi : E \to L$ is given by

$$\pi(\gamma) := \pi_{T^*L}(\gamma(1)).$$

Let $\Phi : E \to \mathbb{R}$ be given by the symplectic action

$$\Phi(\gamma) := \mathcal{A}_H(\gamma) = \int_0^1 \Big(\lambda_{\mathrm{can}}(\dot\gamma(t)) - H_t(\gamma(t))\Big)\, dt.$$

Then the fibre critical points are the solutions $\gamma(t) = \psi_t(\gamma(0))$ of the Hamiltonian differential equation determined by H_t. Moreover, at a fibre critical point the formula (9.4.3) holds with v^* equal to the vertical component of $\gamma(1)$. Hence the Lagrangian submanifold of T^*L generated by Φ is the image of the zero section $L_0 \subset T^*L$ under the time-1 map of the Hamiltonian flow:

$$L_\Phi = \psi_1(L_0).$$

This shows that any exact Lagrangian submanifold of T^*L which is isotopic to the identity by a Hamiltonian isotopy admits a generating variational family, albeit one with infinite-dimensional fibres. Now it is possible to approximate the fibres by finite-dimensional spaces (of *discrete paths*). Thus any such Lagrangian submanifold Λ admits a generating family with finite-dimensional fibres. This idea has been used to great effect by Chaperon [97], Laudenbach and Sikorav [402], and Viterbo [646]. It has been extended to the contact case to derive generating functions for Legendrian submanifolds in 1-jet bundles by Chekanov [101], Théret [619], and Bhupal [63].

However, in some ways it is more natural to remain in the infinite-dimensional context. For example, Oh [508] has used the symplectic action as the foundation of a detailed analysis of the behaviour of Lagrangian submanifolds in the cotangent bundle. His point of view is that the symplectic geometry and topology of the cotangent bundle T^*L is the 'quantization' of the ordinary geometry and topology of the base manifold L. In particular, the thesis that Floer theory in the cotangent bundle is the quantum analogue of Morse theory on L is developed in Fukaya–Oh [248] and Milinković–Oh [480]. □

Exercise 9.4.9 Let E and Φ be as on page 381 and assume that the restriction of $\pi : E \to L$ to $\mathcal{C} = \mathcal{C}(E, \Phi)$ is a diffeomorphism. Prove that $L_\Phi = \mathrm{graph}(dS)$, where $S : L \to \mathbb{R}$ be defined by $S(\pi(c)) := \Phi(c)$ for $c \in \mathcal{C}(E, \Phi)$. Reinterpret Proposition 9.4.2 in this context. □

Exercise 9.4.10 Let (E, Φ) be a transversal variational family. Prove that the map

$$\mathcal{C}(E, \Phi) \to T^*L \times \mathbb{R} : c \mapsto (\pi(c), v^*, \Phi(c)), \qquad v^* \circ d\pi(c) = d\Phi(c),$$

is a Legendrian immersion with respect to the contact form $\lambda_{\mathrm{can}} - dz$ on $T^*L \times \mathbb{R}$. □

The present chapter deals primarily with exact symplectic manifolds (Definition 9.3.2) such as cotangent bundles, which are necessarily noncompact. The closest one can come to such a condition in the closed case is to require that the cohomology class of ω vanishes on $\pi_2(M)$, or, equivalently, that ω lifts to an exact form on the universal cover of M. Such manifolds are called **symplectically aspherical**. Typical examples are tori where the first Chern class $c_1(\omega)$ also vanishes, but there are examples with nonzero $c_1(\omega)$; in the noncompact case this follows from Gromov's theorem 7.3.1, while in the symplectically aspherical case there are examples of Gompf. The generating function methods developed in the present chapter appear to be restricted to such manifolds. They will be used in Chapters 11 and 12 to prove some key theorems in the subject (Gromov nonsqueezing, symplectic rigidity, Arnold conjecture, Weinstein conjecture, nontriviality of the Hofer metric).

Obtaining such results for other symplectic manifolds appears to require the theory of J-holomorphic curves (Section 4.5). This theory has a different character on different types of manifold. For example, a condition diametrically opposite to being symplectically aspherical is that $c_1(\omega) = \lambda[\omega]$ for some $\lambda > 0$. Manifolds satisfying this condition are called **monotone**. (In some situations it is also relevant to consider **spherically monotone** symplectic manifolds, where the above equation only holds on the spherical part of $H_2(M;\mathbb{Z})$.) As we shall see in Sections 14.2 and 14.4, the geometric and dynamical properties of monotone manifolds seem to differ significantly from those of aspherical manifolds.

In this book we will mostly encounter this kind of phenomenon when discussing the properties of Lagrangian submanifolds. In this context also, there are analogues of the exact and monotone conditions (see Definitions 9.3.5 and 3.4.4). The discussions in Section 11.3 and Section 14.3 show that these have both technical and geometric consequences.

10

THE GROUP OF SYMPLECTOMORPHISMS

In this chapter we discuss the basic properties of the group of symplectomorphisms $\mathrm{Symp}(M) = \mathrm{Symp}(M, \omega)$ of a compact connected symplectic manifold (M, ω) and its subgroup $\mathrm{Ham}(M) = \mathrm{Ham}(M, \omega)$ of Hamiltonian symplectomorphisms. We begin by showing that the group $\mathrm{Symp}(M)$ is locally path-connected. We then discuss the flux homomorphism (sometimes also known as the first Calabi homomorphism). The main result here is Theorem 10.2.5, which characterizes the Hamiltonian symplectomorphisms in terms of the flux homomorphism. It states that $\mathrm{Ham}(M)$ is the kernel of a homomorphism from the identity component $\mathrm{Symp}_0(M)$ of $\mathrm{Symp}(M)$ to the quotient of the first cohomology group $H^1(M; \mathbb{R})$ by a countable subgroup Γ. In other words, there is an exact sequence

$$0 \to \mathrm{Ham}(M) \to \mathrm{Symp}_0(M) \to H^1(M; \mathbb{R})/\Gamma \to 0.$$

The proof, which uses elementary and classical methods, first appeared in Banyaga's foundational paper [50]. This result is not used later in the book, but it puts statements such as Arnold's conjectures (which concern the elements of $\mathrm{Ham}(M)$ rather than $\mathrm{Symp}(M)$) in a proper perspective.

Most of the above results also extend to the case of noncompact M, provided that we restrict to symplectomorphisms of compact support. As we shall see in Section 10.3, in this case there is another interesting homomorphism called the Calabi homomorphism that takes values in $\mathbb{R}$ and may be defined on the universal cover of $\mathrm{Ham}_c(M)$. The chapter ends with a brief comparison of the topological properties of the group of symplectomorphisms with those of the group of diffeomorphisms.

10.1 Basic properties

Although many of the results in this chapter have analogues in the noncompact case, for simplicity we assume throughout that M is connected and closed, i.e. compact and without boundary, unless it is explicitly mentioned otherwise. Also, unless otherwise indicated, the group of symplectomorphisms has the C^1-topology.

We first show that two symplectomorphisms ψ_0 and ψ_1 of M which are sufficiently close in the C^1 topology can be joined by a smooth path of symplectomorphisms. The proof is based on the fact that a neighbourhood of the identity in $\mathrm{Symp}(M)$ may be identified with a neighbourhood of zero in the vector space of closed 1-forms on M.

Theorem 10.1.1 (Weinstein [669]) *Let (M, ω) be a closed symplectic manifold. Then the group of symplectomorphisms $\mathrm{Symp}(M, \omega)$ is locally path-connected, i.e. every symplectomorphism has a neighbourhood in $\mathrm{Symp}(M, \omega)$ which is path-connected and open in the C^1 topology.*

Proof: It suffices to show that there is a path-connected neighbourhood $\mathcal{U}_\delta$ of the identity in $\mathrm{Symp}(M)$. Then $\psi\mathcal{U}_\delta$ is a path-connected neighbourhood of ψ for every $\psi \in \mathrm{Symp}(M)$. Now, we saw in Proposition 3.4.14 that there is a bijective correspondence between symplectomorphisms that are C^1-close to the identity and C^1-small closed 1-forms. Hence if we choose a suitable path-connected subspace of closed 1-forms the corresponding subset in $\mathrm{Symp}(M)$ will be the desired neighbourhood of the identity. We now quantify this argument in order to clarify ideas and introduce notation that will be useful later on.

By Theorem 3.4.13 there exist an open neighbourhood $\mathcal{N}(\Delta) \subset M \times M$ of the diagonal $\Delta \subset M \times M$, an open neighbourhood $\mathcal{N}(M_0) \subset T^*M$ of the zero-section $M_0 \subset T^*M$, and a diffeomorphism $\Psi : \mathcal{N}(\Delta) \to \mathcal{N}(M_0)$ such that

$$\Psi^*\omega_{\mathrm{can}} = (-\omega) \times \omega, \qquad \Psi(q,q) = (q,0) \in M_0 \subset T^*M \tag{10.1.1}$$

for every $q \in M$. Now let $\pi : T^*M \to M$ denote the projection of the cotangent bundle and let $\mathrm{pr}_1, \mathrm{pr}_2 : M \times M \to M$ denote the projections on the first and second factor. Choose a Riemannian metric on M, thus obtaining a C^1 norm on the space of 1-forms on M. Because $\mathcal{N}(M_0)$ is open, we may choose a constant $\delta > 0$ such that every 1-form $\sigma \in \Omega^1(M)$ satisfies

$$\|\sigma\|_{C^1} < \delta \implies \begin{array}{l} \mathrm{graph}(\sigma) \subset \mathcal{N}(M_0) \text{ and for i=1,2 the maps} \\ \mathrm{pr}_i \circ \Psi^{-1} \circ \sigma : M \to M \text{ are diffeomorphisms.} \end{array} \tag{10.1.2}$$

Here we identify a 1-form $\sigma \in \Omega^1(M)$ with a smooth map $\sigma : M \to T^*M$, and denote the image of this map by $\mathrm{graph}(\sigma) := \sigma(M)$. Thus by choice of $\delta > 0$ each element σ with $\|\sigma\|_{C^1} < \delta$ defines a diffeomorphism ψ_σ of M given by

$$\psi_\sigma := \mathrm{pr}_2 \circ \Psi^{-1} \circ \sigma \circ \left(\mathrm{pr}_1 \circ \Psi^{-1} \circ \sigma\right)^{-1}. \tag{10.1.3}$$

Further, by construction

$$\Psi\big(\mathrm{graph}(\psi_\sigma)\big) = \{\Psi\big(\mathrm{pr}_1 \circ \Psi^{-1} \circ \sigma(x), \mathrm{pr}_2 \circ \Psi^{-1} \circ \sigma(x)\big) \big| x \in M\} = \mathrm{graph}(\sigma).$$

It now follows from Proposition 3.4.14 that $\psi_\sigma \in \mathrm{Symp}(M)$ exactly if σ is closed. Hence if we define

$$\mathcal{V}_\delta := \big\{\sigma \in \Omega^1(M) \mid \|\sigma\|_{C^1} < \delta, d\sigma = 0\big\},$$

the set

$$\mathcal{U}_\delta := \big\{\psi_\sigma \mid \sigma \in \mathcal{V}_\delta\big\} \tag{10.1.4}$$

is an open neighbourhood of the identity in $\mathrm{Symp}(M)$. But $\mathcal{V}_\delta$ is path-connected. Hence so is $\mathcal{U}_\delta$. This proves Theorem 10.1.1. □

The group of symplectomorphisms of a closed symplectic manifold is an infinite-dimensional Lie group whose Lie algebra is the space $\mathcal{X}(M,\omega)$ of symplectic vector fields (see Proposition 3.1.5). It is one of Cartan's list of indecomposable Lie groups. (See [589, §5.7].) Some other groups on the list are $\mathrm{Diff}(M)$, $\mathrm{Diff}_{\mathrm{vol}}(M)$, and the group of contactomorphisms. These groups have interesting algebraic structures. For example, it was discovered in the 1970s by Epstein and Herman that when M is closed, the identity component $\mathrm{Diff}_0(M)$ of $\mathrm{Diff}(M)$, like that of the group $\mathrm{Homeo}(M)$ of homeomorphisms of M, is a simple group, i.e. it has no nontrivial normal subgroups. By contrast, if M is a closed symplectic manifold with $H^1(M;\mathbb{R}) \neq 0$, then $\mathrm{Ham}(M)$ is a nontrivial normal subgroup of the identity component $\mathrm{Symp}_0(M)$ of $\mathrm{Symp}(M)$ (see Exercise 3.1.14). However a theorem of Banyaga [50] asserts that $\mathrm{Ham}(M)$ is simple. Similarly, in the case of noncompact M, the kernel of the Calabi homomorphism $\mathrm{CAL} : \mathrm{Ham}_c(M) \to \mathbb{R}$ is a nontrival normal subgroup of $\mathrm{Ham}_c(M)$ (Section 10.3), that is simple by another theorem of Banyaga. As we explain in Section 10.3 below, there are analogous results for the group $\mathrm{Diff}_{\mathrm{vol}}(M)$ of volume-preserving diffeomorphisms.

The group $\mathrm{Symp}(M)$ is significantly smaller than $\mathrm{Diff}_{\mathrm{vol}}(M)$. For instance, the Lie algebra of $\mathrm{Symp}(M)$ consists of symplectic vector fields which (at least locally) are determined by a single function on M, while that of $\mathrm{Diff}_{\mathrm{vol}}(M)$ consists of divergence-free vector fields which depend locally on a set of $2n-1$ functions on M. This makes it plausible that the C^0-closure of $\mathrm{Symp}(M)$ in the group of diffeomorphisms is also smaller than $\mathrm{Diff}_{\mathrm{vol}}(M)$, i.e. that rigidity holds. However, this reasoning is nowhere near a proof and, as we mentioned in Chapter 1, it was only in the 1980s that rigidity was established. All proofs of this fact begin with a basic result along the lines of the nonsqueezing theorem. The proofs of Eliashberg and Hofer then use a local argument as in Chapter 12 below. On the other hand, Gromov's original proof took a different, more global approach. Using a difficult implicit function theorem, he showed in his book [288] (see the Maximality Theorem in [288, 3.4.4 (H)]) that the C^0-closure of $\mathrm{Symp}(M)$ in $\mathrm{Diff}(M)$ must either be $\mathrm{Symp}(M)$ or a subgroup of finite codimension in $\mathrm{Diff}_{\mathrm{vol}}(M)$ with no intermediate possibilities. Moreover, this subgroup contains all volume-preserving diffeomorphisms with zero flux (see Exercise 10.2.23) and hence, if M is simply connected, it must be the whole group $\mathrm{Diff}_{\mathrm{vol}}(M)$ itself.

Hamiltonian symplectomorphisms

Let $\mathrm{Symp}_0(M) = \mathrm{Symp}_0(M,\omega)$ denote the connected component of the identity in $\mathrm{Symp}(M)$. It follows from Theorem 10.1.1 that for every $\psi \in \mathrm{Symp}_0(M,\omega)$ there exists a smooth family of symplectomorphisms $\psi_t \in \mathrm{Symp}(M,\omega)$ starting at the identity such that $\psi_1 = \psi$. For such a family of symplectomorphisms there exists a unique family of vector fields $X_t : M \to TM$ that generates the isotopy ψ_t in the sense that

$$\partial_t \psi_t = X_t \circ \psi_t, \qquad \psi_0 = \mathrm{id}. \tag{10.1.5}$$

Recall from Proposition 3.1.5 that since ψ_t is a symplectomorphism for every t, the vector fields X_t are symplectic and so $d\,\iota(X_t)\omega = 0$. If all these 1-forms are exact then there exists a smooth family of Hamiltonian functions $H_t : M \to \mathbb{R}$ such that

$$\iota(X_t)\omega = dH_t. \tag{10.1.6}$$

Thus, for each t, X_t is the Hamiltonian vector field associated to H_t. In this situation ψ_t is said to be the Hamiltonian isotopy generated by the time-dependent Hamiltonian function $[0,1] \times M \to \mathbb{R} : (t,p) \mapsto H_t(p)$. A symplectomorphism $\psi \in \mathrm{Symp}(M)$ is called **Hamiltonian** if there is a Hamiltonian isotopy ψ_t from $\psi_0 = \mathrm{id}$ to $\psi_1 = \psi$. As in (3.1.8) the space of Hamiltonian symplectomorphisms is denoted by $\mathrm{Ham}(M) = \mathrm{Ham}(M,\omega)$. The next proposition summarizes the basic observations of Exercise 3.1.14.

Proposition 10.1.2 *The Hamiltonian symplectomorphisms form a path-connected normal subgroup* $\mathrm{Ham}(M)$ *of* $\mathrm{Symp}(M)$.

Proof: See Exercise 3.1.14. That $\mathrm{Ham}(M)$ is path-connected follows from its very definition. □

Exercise 10.1.3 Another way of seeing that $\mathrm{Ham}(M)$ is a group is to use juxtaposition of paths. In other words, think of the composite $\phi \circ \psi$ as the endpoint of (a smooth reparametrization of) the path

$$\chi_t = \begin{cases} \phi_{2t}, & 0 \le t \le 1/2, \\ \psi_{2t-1} \circ \phi_1, & 1/2 \le t \le 1. \end{cases}$$

Show that given any diffeomorphism ϕ, the paths ψ_t and $\psi_t \circ \phi$ have the same family of generating vector fields X_t. Use this to show that $\mathrm{Ham}(M,\omega)$ is closed under composition. Furthermore, by considering the path

$$\psi_{1-t} \circ \psi_1^{-1}, \qquad 0 \le t \le 1,$$

show that $\mathrm{Ham}(M)$ is closed under inverses. □

Exercise 10.1.4 Consider the 2-torus

$$\mathbb{T}^2 = \mathbb{R}^2/\mathbb{Z}^2$$

with its standard symplectic form. Every diffeomorphism ψ of $\mathbb{T}^2$ which is homotopic to the identity can be written in the form

$$\psi(x,y) = (x + \alpha(x,y), y + \beta(x,y)),$$

where the functions α and β are of period 1 in both variables $x, y \in \mathbb{R}$. Show that if ψ is Hamiltonian then

$$\int_{\mathbb{T}^2} \alpha\, dxdy = \int_{\mathbb{T}^2} \beta\, dxdy = 0.$$

Hint: If ψ_t is a Hamiltonian isotopy, write down expressions for the corresponding functions α_t and β_t in terms of the generating Hamiltonian. This result is generalized in Proposition 10.2.21 below. □

Remark 10.1.5 (Noncompact manifolds) As explained in Remark 3.1.15, the above definitions make perfectly good sense for noncompact manifolds, which we assume here to be without boundary. We can either consider symplectomorphisms with arbitrary support, in which case we will denote the groups by $\mathrm{Symp}(M)$ and $\mathrm{Ham}(M)$ as before, or restrict to the case of compact support in which case we will call the groups $\mathrm{Symp_c}(M) = \mathrm{Symp_c}(M, \omega)$ and $\mathrm{Ham_c}(M) = \mathrm{Ham_c}(M, \omega)$. The former groups are given the topology of C^1-convergence on compact sets, while the latter are topologized as direct limits. Thus,

$$\mathrm{Symp_c}(M) := \bigcup_{\substack{K \subset M \\ K \text{ compact}}} \mathrm{Symp}_K(M),$$

where Symp_K consists of all symplectomorphisms with support in the compact subset K. This means that a set $U \subset \mathrm{Symp_c}(M)$ is open if and only if $U \cap \mathrm{Symp}_K(M)$ is open for all K. One can then show that for any finite continuous path $[0,1] \to \mathrm{Symp_c}(M) : t \mapsto \phi_t$, there is some compact set K which contains the support of all the ϕ_t. The connected component of the identity in $\mathrm{Symp_c}(M)$ will be denoted by $\mathrm{Symp_{c,0}}(M)$. Similarly,

$$\mathrm{Ham_c}(M) := \bigcup_{\substack{K \subset M \\ K \text{ compact}}} \mathrm{Ham}_K(M),$$

where $\mathrm{Ham}_K(M)$ denotes all symplectomorphisms ϕ which are time-1 maps of some Hamiltonian isotopy ϕ_t with support in K. □

Consider an exact symplectic manifold $(M, -d\lambda)$ which is necessarily noncompact. Then for every compactly supported symplectomorphism $\psi \in \mathrm{Symp_{c,0}}(M)$, the 1-form $\psi^*\lambda - \lambda$ is necessarily closed, and hence represents an element of the compactly supported de Rham cohomology group $H^1_c(M; \mathbb{R})$ (see Bott and Tu [78]). We saw in Proposition 9.3.1 that a compactly supported symplectic isotopy $\psi_t : M \to M$ with $\psi_0 = \mathrm{id}$ is a Hamiltonian isotopy if and only if the 1-forms $\psi_t^*\lambda - \lambda = dF_t$ are all exact. Hence it is natural to consider the map

$$\mathrm{Symp_{c,0}}(M) \to H^1_c(M; \mathbb{R}) : \psi \mapsto [\lambda - \psi^*\lambda].$$

It is easy to see that this is a homomorphism. The above considerations show that the group of Hamiltonian symplectomorphisms is contained in the kernel. In fact, one can show that the group $\mathrm{Ham_c}(M)$ agrees with the kernel of this homomorphism. (See Exercise 10.2.6 below.) It follows that there is an exact sequence

$$0 \to \mathrm{Ham_c}(M) \to \mathrm{Symp_{c,0}}(M) \to H^1_c(M; \mathbb{R}) \to 0. \tag{10.1.7}$$

Moreover, by Corollary 9.3.3 every isotopy of symplectomorphisms in $\mathrm{Ham_c}(M)$ is a Hamiltonian isotopy. In the next section we show how these results can be generalized to arbitrary symplectic manifolds by generalizing the above homomorphism to a homomorphism known as the **flux**.

Example 10.1.6 (Cotangent bundles) Consider the case when our exact symplectic manifold is a cotangent bundle $(T^*L, -d\lambda_{\mathrm{can}})$. We shall use the notation (q, v^*) for points of T^*L, where $q \in L$ and $v^* \in T_q^*L$. If σ is a closed 1-form on L, there is an associated vertical diffeomorphism ν_σ given by

$$\nu_\sigma(q, v^*) = (q, v^* + \sigma(q)).$$

It follows easily from Proposition 3.1.18 that

$$\nu_\sigma^* \lambda_{\mathrm{can}} - \lambda_{\mathrm{can}} = \pi^*\sigma,$$

where $\pi : T^*L \to L$ is the projection. Hence ν_σ is a symplectomorphism. Moreover, it is the endpoint of the isotopy $t \mapsto \nu_{t\sigma}$. Hence it follows from Proposition 9.3.1 that ν_σ is Hamiltonian if and only if σ is exact. □

10.2 The flux homomorphism

The flux homomorphism is most naturally defined on the universal cover of $\mathrm{Symp}_{c,0}(M)$. For simplicity we will start by assuming that M is closed. Let $\widetilde{\mathrm{Symp}}_0(M, \omega)$ denote the universal cover of the identity component $\mathrm{Symp}_0(M, \omega)$. A point in $\widetilde{\mathrm{Symp}}_0(M, \omega)$ is a homotopy class of smooth paths $\psi_t \in \mathrm{Symp}_0(M, \omega)$ with fixed endpoints $\psi_0 = \mathrm{id}$ and $\psi_1 = \psi$. We denote such a homotopy class by $\{\psi_t\}$. The group structure can be defined in two equivalent ways, either by composition of symplectomorphisms or by juxtaposition of paths. Thus we may think of $\{\psi_t\} \cdot \{\phi_t\}$ either as the equivalence class of the path $\psi_t \circ \phi_t$ or as the equivalence class of (a smooth reparametrization of) the juxtaposition χ_t defined by

$$\chi_t = \begin{cases} \phi_{2t}, & 0 \le t \le 1/2, \\ \psi_{2t-1} \circ \phi_1, & 1/2 \le t \le 1. \end{cases}$$

It is easy to check that the path $\psi_t \circ \phi_t$ is homotopic to χ_t with fixed endpoints and hence $\{\chi_t\} = \{\psi_t \circ \phi_t\}$.

The **flux homomorphism** $\mathrm{Flux} : \widetilde{\mathrm{Symp}}_0(M, \omega) \to H^1(M; \mathbb{R})$ is defined by

$$\mathrm{Flux}(\{\psi_t\}) = \int_0^1 [\iota(X_t)\omega]\, dt \in H^1(M; \mathbb{R}), \tag{10.2.1}$$

where the vector field X_t is determined by (10.1.5). We must check that this is well-defined, i.e. that it only depends on the homotopy class of the path ψ_t with fixed endpoints. We will use the fact that, under the usual identification of $H^1(M; \mathbb{R})$ with $\mathrm{Hom}(\pi_1(M); \mathbb{R})$, the above cohomology class corresponds to the homomorphism $\pi_1(M) \to \mathbb{R}$ defined by

$$\gamma \mapsto \int_0^1 \int_0^1 \omega(X_t(\gamma(s)), \dot\gamma(s))\, ds dt \tag{10.2.2}$$

for $\gamma : \mathbb{R}/\mathbb{Z} \to M$.

Lemma 10.2.1 *The right-hand side of* (10.2.2) *depends only on the homotopy class of* γ *and on the homotopy class of* ψ_t *with fixed endpoints.*

Proof: Since ψ_t is a family of symplectomorphisms, the 1-forms $\iota(X_t)\omega$ are closed. Hence the right-hand side of (10.2.2) depends only on the homotopy class of γ. Now define the map $\beta : \mathbb{R}/\mathbb{Z} \times [0,1] \to M$ by setting $\beta(s,t) := \psi_t^{-1}(\gamma(s))$. Differentiate the identity $\psi_t(\beta(s,t)) = \gamma(s)$ with respect to s and t to obtain

$$d\psi_t(\beta)\frac{\partial \beta}{\partial s} = \dot{\gamma}(s), \qquad d\psi_t(\beta)\frac{\partial \beta}{\partial t} = -X_t(\gamma(s)).$$

Since $\psi_t^*\omega = \omega$ this gives

$$\mathrm{Flux}(\{\psi_t\})(\gamma) = \int_0^1\int_0^1 \omega\left(\frac{\partial \beta}{\partial s}, \frac{\partial \beta}{\partial t}\right) dsdt = \int_{\mathbb{R}/\mathbb{Z}\times[0,1]} \beta^*\omega.$$

This expression depends only on the homotopy class of β subject to the boundary conditions

$$\beta(s+1,t) = \beta(s,t), \qquad \beta(s,1) = \psi^{-1}(\beta(s,0)),$$

where $\psi = \psi_1$. Geometrically these conditions mean that β is a map of the cylinder $S^1 \times [0,1]$ into M with fixed boundary circles. Hence the right-hand side of (10.2.2) depends only on the homotopy class of ψ_t subject to $\psi_0 = \mathrm{id}$ and $\psi_1 = \psi$. This proves Lemma 10.2.1. □

We have now established that the flux is well-defined. Lemma 10.2.1 also gives a geometric interpretation. The value of $\mathrm{Flux}(\{\psi_t\})$ on the loop γ is simply the symplectic area swept out by the path of γ under the isotopy ψ_t. In fact our definition of β in terms of the inverse ${\psi_t}^{-1}$ was convenient for the purpose of the proof, but the same argument also works with the perhaps more natural map $\beta' : [0,1] \times \mathbb{R}/\mathbb{Z} \to M$ defined by

$$\beta'(t,\theta) := \psi_t(\gamma(\theta)).$$

This map represents the image of the loop γ under the flow ψ_t. (See Fig. 10.1.) The next lemma shows that this gives the same answer.

Lemma 10.2.2

$$\int \beta^*\omega = \int \beta'^*\omega.$$

Proof: It suffices to shows that the maps

$$\psi_1(\beta(s,t)) = \psi_1 \circ \psi_t^{-1}(\gamma(s)), \qquad \beta'(1-t,s) = \psi_{1-t}(\gamma(s))$$

are homotopic with fixed boundary circles (at $t = 0$ and $t = 1$). An explicit homotopy is given by $\beta_\lambda(s,t) := \psi_{1-\lambda t} \circ \psi_{(1-\lambda)t}^{-1}(\gamma(s))$ from $\beta_0 = \psi \circ \beta$ to $\beta_1(s,t) = \beta'(1-t,s)$, and it satisfies $\beta_\lambda(s,0) = \psi(\gamma(s))$ and $\beta_\lambda(s,1) = \gamma(s)$ for all λ and s. This proves Lemma 10.2.2. □

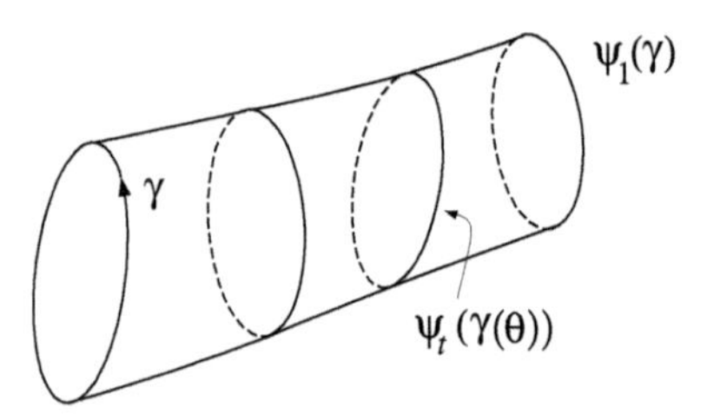

FIG. 10.1. The flux through γ.

Exercise 10.2.3 Prove that $\text{Flux} : \widetilde{\text{Symp}}_0(M,\omega) \to H^1(M;\mathbb{R})$ is a homomorphism. **Hint**: Represent the product by juxtaposition of paths as in Exercise 10.1.3. Alternatively, use Proposition 3.1.5 to prove that if $X \in \mathcal{X}(M,\omega)$ is a symplectic vector field and $\psi \in \text{Symp}_0(M,\omega)$ then $\psi^*X - X$ is a Hamiltonian vector field. Now use the fact that if ϕ_t is generated by X_t and ψ_t is generated by Y_t then $\psi_t \circ \phi_t$ is generated by $Y_t + \psi_{t*}X_t$. □

Exercise 10.2.4 Prove that $\text{Flux} : \widetilde{\text{Symp}}_0(M,\omega) \to H^1(M;\mathbb{R})$ is surjective. **Hint**: If ψ_t is the flow of a (time-independent) symplectic vector field X on the time interval $0 \le t \le 1$ then

$$\text{Flux}(\{\psi_t\}) = [\iota(X)\omega].$$

Use this to construct a right inverse of the flux homomorphism, i.e. a smooth map $s : H^1(M;\mathbb{R}) \to \widetilde{\text{Symp}}_0(M,\omega)$ such that $\text{Flux} \circ s = \text{id}$. Calculate s explicitly when M is the 2-torus $\mathbb{T}^2$. □

The next theorem is the main result of this section. It characterizes the kernel of the flux homomorphism.

Theorem 10.2.5 *Let (M,ω) be a closed connected symplectic manifold and $\psi \in \text{Symp}_0(M,\omega)$. Then ψ is a Hamiltonian symplectomorphism if and only if there exists a symplectic isotopy*

$$[0,1] \to \text{Symp}_0(M,\omega) : [0,1] \mapsto \psi_t$$

such that

$$\psi_0 = \text{id}, \qquad \psi_1 = \psi, \qquad \text{Flux}(\{\psi_t\}) = 0.$$

Moreover, if $\text{Flux}(\{\psi_t\}) = 0$ then $\{\psi_t\}$ is isotopic with fixed endpoints to a Hamiltonian isotopy.

Proof: If ψ is Hamiltonian, it is the endpoint of a Hamiltonian isotopy ψ_t corresponding to some family of Hamiltonian functions $H_t : M \to \mathbb{R}$, and

$$\text{Flux}(\{\psi_t\}) = \int_0^1 [\iota(X_t)\omega]\, dt = \int_0^1 [dH_t]\, dt = 0.$$

Conversely, let $\psi_t \in \text{Symp}_0(M,\omega)$ be a symplectic isotopy from $\psi_0 = \text{id}$ to $\psi_1 = \psi$ such that $\text{Flux}(\{\psi_t\}) = 0$, and define $X_t \in \mathcal{X}(M,\omega)$ by

$$\frac{d}{dt}\psi_t = X_t \circ \psi_t.$$

We know that the integral $\int_0^1 \iota(X_t)\omega\,dt$ is exact, and what we must do is change the isotopy ψ_t so that $\iota(X_t)\omega$ is exact for each t. Equivalently, we must make the integral $\int_0^T \iota(X_t)\omega\,dt$ exact for each $T \in [0,1]$. Thus $\mathrm{Flux}(\{\psi_t\}_{0\le t\le T}) = 0$ for every $T \in [0,1]$.

The first step is to modify ψ_t by a Hamiltonian isotopy such that the 1-form $\int_0^1 \iota(X_t)\omega\,dt$ is zero rather than merely exact. To achieve this, note first that since $\mathrm{Flux}(\{\psi_t\}) = 0$, there exists a function $F : M \to \mathbb{R}$ such that

$$\int_0^1 \iota(X_t)\omega\,dt = dF.$$

Let ϕ_F^s be the Hamiltonian flow of F. Since ϕ_F^s is Hamiltonian for each $s \in \mathbb{R}$, it suffices to prove the theorem for the composition $\phi_F^{-1} \circ \psi$ instead of ψ. But this is the endpoint of the juxtaposition ψ_t' defined by $\psi_t' := \psi_{2t}$ for $0 \le t \le 1/2$, and $\psi_t' := \phi_F^{1-2t} \circ \psi_1$ for $1/2 \le t \le 1$. This isotopy ψ_t' (or a suitable smooth reparametrization) is generated by a smooth family of vector fields X_t' such that $\int_0^1 X_t'\,dt = 0$. Hence, from now on we assume that $\psi = \psi_1$ for some isotopy with

$$\int_0^1 X_t\,dt = 0.$$

Next, for every t, let $\theta_t^s \in \mathrm{Symp}_0(M,\omega)$, $s \in \mathbb{R}$, be the flow generated by the symplectic vector field

$$Y_t := -\int_0^t X_\lambda\,d\lambda.$$

Thus

$$\partial_s\theta_t^s = Y_t \circ \theta_t^s, \qquad \theta_t^0 = \mathrm{id}.$$

Observe that $Y_0 = Y_1 = 0$ and hence $\theta_0^s = \theta_1^s = \mathrm{id}$ for all s.

We claim that

$$\phi_t := \theta_t^1 \circ \psi_t$$

is the desired Hamiltonian isotopy from $\phi_0 = \mathrm{id}$ to $\phi_1 = \psi_1 = \psi$. (See Fig. 10.2.) To see this, note that because Flux is a homomorphism of groups (see Exercise 10.2.3),

$$\begin{aligned}
\mathrm{Flux}(\{\phi_t\}_{0\le t\le T}) &= \mathrm{Flux}(\{\theta_t^1\}_{0\le t\le T}) + \mathrm{Flux}(\{\psi_t\}_{0\le t\le T}) \\
&= \mathrm{Flux}(\{\theta_T^s\}_{0\le s\le 1}) + \int_0^T [\iota(X_t)\omega]\,dt \\
&= [\iota(Y_T)\omega] + \int_0^T [\iota(X_t)\omega]\,dt \\
&= 0.
\end{aligned}$$

Here the second step uses the homotopy invariance of the flux, and the third follows from the fact that θ_T^s is the flow of Y_T. □

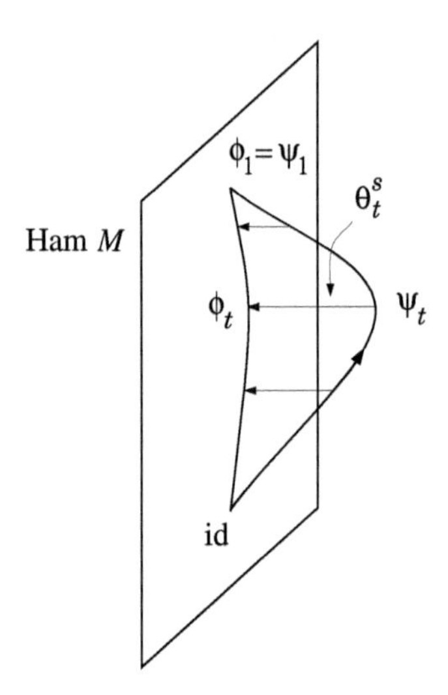

FIG. 10.2. The homotopy to Ham(M).

Exercise 10.2.6 (Noncompact manifolds) Prove that Theorem 10.2.5 extends to compactly supported isotopies of noncompact symplectic manifolds (M, ω). If $\omega = -d\lambda$ is exact, show that the sequence (10.1.7) is exact, i.e. prove that a compactly supported symplectomorphism $\phi \in \mathrm{Symp}_c(M, \omega)$ belongs to the subgroup $\mathrm{Ham}_c(M, \omega)$ if and only if it belongs to the identity component of $\mathrm{Symp}_c(M, \omega)$ and there exists a compactly supported smooth function $F : M \to \mathbb{R}$ such that $\phi^*\lambda - \lambda = dF$. **Hint:** See Proposition 9.3.1 and Remark 10.1.5. When extending Theorem 10.2.5 to the noncompact case, replace $H^1(M; \mathbb{R})$ by the group $H^1_c(M; \mathbb{R})$ of compactly supported 1-cocycles. These are represented by compactly supported 1-forms (see Bott and Tu [78]). In particular, if M is the interior of a compact manifold $\overline{M}$ with boundary $\partial\overline{M}$ then $H^1_c(M; \mathbb{R}) \cong H^1(\overline{M}, \partial\overline{M}; \mathbb{R})$. □

The next lemma rephrases an observation made in Section 10.1 in terms of the flux homomorphism.

Lemma 10.2.7 *If $\omega = -d\lambda$ and $\psi_t : M \to M$ is a compactly supported symplectic isotopy then* $\mathrm{Flux}(\{\psi_t\}) = [\lambda - \psi_1^*\lambda]$.

Proof: If ψ_t is a symplectic isotopy generated by compactly supported symplectic vector fields X_t, then

$$[\iota(X_t)\omega] = -[\psi_t^*\iota(X_t)d\lambda] = -[\psi_t^*\mathcal{L}_{X_t}\lambda] = -\frac{d}{dt}[\psi_t^*\lambda].$$

The lemma follows by integrating this identity from $t = 0$ to $t = 1$. □

Consider the correspondence $\psi \mapsto \sigma = \mathcal{C}(\psi)$ between symplectomorphisms $\psi : M \to M$ which are C^1-close to the identity and closed 1-forms σ on M which was defined in the proof of Theorem 10.1.1. More precisely, $\Psi : \mathcal{N}(\Delta) \to \mathcal{N}(M_0)$ is a symplectomorphism from a neighbourhood of the diagonal in $\Delta \subset M \times M$

to a neighbourhood of the zero section $M_0 \subset T^*M$ such that $\Psi^*\omega_{\mathrm{can}} = (-\omega) \oplus \omega$ and $\Psi(q, q) = (q, 0)$ for $q \in M$. Then $\sigma = \mathcal{C}(\psi) \in \Omega^1(M)$ is defined by

$$\Psi(\mathrm{graph}(\psi)) = \mathrm{graph}(\sigma)$$

whenever $\psi \in \mathrm{Symp}(M)$ is sufficiently close to the identity in the C^1-topology.

Lemma 10.2.8 *If $\psi_t \in \mathrm{Symp}(M)$ is a symplectic isotopy such that ψ_t is sufficiently C^1-close to the identity for every t then*

$$\mathrm{Flux}(\{\psi_t\}) = -[\sigma_1],$$

where $\sigma_t = \mathcal{C}(\psi_t)$. In particular, ψ_t is a Hamiltonian isotopy if and only if the 1-forms σ_t are all exact.

Proof: Consider the local symplectic isotopy $\Phi_t = \Psi \circ (\mathrm{id} \times \psi_t) \circ \Psi^{-1}$ in $\mathcal{N}(M_0) \subset T^*M$. There exists a smooth family of diffeomorphisms $f_t : M \to M$ such that

$$\Phi_t \circ \iota = \sigma_t \circ f_t,$$

where $\iota : M \to T^*M$ is the inclusion of the zero section and we think of σ_t as a map $M \to T^*M$. Now, $\iota_\Delta = \Psi^{-1} \circ \iota : M \to M \times M$ is the inclusion of the diagonal and hence

$$\mathrm{Flux}(\{\psi_t\}) = \iota_\Delta^* \mathrm{Flux}(\{\mathrm{id} \times \psi_t\}) = \iota_\Delta^* \Psi^* \mathrm{Flux}(\{\Phi_t\}) = \iota^* \mathrm{Flux}(\{\Phi_t\}).$$

Note that the isotopy Φ_t on an open neighbourhood of the zero section in T^*M is not compactly supported. However, Lemma 10.2.7 still applies and hence $\mathrm{Flux}(\{\Phi_t\}) = [\lambda_{\mathrm{can}} - \Phi_1^*\lambda_{\mathrm{can}}]$. Thus

$$\iota^* \mathrm{Flux}(\{\Phi_t\}) = -[\iota^*\Phi_1^*\lambda_{\mathrm{can}}] = -[f_1^*\sigma_1^*\lambda_{\mathrm{can}}] = -[\sigma_1].$$

Here we have used the formula $\sigma_1^*\lambda_{\mathrm{can}} = \sigma_1$ and the fact that f_1 is isotopic to the identity. This proves Lemma 10.2.8. □

If ψ_t is a Hamiltonian isotopy such that only the endpoint $\psi = \psi_1$ is C^1-close to the identity then the 1-form $\sigma = \mathcal{C}(\psi)$ need not be exact. Lemma 10.2.11 below shows, however, that the cohomology class of σ must belong to the flux group (sometimes also called the Calabi group), which we define next.

Definition 10.2.9 *The* **flux group** *of a symplectic manifold (M, ω) is the image of the fundamental group of the group of symplectomorphisms $\mathrm{Symp}_0(M, \omega)$ under the flux homomorphism. It is denoted by*

$$\Gamma := \Gamma_\omega := \mathrm{Flux}\big(\pi_1(\mathrm{Symp}_0(M))\big) \subset H^1(M; \mathbb{R}). \tag{10.2.3}$$

Note that $\pi_1(\mathrm{Symp}_0(M, \omega)) \subset \widetilde{\mathrm{Symp}}_0(M, \omega)$. By equation (10.2.2), the flux group Γ_ω is a subgroup of $H^1(M; P_\omega)$, where $P_\omega := [\omega] \cdot H_2(M; \mathbb{Z}) \subset \mathbb{R}$ is the countable group of periods of ω, i.e. the set of values taken by ω on the countable group $H_2(M; \mathbb{Z})$ of integral 2-cycles. Hence Γ is itself a countable group.

Exercise 10.2.10 Let $\mathbb{R}/\mathbb{Z} \times M \to M : (t,p) \mapsto f_t(p)$ be a smooth map and define

$$\alpha_f := \int_0^1 \omega(\partial_t f_t, df_t \cdot)\, dt \in \Omega^1(M).$$

Prove that α_f is closed, and represents the class $[\alpha_f] \in H^1(M;\mathbb{R})$ whose value on the homology class of the loop $\gamma : S^1 \to M$ is $\int \beta'^*\omega$, where $\beta' : \mathbb{T}^2 \to M$ is given by $\beta'(t,\theta) = f_t(\gamma(\theta))$. Prove further that the set

$$\Gamma_{\mathrm{top}} := \big\{[\alpha_f]\,|\, f \in C^\infty(\mathbb{R}/\mathbb{Z} \times M, M)\big\} \subset H^1(M;\mathbb{R})$$

is a subgroup of $H^1(M; P_\omega)$, and that

$$\Gamma_\omega \subset \Gamma_{\mathrm{top}}.$$

Hint: Lemmas 10.2.1 and 10.2.2. □

The next lemma characterizes Hamiltonian symplectomorphisms close to the identity in terms of the flux group Γ_ω. This result has several important consequences which will be discussed in the subsequent propositions.

Lemma 10.2.11 *If $\psi \in \mathrm{Symp}_0(M,\omega)$ is sufficiently close to the identity in the C^1-topology and $\sigma = \mathcal{C}(\psi) \in \Omega^1(M)$ is defined as above, then*

$$\psi \in \mathrm{Ham}(M,\omega) \qquad \Longleftrightarrow \qquad [\sigma] \in \Gamma_\omega.$$

Proof: Consider the symplectic isotopy ψ_t with $\mathcal{C}(\psi_t) = t\sigma$. Then it follows from Lemma 10.2.8 that $\mathrm{Flux}(\{\psi_t\}) = -[\sigma]$. Since ψ is a Hamiltonian symplectomorphism, this extends to a loop $[0,2] \to \mathrm{Symp}_0(M,\omega) : t \mapsto \psi_t$ which for $t \geq 1$ is generated by Hamiltonian vector fields and satisfies $\psi_0 = \psi_2 = \mathrm{id}$. Moreover,

$$\mathrm{Flux}(\{\psi_t\}_{0\leq t\leq 2}) = \mathrm{Flux}(\{\psi_t\}_{0\leq t\leq 1}) = -[\sigma].$$

This shows that $[\sigma] \in \Gamma_\omega$. Conversely, assume $[\sigma] \in \Gamma_\omega$, choose a loop of symplectomorphisms $\psi_t \in \mathrm{Symp}_0(M,\omega)$ with $\mathrm{Flux}(\{\psi_t\}) = [\sigma]$, and extend this to the interval $[0,2]$ by $\mathcal{C}(\psi_t) = [(t-1)\sigma]$ for $1 \leq t \leq 2$. This works whenever σ is sufficiently small in the C^1-topology. The resulting path ψ_t, $0 \leq t \leq 2$, has zero flux and hence, by Theorem 10.2.5, can be deformed to a Hamiltonian isotopy. This shows that $\psi = \psi_2$ is a Hamiltonian symplectomorphism. This proves Lemma 10.2.11. □

Proposition 10.2.12 *Every smooth path $\psi_t \in \mathrm{Ham}(M)$ is generated by Hamiltonian vector fields.*

Proof: Assume $\psi_0 = \mathrm{id}$. Then, by Lemma 10.2.11, the cohomology class of $\sigma_t = \mathcal{C}(\psi_t)$ is in $\Gamma_\omega = \mathrm{Flux}\big(\pi_1(\mathrm{Symp}_0(M))\big)$ for t sufficiently small. Since Γ_ω is a countable set, the smooth map $t \mapsto [\sigma_t]$ is constant. Since $\sigma_0 = 0$ it follows that σ_t is exact for every sufficiently small t. By Lemma 10.2.8, this implies that $\mathrm{Flux}(\{\psi_t\}_{0\leq t\leq\varepsilon}) = 0$ for $\varepsilon > 0$ sufficiently small, and hence ψ_t is a Hamiltonian isotopy for t near zero. This proves Proposition 10.2.12. □

Proposition 10.2.13 *Let (M, ω) be a closed, connected symplectic manifold.*
(i) *There is an exact sequence of simply connected Lie groups*

$$0 \to \widetilde{\mathrm{Ham}}(M, \omega) \to \widetilde{\mathrm{Symp}}_0(M, \omega) \to H^1(M; \mathbb{R}) \to 0,$$

where $\widetilde{\mathrm{Ham}}(M, \omega)$ is the universal cover of $\mathrm{Ham}(M, \omega)$ and the third arrow is the flux homomorphism.
(ii) *There is an exact sequence of Lie algebras*

$$0 \to \mathbb{R} \to C^\infty(M) \to \mathcal{X}(M, \omega) \to H^1(M; \mathbb{R}) \to 0.$$

Here the third map is $H \mapsto X_H$ and the fourth map is $X \mapsto [\iota(X)\omega]$.
(iii) *There is an exact sequence of groups*

$$0 \to \pi_1(\mathrm{Ham}(M, \omega)) \to \pi_1(\mathrm{Symp}_0(M, \omega)) \to \Gamma_\omega \to 0,$$

where Γ_ω is the flux group in (10.2.3).
(iv) *There is an exact sequence of groups*

$$0 \to \mathrm{Ham}(M, \omega) \to \mathrm{Symp}_0(M, \omega) \xrightarrow{\rho} H^1(M; \mathbb{R})/\Gamma_\omega \to 0,$$

where ρ is the map induced by Flux. *Thus $\mathrm{Symp}_0(M, \omega)/\mathrm{Ham}(M, \omega)$ is isomorphic to $H^1(M; \mathbb{R})/\Gamma_\omega$.*

Proof: By Proposition 10.2.12, every smooth path $\psi_t \in \mathrm{Ham}(M)$ which starts at the identity is a Hamiltonian isotopy and hence has zero flux. This shows that $\widetilde{\mathrm{Ham}}(M) \subset \ker(\mathrm{Flux})$. Conversely, Theorem 10.2.5 shows that if $\mathrm{Flux}(\{\psi_t\}) = 0$ then the path ψ_t is homotopic, with fixed endpoints, to a Hamiltonian isotopy, and hence $\{\psi_t\} \in \widetilde{\mathrm{Ham}}(M)$. This shows that

$$\widetilde{\mathrm{Ham}}(M) = \ker(\mathrm{Flux}).$$

The first statement now follows from the surjectivity of Flux. The second statement is obvious. The third statement is also clear except for the fact that $\pi_1(\mathrm{Ham}(M))$ injects into $\pi_1(\mathrm{Symp}_0(M))$. To see this, it suffices to show that any path $[0, 1] \to \widetilde{\mathrm{Symp}}_0(M)$ with endpoints in $\widetilde{\mathrm{Ham}}(M)$ is isotopic with fixed endpoints to a path in $\widetilde{\mathrm{Ham}}(M) = \ker(\mathrm{Flux})$. This is a parametrized version of (i). The last statement is obvious. □

Remark 10.2.14 When (M, ω) is a compact, connected, symplectic manifold there is an exact sequence of Lie algebras

$$0 \to C^\infty_{\mathrm{mvz}}(M) \to \mathcal{X}(M, \omega) \to H^1(M; \mathbb{R}) \to 0,$$

where $C^\infty_{\mathrm{mvz}}(M)$ is the space of smooth functions on M with mean value zero. (See part (ii) of Proposition 10.2.13 and Remark 3.1.11.) □

Exercise 10.2.15 Show that the inclusion $\mathrm{Ham}(M) \to \mathrm{Symp}_0(M)$ induces an isomorphism on π_k for $k > 1$. □

The flux conjecture

The flux group $\Gamma_\omega \subset H^1(M;\mathbb{R})$ in Definition 10.2.9 has the following geometric significance.

Proposition 10.2.16 *Let (M,ω) be a closed symplectic manifold and denote by $\Gamma = \Gamma_\omega \subset H^1(M;\mathbb{R})$ the image of $\pi_1(\mathrm{Symp}_0(M,\omega))$ under the flux homomorphism. Then the following are equivalent.*

(i) *Γ_ω is discrete.*
(ii) $\mathrm{Ham}(M,\omega)$ *is a submanifold of* $\mathrm{Symp}(M,\omega)$.
(iii) $\mathrm{Ham}(M,\omega)$ *is C^∞-closed in* $\mathrm{Symp}(M,\omega)$.
(iv) $\mathrm{Ham}(M,\omega)$ *is C^1-closed in* $\mathrm{Symp}(M,\omega)$.

Proof: We prove that (i) is equivalent to (ii). Let $\mathcal{U}_\delta \subset \mathrm{Symp}_0(M,\omega)$ be the C^1-neighbourhood of the identity defined by equation (10.1.4) in the proof of Theorem 10.1.1, and denote by

$$\mathcal{C} : \mathcal{U}_\delta \to \mathcal{V}_\delta \subset (\ker d) \cap \Omega^1(M)$$

the diffeomorphism $\psi_\sigma \mapsto \sigma$, where ψ_σ is as in (10.1.2). By Lemma 10.2.11 the diffeomorphism $\mathcal{C} : \mathcal{U}_\delta \to \mathcal{V}_\delta$ maps the Hamiltonian symplectomorphisms onto the set of elements of $\mathcal{V}_\delta$ whose cohomology classes lie in Γ, i.e.

$$\mathcal{C}\big(\mathcal{U}_\delta \cap \mathrm{Ham}(M,\omega)\big) = \{\sigma \in \mathcal{V}_\delta \,|\, [\sigma] \in \Gamma\} =: \mathcal{V}_{\delta,\Gamma}.$$

Now Γ is discrete if and only if $\mathcal{V}_{\delta,\Gamma}$ consists entirely of exact 1-forms for $\delta > 0$ sufficiently small, and in this case $\mathrm{Ham}(M,\omega) \cap \mathcal{U}_\delta$ is mapped to a linear subspace of $\mathcal{V}_\delta$ under the coordinate chart $\mathcal{C}$ of $\mathrm{Symp}_0(M,\omega)$. This shows that (i) is equivalent to (ii).

We prove that (iii) implies (i). Thus assume that $\mathrm{Ham}(M,\omega)$ is C^∞-closed in $\mathrm{Symp}(M,\omega)$ and, by contradiction, that Γ is not discrete. Since Γ is a countable subgroup of the real vector space $H^1(M;\mathbb{R})$, it follows that Γ is not closed and, moreover, the set $\mathrm{cl}(\Gamma) \setminus \Gamma$ contains arbitrarily small elements. Hence there exists a closed 1-form $\sigma \in \mathcal{V}_\delta \setminus \mathcal{V}_{\delta,\Gamma}$ which can be approximated by a sequence $\sigma_\nu \in \mathcal{V}_{\delta,\Gamma}$ in the C^∞-topology. Choose $\psi, \psi_\nu \in \mathcal{U}_\delta \subset \mathrm{Symp}_0(M,\omega)$ such that $\mathcal{C}(\psi) = \sigma$ and $\mathcal{C}(\psi_\nu) = \sigma_\nu$. Then, by Lemma 10.2.11, $\psi_\nu \in \mathrm{Ham}(M,\omega)$ and $\psi \in \mathrm{Symp}_0(M,\omega) \setminus \mathrm{Ham}(M,\omega)$. Since ψ_ν converges to ψ in the C^∞ topology, this contradicts (iii).

We prove that (i) implies (iv). Assume Γ is a discrete subgroup of $H^1(M;\mathbb{R})$ and, by contradiction, that $\mathrm{Ham}(M,\omega)$ is not C^1-closed in $\mathrm{Symp}_0(M,\omega)$. Then there exists a symplectomorphism $\psi \in \mathrm{Symp}(M,\omega) \setminus \mathrm{Ham}(M,\omega)$ which can be approximated by a sequence $\psi_\nu \in \mathrm{Ham}(M,\omega)$ in the C^1-topology. Choose an integer k such that $\psi_k^{-1} \circ \psi \in \mathcal{U}$. Then $\psi_k^{-1} \circ \psi \in \mathcal{U} \setminus \mathrm{Ham}(M,\omega)$ and hence, by Lemma 10.2.11, we have $[\mathcal{C}(\psi_k^{-1} \circ \psi)] \notin \Gamma$ and $\gamma_\nu = [\mathcal{C}(\psi_k^{-1} \circ \psi_\nu)] \in \Gamma$ for every (sufficiently large) ν. Hence Γ is not closed, so that Γ is not discrete, in contradiction to (i).

Since (iv) immediately implies (iii), this proves Proposition 10.2.16. □

Proposition 10.2.16 shows that if Γ_ω is discrete, then $\mathrm{Ham}(M,\omega)$ is a Lie subgroup of $\mathrm{Symp}(M,\omega)$ whose Lie algebra is the space of Hamiltonian vector fields. If it is not, one should think of $\mathrm{Ham}(M,\omega)$ as a leaf of a codimension-k foliation of $\mathrm{Symp}_0(M,\omega)$, where

$$k := \operatorname{rank} H^1(M;\mathbb{R})$$

is the first Betti number of M.

Recall from the discussion after Definition 10.2.9 that Γ_ω is a subgroup of $H^1(M;P_\omega)$, where

$$P_\omega := [\omega]\cdot H_2(M;\mathbb{Z}) \subset \mathbb{R}$$

is the countable group of periods of ω. Thus Γ_ω is obviously discrete when P_ω itself is discrete, for example if $[\omega] \in H^2(M;\mathbb{Q})$. When M is noncompact and ω is exact, it follows from the exact sequence (10.1.7) that Γ_ω is the trivial group and, in particular, is discrete. As pointed out by Banyaga [50], another easy case in which Γ_ω is discrete is when (M,ω) is Kähler, or more generally when (M,ω) is a **Lefschetz manifold**, i.e. the homomorphism

$$H^1(M;\mathbb{R}) \to H^{2n-1}(M;\mathbb{R}) : [\alpha] \mapsto [\alpha \wedge \omega^{n-1}]$$

is an isomorphism. (Here $2n = \dim M$.) These results are 'soft', in that they follow from simple cohomological arguments. In [397] Lalonde, McDuff, and Polterovich proved that Γ_ω is discrete whenever $[\omega]$ takes integral values on the image of the Hurewicz homomorphism $\pi_2(M) \to H_2(M;\mathbb{Z})$. Their result is 'hard', in that its proof is based on the fact that the homomorphism

$$\pi_1(\mathrm{Ham}(M)) \to \pi_1(M)$$

induced by the evaluation map $\phi_t \mapsto \phi_t(x)$ has trivial image. To date, this has been proved only by using the deep pseudoholomorphic curve methods in modern symplectic topology (see the discussion after Proposition 10.2.19). The **flux conjecture** asserts that the group Γ_ω is always discrete. It is discussed in depth in the aforementioned paper [397] by Lalonde, McDuff, and Polterovich, which also contains several other partial results towards it. The flux conjecture was settled in 2006 by Kaoru Ono, again by using pseudoholomorphic curves.

Theorem 10.2.17. (Ono) *Let (M,ω) be a closed symplectic manifold. Then the group $\mathrm{Ham}(M,\omega)$ is C^1-closed in $\mathrm{Symp}_0(M,\omega)$.*

Proof: See Ono [516]. □

A refinement of the flux conjecture asserts that the group $\mathrm{Ham}(M,\omega)$ is a closed subset of $\mathrm{Symp}_0(M,\omega)$ in the C^0-topology.

Conjecture 10.2.18 (The C^0 flux conjecture) *For every closed symplectic manifold (M,ω) the group $\mathrm{Ham}(M,\omega)$ is C^0-closed in $\mathrm{Symp}_0(M,\omega)$.*

The rigidity theorem of Eliashberg and Gromov asserts that the group of symplectomorphisms $\mathrm{Symp}(M, \omega)$ is C^0-closed in the group of all diffeomorphisms $\mathrm{Diff}(M)$ (see Theorem 12.2.1 below). However, it is an open question whether its identity component $\mathrm{Symp}_0(M, \omega)$ is also C^0-closed in $\mathrm{Diff}(M)$. To avoid this difficulty the C^0 flux conjecture is usually stated in the above form. It was confirmed by Lalonde–McDuff–Polterovch [397] under the assumption that the flux group Γ_ω in Definition 10.2.9 agrees with the group Γ_{top} in Exercise 10.2.10 and (M, ω) is a Lefschetz manifold. The Lefschetz assumption was removed by Buhovsky [83] who established the C^0 flux conjecture under the assumption

$$\Gamma_\omega = \Gamma_{\mathrm{top}}.$$

In full generality the conjecture is still open.

It is quite possible that the group $\mathrm{Ham}(M, \omega)$ is not always C^0-closed in the full group of symplectomorphisms $\mathrm{Symp}(M, \omega)$, though no specific counterexample is known. Similarly, it is not known whether the identity component $\mathrm{Symp}_0(M, \omega)$ is C^0-closed in $\mathrm{Symp}(M, \omega)$. Moreover, the group $\mathrm{Symp}_0(M, \omega)$ may well not be open in the C^0 topology on $\mathrm{Symp}(M, \omega)$. For example, any compactly supported symplectomorphism ϕ of $\mathbb{R}^{2n}$ can be conformally rescaled to have support in an arbitrarily small neighbourhood of a point. It follows that $\mathrm{Symp}_c(\mathbb{R}^{2n})$ is contractible in the C^0-topology. Hence if $\mathrm{Symp}_c(\mathbb{R}^{2n})$ is not C^1 path-connected for some n, we can conclude that $\mathrm{Symp}_{c,0}(\mathbb{R}^{2n})$ is not C^0-open. By Darboux's theorem, such a result might also transfer to any symplectic manifold. However, one of the few results known in this connection is that $\mathrm{Symp}_c(\mathbb{R}^4)$ is contractible, and hence connected. This is a deep result proved by Gromov [287]. An exposition of the proof may be found in Lalonde–McDuff [394] and in McDuff–Salamon [470, Theorem 9.5.2].

There are also interesting questions concerning the extension of the flux homomorphism. Above it was defined on the identity component $\mathrm{Symp}_0(M)$ of $\mathrm{Symp}(M)$. If it could be extended to the full group $\mathrm{Symp}(M)$ in some way, then each component of $\mathrm{Symp}(M)$ would contain a distinguished set of elements that would be compatible with the group operations. One could also extend the characterization of Hamiltonian fibrations given in Theorem 6.5.3 to more general symplectic fibrations. Such questions are discussed in Kędra–Kotschick–Morita [363] and McDuff [461]. Among other results, these papers show that the total space of every symplectic M-bundle carries a cohomology class that restricts to the symplectic class on the fibre exactly if $\Gamma_\omega = 0$ and the flux homomorphism extends to a crossed homomorphism

$$f : \mathrm{Symp}(M) \to H^1(M; \mathbb{R}),$$

i.e. a map f that satisfies the identity $f(\phi\psi) = f(\phi) + \phi_*(f(\psi))$, where ϕ acts on the class $[\alpha] \in H^1(M; \mathbb{R})$ represented by the 1-form α by $\phi_*([\alpha]) = [(\phi^{-1})^*(\alpha)]$. The former paper also finds interesting topological conditions under which the flux group Γ_ω must vanish, while the latter gives some extension formulas.

The flux homomorphism and the evaluation map

Fix a point $x \in M$ and define the evaluation map $\mathrm{ev} : \mathrm{Ham}(M, \omega) \to M$ by $\mathrm{ev}(\phi) := \phi(x)$. The next proposition shows that the composition of the induced map $\mathrm{ev}_* : \pi_1(\mathrm{Ham}(M, \omega)) \to \pi_1(M)$ on fundamental groups with the homomorphism $\pi_1(M) \to H_1(M; \mathbb{Z})$ is trivial.

Proposition 10.2.19 *If (M, ω) is a closed symplectic manifold and $\{\phi_t\}_{t \in \mathbb{R}/\mathbb{Z}}$ is a loop in* $\mathrm{Ham}(M, \omega)$, *then the loops $t \mapsto \phi_t(x)$ are null-homologous.*

Proof: We examine the Poincaré dual of the homology class of the loop

$$\gamma : \mathbb{R}/\mathbb{Z} \to M, \qquad \gamma(t) := \phi_t(p_0).$$

Let Q be a closed oriented $(2n-1)$-manifold and $f : Q \to M$ be a smooth map. Then the intersection number of γ with $f(Q)$ is equal to the degree of the map $\beta : Q \times S^1 \to M$ given by

$$\beta(q, t) := \phi_t^{-1}(f(q)).$$

The degree is given by

$$\begin{aligned}
\gamma \cdot f(Q) &= \deg(\beta) \\
&= \frac{1}{\mathrm{Vol}(M)} \int_{Q \times S^1} \beta^* \frac{\omega^n}{n!} \\
&= \frac{1}{\mathrm{Vol}(M)} \int_0^1 \int_Q f^* \iota(X_t) \frac{\omega^n}{n!} \\
&= \frac{1}{\mathrm{Vol}(M)} \int_0^1 \int_Q f^* \left((\iota(X_t)\omega) \wedge \frac{\omega^{n-1}}{(n-1)!} \right) \\
&= \frac{1}{\mathrm{Vol}(M)} \int_Q f^* \left(\mathrm{Flux}(\{\phi_t\}) \wedge \frac{\omega^{n-1}}{(n-1)!} \right).
\end{aligned}$$

Here the third identity follows as in the proof of Lemma 10.2.1. This shows that the homology class of γ is Poincaré dual to the class

$$\frac{1}{\mathrm{Vol}(M)} \left[\mathrm{Flux}(\{\phi_t\}) \wedge \frac{\omega^{n-1}}{(n-1)!} \right] \in H^{2n-1}(M; \mathbb{R}).$$

Hence γ is homologous to zero whenever $\mathrm{Flux}(\{\phi_t\}) = 0$. This proves the proposition. □

Remark 10.2.20 Proposition 10.2.19 is closely connected to Theorem 5.1.6, and could also be proved by adapting that argument. The underlying phenomenon here is the connection between the symplectic and volume-preserving flux that is explored in Exercise 10.2.23. □

The stronger statement that the loops $t \mapsto \phi_t(x)$ are contractible in M for every loop $t \mapsto \phi_t$ of Hamiltonian symplectomorphisms is a much deeper theorem. It holds for all closed symplectic manifolds and can be derived as a consequence of Floer's proof of the Arnold conjecture (see Remark 11.4.2). It would be interesting if one could prove this with direct topological arguments.

Below we give another proof of Proposition 10.2.19 for the standard symplectic structure on the torus $M = \mathbb{T}^{2n}$. Note that in this case a loop in M is contractible if and only if it is null-homologous.

The flux homomorphism for the torus

We now look in detail at the case of the torus $\mathbb{T}^{2n} = \mathbb{R}^{2n}/\mathbb{Z}^{2n}$ with its standard symplectic form. In this case we shall see that

$$\Gamma_\omega = H^1(\mathbb{T}^{2n}; \mathbb{Z}).$$

Let $\phi \in \mathrm{Symp}(\mathbb{T}^{2n}, \omega_0)$ and choose a lift $\widetilde{\phi} : (\mathbb{R}^{2n}, \omega_0) \to (\mathbb{R}^{2n}, \omega_0)$. Then there is an integer matrix $A \in \mathrm{GL}(2n, \mathbb{Z})$ such that

$$\widetilde{\phi}(w + \ell) = \widetilde{\phi}(w) + A\ell$$

for $w \in \mathbb{R}^{2n}$ and $\ell \in \mathbb{Z}^{2n}$. We claim that A is symplectic. To see this, note that the matrix A represents the map induced by ϕ on the fundamental group $\pi_1(\mathbb{T}^{2n}) = \mathbb{Z}^{2n}$. Hence its transpose A^{T} gives the induced action on $H^1(\mathbb{T}^{2n}; \mathbb{R})$ in terms of the basis $dx_1, dy_1, \dots, dx_n, dy_n$. But this is a symplectic vector space with the bilinear form

$$\Omega([\alpha], [\beta]) = \int_{\mathbb{T}^{2n}} \alpha \wedge \beta \wedge \omega^{n-1},$$

for two closed 1-forms $\alpha, \beta \in \Omega^1(\mathbb{T}^{2n})$. Since $\phi^*\omega = \omega$, the action of A^{T} preserves Ω. Moreover, the given basis identifies $(H^1(\mathbb{T}^{2n}; \mathbb{R}), \Omega)$ with $(\mathbb{R}^{2n}, \omega_0)$. It follows that A^{T} and hence also A is symplectic. Note that $A = \mathbb{1}$ if and only if ϕ acts trivially on homology.

Proposition 10.2.21 *Let $\phi_t \in \mathrm{Symp}_0(\mathbb{T}^{2n}, \omega_0)$ be a symplectic isotopy with a lift $\widetilde{\phi}_t : \mathbb{R}^{2n} \to \mathbb{R}^{2n}$ such that*

$$\widetilde{\phi}_t(w + \ell) = \widetilde{\phi}_t(w) + \ell, \qquad \widetilde{\phi}_0(w) = w$$

for $w \in \mathbb{R}^{2n}$ and $\ell \in \mathbb{Z}^{2n}$. Then

$$\mathrm{Flux}(\{\phi_t\}) = \left[\sum_{j=1}^{2n} a_j dw_j\right], \quad a = (a_1, \dots, a_{2n}) = J_0 \int_{\mathbb{T}^{2n}} (\widetilde{\phi}_1(w) - w)\, dw.$$

Hence the flux homomorphism for the torus $(\mathbb{T}^{2n}, \omega_0)$ descends to a homomorphism $\rho : \mathrm{Symp}_0(\mathbb{T}^{2n}, \omega_0) \to H^1(\mathbb{T}^{2n}; \mathbb{R})/H^1(\mathbb{T}^{2n}; \mathbb{Z})$ and $\Gamma_\omega = H^1(\mathbb{T}^{2n}; \mathbb{Z})$.

Proof: Let H_t be a smooth family of generating Hamiltonians for ϕ_t, so that

$$\frac{d}{dt}\widetilde{\phi}_t = -J_0 \nabla H_t \circ \widetilde{\phi}_t.$$

We claim that there are functions $h_j(t)$ such that

$$H_t(w+\ell) = H_t(w) + \sum_{j=1}^{2n} h_j(t)\ell_j$$

for $w \in \mathbb{R}^{2n}$, $\ell \in \mathbb{Z}^{2n}$, $0 \le t \le 1$. To see this, define $h_j(t)$ so that the equation holds when ℓ is the jth vector of the standard basis, and then use linearity with respect to ℓ. Hence

$$\begin{aligned}
a &= J_0 \int_{\mathbb{T}^{2n}} (\widetilde{\phi}_1(w) - w)\, dw \\
&= J_0 \int_0^1 \int_{\mathbb{T}^{2n}} \frac{d}{dt}\widetilde{\phi}_t(w)\, dwdt \\
&= \int_0^1 \int_{\mathbb{T}^{2n}} \nabla H_t \circ \widetilde{\phi}_t(w)\, dwdt \\
&= \int_0^1 \int_{\mathbb{T}^{2n}} \nabla H_t(w)\, dwdt \\
&= \int_0^1 h(t)\, dt.
\end{aligned}$$

Furthermore, although the functions H_t are not defined on $\mathbb{T}^{2n}$, both dH_t and X_{H_t} descend to $\mathbb{T}^{2n}$. Thus we find that

$$\begin{aligned}
\mathrm{Flux}(\{\phi_t\}) &= \int_0^1 [\iota(X_{H_t})\omega_0]\, dt \\
&= \int_0^1 [dH_t]\, dt \\
&= \left[\sum_{j=1}^{2n} \left(\int_0^1 h_j(t)\, dt\right) dw_j\right] \\
&= \left[\sum_{j=1}^{2n} a_j dw_j\right].
\end{aligned}$$

This proves Proposition 10.2.21. □

The vector $\rho(\phi) = a$ in Proposition 10.2.21 can be thought of as the image under J_0 of the amount by which the diffeomorphism ϕ moves the centre of mass of the torus.

Proof of Proposition 10.2.19 for the torus: Let $\{\phi_t\}_{t\in\mathbb{R}/\mathbb{Z}}$ be a loop in $\mathrm{Ham}(\mathbb{T}^{2n},\omega_0)$ such that $\phi_0=\mathrm{id}$ and choose a lift $\widetilde{\phi}_t:\mathbb{R}^{2n}\to\mathbb{R}^{2n}$ with $\widetilde{\phi}_0=\mathrm{id}$. Then there is an integer vector $\ell\in\mathbb{Z}^{2n}$ such that $\widetilde{\phi}_1(w)=w+\ell$ for every $w\in\mathbb{R}^{2n}$. Since ϕ_t is a Hamiltonian isotopy, it follows from Proposition 10.2.21 that

$$0=\mathrm{Flux}(\{\phi_t\})=\left[\sum_{j=1}^{2n}a_j dw_j\right],\qquad a=J_0\ell.$$

Hence $\ell=0$ and so the loops $t\mapsto\phi_t(q)$ are contractible. □

A symplectomorphism ϕ of the torus $(\mathbb{T}^{2n},\omega_0)$ is called **exact**[36] if it admits a lift $\widetilde{\phi}:\mathbb{R}^{2n}\to\mathbb{R}^{2n}$ such that

$$\widetilde{\phi}(w+k)=\widetilde{\phi}(w)+k,\qquad \int_{\mathbb{T}^{2n}}(\widetilde{\phi}(w)-w)\,dw=0.$$

Geometrically, this can be interpreted as the condition that $\widetilde{\phi}$ acts trivially on homology and preserves the centre of mass. It follows from Proposition 10.2.21 and Theorem 10.2.5 that a symplectomorphism ϕ of $(\mathbb{T}^{2n},\omega_0)$ is Hamiltonian if and only if it is exact *and* isotopic to the identity. In the 2-dimensional case, one can show using complex variable theory that the identity component of the group $\mathrm{Diff}(\mathbb{T}^2)$ consists of all diffeomorphisms which act trivially on homology.

Exercise 10.2.22 Prove that a symplectomorphism $\phi\in\mathrm{Symp}(\mathbb{T}^2,\omega_0)$ is Hamiltonian if and only if it is exact. **Hint:** Use the preceding discussion and Exercise 3.2.8. □

For $n>1$ it is unknown whether every symplectomorphism of $(\mathbb{T}^{2n},\omega_0)$ which acts trivially on homology is isotopic to the identity in $\mathrm{Diff}(\mathbb{T}^{2n})$. Even if it were, it might not be isotopic to the identity in $\mathrm{Symp}(\mathbb{T}^{2n})$. In other words, it is an open question whether or not the assertion of Exercise 10.2.22 extends to higher-dimensional tori.

Volume-preserving diffeomorphisms

Let M be a closed oriented m-dimensional manifold with volume form σ and volume $\int_M\sigma=1$. Denote by $\mathrm{Diff}_{\mathrm{vol},0}(M)$ the identity component of the group $\mathrm{Diff}_{\mathrm{vol}}(M)$ of volume-preserving diffeomorphisms and by

$$\mathcal{X}_{\mathrm{vol}}(M):=\{X\in\mathcal{X}(M)\,|\,d\iota(X)\sigma=0\}$$

the space of divergence-free vector fields. Call a divergence-free vector field X **exact** if $\iota(X)\sigma$ is exact and denote the space of such vector fields by $\mathcal{X}_{\mathrm{vol}}^{\mathrm{ex}}(M)$. Call a volume-preserving diffemorphism ψ **exact** if it is generated by a smooth

[36] Warning: some authors use the word 'exact' to mean 'Hamiltonian'. Our usage is consistent with the notion of exact Lagrangian submanifolds of cotangent bundles where the existence of a Hamiltonian isotopy is not required.

family of vector fields $X_t \in \mathcal{X}^{\mathrm{ex}}_{\mathrm{vol}}(M)$ via $\partial_t\psi_t = X_t \circ \psi_t$, $\psi_0 = \mathrm{id}$, $\psi_1 = \psi$, and denote the space of such diffeomorphisms by $\mathrm{Diff}^{\mathrm{ex}}_{\mathrm{vol}}(M)$. Define the map

$$\mathrm{Flux}_{\mathrm{vol}} : \widetilde{\mathrm{Diff}}_{\mathrm{vol},0}(M) \to H^{m-1}(M;\mathbb{R})$$

by

$$\mathrm{Flux}_{\mathrm{vol}}(\{\psi_t\}) := \left[\int_0^1 \iota(X_t)\sigma\right] \in H^{m-1}(M), \tag{10.2.4}$$

where the vector fields $X_t \in \mathcal{X}_{\mathrm{vol}}(M)$ generate the volume-preserving isotopy ψ_t.

Exercise 10.2.23 **(i)** Prove that $\mathrm{Flux}_{\mathrm{vol}}$ is a surjective homomorphism.

(ii) Prove that $\mathrm{Diff}^{\mathrm{ex}}_{\mathrm{vol}}(M)$ is a normal subgroup of $\mathrm{Diff}_{\mathrm{vol},0}(M)$.

(iii) Prove that a volume-preserving isotopy ψ_t with $\psi_0 = \mathrm{id}$ can be deformed to an exact isotopy with fixed endpoints if and only if

$$\mathrm{Flux}_{\mathrm{vol}}(\{\psi_t\}) = 0.$$

(This is the analogue of Theorem 10.2.5.)

(iv) Let Q be a closed oriented $(m-1)$-manifold, and $\iota : Q \to M$ be a smooth map. If $\psi_t = \psi_{t+1}$ is a loop in $\mathrm{Diff}_{\mathrm{vol},0}(M)$ with $\psi_0 = \mathrm{id}$, prove that $\int_Q \iota^*\mathrm{Flux}_{\mathrm{vol}}(\{\psi_t\})$ is the degree of the map $\mathbb{R}/\mathbb{Z} \times Q \to M : (t,q) \mapsto \psi_t \circ \iota(q)$.

(v) Define the **volume flux group** by

$$\Gamma_{\mathrm{vol}} := \mathrm{Flux}_{\mathrm{vol}}(\pi_1(\mathrm{Diff}_{\mathrm{vol},0}(M))) \subset H^{m-1}(M;\mathbb{R}).$$

Prove that Γ_σ is the image of the composition

$$\pi_1(\mathrm{Diff}_0(M)) \xrightarrow{\mathrm{ev}} \pi_1(M) \longrightarrow H_1(M;\mathbb{Z}) \xrightarrow{\mathrm{PD}} H^{m-1}(M;\mathbb{Z}) \longrightarrow H^{m-1}(M;\mathbb{R}).$$

Hint: Use Moser isotopy for volume forms to show that the inclusion of $\mathrm{Diff}_{\mathrm{vol},0}(M)$ into $\mathrm{Diff}_0(M)$ induces an isomorphism of fundamental groups. (It is actually a homotopy equivalence.) Then use part (iv).

(vi) Prove that $\mathrm{Flux}_{\mathrm{vol}}$ descends to a group isomorphism

$$\mathrm{Diff}_{\mathrm{vol},0}(M)/\mathrm{Diff}^{\mathrm{ex}}_{\mathrm{vol}}(M) \cong H^{m-1}(M;\mathbb{R})/\Gamma_{\mathrm{vol}}.$$

(This requires the analogue of Lemma 10.2.11.)

(vii) When ω is a symplectic form and

$$\sigma := \frac{\omega^n}{n!},$$

what is the relationship between Flux and $\mathrm{Flux}_{\mathrm{vol}}$? □

In [363], Kędra–Kotschick–Morita show that in many cases Γ_{vol} must vanish. For example, if this group is nontrivial, then $\pi_1(M)$ must have infinite centre.

10.3 The Calabi homomorphism

We next discuss the algebraic structure of the group of symplectomorphisms. Further references and proofs of the results mentioned here can be found in Banyaga's celebrated paper [50][37] as well as his book [51].

We begin by recalling some facts about the group $\mathrm{Diff}(M)$ of all diffeomorphisms of the compact manifold M, and its subgroup $\mathrm{Diff}_{\mathrm{vol}}(M)$ of volume-preserving diffeomorphisms. Denote by $\mathrm{Diff}_0(M)$ and $\mathrm{Diff}_{\mathrm{vol},0}(M)$ the identity components of these groups. Since these groups are both locally path-connected, these components consist of all elements which are isotopic to the identity.

One can prove that the commutator subgroup $[\mathrm{Diff}_0(M), \mathrm{Diff}_0(M)]$, which is generated by the commutators $\phi\psi\phi^{-1}\psi^{-1}$ with $\phi, \psi \in \mathrm{Diff}_0(M)$, is simple, i.e. it has no normal subgroups. This follows from a very general argument due to Epstein [207] which applies also in the volume-preserving and symplectic cases. A much more delicate result is that $\mathrm{Diff}_0(M)$ is perfect, that is,

$$\mathrm{Diff}_0(M) = [\mathrm{Diff}_0(M), \mathrm{Diff}_0(M)].$$

One proves this first for the torus by some hard analysis due to Herman, and then transfers this result to an arbitrary manifold using a geometric technique developed by Thurston. Putting these result together, one finds that $\mathrm{Diff}_0(M)$ itself is simple. In other words, there is no nontrivial homomorphism from $\mathrm{Diff}_0(M)$ onto any group. When M is noncompact, the group of all diffeomorphisms of M does have some obvious normal subgroups, for example the group $\mathrm{Diff}_{\mathrm{c}}(M)$ of compactly supported diffeomorphisms. However, the identity component $\mathrm{Diff}_{\mathrm{c},0}(M)$ of $\mathrm{Diff}_{\mathrm{c}}(M)$ is again a simple group.

As we remarked in Exercise 10.2.23, there is an analogue of the flux homomorphism in the volume-preserving case, which induces a surjective homomorphism

$$\rho_{\mathrm{vol}} : \mathrm{Diff}_{\mathrm{vol},0}(M) \to H^{m-1}(M;\mathbb{R})/\Gamma_{\mathrm{vol}}$$

for a suitable subgroup $\Gamma_{\mathrm{vol}} \subset H^{m-1}(M;\mathbb{R})$ (compare Proposition 10.2.13). The group Γ_{vol} is always discrete, and so the kernel of ρ_{vol} is closed. Furthermore, because $H^{m-1}(M;\mathbb{R})$ is abelian, this subgroup lies in the commutator subgroup of $\mathrm{Diff}_{\mathrm{vol},0}(M)$. Thurston showed in [623] that this kernel is its own commutator subgroup, and hence is simple. A similar result holds in the symplectic case.

Theorem 10.3.1 (Banyaga) *If (M,ω) is a closed connected symplectic manifold then the group $\mathrm{Ham}(M,\omega)$ is simple, i.e. it contains no nontrivial normal subgroups.*

Proof: See Banyaga [50, 51]. □

Banyaga's proof closely follows Thurston's line of argument. Since Herman's analysis extends to both the volume-preserving and the symplectic case, the

[37] This paper also contains proofs of most of the results of Section 10.2, and our proof of Theorem 10.2.5 is basically taken from it.

problem is to make the geometric part of the proof go through. The main step is a **fragmentation lemma**, which shows that any element of $\mathrm{Ham}(M)$ can be written as a finite product of Hamiltonian symplectomorphisms each of which is the time-1 map of some Hamiltonian isotopy with support in a Darboux chart.

Remark 10.3.2 The existence and properties of fragmentations have many other significant consequences for the algebraic structure of symplectomorphism groups; see for example Burago–Ivanov–Polterovich [87] and Leroux [413]. The latter paper also explains Thurston's elegant version of the argument that perfection implies simplicity. □

Let us now consider the noncompact case. Thurston showed that the group of compactly supported volume-preserving diffeomorphisms which are in the kernel of the flux homomorphism is simple, provided that $m > 2$. The case $m = 2$ coincides with the symplectic case, of course. In this case the kernel of the flux is no longer simple: as we shall see, another homomorphism comes into play. We will follow customary usage by calling this the Calabi homomorphism, though some authors call this the second Calabi homomorphism, the first such being the flux. (Both homomorphisms were described by Calabi [92].)

The Calabi homomorphism is easiest to understand for exact manifolds such as $\mathbb{R}^{2n}$. Thus let (M, ω) be a (necessarily noncompact) exact symplectic manifold and choose a 1-form $\lambda \in \Omega^1(M)$ such that

$$\omega = -d\lambda.$$

Denote by $\mathrm{Symp}_{c,0}(M, \omega)$ the identity component of the group of compactly supported symplectomorphisms (with respect to the direct limit topology described above) and by $\mathrm{Ham}_c(M, \omega)$ the group of compactly supported Hamiltonian symplectomorphisms. By Proposition 9.3.1 and Exercise 10.2.6, an element $\phi \in \mathrm{Symp}_{c,0}(M)$ belongs to the subgroup $\mathrm{Ham}_c(M, \omega)$ if and only if there exists a compactly supported smooth function $F = F_\phi : M \to \mathbb{R}$ such that

$$\phi^*\lambda - \lambda = dF_\phi. \tag{10.3.1}$$

Given $\phi \in \mathrm{Ham}_c(M)$, we define

$$\mathrm{CAL}(\phi) = -\frac{1}{n+1} \int_M F_\phi \omega^n. \tag{10.3.2}$$

Lemma 10.3.3 *The number* $\mathrm{CAL}(\phi)$ *is independent of the* 1*-form* $\lambda \in \Omega^1(M)$ *with* $\omega = -d\lambda$ *used to define it. Moreover, the map*

$$\mathrm{CAL} : \mathrm{Ham}_c(M) \to \mathbb{R}$$

is a homomorphism. It is called the **Calabi homomorphism**.

Proof: Let $\lambda + \alpha$ be any other such 1-form with $d\alpha = 0$. Then, since ϕ induces the identity map on compactly supported cohomology, there exists a compactly supported smooth function $G : M \to \mathbb{R}$ such that $\phi^*\alpha - \alpha = dG$. We must prove that

$$\int G\omega^n = 0.$$

But this holds because

$$\begin{aligned}
\int G\omega^n &= -\int G\,d\lambda \wedge \omega^{n-1} \\
&= \int \big(dG \wedge \lambda - d(G\lambda)\big) \wedge \omega^{n-1} \\
&= \int (\phi^*\alpha - \alpha) \wedge \lambda \wedge \omega^{n-1} \\
&= \int \phi^*\alpha \wedge \lambda \wedge \omega^{n-1} - \int \phi^*(\alpha \wedge \lambda \wedge \omega^{n-1}) \\
&= \int \phi^*\alpha \wedge (\lambda - \phi^*\lambda) \wedge \omega^{n-1}.
\end{aligned}$$

The last term vanishes because α is closed and $\lambda - \phi^*\lambda$ is exact. Now use the identity $\phi^*\psi^*\lambda - \lambda = \phi^*(\psi^*\lambda - \lambda) + \phi^*\lambda - \lambda$ to prove that CAL is a group homomorphism. □

Here we have given a direct proof of the existence of the Calabi homomorphism for exact manifolds. An alternative argument follows from Proposition 9.3.1, which asserts that the unique compactly supported function F which satisfies the equation $\phi^*\lambda - \lambda = dF$ is given by the symplectic action

$$F(z) = \int_0^1 (\iota(X_s)\lambda - H_s) \circ \phi_s(z)\,ds \tag{10.3.3}$$

with respect to λ, where $\phi_s \in \mathrm{Ham}_c(M)$ is any Hamiltonian isotopy from $\phi_0 = \mathrm{id}$ to $\phi_1 = \phi$, i.e.

$$\frac{d}{dt}\phi_t = X_t \circ \phi_t, \qquad \phi_0 = \mathrm{id}, \qquad \iota(X_t)\omega = dH_t,$$

where H_t has compact support. Thus we may think of $\mathrm{CAL}(\phi)$ as the (spacewise) **average symplectic action** of ϕ. Fathi [211] and Gambaudo–Ghys [257] point out that in the two-dimensional case, $\mathrm{CAL}(\phi)$ can also be interpreted as a mean rotation number; see Shelukhin [581] for a short proof. We shall now give a formula for the Calabi homomorphism which is evidently independent of the form λ.

Lemma 10.3.4 *Let $\phi \in \mathrm{Ham}_c(M, -d\lambda)$. Then, using the above notation, we have*

$$\mathrm{CAL}(\phi) = \int_0^1 \int_M H_t\omega^n\,dt \tag{10.3.4}$$

and

$$\mathrm{CAL}(\phi) = -\frac{1}{n+1}\int_M \phi^*\lambda \wedge \lambda \wedge \omega^{n-1}. \tag{10.3.5}$$

Proof: For each t let $F_t : M \to \mathbb{R}$ be the unique compactly supported function such that

$$\phi_t^*\lambda - \lambda = dF_t.$$

Then, by a simple calculation

$$\begin{aligned}
\int_M \phi_t^*\lambda \wedge \lambda \wedge \omega^{n-1} &= \int_M (\phi_t^*\lambda - \lambda) \wedge \lambda \wedge \omega^{n-1} \\
&= \int_M (dF_t) \wedge \lambda \wedge \omega^{n-1} \\
&= \int_M (d(F_t\lambda) - F_t d\lambda) \wedge \omega^{n-1} \\
&= \int_M d(F_t\lambda \wedge \omega^{n-1}) - \int_M F_t(d\lambda) \wedge \omega^{n-1} \\
&= \int_M F_t\omega^n.
\end{aligned}$$

This proves equation (10.3.5). To prove (10.3.4), recall from (10.3.3) that

$$\frac{d}{dt}\int_M F_t\omega^n = \int_M (\iota(X_t)\lambda - H_t)\omega^n.$$

Since every $(2n+1)$-form on M is zero, we have

$$\begin{aligned}
0 &= \iota(X_t)(\lambda \wedge \omega^n) \\
&= (\iota(X_t)\lambda)\omega^n - \lambda \wedge \iota(X_t)(\omega^n) \\
&= (\iota(X_t)\lambda)\omega^n - n\lambda \wedge \iota(X_t)\omega \wedge \omega^{n-1} \\
&= (\iota(X_t)\lambda)\omega^n + n(dH_t) \wedge \lambda \wedge \omega^{n-1} \\
&= (\iota(X_t)\lambda)\omega^n + nd(H_t\lambda) \wedge \omega^{n-1} - nH_t d\lambda \wedge \omega^{n-1} \\
&= (\iota(X_t)\lambda + nH_t)\omega^n + nd(H_t\lambda \wedge \omega^{n-1}).
\end{aligned}$$

Integrate this identity over M to obtain

$$\int_M (\iota(X_t)\lambda + nH_t)\omega^n = 0,$$

and hence

$$\frac{d}{dt}\int_M F_t\omega^n = \int_M (\iota(X_t)\lambda - H_t)\omega^n = -(n+1)\int_M H_t\omega^n.$$

This proves equation (10.3.4) and Lemma 10.3.4. □

Equation (10.3.4) gives another proof that the number $\mathrm{CAL}(\phi)$ is independent of λ. Equation (10.3.5) shows that the Calabi invariant extends to a map

$$\mathrm{CAL} : \mathrm{Symp}_{\mathrm{c}}(M, -d\lambda) \to \mathbb{R}.$$

It was noted by Morita and Kotschick in [495] that when restricted to the identity component, the extended map satisfies the identity

$$\mathrm{CAL}(\phi\psi) = \mathrm{CAL}(\phi) + \mathrm{CAL}(\psi) + \frac{1}{n+1}\int_M \mathrm{Flux}(\phi) \wedge \mathrm{Flux}(\psi) \wedge \omega^{n-1}. \quad (10.3.6)$$

Since the last term above vanishes when $2n \geq 4$, on an exact symplectic manifold (M, ω) of dimension $2n \geq 4$ there is a homomorphism

$$\mathrm{Flux} \oplus \mathrm{CAL} : \mathrm{Symp}_{\mathrm{c},0}(M, \omega) \longrightarrow H^1_{\mathrm{c}}(M; \mathbb{R}) \oplus \mathbb{R}.$$

Exercise 10.3.5 Verify equation (10.3.6). Deduce that $\mathrm{CAL} : \mathrm{Symp}_{\mathrm{c},0}(M, \omega) \to \mathbb{R}$ is a surjective homomorphism for $\dim M \geq 4$. Show that $\mathrm{Flux} \oplus \mathrm{CAL}$ is a surjective homomorphism when $\dim M \geq 4$. **Hint:** Since Flux is surjective, it suffices to find an element $\phi \in \mathrm{Ham}_{\mathrm{c}}(M)$ with nonzero Calabi invariant. □

Exercise 10.3.6 (Flux for the annulus) Let M be the annulus given in polar coordinates by

$$M := \left\{ re^{i\theta} \in \mathbb{C} \,|\, \tfrac{1}{2} < r < 1 \right\}.$$

Consider a compactly supported diffeomorphism ϕ of the form

$$\phi(re^{i\theta}) = re^{i(\theta + f(r))}.$$

Take $\lambda = \frac{1}{2} r^2 d\theta$ and calculate $\phi^*\lambda - \lambda$. According to the compactly supported analogue of Lemma 10.2.1, its integral over the closed noncompact arc $\gamma := \{(r, 0) \,|\, \frac{1}{2} < r < 1\}$ should be the geometric area swept out by γ under any isotopy from id to ϕ. Check this. Moreover, calculate $\mathrm{CAL}(\phi)$. Show that $\mathrm{CAL}(\phi)$ remains unchanged if instead you consider ϕ to be an element of the group $\mathrm{Ham}_{\mathrm{c}}(\mathbb{D})$, where $\mathbb{D}$ is the open 2-disc. □

In the general case, where M is still noncompact but ω need not be exact, the natural place to define the Calabi homomorphism is the universal cover $\widetilde{\mathrm{Ham}}_{\mathrm{c}}(M)$ of $\mathrm{Ham}_{\mathrm{c}}(M)$. It is given by the formula

$$\widetilde{\mathrm{CAL}}(\{\phi_t\}) := \int_0^1 \int_M H_t \, \omega^n \, dt,$$

where H_t is the unique compactly supported Hamiltonian which generates the isotopy ϕ_t. The proof that this is a well-defined surjective homomorphism is left to the reader. If Λ denotes the image of $\pi_1(\mathrm{Ham}_{\mathrm{c}}(M))$ under $\widetilde{\mathrm{CAL}}$, we obtain a homomorphism $\mathrm{CAL} : \mathrm{Ham}_{\mathrm{c}}(M, \omega) \to \mathbb{R}/\Lambda$, induced by $\widetilde{\mathrm{CAL}}$. The following theorem is a noncompact analogue of Theorem 10.3.1.

Theorem 10.3.7 (Banyaga) *Let (M, ω) be a noncompact connected symplectic manifold without boundary. Then the kernel of the Calabi homomorphism $\mathrm{CAL} : \mathrm{Ham}_{\mathrm{c}}(M) \to \mathbb{R}/\Lambda$ is a simple group.*

Proof: See [50]. □

Very little is known about the group Λ in general, for example conditions under which it is discrete or trivial; cf. McDuff [465, Remark 3.11]. One case in which we know a little is when U is an open displaceable[38] subset of a closed symplectic manifold (M, ω) with the induced symplectic form. In this case the Calabi homomorphism on $\widetilde{\mathrm{Ham}}_{\mathrm{c}}(U, \omega)$ vanishes on every loop in $\mathrm{Ham}_{\mathrm{c}}(U, \omega)$ that contracts in $\mathrm{Ham}_{\mathrm{c}}(M, \omega)$, and hence descends to a homomorphism defined on the image $\mathcal{G}(U)$ of $\widetilde{\mathrm{Ham}}_{\mathrm{c}}(U, \omega)$ in $\widetilde{\mathrm{Ham}}_{\mathrm{c}}(M, \omega)$. For a proof see Corollary 4.3.7 in the book [532] by Polterovich and Rosen.

This lovely book explains the Polterovich–Entov theory of Calabi quasimorphisms for closed symplectic manifolds M.[39] Quasimorphisms are continuous maps $\mu : \mathrm{Ham}(M) \to \mathbb{R}$ that are a bounded distance from being a homomorphism. In other words, there is a constant $K > 0$ such that

$$|\mu(\phi\psi) - \mu(\phi) - \mu(\phi)| \leq K, \quad \phi, \psi \in \mathrm{Ham}(M).$$

Further, μ is called a **Calabi quasimorphism** if μ restricts on $\mathrm{Ham}_{\mathrm{c}}(U)$ to CAL for every displaceable open subset $U \subset M$. Note that because $\mathrm{Ham}(M)$ is simple, we must have $K > 0$ whenever μ is not identically zero. The existence of Calabi quasimorphisms is highly nontrivial even for $\mathbb{C}\mathrm{P}^n$, being based on the properties of Floer-theoretic spectral invariants. Moreover, often one can only define them on the universal cover $\widetilde{\mathrm{Ham}}(M)$. This theory has had many applications, some in Hofer geometry (see Section 12.3), and others that relate to other algebraic properties of the group $\mathrm{Ham}(M)$ such as the behaviour of the commutator length. There are also quasimorphisms on $\widetilde{\mathrm{Ham}}(M)$ of other types, i.e. that restrict on the subgroups $\mathcal{G}(U)$ for displaceable U to other homomorphisms; cf. the foundational paper by Barge–Ghys [52] that constructs the "average Maslov quasimorphism", and the paper [580] by Shelukhin that constructs a quasimorphism μ_S on $\widetilde{\mathrm{Ham}}(M)$ of Calabi–Maslov local type. Indeed, Shelukhin defines μ_S for any closed symplectic manifold so that for any displaceable open set $U \subset M$, its restriction to $\mathcal{G}(U)$ is a linear combination of the Calabi homomorphism with the average Maslov quasimorphism of Barge–Ghys. His construction uses the infinite-dimensional reduction techniques outlined in Example 5.3.19.

10.4 The topology of symplectomorphism groups

Here are two interesting and fundamental questions about the relationship between diffeomorphisms and symplectomorphisms.

> *Which diffeomorphisms of a given symplectic manifold are isotopic to a symplectomorphism? If two symplectomorphisms are smoothly isotopic are they also symplectically isotopic?*

For many symplectic manifolds these questions are wide open. We now briefly discuss some examples and known results. First observe that every symplectomorphism preserves the cohomology class of the symplectic form, and so the

[38] A subset $U \subset M$ is called **displaceable** if there is $\phi \in \mathrm{Ham}(M)$ such that $\phi(U) \cap U = \emptyset$.

[39] See Lanzat [399] for an extension to the noncompact case.

image of $\pi_0(\mathrm{Symp}(M,\omega))$ in $\pi_0(\mathrm{Diff}(M))$ is contained in the subgroup consisting of elements that fix the class $[\omega]$.

Example 10.4.1 (Symplectomorphisms of the torus) Consider the torus $\mathbb{T}^{2n} = \mathbb{R}^{2n}/\mathbb{Z}^{2n}$ with the standard symplectic form ω_0. Every matrix

$$\Psi \in \mathrm{SL}(2n, \mathbb{Z})$$

gives rise to a diffeomorphism of $\mathbb{T}^{2n}$, and this diffeomorphism preserves the cohomology class of the standard symplectic structure if and only if Ψ is a symplectic matrix (see the remarks just before Proposition 10.2.21). Since every diffeomorphism of $\mathbb{T}^{2n}$ is homotopic to such a linear map Ψ, every diffeomorphism which preserves the cohomology class $[\omega_0]$ is homotopic to a symplectomorphism. But, for $n \geq 2$, it is not known whether any diffeomorphism of $\mathbb{T}^{2n}$ which preserves $[\omega_0]$ is isotopic to a symplectomorphism. Likewise, it is not known whether every symplectomorphism of $\mathbb{T}^{2n}$ is symplectically isotopic to a linear symplectomorphism Ψ. □

Example 10.4.2 (i) (Symplectomorphisms of $\mathbb{C}\mathrm{P}^2$) A theorem of Gromov asserts that the symplectomorphism group of $\mathbb{C}\mathrm{P}^2$ with the Fubini–Study form retracts onto the isometry group PU(3) (see [287] and also [470, Theorem 9.5.3]). In particular, the group $\mathrm{Symp}(\mathbb{C}\mathrm{P}^2, \omega_{\mathrm{FS}})$ is connected.

(ii) (Symplectomorphisms of $\mathbb{C}\mathrm{P}^2\#k\overline{\mathbb{C}\mathrm{P}}^2$) A theorem of Abreu–McDuff [8, 12] asserts that the symplectomorphism group of the one-point blowup of $\mathbb{C}\mathrm{P}^2$ (with any symplectic form) is connected. A theorem of Li–Li–Wu [417] asserts that $\mathrm{Symp}_h(M,\omega)$ is connected for all blowups of $\mathbb{C}\mathrm{P}^2$ at up to four points.

(iii) (Symplectomorphisms of $S^2 \times S^2$) Consider the product $M := S^2 \times S^2$ with the symplectic form $\omega_\lambda := \lambda\pi_1^*\sigma + \pi_2^*\sigma$, where σ is an area form on S^2, the maps $\pi_1, \pi_2 : S^2 \times S^2 \to S^2$ are the projections onto the two factors, and $\lambda \geq 1$. For $\lambda = 1$ Gromov [287] proved that the group of symplectomorphisms retracts onto the isometry group $\mathbb{Z}/2\mathbb{Z} \times \mathrm{SO}(3) \times \mathrm{SO}(3)$. In particular, a symplectomorphism is smoothly isotopic to the identity if and only it is symplectically isotopic to the identity if and only if it induces the identity on homology. This assertion remains valid for $\lambda > 1$ by a theorem of Abreu and McDuff [8, 12]. In that case every symplectomorphism induces the identity on homology and Abreu–McDuff proved that the symplectomorphism group

$$\mathcal{G}_\lambda := \mathrm{Symp}(S^2 \times S^2, \omega_\lambda)$$

is connected for $\lambda > 1$. They also calculated the rational higher homotopy groups of $\mathcal{G}_\lambda$. It was already noted by Gromov [287] that the fundamental group of $\mathcal{G}_\lambda$ is infinite for $\lambda > 1$. Abreu–McDuff [8, 12] proved that

$$\pi_1(\mathcal{G}_\lambda) \cong \mathbb{Z}/2\mathbb{Z} \oplus \mathbb{Z}/2\mathbb{Z} \oplus \mathbb{Z}$$

for $\lambda > 1$ and that the element of infinite order is the loop $\mathbb{R}/\mathbb{Z} \to \mathcal{G}_\lambda : t \mapsto \psi_t$ defined by $\psi_t(z_1, z_2) := (z_1, \phi_{t,z_1}(z_2))$, where $\phi_{t,z_1} \in \mathrm{SO}(3) \subset \mathrm{Diff}(S^2)$ denotes the rotation about the axis determined by $z_1 \in S^2$ through the angle $2\pi t$. (This loop also appears in Example 13.2.9). But $\pi_1(\mathrm{Diff}(S^2 \times S^2))$ has rank at least two because it contains both this loop and its conjugate by the involution that interchanges the two S^2 factors. Hence the above map $\pi_1(\mathcal{G}_\lambda) \to \pi_1(\mathrm{Diff}(S^2 \times S^2))$ is never surjective. See Anjos–Granja [22], Abreu–Granja–Kitchloo [10] and Kędra–McDuff [364] for different generalizations of the above results. □

The proofs of the above results rely heavily on the theory of J-holomorphic spheres, and do not always work even for manifolds as simple as a k-fold blowup of $\mathbb{C}\mathrm{P}^2$ or for products $\Sigma_g \times S^2$ where Σ_g is a Riemann surface of genus g. For example, if ω_λ is the product form on $\Sigma_g \times S^2$ as above, it is not known whether the group $\mathrm{Symp}_h(\Sigma_g \times S^2, \omega_\lambda)$ is connected for all $g > 0$ and $\lambda > 0$; cf. the discussion in Example 13.4.3 (v). On the other hand, the squares of the generalized Dehn twists described in Exercise 6.3.7 provide an important class of examples of symplectomorphisms that are not usually symplectically isotopic to the identity and yet are smoothly isotopic to the identity (and in particular act trivially on homotopy and homology groups).

Example 10.4.3 (Generalized Dehn twists) In his PhD thesis, Seidel constructed symplectomorphisms on many symplectic four-manifolds (M, ω) that are smoothly, but not symplectically, isotopic to the identity. He proved the following two theorems (see [570] and [575, Cor 2.9 & Thm 0.5]).

Theorem A. *Let M be a closed symplectic four-manifold with Betti numbers*

$$b_1 := \dim H^1(M; \mathbb{R}) = 0, \qquad b_2 := \dim H^2(M; \mathbb{R}) \geq 3.$$

Assume M is minimal (i.e. it does not contain a symplectically embedded two-sphere with self-intersection number minus one). If $\Lambda \subset M$ is a Lagrangian sphere, then the square $\phi = \tau_\Lambda \circ \tau_\Lambda$ of the generalized Dehn twist determined by Λ is smoothly, but not symplectically, isotopic to the identity.

Theorem B. *Let M be an algebraic surface and a complete intersection, not diffeomorphic to $\mathbb{C}\mathrm{P}^2$ or $\mathbb{C}\mathrm{P}^1 \times \mathbb{C}\mathrm{P}^1$. Then there exists a symplectomorphism $\phi : M \to M$ that is smoothly, but not symplectically, isotopic to the identity.*

Here are some remarks. For a much more detailed and wide ranging discussion of examples and ramifications of these theorems, see Seidel [575].

(i) The assumptions of Theorem A allow for examples of symplectic 4-manifolds that do not admit Kähler structures (see Gompf–Mrowka [279] and Fintushel–Stern [219,222]). The assumptions of Theorem B allow for examples that are not minimal, such as the six-fold blowup of the projective plane (cubics in $\mathbb{C}\mathrm{P}^3$).

(ii) The symplectomorphism $\phi = \tau_\Lambda^2$ in Theorem A is smoothly isotopic to the identity by an isotopy localized near Λ. Seidel computed the Floer cohomology group $\mathrm{HF}^*(\tau_\Lambda)$ with its module structure over the quantum cohomology ring, via his exact sequence [570, 573]. As a result he was able to show that, under the assumptions of minimality and $b_2 \geq 3$, the Floer cohomology group $\mathrm{HF}^*(\tau_\Lambda)$

is not isomorphic to the Floer homology group $\mathrm{HF}_*(\tau_\Lambda) = \mathrm{HF}^*(\tau_\Lambda^{-1})$ by an isomorphism of modules. Hence τ_Λ and τ_Λ^{-1} are not Hamiltonian isotopic. When $b_1 = 0$ it then follows that they are not symplectically isotopic.

(iii) The assumption that $b_2 \geq 3$ cannot be removed in Theorem A. An example is the product $M = \mathbb{C}\mathrm{P}^1 \times \mathbb{C}\mathrm{P}^1$ with its monotone symplectic structure and with Λ equal to the anti-diagonal. In this example $\phi = \tau_\Lambda^2$ *is* symplectically isotopic to the identity, by Gromov's theorem in [287] (see Example 10.4.2 (iii)).

(iv) The assumption of minimality cannot be removed in Theorem A. Examples are the blowups $M_k := \mathbb{C}\mathrm{P}^2 \# k\overline{\mathbb{C}\mathrm{P}}^2$ for $2 \leq k \leq 4$. On these manifolds there exist symplectic forms (monotone in the cases $k = 3, 4$) and Lagrangian spheres such that the squares of the generalized Dehn twists *are* symplectically isotopic to the identity (see [575, Examples 1.10 and 1.12]). In contrast, for $5 \leq k \leq 8$, the square of a generalized Dehn twist in the k-fold blowup of the projective plane with its monotone symplectic form is *never* symplectically isotopic to the identity (see [575, Example 2.10]).

(v) Seidel showed that by an arbitrarily small perturbation of the cohomology class of ω, the square of the generalized Dehn twist deforms to a symplectomorphism that *is* symplectically isotopic to the identity. □

Recall from Example 10.4.2 (iii) that there are symplectic manifolds (M, ω) for which the inclusion induced homomorphism

$$\pi_1(\mathrm{Symp}(M, \omega)) \to \pi_1(\mathrm{Diff}(M))$$

is not surjective. In that example the manifold is simply connected and the symplectomorphism group is connected so that $\mathrm{Symp}(M, \omega) = \mathrm{Ham}(M, \omega)$. There are other kinds of reasons to expect that $\pi_k(\mathrm{Symp}(M, \omega))$ may be significantly smaller than $\pi_k(\mathrm{Diff}(M))$ which are related to the properties of the evaluation map $\pi_k(\mathrm{Ham}(M, \omega)) \to \pi_k(M)$, or, more generally, of the natural action of $\pi_k(\mathrm{Ham}(M, \omega))$ on the homology $H_*(M)$ of M. Observe that this statement involves the group $\pi_k(\mathrm{Ham}(M, \omega))$ rather than $\pi_k(\mathrm{Symp}_0(M, \omega))$. When $k > 1$ this makes no difference, as we noted in Exercise 10.2.15. However, when $k = 1$ there may be a significant difference.

We saw in Proposition 10.2.19 and the subsequent discussion that the evaluation map on $\pi_1(\mathrm{Ham}(M, \omega))$ is trivial for every closed symplectic manifold (M, ω), although in many cases it is nontrivial on $\pi_1(\mathrm{Symp}_0(M, \omega))$ (for example in the case of the standard symplectic torus $(\mathbb{T}^{2n}, \omega_0)$). Its image lies in the **evaluation subgroup** of $\pi_1(M)$. This is the image of $\pi_1(\mathrm{Map}(M, M))$ under the evaluation map $\mathrm{Map}(M, M) \to M : \phi \mapsto \phi(q)$, from the space Map(M, M) of continuous self-maps of M to M. It lies in the centre of $\pi_1(M)$ and has various interesting properties; cf Gottlieb [283]. Much less is known about the corresponding maps on homotopy for $k > 1$, though the action of $\pi_k(\mathrm{Ham}(M))$ on the homology $H_*(M; \mathbb{Q})$ vanishes by Lalonde–McDuff [396]. We refer the reader to the papers by Seidel [571], Lalonde–McDuff–Polterovich [398], and Kędra [362] for further information.

PART IV

SYMPLECTIC INVARIANTS

11

THE ARNOLD CONJECTURE

This chapter takes up the study of the geometric behaviour of symplectomorphisms which was started in Chapter 8. A natural generalization of the Poincaré–Birkhoff theorem concerns the existence of fixed points of a symplectomorphism $\psi \in \mathrm{Symp}(M, \omega)$ of a compact symplectic manifold. The question of the existence of such fixed points had already been raised by Birkhoff. In 1927 he wrote with reference to the Poincaré–Birkhoff theorem:

> Up to the present time no proper generalization of this theorem to higher dimensions has been found, so that its application remains limited to dynamical systems with two degrees of freedom. [74, page 150]

In general we cannot expect symplectomorphisms For example a rotation on the 2-torus is a symplectomorphism without fixed points. If, however, the symplectomorphism of $\mathbb{T}^2$ is the time-1 map of a (time-independent) Hamiltonian flow then it must have at least 3 fixed points. (These correspond to the critical points of the generating Hamiltonian.) More generally, if a symplectomorphism is sufficiently close to the identity in the C^1-topology and is the time-1 map of a time-dependent Hamiltonian flow, then the fixed points also correspond to the critical points of a generating function (see Proposition 9.3.17). Based on this observation and on the Poincaré–Birkhoff theorem, Arnold formulated in the 1960s his famous conjecture:

> A symplectomorphism that is generated by a time-dependent Hamiltonian vector field should have at least as many fixed points as a function on the manifold must have critical points.

This simple and compelling conjecture has proved to be a powerful motivating force in the development of the modern theory. Symplectic topology aims to understand global symplectic phenomena, and this transition from a symplectomorphism near the identity to an arbitrary one is a particularly transparent example of global versus local symplectic topology.

The Arnold conjecture was first proved by Eliashberg [174] for Riemann surfaces. His methods are strictly for the 2-dimensional case. For tori of arbitrary dimension the Arnold conjecture was proved in the celebrated paper by Conley and Zehnder [124] via a finite-dimensional approximation of the symplectic action functional on the loop space. This work has been extended by several authors (see Golé [276], for example). The most important breakthrough was Floer's proof of the Arnold conjecture for monotone symplectic manifolds [230]. In his proof Floer developed a new approach to infinite-dimensional Morse theory, called **Floer homology**, with far reaching consequences and many applications in various fields of mathematics. *Hamiltonian Floer homology* is the version developed by Floer for the study of periodic orbits of Hamiltonian systems. All work on the Arnold conjecture since then has built on Floer's ideas.

In the present chapter we first give a proof of the Arnold conjecture for the standard torus, which is based on the discrete symplectic action. Our proof is a reformulation of Chaperon's method of *broken geodesics* [97] which appeared shortly after the work of Conley and Zehnder. We were also influenced by Givental's approach to this problem in [273]. The symplectic part of this proof is very easy. However, to make it complete we devote Section 11.2 to a fairly detailed discussion of the relevant index theory and of Ljusternik–Schnirelmann theory.

Section 11.3 concerns the closely related Lagrangian intersection problem. We outline a proof of Arnold's conjecture for cotangent bundles that again uses the discrete symplectic action, this time to construct generating functions for Lagrangian submanifolds. Here we follow the approach of Bhupal [63]. The chapter ends with a brief outline of the construction and applications of Floer homology.

11.1 Symplectic fixed points

Let M be a smooth manifold, let $\phi : M \to M$ be a diffeomorphism, and denote the set of fixed points of ϕ by

$$\mathrm{Fix}(\phi) := \{p \in M \mid \phi(p) = p\}.$$

When estimating the number of fixed points, it is important to distinguish the nondegenerate from the degenerate ones. A fixed point $p = \phi(p)$ is called **nondegenerate** if the graph of ϕ intersects the diagonal in $M \times M$ transversally at (p, p), and is called **degenerate** otherwise. Thus p is nondegenerate if and only if no eigenvalue of $d\phi(p) : T_pM \to T_pM$ is equal to one, i.e.

$$\det\left(d\phi(p) - \mathbb{1}\right) \neq 0.$$

This distinction corresponds to the notion of nondegeneracy for critical points of a function. For example, if (M, ω) is a symplectic manifold and

$$\phi : M \to M$$

is a symplectomorphism that admits a generating function

$$F : M \to \mathbb{R}$$

as in Proposition 9.3.17, then the fixed points of ϕ are the critical points of F. The reader may verify that in this situation p is a nondegenerate fixed point of ϕ if and only if it is a nondegenerate critical point of F (i.e. the Hessian of F at p is nonsingular). Thus the fixed points of ϕ are all nondegenerate if and only if $F : M \to \mathbb{R}$ is a **Morse function**, i.e. all its critical points are nondegenerate.

Likewise, if $\phi = \phi_H$ is a Hamiltonian symplectomorphism generated by a time-independent Hamiltonian function $H : M \to \mathbb{R}$ with small C^2-norm, then the fixed points of ϕ are the critical points of H and they are all nondegenerate if and only if H is a Morse function.

The proof of the Arnold conjecture for the torus presented here is based on the existence of a generating function for symplectomorphisms which are C^1-close to the identity and are generated by a Hamiltonian differential equation. Such generating functions were constructed for arbitrary manifolds in Proposition 9.3.17. Because the universal cover of the torus is the Euclidean space $\mathbb{R}^{2n}$, there is an alternative and more explicit construction of these generating functions along the lines of Lemma 9.2.1. We shall use the fact that by Proposition 10.2.21, every Hamiltonian symplectomorphism of $\mathbb{T}^{2n}$ is exact in the sense that

$$\int_{\mathbb{T}^{2n}} (\psi(w) - w)\, dw = 0 \tag{11.1.1}$$

for any lift $\psi : \mathbb{R}^{2n} \to \mathbb{R}^{2n}$ (see the remarks preceding Exercise 10.2.22).

Lemma 11.1.1 *Let $\psi : \mathbb{R}^{2n} \to \mathbb{R}^{2n}$ be the lift of an exact symplectomorphism of $\mathbb{T}^{2n}$ which is sufficiently close to the identity in the C^1-topology. Then there exists a smooth function $V : \mathbb{R}^n \times \mathbb{R}^n \to \mathbb{R}$ which satisfies*

$$V(x + k, y + \ell) = V(x, y)$$

for $x, y \in \mathbb{R}^n$ and $k, \ell \in \mathbb{Z}^n$ and generates the symplectomorphism ψ in the sense that $(x_1, y_1) = \psi(x_0, y_0)$ if and only if

$$x_1 - x_0 = \frac{\partial V}{\partial y}(x_1, y_0), \qquad y_1 - y_0 = -\frac{\partial V}{\partial x}(x_1, y_0). \tag{11.1.2}$$

Proof: A lift $\psi : \mathbb{R}^{2n} \to \mathbb{R}^{2n}$ of a diffeomorphism of the $2n$-torus which is homotopic to the identity can be written in the form

$$\psi(x, y) = (x + p(x, y), y + q(x, y)),$$

where p and q are of period 1 in all variables. Since ψ is exact symplectic, equation (11.1.1) implies that

$$\int_{\mathbb{T}^{2n}} p\, dxdy = \int_{\mathbb{T}^{2n}} q\, dxdy = 0.$$

Since ψ is C^1 close to the identity it is symplectically isotopic to the identity by Theorem 10.1.1. Hence ψ is Hamiltonian, with C^1 small Hamiltonian functions, by Theorem 10.2.5 and Proposition 10.2.21. Moreover, by Lemma 9.2.1, there is a generating function $V : \mathbb{R}^n \times \mathbb{R}^n \to \mathbb{R}$ such that the equation $(x_1, y_1) = \psi(x_0, y_0)$ is equivalent to (11.1.2). Hence

$$p(x, y) = \frac{\partial V}{\partial y}(x + p(x, y), y), \qquad q(x, y) = -\frac{\partial V}{\partial x}(x + p(x, y), y).$$

We must prove that V descends to a function on the torus, i.e. that it satisfies the periodicity condition $V(x + k, y + \ell) = V(x, y)$ for $x, y \in \mathbb{R}^n$ and $k, \ell \in \mathbb{Z}^n$.

We first prove that $\partial V/\partial x$ and $\partial V/\partial y$ descend to the torus. Let $x_1, y_0 \in \mathbb{R}^n$ and define $x_0, y_1 \in \mathbb{R}^n$ by

$$x_0 := x_1 - \frac{\partial V}{\partial y}(x_1, y_0), \qquad y_1 := y_0 - \frac{\partial V}{\partial x}(x_1, y_0).$$

Then $(x_1, y_1) = \psi(x_0, y_0)$. Now let $k, \ell \in \mathbb{Z}^n$. Then

$$(x_1 + k, y_1 + \ell) = \psi(x_0 + k, y_0 + \ell),$$

and hence

$$\frac{\partial V}{\partial y}(x_1 + k, y_0 + \ell) = (x_1 + k) - (x_0 + k) = \frac{\partial V}{\partial y}(x_1, y_0)$$

and

$$\frac{\partial V}{\partial x}(x_1 + k, y_0 + \ell) = (y_0 + \ell) - (y_1 + \ell) = \frac{\partial V}{\partial x}(x_1, y_0).$$

This shows that $\partial V/\partial x$ and $\partial V/\partial y$ descend to the torus. Hence the formula

$$\beta(x_1, y_0) := \frac{\partial V}{\partial x}(x_1, y_0) \cdot dx_1 + \frac{\partial V}{\partial y}(x_1, y_0) \cdot dy_0$$

defines a closed 1-form on $\mathbb{T}^n \times \mathbb{T}^n$.

We must prove that β is exact. Define the map $f : \mathbb{T}^{2n} \to \mathbb{T}^n \times \mathbb{T}^n$ by

$$f(x_0, y_0) := (x_0 + p(x_0, y_0), y_0).$$

This map is a diffeomorphism and its inverse is given by

$$f^{-1}(x_1, y_0) = (x_1 - \partial_y V(x_1, y_0), y_0).$$

The pullback of β under f is the 1-form

$$\begin{aligned} f^*\beta &= \frac{\partial V}{\partial x}(x_0 + p(x_0, y_0), y_0) \cdot f^*dx_1 + \frac{\partial V}{\partial y}(x_0 + p(x_0, y_0), y_0) \cdot dy_0 \\ &= -q \cdot (dx_0 + dp) + p \cdot dy_0 \\ &= -gr_\psi^* \alpha. \end{aligned}$$

Here $gr_\psi : \mathbb{T}^{2n} \to \mathbb{T}^{2n} \times \mathbb{T}^{2n}$ denotes the embedding

$$gr_\psi(x_0, y_0) := (x_0, y_0, x_1, y_1) := (x_0, y_0, x_0 + p(x_0, y_0), y_0 + q(x_0, y_0))$$

and α denotes the 1-form

$$\alpha := (y_1 - y_0) \cdot dx_1 + (x_0 - x_1) \cdot dy_0 \tag{11.1.3}$$

on a neighbourhood of the diagonal in $\mathbb{T}^{2n} \times \mathbb{T}^{2n}$. Then

$$-d\alpha = dx_1 \wedge dy_1 - dx_0 \wedge dy_0,$$

and α vanishes on the diagonal in $\mathbb{T}^{2n} \times \mathbb{T}^{2n}$. Hence it follows from Proposition 9.3.17 that $gr_\psi^*\alpha$ is exact. This implies that β is exact and hence $V : \mathbb{R}^{2n} \to \mathbb{R}$ descends to a function on the torus as claimed. This proves Lemma 11.1.1. □

Exercise 11.1.2 Find a neighbourhood of the diagonal in $\mathbb{T}^{2n} \times \mathbb{T}^{2n}$ on which the 1-form α in (11.1.3) is well-defined. □

Exercise 11.1.3 Here is an attempt at an alternative proof of the fact that the function $V : \mathbb{R}^{2n} \to \mathbb{R}$ in the proof of Lemma 11.1.1 descends to the torus. Since the partial derivatives of V have period one in all variables, there exist vectors $a, b \in \mathbb{R}^n$ and a function $W : \mathbb{T}^{2n} \to \mathbb{R}$ such that

$$V(x, y) = \langle a, x\rangle + \langle b, y\rangle + W(x, y).$$

Differentiate this identity to obtain

$$a + \frac{\partial W}{\partial x}(x + p(x, y), y) = -q(x, y), \qquad b + \frac{\partial W}{\partial y}(x + p(x, y), y) = p(x, y).$$

Differentiate the second identity again with respect to x to obtain

$$\mathbb{1} - \frac{\partial^2 W}{\partial y \partial x}(x + p(x, y), y) = \left(\mathbb{1} + \frac{\partial p}{\partial x}(x, y)\right)^{-1}.$$

Since $q(x, y)$ has mean value zero, the vector a is given by

$$\begin{aligned} a &= -\int_{\mathbb{T}^{2n}} \frac{\partial W}{\partial x}(x + p(x, y), y)\, dxdy \\ &= -\int_{\mathbb{T}^{2n}} \frac{\partial W}{\partial x}(x, y) \det\left(\mathbb{1} - \frac{\partial^2 W}{\partial x \partial y}(x, y)\right) dxdy. \end{aligned}$$

Here the second identity follows by change of variables. The task at hand is to prove that the last integral vanishes. This is easy to see when $n = 1$ but becomes quite technical for $n \geq 2$. The trick in the proof of Lemma 11.1.1 is to avoid this calculation by using Proposition 9.3.17. □

Exercise 11.1.4 Let $\psi : \mathbb{T}^{2n} \to \mathbb{T}^{2n}$ be an exact symplectomorphism which is sufficiently close to the identity in the C^1-topology and let $V : \mathbb{T}^{2n} \to \mathbb{R}$ be the generating function constructed in Lemma 11.1.1. Prove that the fixed points of ψ are the critical points of V. Prove that an element $z \in \mathrm{Fix}(\psi) = \mathrm{Crit}(V)$ is nondegenerate as a fixed point of ψ if and only if it is nondegenerate as a critical point of V. □

Proposition 11.1.5 *Let $\psi : \mathbb{T}^{2n} \to \mathbb{T}^{2n}$ be an exact symplectomorphism which is sufficiently close to the identity in the C^1-topology. Then ψ has at least $2n+1$ fixed points. If the fixed points of ψ are all nondegenerate then ψ has at least 2^{2n} fixed points.*

Proof: Let $V : \mathbb{T}^{2n} \to \mathbb{R}$ be the generating function of Lemma 11.1.1. Then the critical points of V are the fixed points of ψ. But any function on the torus must have at least $2n + 1$ critical points. (If a function on a manifold has ℓ critical points then the manifold can be covered by ℓ contractible open sets. Thus the result follows from Lemma 11.2.8 below.) Moreover, a point (x, y) is nondegenerate as a fixed point of ψ if and only if it is nondegenerate as a critical point of V. Hence it follows from Morse theory (see, for example, Milnor [483]) that in the nondegenerate case ψ must have at least 2^{2n} fixed points. □

Example 11.1.6 The function

$$H(x,y) = \sin \pi x \, \sin \pi y \, \sin \pi(x+y)$$

on the 2-torus $\mathbb{T}^2 = \mathbb{R}^2/\mathbb{Z}^2$ has precisely three critical points. Hence the time-1 map $\phi = \phi_{\varepsilon H}$ of the Hamiltonian flow with the Hamiltonian function εH has precisely three fixed points for $\varepsilon > 0$ sufficiently small. The function

$$H(x,y) = \sin 2\pi x \, + \, \sin 2\pi y$$

has precisely four critical points and these are all nondegenerate. □

It was conjectured by Arnold [23] in the 1960s that Proposition 11.1.5 should remain valid for arbitrary Hamiltonian symplectomorphisms of general compact symplectic manifolds which are not necessarily close to the identity.

Conjecture 11.1.7 (Arnold) *Let $\psi : M \to M$ be a Hamiltonian symplectomorphism of a compact symplectic manifold (M, ω). Then ψ must have at least as many fixed points as a function on M must have critical points. If the fixed points are all nondegenerate then the number of fixed points is at least the minimal number of critical points of a Morse function on M.*

Arnold first formulated his conjecture in the 1966 Congress in Moscow for the 2-torus, the extension to arbitrary symplectic manifolds M only appearing later. When M is arbitrary it is very hard to estimate the minimal number of critical points of a function on M, even if one restricts to Morse functions. Therefore, one often considers a slightly weaker version of the conjecture in which these bounds are replaced by appropriate numbers that depend only on the cohomology of M. This weaker version is called the **homological Arnold conjecture**.

In the case when the critical points are allowed to be degenerate, the correct number to use is the **cuplength** $cl\,(M)$. This is defined to be the minimal number N such that for any cohomology classes $a_1, \dots, a_N \in H^*(M)$ with $\deg(a_j) \geq 1$, the cup product $a_1 \cup \cdots \cup a_N = 0$. As we shall see in Lemma 11.2.8 below, it follows from Ljusternik–Schnirelmann theory that the number of critical points of a function $f : M \to \mathbb{R}$ is bounded below by $cl\,(M)$. So the homological Arnold conjecture asserts that every Hamiltonian symplectomorphism ψ of M satisfies

$$\#\mathrm{Fix}(\psi) \geq cl\,(M).$$

In the nondegenerate case it follows from finite-dimensional Morse theory (see e.g. Milnor [483]) that the number of critical points of a Morse function is bounded below by the sum of the Betti numbers $b_k(M) = \dim H_k(M)$. So the nondegenerate homological Arnold conjecture asserts that

$$\#\mathrm{Fix}(\psi) \geq \sum_{k=0}^{2n} b_k(M),$$

where $2n := \dim M$. Here we have not specified the coefficient ring used in the definition of $cl\,(M)$ and $b_k(M)$. The homological Arnold conjecture should in fact hold for any coefficient ring.

The nondegenerate homological Arnold conjecture is much stronger than the Lefschetz fixed point theorem, which would only give the alternating sum of the Betti numbers as a lower bound. For example, in the case of the torus the latter number is zero.

Most of the known results on the Arnold conjecture concern the weaker homological forms of the estimates. Of course there are some manifolds (projective space, for example) where both estimates agree, but this is usually not the case when the fundamental group of M is nontrivial. However, some progress has been made with the original conjecture by Rudyak [551], using a development of Ljusternik–Schnirelmann theory called category weight. Using these ideas, Rudyak and Oprea [517] proved that the original conjecture holds (in the case where degenerate fixed points are allowed) whenever both cohomology classes $[\omega]$ and c_1 vanish on $\pi_2(M)$.

As pointed out in the beginning of the present section, there are two obvious cases in which the Arnold conjecture holds on an arbitrary manifold. The first is where the Hamiltonian function is independent of t, and the second, slightly less obvious, case is where the exact symplectomorphism is C^1 close to the identity and the result reduces to generating functions as in the case of the torus.

Proposition 11.1.8 *The Arnold conjecture holds for every Hamiltonian symplectomorphism ψ of a closed symplectic manifold (M, ω) which is sufficiently close to the identity in the C^1-topology.*

Proof: Recall from Proposition 3.4.14 that the graph of ψ is a Lagrangian submanifold of $M \times M$ which is close to the diagonal Δ. Also a neighbourhood of Δ may be identified with a neighbourhood of the zero section in T^*M via a symplectomorphism Ψ. With this identification the graph of ψ corresponds to the graph of a closed 1-form $\sigma : M \to T^*M$, i.e. $\mathrm{graph}(\psi) = \Psi(\mathrm{graph}(\sigma))$. We showed in Proposition 9.3.17 that Hamiltonian symplectomorphisms corresponded precisely to exact 1-forms. Hence there exists a function $F : M \to \mathbb{R}$ such that

$$\mathrm{graph}(\psi) = \Psi(\mathrm{graph}(dF)).$$

It follows that the fixed points of ψ are the critical points of F, and that an element $p \in \mathrm{Fix}(\psi) = \mathrm{Crit}(F)$ is nondegenerate as a fixed point of ψ if and only if it is nondegenerate as a critical point of F. This proves Proposition 11.1.8. □

In 1983, Conley and Zehnder proved the Arnold conjecture for the torus with the standard symplectic structure [124]. This is a much deeper result than Proposition 11.1.5. Their methods exploit the linear structure of the covering space $\mathbb{R}^{2n}$ of $\mathbb{T}^{2n}$ and were a direct inspiration for the variational argument which we present in Section 12.5 below. Shortly afterwards Chaperon gave an alternative proof of the Conley–Zehnder theorem using the method of *broken geodesics* [97]. Givental [273] reformulated Chaperon's argument in terms of generating functions. The argument we give here is a further reformulation in terms of the discrete symplectic action functional (9.2.4) which was used by Robbin and Salamon [540, 543] and was discussed in Chapter 9.

Theorem 11.1.9 (Conley–Zehnder) *Every Hamiltonian symplectomorphism of the standard torus $\mathbb{T}^{2n}$ has at least 2^{2n} geometrically distinct fixed points provided that these are all nondegenerate. In the degenerate case the number of fixed points is at least $2n+1$.*

Proof: See pages 430 and 434. □

Let $\psi_H^{t,t_0} : \mathbb{R}^{2n} \to \mathbb{R}^{2n}$ be a lift of a time-dependent Hamiltonian flow of $\mathbb{T}^{2n}$ generated by a smooth family of Hamiltonian functions $H_t : \mathbb{R}^{2n}/\mathbb{Z}^{2n} \to \mathbb{R}$ which is 1-periodic in t. (This step is justified in Exercise 11.1.11 below.) Fix a sufficiently large integer N and denote

$$\psi = \psi_H^{1,0}, \qquad \psi_j = \psi_H^{(j+1)/N, j/N}$$

so that

$$\psi = \psi_{N-1} \circ \psi_{N-2} \circ \cdots \circ \psi_0.$$

For large N all the symplectomorphisms ψ_j will satisfy the requirements of Lemma 11.1.1. Hence for every j there is a smooth function $V_j : \mathbb{R}^{2n}/\mathbb{Z}^{2n} \to \mathbb{R}$ which is sufficiently small in the C^2-norm and generates the symplectomorphism $(x_{j+1}, y_{j+1}) = \psi_j(x_j, y_j)$ via the discrete Hamiltonian equations

$$x_{j+1} - x_j = \frac{\partial V_j}{\partial y}(x_{j+1}, y_j), \qquad y_{j+1} - y_j = -\frac{\partial V_j}{\partial x}(x_{j+1}, y_j). \tag{11.1.4}$$

In view of Lemma 9.2.6, the fixed points of ψ correspond to the critical points of the discrete symplectic action functional as follows.

Denote by $X \simeq \mathbb{R}^{2nN}$ the space of sequences $\mathbf{z} = \{z_j\}_{j\in\mathbb{Z}}$ in $\mathbb{R}^{2n}$ which satisfy

$$z_j = z_{j+N}$$

for every $j \in \mathbb{Z}$. As before we use the notation $\mathbf{z} = (\mathbf{x}, \mathbf{y})$, where $\mathbf{x} = \{x_j\}_{j\in\mathbb{Z}}$ and $\mathbf{y} = \{y_j\}_{j\in\mathbb{Z}}$ are sequences in $\mathbb{R}^n$ which satisfy the same periodicity condition. The lattice $\mathbb{Z}^{2n}$ acts on the space X by $\{z_j\}_j \mapsto \{z_j + k\}_j$ for $k \in \mathbb{Z}^{2n}$. Consider the discrete symplectic action function $\Phi : X \to \mathbb{R}$, defined by

$$\Phi(\mathbf{z}) = \sum_{j=0}^{N-1} \Big(\langle y_j, x_{j+1} - x_j \rangle - V_j(x_{j+1}, y_j) \Big). \tag{11.1.5}$$

This function is invariant under the action of $\mathbb{Z}^{2n}$ on X and hence descends to the quotient $X/\mathbb{Z}^{2n}$. By Lemma 9.2.6, a sequence $\mathbf{z}$ is a critical point of Φ if and only if $z_{j+1} = \psi_j(z_j)$ for $j = 0, \dots, N-1$. Hence there is a one-to-one correspondence of the critical points of Φ on $X/\mathbb{Z}^{2n}$ with the fixed points of ψ. Moreover, a critical point $\mathbf{z}$ of Φ is nondegenerate if and only if the corresponding fixed point z_0 of ψ is nondegenerate. Two critical points $\mathbf{z}$ and $\mathbf{z}'$ of Φ are called *geometrically distinct* if they are not related by the action of $\mathbb{Z}^{2n}$, or equivalently, z_0 and z_0' represent different fixed points of ψ on $\mathbb{T}^{2n}$.

It remains to prove that if Φ is a Morse function it must have at least 2^{2n} critical points on $X/\mathbb{Z}^{2n}$. We will do this in the next section with an argument based on the Conley index. To prove the assertion about degenerate fixed points we will use Ljusternik–Schnirelmann theory to show that the number of critical points of Φ is always bounded below by $2n+1$.

Remark 11.1.10 Givental [273] has suggested the following approach. With ψ and ψ_j as above, consider the space $X = \mathbb{R}^{2nN}$ with the maps $\sigma : X \to X$ and $\Psi : X \to X$ defined by

$$\sigma(z_0, \dots, z_{N-1}) = (z_1, \dots, z_{N-1}, z_0),$$

$$\Psi(z_0, \dots, z_{N-1}) = (\psi_0(z_0), \dots, \psi_{N-1}(z_{N-1})).$$

Then the fixed points of ψ correspond to the solutions $\mathbf{z} \in X$ of the equation

$$\sigma(\mathbf{z}) = \Psi(\mathbf{z}).$$

Now both these symplectomorphisms have generating functions. That is, after a suitable change of coordinates which identifies $X \times X$ with T^*X, we may represent the graphs of σ and Ψ by exact 1-forms dG and dQ, where G and Q are real valued functions on X. Hence the fixed points of ψ correspond to the solutions of the equation $dG(\mathbf{z}) = dQ(\mathbf{z})$. Thus the problem is reduced to finding critical points of the function $G - Q : X \to \mathbb{R}$. It turns out that this function is precisely the discrete symplectic action (11.1.5). (Exercise: Prove this.) □

Exercise 11.1.11 Let $\{\phi_t\}_{0\le t\le 1}$ be a Hamiltonian isotopy generated by a smooth time-dependent family of Hamiltonian functions $H_t : M \to \mathbb{R}$. If $\beta : [0,1] \to [0,1]$ is any smooth function such that $\beta' \ge 0$, verify that the reparametrized path $\phi_{\beta(t)}$ is generated by the Hamiltonian functions $\beta'(t)H_{\beta(t)}$. Deduce that there exists a reparametrization $t \mapsto \phi_{\beta(t)}$ of the Hamiltonian isotopy ϕ_t whose generating Hamiltonian extends smoothly to a 1-periodic function on $\mathbb{R} \times M$. □

11.2 Morse theory and the Conley index

To estimate the number of geometrically distinct critical points of the discrete symplectic action $\Phi : X/\mathbb{Z}^{2n} \to \mathbb{R}$ of (11.1.5) we need a localized version of Morse theory, since the manifold $X/\mathbb{Z}^{2n}$ is not compact. Conley developed the following very elegant approach, which applies to general dynamical systems. Assume that $\phi^t : M \to M$ is a flow on a locally compact metric space M, i.e.

$$\phi^{t+s} = \phi^t \circ \phi^s, \qquad \phi^0 = \mathrm{id},$$

for $s, t \in \mathbb{R}$. The model example is the gradient flow of the discrete symplectic action functional Φ on $M = X/\mathbb{Z}^{2n}$. A set $\Lambda \subset M$ is called **invariant** if $\phi^t(\Lambda) = \Lambda$ for every t. A compact invariant set Λ is called **isolated** if there exists a neighbourhood N of Λ such that

$$\Lambda = I(N) := \bigcap_{t\in\mathbb{R}} \phi^t(N).$$

Definition 11.2.1 An **index pair** for an isolated invariant set Λ is a pair of compact sets $L \subset N$ satisfying the following conditions.

(i) The closure of $N \setminus L$ is an isolating neighbourhood of Λ and $N \setminus L$ is a neighbourhood of Λ, i.e.

$$\Lambda = I(\mathrm{cl}(N \setminus L)) \subset \mathrm{int}(N \setminus L).$$

(ii) L is **positively invariant** in N, i.e.

$$x \in L, \quad \phi^{[0,t]}(x) \subset N \qquad \Longrightarrow \qquad \phi^t(x) \in L.$$

(iii) Every orbit which leaves N must go through L first, i.e.

$$x \in N \setminus L \qquad \Longrightarrow \qquad \exists\, t > 0 \text{ such that } \phi^{[0,t]}(x) \subset N.$$

The set L is called the **exit set** (see Fig. 11.1).

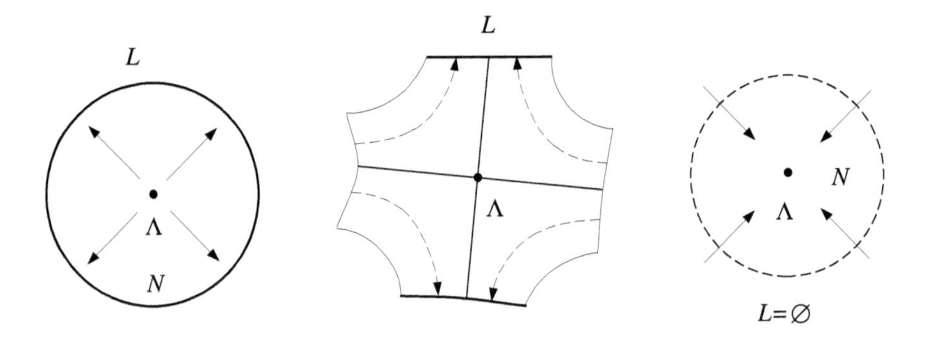

FIG. 11.1. Index pairs.

In [123], [125] and [539] it is shown that every isolated invariant set admits an index pair and, if M is a manifold, that the index pair can be chosen such that the topological quotient N/L has the homotopy type of a finite polyhedron. Moreover, the homotopy type of N/L is independent of the choice of the index pair (cf. [123, 552, 539]).

Lemma 11.2.2 *If (N_α, L_α) and (N_β, L_β) are index pairs for Λ then the index spaces N_α/L_α and N_β/L_β are homotopy equivalent.*

Proof: The proof is an elementary homotopy argument. For each $t \geq 0$, the flow determines a (possibly discontinuous) map $\phi^t_{\beta\alpha} : N_\alpha/L_\alpha \to N_\beta/L_\beta$ defined by

$$\phi^t_{\beta\alpha}(x) := \begin{cases} \phi^t(x), & \text{if } \phi^{[0,2t/3]}(x) \subset N_\alpha \setminus L_\alpha \\ & \text{and } \phi^{[t/3,t]}(x) \subset N_\beta \setminus L_\beta, \\ \star, & \text{otherwise.} \end{cases}$$

This map is continuous for $t > t_{\alpha\beta}$, where $t_{\alpha\beta} \geq 0$ is the infimum over all $t \geq 0$ such that

$$\phi^{[-t,t]}(x) \subset N_\alpha \setminus L_\alpha \quad \Longrightarrow \quad \phi^{[-2t/3,2t/3]}(x) \subset N_\beta \setminus L_\beta,$$
$$\phi^{[-t,t]}(x) \subset N_\beta \setminus L_\beta \quad \Longrightarrow \quad \phi^{[-2t/3,2t/3]}(x) \subset N_\alpha \setminus L_\alpha.$$

In particular, $t_{\alpha\alpha} = 0$, and the induced semi-dynamical system

$$[0,\infty) \times N_\alpha/L_\alpha \to N_\alpha/L_\alpha : (t,x) \mapsto \phi^t_{\alpha\alpha}(x), \tag{11.2.1}$$

given by

$$\phi^t_{\alpha\alpha}(x) := \begin{cases} \phi^t(x), & \text{if } \phi^{[0,t]}(x) \subset N_\alpha \setminus L_\alpha, \\ \star, & \text{otherwise}, \end{cases}$$

is continuous. (It is shown in [539, Theorem 4.2] that the map (11.2.1) is continuous if and only if the pair (N_α, L_α) satisfies conditions (ii) and (iii) in 11.2.1.) Moreover, $t_{\alpha\gamma} \leq t_{\alpha\beta} + t_{\beta\gamma}$ and

$$\phi^t_{\gamma\beta} \circ \phi^s_{\beta\alpha} = \phi^{t+s}_{\gamma\alpha}, \qquad \phi^0_{\alpha\alpha} = \mathrm{id},$$

for $s > t_{\alpha\beta}$ and $t > t_{\beta\gamma}$. Hence $\phi^t_{\alpha\beta}$ is a homotopy inverse of $\phi^s_{\beta\alpha}$. □

The **Conley index** of Λ is the homotopy type of the pointed space N/L. An index pair is called **regular** if the exit set L is a neighbourhood deformation retract in N. In this case the homology of the pair (N,L) agrees with the homology of the index space N/L and is therefore an invariant of the isolated invariant set Λ. If the homology is finite-dimensional (e.g. if M is a manifold) it can be characterized by the index polynomial

$$p_\Lambda(s) = \sum_k \dim H_k(N,L) s^k.$$

The Conley index is additive in the sense that $p_\Lambda(s) = p_{\Lambda_1}(s) + p_{\Lambda_2}(s)$ whenever Λ is the disjoint union of the isolated invariant sets Λ_1 and Λ_2.

Example 11.2.3 As an example consider a hyperbolic fixed point $x = 0$ of a differential equation

$$\dot{x} = v(x)$$

in $\mathbb{R}^n$. Denote by E^s and E^u the stable and unstable subspaces of the linearized system $\dot{\xi} = dv(0)\xi$. Thus E^s is the space spanned by the eigenvectors whose eigenvalues have negative real part and E^u the corresponding space for eigenvalues with positive real part. Then an index pair for the isolated invariant set $\Lambda = \{0\}$ of the nonlinear flow is given by

$$N = \{x_u + x_s \,|\, x_u \in E^u,\, x_s \in E^s,\, |x_u| \leq \varepsilon,\, |x_s| \leq \varepsilon\},$$
$$L = \{x_u + x_s \in N \,|\, |x_u| = \varepsilon\}$$

for $\varepsilon > 0$ sufficiently small (see Fig. 11.2). It follows that N/L has the homotopy type of a pointed k-sphere, where $k = \dim E^u$ is the index of the hyperbolic fixed point and the index polynomial is $p_\Lambda(s) = s^k$. □

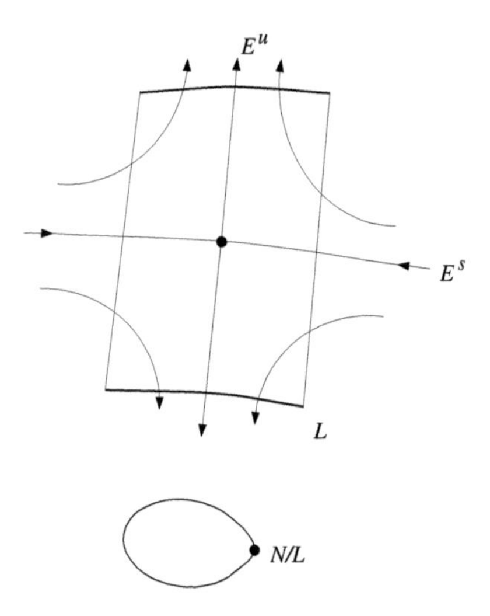

FIG. 11.2. A hyperbolic fixed point.

We are now in a position to prove a localized version of the Morse inequalities. Let $\Phi : M \to \mathbb{R}$ be a Morse function on an n-dimensional Riemannian manifold M and consider the gradient flow

$$\dot{x} = -\nabla\Phi(x). \tag{11.2.2}$$

Assume that the gradient field of Φ is complete and denote by ϕ^t the flow of (11.2.2). Since Φ is a Morse function every critical point x of Φ is a hyperbolic equilibrium of (11.2.2) and hence its unstable set

$$W^u(x) = \left\{ y \in M \,\middle|\, \lim_{t\to-\infty} \phi^t(y) = x \right\}$$

is a submanifold of M. Its dimension is the **Morse index** of x:

$$\mathrm{ind}(x) = \dim W^u(x).$$

Equivalently, the index can be defined as the number of negative eigenvalues of the Hessian $d^2\Phi(x)$. If $\Lambda \subset M$ is a compact isolated invariant set for the gradient flow of Φ, then the Morse inequalities relate the numbers

$$c_k(\Lambda) = \#\{x \in \Lambda \,|\, d\Phi(x) = 0,\, \mathrm{ind}(x) = k\}$$

of critical points in Λ to the **Conley–Betti numbers**

$$b_k(\Lambda) = \dim H_k(N, L),$$

where (N, L) is a regular index pair for Λ.

Theorem 11.2.4 (Morse inequalities) *For $k = 0, \dots, n$,*

$$c_k(\Lambda) - c_{k-1}(\Lambda) + \cdots \pm c_0(\Lambda) \geq b_k(\Lambda) - b_{k-1}(\Lambda) + \cdots \pm b_0(\Lambda)$$

and equality holds for $k = n$. In particular, the number of critical points in Λ is bounded below by the sum of the Conley–Betti numbers $b_k(\Lambda)$.

Proof: Fix an index pair (N, L) for Λ. For every regular value $a \in \mathbb{R}$ of $\Phi|_N$ define

$$N^a = \{x \in N \mid \Phi(x) \leq a\} \cup L.$$

Then for every critical value c of $\Phi|_N$ the set of critical points $x \in \Lambda$ with $\Phi(x) = c$ is an isolated invariant set and an index pair is given by $L' = N^a$ and $N' = N^b$, where $a < c < b$ are chosen such that there is no other critical value of $\Phi|_N$ in the interval $[a, b]$. Hence it follows from the additivity of the Conley index and Lemma 11.2.2 that in this case

$$\sum_k \dim H_k(N^b, N^a) s^k = \sum_{d\Phi(x)=0,\, \Phi(x)=c} s^{\mathrm{ind}(x)}.$$

Now define $b_k^a = b_k^a(\Lambda) = \dim H_k(N^a, L)$ and let $c_k^a = c_k^a(\Lambda)$ denote the number of critical points of S in $\Lambda \cap N^a$ with index k. Then the above identity can be restated in the form

$$\dim H_k(N^b, N^a) = c_k^b - c_k^a.$$

The homology exact sequence of the triple (N^b, N^a, L) shows that

$$d_{k-1}^{ab} + d_k^{ab} = c_k^b - c_k^a - b_k^b + b_k^a,$$

where d_k^{ab} is the rank of the boundary operator $H_{k+1}(N^b, N^a) \to H_k(N^a, L)$. Equivalently

$$p_{\mathrm{crit}}^b(s) - p_\Lambda^b(s) = p_{\mathrm{crit}}^a(s) - p_\Lambda^a(s) + (1+s)q^{ab}(s),$$

where

$$p_{\mathrm{crit}}^a(s) = \sum_{k=0}^n c_k^a s^k, \qquad p_\Lambda^a(s) = \sum_{k=0}^n b_k^a s^k, \qquad q^{ab}(s) = \sum_{k=0}^n d_k^{ab} s^k.$$

In particular, $q^{ab}(s)$ is a polynomial with nonnegative coefficients. It follows by induction that

$$p_{\mathrm{crit}}^a(s) - p_\Lambda^a(s) = (1+s)q^a(s),$$

where $q^a(s)$ is a polynomial with nonnegative coefficients. The coefficients can be written in the form

$$d_k^a = \sum_{j=0}^k (-1)^j \left(c_{k-j}^a - b_{k-j}^a\right) \geq 0.$$

For $a > \sup_N \Phi$ these are the required Morse inequalities. □

Remark 11.2.5 An alternative proof of the Morse inequalities as stated in Theorem 11.2.4 can be given using the Morse complex as explicitly formulated in Milnor's book on the h-cobordism theorem [485] and in Witten's paper [688]. The idea is to construct a chain complex generated by the critical points of Φ in M. This chain complex is graded by the Morse index, and the entries of the boundary operator are given by counting the number of connecting gradient trajectories. Under suitable transversality conditions these numbers are finite, and if the orbits are counted with suitably defined signs then it turns out that the square of this operator is zero. The homology of the resulting chain complex is invariant under deformation and in the compact case it is isomorphic to the homology of the manifold M. In the localized version considered by Conley this complex computes the homology of an index pair of the isolated invariant set. It follows that the number of critical points (the dimension of the chain complex) is bounded below by the dimension of the homology. Conley's idea that one could get information from a partially defined flow greatly influenced Floer in the development of his homology theory. □

Morse theory for the discrete symplectic action

We shall now return to the case where $X = \mathbb{R}^{2nN}$, $M = X/\mathbb{Z}^{2n}$, and the functional $\Phi : M \to \mathbb{R}$ is the discrete symplectic action.

Proof of Theorem 11.1.9 'The nondegenerate case': Let

$$\Phi : M = \mathbb{R}^{2nN}/\mathbb{Z}^{2n} \to \mathbb{R}$$

be the discrete symplectic action functional defined by (11.1.5), where the functions $V_j : \mathbb{R}^{2n} \to \mathbb{R}$ all descend to $\mathbb{T}^{2n} = \mathbb{R}^{2n}/\mathbb{Z}^{2n}$. We have seen that the V_j can be chosen such that the critical points of Φ are in one-to-one correspondence with the fixed points of our Hamiltonian symplectomorphism of $\mathbb{T}^{2n}$ and that if these fixed points are all nondegenerate, then Φ is a Morse function. We must prove that Φ has at least 2^{2n} geometrically distinct critical points.

We identify the quotient $\mathbb{R}^{2nN}/\mathbb{Z}^{2n}$ with the space $\mathcal{X} = \mathbb{T}^{2n} \times \mathbb{R}^{2n(N-1)}$ by choosing coordinates $(z_0, \zeta_1, \dots, \zeta_{N-1}) \in \mathbb{T}^{2n} \times \mathbb{R}^{2n(N-1)}$ such that

$$\zeta_j = z_j - z_{j-1}, \qquad j = 1, \dots, N-1.$$

Then a simple calculation shows that

$$\Phi(\mathbf{z}) = \Psi(z_0, \zeta) = \tfrac{1}{2}\langle \zeta, P\zeta \rangle + W(z_0, \zeta),$$

where $\zeta := (\zeta_1, \dots, \zeta_{N-1})$,

$$P := \begin{pmatrix} 0 & -B \\ -B^T & 0 \end{pmatrix}, \qquad B := \begin{pmatrix} \mathbb{1}_n & \cdots & \cdots & \mathbb{1}_n \\ 0 & \ddots & & \vdots \\ \vdots & \ddots & \ddots & \vdots \\ 0 & \cdots & 0 & \mathbb{1}_n \end{pmatrix},$$

and the function $W : \mathbb{R}^{2nN} \to \mathbb{R}$ descends to the torus $\mathbb{T}^{2nN} = \mathbb{R}^{2nN}/\mathbb{Z}^{2nN}$.

Since B is nonsingular the matrix P has signature 0 and there is a splitting

$$\mathbb{R}^{2n(N-1)} = E^- \oplus E^+$$

into the negative and positive eigenspaces of P. Now the gradient flow of Ψ takes the form

$$\dot\zeta = -P\zeta - \frac{\partial W}{\partial \zeta}(z_0,\zeta), \qquad \dot z_0 = -\frac{\partial W}{\partial z_0}(z_0,\zeta).$$

The set Λ of all bounded orbits is an isolated invariant set in the sense of Conley and it contains all the critical points of Ψ. Since Ψ is a bounded perturbation of the nondegenerate quadratic function $\zeta \mapsto \frac{1}{2}\langle\zeta, P\zeta\rangle$, the gradient flow of Φ is essentially the same as $\dot\zeta = -P\zeta$ outside a large compact set. Hence an index pair for the set Λ of bounded solutions is given by

$$N := \left\{(z_0, \zeta^- + \zeta^+)\,|\,\zeta^\pm \in E^\pm,\ |\zeta^\pm| \le R\right\},$$

$$L := \left\{(z_0, \zeta^- + \zeta^+) \in N\,|\,|\zeta^-| = R\right\},$$

provided that R is sufficiently large. Now $\dim E^\pm = n(N-1) =: m$ and so the space N/L is homotopy-equivalent to the quotient $\mathbb{T}^{2n} \times B^m/\mathbb{T}^{2n} \times S^{m-1}$. Hence $H_{k+m}(N,L) \cong H_k(\mathbb{T}^{2n})$ and this implies that

$$b_{k+m}(\Lambda) = \binom{2n}{k}.$$

Thus the sum of the Conley–Betti numbers of Λ is 2^{2n} and this proves Theorem 11.1.9 in the nondegenerate case. □

Ljusternik–Schnirelmann theory

Let M be a compact metric space. A **category** on M is a map $\nu : 2^M \to \mathbb{N}$ which assigns an integer $\nu(A) \ge 0$ to every subset $A \subset M$ and satisfies the following axioms:

(continuity) For every subset $A \subset M$ there exists an open set $U \subset M$ such that $A \subset U$ and $\nu(A) = \nu(U)$.

(monotonicity) If $A \subset B$ then $\nu(A) \le \nu(B)$.

(subadditivity) $\nu(A \cup B) \le \nu(A) + \nu(B)$.

(naturality) If $\phi : M \to M$ is a homeomorphism then $\nu(\phi(A)) = \nu(A)$.

(normalization) $\nu(\emptyset) = 0$ and if $A = \{x_0, \dots, x_N\}$ is a finite set then $\nu(A) = 1$.

Remark 11.2.6 If M is a connected manifold then the normalization axiom reduces to the assumption that $\nu(\emptyset) = 0$ and $\nu(\{x_0\}) = 1$ for every $x_0 \in M$. This implies that every finite set $A = \{x_0, x_1, \dots, x_N\}$ has category $\nu(A) = 1$. To see this choose an embedding $\gamma : [0,1] \to M$ such that $\gamma(j/N) = x_j$. (If M is a 1-manifold the points should first be reordered.) Consider the set T of all $t \in [0,1]$ such that $\nu(\gamma([0,t])) = 1$. Then the normalization axiom shows that $0 \in T$. By the continuity and monotonicity axioms, T is open and, by the naturality axiom, T is closed. Hence $T = [0,1]$. This implies that $\nu(\gamma([0,1])) = 1$ and the monotonicity axiom then shows that $\nu(A) = 1$. □

Now fix a cohomology functor H^* with coefficient ring R. Call an open set $U \subset M$ **cohomologically trivial** if the induced map $H^k(M) \to H^k(U)$ on cohomology is zero for $k \geq 1$. If M is a compact manifold and H^* denotes the de Rham cohomology then this means that every closed form $\alpha \in \Omega^k(M)$ of degree $k \geq 1$ is exact on U. Define the **Ljusternik–Schnirelmann category**

$$\nu_{\mathrm{LS}}(A)$$

of a set $A \subset M$ to be the minimal number of cohomologically trivial sets $U_1, \ldots, U_N$ required to cover A:

$$A \subset U_1 \cup \cdots \cup U_N.$$

This function $\nu_{\mathrm{LS}} : 2^M \to \mathbb{N}$ obviously satisfies the above axioms.

Example 11.2.7 The Ljusternik–Schnirelmann category of an n-dimensional smooth manifold M is less than or equal to $n+1$. To see this, triangulate M and then consider the first barycentric subdivision $\mathcal{T}_1$ of this triangulation $\mathcal{T}$. This is the triangulation whose vertices consist of all the vertices in $\mathcal{T}$ plus the barycentres of all the faces in $\mathcal{T}$. Let $\mathcal{V}_j$ denote the set of vertices in $\mathcal{T}'$ corresponding to the j-dimensional faces of $\mathcal{T}$. So $\mathcal{V}_0$ is the set of vertices in $\mathcal{T}$. Recall that the **open star** $\mathrm{St}(v)$ of a vertex v is the union of all open simplices whose closure contains v. It is easy to see that the open stars in $\mathcal{T}'$ of any two vertices in $\mathcal{V}_j$ are disjoint. (See Fig. 11.3.) Thus the union U_j of all such open stars is cohomologically trivial. Furthermore, they cover M. □

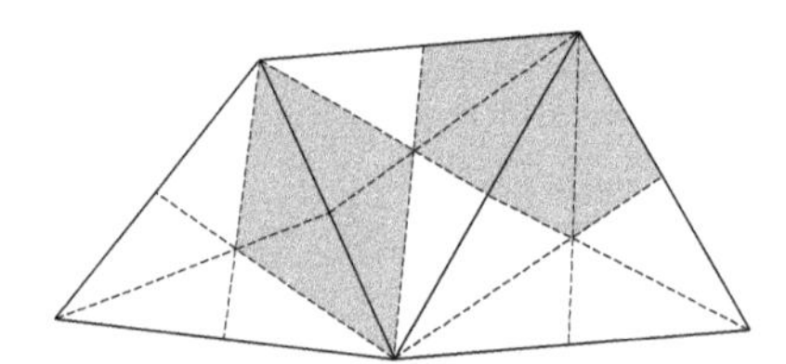

FIG. 11.3. Open stars of the first barycentric subdivision.

The **cuplength** $cl\,(M)$ is the minimal integer $N \geq 1$ such that for any set of cohomology classes

$$\alpha_j \in H^{k_j}(M), \qquad j = 1, \ldots, N,$$

of degrees $k_j \geq 1$, the class $\alpha_1 \cup \cdots \cup \alpha_N$ is zero. If M is a manifold and H^* denotes the de Rham cohomology then this means that for any set of closed differential forms $\alpha_j \in \Omega^{k_j}(M)$ of degrees $k_j \geq 1$ the exterior product $\alpha_1 \wedge \cdots \wedge \alpha_N$ is exact. Obviously, in this case

$$cl\,(M) \leq \dim M + 1.$$

In general, the cuplength of M is a lower bound for the Ljusternik–Schnirelmann category.

Lemma 11.2.8 *For every compact metric space M*

$$\nu_{\mathrm{LS}}(M) \geq cl\,(M).$$

Proof: We prove the lemma in the case where M is a manifold and H^* denotes the de Rham cohomology. The argument can be easily adapted to the general case. Assume that

$$M \subset U_1 \cup \cdots \cup U_N,$$

where the sets U_j are all cohomologically trivial. We must prove that for any set of closed forms $\alpha_j \in \Omega^{k_j}(M)$ of degree $k_j \geq 1$, the form $\alpha_1 \wedge \cdots \wedge \alpha_N$ is exact. We prove by induction that $\alpha_1 \wedge \cdots \wedge \alpha_k$ is exact on $U_1 \cup \cdots \cup U_k$. Our argument is a typical use of the Mayer–Vietoris principle: see Bott and Tu [78, Chapter 1].

For $k = 1$ this is obvious by assumption. If this holds for $k < N$, apply the following claim to $U := U_1 \cup \cdots \cup U_k$, $V := U_{k+1}$, $\alpha := \alpha_1 \wedge \cdots \wedge \alpha_k$, $\beta := \alpha_{k+1}$.

Claim: *Assume $\alpha \in \Omega^k(M)$ is exact on U and on V while $\beta \in \Omega^\ell(U \cup V)$ is closed on U and exact on V. Then $\alpha \wedge \beta$ is exact on $U \cup V$.*

To prove the claim, choose $\sigma_U \in \Omega^{k-1}(U)$, $\sigma_V \in \Omega^{k-1}(V)$, and $\tau_V \in \Omega^{\ell-1}(V)$ such that

$$\alpha|_U = d\sigma_U, \qquad \alpha|_V = d\sigma_V, \qquad \beta|_V = d\tau_V.$$

Now extend σ_U to a $(k-1)$-form on $U \cup V$ and define $\rho \in \Omega^{k+\ell-1}(U \cup V)$ by

$$\rho|_U = \sigma_U \wedge \beta, \qquad \rho|_V = \sigma_V \wedge \beta + (-1)^{k-1} d((\sigma_U - \sigma_V) \wedge \tau_V).$$

Note that both definitions of ρ agree on $U \cap V$ and, since β is closed, $d\rho = \alpha \wedge \beta$. This proves the claim and the lemma. □

Now let $\phi^t : M \to M$ be a flow on the compact metric space M, i.e.

$$\phi^{t+s} = \phi^t \circ \phi^s, \qquad \phi^0 = \mathrm{id}.$$

This flow is called **gradient-like** if there exists a (Lyapunov) function

$$\Phi : M \to \mathbb{R}$$

which is strictly decreasing along the nonconstant orbits of ϕ^t. This means that

$$t \geq 0 \qquad \Longrightarrow \qquad \Phi(\phi^t(x)) \leq \Phi(x)$$

and

$$\Phi(\phi^s(x)) = \Phi(x) \text{ for some } s > 0 \quad \Longrightarrow \quad \phi^t(x) \equiv x \text{ for all } t.$$

Any constant orbit $x \equiv \phi^t(x)$ is called a **critical point** of Φ. Note that any point at which Φ attains its maximum or minimum is a critical point. A **critical level** of Φ is a number c such that the set $\Phi^{-1}(c) \subset M$ contains a critical point. The set of critical points is closed and so is the set of critical levels.

Theorem 11.2.9 (Ljusternik–Schnirelmann) *Let M be a compact metric space and denote by $cl\,(M)$ the cuplength of M with respect to any cohomology theory. Then a gradient-like flow on M has at least $cl\,(M)$ constant solutions.*

Proof: Let ν be any category on M. We prove that a gradient-like flow on M has at least $\nu(M)$ constant orbits. Then the theorem follows from Lemma 11.2.8.

Given any number $c > 0$ define

$$M^c = \{x \in M \,|\, \Phi(x) \leq c\}.$$

If c is not a critical level of Φ then there exist numbers $\varepsilon > 0$ and $t > 0$ such that $\phi^t(M^{c+\varepsilon}) \subset M^{c-\varepsilon}$. More generally, if c is a critical level and U is any neighbourhood of the critical subset of $\Phi^{-1}(c)$, then there exist constants $\varepsilon > 0$ and $t > 0$ such that

$$\Phi^t(M^{c+\varepsilon} \setminus U) \subset M^{c-\varepsilon}. \tag{11.2.3}$$

Now for $j = 1, \dots, \nu(M)$ denote

$$c_j = \sup_{\nu(M^c) < j} c.$$

Obviously, $c_1 = \min_M \Phi$ and $c_N = \max_M \Phi$. It follows from (11.2.3) that c_j is a critical level of Φ for every j and, by definition,

$$c_1 \leq c_2 \leq \cdots \leq c_N.$$

We prove that for every j either $c_j < c_{j+1}$ or the set $\Phi^{-1}(c_j)$ contains infinitely many critical points. Suppose that there are only finitely many critical points $x_1, \dots, x_k$ with $\Phi(x_i) = c_j$. By the normalization axiom, there exists a neighbourhood U of $\{x_1, \dots, x_k\}$ such that $\nu(U) = 1$. By (11.2.3) and the monotonicity and naturality axioms of ν, we have

$$\nu(M^{c_j+\varepsilon}) \leq \nu(M^{c_j+\varepsilon} \setminus U) + 1 \leq \nu(M^{c_j-\varepsilon}) + 1 < j + 1.$$

This implies $c_{j+1} \geq c_j + \varepsilon > c_j$. □

Proof of Theorem 11.1.9 'The degenerate case': Let $\Phi : \mathbb{R}^{2nN} \to \mathbb{R}$ be the discrete symplectic action defined by (11.1.5) whose critical points agree with the fixed points of our Hamiltonian symplectomorphism of $\mathbb{T}^{2n}$. We must prove that Φ has at least $2n+1$ geometrically distinct critical points. We only sketch the main point. Denote by $\Lambda \subset \mathbb{R}^{2nN}/\mathbb{Z}^{2n}$ the set of points $\mathbf{z}$ such that the corresponding gradient flow line of Φ is bounded. Consider the projection $\pi : \Lambda \to \mathbb{T}^{2n}$ induced by $\mathbf{z} \mapsto z_0$. Then the induced map on Čech cohomology

$$\pi^* : \check{H}^*(\mathbb{T}^{2n}) \to \check{H}^*(\Lambda)$$

is injective and hence

$$cl\,(\Lambda) \geq cl\,(\mathbb{T}^{2n}) = 2n + 1.$$

That π^* is injective is obvious in the case where $V_j \equiv 0$ for every j (see (11.1.4)) and hence the set of bounded solutions is diffeomorphic to $\mathbb{T}^{2n}$. To prove it in the general case one uses a homotopy argument. For more details we refer to Floer's work on the Conley index for normally hyperbolic invariant sets [225]. □

11.3 Lagrangian intersections

In the proof of Proposition 11.1.8 we reduced the symplectic fixed point problem to that of the intersection of the Lagrangian submanifolds Δ and $\mathrm{graph}(\psi)$ in $M \times M$. There is in fact a more general version of the Arnold conjecture for Lagrangian submanifolds [24], which is known as the **Arnold–Givental conjecture** and concerns Lagrangian submanifolds that arise as fixed point sets of anti-symplectic involutions. Here, we call a diffeomorphism $\tau : M \to M$ of a symplectic manifold (M, ω) an **anti-symplectic involution** if

$$\tau \circ \tau = \mathrm{id}, \qquad \tau^*\omega = -\omega.$$

Thus the fixed point set $L := \mathrm{Fix}(\tau)$ of an anti-symplectic involution is a (possibly empty) Lagrangian submanifold of M.

Conjecture 11.3.1. (Arnold–Givental) *Let (M, ω) be a closed symplectic manifold, let $L \subset M$ be the fixed point set of an anti-symplectic involution, and let $\psi : M \to M$ be a Hamiltonian symplectomorphism. Then*

$$\#(L \cap \psi(L)) \geq \mathrm{Crit}(L).$$

If L and $\psi(L)$ intersect transversally then $\#(L \cap \psi(L)) \geq \mathrm{Crit}_{\mathrm{Morse}}(L)$. Here $\mathrm{Crit}(L)$ *denotes the minimum number of critical points of a function on the manifold L, while* $\mathrm{Crit}_{\mathrm{Morse}}(L)$ *is the corresponding minimum for Morse functions.*

The assumption that L is the fixed point set of an anti-symplectic involution in Conjecture 11.3.1 cannot be removed without replacement. For example, if $L \subset \mathbb{T}^2$ is a contractible loop in the 2-torus it is the boundary of an embedded disc, by a theorem of Epstein [206]. If the area of this disc is less than half of the area of the 2-torus it is easy to construct a Hamiltonian symplectomorphism $\psi : \mathbb{T}^2 \to \mathbb{T}^2$ such that $L \cap \psi(L) = \emptyset$ (see Fig 11.4).

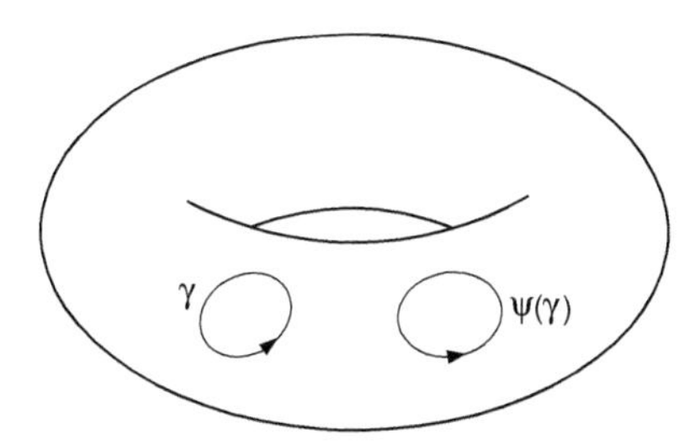

FIG. 11.4. Exactly isotopic circles on the torus.

Remark 11.3.2 (i) It is customary to replace the lower bounds in Conjecture 11.3.1 by the cuplength $cl(L) \leq \mathrm{Crit}(L)$, respectively the sum of the Betti numbers $\dim H_*(L) \leq \mathrm{Crit}_{\mathrm{Morse}}(L)$. The resulting weaker conjecture is called the **homological Arnold–Givental conjecture**.

(ii) The diagonal in $(M\times M,(-\omega)\times\omega)$ is the fixed point set of the anti-symplectic involution $(p,q)\mapsto(q,p)$. Hence Conjecture 11.1.7 is a special case of Conjecture 11.3.1. □

The nondegenerate version of the homological Arnold–Givental conjecture with $\mathbb{Z}_2$ coefficients was proved by Floer [226] under the assumption $\pi_2(M,L)=0$ (instead of L being the fixed point set of an anti-symplectic involution). Floer's proof was extended by Oh [506] to the so-called **monotone case**, where there exists a constant $\lambda>0$ such that

$$\mu(u)=\lambda\int_{\mathbb{D}}u^*\omega$$

for every smooth map $u:(\mathbb{D},\partial\mathbb{D})\to(M,L)$. Here, $\mu(u)$ denotes the Maslov index of the loop of Lagrangian subspaces $\Lambda(t):=T_{u(e^{2\pi it})}L$ in a symplectic trivialization of the vector bundle u^*TM over the disc. In Oh's result the condition that L is the fixed point set of an anti-symplectic involution cannot be removed, as the example in Figure 11.4 shows. The nondegenerate homological Arnold–Givental conjecture was extended to a class of non-monotone Lagrangian submanifolds by Fukaya–Oh–Ohta–Ono [249]. It was proved by Frauenfelder [240] for Lagrangian submanifolds of Marsden–Weinstein quotients of symplectically aspherical manifolds that arise as fixed point sets of anti-symplectic involutions of the ambient manifold.

Exercise 11.3.3 Prove that the fixed point set of an anti-symplectic involution is a Lagrangian submanifold. □

Exercise 11.3.4 Consider the Riemann sphere $\overline{\mathbb{C}}=\mathbb{C}\cup\{\infty\}$ with the Fubini–Study form $\omega_{\mathrm{FS}}:=(1+x^2+y^2)^{-2}dx\wedge dy$. Define the map $\tau:\overline{\mathbb{C}}\to\overline{\mathbb{C}}$ by

$$\tau([z_0:z_1]):=\frac{a\bar{z}+b}{c\bar{z}+d},\qquad ad-bc=1.$$

Assume that τ is not the antipodal map $z\mapsto-1/\bar{z}$. Prove that τ is an anti-symplectic involution if and only if $b=c$, $|a|^2=|d|^2=1-|b|^2$, and $a\bar{b}+b\bar{d}=0$. Prove that the fixed point set of τ is the image of a great circle under the stereographic projection. □

Exercise 11.3.5 Prove that an embedded circle $L\subset S^2$ is the fixed point set of an anti-symplectic involution if and only if the two connected components of $S^2\setminus L$ have equal area. **Hint:** A theorem of Epstein [206] asserts that two embedded circles in a 2-manifold are homotopic if and only if they are isotopic. Thus any two embedded circles in the 2-sphere are isotopic. Show that they are Hamiltonian isotopic whenever they divide the 2-sphere into two discs of equal area. □

Exercise 11.3.6 Prove that an embedded circle $L\subset\mathbb{T}^2$ is the fixed point set of an anti-symplectic involution if and only if its complement is connected. □

Exercise 11.3.7 Let (M,ω) be a closed symplectic 2-manifold and $L\subset M$ be a separating embedded circle. Show that L is the fixed point set of an anti-symplectic involution if and only if the two connected components of $M\setminus L$ have the same genus and the same area. Find assumptions under which M admits a Lagrangian submanifold $L\subset M$ with $\pi_2(M,L)=0$ that is not the fixed point set of anti-symplectic involution. □

Although the Arnold–Givental conjecture is stated for closed symplectic manifolds, the techniques developed by Floer also extend to certain noncompact manifolds and it is often sufficient to assume that the Lagrangian submanifold is compact. An example is the cotangent bundle $M = T^*L$ of a compact manifold L. Identify L with the zero section of T^*L, and note that L is both the fixed point set of an anti-symplectic involution of T^*L and satisfies $\pi_2(T^*L, L) = 0$. The homological Arnold–Givental conjecture indeed does hold for the zero section of a cotangent bundle.

Theorem 11.3.8 *Let L be a closed manifold and let $\psi : T^*L \to T^*L$ be a Hamiltonian symplectomorphism. Then $\#(\psi(L) \cap L) \geq cl(L)$. If $\psi(L)$ intersects L transversally then $\#(\psi(L) \cap L) \geq \sum_k b_k(L)$. Here $b_k(L)$ denotes the kth Betti number of L with coefficients in any field.*

Proof: See page 439. □

This result was proved by Chaperon [96] for the cotangent bundle of the torus, and for general cotangent bundles by Hofer [312] and Laudenbach–Sikorav [402]. The papers by Chaperon and Laudenbach–Sikorav use 'elementary' variational methods that are available because of the linear structure of the cotangent bundle T^*L. Laudenbach and Sikorav show that the Lagrangian submanifold $\psi(L)$ has a generating function Φ, as described in Proposition 9.4.4, that is *quadratic at infinity*. We explain this condition below. It implies that one can find critical points of Φ by the methods of Theorem 11.2.4, and hence find intersection points in $\psi(L) \cap L$ via Remark 9.4.7. Later, Viterbo [643] showed that this generating function Φ is essentially unique, thereby constructing symplectic invariants of $\psi(L)$. Below we explain a variant of the Laudenbach–Sikorav construction, constructing a generating function by means of the discrete symplectic action. These techniques were extended to the contact case by Chekanov [101], Théret [619], and Bhupal [63], who constructed generating functions for Legendrian submanifolds in 1-jet bundles. In our exposition we follow the approach of Bhupal [63].

Before going into the details of the proof, we mention a generalization of Theorem 11.3.8 to exact Lagrangians. Consider the symplectic manifold $M = T^*L$ with the standard symplectic structure $\omega = -d\lambda$ associated to the canonical 1-form $\lambda = \lambda_{\mathrm{can}}$. Recall that a Lagrangian submanifold $\Lambda \subset T^*L$ is called **exact** if the restriction of λ to Λ is exact. In particular, the zero section L is an exact Lagrangian submanifold. More generally, if ψ_t is a Hamiltonian isotopy given by

$$\frac{d}{dt}\psi_t = X_t \circ \psi_t, \qquad \psi_0 = \mathrm{id}, \qquad \iota(X_t)\omega = dH_t,$$

and $\psi := \psi_1$, then, by Proposition 9.3.1,

$$\psi^*\lambda - \lambda = dF, \qquad F := \int_0^1 (\iota(X_t)\lambda - H_t) \circ \psi_t \, dt,$$

and hence $\psi(L)$ is an exact Lagrangian submanifold (compare Proposition 9.4.2).

Remark 11.3.9 Arnold's **nearby Lagrangian conjecture** asserts that every compact exact Lagrangian submanifold

$$\Lambda \subset T^*L$$

without boundary in the cotangent bundle of a closed manifold L is the image of the zero section under a Hamiltonian symplectomorphism. This is easy to prove for $L = S^1$, is known for $L = S^2$ by a theorem of Hind [309], and is an open question for any other closed connected manifold L. Important progress on this conjecture has been made by Fukaya–Seidel–Smith [253, 254], who proved that when the Maslov class of Λ vanishes and Λ is orientable, the projection

$$\pi : \Lambda \to L$$

induces an isomorphism on homology (and in particular has degree one). This result was strengthened by Abouzaid [2] who showed, still under the assumption of vanishing Maslov class, that $\pi : \Lambda \to L$ is a homotopy equivalence. In [3] he showed that an exotic homotopy n-sphere, that does not bound a parallelizable manifold, does not admit a Lagrangian embedding into the cotangent bundle of the standard n-sphere. However, the full nearby Lagrangian conjecture appears to be out of reach at the time of writing (in 2016). □

Gromov [287] has proved the existence of at least one intersection point of any closed exact Lagrangian submanifold in T^*L with the zero-section. As we explain in Section 13.2, a different version of this result is the basis of his construction of an exotic symplectic structure on $\mathbb{R}^{2n}$. In both situations, the exactness property is crucial.

Theorem 11.3.10 (Gromov) *Let L be a closed n-manifold and $\Lambda \subset T^*L$ be a compact exact Lagrangian submanifold without boundary. Then*

$$\Lambda \cap L \neq \emptyset$$

and

$$\Lambda \cap \psi(\Lambda) \neq \emptyset$$

for every Hamiltonian symplectomorphism $\psi : T^*L \to T^*L$.

Proof: See Gromov [287], Audin–Lalonde–Polterovich [35], and McDuff–Salamon [470, Theorem 9.2.16 and Corollary 9.2.17]. All known proofs of this result use J-holomorphic curves or Floer homology and go beyond the scope of this book. □

Remark 11.3.11 Such an existence result cannot hold for exact Lagrangian immersions. Consider for example the cotangent bundle of $L = S^1$. Any immersion $\iota : S^1 \to T^*S^1$ is Lagrangian. It is exact if and only if the total area between S^1 and $\iota(S^1)$ is zero. An example where $\iota(S^1)$ does not intersect the zero section is illustrated in Fig. 11.5. Note, however, that there cannot be a Hamiltonian isotopy connecting $\iota(L)$ to the zero section unless ι is an embedding. □

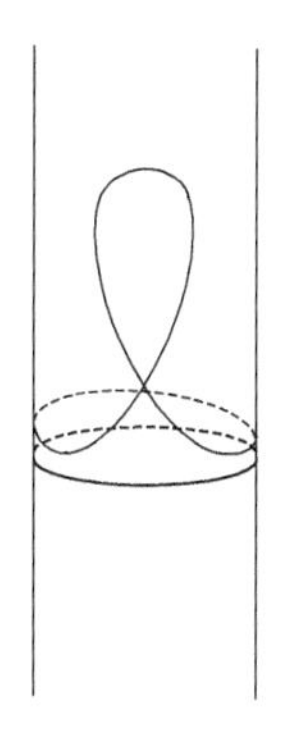

FIG. 11.5. A disjoint exact immersed Lagrangian.

Proof of Theorem 11.3.8: Let $\psi^L : T^*L \to T^*L$ be a Hamiltonian symplectomorphism. As explained above, the main idea is to construct a generating function for $\psi^L(L_0)$ via the discrete symplectic action, as in Bhupal's thesis [63]. The reduction argument in Step 3 and Exercise 11.3.15 below is due to Chekanov.

Let $\Phi : L \times \mathbb{R}^N \to \mathbb{R}$ be a smooth function. Recall from Section 9.4 that the set of fibre critical points is defined by

$$\mathcal{C}_\Phi = \{(x, \xi) \,|\, \partial_\xi \Phi(x, \xi) = 0\},$$

and that there is a natural map $\iota_\Phi : \mathcal{C}_\Phi \to T^*L$ defined by

$$\iota_\Phi(x, \xi) = (x, \partial_x \Phi(x, \xi))$$

for $(x, \xi) \in \mathcal{C}_\Phi$. Recall further that ι_Φ is a Lagrangian immersion whenever Φ is **transverse** in the sense that the graph of $d\Phi$ intersects the conormal bundle $T^*L \times \mathbb{R}^N \times \{0\}$ transversally:

$$\mathrm{graph}(d\Phi) \quad \pitchfork \quad T^*L \times \mathbb{R}^N \times \{0\}.$$

Here both spaces are to be understood as submanifolds of the cotangent bundle $T^*(L \times \mathbb{R}^N) = T^*L \times \mathbb{R}^N \times \mathbb{R}^N$. The goal is to prove that there exists a generating function

$$\Phi : L \times \mathbb{R}^N \to \mathbb{R}$$

which satisfies the following conditions.

(I) Φ is transverse.

(II) The map $\iota_\Phi : \mathcal{C}_\Phi \to T^*L$ is injective.

(III) $\iota_\Phi(\mathcal{C}_\Phi) = \psi^L(L_0)$.

(IV) There exist a $c > 0$ and a nonsingular matrix $P = P^T \in \mathbb{R}^{N \times N}$ such that

$$|\xi| \geq c \qquad \Longrightarrow \qquad \Phi(x, \xi) = \frac{1}{2}\langle \xi, P\xi \rangle$$

for every pair $(x, \xi) \in L \times \mathbb{R}^N$.

Condition (IV) says that Φ is quadratic at infinity. The methods discussed in the proof of Theorem 11.2.4 can be used to find critical points of these functions. Since these critical points correspond to intersection points in $L \cap \psi(L)$, by Remark 9.4.7, the theorem will follow once Φ is constructed.

The generating function can be constructed in three steps. The first step reduces the problem to the case $L = \mathbb{R}^k$.

Step 1: *Suppose that L is a submanifold of some Euclidean space $\mathbb{R}^k$. Choose a Hamiltonian isotopy $\psi_t^L : T^*L \to T^*L$ such that*

$$\psi_0^L = \mathrm{id}, \qquad \psi_1^L = \psi^L,$$

and let $H_t^L : T^*L \to \mathbb{R}$ denote the corresponding Hamiltonian functions. Next choose Hamiltonian functions $H_t : \mathbb{R}^{2k} \to \mathbb{R}$ such that

$$H_t|_{L\times\mathbb{R}^k} = H_t^L \circ \pi^L$$

for every t, where $\pi^L : L \times \mathbb{R}^k \to T^*L$ denotes the projection of Exercise 11.3.13 below. Then

$$\psi^L(L_0) = \pi^L \circ \psi_1(L \times \{0\}).$$

This follows from Exercise 11.3.16 below.

Step 2: *Consider the Lagrangian submanifold*

$$\Lambda = \psi_1(\mathbb{R}^k \times 0) \subset \mathbb{R}^k \times \mathbb{R}^k.$$

There exists a generating function

$$\Phi : \mathbb{R}^k \times \mathbb{R}^N \to \mathbb{R}$$

which satisfies conditions (I), (II), (III), *and* (IV) *above, with $\psi^L(L_0)$ replaced by Λ.*

The basic model for Φ is the symplectic action in Example 9.4.8. This action functional is defined on a bundle with infinite-dimensional fibres. To obtain a finite-dimensional generating function we replace the symplectic action by the discrete symplectic action. To begin with we observe that our Hamiltonian functions H_t do not have compact support. However, we will only be interested in the image of the compact set $L \times \{0\}$ under ψ_t, and hence we can modify H_t by multiplication with a cutoff function which is equal to 1 on a ball containing the set $\{\psi_t(x,0) \,|\, x \in L,\, 0 \le t \le 1\}$. The modified function, still denoted by H_t, has compact support and leaves the image $\psi_t(L \times \{0\})$ unchanged. Of course this modified function will only satisfy equation (11.3.1) in Exercise 11.3.16 below for y and y' in some bounded domain.

With H_t of compact support let us denote by $\psi_H^{t,t_0} : \mathbb{R}^{2k} \to \mathbb{R}^{2k}$ the flow of the corresponding time-dependent Hamiltonian system, and write $\psi = \psi_H^{1,0}$ and $\psi_j = \psi^{(j+1)/N,j/N}$ for $j = 0, \dots, N-1$. Then

$$\psi = \psi_{N-1} \circ \psi_{N-2} \circ \cdots \circ \psi_0.$$

For large N the symplectomorphisms ψ_j are all C^1-close to the identity and hence, by Lemma 9.2.1, admit generating functions $V_j : \mathbb{R}^{2k} \to \mathbb{R}$ that have compact support and generate $(x_{j+1}, y_{j+1}) = \psi_j(x_j, y_j)$ via the difference equation

$$x_{j+1} - x_j = \frac{\partial V_j}{\partial y}(x_j, y_{j+1}), \qquad y_{j+1} - y_j = -\frac{\partial V_j}{\partial x}(x_j, y_{j+1}).$$

Note here that the pair (x_{j+1}, y_j) in Lemma 9.2.1 has been replaced by the pair (x_j, y_{j+1}). This change is necessary in order to deal with the boundary condition $y_0 = 0$. Let $E^N \cong \mathbb{R}^{(2N+1)k}$ denote the space of discrete paths

$$c^N = (x_0, x_1, \dots, x_N, y_1, \dots, y_N)$$

with $x_j, y_j \in \mathbb{R}^k$ for all j. Define $\pi^N : E^N \to \mathbb{R}^k$ and $\Phi^N : E^N \to \mathbb{R}$ by

$$\pi^N(c^N) = x_N, \qquad \Phi^N(c^N) = \sum_{j=1}^{N} \langle y_j, x_j - x_{j-1} \rangle - V_{j-1}(x_{j-1}, y_j).$$

This is the discrete symplectic action. The fibre critical points are the discrete paths c^N which satisfy $z_{j+1} = \psi_j(z_j)$, where $z_j = (x_j, y_j)$ and $y_0 = 0$. Moreover,

$$\frac{\partial \Phi^N}{\partial x_N}(c^N) = y_N$$

and hence $\iota_{\Phi^N}(c^N) = \psi(x_0, 0)$. This shows that the map $\iota_{\Phi^N} : C_{\Phi^N} \to \mathbb{R}^{2k}$ is injective and that its image is equal to $\psi(\mathbb{R}^k \times \{0\})$. Transversality is left as an exercise. This generating function does not, however, satisfy (IV) and it is necessary to make a further modification. For this it is useful to introduce the new variables

$$x = x_N, \qquad \xi_j = x_j - x_{j-1},$$

for $j = 1, \dots, N$. Abbreviate $\gamma := (\xi_1, \dots, \xi_N, y_1, \dots, y_N)$. Then

$$\Phi^N(x, \gamma) = \frac{1}{2}\langle \gamma, P\gamma \rangle + V(x, \gamma),$$

where

$$P := \begin{pmatrix} 0 & \mathbb{1} \\ \mathbb{1} & 0 \end{pmatrix}, \qquad V(x, \gamma) := \sum_{j=1}^{N} V_{j-1}(x - \xi_N - \cdots - \xi_j, y_j).$$

Here the functions V_{j-1} have compact support but V does not have compact support. However, V is uniformly bounded and so are its first derivatives. Moreover, the set of fibre critical points with $x_N \in L$ is compact. Hence we can modify V by multiplication with a suitable cutoff function without changing the set of fibre critical points. This modified function satisfies the requirements of Step 2.

Step 3: *The generating function* $\Phi^L = \Phi|_{L\times\mathbb{R}^N} : L \times \mathbb{R}^N \to \mathbb{R}$ *satisfies the conditions* (I), (II), (III), *and* (IV) *above.*

This follows from Exercises 11.3.15 and 11.3.16 below. Step 3 shows that the critical points of Φ^L are in one-to-one correspondence with the intersection points $L_0 \cap \psi^L(L_0)$. As in Lemma 9.2.6, transversality of the intersections implies that Φ is a Morse function. Hence the estimates for the number of intersection points follow from Theorem 11.2.4 (in the transverse case) and Theorem 11.2.9 (in the nontransverse case). The details are as in the proof of Theorem 11.1.9 and will not be repeated here. □

Remark 11.3.12 The second assertion in Theorem 11.3.10 (that $\psi(\Lambda) \cap \Lambda \neq \emptyset$ for every Hamiltonian symplectomorphism ψ) follows from Theorem 11.3.8 in the case where $\Lambda = L$ is the zero section. All known proofs in the general case require J-holomorphic curves and therefore lie beyond the scope of this book. However, it is easy to see that the second assertion implies the first. Namely, suppose by contradiction that $\Lambda \cap L = \emptyset$ and consider the embedding

$$\iota_t := e^t\iota : \Lambda \to T^*L,$$

where $\iota : \Lambda \to T^*L$ is the inclusion. Then $\iota_t(\Lambda) \cap \Lambda = \emptyset$ for large t. Moreover, ι_t is an exact Lagrangian embedding for every t. Hence, by Exercise 11.3.17 below, there exists a Hamiltonian isotopy $\psi_t : T^*L \to T^*L$ such that $\psi_t|_\Lambda = \iota_t$ for all t. Since $\iota_t(\Lambda) \cap \Lambda = \emptyset$ for large t, this contradicts the assertion that the image of Λ under any Hamiltonian symplectomorphism must intersect Λ. □

Exercise 11.3.13 Consider the manifold

$$\mathbb{R}^k \times \mathbb{R}^k = T^*\mathbb{R}^k$$

with the standard symplectic structure. Let $L \subset \mathbb{R}^k$ be a smooth submanifold. Prove that $L \times \mathbb{R}^k$ is a coisotropic submanifold of $\mathbb{R}^k \times \mathbb{R}^k$ with isotropic leaves $x \times T_xL^\perp$ for $x \in L$. Prove that the cotangent bundle T^*L can be naturally identified with the quotient $L \times \mathbb{R}^k/{\sim}$, where

$$(x, y) \sim (x', y') \qquad \Longleftrightarrow \qquad x = x', \quad y' - y \perp T_xL.$$

More precisely, let $\pi^L : L \times \mathbb{R}^k \to T^*L$ denote the projection which sends a pair $(x, y) \in L \times \mathbb{R}^k$ to the $(x, \pi_x(y))$, where $\pi_x(y)$ denotes the restriction of the linear functional $v \mapsto \langle y, v\rangle$ to T_xL. Prove that π^L descends to a symplectomorphism from the symplectic quotient $L \times \mathbb{R}^k/\sim$ to T^*L. □

Exercise 11.3.14 Let $L \subset \mathbb{R}^k$ be as in Exercise 11.3.13. Suppose that $\Lambda \subset \mathbb{R}^k \times \mathbb{R}^k$ is a Lagrangian submanifold, transverse to $L \times \mathbb{R}^k$. Prove that $\Lambda^L := \pi^L(\Lambda \cap (L \times \mathbb{R}^k))$ is a Lagrangian submanifold of T^*L. Prove that if Λ is transverse to the zero section $\mathbb{R}^k \times \{0\}$, then Λ^L is transverse to the zero section in T^*L. □

Exercise 11.3.15 Let $\Phi : \mathbb{R}^k \times \mathbb{R}^N \to \mathbb{R}$ be a transverse generating function such that $\iota_\Phi : \mathcal{C}_\Phi \to \mathbb{R}^k \times \mathbb{R}^k$ is injective. Suppose that the Lagrangian submanifold

$$\Lambda_\Phi = \iota_\Phi(\mathcal{C}_\Phi) \subset \mathbb{R}^k \times \mathbb{R}^k$$

is transverse to $L \times \mathbb{R}^k$. Prove that the function

$$\Phi^L = \Phi|_{L\times\mathbb{R}^N} : L \times \mathbb{R}^N \to \mathbb{R}$$

is transverse and that it generates the Lagrangian submanifold

$$\Lambda_{\Phi^L} = \iota_{\Phi^L}(\mathcal{C}_{\Phi^L}) = \pi^L(\Lambda_\Phi \cap (L \times \mathbb{R}^k)) \subset T^*L.$$

Hint: Show that $\mathcal{C}_{\Phi^L} = {\iota_\Phi}^{-1}(L \times \mathbb{R}^k)$ and $\iota_{\Phi^L} = \pi^L \circ \iota_\Phi$. □

Exercise 11.3.16 Suppose that $L \subset \mathbb{R}^k$ is a smooth submanifold. Let $H_t : \mathbb{R}^k \times \mathbb{R}^k \to \mathbb{R}$ be a time-dependent Hamiltonian function which satisfies

$$x \in L, \quad y' - y \perp T_xL \qquad \Longrightarrow \qquad H_t(x, y) = H_t(x, y'). \tag{11.3.1}$$

Let $\psi_t : \mathbb{R}^k \times \mathbb{R}^k$ denote the Hamiltonian isotopy generated by H_t. Prove that $L \times \mathbb{R}^k$ is invariant under ψ_t for every t. Let $\psi_t^L : T^*L \to T^*L$ denote the Hamiltonian isotopy generated by the induced Hamiltonian functions $H_t^L : T^*L \to \mathbb{R}$ defined by

$$H_t = H_t^L \circ \pi^L.$$

Prove that $\psi_t^L \circ \pi^L = \pi^L \circ \psi^t$. Prove that $\psi_t(\mathbb{R}^k \times \{0\})$ is transverse to $L \times \mathbb{R}^k$ and

$$\psi_t^L(L_0) = \pi^L \circ \psi_t(L \times \{0\})$$

for all t, where $L_0 \subset T^*L$ denotes the zero section. □

Exercise 11.3.17 Let $(M, \omega = -d\lambda)$ be an exact symplectic manifold of dimension $2n$ and Λ be a compact n-manifold. Suppose that $\iota_t : \Lambda \to M$ is a smooth family of embeddings and $F_t : \Lambda \to \mathbb{R}$ is a smooth family of functions such that

$${\iota_t}^*\lambda = dF_t$$

for every t. Prove that there exists a Hamiltonian isotopy $\psi_t : M \to M$ such that $\psi_0 = \mathrm{id}$ and

$$\iota_t = \psi_t \circ \iota_0$$

for every t. **Hint:** Choose the Hamiltonian functions H_t such that

$$H_t \circ \iota_t = \lambda(\partial_t\iota_t) - \partial_t F_t$$

and

$$dH_t(\iota_t(\xi))v = \omega(\partial_t\iota_t(\xi), v)$$

for $\xi \in \Lambda$ and $v \perp \operatorname{im} d\iota_t(\xi)$. Prove that the last condition continues to hold for all $v \in T_{\iota_t(\xi)}M$. Deduce that $\partial_t\iota_t = X_{H_t} \circ \iota_t$. □

Rigidity

There is a beautiful and surprising consequence of Theorem 11.3.10 which was discovered by Benci and Sikorav [585]. This result illustrates the notion of **symplectic rigidity** in the cotangent bundle $T^*\mathbb{T}^n = \mathbb{T}^n \times \mathbb{R}^n$ of the torus $\mathbb{T}^n = \mathbb{R}^n/\mathbb{Z}^n$.

Theorem 11.3.18 (Benci–Sikorav) *Let $A, B \subset \mathbb{R}^n$ be two connected open sets. Suppose that there exists a symplectomorphism $\psi : T^*\mathbb{T}^n \to T^*\mathbb{T}^n$ such that*

$$\psi(A \times \mathbb{T}^n) = B \times \mathbb{T}^n.$$

Then there exist a unimodular matrix $\Psi \in \mathbb{Z}^{n\times n}$ and a vector $x \in \mathbb{R}^n$ such that

$$B = \Psi A + x.$$

Proof: On $T^*\mathbb{T}^n$ we introduce the coordinates (x, y), where $x \in \mathbb{T}^n = \mathbb{R}^n/\mathbb{Z}^n$ and $y \in \mathbb{R}^n$. Then the symplectomorphism ψ can be lifted to a map

$$\mathbb{R}^n \times \mathbb{R}^n \to \mathbb{R}^n \times \mathbb{R}^n : (x, y) \mapsto (u(x, y), v(x, y)).$$

Now $u(x+j, y) - u(x, y) \in \mathbb{Z}^n$ for every integer vector $j \in \mathbb{Z}^n$. Hence there exists a unimodular matrix $\Psi \in \mathbb{Z}^{n\times n}$ such that

$$u(x + j, y) = u(x, y) + \Psi j.$$

Composing ψ with the linear symplectomorphism

$$\begin{pmatrix} \Psi^{-1} & 0 \\ 0 & \Psi^{\mathrm{T}} \end{pmatrix} \in \mathbb{Z}^{2n\times 2n}$$

we may assume without loss of generality that when $j \in \mathbb{Z}^n$

$$u(x + j, y) = u(x, y) + j, \qquad v(x + j, y) = v(x, y). \tag{11.3.2}$$

In other words, the induced map ψ_* on H_1 is the identity. This implies that

$$\int_{\mathbb{T}^n} \frac{\partial u}{\partial x}\, dx = \mathbb{1}.$$

Combining this with the identity

$$\frac{\partial u}{\partial x}^{\mathrm{T}} \frac{\partial v}{\partial y} - \frac{\partial v}{\partial x}^{\mathrm{T}} \frac{\partial u}{\partial y} = \mathbb{1}$$

we obtain

$$\frac{\partial}{\partial y} \int_{\mathbb{T}^n} \frac{\partial u}{\partial x}^{\mathrm{T}} (v - y)\, dx = \int_{\mathbb{T}^n} \frac{\partial}{\partial x} \left(\frac{\partial u}{\partial y}^{\mathrm{T}} (v - y) \right) dx = 0.$$

Adding a constant vector to $v(x, y)$ we may assume that

$$\int_{\mathbb{T}^n} \left(\frac{\partial u}{\partial x}(x, 0) \right)^{\mathrm{T}} v(x, 0)\, dx = 0,$$

and by what we have just proved, this implies that

$$\int_{\mathbb{T}^n} \left(\frac{\partial u}{\partial x}(x, y) \right)^{\mathrm{T}} (v(x, y) - y)\, dx = 0 \tag{11.3.3}$$

for all $y \in \mathbb{R}^n$.

Now let $a \in A$. Then the map $\iota : \mathbb{T}^n \to \mathbb{T}^n \times \mathbb{R}^n$ defined by

$$\iota(x) = (u(x, a), v(x, a) - a)$$

is a Lagrangian embedding. Since the canonical 1-form on $\mathbb{T}^n \times \mathbb{R}^n$ is

$$\lambda = \sum_{j=1}^{n} y_j dx_j,$$

the 1-form $\iota^* \lambda$ is given by

$$\iota^* \lambda = \sum_{j,k=1}^{n} (v_j(x, a) - a_j) \frac{\partial u_j}{\partial x_k}(x, a)\, dx_k.$$

This is a closed 1-form on $\mathbb{R}^n$ and hence there exists a function $h : \mathbb{R}^n \to \mathbb{R}$ such that

$$\frac{\partial h}{\partial x_k} = \sum_{j=1}^{n} (v_j(x, a) - a_j) \frac{\partial u_j}{\partial x_k}(x, a).$$

It follows from Equation (11.3.2) that the gradient of h is of period 1 in all variables. Moreover, Equation (11.3.3) shows that the gradient of h is of mean value zero. Hence h itself is of period 1 and is therefore a function on $\mathbb{T}^n$. Hence the 1-form $\iota^* \lambda$ is exact:

$$\iota^* \lambda = dh, \qquad h : \mathbb{T}^n \to \mathbb{R}.$$

Hence it follows from Theorem 11.3.10 that the embedded torus $\iota(\mathbb{T}^n)$ must intersect the zero section. Hence $v(x, a) = a$ for some $x \in \mathbb{R}^n$. Hence $a \in B$. Thus we have shown that $A \subset B$. Interchanging the roles of A and B we obtain $A = B$. This proves Theorem 11.3.18. □

This theorem inspired Eliashberg in his discovery of new invariants for symplectic and contact manifolds [180]. Another very simple and appealing application of the Lagrangian intersection theorem is Polterovich's proof [530] that the Hofer diameter of the group $\mathrm{Ham}(S^2)$ of Hamiltonian symplectomorphisms of S^2 is infinite.

11.4 Floer homology

We end this chapter with a brief outline of Floer's proof of the Arnold conjecture. Let (M, ω) be a closed symplectic manifold, let $H : \mathbb{R}/\mathbb{Z} \times M \to \mathbb{R}$ be a smooth 1-periodic Hamiltonian function, abbreviate $H_t := H(t, \cdot) = H_{t+1}$ for $t \in \mathbb{R}$, and denote by $\mathbb{R} \to \mathrm{Ham}(M, \omega) : t \mapsto \phi_t$ the Hamiltonian isotopy generated by H via $\partial_t \phi_t = X_{H_t} \circ \phi_t$ and $\phi_0 = \mathrm{id}$. The (homological) Arnold Conjecture 11.1.7 is concerned with lower bounds for the number of periodic solutions of Hamilton's equation

$$\dot{z}(t) = X_{H_t}(z(t)), \qquad z(t+1) = z(t), \tag{11.4.1}$$

or, equivalently, with lower bounds for the number of fixed points of ϕ_1. Floer's proof actually establishes lower bounds for the number of *contractible* solutions of equation (11.4.1). The space

$$\mathcal{L}M := \{\gamma : \mathbb{R}/\mathbb{Z} \to M \,|\, \gamma \text{ is smooth and contractible}\}$$

of contractible loops is an infinite-dimensional Fréchet manifold whose tangent space at γ is the space $T_\gamma \mathcal{L}M = \Omega^0(\mathbb{R}/\mathbb{Z}, \gamma^*TM)$ of vector fields along γ. The loop space carries a natural 1-form $\delta\mathcal{A}_H$ given by

$$\delta\mathcal{A}_H(\gamma)\widehat{\gamma} := \int_0^1 \omega\big(\dot{\gamma}(t) - X_{H_t}(\gamma(t)), \widehat{\gamma}(t)\big)\, dt \tag{11.4.2}$$

for $\gamma \in \mathcal{L}M$ and $\widehat{\gamma} \in T_\gamma\mathcal{L}M$. This 1-form is closed and its zeros are the contractible solutions of equation (11.4.1). The 1-form (11.4.2) is exact if and only if (M, ω) is **symplectically aspherical**, i.e.

$$\int_{S^2} v^*\omega = 0 \tag{11.4.3}$$

for every smooth function $v : S^2 \to M$. In this case there exists a function $\mathcal{A}_H : \mathcal{L}M \to \mathbb{R}$ whose differential is the 1-form (11.4.2). It is defined by

$$\mathcal{A}_H(\gamma) := -\int_{\mathbb{D}} u^*\omega - \int_0^1 H_t(\gamma(t))\, dt \tag{11.4.4}$$

for $\gamma \in \mathcal{L}M$, where $\mathbb{D} \subset \mathbb{C}$ denotes the closed unit disc in the complex plane and $u : \mathbb{D} \to M$ is a smooth map such that $u(e^{2\pi i t}) = \gamma(t)$ for all $t \in \mathbb{R}$. The map u exists because γ is contractible, and the right hand side of (11.4.4) is independent of the choice of u by (11.4.3). Thus the periodic solutions of Hamilton's equation appear as the critical points of the **symplectic action functional** $\mathcal{A}_H : \mathcal{L}M \to \mathbb{R}$. While this had been known for a long time, before the 1980s no one attempted to develop a Morse theory for $\mathcal{A}_H$ because it does not have a well-defined L^2 gradient flow, and also each critical point has infinite Morse index and coindex. It was Floer's ingenious idea to make the Morse complex work in this setting despite these obstacles, using the fact that the L^2 gradient flow of $\mathcal{A}_H$ is an elliptic equation.

To be more precise, one first chooses an L^2 metric on $\mathcal{L}M$ that is compatible with the symplectic structure. Such a metric is determined by a 1-periodic smooth family $J_t = J_{t+1} \in \mathcal{J}(M, \omega)$ of ω-compatible almost complex structures on M. Then the L^2 inner product on $T_\gamma \mathcal{L}M = \Omega^0(\mathbb{R}/\mathbb{Z}, \gamma^*TM)$ is given by

$$\langle \widehat{\gamma}_1, \widehat{\gamma}_2 \rangle_{L^2} := \int_0^1 \langle \widehat{\gamma}_1(t), \widehat{\gamma}_2(t) \rangle_{J_t} \, dt = \int_0^1 \omega\big(\widehat{\gamma}_1(t), \, J_t(\gamma(t))\widehat{\gamma}_2(t)\big) \, dt.$$

By equation (11.4.2) the gradient of $\mathcal{A}_H$ at $\gamma \in \mathcal{L}M$ with respect to this inner product is given by $\mathrm{grad}\mathcal{A}_H(\gamma) = J_t(\gamma)(\dot{\gamma} - X_{H_t}(\gamma))$. Thus a negative gradient flow line of $\mathcal{A}_H$ is a contractible smooth map $u : \mathbb{R} \times \mathbb{R}/\mathbb{Z} \to M$, understood as a smooth path $\mathbb{R} \to \mathcal{L}M : s \mapsto u(s, \cdot)$, that satisfies the partial differential equation

$$\partial_s u + J_t(u) \left(\partial_t u - X_{H_t}(u) \right) = 0. \tag{11.4.5}$$

This is the **Floer equation**. The solutions of (11.4.5) with $H = 0$ are J-holomorphic curves, the solutions with $\partial_s u = 0$ are periodic orbits of the Hamiltonian system associated to H, and the solutions with $\partial_t u = 0$ (in the case where $J_t \equiv J$ and $H_t \equiv H$ are also independent of t) are gradient flow lines of H. In particular, the interplay of the Floer equation with Gromov's pseudoholomorphic curves guarantees that the elliptic theory outlined in Section 4.5 can be used to understand the Fredholm theory and compactness properties of Floer trajectories.

Let us now assume that all the contractible periodic solutions of Hamilton's equation (11.4.1) are nondegenerate. Then the symplectic action $\mathcal{A}_H : \mathcal{L}M \to \mathbb{R}$ is a Morse function and its critical points form a finite set, denoted by

$$\mathrm{Per}(H) := \{ z \in \mathcal{L}M \,|\, z \text{ satisfies } (11.4.1) \}.$$

As in Morse theory, the energy of a solution $u : \mathbb{R} \times \mathbb{R}/\mathbb{Z} \to M$ of the Floer equation (11.4.5) is defined by

$$E(u) := \int_{-\infty}^{\infty} \int_0^1 |\partial_s u|^2 \, dt ds.$$

If u is a contractible finite energy solution of (11.4.5), then the limits

$$\lim_{s \to \pm\infty} u(s, t) = z^{\pm}(t) \tag{11.4.6}$$

exist and are contractible periodic solutions of Hamilton's equation (11.4.1). Conversely, every solution of (11.4.5) and (11.4.6) has finite energy and satisfies the **energy identity**

$$E(u) = \mathcal{A}_H(z^-) - \mathcal{A}_H(z^+).$$

Given $z^{\pm} \in \mathrm{Per}(H)$, denote the set of Floer trajectories connecting z^- to z^+ by

$$\mathcal{M}(z^-, z^+; H, J) := \{ u : \mathbb{R} \times \mathbb{R}/\mathbb{Z} \to M \,|\, u \text{ satisfies } (11.4.5) \text{ and } (11.4.6) \}.$$

The goal is to show that these spaces are finite-dimensional manifolds, for a generic choice of J, whose properties are similar to those of the spaces of connecting orbits in Morse–Smale flows.

The details are carried out in a series of papers by Floer [226,227,228,229,230]. The construction of Floer homology has three main technical ingredients. The first is the Fredholm theory needed to establish that the moduli spaces are finite-dimensional, transversally cut out manifolds of the appropriate dimensions. Even though $\mathcal{A}_H$ has no well-posed gradient flow and its critical points have infinite Morse index and coindex, the theory works because the Floer equation is elliptic. Moreover Floer circumvented the absence of a finite Morse index, by introducing a **relative Morse index** $\mu(u)$ along a Floer trajectory u which coincides with the dimension of the moduli space $\mathcal{M}(z^-, z^+; H, J)$ near u (see Floer [227]). It also agrees with the difference of the Conley–Zehnder indices in a suitable trivialization of the pullback tangent bundle (see Salamon–Zehnder [563] and Robbin–Salamon [541,542]). With this understood the transversality theory goes through much as in finite-dimensional Morse theory (see Floer [227, 235]).

The second technical ingredient is a Floer–Gromov compactness theorem. This is where the main differences to finite-dimensional Morse theory may arise, since solutions of the Floer equation behave, roughly speaking, like J-holomorphic curves, so that in general J-holomorphic spheres may *bubble off*. However, in the symplectically aspherical case, where (11.4.3) holds, this cannot happen, and the compactness behaviour in Floer theory is exactly as in finite-dimensional Morse theory. The upshot is that in this case, for generic J, the spaces of index-1 Floer gradient trajectories are finite sets and their count can be used to define the Floer boundary operator $\partial = \partial^{H,J}$ on the Floer chain complex (see Floer [226]).

The third technical ingredient is the Floer gluing theorem, which is needed to prove that the Floer boundary operator indeed satisfies $\partial \circ \partial = 0$. While this is a highly nontrivial piece of analysis, the Floer gluing theorem is strictly analogous to the corresponding theorem in Morse theory, and so does not pose any obstacle to developing the theory.

With this understood one can define the Floer chain complex by

$$\mathrm{CF}_*(H) := \bigoplus_{z \in \mathrm{Per}(H)} \mathbb{Z}/2\mathbb{Z}\langle z \rangle, \qquad \partial^{H,J}\langle z \rangle := \sum_{w \in \mathrm{Per}(H)} n_2(z, w)\langle w \rangle.$$

Here $n_2(z, w) \in \{0, 1\}$ is the parity of the number of index-1 solutions of the Floer equation from z to w modulo time shift, i.e.

$$n_2(z, w) := \#_2 \mathcal{M}^1(z, w; H, J)/\mathbb{R},$$

where

$$\mathcal{M}^1(z, w; H, J) := \{u \in \mathcal{M}(z, w; H, J) \mid \mu(u) = 1\}.$$

Here we have used mod 2 coefficients to avoid addressing problems of orientation. However, Floer's construction extends naturally to integer coefficients. This setup should be compared to the Morse complex in finite dimensions, as constructed by Thom, Smale, Milnor, and Witten. (See Milnor's book on the h-cobordism theorem [485], Witten [688], Floer [229], and the discussion in Remark 11.2.5.)

It follows from the transversality, compactness, and gluing analysis developed by Floer that this construction is well-defined and gives rise to a chain complex. The resulting homology groups

$$\mathrm{HF}_*(H,J) := H_*\left(\mathrm{CF}_*(H), \partial^{H,J}\right).$$

are called the **Floer homology of the pair** (H,J). The grading is by the Conley–Zehnder index and is well-defined modulo $2N$, where N is the so-called **minimal Chern number**. It is defined by

$$N := \min\left\{\langle c_1,[v]\rangle \,|\, v : S^2 \to M,\ \langle c_1,[v]\rangle > 0\right\}, \tag{11.4.7}$$

when the first Chern class $c_1 := c_1(TM,J) \in H^2(M;\mathbb{Z})$ for $J \in \mathcal{J}(M,\omega)$ does not vanish on $\pi_2(M)$, and by $N := \infty$ otherwise. Thus Floer homology is integer graded whenever $N = \infty$.

Floer then proved that his homology groups are independent of the choices of H and J up to natural isomorphism. This proof is based on the same three technical ingredients (transversality, compactness, gluing) as the very construction of the Floer homology groups. The final step in his proof of the Arnold conjecture is to show that the Floer homology groups $\mathrm{HF}_*(H,J)$ are naturally isomorphic to the singular homology $H_*(M;\mathbb{Z}/2\mathbb{Z})$ of the underlying manifold M with coefficients in $\mathbb{Z}/2\mathbb{Z}$ and with the grading *rolled up modulo* $2N$. The proof is based, heuristically, on the observation that, in the case of time-independent Floer data $J_t = J$ and C^2-small $H_t = H$, the relevant Floer gradient trajectories are gradient flow lines of the function $-H$, and hence the Floer chain complex reduces to the Morse complex of $-H$. Once all these results have been established, it follows that the dimension of $H_*(M;\mathbb{Z}/2\mathbb{Z})$ is a lower bound for the number of critical points of $\mathcal{A}_H$. This implies the nondegenerate homological Arnold conjecture with coefficients in $\mathbb{Z}/2\mathbb{Z}$ for symplectically aspherical manifolds.

In [230] Floer extended his proof of the Arnold conjecture to **monotone** symplectic manifolds (M,ω). In this case the homomorphisms from $\pi_2(M)$ to $\mathbb{R}$ determined by the cohomology classes $[\omega]$ and c_1 agree up to a positive factor, i.e. there is a number $\lambda > 0$ such that

$$\langle c_1(v^*TM),[S^2]\rangle = \lambda \int_{S^2} v^*\omega \tag{11.4.8}$$

for every smooth map $v : S^2 \to M$. In this case J-holomorphic spheres can bubble off, and so the compactness problem is more subtle. Floer proved that for a generic family of almost complex structures bubbling does not occur for Floer trajectories of index less than or equal to two, so that the construction of Floer homology and the proof of the Arnold conjecture carry over to this class of symplectic manifolds. This is a rough sketch of Floer's proof of the nondegenerate homological Arnold conjecture in the monotone case. More detailed expositions of his arguments can be found in McDuff [448], Salamon–Zehnder [563], Salamon [555], and in the book by Audin–Damian [33]. Further extensions of the story in the existing literature are discussed at the end of this section.

Exercise 11.4.1 Let (M, ω) be a closed symplectic 2-manifold of genus $g \geq 1$. Then M is symplectically aspherical and hence the action 1-form (11.4.2) is exact on the space of contractible loops. When $g = 1$ show that it is not exact on any other connected component of the free loop space. When $g \geq 2$ show that it is exact on every connected component of the free loop space. □

Remark 11.4.2 Let (M, ω) be a closed monotone symplectic manifold. Then the evaluation map $\pi_1(\mathrm{Ham}(M, \omega)) \to \pi_1(M) : [\{\phi_t\}] \mapsto [\{\phi_t(p_0)\}]$ is trivial for every $p_0 \in M$. To see this, note first that Floer's theorem states that if all the contractible 1-periodic solutions of Hamilton's equation are nondegenerate then $\#\mathrm{Per}(H) \geq \dim H_*(M; \mathbb{Z}/2\mathbb{Z})$. Since the hypothesis holds trivially if $\mathrm{Per}(H) = \emptyset$, it follows that $\mathrm{Per}(H)$ is never empty. The assertion remains valid without the monotonicity assumption, but the proof is much harder. □

Remark 11.4.3 Our generating function proof of Theorem 11.1.9 can be viewed as a discretization of Floer's proof of the Arnold conjecture. Namely, consider the construction of Floer homology for $M = \mathbb{R}^{2n}$, equipped with the standard symplectic form ω_0, the standard complex structure J_0, and the standard Riemannian metric. Then the symplectic action functional in (11.4.4) agrees with the one in (1.1.8) and the Floer equation (11.4.5) has the form

$$\frac{\partial u}{\partial s} + J_0 \frac{\partial u}{\partial t} - \nabla H_t(u) = 0. \tag{11.4.9}$$

On the other hand, a gradient flow line of the discrete symplectic action (11.1.5) is a solution $\mathbf{z}(s) = (x_1(s), \ldots, y_N(s))$ of the ordinary differential equation

$$\begin{aligned} \dot{x}_j &= y_j - y_{j-1} + \frac{\partial V_{j-1}}{\partial x}(x_j, y_{j-1}), \\ \dot{y}_j &= x_j - x_{j+1} + \frac{\partial V_j}{\partial y}(x_{j+1}, y_j). \end{aligned}$$

Now think of the sequence $\mathbf{x} = (x_0, \ldots, x_N)$ as a discretized loop $t \mapsto x(s,t)$ and replace $x_{j+1} - x_j$ by $\partial x/\partial t$. Similarly, replace $y_j - y_{j-1}$ by $\partial y/\partial t$ and $V_j(x_{j+1}, y_j)$ by $H_t(x, y)$ to obtain

$$\frac{\partial x}{\partial s} = \frac{\partial y}{\partial t} + \frac{\partial H_t}{\partial x}(x, y), \qquad \frac{\partial y}{\partial s} = -\frac{\partial x}{\partial t} + \frac{\partial H_t}{\partial y}(x, y).$$

This is the Floer equation (11.4.9) in the notation $u(s,t) = (x(s,t), y(s,t))$. Thus the gradient flow lines of the discrete symplectic action can be viewed as discretized J-holomorphic curves. Observe that the Hessian of the discrete symplectic action has very large dimensional positive and negative eigenspaces, each of approximately half the dimension of the *discretized loop space*, and so the Morse index will diverge to infinity as the mesh size $1/N$ goes to zero. However, the signature stabilizes and converges to the Conley–Zehnder index (see Remark 9.2.7). In Floer homology it is the Conley–Zehnder index which plays the role of the relative Morse index. □

Comments on the literature

Floer homology has become a vast area of research with many new directions emerging over the last quarter of a century. It is not possible to do justice to all these developments in a brief discussion. We focus on selected topics to illustrate the rich variety of applications of Floer theory in many fields of mathematics.

Hamiltonian Floer homology

Floer's proof of the nondegenerate homological Arnold conjecture was extended to the symplectic analogue of Calabi–Yau manifolds (where the first Chern class vanishes on $\pi_2(M)$) and to symplectic $2n$-manifolds with minimal Chern number $N \geq n-2$ (see equation (11.4.7)) by Hofer–Salamon [318] and Ono [515]. The extension to these cases requires the introduction of Novikov rings, as already pointed out by Floer in [230]. While, in principle, it has been understood since the spring of 1996 how to extend Floer's proof of the nondegenerate homological Arnold conjecture with rational coefficients to all closed symplectic manifolds, the state of the literature on this subject is still somewhat unsatisfactory. A number of papers appeared on this subject in the 1990s, e.g. by Fukaya–Ono [252], Liu–Tian [432], Ruan [548], Siebert [582] and Li–Tian [428]. However, some of the details were not fully carried out and many nontrivial technical difficulties appeared upon closer examination. Research is still ongoing to put some of the fine points on a rigorous mathematical footing, on the one hand via the polyfold approach of Hofer–Wysocki–Zehnder [320,321,322,323] and on the other via a more careful exposition of the approach through finite-dimensional reduction (see Pardon [523] and also McDuff–Wehrheim [473], Fukaya–Oh–Ohta–Ono [251], Chen–Li–Wang [104], Joyce [352].)

Lagrangian Floer homology

Floer's original theory in [226,227,228,229] includes the case of Lagrangian intersections so as to prove the Arnold–Givental conjecture. In this variant of the theory the loop space is replaced by the space of paths joining two Lagrangian submanifolds L_0 and L_1, the periodic solutions of Hamilton's equation are replaced by the Lagrangian intersections in $L_0 \cap \phi_H^{-1}(L_1)$, and periodicity in the Floer equation (11.4.5) is replaced by the boundary conditions

$$u(s,0) \in L_0, \qquad u(s,1) \in L_1.$$

In the simplest case of embedded loops in oriented 2-manifolds, Lagrangian Floer theory has a combinatorial description (see de Silva–Robbin–Salamon [135] for an exposition). In good cases the theory goes through as in the Hamiltonian setting and the resulting Floer homology groups $\mathrm{HF}_*(L_0, L_1)$ are independent of the choice of H and J up to natural isomorphism. Floer developed his theory in the case where $L = L_0 = L_1$ satisfies $\pi_2(M, L) = 0$ and proved that in this case the Floer homology groups are isomorphic to $H_*(L; \mathbb{Z}/2\mathbb{Z})$, which implies the Arnold–Givental conjecture.

Lagrangian Floer theory was extended by Oh [506] to the monotone case. In more general cases Lagrangian Floer homology may not always be well-defined. There is a remarkable obstruction theory, developed by Fukaya–Oh–Ohta–Ono [249], with many applications in symplectic topology.

Product structures

Both in Hamiltonian and in Lagrangian Floer theory there are natural associative product structures. Hamiltonian Floer theory carries the **pair-of-pants product** (see Schwarz [567]), that is isomorphic to quantum cohomology (see Piunikhin–Salamon–Schwarz [525]), while Lagrangian Floer homology has the **Donaldson triangle product**

$$\mathrm{HF}^*(L_0, L_1) \otimes \mathrm{HF}^*(L_1, L_2) \to \mathrm{HF}^*(L_0, L_2)$$

(see de Silva [134]). The latter gives rise to the **Donaldson category** of (M, ω) with (monotone) Lagrangian submanifolds as objects, the Floer cohomology groups $\mathrm{HF}^*(L_0, L_1)$ as spaces of morphisms, and the Donaldson triangle product as composition. In [664, 665, 666, 667, 668], Wehrheim and Woodward proved via **quilted Floer homology** that Lagrangian correspondences induce morphisms between (suitably extended) Donaldson categories.

An important chain level refinement of the Donaldson category is **Fukaya's** A^∞ **category** and the **derived Fukaya category** of M (see the papers by Fukaya [245, 249] and Seidel [572, 574]). Applications of this theory include the results by Fukaya–Seidel–Smith [253, 254] and Abouzaid [3, 2] about exact Lagrangian submanifolds in cotangent bundles, discussed in Remark 11.3.9.

Floer homology of symplectomorphisms

Another version of symplectic Floer homology is associated to a symplectomorphism $\phi : M \to M$ (see [164]). In this theory the free loop space is replaced by the space of twisted loops $\gamma : \mathbb{R} \to M$ that satisfy the condition $\gamma(t) = \phi(\gamma(t+1))$, the periodic solutions of Hamilton's equation are replaced by solutions of Hamilton's equation $\partial_t x(t) = X_t(x)$ that satisfy the condition $x(0) = \phi(x(1)) = \phi(\phi_H(x(0)))$, and so correspond to fixed points of $\phi \circ \phi_H$, and periodicity in the Floer equation (11.4.5) is replaced by the twisted periodicity condition $u(s,t) = \phi(u(s, t+1))$. The resulting Floer homology group $\mathrm{HF}_*(\phi)$ is independent of the choice of the Hamiltonian perturbation and the family of almost complex structures up to canonical isomorphism. It is naturally isomorphic to the Lagrangian Floer homology $\mathrm{HF}_*(\Delta, \mathrm{graph}(\phi))$, associated to the diagonal and the graph of ϕ in $(M \times M, (-\omega) \oplus \omega)$. Moreover, it is isomorphic to Hamiltonian Floer homology in the case $\phi = \mathrm{id}$. The Floer cohomology group $\mathrm{HF}^*(\phi)$ is a module over the quantum cohomology ring of M via the pair-of-pants product.

This version of Floer cohomology was used by Seidel to prove that many symplectic 4-manifolds admit symplectomorphisms that are smoothly, but not symplectically, isotopic to the identity [570, 575]. The symplectomorphism in question arises in each case as the square $\phi = \psi_L \circ \psi_L$ of a generalized Dehn twist ψ_L along a Lagrangian sphere L. Seidel's proof is based on a computation of the Floer cohomology group $\mathrm{HF}^*(\psi_L)$. The computation uses Seidel's exact triangle [570, 573] which relates the Floer cohomology of the symplectomorphism ψ_L to the Lagrangian Floer cohomology group $\mathrm{HF}^*(L, L)$ and to Hamiltonian Floer cohomology, and which respects the module structure over the (quantum) cohomology ring of M. The computation shows that, as a module, the Floer cohomology group $\mathrm{HF}^*(\psi_L^{-1}) \cong \mathrm{HF}_*(\psi_L)$ is not isomorphic to $\mathrm{HF}^*(\psi_L)$. Thus ψ_L^{-1} is not *Hamiltonian isotopic* to ψ_L, and hence in the simply connected case it is not *symplectically isotopic* to ψ_L. (See also Example 10.4.3.)

Instanton Floer homology

The Chern–Simons functional on the space of connections over a 3-manifold Y can be viewed as a gauge theoretic analogue of the symplectic action functional. In [231] Floer used it to define instanton Floer homology groups $\mathrm{HF}_*(Y)$ of a 3-manifold in the cases where either Y is an integral homology 3-sphere equipped with the trivial SU(2)-bundle or Y is equipped with a nontrivial SO(3)-bundle. In this theory the periodic orbits of Hamilton's equation are replaced by flat connections over the 3-manifold Y and the solutions of the Floer equation (11.4.5) are replaced by anti-self-dual instantons over the tube $\mathbb{R} \times Y$. The instanton Floer homology groups are closely related to the Donaldson invariants of smooth 4-manifolds, in analogy to the relation between symplectic Floer homology and Gromov–Witten theory. They have played an important role in three- and four-dimensional topology (see Donaldson's book [150] and the references therein).

In [232, 233] Floer discovered a long exact sequence associated to surgery, which was a model for Seidel's exact sequence in symplectic Floer theory. There is in fact a close connection between instanton Floer homology for three-manifolds and symplectic Floer theory. It is based on the observation that the moduli space of flat connections over a Riemann surface Σ is a symplectic manifold $\mathcal{M}_\Sigma$ (see Example 5.3.18 and Exercise 5.4.23), and that flat connections over Σ that extend over a handlebody Y with boundary Σ give rise to a Lagrangian submanifold $\mathcal{L}_Y \subset \mathcal{M}_\Sigma$. Thus, for a Heegaard splitting $Y = Y_0 \cup_\Sigma Y_1$, the instanton Floer homology groups $\mathrm{HF}_*(Y)$ and the Lagrangian Floer homology groups $\mathrm{HF}_*(\mathcal{L}_{Y_0}, \mathcal{L}_{Y_1})$ are based on the same chain complex, with different boundary operators. Atiyah and Floer conjectured that these Floer homology groups are isomorphic. At the time of writing (in 2016) the **Atiyah–Floer conjecture** is still open, although important progress has been made by Wehrheim [660, 661, 662, 663], leading to the definition of instanton Floer homology for 3-manifolds with Lagrangian boundary conditions in Salamon–Wehrheim [561]. Another version of this conjecture, relating the instanton Floer homology of a 3-dimensional mapping torus to the symplectic Floer homology of a corresponding symplectomorphism of $\mathcal{M}_\Sigma$, was settled in the early 1990s by Dostoglou–Salamon [164, 165, 166]. For other work related to the Atiyah–Floer conjecture see Fukaya [246], Duncan [169] and Lipyanskiy [430].

Seiberg–Witten Floer homology

Another version of Floer homology for 3-manifolds was found as a result of the discovery of the Seiberg–Witten invariants in late 1994. A full exposition of Seiberg–Witten Floer homology is given in the book [382] by Kronheimer–Mrowka. The Seiberg–Witten invariants were used by Ozsvath–Szabó [519] to settle the **symplectic Thom conjecture**, which asserts that every 2-dimensional connected symplectic submanifold of a closed symplectic 4-manifold minimizes the genus in its homology class (see Theorem 13.3.17). For surfaces with nonnegative self-intersection the symplectic Thom conjecture was settled earlier by Kronheimer and Mrowka, first for Kähler surfaces using Donaldson theory and then for all closed symplectic four-manifolds using Seiberg–Witten theory.

Important applications of Seiberg–Witten Floer homology were discovered by Taubes by extending his correspondence between the 4-dimensional Seiberg–Witten invariants and pseudoholomorphic curves in [608,609,610,611] to the 3-dimensional setting. In [612] he used these ideas to settle the Weinstein conjecture (every contact form has a closed Reeb orbit) in dimension three; see also Hutchings [336]. In [342], Hutchings and Taubes extended the Weinstein conjecture to so-called **stable Hamiltonian structures** on 3-manifolds. In [343, 344] Hutchings and Taubes settled **Arnold's chord conjecture** (for every Legendrian submanifold and every contact form there exists a Reeb chord) in dimension three. Their work is based on Hutchings' **embedded contact homology** [337,340], another variant of 3-dimensional Floer homology that was shown by Taubes [613, 614, 615, 616, 617] to be isomorphic to Seiberg–Witten Floer homology. Other applications of embedded contact homology include the existence of a second closed Reeb orbit in dimension three (see the paper [130] by Cristofaro-Gardiner and Hutchings) and sharp estimates for symplectic embeddings of ellipsoids and polydiscs in dimension four (see Hutchings [337,338,339,341], McDuff [467], and McDuff–Schlenk [471]).

Heegaard–Floer homology

The above discussion shows that there have been many applications of the new invariants of low-dimensional manifolds in symplectic topology. In the reverse direction ideas from symplectic topology have also impacted on three- and four-dimensional topology. One instance of this is the **Heegaard–Floer homology** introduced by Ozsvath–Szabó [520, 521]. This is a variant of Lagrangian Floer homology in the g-fold symmetric product $\mathrm{Sym}^g(\Sigma)$ of a closed Riemann surface of genus g. The starting point is the observation that a handlebody with boundary Σ gives rise to a union α of g pairwise disjoint embedded loops in Σ whose complement is connected. Any such set determines a monotone Lagrangian torus $T_\alpha \subset \mathrm{Sym}^g(\Sigma)$ for a suitable choice of a symplectic structure. Two handlebodies give rise to two monotone tori $T_\alpha, T_\beta \subset \mathrm{Sym}^g(\Sigma)$ and hence to a Floer homology group $\mathrm{HF}_*(T_\alpha, T_\beta)$. While this group is not particularly interesting, the story can be refined by including the winding number about a marked point in $\Sigma \setminus (\alpha \cup \beta)$ in the count of the holomorphic strips, and this leads to the definition of Heegaard–

Floer homology. It turns out that this is an invariant of the 3-manifold obtained as the union of the two handlebodies associated to α and β, that there is an analogous invariant of smooth 4-manifolds, and that these invariants have interesting applications, complementing the results obtained via Donaldson and Seiberg–Witten theory. It was conjectured from the outset that Heegard–Floer theory is equivalent to Seiberg–Witten Floer theory. This equivalence was then established by Kutluhan–Lee–Taubes [384, 385, 386, 387, 388]. Combining their results with those of Taubes in [613, 614, 615, 616, 617], one obtains that the Seiberg–Witten Floer homology of a contact 3-manifold is isomorphic to its embedded contact homology. This was proved directly by Colin–Ghigghini–Honda [120, 121, 122].

Donaldson–Thomas theory

In [163] Donaldson and Thomas extended the $(3+1)$-dimensional Floer–Donaldson theory to higher-dimensional manifolds with special holonomy. (See the books by Dominic Joyce [349, 350] for Riemannian manifolds with special holonomy.) There are many variants of this story; further ideas are explained in Donaldson–Segal [161]. While the algebro-geometric version of this theory has led to the rigorous introduction of new invariants by Richard Thomas, the gauge theoretic definition of the invariants is still largely conjectural. In one version 3-manifolds are replaced by 7-manifolds equipped with G_2-structures and smooth 4-manifolds are replaced by 8-manifolds with Spin(7)-structures. In particular, there is a conjectural Donaldson–Thomas Floer theory for closed 7-dimensional G_2-manifolds. As pointed out by Donaldson–Thomas [163], there is also a version of the Atiyah–Floer conjecture in this setting. The analogue of Hamilton's equation is the three-dimensional **Fueter equation** for maps from a 3-manifold equipped with a divergence-free frame (instead of a circle) into a hyperkähler manifold (instead of a symplectic manifold). The analogue of the Floer equation (11.4.5) in this setting is the four-dimensional Fueter equation. This equation was introduced by Rudolph Fueter [242, 243] in the 1930s in his study of analytic functions of one quaternionic variable. In the simplest case, where the hyperkähler target manifold is flat (i.e. a finite quotient of a hyperkähler torus), the **hyperkähler Floer homology groups** were defined by Hohloch–Noetzel–Salamon [326, 327], under the assumption that the source is the standard 3-sphere or the standard 3-torus. The theory extends to all closed 3-manifolds with *regular divergence-free frames* (see Salamon [559]). As a result, one obtains an analogue of the nondegenerate Arnold conjecture for the perturbed Fueter equation with values in flat hyperkähler target manifolds [326, 327, 559]. The **hyperkähler Arnold conjecture** was extended to flat target manifolds with arbitrary Clifford pencils and to the degenerate case by Ginzburg–Hein [267, 268]. They circumvented the Floer theory argument by a Conley–Zehnder type finite-dimensional reduction. The relation with the Fueter equation and Atiyah–Floer type adiabatic limit arguments was used by Walpuski to prove existence theorems for G_2-instantons in dimension seven [656, 657] and Spin(7)-instantons in dimension eight [658]. Walpuski also made important progress towards understanding the highly nontrivial compactness problem for solutions of the Fueter equation with values in non-flat hyperkähler target manifolds [659], and in ongoing research, in part jointly with Andriy Haydys [305], is developing a program for a rigorous definition of the G_2-invariants.

12

SYMPLECTIC CAPACITIES

In this chapter we return to the problems which were formulated in Chapter 1, namely the Weinstein conjecture, the Nonsqueezing Theorem, and symplectic rigidity. As we shall see, these questions are all related to the existence and properties of certain symplectic invariants. Roughly speaking these invariants measure the 2-dimensional size of a symplectic manifold just as the length spectra determined by geodesics measure the 1-dimensional size of a Riemannian manifold. They are very different from the volume. Gromov has suggested the terms *symplectic area* or *symplectic width*. However, the term **symplectic capacity**, even though intuitively it is closely related to volume, has now become widely used for this type of invariant and we shall adopt this terminology.

We begin the chapter by discussing some of the many interesting consequences which follow from the existence of capacities. In particular, we show in Section 12.2 that rigidity holds. We then discuss the relation between capacities and Hofer's invariant metric on the group of Hamiltonian symplectomorphisms. In Section 12.4, we define a particular capacity, called the Hofer–Zehnder capacity, and show that its existence gives rise to a proof of the Weinstein conjecture for hypersurfaces of Euclidean space. Finally, in Section 12.5 we prove that the Hofer–Zehnder capacity satisfies the required axioms. Here we combine Hofer's method from the calculus of variations with the finite-dimensional techniques of the discrete symplectic action. One advantage of our approach is that it bridges the gap between the proof given by Hofer and Zehnder [325], and the finite-dimensional approach of Viterbo [646] which is based on generating functions. Readers may also find it useful to consult Sikorav [586], where they can find a brief exposition of many of the ideas in this chapter using the original variational approach of Ekeland, Hofer, and Zehnder.

12.1 Nonsqueezing and capacities

Gromov's celebrated Nonsqueezing Theorem is a precursor to the whole development described in this chapter. It asserts that the closed ball $B^{2n}(r)$ cannot be symplectically embedded into a cylinder

$$Z^{2n}(R) := \left\{(x, y) \in \mathbb{R}^{2n} \mid x_1^2 + y_1^2 \le R^2\right\}$$

of smaller radius.

Theorem 12.1.1 (Nonsqueezing theorem) *If there exists a symplectic embedding of $(B^{2n}(r), \omega_0)$ into $(Z^{2n}(R), \omega_0)$ then $r \le R$.*

Proof: See page 484. □

Remark 12.1.2 There is no analogue of this result when the splitting is not symplectic. For example consider the Lagrangian splitting $\mathbb{R}^{2n} = \mathbb{R}^n \times \mathbb{R}^n$. Then for every $\delta > 0$ the map $(x, y) \mapsto (\delta x, \delta^{-1} y)$ is a symplectic embedding of $B^{2n}(1)$ into $B^n(\delta) \times \mathbb{R}^n$. Further, as noted by [301], there is a symplectic linear map that takes a large ball to an ellipsoid that intersects each 2-plane parallel to the (x_1, y_1)-axis in a disc of small area. This does not contradict nonsqueezing because the other axes of the ellipsoid are not symplectically orthogonal to the (x_1, y_1)-plane. □

Theorem 12.1.1 not only inspired the ideas about rigidity which are discussed in this chapter. It may also be considered as the most basic geometric expression of this rigidity. Indeed, Weinstein made the point that it can be considered as a geometric expression of the uncertainty principle. Given a point $(x_1, y_1, \ldots, x_n, y_n)$ of $\mathbb{R}^{2n}$, think of x_i as the ith position coordinate and y_i as the ith momentum coordinate of some Hamiltonian system. By taking a measurement, we might find out that the state of the system lies somewhere in a subset U of $\mathbb{R}^{2n}$. (In a more sophisticated model, we would be able to determine the probability of the state lying in U.) Suppose that U is (or contains) a ball of radius r. Then the range of uncertainty in our knowledge of the values of a conjugate pair (x_i, y_i) is measured by the area πr^2, and the Nonsqueezing Theorem can be thought of as saying that no matter how the system is transformed this range of uncertainty can never be made smaller.

Gromov's original proof of Theorem 12.1.1 uses J-holomorphic curves and will not be discussed here. Perhaps the simplest proof uses generating functions, and is due to Viterbo [646]. In the last section of this chapter, we will give a proof due to Hofer and Zehnder which is based on the calculus of variations.

Remark 12.1.3 Theorem 12.1.1 has been extended by Lalonde–McDuff [392] to arbitrary symplectic manifolds (M, ω) of dimension $2n$. Namely, if there exists a symplectic embedding of $(B^{2n+2}(r), \omega_0)$ into $(S^2 \times M, \sigma \oplus \omega)$ then $\pi r^2 \le \int_{S^2} \sigma$. The proof uses J-holomorphic curves, together with techniques of embedding balls similar to those discussed in Section 12.2 below. It follows that if there exists a symplectic embedding of $B^{2n+2}(r)$ into $B^2(R) \times M$ then $r \le R$. □

Remark 12.1.4 (i) In Theorem 12.1.1 the 2-sphere cannot be replaced by the 2-torus $\mathbb{T}^2 = \mathbb{R}^2/\mathbb{Z}^2$ with its standard symplectic form. Polterovich observed that for every $r > 0$ one can construct a symplectic embedding of $B^{2n+2}(r)$ into $\mathbb{T}^2 \times \mathbb{R}^{2n}$ as follows. Find a linear Lagrangian subspace L of $\mathbb{R}^2 \times \mathbb{R}^{2n}$ whose ε-neighbourhood L_ε projects injectively into $\mathbb{R}^2/\mathbb{Z}^2 \times \mathbb{R}^{2n} = \mathbb{T}^2 \times \mathbb{R}^{2n}$. Then consider a composition $B^{2n+2}(r) \hookrightarrow L_\varepsilon \hookrightarrow \mathbb{T}^2 \times \mathbb{R}^{2n}$.

(ii) Lalonde [389] proved that there exists an embedding of $B^4(r)$ into a symplectic product $(\mathbb{T}^2 \times S^2, \tau \oplus \sigma)$ if and only if the volume of the ball is smaller than the volume of $\mathbb{T}^2 \times S^2$ and $\pi r^2 < \int_{S^2} \sigma$. Similar results hold with $\mathbb{T}^2$ replaced by closed symplectic 2-manifolds of genus greater than one. □

Remark 12.1.5 Let ω be a symplectic form on $M = \mathbb{T}^4$ with constant coefficients. A theorem of Latschev–McDuff–Schlenk [412] asserts that $B^4(r)$ admits a symplectic embedding into $(\mathbb{T}^4, \omega)$ if and only if its volume $\pi^2 r^4/2$ is smaller than the volume of $(\mathbb{T}^4, \omega)$. This gives rise to another proof of the assertion in Remark 12.1.4 (i). □

Gromov's Nonsqueezing Theorem gave rise to the following definition which is due to Ekeland and Hofer [171]. A **symplectic capacity** is a functor c which assigns to every symplectic manifold (M, ω) a nonnegative (possibly infinite) number $c(M, \omega)$ and satisfies the following conditions.

(monotonicity) If there is a symplectic embedding $(M_1, \omega_1) \hookrightarrow (M_2, \omega_2)$ and $\dim M_1 = \dim M_2$ then $c(M_1, \omega_1) \le c(M_2, \omega_2)$.
(conformality) $c(M, \lambda\omega) = \lambda c(M, \omega)$.
(nontriviality) $c(B^{2n}(1), \omega_0) > 0$ and $c(Z^{2n}(1), \omega_0) < \infty$.

The first axiom implies naturality: if two symplectic manifolds (M_1, ω_1) and (M_2, ω_2) are symplectomorphic then $c(M_1, \omega_1) = c(M_2, \omega_2)$.

The key to understanding symplectic capacities is the observation that the nontriviality axiom makes it impossible for the nth root of the volume of M to be a capacity. Indeed, the requirement that $c(Z^{2n}(1), \omega_0)$ be finite means that these capacities are essentially 2-dimensional invariants.

A priori it is not at all clear that capacities exist. As a matter of fact the existence of a symplectic capacity with

$$c(B^{2n}(1), \omega_0) = c(Z^{2n}(1), \omega_0) = \pi \tag{12.1.1}$$

is equivalent to Gromov's Nonsqueezing Theorem. If there is such a capacity then, by the monotonicity and conformality axioms, the ball $B^{2n}(1)$ cannot be symplectically embedded in $Z^{2n}(R)$ unless $R \ge 1$. Conversely, define the **Gromov width** of a symplectic manifold (M, ω) by

$$w_G(M, \omega) = w_G(M) = \sup\{\pi r^2 \mid B^{2n}(r)\,\text{embeds symplectically in}\, M\}. \tag{12.1.2}$$

The Gromov width clearly satisfies the monotonicity and conformality axioms and satisfies the nontriviality axiom by the Nonsqueezing Theorem 12.1.1.

Example 12.1.6 For $a > 0$ denote by $S^2(a)$ the 2-sphere of area a and by $\mathbb{T}^2(a)$ the 2-torus of area a. It follows from Gromov's proof of the Nonsqueezing Theorem and from Remark 12.1.3 that for $r \le R$

$$w_G(S^2(\pi r^2) \times \mathbb{T}^{2n}(\pi R^2)) = \pi r^2.$$

Moreover, when $n = 1$ it follows from part (ii) of Remark 12.1.4 that we may even choose $R = r/\sqrt{2}$ without decreasing the capacity:

$$w_G(S^2(\pi r^2) \times \mathbb{T}^2(\pi r^2/2)) = \pi r^2,$$

while part (i) of this remark can be rephrased as

$$w_G(\mathbb{R}^2 \times \mathbb{T}^2(\pi/2)) = \infty.$$

It also follows from Gromov's proof of the Nonsqueezing Theorem that

$$w_G(S^2(\pi R_1^2) \times \cdots \times S^2(\pi R_n^2)) = \pi R_1^2$$

for $0 < R_1 \le \cdots \le R_n$. □

We shall now restrict the discussion to subsets of $\mathbb{R}^{2n}$. These subsets are not required to be open, i.e. they are not required to be manifolds. A symplectic embedding $\psi : A \to \mathbb{R}^{2n}$ defined on an arbitrary subset $A \subset \mathbb{R}^{2n}$ is by definition a map which extends to a symplectic embedding of an open neighbourhood of A. Now a **symplectic capacity** c on $\mathbb{R}^{2n}$ assigns a number $c(A) \in [0, \infty]$ to every subset $A \subset \mathbb{R}^{2n}$ such that the following holds:

(monotonicity) If there exists a symplectic embedding $\psi : A \to \mathbb{R}^{2n}$ such that $\psi(A) \subset B$ then $c(A) \le c(B)$.

(conformality) $c(\lambda A) = \lambda^2 c(A)$.

(nontriviality) $c(B^{2n}(1)) > 0$ and $c(Z^{2n}(1)) < \infty$.

Such a capacity is **intrinsic** in the sense that it depends only on the symplectic structure of A and not on the way in which it is embedded in the ambient manifold $\mathbb{R}^{2n}$. There are several interesting invariants which do depend on the embedding, for example the Ekeland–Hofer capacity defined in [171] and the displacement energy discussed below. These are invariants of the pair $(\mathbb{R}^{2n}, A)$ and only satisfy the following weaker version of monotonicity as well as the conformality and nontriviality axioms above.

(relative monotonicity) If there exists a symplectomorphism ψ of $\mathbb{R}^{2n}$ such that $\psi(A) \subset B$ then $c(A) \le c(B)$.

We will always use the word capacity to refer to an intrinsic capacity which satisfies the full monotonicity axiom. Capacities which satisfy relative monotonicity will be called **relative** or **nonintrinsic**.

For every subset $A \subset \mathbb{R}^{2n}$, define

$$\overline{w}_G(A) = \inf\{\pi r^2 \,|\, A \,\text{embeds symplectically in}\, Z^{2n}(r)\}.$$

It follows again from Gromov's Nonsqueezing Theorem that $\overline{w}_G$ satisfies the axioms of a symplectic capacity on $\mathbb{R}^{2n}$. If c is any other capacity on $\mathbb{R}^{2n}$ such that $c(B^{2n}(1)) = c(Z^{2n}(1)) = \pi$ then, by the monotonicity axiom, we have

$$w_G(A) \le c(A) \le \overline{w}_G(A)$$

for every subset $A \subset \mathbb{R}^{2n}$.

Example 12.1.7 (Capacity of ellipsoids) Recall from Lemma 2.4.6 that given any ellipsoid

$$E = \left\{ w \in \mathbb{R}^{2n} \,\middle|\, \sum_{j,k=1}^{2n} a_{jk} w_j w_k \le 1 \right\}$$

there is a linear symplectomorphism $\Psi \in \mathrm{Sp}(2n)$ which takes E to an ellipsoid of the form

$$\Psi E = E(r) := \left\{ z \in \mathbb{C}^n \,\middle|\, \sum_{j=1}^{n} \left| \frac{z_j}{r_j} \right|^2 \le 1 \right\}, \qquad 0 < r_1 \le r_2 \le \cdots \le r_n.$$

Recall also that the n-tuple $r = (r_1, \ldots, r_n)$ is uniquely determined by E and is called the spectrum of E. Since

$$B^{2n}(r_1) \subset \Psi E \subset Z^{2n}(r_1),$$

it follows that

$$c(E) = \pi {r_1}^2 = w_L(E)$$

for every relative symplectic capacity c which satisfies (12.1.1). Here w_L denotes the linear symplectic width introduced in Chapter 2 and the last equation follows from Theorem 2.4.8. □

Exercise 12.1.8 Let c be a capacity on $\mathbb{R}^{2n}$. Prove that the restriction of c to compact convex sets is continuous with respect to the Hausdorff metric. Recall that this is a metric on the space of closed subsets of $\mathbb{R}^{2n}$ and is defined by

$$d(A, B) = \sup_{x \in A} d(x, B) + \sup_{y \in B} d(y, A)$$

for $A, B \subset \mathbb{R}^{2n}$. Let $A \subset \mathbb{R}^{2n}$ be a compact convex set. Prove that

$$\lim_{\substack{B \text{ convex} \\ B \to A}} c(B) = c(A),$$

where the convergence is over compact convex subsets of $\mathbb{R}^{2n}$ in the Hausdorff metric. In other words, for every $\varepsilon > 0$ there exists a $\delta > 0$ such that for every compact convex set $B \subset \mathbb{R}^{2n}$,

$$d(A, B) < \delta \qquad \Longrightarrow \qquad |c(A) - c(B)| < \varepsilon.$$

Hint: Distinguish the cases $\mathrm{int}(A) = \emptyset$ and $\mathrm{int}(A) \ne \emptyset$. If $0 \in \mathrm{int}(A)$ prove that for every $\lambda > 1$ there is a constant $\delta > 0$ such that for every compact convex set $B \subset \mathbb{R}^{2n}$,

$$d(A, B) < \delta \qquad \Longrightarrow \qquad \lambda^{-1} A \subset B \subset \lambda A.$$

If $\mathrm{int}(A) = \emptyset$ prove that A is contained in some hyperplane $W \subset \mathbb{R}^{2n}$ and deduce that $c(A) = 0$. More precisely, show that for every $\delta > 0$, there exists a symplectic matrix $\Psi \in \mathrm{Sp}(2n)$ such that $\Psi A \subset Z^{2n}(\delta)$. □

Remark 12.1.9 The following examples, which are due to Sean Bates [58], show that the above continuity property does not extend to the nonconvex case. For example, one can approximate a 2-disc A arbitrarily closely in the Hausdorff metric by sets B which are disjoint unions of thin annuli. If $c = w_G$ is the Gromov width then $c(B)$ is the maximum of the areas of its connected components and so may be much smaller than $c(A)$. It is also easy to see that for any $r < 1$ the $2n$-ball $A = B^{2n}(1)$ may be approximated arbitrarily closely in the Hausdorff metric by an embedded ball B of radius r. To see this let B be the image of the standard ball B' of radius r by the time-1 map of H, where H is a Hamiltonian with support in A whose flow takes some subset of N points in B' to a finite set in A whose ε-neighbourhood covers A. □

Since capacities are symplectic invariants, they can be used to distinguish different symplectic manifolds. Such applications will be discussed in Chapter 13. In this chapter, we give applications involving symplectomorphisms. In particular, capacities are closely related to an invariant metric on the group of compactly supported symplectomorphisms of $\mathbb{R}^{2n}$ which is due to Hofer. This is discussed in Section 12.3.

There are many possible definitions of symplectic capacities with different properties. One of the most important is the Hofer–Zehnder capacity. It is based on properties of periodic orbits of Hamiltonian systems and gives rise to a proof of the Weinstein conjecture for hypersurfaces of Euclidean space. Its general properties will be discussed in Section 12.4 and in Section 12.5 we shall prove that it does satisfy all the axioms. In particular, as we have seen above, the existence of this capacity implies Gromov's Nonsqueezing Theorem.

In the next section we shall assume that capacities exist and show how they can be used to prove symplectic rigidity.

12.2 Rigidity

The goal of this section is to prove that the group of symplectomorphisms is closed in the group of all diffeomorphisms with respect to the C^0-topology. This is the content of the following theorem which is due to Eliashberg [175,176] and Gromov [288]. Another proof was given by Ekeland and Hofer [171]. The proof given below uses the existence of a symplectic capacity.

Theorem 12.2.1 *For every symplectic manifold (M, ω) the group of symplectomorphisms of (M, ω) is C^0-closed in the group of all diffeomorphisms of M.*

Proof: See page 464 for a proof based on Theorem 12.1.1. □

A symplectomorphism of a symplectic manifold (M, ω) is a diffeomorphism ϕ such that $\phi^*\omega = \omega$. This definition involves the first derivatives of ϕ and so cannot be generalized in an obvious way to homeomorphisms. This is in contrast to the volume-preserving case: a diffeomorphism preserves a volume form if and only if it preserves the corresponding measure. Eliashberg in [175,176] and, independently, Ekeland–Hofer in [171], realized that one can use capacities in a similar way to give an alternative definition of a symplectomorphism which does not involve derivatives. Their observation is summarized in the following proposition.

Proposition 12.2.2 *Let $\psi : \mathbb{R}^{2n} \to \mathbb{R}^{2n}$ be a diffeomorphism and let c be a symplectic capacity on $\mathbb{R}^{2n}$ which satisfies* (12.1.1). *Then the following are equivalent.*

(i) *ψ preserves the capacity of ellipsoids, i.e. it satisfies $c(\psi(E)) = c(E)$ for every ellipsoid E in $\mathbb{R}^{2n}$.*

(ii) *ψ is either a symplectomorphism or an anti-symplectomorphism, i.e. it satisfies $\psi^*\omega_0 = \pm\omega_0$.*

Proof: That (ii) implies (i) is obvious. The converse is proved on page 464. □

Here we consider ellipsoids with arbitrary centre. It follows from the definition of a capacity that every symplectomorphism and every anti-symplectomorphism preserves the capacity of ellipsoids. (For anti-symplectomorphisms one needs the additional elementary fact that for every ellipsoid there exists an anti-symplectomorphism which maps this ellipsoid to itself.) In Section 2.4 we have shown that conversely, every linear map which preserves the linear symplectic width of ellipsoids, is either symplectic or anti-symplectic. Proposition 12.2.2 is the nonlinear version of this result. As we now show, the proof is elementary once given the existence of a symplectic capacity.

Lemma 12.2.3 *Let c be a symplectic capacity on $\mathbb{R}^{2n}$ which satisfies* (12.1.1). *Let $\psi_\nu : \mathbb{R}^{2n} \to \mathbb{R}^{2n}$ be a sequence of continuous maps converging to a homeomorphism $\psi : \mathbb{R}^{2n} \to \mathbb{R}^{2n}$, uniformly on compact sets. Assume that ψ_ν preserves the capacity of ellipsoids for every ν. Then ψ preserves the capacity of ellipsoids.*

Proof: Without loss of generality we consider only ellipsoids centred at zero. We first prove that for every ellipsoid E and every positive number $\lambda < 1$ there exists a $\nu_0 > 0$ such that

$$\psi_\nu(\lambda E) \subset \psi(E) \subset \psi_\nu(\lambda^{-1}E) \tag{12.2.1}$$

for $\nu \geq \nu_0$. To see this, abbreviate $f_\nu := \psi^{-1} \circ \psi_\nu$. Then f_ν converges to the identity, uniformly on compact sets. So the inclusion $f_\nu(\lambda E) \subset E$ is obvious for large ν. However, because we do not assume that the ψ_ν are homeomorphisms, the inclusion $\psi(E) \subset \psi_\nu(\lambda^{-1}E)$ requires proof. To this end, we choose ν_0 so large that for $\nu > \nu_0$

$$x \in \lambda^{-1}\partial E \qquad \Longrightarrow \qquad f_\nu(x) \notin E$$

and the map $\phi_\nu : \lambda^{-1}\partial E \to S^{2n-1}$ defined by

$$\phi_\nu(x) := \frac{f_\nu(x)}{|f_\nu(x)|}$$

has degree 1. Now let $y_0 \in E$ and suppose that $f_\nu(x) \neq y_0$ for all $x \in \lambda^{-1}E$. Then the map $\phi'_\nu : \lambda^{-1}\partial E \to S^{2n-1}$ defined by

$$\phi'_\nu(x) := \frac{f_\nu(x) - y_0}{|f_\nu(x) - y_0|}$$

extends to $\lambda^{-1}E$, and therefore must have degree zero. On the other hand, ϕ'_ν is homotopic to ϕ_ν and therefore must have degree 1. This contradiction implies that $E \subset f_\nu(\lambda^{-1}E)$ for $\nu \geq \nu_0$ and this proves (12.2.1).

Equation (12.2.1) now implies that $\lambda^2 c(E) \leq c(\psi(E)) \leq \lambda^{-2}c(E)$. Since $\lambda < 1$ was chosen arbitrarily close to 1 it follows that ψ preserves the capacity of ellipsoids. This proves Lemma 12.2.3. □

Proof of Proposition 12.2.2 '(i) implies (ii)': Let c be a symplectic capacity on $\mathbb{R}^{2n}$ which satisfies (12.1.1) and let $\psi : \mathbb{R}^{2n} \to \mathbb{R}^{2n}$ be a diffeomorphism that preserves the capacity of ellipsoids. By postcomposing with a translation we may assume that $\psi(0) = 0$. Then the maps $\psi_t(z) := \frac{1}{t}\psi(tz)$ are diffeomorphisms of $\mathbb{R}^{2n}$ which preserve the capacity of ellipsoids, and they converge, uniformly on compact sets, to the linear map $\Psi := d\psi(0)$:

$$\Psi z = \lim_{t \to 0} \psi_t(z).$$

Hence, by Lemma 12.2.3, Ψ preserves the capacity of ellipsoids. By Example 12.1.7, the capacity of an ellipsoid agrees with its linear symplectic width. Hence it follows from Theorem 2.4.4 that $\Psi^*\omega_0 = \pm\omega_0$. The same holds with Ψ replaced by $d\psi(z)$ for any $z \in \mathbb{R}^{2n}$ and, by continuity, the sign is independent of z. This proves Proposition 12.2.2. □

Proof of Theorem 12.2.1, assuming Theorem 12.1.1: This is a local statement and so it suffices to prove it for $M = \mathbb{R}^{2n}$. Let $\psi_\nu : \mathbb{R}^{2n} \to \mathbb{R}^{2n}$ be a sequence of symplectomorphisms converging in the C^0-topology to $\psi : \mathbb{R}^{2n} \to \mathbb{R}^{2n}$ and assume that ψ is a diffeomorphism. By Theorem 12.1.1 there exists a symplectic capacity on $\mathbb{R}^{2n}$ satisfying (12.1.1), namely the Gromov width $c = w_G$. By definition ψ_ν preserves the Gromov width of ellipsoids for every ν. Hence, by Lemma 12.2.3, ψ preserves the Gromov width of ellipsoids. Hence, by Proposition 12.2.2, ψ is either a symplectomorphism or an anti-symplectomorphism. We claim that $\psi^*\omega_0 = \omega_0$. Suppose otherwise that $\psi^*\omega_0 = -\omega_0$. Then the sequence $\phi_\nu := \psi_\nu \times \mathrm{id} \in \mathrm{Diff}(\mathbb{R}^{2n} \times \mathbb{R}^{2n})$ preserves the symplectic structure $\omega := \omega_0 \times \omega_0$ while its C^0-limit $\phi := \psi \times \mathrm{id}$ satisfies $\phi^*\omega = (-\omega_0) \times \omega_0 \neq \pm\omega$, in contradiction to what we have just proved. This contradiction shows that $\psi^*\omega_0 = \omega_0$, as claimed, and this proves Theorem 12.2.1. □

Proposition 12.2.2 suggests the following definition. For each n fix a symplectic capacity c on $\mathbb{R}^{2n}$ that satisfies (12.1.1). Let n be odd. Call a map $\phi : \mathbb{R}^{2n} \to \mathbb{R}^{2n}$ a **symplectic homeomorphism** if it is an orientation-preserving homeomorphism and every point in $\mathbb{R}^{2n}$ has a neighbourhood U such that $c(\phi(V)) = c(V)$ for every open subset $V \subset U$. If n is even, call $\phi : \mathbb{R}^{2n} \to \mathbb{R}^{2n}$ a **symplectic homeomorphism** if $\phi \times \mathrm{id} : \mathbb{R}^{2n+2} \to \mathbb{R}^{2n+2}$ is a symplectic homeomorphism. This rather cumbersome definition is needed to distinguish between symplectic and anti-symplectic homeomorphisms. Note also that by Lemma 12.2.3, the group of symplectic homeomorphisms is closed in the group of all homeomorphisms with respect to the C^0-topology. Moreover, by Proposition 12.2.2, every smooth symplectic homeomorphism is a symplectomorphism in the usual sense.

One can translate this definition to an arbitrary manifold using Darboux's theorem. But there are many open questions. For example, if ϕ preserves the capacity of all small open sets, must it also preserve the capacity of large open sets? Must these symplectic homeomorphisms preserve volume? Is there a similar C^0-characterization of contactomorphisms? For further discussion see Section 14.7.

12.3 The Hofer metric

Let G be a finite-dimensional Lie group with Lie algebra $\mathfrak{g}$. A norm $|\cdot|$ on $\mathfrak{g}$ is called **invariant** if it is invariant under the adjoint action of G, i.e.

$$|\xi| = |g^{-1}\xi g|$$

for every $\xi \in \mathfrak{g}$ and every $g \in \mathrm{G}$. (Here $g^{-1}\xi g$ is defined as the derivative of the curve $\mathbb{R} \to \mathrm{G} : t \mapsto g^{-1}\exp(t\xi)g$ at $t = 0$.) Any such norm gives rise to a metric on the Lie group G via

$$d(g_0, g_1) = \inf_g \int_0^1 \left|\dot{g}(t)g(t)^{-1}\right| \, dt$$

for $g_0, g_1 \in \mathrm{G}$. Here the infimum is over all paths $g : [0,1] \to \mathrm{G}$ connecting $g_0 = g(0)$ to $g_1 = g(1)$. Such invariant metrics play an important role in the theory of finite-dimensional Lie groups. The **Hofer metric** is an analogous invariant metric on the infinite-dimensional *'Lie group'* of compactly supported Hamiltonian symplectomorphisms (cf. [314]). In [314] the distance of a symplectomorphism to the identity in this metric is called the **symplectic energy**. It turns out that the Hofer metric is closely related to symplectic capacities. The main results proved in this section all rely on the existence of a symplectic capacity and hence on Gromov's Nonsqueezing Theorem 12.1.1.

Let (M, ω) be a connected symplectic manifold without boundary. We do not assume that M is compact. A symplectomorphism $\psi : M \to M$ is called compactly supported if there exists a compact subset $K \subset M$ such that $\psi(p) = p$ for every $p \in M \setminus K$. As we saw in Chapter 10, the compactly supported symplectomorphisms form a group $\mathrm{Symp}_{\mathrm{c}}(M, \omega)$ whose Lie algebra is the space $\mathcal{X}_{\mathrm{c}}(M, \omega)$ of compactly supported symplectic vector fields. Likewise, the Lie algebra of the group $\mathrm{Ham}_{\mathrm{c}}(M, \omega)$ of compactly supported Hamiltonian symplectomorphisms is the space of compactly supported Hamiltonian vector fields. If M is noncompact then this space can be identified with the space $C_0^\infty(M)$ of compactly supported functions on M; otherwise the Hamiltonian function is only unique up to an additive constant. Recall from Chapter 3 that a smooth time-dependent compactly supported Hamiltonian function $H : [0,1] \times M \to \mathbb{R}$ generates a Hamiltonian isotopy $\phi_t \in \mathrm{Ham}_c(M,\omega)$ with uniform compact support in $[0,1] \times M$ via

$$\partial_t \phi_t = X_{H_t} \circ \phi_t, \qquad \phi_0 = \mathrm{id}, \qquad \iota(X_{H_t})\omega = dH_t, \tag{12.3.1}$$

where $H_t := H(t,\cdot)$. The time-1 map of this isotopy is called the **Hamiltonian symplectomorphism generated by** H and will be denoted by

$$\phi_H := \phi_1.$$

Since $\psi^* X_H = X_{H\circ\psi}$ for every symplectomorphism ψ, the adjoint action of the group of symplectomorphisms on the Lie algebra of compactly supported Hamiltonian functions is given by $H \mapsto H \circ \psi$.

The **Hofer norm** on the Lie algebra of compactly supported Hamiltonian functions is defined by

$$\|H\| := \max_M H - \min_M H \tag{12.3.2}$$

for $H \in C_0^\infty(M)$. This is only a norm on $C_0^\infty(M)$ when M is noncompact. In the compact case, the constant functions have norm zero and so equation (12.3.2) defines a norm on the quotient space $C^\infty(M)/\mathbb{R}$ or, equivalently, on the space of Hamiltonian vector fields. This norm is obviously invariant under the adjoint action, i.e. under the composition $H \mapsto H \circ \psi$ with a symplectomorphism $\psi \in \mathrm{Symp}_c(M,\omega)$. The corresponding length functional on the space of Hamiltonian isotopies $\{\phi_t\}_{0\le t\le 1}$ with compact support in $[0,1] \times M$ is the number

$$\mathcal{L}(\{\phi_t\}_{0\le t\le 1}) := \int_0^1 \|H_t\|\,dt = \int_0^1 \left(\sup_M H_t - \min_M H_t\right) dt, \tag{12.3.3}$$

where $H : [0,1] \times M \to \mathbb{R}$ is the generating Hamiltonian with compact support in $[0,1] \times M$ and $H_t := H(t,\cdot)$. This gives rise to a distance function on the Lie group $\mathrm{Ham}_c(M,\omega)$, which assigns to a pair of Hamiltonian symplectomorphisms the infimum of the lengths of Hamiltonian isotopies joining them, i.e. the **Hofer distance** of two Hamiltonian symplectomorphism $\phi_0, \phi_1 \in \mathrm{Ham}_c(M,\omega)$ is defined by

$$\rho(\phi_0,\phi_1) := \inf_{\phi_H = \phi_1\circ\phi_0^{-1}} \int_0^1 \|H_t\|\,dt, \tag{12.3.4}$$

where the infimum is taken over all time-dependent compactly supported smooth Hamiltonian functions $H : [0,1] \times M \to \mathbb{R}$ that generate the Hamiltonian symplectomorphism $\phi_1 \circ \phi_0^{-1}$. Sometimes it is useful to consider only those Hamiltonian isotopies ϕ_t which are constant for t near the endpoints 0 and 1. But this can always be arranged by a suitable reparametrization with respect to t and therefore does not alter the definition of the metric ρ.

Proposition 12.3.1 **(i)** *The Hofer distance is symmetric, i.e.*

$$\rho(\phi_0,\phi_1) = \rho(\phi_1,\phi_0)$$

for all $\phi_0, \phi_1 \in \mathrm{Ham}_c(M,\omega)$

(ii) *The Hofer distance satisfies the triangle inequality*

$$\rho(\phi_0, \phi_2) \le \rho(\phi_0, \phi_1) + \rho(\phi_1, \phi_2)$$

for all $\phi_0, \phi_1, \phi_2 \in \mathrm{Ham_c}(M, \omega)$

(iii) *The Hofer distance is bi-invariant, i.e.*

$$\rho(\phi_0 \circ \psi, \phi_1 \circ \psi) = \rho(\psi \circ \phi_0, \psi \circ \phi_1) = \rho(\phi_0, \phi_1)$$

for all $\phi_0, \phi_1, \psi \in \mathrm{Ham_c}(M, \omega)$ *and*

$$\rho(\psi^{-1} \circ \phi_0 \circ \psi, \psi^{-1} \circ \phi_1 \circ \psi) = \rho(\phi_0, \phi_1)$$

for all $\phi_0, \phi_1 \in \mathrm{Ham_c}(M, \omega)$ *and all* $\psi \in \mathrm{Symp_c}(M, \omega)$.

Proof: Abbreviate $\phi\psi := \phi \circ \psi$ for two diffeomorphism ϕ, ψ of M. Choose compactly supported Hamiltonians $F, G : [0,1] \times M \to \mathbb{R}$ such that $\phi_F = \phi_1\phi_0^{-1}$ and $\phi_G = \phi_2\phi_1^{-1}$, and denote the Hamiltonian isotopies generated by F and G by $\phi_{F,t}$ and $\phi_{G,t}$, respectively. Then the Hamiltonian $H_t := G_t + F_t \circ \phi_{G,t}^{-1}$ generates $\phi_2\phi_0^{-1}$ (see Exercise 3.1.14). Since $\|H_t\| \le \|F_t\| + \|G_t\|$, the triangle inequality follows by taking the infimum over all F and G. Secondly, the Hamiltonian $K_t := -F_t \circ \phi_{F,t}$ generates $\phi_0\phi_1^{-1}$ (see Exercise 3.1.14) and this implies the symmetry of the Hofer distance. Thirdly, the Hamiltonian $F_t \circ \psi$ generates $\psi^{-1}\phi_1\phi_0^{-1}\psi$ and this shows that $\rho(\psi^{-1}\phi_0\psi, \psi^{-1}\phi_1\psi) = \rho(\phi_0, \phi_1)$. The equation $\rho(\phi_0\psi, \phi_1\psi) = \rho(\phi_0, \phi_1)$ follows directly from the definition of the Hofer distance and it implies the equation $\rho(\psi\phi_0, \psi\phi_1) = \rho(\phi_0, \phi_1)$. This proves Proposition 12.3.1. □

Proposition 12.3.1 asserts that the Hofer distance is a bi-invariant pseudo-metric on the group $\mathrm{Ham_c}(M, \omega)$. The missing property is that the distance between two distinct elements of the group $\mathrm{Ham_c}(M, \omega)$ is positive. Note that the compactly supported Hamiltonian symplectomorphisms with zero distance to the identity form a normal subgroup of $\mathrm{Ham_c}(M, \omega)$. Hence Banyaga's Theorem 10.3.1 implies in the compact case that every bi-invariant pseudo-metric is either identically zero or is a metric. Thus, to prove it is a metric for a given compact manifold M, one just has to find one symplectomorphism $\phi \ne \mathrm{id}$ of M with $\rho(\phi, \mathrm{id}) > 0$. In the noncompact case there do exist nontrivial pseudo-metrics. Examples are discussed below.

Remark 12.3.2 **(i)** That the Hofer distance ρ is a metric is highly nontrivial. This is due to the fact that the Hofer norm does not involve the derivatives of H while the Hamiltonian vector field does depend on dH. So a priori it is not clear that two different Hamiltonian symplectomorphisms cannot be connected by a Hamiltonian isotopy with arbitrarily small Hamiltonian functions. On the other hand, an invariant metric cannot involve the derivatives of H. For example the C^1-norm of H is not invariant under the adjoint action $H \mapsto H \circ \psi$.

(ii) Another important point is that it is not enough to prove that ρ is a metric just for one manifold. For example, suppose that we have proved that ρ is a metric when $M = \mathbb{R}^{2n}$. Then one knows that $\rho(\phi, \mathrm{id}) > 0$ for any nontrivial element $\phi \in \mathrm{Ham}_c(\mathbb{R}^{2n})$. If the support of ϕ is sufficiently small, we may transfer ϕ to an arbitrary manifold by conjugating it by a Darboux chart. However, to calculate $\rho_M(\phi, \mathrm{id})$ we have to look at all Hamiltonian isotopies of M, and there is no obvious way to rule out the possibility that there is a cunning way to move M so as to reach ϕ with an isotopy which has smaller energy than is needed in $\mathbb{R}^{2n}$. □

The following theorem was first proved for $M = \mathbb{R}^{2n}$ by Hofer in [314]. It was generalized to certain other manifolds (basically, those for which the cohomology class $[\omega]$ is rational) by Polterovich [527] and then proved for all manifolds by Lalonde and McDuff [390].

Theorem 12.3.3 *Let* $\phi, \psi \in \mathrm{Ham}_c(M, \omega)$. *Then*

$$\rho(\phi, \psi) = 0 \quad \Longrightarrow \quad \phi = \psi.$$

Proof: See page 475 for $M = \mathbb{R}^{2n}$. The proof assumes Theorem 12.1.1. □

All three proofs use Hofer's idea of **displacement energy** which we introduce below. Hofer's proof uses a variant of the variational argument which we give in Section 12.5. His argument is elegant, but it is not very obvious why it works. The other two proofs are similar in spirit, and use geometric arguments to relate this result to the Nonsqueezing Theorem. Polterovich's argument is based on embeddings of Lagrangians, and has been generalized and extended by Chekanov [100] and Oh [508]. Lalonde and McDuff use symplectic balls. The basic idea of their proof is that if there were a nontrivial symplectomorphism ϕ of M with zero norm, one could construct an embedding of a large ball into a thin cylinder $B^2(r) \times M$. This is explained in the case of $M = \mathbb{R}^{2n}$ in Theorem 12.3.9 below.

Invariant pseudo-metrics

Assume M is noncompact. Then another bi-invariant pseudo-metric on the group $\mathrm{Ham}_c(M, \omega)$ is given by

$$\rho_p(\phi, \psi) = \inf_{\phi_H = \phi_0^{-1} \circ \phi_1} \int_0^1 \left(\int_M |H|^p \, \omega^n \right)^{1/p} dt$$

for $\phi_0, \phi_1 \in \mathrm{Ham}_c(M, \omega)$, where the infimum is taken over all time-dependent compactly supported smooth Hamiltonian functions $H : [0,1] \times M \to \mathbb{R}$ that generate the Hamiltonian symplectomorphism $\phi_0^{-1} \circ \phi_1$. This formula only defines a pseudo-metric on $\mathrm{Ham}_c(M, \omega)$, as was proved by Eliashberg and Polterovich [193]. In other words, ρ_p satisfies the requirements of Proposition 12.3.1 but not the assertion of Theorem 12.3.3.

It follows from Banyaga's Theorem 10.3.7 that any pseudo-metric which is not a metric must be a function of the Calabi invariant. We now give an explicit formula for d_p when $\omega = -d\lambda$. Recall from Chapter 10 that in this case the Calabi invariant of a Hamiltonian symplectomorphism $\phi \in \mathrm{Ham}^c(M, \omega)$ is given by

$$\mathrm{CAL}(\phi) = \int_0^1 \int_M H_t \omega^n \, dt,$$

where H_t is a smooth family of compactly supported Hamiltonians generating $\phi = \phi_H^1$. Eliashberg and Polterovich proved that for every integer $p \in [1, \infty)$,

$$\rho_p(\phi, \mathrm{id}) = \mathrm{Vol}(M)^{(1-p)/p} |\mathrm{CAL}(\phi)|.$$

If M has infinite volume the factor is $(+\infty)^0 = 1$ for $p = 1$ or $(+\infty)^{-\mu} = 0$ for $p > 1$. So, in particular, $\rho_1(\phi, \mathrm{id}) = |\mathrm{CAL}(\phi)|$.

The energy–capacity inequality

Following Hofer we define the **displacement energy** $e(K)$ of a compact subset $K \subset M$ by

$$e(K) = \inf_{\substack{\phi \in \mathrm{Ham}^c(M,\omega) \\ K \cap \phi(K) = \emptyset}} \rho(\phi, \mathrm{id}).$$

In Hofer's terminology this is the minimum of the energy $E(\phi) = \rho(\phi, \mathrm{id})$ of a Hamiltonian symplectomorphism ϕ that *disjoins* K from itself. If there are no disjoining Hamiltonian symplectomorphisms we define $e(K) = \infty$. If $A \subset M$ is not compact, define

$$e(A) = \sup_{\substack{K \text{ compact} \\ K \subset A}} e(K).$$

We shall see that the displacement energy is a new capacity for the subsets A of a symplectic manifold M.

Theorem 12.3.4 *The displacement energy is a relative symplectic capacity for subsets of* $\mathbb{R}^{2n}$ *and it satisfies*

$$e(B^{2n}(r)) = e(Z^{2n}(r)) = \pi r^2$$

for $r > 0$. *In particular,* $w_G(A) \le e(A) \le \overline{w}_G(A)$ *for every subset* $A \subset \mathbb{R}^{2n}$.

Proof: See page 475. The proof assumes Theorem 12.1.1. □

The proof of relative monotonicity is a straightforward exercise which we leave to the reader. (By considering annuli in $\mathbb{R}^2$ it is easy to see that the monotonicity property does not hold.) The proof of conformality and of the inequality $e(Z^{2n}(r)) \le \pi r^2$ are easy exercises with hints given below. The hard part of the theorem is the estimate $e(B^{2n}(r)) \ge \pi r^2$ and we will give a proof which is due to Lalonde and McDuff [390].

Remark 12.3.5 In his original work, Hofer proved the **energy–capacity inequality**

$$e(A) \geq c_{\mathrm{HZ}}(A)$$

for $A \subset \mathbb{R}^{2n}$ (cf. [314, 316, 527]). Here c_{HZ} is the Hofer–Zehnder capacity introduced in the next section. Since this capacity satisfies (12.1.1) we have

$$c_{\mathrm{HZ}}(A) \geq w_G(A),$$

and this implies the nontriviality axiom for the displacement energy. As we mentioned above, Hofer's original proof of the inequality $e(A) \geq c_{\mathrm{HZ}}(A)$ is based on variational techniques. These use the linear structure of $\mathbb{R}^{2n}$ and so do not extend to arbitrary manifolds. In contrast, the proof which we give below of the slightly weaker inequality $e(A) \geq w_G(A)$ is based on a geometric embedding argument. As the reader will note, some special properties of Euclidean space are used in Step 1 of the proof, and for arbitrary manifolds, these arguments only show that $e(A) \geq \frac{1}{2} w_G(A)$. (See Exercise 12.3.10.) □

Exercise 12.3.6 Let $\phi \in \mathrm{Ham}_c(\mathbb{R}^{2n}, \omega_0)$ and define $\phi_\lambda(z) = \lambda\phi(\lambda^{-1}z)$ for $\lambda > 0$. Prove that $\phi_\lambda \in \mathrm{Ham}^c(\mathbb{R}^{2n}, \omega_0)$ and

$$\rho(\phi_\lambda, \mathrm{id}) = \lambda^2 \rho(\phi, \mathrm{id}).$$

Deduce that the displacement energy satisfies the conformality axiom

$$e(\lambda A) = \lambda^2 e(A)$$

for $A \subset \mathbb{R}^{2n}$ and $\lambda > 0$. **Hint:** Consider the Hamiltonian functions $H_\lambda(z) := \lambda^2 H(\lambda^{-1}z)$ for $\lambda > 0$. □

Exercise 12.3.7 Prove that the displacement energy of a cylinder

$$Z^{2n}(r) = B^2(r) \times \mathbb{R}^{2n-2}$$

in $\mathbb{R}^{2n}$ is bounded above by its Gromov width, i.e.

$$e(Z^{2n}(r)) \leq w_G(Z^{2n}(r)) = \pi r^2.$$

Hint: Identify the open ball $B^2(r)$ with a square and calculate the energy $\rho(\phi, \mathrm{id})$ of a translation disjoining $B^2(r) \times K$ from itself, where $K \subset \mathbb{R}^{2n-2}$ is a compact set. □

These exercises show that the displacement energy satisfies all the axioms of a relative capacity except for the crucial energy–capacity inequality

$$e(A) \geq w_G(A).$$

In view of the definition of w_G and the monotonicity property of the displacement energy it suffices to prove this inequality for balls and this is the statement of Theorem 12.3.9 below. It then follows that the displacement energy of a ball $B^{2n}(r)$ and of the cylinder $Z^{2n}(r)$ in $\mathbb{R}^{2n}$ are exactly equal to their Gromov width πr^2.

The proof of the energy–capacity inequality is based on the explicit construction of a symplectic embedding and on Gromov's Nonsqueezing Theorem. This requires some preparation. We assume that $[0,1] \times M \to \mathbb{R} : (t,z) \mapsto H_t(z)$ is a compactly supported Hamiltonian on an exact symplectic manifold (M,ω) and consider the **graph**

$$\Gamma_H := \{(z, H_t(z), t) \,|\, z \in M, 0 \le t \le 1\}.$$

This graph is a hypersurface in the symplectic manifold $M \times \mathbb{R} \times [0,1]$ with symplectic form

$$\Omega := \omega + dh \wedge dt.$$

Note that this is the differential form of Cartan which we have already encountered in Section 8.1 in the proof of Lemma 8.1.2. The kernel of Cartan's form on the graph of Γ_H is given by the Hamiltonian vector field and so the leaves of the characteristic foliation are precisely the curves

$$t \mapsto (\phi_t(z), H_t(\phi_t(z)), t)$$

which correspond to the Hamiltonian isotopy determined by equation (12.3.1). One can either prove this by thinking of Γ_H as the level set $h - H_t(z) = 0$ or use the following lemma. The proof is an easy exercise which we leave to the reader.

Lemma 12.3.8 *Let $\mathbb{R} \times M \to \mathbb{R} : (t,z) \mapsto H_t(z)$ be a compactly supported Hamiltonian function and denote by $\{\phi_t\}_{t\in\mathbb{R}}$ the Hamiltonian isotopy determined by H via equation* (12.3.1). *Then the map*

$$\psi_H : M \times \mathbb{R}^2 \to M \times \mathbb{R}^2,$$

defined by

$$\psi_H(z,h,t) := (\phi_t(z), h + H_t(\phi_t(z)), t),$$

is a symplectomorphism with respect to the symplectic form $\Omega = \omega + dh \wedge dt$.

Proof: Exercise. □

Theorem 12.3.9 *Let $B = B^{2n}(r) \subset \mathbb{R}^{2n}$ be the closed ball of radius r, centred at the origin. Then*

$$e(B) \ge w_G(B) = \pi r^2.$$

Proof, assuming Theorem 12.1.1: Assume that ϕ_1 is a compactly supported Hamiltonian symplectomorphism which satisfies $\phi_1(B) \cap B = \emptyset$ and denote

$$c := w_G(B) = \pi r^2, \qquad e := \rho(\phi_1, \mathrm{id}).$$

We must prove that $e \ge c$. To see this we shall first construct another symplectomorphism $\phi = \psi \circ \phi_1 \circ \psi^{-1}$ such that $\phi(B) \cap \sqrt{2}B = \emptyset$. This symplectomorphism has the same distance to the identity as ϕ_1. Using ϕ we shall then construct, for every $\varepsilon > 0$, an embedding of the ball $B^{2n+2}(\sqrt{2}r)$ with Gromov width $2c$ into the cylinder $Z^{2n+2}(R) = B^2(R) \times \mathbb{R}^{2n}$ with Gromov width $\pi R^2 = c + e + \varepsilon$. By Gromov's Nonsqueezing Theorem 12.1.1, we must have $2c \le \pi R^2$ and therefore $c \le \pi R^2 - c = e + \varepsilon$. Since $\varepsilon > 0$ can be chosen arbitrarily small the result follows. We construct ϕ and the embedding $B^{2n+2}(\sqrt{2}r) \hookrightarrow Z^{2n+2}(R)$ in three steps.

Step 1 *There exists a symplectomorphism* $\phi \in \mathrm{Ham}_c(\mathbb{R}^{2n}, \omega_0)$ *such that*

$$\phi(B^{2n}(r)) \cap B^{2n}(\sqrt{2}r) = \emptyset, \qquad \rho(\phi, \mathrm{id}) = e.$$

We will construct a symplectomorphism ψ which is the identity on B and which takes $\phi_1(B)$ to a ball which is disjoint from $\sqrt{2}B$. Since energy is invariant under conjugation, we may then take $\phi = \psi \circ \phi_1 \circ \psi^{-1}$.

Define $\phi_\lambda(z) = \lambda\phi_1(z/\lambda)$ for $1 \le \lambda \le 2$ and observe that

$$\phi_\lambda(B^{2n}(r)) \cap \lambda B^{2n}(r) = \emptyset, \qquad 1 \le \lambda \le 2.$$

Since $\lambda \mapsto \phi_\lambda$ is a Hamiltonian isotopy, there exist compactly supported Hamiltonian functions $H_\lambda \in C_0^\infty(\mathbb{R}^{2n})$ for $1 \le \lambda \le 2$ such that

$$\frac{d}{d\lambda}\phi_\lambda = X_{H_\lambda} \circ \phi_\lambda.$$

Now multiply H_λ by a cutoff function $\beta : \mathbb{R}^{2n} \to \mathbb{R}$ which vanishes on $B^{2n}(r)$ and is equal to 1 on $\phi_\lambda(B^{2n}(r))$ for every $\lambda \in [1, 2]$. Consider the Hamiltonian isotopy ψ_λ generated by βH_λ via

$$\frac{d}{d\lambda}\psi_\lambda = X_{\beta H_\lambda} \circ \psi_\lambda, \qquad \psi_1 = \mathrm{id}.$$

The symplectomorphism $\psi = \psi_2 \in \mathrm{Ham}^c(\mathbb{R}^{2n})$ satisfies

$$\psi|_{B^{2n}(r)} = \mathrm{id}, \qquad \psi|_{\phi_1(B^{2n}(r))} = \phi_2 \circ {\phi_1}^{-1}.$$

Hence the symplectomorphism

$$\phi = \psi \circ \phi_1 \circ \psi^{-1}$$

maps $B^{2n}(r)$ to $\phi_2(B^{2n}(r))$ and this set is disjoint from $B^{2n}(2r)$. By Proposition 12.3.1, $\rho(\phi, \mathrm{id}) = \rho(\phi_1, \mathrm{id}) = e$. This proves Step 1.

Step 2 *Let* $U_1, U_2 \subset \mathbb{R}^2$ *be two closed squares of area* c *joined by a line segment* L *and denote*

$$Y := U_1 \cup L \cup U_2.$$

Then there exists a symplectic embedding of the ball $B = B^{2n+2}(\sqrt{2}r)$ *of capacity* $2c$ *into an arbitrarily small neighbourhood of the set*

$$Z_{2c} = U_1 \times B^{2n}(\sqrt{2}r) \;\cup\; L \times B^{2n}(r) \;\cup\; U_2 \times B^{2n}(r).$$

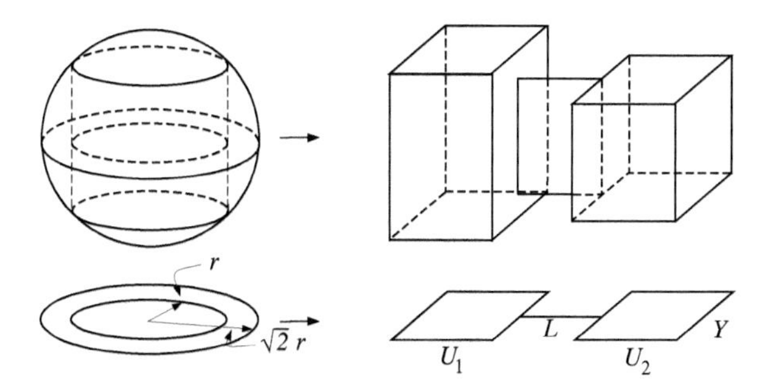

FIG. 12.1. Fibring the ball.

Let $\pi : \mathbb{R}^{2n+2} \to \mathbb{R}^2$ denote the projection associated to the symplectic splitting $\mathbb{R}^{2n+2} = \mathbb{R}^2 \times \mathbb{R}^{2n}$ with coordinates (z_0, z). Restricting this to the ball B, we see that this displays B as a fibred space, with fibres which consist of $2n$-dimensional balls of varying radii. In fact, it is easy to check that $|z_0| < r$ if and only if $w_G\left(\pi^{-1}(z_0) \cap B)\right) > c$.

Now let $\mathcal{N}(Y)$ denote a small neighbourhood of Y, and choose an area-preserving embedding $f : B^2(\sqrt{2}r) \to \mathcal{N}(Y)$ such that $B^2(r)$ is mapped to a small neighbourhood of U_1, a small neighbourhood of the arc $(-\sqrt{2}r, -r)$ is mapped to a neighbourhood of $\partial U_1 \cup L$, and almost all the remaining part of the outer annulus $B^2(\sqrt{2}r) \setminus B^2(r)$ is mapped to U_2. (This is possible because of the great flexibility of area-preserving maps: see Exercise 3.2.8.) The resulting map

$$B \to \mathbb{R}^{2n+2} : (z_0, z) \to (f(z_0), z)$$

is as required. See Fig. 12.1.

Step 3 *For every $\varepsilon > 0$ there exists a symplectic embedding*

$$\mathcal{N}(Z_{2c}) \hookrightarrow B^2(R) \times \mathbb{R}^{2n},$$

where $\pi R^2 = c + e + \varepsilon$.

By Step 1, there exists a Hamiltonian isotopy ϕ_t from $\phi_0 = \mathrm{id}$ to $\phi_1 = \phi$ whose generating time-dependent Hamiltonian $[0,1] \times M \to \mathbb{R} : (t, z) \mapsto H_t(z)$ satisfies

$$\int_0^1 \|H_t\| \, dt \le e + \varepsilon/2.$$

We may assume without loss of generality that $H_t = 0$ for t near 0 and 1 so that the isotopy may be smoothly extended to $t \in \mathbb{R}$ by setting $\phi_t = \mathrm{id}$ for $t \le 0$ and $\phi_t = \phi$ for $t \ge 1$. Define $Q \subset \mathbb{R}^{2n+2}$ to be the graph of H_t:

$$Q = \left\{(H_t(z), t, z) \,|\, t \in [0,1],\, z \in \mathbb{R}^{2n}\right\}.$$

Then the hypersurface Q is contained in the product $A \times \mathbb{R}^{2n}$, where

$$A = \{(h, t) \,|\, \inf H_t \le h \le \sup H_t\}.$$

By assumption on H_t this set has area $e \le \mathrm{area}(A) \le e + \varepsilon/2$. See Fig. 12.2.

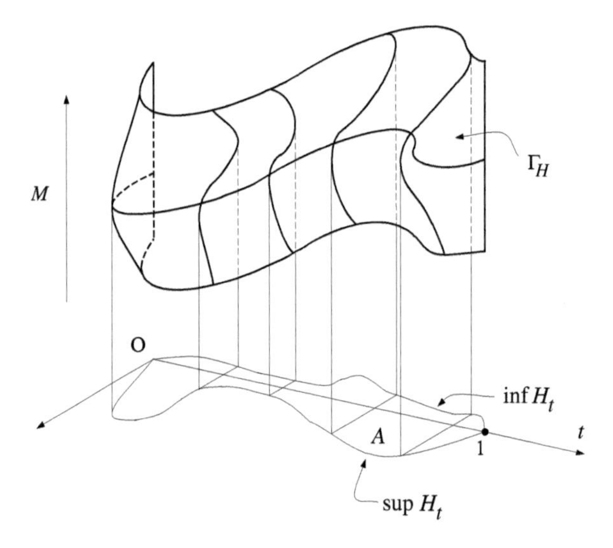

FIG. 12.2. The graph projects onto a set of area $A = e$.

The essential geometric point is the following. By Lemma 12.3.8, the characteristic flow lines on Q are the arcs $t \mapsto (H_t(\phi_t(w)), t, \phi_t(w))$ for $w \in \mathbb{R}^{2n}$. These flow lines begin at the point w at time 0 and end at the point $\phi(w)$ at time 1. We now show how to exploit the fact that ϕ moves the ball $B^{2n}(r)$ to a position disjoint from $B^{2n}(\sqrt{2}r)$ in order to wrap the set Z_{2c} around itself so that it fits into a thin cylinder.

To see this, consider the symplectomorphism $\psi_H : \mathbb{R}^{2n+2} \to \mathbb{R}^{2n+2}$ defined by $\psi_H(h, t, z) := (h + H_t(\phi_t(z)), t, \phi_t(z))$ as in Lemma 12.3.8, where the Hamiltonian H_t is chosen to be identically zero except for t strictly between 0 and 1. We may suppose that the set $Y = U_1 \cup L \cup U_2$ is embedded in $\mathbb{R}^2$ so that the segment L is the line $\{0\} \times [0,1] \subset A$. Then the restriction of ψ_H to Z_{2c} gives rise to an embedding $\psi_H : Z_{2c} \to (U_1 \cup A \cup U_2) \times \mathbb{R}^{2n}$ (denoted by the same symbol). It has the following properties.

(i) The restriction of ψ_H to $U_1 \times B^{2n}(\sqrt{2}r)$ is the identity.
(ii) ψ_H maps $L \times B^{2n}(r)$ to the hypersurface $Q \subset A \times \mathbb{R}^{2n}$.
(iii) The restriction of ψ_H to $U_2 \times B^{2n}(r)$ is the map $\mathrm{id} \times \phi$.

This map obviously extends to a smooth symplectomorphism from some neighbourhood $\mathcal{N}(Z_{2c})$ into a neighbourhood $\mathcal{N}(U_1 \cup A \cup U_2) \times \mathbb{R}^{2n}$. See Fig. 12.3.

Now we shall construct the required embedding $\mathcal{N}(Z_{2c}) \to B^2(R) \times \mathbb{R}^{2n}$. To do this, let X be the annulus obtained from $U_1 \cup A \cup U_2$ by identifying the two squares U_1 and U_2 by a translation. By construction, X has area bounded by $c + e + \varepsilon/2$ and so a neighbourhood can be embedded into the disc $B^2(R)$, where $\pi R^2 = c + e + \varepsilon$. Thus we have a symplectic covering map

$$G : \mathcal{N}(U_1 \cup A \cup U_2) \times \mathbb{R}^{2n} \to B^2(R) \times \mathbb{R}^{2n}$$

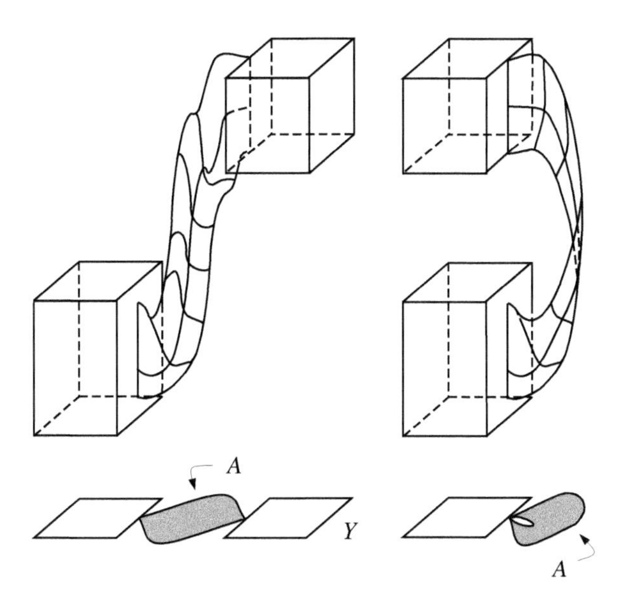

FIG. 12.3. Embedding the ball.

with $G(U_1 \times \mathbb{R}^{2n}) = G(U_2 \times \mathbb{R}^{2n})$. Since, by Step 1, the ball $B^{2n}(\sqrt{2}r)$ is disjoint from $\phi(B^{2n}(r))$, the restriction of G to the image of ψ_H is injective and thus the composition $G \circ \psi_H : \mathcal{N}(Z_{2c}) \to B^2(R) \times \mathbb{R}^{2n}$ satisfies the requirements of Step 3. Combine this with Step 2 to obtain the required symplectic embedding of $B^{2n+2}(\sqrt{2}r)$ into $B^2(R) \times \mathbb{R}^{2n}$. This proves Theorem 12.3.9. □

Proof of Theorem 12.3.4, assuming Theorem 12.1.1: That the displacement energy satisfies the monotonicity axiom is obvious, the conformality axiom is proved in Exercise 12.3.6, the inequality $e(Z^{2n}(r)) \le \pi r^2$ is proved in Exercise 12.3.7, and the inequality $\pi r^2 \le e(B^{2n}(r))$ is proved in Theorem 12.3.9 (assuming Theorem 12.1.1). This proves Theorem 12.3.4. □

Proof of Theorem 12.3.3 for $M = \mathbb{R}^{2n}$, assuming Theorem 12.1.1: Let ϕ be a symplectomorphism of $(\mathbb{R}^{2n}, \omega_0)$, not equal to the identity. Then there exists a ball B of radius r such that $\phi(B) \cap B = \emptyset$. Hence it follows from Theorem 12.3.4 that $\rho(\phi, \mathrm{id}) \ge e(B) = \pi r^2 > 0$. This proves Theorem 12.3.3 for $M = \mathbb{R}^{2n}$. □

We emphasize that the proofs of the main theorems in this section rely on Gromov's Nonsqueezing Theorem which will only be proved at the end of this chapter with the help of the Hofer–Zehnder capacity. The reader should take care to avoid circular arguments. For example, Theorem 12.3.4 also implies Gromov's Nonsqueezing Theorem.

Exercise 12.3.10 Let (M, ω) be a symplectic $2n$-manifold without boundary and let ϕ be a compactly supported Hamiltonian symplectomorphism of (M, ω) that displaces a $2n$-dimensional symplectically embedded ball $B \subset M$ of Gromov width $w_G(B) = \pi r^2$. If $\pi R^2 > \rho(\phi, \mathrm{id}) + c/2$ construct a symplectic embedding $B^{2n+2}(r) \hookrightarrow B^2(R) \times M$.

Use the generalization of the Nonsqueezing Theorem in Remark 12.1.3 to deduce that $e(B) \geq \frac{1}{2}w_G(B)$. Hence show that Theorem 12.3.3 holds for (M, ω). □

Geodesics

The Hofer metric on a neighbourhood of the identity in $\mathrm{Ham}_c(\mathbb{R}^{2n}, \omega_0)$ can be characterized in terms of generating functions. Recall from Lemma 9.2.1 that every compactly supported smooth function $V : \mathbb{R}^{2n} \to \mathbb{R}$ with sufficiently small second derivatives determines a symplectomorphism $\psi_V : \mathbb{R}^{2n} \to \mathbb{R}^{2n}$ via

$$(x_1, y_1) = \psi_V(x_0, y_0) \qquad \iff \qquad \begin{aligned} x_1 - x_0 &= \partial_y V(x_1, y_0), \\ y_1 - y_0 &= -\partial_x V(x_1, y_0). \end{aligned} \tag{12.3.5}$$

The following theorem is due to Bialy–Polterovich [64]. It shows that the Hofer metric is *flat* on a neighbourhood of the identity in $\mathrm{Ham}_c(\mathbb{R}^{2n}, \omega_0)$.

Theorem 12.3.11. (Bialy–Polterovich) *Let $V_1, V_2 : \mathbb{R}^{2n} \to \mathbb{R}$ be compactly supported smooth functions with sufficiently small second derivatives. Then*

$$\rho(\psi_{V_1}, \psi_{V_2}) = \|V_1 - V_2\| \,.$$

Proof: See Bialy–Polterovich [64]. □

We do not prove Theorem 12.3.11 in this book. However, we show in a series of exercises how it can be reduced to a result about geodesics in the Hofer metric (Theorem 12.3.13, which also will not be proved in this book). Theorem 12.3.11 was extended to other symplectic manifolds by Lalonde–McDuff [391].

Let $V : \mathbb{R}^{2n} \to \mathbb{R}$ be a compactly supported smooth function with sufficiently small second derivatives and consider the Hamiltonian isotopy $t \mapsto \psi_{tV}$, which connects ψ_V to the identity. It is easy to see that this path has length $\|V\|$ (Exercise 12.3.15) and that any other path from $\psi_0 = \mathrm{id}$ to $\psi_1 = \psi_V$ which remains C^1-close to the identity for all t has length greater than or equal to $\|V\|$ (Exercise 12.3.14). In other words, the path $\{\psi_{tV}\}_{0 \leq t \leq 1}$ is **locally** length minimizing among all paths from id to ψ_V. However, it is much harder to prove that this path is also **globally** length minimizing. This means that every path $\{\psi_t\}_{0 \leq t \leq 1}$ in $\mathrm{Ham}^c(\mathbb{R}^{2n})$ with $\psi_0 = \mathrm{id}$ and $\psi_1 = \psi_V$ must have length greater than or equal to $\|V\|$, regardless of whether or not ψ_t admits a generating function for every t. That this is true is a deep theorem, and its proof relies on subtle symplectic invariants. The original proof by Bialy–Polterovich [64] uses variational techniques and Hofer's results about the action spectrum. The proof by Lalonde–McDuff [391] uses a generalization of Gromov's Nonsqueezing Theorem which so far has only been proved with the help of J-holomorphic curves. Both proofs of Theorem 12.3.11 rely on the notion of geodesics in the Hofer metric.

There are various possible ways to define the notion of geodesic. In particular, there is a variational theory due to Ustilovsky [636] that meshes well with the definitions given below under suitable nondegeneracy conditions; see also [391]. Here we adopt a modified version of the Bialy–Polterovich approach that applies to all manifolds.

Definition 12.3.12 *Let (M, ω) be a connected (possibly noncompact) symplectic manifold without boundary. A smooth Hamiltonian isotopy*

$$[t_0, t_1] \times M \to M : (t, p) \mapsto \phi_t(p)$$

with compact support is called

- **regular** *if $\frac{d}{dt}\phi_t \neq 0$ for all $t \in [t_0, t_1]$;*
- *a* **minimal Hofer geodesic** *if $\mathcal{L}(\{\phi_t\}_{t_0 \le t \le t_1})$ is the minimum of the Hofer lengths of all compactly supported Hamiltonian isotopies from ϕ_{t_0} to ϕ_{t_1} that are isotopic to $\{\phi_t\}_{t_0 \le t \le t_1}$ with fixed endpoints;*
- *a* **Hofer geodesic** *if each $t \in [t_0, t_1]$ has a neighbourhood I such that $\{\phi_t\}_{t \in I}$ is a minimal geodesic;*
- **quasi-autonomous** *if there exist points $z^\pm \in M$ such that*

$$H_t(z^+) = \max_M H_t, \qquad H_t(z^-) = \min_M H_t \tag{12.3.6}$$

  for every $t \in [t_0, t_1]$;
- **locally quasi-autonomous** *if each $t \in [t_0, t_1]$ has a neighbourhood I such that $\{\phi_t\}_{t \in I}$ is quasi-autonomous.*

Theorem 12.3.13 *Let (M, ω) be either $(\mathbb{R}^{2n}, \omega_0)$ or a closed symplectic manifold, and let $[t_0, t_1] \times M \to M : (t, z) \mapsto \phi_t(z)$ be a compactly supported regular Hamiltonian isotopy with generating Hamiltonian*

$$[t_0, t_1] \times M \to \mathbb{R} : (t, z) \mapsto H_t(z).$$

Then the following holds.

(i) *The Hamiltonian isotopy $\{\phi_t\}$ is a geodesic for the Hofer metric if and only if it is locally quasi-autonomous.*

(ii) *If the Hamiltonian isotopy $\{\phi_t\}$ is a geodesic for the Hofer metric then it is a global minimum of the length functional on sufficiently small time intervals, i.e. there exists an $\varepsilon > 0$ such that*

$$\mathcal{L}(\{\psi_s\}_{t \le s \le t+\varepsilon}) \ge \mathcal{L}(\{\phi_s\}_{t \le s \le t+\varepsilon})$$

for every $t \in [t_0, t_1 - \varepsilon]$ and every compactly supported Hamiltonian isotopy $[t, t+\varepsilon] \to \mathrm{Ham}_c(M, \omega) : s \mapsto \psi_s$ connecting $\psi_t = \phi_t$ to $\psi_{t+\varepsilon} = \phi_{t+\varepsilon}$.

Proof: See Bialy–Polterovich [64] for the case $M = \mathbb{R}^{2n}$, Lalonde–McDuff [391] for the proof of (i) for general closed M, and McDuff [460] for (ii). □

Explicit curve shortening procedures show that every regular geodesic must be locally quasi-autonomous. The converse statement that a sufficiently short quasi-autonomous path minimizes length within its homotopy class is much harder to prove. The argument used by Bialy–Polterovich in the case $M = \mathbb{R}^{2n}$ in fact also proves that such paths are globally length minimizing in the sense of (ii). However, extra arguments are needed to prove (ii) for general closed M; McDuff [460] uses variants of the Nonsqueezing Theorem, while Oh [509] bases his arguments on chain level Floer theory.

Once Theorem 12.3.13 is established, Theorem 12.3.11 is a fairly easy consequence. The key observation is that for every Hamiltonian isotopy ϕ_t which is sufficiently close to the identity in the C^1 topology, the generating functions V_t in (12.3.5) are related to Hamiltonian functions H_t in (12.3.1) via the Hamilton–Jacobi equation. It then follows that the critical points of H_t agree with the critical points of the time derivative $\partial_t V_t$. This is the content of Exercise 12.3.14. The next exercises show how to deduce Theorem 12.3.11 from Theorem 12.3.13.

Exercise 12.3.14 Let $[0,1] \to \mathrm{Ham}_c(\mathbb{R}^{2n}, \omega_0) : t \mapsto \phi_t$ be a Hamiltonian isotopy starting at the identity $\phi_0 = \mathrm{id}$ and generated by a compactly supported Hamiltonian function $[0,1] \times \mathbb{R}^{2n} \to \mathbb{R} : (t,z) \mapsto H_t(z)$. Assume $\|d\phi_t(z) - \mathbb{1}\| \le \frac{1}{2}$ for all z and t and denote by $V_t \in C_0^\infty(R^{2n})$ the generating function of ϕ_t in Lemma 9.2.1. Thus the vectors $x_0, y_0, x_t, y_t \in \mathbb{R}^n$ satisfy $(x_t, y_t) = \phi_t(x_0, y_0)$ if and only if

$$x_t - x_0 = \partial_y V_t(x_t, y_0), \qquad y_t - y_0 = -\partial_x V_t(x_t, y_0). \tag{12.3.7}$$

Prove that V_t satisfies the Hamilton–Jacobi equation

$$\partial_t V_t(x,y) = H_t(x, y - \partial_x V_t(x,y)), \qquad V_0 = 0. \tag{12.3.8}$$

Deduce that $\|H_t\| = \|\partial_t V_t\|$. Prove that the length of the path $s \mapsto \psi_s$ o the interval $[0,t]$ is bounded below by the Hofer norm of V_t, i.e.

$$\mathcal{L}(\{\phi_s\}_{0 \le s \le t}) \ge \|V_t\|. \tag{12.3.9}$$

This proof of the estimate (12.3.9) works only if ψ_s is C^1-close to the identity for all s. **Hint:** Prove that the generating functions V_t are related to the symplectic action by the formula

$$\langle y_0, x_t - x_0 \rangle - V_t(x_t, y_0) = \int_0^t \Big(\langle y_s, \partial_s x_s \rangle - H_s(x_s, y_s) \Big)\, ds, \tag{12.3.10}$$

where $(x_s, y_s) := \phi_s(x_0, y_0)$. Namely, use equation (12.3.10) to define V_t and show that this function satisfies (12.3.7). (See also Lemma 12.5.4 below.) Differentiate equation (12.3.10) with respect to t. □

Exercise 12.3.15 Consider the path $[0,1] \to \mathrm{Ham}^c(\mathbb{R}^{2n}, \omega_0) : t \mapsto \psi_t$ with generating functions $V_t = tV$. Prove that

$$\mathcal{L}(\{\phi_t\}_{0 \le t \le 1}) = \|V\|.$$

Hint: The Hamilton–Jacobi equation takes the form $V(x_t, y_0) = H_t(x_t, y_t)$. Alternatively, use Corollary 9.3.14 to prove that the maxima and minima of V agree with the maxima and minima of H_t for every t. □

Exercise 12.3.16 Let ϕ_1, ϕ_2 and V_1, V_2 be as in Theorem 12.3.11. Denote by V_{21} the generating function of $\phi_{21} := \phi_2 \circ \phi_1^{-1}$. Prove that

$$V_{21}(x_2, y_1) = V_2(x_2, y_0) - V_1(x_1, y_0) + \langle y_1 - y_0, x_2 - x_1 \rangle,$$

where $(x_1, y_1) := \phi_1(x_0, y_0)$ and $(x_2, y_2) := \phi_2(x_0, y_0)$. By examining the critical points of the functions V_{21} and $V_2 - V_1$, prove that

$$\|V_{21}\| = \|V_2 - V_1\|.$$

Deduce that Theorem 12.3.11 is equivalent to the assertion

$$\rho(\mathrm{id}, \phi) = \|V\|,$$

whenever V is the generating function of ϕ. □

Comments on the literature

Hameomorphisms

The Hofer metric ρ provides a kind of C^{-1} topology on the group of Hamiltonian symplectomorphisms; it involves the sup-norm of the time-dependent Hamiltonian H while the symplectomorphism ϕ_H depends on the first derivatives. However, the Hofer metric on $\mathrm{Ham}_c(M, \omega)$ is not controlled by its C^0 topology (i.e. C^0 convergence does not imply convergence in the Hofer metric). Thus it is interesting to examine the completion of the group $\mathrm{Ham}_c(M, \omega)$ with respect to the sum of the sup-norm and the Hofer distance.

Definition 12.3.17 *Let (M, ω) be a compact connected symplectic manifold. A continuous map $[0,1] \times M \to M : (t, p) \mapsto \phi_t(p)$ is called a* **continuous Hamiltonian isotopy** *if there exists a sequence of smooth Hamiltonian isotopies $[0,1] \times M \to M : (t,p) \mapsto \phi_t^\nu(p)$ with $\phi_0^\nu = \mathrm{id}$ and a measurable function $[0,1] \times M \to \mathbb{R} : (t,p) \mapsto H_t(p)$ such that H_t is continuous for each t, ϕ_t is a homeomorphism for each t, and*

$$\lim_{\nu \to \infty} \max_{t,p} d(\phi_t^\nu(p), \phi_t(p)) = 0, \qquad \lim_{\nu \to \infty} \int_0^1 \|H_t^\nu - H_t\| \, dt = 0. \qquad (12.3.11)$$

Here, $[0,1] \times M \to M : (t,p) \mapsto H_t^\nu(p)$ denotes the unique Hamiltonian function that generates ϕ_t^ν such that the H_t^ν have mean value zero. The term $\|H_t^\nu - H_t\|$ in (12.3.11) denotes the Hofer norm introduced in (12.3.2) and d denotes the distance function of a Riemannian metric on M.

This definition is due to Oh and Müller [510] who raised the question whether H is uniquely determined by ϕ. In [649], Viterbo proved uniqueness under the assumption that convergence of H^ν in the Hofer norm is replaced by convergence in the sup-norm (and so H is continuous). In [85], Buhovski and Seyfaddini settled the uniqueness question for H in the affirmative. They also give an example of a continuous function $H : [0,1] \times M \to \mathbb{R}$ that does not generate a continuous

Hamiltonian isotopy and an example of a continuous isotopy that is the C^0 limit of smooth Hamiltonian isotopies but is not a continuous *Hamiltonian* isotopy.

Call a homeomorphism $\phi : M \to M$ a **hameomorphism** if there exists a continuous Hamiltonian isotopy $\{\phi_t\}_{0 \le t \le 1}$ such that $\phi_1 = \phi$. It was noted by Oh and Müller that hameomorphisms form a group. Denote this by $\mathrm{Hameo}(M, \omega)$. When M has dimension two, $\mathrm{Hameo}(M, \omega)$ is a normal subgroup of the group of area and orientation-preserving homeomorphisms. It is an open question whether these two groups agree. An example of an area and orientation-preserving homeomorphism of the open unit disc $\mathbb{D} \subset \mathbb{C}$ is the map $\phi(re^{i\theta}) := re^{i\theta + f(r)}$, where $f : (0, 1] \to \mathbb{R}$ is a smooth function such that $\lim_{r \to 0} f(r) = \infty$ and $f(r) = 0$ for r near 1. This map exhibits *'infinite rotation'* near the origin and it is an open question whether it is a hameomorphism. If it is not, then the group of compactly supported area-preserving homeomorphisms of $\mathbb{D}$ is not simple. Note that the corresponding group of smooth area-preserving maps is not simple because the Calabi homomorphism is nontrivial (Exercise 10.3.6).

Loops of Hamiltonian symplectomorphisms

In [528], Polterovich examined the Hofer lengths of loops of Hamiltonian symplectomorphisms. He related the minimal Hofer length to curvature estimates for symplectic connections of an associated Hamiltonian fibration over the 2-sphere (see Section 6.3) and found lower bounds via J-holomorphic sections of this fibration. Similar results were independently obtained by Seidel [569, 571].

The Hofer diameter

The **Hofer diameter conjecture** asserts that $\mathrm{Ham}_c(M, \omega)$ has infinite diameter with respect to Hofer's metric for every nonempty symplectic manifold (M, ω) of positive dimension. For $(M, \omega) = (\mathbb{R}^{2n}, \omega_0)$ this follows from the energy–capacity inequality of Theorem 12.3.9. On the other hand, Sikorav [586] proved that if $U \subset \mathbb{R}^{2n}$ is a star-shaped domain and $\phi \in \mathrm{Ham}_c(\mathbb{R}^{2n}, \omega_0)$ is generated by a Hamiltonian function with support in $[0, 1] \times U$, then the Hofer distance $\rho(\phi, \mathrm{id})$ can be estimated above by a constant times the displacement energy $e(U)$. For closed symplectic manifolds of dimension $2n \ge 2$ the conjecture was confirmed by Polterovich [530] in the case of the 2-sphere, by Schwarz [568] in the symplectically aspherical case, by McDuff [464] in the case of minimal Chern number $N \ge n + 1$ and in the case of vanishing spherical 3-point Gromov–Witten invariants (unless the manifold is monotone with $N \le n$ and $\mathrm{rank} H_2(M) = 1$), and by Usher [634] for symplectic manifolds that admit nonconstant Hamiltonian functions all of whose contractible periodic orbits are constant. In [633], Usher found many examples that satisfy this condition, including positive genus surfaces, many symplectic 4-manifolds (such as K3-surfaces and Gompf's examples in Section 7.2), and products of these with other symplectic manifolds.

The Hofer metric for Lagrangian submanifolds

There is an analogue of the Hofer length for exact paths of Lagrangian submanifolds. In [102], Chekanov proved that the analogue of the Hofer metric for compact Lagrangian submanifolds of (M, ω) is nondegenerate. When (M, ω) is noncompact this requires the condition that the symplectic form be **tame**, i.e. there exists an ω-compatible almost complex structure $J \in \mathcal{J}(M, \omega)$ such that the Riemannian metric $\omega(\cdot, J\cdot)$ is geometrically bounded in the sense that the sectional curvature is bounded above and the injectivity radius is bounded below. (Otherwise there are counterexamples: see Chekanov [102].) In [481], Milinković proved a Lagrangian analogue of Theorem 12.3.11. In [18], Akveld and Salamon extended Polterovich's ideas [528] to the Lagrangian setting. They used J-holomorphic discs to find lower bounds for the Hofer lengths of some exact Lagrangian loops; e.g. they showed that a half rotation of the equator in S^2 minimizes the Hofer length in its homotopy class. In [368], Khanevsky proved that the space of *diameters of the disc* (i.e. arcs that divide the disc into two regions of equal area) has infinite Hofer diameter. It is an open question whether the space of *equators of the* 2*-sphere* (i.e. embedded circles that divide S^2 into two regions of equal area) has infinite Hofer diameter: see Problem 32 in Chapter 14.

12.4 The Hofer–Zehnder capacity

In this section we introduce the Hofer–Zehnder capacity and show how it implies Gromov's nonsqueezing theorem 12.1.1 and the Weinstein conjecture for hypersurfaces in $\mathbb{R}^{2n}$ (see Theorem 12.4.6 below). The Hofer–Zehnder capacity was defined by Hofer and Zehnder in [324]. It is based on the properties of periodic orbits of Hamiltonian systems, and the main technical ingredient in establishing its nontriviality is an existence theorem for nonconstant periodic orbits for a certain class of time-independent Hamiltonian differential equations on $\mathbb{R}^{2n}$ (Theorem 12.4.3). The proof of Theorem 12.4.3 is deferred to Section 12.5.

Let (M, ω) be a connected symplectic manifold without boundary, and denote the set of all nonnegative Hamiltonian functions whose support is a proper compact subset of M and which attain their maximum on some open set by

$$\mathcal{H}(M) := \left\{ H \in C_0^\infty(M) \,\middle|\, \begin{array}{l} H \geq 0,\ \mathrm{supp}(H) \subsetneq M, \\ H|_U = \sup H \text{ for some open set } U \end{array} \right\}.$$

For $H \in \mathcal{H}(M)$ let $\phi_H^t \in \mathrm{Symp}^c(M, \omega)$ denote the flow of the Hamiltonian vector field X_H defined by $\iota(X_H)\omega = dH$. Thus $\partial_t \phi_H^t = X_H \circ \phi_H^t$ and $\phi_H^0 = \mathrm{id}$. A solution $x(t) = \phi_H^t(x_0)$ of Hamilton's equation is called a T**-periodic orbit** if $x(t+T) = x(t)$ for every $t \in \mathbb{R}$. Call a function $H \in \mathcal{H}(M)$ **admissible** if the corresponding Hamiltonian flow has no nonconstant T-periodic orbits with period $T \leq 1$. In other words, every nonconstant periodic orbit has period greater than one. Denote the set of admissible Hamiltonian functions by

$$\mathcal{H}_{\mathrm{ad}}(M, \omega) := \left\{ H \in \mathcal{H}(M) \,|\, \text{if } 0 < T \leq 1 \text{ and } \phi_H^T(x_0) = x_0 \text{ then } X_H(x_0) = 0 \right\}.$$

The following lemma shows that for every $H \in \mathcal{H}(M)$ the function εH is admissible for $\varepsilon > 0$ sufficiently small. Roughly speaking, if a vector field is small then the solutions are slow and hence the periods must be long.

Lemma 12.4.1 *Let $f : \mathbb{R}^m \to \mathbb{R}^m$ be continuously differentiable and assume $x(t) = x(t+T) \in \mathbb{R}^m$ is a periodic solution of the differential equation $\dot{x} = f(x)$. If $T \cdot \sup_x \|df(x)\| < 1$ then $x(t)$ is constant.*

Proof: Since $x(0) = x(T)$, an easy calculation shows that

$$\dot{x}(t) = \int_0^t \frac{s}{T}\ddot{x}(s)\,ds + \int_t^T \frac{s-T}{T}\ddot{x}(s)\,ds.$$

This implies that

$$|\dot{x}(t)| \le \int_0^T |\ddot{x}(s)|\;ds \le \sqrt{T}\,\|\ddot{x}\|_{L^2[0,T]}\,,$$

and hence $\|\dot{x}\|_{L^2} \le T\,\|\ddot{x}\|_{L^2}$. Now denote $\varepsilon := \sup_x \|df(x)\|$ and observe that $|\ddot{x}| \le \|df(x)\| \cdot |\dot{x}| \le \varepsilon\,|\dot{x}|$. Hence

$$\|\ddot{x}\|_{L^2[0,T]} \le \varepsilon\,\|\dot{x}\|_{L^2[0,T]} \le \varepsilon T\,\|\ddot{x}\|_{L^2[0,T]}\,.$$

Since $\varepsilon T < 1$ it follows that $\ddot{x}(t) \equiv 0$. Hence $\dot{x}(t)$ is constant and, since $x(t)$ is periodic, it follows that $x(t)$ is constant. This proves Lemma 12.4.1. □

The **Hofer–Zehnder capacity** of (M, ω) is defined by

$$c_{\mathrm{HZ}}(M,\omega) := \sup_{H\in\mathcal{H}_{\mathrm{ad}}(M,\omega)} \|H\|\,. \tag{12.4.1}$$

Here $\|H\|$ is the Hofer norm, introduced in equation (12.3.2) in Section 12.3. The Hofer–Zehnder capacity of a manifold with boundary is understood as the Hofer–Zehnder capacity of its interior $M\backslash\partial M$. Lemma 12.4.1 shows that $c_{\mathrm{HZ}}(M,\omega) > 0$ for every nonempty symplectic manifold (M, ω) of positive dimension. The following theorem is due to Hofer and Zehnder [324].

Theorem 12.4.2 *The map $(M,\omega) \mapsto c_{\mathrm{HZ}}(M,\omega)$ satisfies the monotonicity, conformality, and normalization axioms of a symplectic capacity. Moreover,*

$$c_{\mathrm{HZ}}(B^{2n}(r)) = c_{\mathrm{HZ}}(Z^{2n}(r)) = \pi r^2$$

for every $r > 0$.

Proof: See page 483. □

The proof of Theorem 12.4.2 rests on the following existence result for periodic orbits of Hamiltonian differential equations in $\mathbb{R}^{2n}$.

Theorem 12.4.3 *Assume $H \in \mathcal{H}(Z^{2n}(1))$ with $\sup H > \pi$. Then the Hamiltonian flow of H has a nonconstant periodic orbit of period 1.*

Proof: See Section 12.5, page 501. □

The 1-periodic orbits of H are the critical points of the symplectic action functional

$$\mathcal{A}_H(z) = \int_0^1 \Big(\langle y, \dot{x} \rangle - H(x,y) \Big)\, dt$$

on the loop space of $\mathbb{R}^{2n}$. (See Lemma 1.1.8.) Here we think of a loop as a smooth map $z : \mathbb{R} \to \mathbb{R}^{2n}$ such that $z(t+1) = z(t)$. Hofer and Zehnder [324] use techniques from the calculus of variations to establish the existence of a nontrivial critical point (and hence of a nonconstant 1-periodic solution) under the above assumptions. Their method is an adaptation and simplification of an argument due to Viterbo [646]. We shall postpone the proof to Section 12.5, where we explain a modified version of the argument by Hofer and Zehnder which is based on the discrete symplectic action.

Theorem 12.4.3 is the main technical result of the present chapter and the main ingredient in the proof that c_{HZ} satisfies the axioms of a symplectic capacity. It therefore gives rise to a proof of Gromov's Nonsqueezing Theorem 12.1.1. We will see that the Weinstein conjecture for hypersurfaces, as stated in Section 1.2, is also an immediate corollary (Theorem 12.4.6). Thus there is a deep interconnection between symplectic topology and Hamiltonian dynamics, which is yet to be fully understood.

Proof of Theorem 12.4.2, assuming Theorem 12.4.3: We prove monotonicity. Let (M_1, ω_1) and (M_2, ω_2) be symplectic manifolds of the same dimension $2n$ and let $\psi : M_1 \to M_2$ be a symplectic embedding. If $H_1 : M_1 \to \mathbb{R}$ is a compactly supported Hamiltonian function, then there is a unique compactly supported function $H_2 : M_2 \to \mathbb{R}$ such that H_2 vanishes on $M_2 \setminus \psi(M_1)$ and $H_1 = H_2 \circ \psi$. Since H_1 is compactly supported, the function H_2 is smooth. Since ψ intertwines the Hamiltonian flows of H_1 and H_2 there is a one-to-one correspondence of nonconstant periodic orbits of these flows. Hence

$$\begin{aligned} c_{\mathrm{HZ}}(M_1, \omega_1) &= \sup_{H_1 \in \mathcal{H}_{\mathrm{ad}}(M_1, \omega_1)} \|H_1\| \\ &= \sup_{\substack{H_2 \in \mathcal{H}_{\mathrm{ad}}(M_2, \omega_2) \\ \mathrm{supp}(H_2) \subset \psi(M_1)}} \|H_2\| \\ &\le c_{\mathrm{HZ}}(M_2, \omega_2). \end{aligned}$$

This proves monotonicity. Conformality follows from the identity

$$\mathcal{H}_{\mathrm{ad}}(M, \lambda\omega) = \{\lambda H \,|\, H \in \mathcal{H}_{\mathrm{ad}}(M, \omega)\}\,.$$

In fact the Hamiltonian vector field of H with respect to ω agrees with the Hamiltonian vector field of λH with respect to $\lambda\omega$.

We shall now prove the inequality $c_{\mathrm{HZ}}(B^{2n}(1)) \ge \pi$. Let $\varepsilon > 0$ and choose a smooth function $f : [0,1] \to \mathbb{R}$ such that

$$\begin{aligned} -\pi < f'(r) \le 0, &\quad \text{for all } r, \\ f(r) = \pi - \varepsilon, &\quad \text{for } r \text{ near } 0, \\ f(r) = 0, &\quad \text{for } r \text{ near } 1. \end{aligned}$$

Define $H(z) = f(|z|^2)$ for $z \in B^{2n}(1)$. Then $H \in \mathcal{H}(B^{2n}(1))$ and $\|H\| = \pi - \varepsilon$. We must prove that H is admissible. But the orbits of the Hamiltonian flow are easy to calculate explicitly. The Hamiltonian differential equation of H is of the form

$$\dot{x} = 2f'(|z|^2)y, \qquad \dot{y} = -2f'(|z|^2)x,$$

and it follows that $r = |z(t)|^2$ is constant along the solutions. In complex notation $z = x+iy$ the differential equation is of the form $\dot{z} = -2if'(r)z$ and the solutions $z(t) = e^{-2if'(r)t}z_0$ are all periodic. They are nonconstant whenever $f'(r) \neq 0$ and in this case the period is $T = \pi/|f'(r)| > 1$. Hence for every $\varepsilon > 0$ there is an admissible Hamiltonian function $H \in \mathcal{H}(B^{2n}(1))$ with $\|H\| = \pi - \varepsilon$ and thus

$$c_{\mathrm{HZ}}(B^{2n}(1)) \geq \pi.$$

Now, Theorem 12.4.3 asserts that for every $H \in \mathcal{H}(Z^{2n}(1))$ with $\|H\| > \pi$ the corresponding Hamiltonian flow has a nonconstant periodic orbit of period 1. Hence no such function is admissible, which implies $c_{\mathrm{HZ}}(Z^{2n}(1)) \leq \pi$. By the monotonicity axiom, we have $c_{\mathrm{HZ}}(B^{2n}(1)) = c_{\mathrm{HZ}}(Z^{2n}(1)) = \pi$, and this proves Theorem 12.4.2. □

An immediate corollary to the fact that c_{HZ} satisfies the axioms of a capacity is the following proof of Gromov's Nonsqueezing Theorem.

Proof of Theorem 12.1.1, assuming Theorem 12.4.3: We have shown in Theorem 12.4.2, using Theorem 12.4.3, that c_{HZ} is a symplectic capacity satisfying

$$c_{\mathrm{HZ}}(B^{2n}(r)) = c_{\mathrm{HZ}}(Z^{2n}(r)) = \pi r^2$$

for every $r > 0$. Assume that there exists a symplectic embedding

$$\psi : B^{2n}(r) \hookrightarrow Z^{2n}(R).$$

Then

$$\pi r^2 = c_{\mathrm{HZ}}(B^{2n}(r)) = c_{\mathrm{HZ}}(\psi(B^{2n}(r))) \leq c_{\mathrm{HZ}}(Z^{2n}(R)) = \pi R^2.$$

The first equality follows from the normalization axiom for c_{HZ} and the (third) inequality follows from the monotonicity axiom. We conclude that $r \leq R$. This proves Theorem 12.1.1 □

Another consequence of Theorem 12.4.2 is an 'almost existence' theorem for periodic orbits of the characteristic flow on a hypersurface Q in $\mathbb{R}^{2n}$. This theorem is due to Hofer and Zehnder [324]. We begin with a brief discussion of hypersurfaces in Euclidean space.

Lemma 12.4.4 *Let Q be a compact connected hypersurface in $\mathbb{R}^m$. Then Q is oriented and $\mathbb{R}^m \setminus Q$ has two connected components. One of these connected components is bounded.*

Proof: The proof is by standard arguments in differential topology and we only sketch the main points. The idea is to define, for every point $x \notin Q$, a map

$$f_x : Q \to S^{m-1}$$

by

$$f_x(\xi) := \frac{\xi - x}{|\xi - x|}$$

and denote

$$U_0 := \{x \in \mathbb{R}^m \setminus Q \mid \deg(f_x) = 0 \pmod 2\},$$
$$U_1 := \{x \in \mathbb{R}^m \setminus Q \mid \deg(f_x) = 1 \pmod 2\}.$$

Both sets are nonempty and connected. To see the former, choose a curve crossing Q and prove that it passes from U_0 to U_1 or vice versa. To prove connectedness, prove first that Q divides a tubular neighbourhood V of Q into two parts

$$V_0 := V \cap U_0, \qquad V_1 := V \cap V_1,$$

which are path-connected because Q is path-connected. Now any point in U_0 can obviously be moved to a point in V_0 and similarly for U_1. This proves that $\mathbb{R}^m \setminus Q$ has two components. Note in fact that U_1 is bounded and U_0 is unbounded. Finally, prove that Q is oriented by choosing an outward pointing normal vector field. □

Theorem 12.4.5 *Let Q be a compact $(2n-1)$-dimensional manifold without boundary, and let $\iota : Q \times [0,1] \to \mathbb{R}^{2n}$ be any embedding. Then for a dense set of parameters $s \in [0,1]$ the Hamiltonian flow on $\iota(Q \times s)$ has a periodic orbit.*

Proof, assuming Theorem 12.4.3: By Lemma 12.4.4 the set $\mathbb{R}^{2n} \setminus \operatorname{Im} \iota$ has two connected components U_0 and U_1. Denote by U_0 the unbounded component and by U_1 the bounded component and assume without loss of generality that

$$\partial U_0 = \iota(Q \times 0), \qquad \partial U_1 = \iota(Q \times 1).$$

By Theorem 12.4.2, the bounded open set

$$W_1 = \mathbb{R}^{2n} \setminus \overline{U_0} = U_1 \cup \iota(Q \times (0,1])$$

has finite Hofer–Zehnder capacity $c_{\mathrm{HZ}}(W_1) < \infty$. (Note that this involves the hard part of Theorem 12.4.2, namely the assertion $c_{\mathrm{HZ}}(Z^{2n}(r)) < \infty$, which uses Theorem 12.4.3, which will only be proved in Section 12.5.) Choose a function

$$f : [0,1] \to \mathbb{R}$$

such that

$$\begin{array}{ll} f(s) \geq 0, & \text{for all } s, \\ f(s) = 0, & \text{for } s \text{ near } 0, \\ f(s) = c_1 > c_{\mathrm{HZ}}(W_1), & \text{for } s \text{ near } 1. \end{array}$$

Define $H : \mathbb{R}^{2n} \to \mathbb{R}$ by

$$H(z) = \begin{cases} 0, & \text{for } z \in U_0, \\ f(s), & \text{for } z \in \iota(Q \times s), \\ c_1, & \text{for } z \in U_1. \end{cases}$$

Then H is nonnegative with compact support $\operatorname{supp}(H) \subset W_1$, and hence

$$H \in \mathcal{H}(W_1).$$

But since $\|H\| > c_{\mathrm{HZ}}(W_1)$ it follows from the definition of the Hofer–Zehnder capacity that the Hamiltonian flow of X_H has a nonconstant periodic orbit (of period $T \leq 1$). Any such orbit must lie on a level set of H, where $H \neq 0$ and $H \neq c_1$. These level sets are given by $\iota(Q \times s)$ and hence one of these must carry a periodic orbit. By varying f this argument gives the existence of a periodic orbit in $\iota(Q \times [s_0, s_1])$ for any $s_0 < s_1$. This proves Theorem 12.4.5. □

Theorem 12.4.5 was strengthened by Struwe [602], who proved the existence of a periodic orbit for a set of parameters s of measure 1. As a corollary to Theorem 12.4.5 we obtain a proof of the Weinstein conjecture for hypersurfaces of Euclidean space. The result is due to Viterbo [643] and the proof via the Hofer–Zehnder capacity is, of course, due to Hofer and Zehnder [324]. Counterexamples by Herman [307] and Ginzburg [263,264,265] show that the contact condition in Theorem 12.4.6 cannot be removed when $2n \geq 6$. A C^2 counterexample in $\mathbb{R}^4$ was constructed by Ginzburg–Gurel [266].

Theorem 12.4.6 (Weinstein conjecture) *If $Q \subset \mathbb{R}^{2n}$ is a hypersurface of contact type then its characteristic foliation has a closed orbit.*

Proof, assuming Theorem 12.4.3: By Proposition 3.5.31, there exists a vector field $X : V \to \mathbb{R}^{2n}$ defined on a neighbourhood of Q such that X is transverse to Q and $\mathcal{L}_X\omega_0 = \omega_0$. Denote by $\phi_s : V_s \to V$ the flow of X and consider the foliation of a neighbourhood of Q by the surfaces

$$Q_s := \phi_s(Q), \qquad -\delta < s < \delta.$$

Since $\phi_s^*\omega_0 = e^s\omega_0$ we have, for every $\zeta \in T_zQ$,

$$\omega_0(\zeta, \eta) = 0 \; \forall \, \eta \in T_zQ \quad \Longleftrightarrow \quad \omega_0(d\phi_s(z)\zeta, \eta) = 0 \; \forall \, \eta \in T_{\phi_s(z)}Q_s.$$

In other words, ϕ_s maps the characteristic foliation of Q onto the characteristic foliation of Q_s. Now it follows from Theorem 12.4.5 that for some $s \in (-\delta, \delta)$ the hypersurface Q_s carries a closed characteristic $z(t) = z(t+1)$. The curve $t \mapsto \phi_{-s}(z(t))$ is the corresponding closed characteristic on Q. □

Exercise 12.4.7 Show, by an elementary argument which does not rely on Theorem 12.4.3, that for every closed symplectic 2-manifold (Σ, σ) the Hofer–Zehnder capacity agrees with the area, i.e.

$$\mathrm{c_{HZ}}(\Sigma, \sigma) = \int_\Sigma \sigma.$$

Hint: We showed in Theorem 12.4.2 that $\mathrm{c_{HZ}}(B^2(r)) \geq \pi r^2$. Hence the inequality $\mathrm{c_{HZ}}(\Sigma) \geq \mathrm{area}(\Sigma)$ follows from the monotonicity axiom and the existence of a suitable embedding $\mathrm{int}\, B^2(r) \hookrightarrow \Sigma$, where $\pi r^2 = \mathrm{area}(\Sigma)$. To prove the other inequality one must show that for every Hamiltonian H with $\|H\| > \mathrm{area}(\Sigma)$ the corresponding Hamiltonian flow has a nonconstant T-periodic orbit for some $T \leq 1$. To see this write ω in the form $dH \wedge \alpha$ on the set of regular points of H, where α is a suitable 1-form, and estimate the area. Note that the level sets of H are compact, and so all orbits are periodic: it is just a question of estimating their periods. □

Exercise 12.4.8 Find a symplectic form ω on the 4-torus $M := \mathbb{T}^4 = \mathbb{R}^4/\mathbb{Z}^4$ and a nonconstant smooth Hamiltonian function $H : \mathbb{T}^4 \to \mathbb{R}$ whose Hamiltonian flow does not have any nonconstant periodic solutions. Deduce that $(\mathbb{T}^4, \omega)$ has infinite Hofer–Zehnder capacity. □

Exercise 12.4.9 For connected symplectic manifolds (M, ω) that are not simply connected the Hofer–Zehnder capacity can be modified as follows. Call a Hamiltonian function $H \in \mathcal{H}(M)$ **weakly admissible** if every *contractible* periodic solution with period $T \leq 1$ is constant. Denote the space of weakly admissible Hamiltonian functions by $\mathcal{H}_{\mathrm{ad},0}(M, \omega)$. Define the **contractible Hofer–Zehnder capacity** of (M, ω) by

$$\mathrm{c_{HZ,0}}(M, \omega) := \sup_{H \in \mathcal{H}_{\mathrm{ad},0}(M,\omega)} \|H\| \geq \mathrm{c_{HZ}}(M, \omega).$$

Prove that $\mathrm{c_{HZ,0}}(\Sigma, \sigma) = \infty$ for every closed symplectic 2-manifold (Σ, σ) of genus $g \geq 1$.
Hint: Construct a Hamiltonian function whose nonconstant periodic solutions are all noncontractible. This observation is due to Usher [633]. He noted that the existence of such Hamiltonian functions implies that the group of Hamiltonian symplectomorphisms has infinite Hofer diameter. □

Other capacities

There are many other definitions of symplectic capacities: see for example Ekeland and Hofer [171, 172], Hofer and Zehnder [324], Cieliebak–Hofer–Latschev–Schlenk [116], and Hutchings [338]. In particular, Ekeland and Hofer construct a sequence of capacities c_k for integers $k \geq 1$ which are defined for subsets of $\mathbb{R}^{2n}$ and have the following properties.

(i) For each $U \subset \mathbb{R}^{2n}$ the sequence $c_k(U)$ is nondecreasing, i.e.

$$c_1(U) \leq c_2(U) \leq c_3(U) \leq \cdots.$$

(ii) If U is open and convex then $c_1(U) = \mathrm{c_{HZ}}(U)$ and $c_k(U) = c_k(\partial U)$ for every k. In other words, $c_k(U)$ is determined by the boundary of U.

(iii) If U is open and convex then each $c_k(U)$ is the action $\mathcal{A}(z)$ of a closed characteristic on ∂U. Moreover, $c_1(U)$ is the minimal such action.

Here the symplectic action of a loop $z : \mathbb{R}/\mathbb{Z} \to \mathbb{R}^{2n}$ is the integral

$$\mathcal{A}(z) = \int_0^1 \langle y, \dot{x} \rangle \, dt.$$

In other words, $\mathcal{A}(z)$ is the integral of the 1-form

$$\lambda = \sum_{j=1}^n y_j dx_j$$

over the loop z. This form can be replaced by any other 1-form which satisfies $d\lambda = -\omega_0$. Equivalently, $\mathcal{A}(z)$ can be expressed as the integral of $-\omega_0$ over any disc with boundary z. This action functional $\mathcal{A}$ corresponds to the case $H = 0$ in the previous context and its critical points are the constant paths. If, however, the functional is restricted to the loop space of $Q = \partial U$ then the critical points of $\mathcal{A}$ are precisely the closed characteristics. The analysis used by Ekeland and Hofer to find critical points of $\mathcal{A}$ is very similar to that in Section 12.5 below. However, in order to find a countable family of such critical points they use an index theory which is more subtle than the simple linking argument explained below and exploits the natural S^1-action on the loop space of Q. This is reminiscent of the construction of the length spectra of closed geodesics in Riemannian geometry (see Gromov's paper [290]). As we discuss further in Section 14.8, these capacities give less information in dimension four than do Hutchings' capacities. However, in contrast to the latter, they are defined in all dimensions.

Exercise 12.4.10 Prove that the critical points of the symplectic action functional $\mathcal{A}$ on the loop space of a hypersurface $Q \subset \mathbb{R}^{2n}$ are the closed orbits of the characteristic flow on Q. □

An alternative approach to the theory of capacities was proposed by Viterbo in [646,647]. His work is based on generating functions and gives rise to invariants of Lagrangian submanifolds of cotangent bundles. Now the graph of a compactly supported Hamiltonian symplectomorphism $\psi \in \mathrm{Ham}^c(\mathbb{R}^{2n})$ can be identified with a Lagrangian submanifold of T^*S^{2n} and its Viterbo invariants determine a bi-invariant metric on $\mathrm{Ham}^c(\mathbb{R}^{2n})$ called the **Viterbo metric**. The values of this metric may be interpreted in terms of the fixed points of the symplectomorphism. Such an interpretation is also possible for Hofer's metric, at least in the case of $\mathbb{R}^{2n}$ (see Hofer and Zehnder [325]). As in Section 12.3, Viterbo's metric gives rise to a displacement energy and Viterbo proves that this energy satisfies the axioms of a relative capacity. Thus his approach gives rise to an alternative proof of many of the main theorems in this chapter. It is an open question how the Viterbo metric and the Hofer metric (and hence the corresponding displacement energies) are related in general, though it is now known that they agree on a C^2-small neighbourhood of the identity in $\mathrm{Ham}^c(\mathbb{R}^{2n})$: see Bialy–Polterovich [64].

12.5 A variational argument

The purpose of this section is to give a proof of Theorem 12.4.3. It has been known for a long time that the problem of finding a periodic solution of a Hamiltonian differential equation on $\mathbb{R}^{2n}$ is equivalent to that of finding a critical point of the *symplectic action functional*

$$\mathcal{A}_H(z) = \int_0^1 \Big(\langle y, \dot{x} \rangle - H(x, y) \Big) \, dt$$

on the space of loops in $\mathbb{R}^{2n}$. Various sophisticated techniques for solving this kind of problem were developed by Rabinowitz, Weinstein, Ekeland, Amman, Zehnder, Floer, and Hofer, among others, over the past two decades. Even though these techniques originated from the study of Hamiltonian dynamical systems, they have played an increasingly important role in symplectic topology. Some of the high points of these developments are Rabinowitz's existence proof for closed characteristics on star-shaped hypersurfaces in $\mathbb{R}^{2n}$ (1979), the proof of the Arnold conjecture for the standard torus by Conley and Zehnder (1983), Floer's proof of the Arnold conjecture for monotone symplectic manifolds (1987), Viterbo's proof of the Weinstein conjecture for contact hypersurfaces of $\mathbb{R}^{2n}$ (1987), and Hofer's bi-invariant metric on the group of Hamiltonian symplectomorphisms (1990). Some of these developments are described in Viterbo's Bourbaki seminar [644]. Here we will present a modified version of the argument by Hofer and Zehnder for the proof of Theorem 12.4.3. The crucial difference is that we shall use the *discrete symplectic action functional* on the (finite-dimensional) space of *discrete loops* in $\mathbb{R}^{2n}$ as opposed to the symplectic action on the (infinite-dimensional) loop space.

An apparent obstacle for the proof is the fact that both the nonconstant 1-periodic solutions and the fixed points appear as critical points of the action functional and, in particular, the set of critical points is not compact. This difficulty can be overcome by means of an extension of the Hamiltonian function, due to Hofer and Zehnder, after which the set of critical points of $\mathcal{A}_H$ is compact and all the irrelevant critical points have action less than or equal to zero. It then remains to establish the existence of a critical point with $\mathcal{A}_H > 0$. This will be done in three steps. The first step is to replace $\mathcal{A}_H$ by a *discrete symplectic action functional* $\mathcal{A}_H^\tau$ which is defined on a space of *discrete loops* in $\mathbb{R}^{2n}$. There is a one-to-one correspondence between the critical points of $\mathcal{A}_H$ and those of $\mathcal{A}_H^\tau$ and both functionals have the same values on corresponding critical points. The second step is to prove that $\mathcal{A}_H^\tau$ satisfies the Palais–Smale condition. The final step is a linking argument which proves the existence of a critical point with $\mathcal{A}_H^\tau > 0$.

Extending the Hamiltonian function

This is the point in the proof where the geometry and analysis meet, and where the exact conditions both on the support of H and on its size are of crucial importance, since both of these affect the possible choices of a quadratic extension.

We begin by changing slightly the normalization of H. Instead of considering functions $H \geq 0$ which attain their maximum M at an interior point and are 0 near the boundary, we consider functions of the form $M - H$. Thus, let

$$H_0 : Z^{2n}(1) \to \mathbb{R}$$

be a Hamiltonian function which satisfies the following conditions.

(I) $\pi < M = \sup H_0 < \infty$ and $0 \leq H_0(z) \leq M$ for all $z \in Z^{2n}(1)$.

(II) There exists a compact set $K \subset \operatorname{int} Z^{2n}(1)$ such that $H_0(z) = M$ for $z \notin K$.

(III) There exists an open set $U \subset K$ such that $H_0(z) = 0$ for $z \in U$.

We shall prove that the Hamiltonian flow of every function H_0 which satisfies (I), (II), (III) has a nonconstant 1-periodic orbit. This assertion is equivalent to Theorem 12.4.3.

Remark 12.5.1 In (III) we may assume without loss of generality that $0 \in U$. To see this choose a symplectomorphism $\psi : Z^{2n}(1) \to Z^{2n}(1)$ such that $\psi(0) \in U$ and $\psi(z) = z$ outside a compact set. Then replace H_0 by $H_0 \circ \psi$. For example, ψ can be chosen as the time-1-map of a Hamiltonian flow where the Hamiltonian is a product of a linear function with a cutoff function. The Hamiltonian flow of such a function has the form $\dot{z} = \text{const}$ in the region where the cutoff function is equal to 1. □

A Hamiltonian function $H : \mathbb{R}^{2n} \to \mathbb{R}$ is said to have **quadratic growth** if there exists a constant $c > 0$ such that

$$\left\| d^2 H(z) \right\| \leq c \tag{12.5.1}$$

for all $z \in \mathbb{R}^{2n}$. Throughout we shall denote by $\phi_H^t : \mathbb{R}^{2n} \to \mathbb{R}^{2n}$ the Hamiltonian flow of H and by $\operatorname{Per}(H)$ the set of 1-periodic solutions of $\dot{z} = X_H(z)$. Note that this set can be identified with the set $\operatorname{Fix}(\phi_H^1)$ of fixed points of the time-1-map.

Exercise 12.5.2 Let $H : \mathbb{R}^{2n} \to \mathbb{R}$ be a Hamiltonian function with quadratic growth. If $|\nabla H(0)| \leq c$ and $|H(0)| \leq c/2$, prove that

$$|H(z)| \leq c(|z|^2 + 1), \qquad |\nabla H(z)| \leq c(|z| + 1).$$

Prove that the solutions of the Hamiltonian differential equation $\dot{z} = X_H(z)$ exist for all time. Prove that for all $t \in \mathbb{R}$ and $z \in \mathbb{R}^{2n}$,

$$\left\| d\phi_H^t(z) - \mathbb{1} \right\| \leq e^{c|t|} - 1,$$

where c is the constant of (12.5.1). □

Hofer and Zehnder [324] found an ingenious extension of H_0 to a Hamiltonian function on $\mathbb{R}^{2n}$ with quadratic growth. This was also the key point in Viterbo's earlier proof of the Weinstein conjecture in [643]. This extension is explained in the proof of the following lemma.

Lemma 12.5.3 *Suppose that* $H_0 : Z^{2n}(1) \to \mathbb{R}$ *satisfies* (I), (II), (III). *Then there exists a Hamiltonian function* $H : \mathbb{R}^{2n} \to \mathbb{R}$ *which satisfies the following conditions.*

(i) *There exists a constant* $R > 0$ *such that* $K \subset B^{2n}(R)$ *and* $H(z) = H_0(z)$ *for* $z \in Z^{2n}(1) \cap B^{2n}(R)$.

(ii) H *has quadratic growth.*

(iii) *There exists a constant* $c > 0$ *such that*

$$|z| > c \qquad \Longrightarrow \qquad |\phi_H^1(z) - z| \geq 1,$$

where ϕ_H^1 *denotes the time-1-map of the Hamiltonian flow of* H. *In particular, the set* $\mathrm{Per}(H) \cong \mathrm{Fix}(\phi_H^1)$ *is compact.*

(iv) *If* $z \in \mathrm{Per}(H)$ *with* $\mathcal{A}_H(z) > 0$ *then* $z(t)$ *is nonconstant and* $z(t) \in K$ *for all* t.

Proof: First choose a number $0 < \varepsilon < \pi/2$ such that $M > \pi + \varepsilon$. Then there exists a smooth function $f : [0, \infty) \to \mathbb{R}$ such that

$$\begin{aligned} f(s) &= M, && \text{for } 0 \leq s \leq 1, \\ f(s) &= (\pi + \varepsilon)s, && \text{for } s \geq M, \\ f(s) &\geq (\pi + \varepsilon)s, && \text{for all } s, \\ 0 \leq f'(s) &\leq \pi + \varepsilon, && \text{for all } s. \end{aligned}$$

Next choose $R \geq 1$ such that $H(z) = M$ for $z \notin B^{2n}(R)$. Then choose a smooth function $g : [0, \infty) \to \mathbb{R}$ such that

$$\begin{aligned} g(s) &= 0, && \text{for } 0 \leq s \leq R^2, \\ g(s) &= \pi s/2, && \text{for } s \geq 3R^2, \\ 0 \leq g'(s) &< \pi, && \text{for all } s. \end{aligned}$$

Denote $z = (z_1, w) \in \mathbb{R}^{2n}$, where $z_1 \in \mathbb{R}^2$ and $w = (z_2, \ldots, z_n) \in \mathbb{R}^{2n-2}$. In this notation the cylinder $Z^{2n}(1)$ is the set $\{z \in \mathbb{R}^{2n} \,|\, |z_1| < 1\}$. The required extension H of H_0 is now given by

$$H(z) = \begin{cases} H_0(z), & \text{if } z \in Z^{2n}(1) \cap B^{2n}(R), \\ f(|z_1|^2) + g(|w|^2), & \text{otherwise.} \end{cases}$$

This function obviously satisfies the conditions (i) and (ii).

We prove that H satisfies (iii). Note first that in the domain where either $|z_1| \geq 1$ or $|w| \geq R$, the Hamiltonian differential equation $\dot{z} = -J_0 \nabla H(z)$ splits into the two independent equations

$$\dot{z}_1 = -2if'(|z_1|^2)z_1, \qquad \dot{w} = -2ig'(|w|^2)w.$$

Here we have identified $\mathbb{R}^{2n}$ with $\mathbb{C}^n$ in the obvious way. It follows that the functions $z \mapsto |z_1|^2$ and $z \mapsto |w|^2$ are integrals of the Hamiltonian flow. In particular, the domain $\{\max\{|z_1|, |w|/R\} \geq 1\}$ is invariant under the Hamiltonian flow and, in this domain, the time-1 map is given by

$$\phi_H^1(z_1, w) = (e^{-2if'(|z_1|^2)}z_1, e^{-2ig'(|w|^2)}w).$$

Now suppose that

$$|z|^2 = |z_1|^2 + |w|^2 \geq c^2 \geq 2\max\{M, 3R^2, |1 - e^{-2i\varepsilon}|^{-2}\}.$$

Then we have either $|z_1|^2 \geq c^2/2 \geq M$, and hence $e^{-2if'(|z_1|^2)} = e^{-2i\varepsilon} \neq 1$, or $|w|^2 \geq c^2/2 \geq 3R^2$, and hence $e^{-2ig'(|w|^2)} = -1$. In the first case we obtain $|\phi_H^1(z) - z| \geq |1 - e^{-2i\varepsilon}|\,|z_1| \geq |1 - e^{-2i\varepsilon}|\,c/\sqrt{2} \geq 1$ and in the second case, $|\phi_H^1(z) - z| \geq 2\,|w| \geq 1$. This shows that $|\phi_H^1(z) - z| \geq 1$ whenever $|z| \geq c$.

We prove that H satisfies (iv). Hence let $z \in \mathrm{Per}(H)$ with $\mathcal{A}_H(z) > 0$. Since the symplectic action of a constant solution $z(t) \equiv z_0$ is $\mathcal{A}_H(z) = -H(z_0) \leq 0$ it follows that our solution must be nonconstant. Since the level set $H^{-1}(M)$ consists entirely of critical points this implies that $H(z(t)) \neq M$ for all t. Now suppose, by contradiction, that $z(t) \notin K$ for some t. Then $H(z(t)) > M$ for this and hence all values of t. This means that our periodic orbit $z(t) = (z_1(t), w(t))$ lies in a region of $\mathbb{R}^{2n} = \mathbb{C}^n$ where the Hamiltonian function H is given by the formula $H(z) = f(|z_1|^2) + g(|w|^2)$. Hence, as above,

$$z_1(t) = e^{-2if'(|z_1|^2)}z_1(0), \qquad w(t) = e^{-2ig'(|w|^2)}w(0).$$

Since $0 \leq 2g'(|w|^2) < 2\pi$ the function $w(t)$ cannot be of period 1 unless either $w(t) \equiv 0$ or $g'(|w|^2) = 0$. In either case $w(t)$ is constant and $g(w(t)) \equiv 0$. Hence $z_1(t)$ must be nonconstant. This means that $|z_1| > 1$ and

$$0 < 2f'(|z_1|^2) \leq 2\pi + 2\varepsilon \leq 3\pi.$$

Since $z_1(t) = z_1(t+1)$, we must have $2f'(|z_1|^2) = 2\pi$. Hence the symplectic action of our solution $z(t)$ is given by

$$\begin{aligned}
\mathcal{A}_H(z) &= \int_0^1 \left(\frac{1}{2}y_1\dot{x}_1 - \frac{1}{2}x_1\dot{y}_1 - f(|z_1|^2)\right) dt \\
&= f'(|z_1|^2)|z_1|^2 - f(|z_1|^2) \\
&= \pi|z_1|^2 - f(|z_1|^2) \\
&\leq -\varepsilon|z_1|^2.
\end{aligned}$$

This contradicts the condition $\mathcal{A}_H(z) > 0$ and thus our assumption that $z(t) \notin K$ must have been wrong. Hence H satisfies (iv), which proves Lemma 12.5.3. □

Discrete symplectic action

Let H be a Hamiltonian function which satisfies the quadratic growth condition (12.5.1). It follows from Lemma 9.2.1 that for τ sufficiently small, the symplectomorphism ϕ_H^τ admits a generating function of type V. We give below an alternative proof of this fact, and establish further properties of this generating function in the time-independent case. Note also the presence of the normalizing constant τ in the equations which we shall use to define the generating function $V_\tau = V_\tau(x,y)$. This is inserted so that we have the proper convergence as $\tau \to 0$.

Lemma 12.5.4 *Suppose that $H : \mathbb{R}^{2n} \to \mathbb{R}$ satisfies the quadratic growth condition (12.5.1). Then, for $\tau > 0$ sufficiently small, there exists a unique function $V_\tau : \mathbb{R}^{2n} \to \mathbb{R}$ which satisfies the following conditions.*

(i) $(x_1, y_1) = \phi_H^\tau(x_0, y_0)$ *if and only if*

$$\frac{x_1 - x_0}{\tau} = \frac{\partial V_\tau}{\partial y}(x_1, y_0), \qquad \frac{y_1 - y_0}{\tau} = -\frac{\partial V_\tau}{\partial x}(x_1, y_0). \tag{12.5.2}$$

(ii) *V_τ converges to H in the C^∞-topology as $\tau \to 0$. Moreover, V_τ has quadratic growth and there exists a constant $c > 0$ such that*

$$\sup_{z \in \mathbb{R}^{2n}} |V_\tau(z) - H(z)| \le c\tau(|z|^2 + 1).$$

(iii) $\mathrm{Crit}(H) = \mathrm{Crit}(V_\tau) = \mathrm{Fix}(\phi_H^\tau)$ *and $H(z) = V_\tau(z)$ for every critical point z of H.*

(iv) *For all $x_1, y_0 \in \mathbb{R}^n$ there exists a unique solution*

$$z(t) = (x(t), y(t)) = z(t; x_1, y_0, \tau)$$

of the boundary value problem

$$\dot{z} = X_H(z), \qquad x(\tau) = x_1, \qquad y(0) = y_0. \tag{12.5.3}$$

The action of this solution on the interval $[0,\tau]$ is given by

$$\mathcal{A}_H^{[0,\tau]}(z) = \langle y_0, x_1 - x_0 \rangle - \tau V_\tau(x_1, y_0), \tag{12.5.4}$$

where $x_0 := x(0)$.

Proof: That the boundary value problem (12.5.3) has a unique solution for all x_1 and y_0 whenever τ is sufficiently small follows from the implicit function theorem and Exercise 12.5.2. Hence equation (12.5.4) can be used to define V_τ. We prove that V_τ converges to H in the C^∞-topology as τ tends to 0. Denote

$$z_\tau(s; x_1, y_0) = (x_\tau(s), y_\tau(s)) = z(\tau s; x_1, y_0, \tau)$$

for $0 \le s \le 1$ and $\tau > 0$. This means that z_τ is the solution of the boundary value problem $\dot{z}_\tau = \tau X_H(z_\tau)$ with $x_\tau(1) = x_1$ and $y_\tau(0) = y_0$. In particular, for $\tau = 0$, we have $z_0(s; x_1, y_0) = (x_1, y_0)$ and $V_0 = H$. Then

$$V_\tau(x_1, y_0) = \int_0^1 H(x_\tau(s), y_\tau(s))\, ds + \int_0^1 \int_0^s \langle \partial_x H(x_\tau(r), y_\tau(r)), \dot{x}_\tau(s) \rangle\, dr ds$$

for $\tau \ge 0$. Hence the C^∞-convergence of V_τ to $H = V_0$ follows from the fact that the function $(s, \tau, x_1, y_0) \mapsto z_\tau(s; x_1, y_0)$ is smooth in the region $\tau \ge 0$. The uniform quadratic estimate for the difference $H - V_\tau$ follows by differentiating the above formula for V_τ with respect to τ. The quadratic growth property of V_τ is left as an exercise. Thus we have proved that V_τ satisfies (ii) and (iv).

We prove that V_τ satisfies (i). Fix τ and vectors $x_1, y_0 \in \mathbb{R}^n$ and denote by $z(t) = (x(t), y(t))$ the solution of (12.5.3). We must prove that the partial derivatives of V_τ are given by (12.5.2), where $x_0 = x(0)$ and $y_1 = y(\tau)$. To see this, fix vectors $\xi_1, \eta_0 \in \mathbb{R}^n$ and denote

$$(x_s(t), y_s(t)) = z_s(t) = z(t; x_1 + s\xi_1, y_0 + s\eta_0, \tau),$$

$$\xi(t) = \left.\frac{d}{ds}\right|_{s=0} x_s(t), \quad \eta(t) = \left.\frac{d}{ds}\right|_{s=0} y_s(t), \quad \eta_1 = \eta(\tau), \quad \xi_0 = \xi(0).$$

Abbreviate $\partial_x V = \partial_x V(x_1, y_0)$ and $\partial_y V = \partial_y V(x_1, y_0)$. Then

$$\begin{aligned}
\langle \partial_x V, \xi_1 \rangle + \langle \partial_y V, \eta_0 \rangle &= \left.\frac{d}{ds}\right|_{s=0} V_\tau(x_1 + s\xi_1, y_0 + s\eta_0) \\
&= \left.\frac{d}{ds}\right|_{s=0} \frac{1}{\tau}\langle y_s(0), x_s(\tau) - x_s(0)\rangle \\
&\quad - \left.\frac{d}{ds}\right|_{s=0} \frac{1}{\tau}\int_0^\tau \Big(\langle y_s, \dot{x}_s \rangle - H(x_s, y_s) \Big)\, dt \\
&= \frac{1}{\tau}\langle \eta_0, x_1 - x_0 \rangle + \frac{1}{\tau}\langle y_0, \xi_1 - \xi_0 \rangle \\
&\quad - \frac{1}{\tau}\int_0^\tau \Big(\langle \eta, \dot{x} \rangle + \langle y, \dot{\xi} \rangle - \langle \partial_x H, \xi \rangle - \langle \partial_y H, \eta \rangle \Big)\, dt \\
&= \frac{1}{\tau}\langle \eta_0, x_1 - x_0 \rangle + \frac{1}{\tau}\langle y_0 - y_1, \xi_1 \rangle.
\end{aligned}$$

This shows that $\partial_x V_\tau = \tau^{-1}(y_0 - y_1)$ and $\partial_y V_\tau = \tau^{-1}(x_1 - x_0)$. Hence V_τ satisfies (i). The critical points of V_τ agree with the fixed points of ϕ_H^τ by (i), and $\mathrm{Crit}(H) = \mathrm{Fix}(\phi_H^\tau)$ by Lemma 12.4.1. Now use (iv) to conclude that V_τ and H agree on the set of critical points. This proves Lemma 12.5.4. □

Exercise 12.5.5 Suppose that H is the quadratic function $H(z) = \frac{1}{2}\langle z, Pz \rangle$, where $P = P^{\mathrm{T}}$. So its flow is given by $\phi_H^t(z) = \Psi(t)z$, where $\Psi(t) = e^{-J_0 P t}z$. For $t = \tau$ sufficiently small let $V_\tau(x, y)$ be defined as above. Check explicitly that

$$\lim_{\tau \to 0} V_\tau = H.$$

Hint: Write $\Psi(\tau)$ in block form

$$\Psi(\tau) = \begin{pmatrix} A(\tau) & B(\tau) \\ C(\tau) & D(\tau) \end{pmatrix} \in \mathrm{Sp}(2n).$$

Prove that for τ sufficiently small,

$$V_\tau(x_1, y_0) = -\frac{1}{2\tau}\langle x_1, CA^{-1}x_1 \rangle + \frac{1}{\tau}\langle y_0, (\mathbb{1} - A^{-1})x_1 \rangle + \frac{1}{2\tau}\langle y_0, A^{-1}By_0 \rangle.$$

Compare this with Example 9.2.3. □

Now choose $\tau = 1/N$ for a sufficiently large integer N, identify $\mathcal{P} = \mathbb{R}^{2nN}$ with the space of *discrete loops* $\mathbf{z} = (z_j)_{j\in\mathbb{Z}}$ such that $z_{j+N} = z_j$ for all j, and define the discrete symplectic action

$$\mathcal{A}_H^\tau : \mathbb{R}^{2nN} \to \mathbb{R}$$

by

$$\mathcal{A}_H^\tau(\mathbf{z}) = \sum_{j=0}^{N-1}\Big(\langle y_j, x_{j+1} - x_j\rangle - \tau V_\tau(x_{j+1}, y_j)\Big). \qquad (12.5.5)$$

Recall from Chapter 9 that the critical points of $\mathcal{A}_H^\tau$ are the sequences $z_j \in \mathbb{R}^{2n}$ which satisfy

$$z_{j+1} = \phi_H^\tau(z_j), \qquad z_{j+N} = z_j.$$

Since $\tau = 1/N$ these correspond to the fixed points of ϕ_H^1. The group $\mathbb{Z}/\mathbb{Z}_N$ acts on the *discrete loop space* $\mathbb{R}^{2nN}$ and $\mathcal{A}_H^\tau$ is invariant under this action. This corresponds to the action of S^1 on the loop space. But here we shall not use this fact. In the following we denote by $\mathrm{Per}(H)$ the set of 1-periodic solutions of the Hamiltonian system $\dot{z} = X_H(z)$. Note that these are the critical points of the symplectic action $\mathcal{A}_H$.

Corollary 12.5.6 *The map $C^\infty(\mathbb{R}/\mathbb{Z}, \mathbb{R}^{2n}) \to \mathbb{R}^{2nN} : z \mapsto \{z(j/N)\}_{j\in\mathbb{Z}}$ identifies the critical points of the symplectic action $\mathcal{A}_H$ with the critical points of the discrete symplectic action $\mathcal{A}_H^\tau$ for $\tau = 1/N$. Moreover,*

$$\mathcal{A}_H(z) = \mathcal{A}_H^\tau(\{z(j/N)\}_{j\in\mathbb{Z}})$$

for every $z \in \mathrm{Per}(H)$.

Proof: Lemma 12.5.4. □

The Palais–Smale condition

Let X be a (not necessarily compact) Riemannian manifold and

$$\Phi : X \to \mathbb{R}$$

be a smooth function whose gradient vector field $\mathrm{grad}\,\Phi : X \to TX$ is complete. Denote by

$$\phi_t : X \to X$$

the gradient flow of Φ. This means that for every point $x_0 \in X$ the function $x(t) := \phi_t(x_0)$ is the unique solution of the ordinary differential equation

$$\dot{x} = -\mathrm{grad}\,\Phi(x)$$

with initial condition $x(0) = x_0$. We say that Φ satisfies the **Palais–Smale condition** if every sequence $x_\nu \in X$ with

$$\lim_{\nu\to\infty} \|\mathrm{grad}\,\Phi(x_\nu)\| = 0$$

has a convergent subsequence. In particular, this implies that the set of critical points of Φ is compact.

Lemma 12.5.7 *Assume that Φ satisfies the Palais–Smale condition and let $c \in \mathbb{R}$ be a regular value of Φ. Then for every $T > 0$ there exists a number $\delta > 0$ such that*

$$\Phi(x) \le c + \delta \qquad \Longrightarrow \qquad \Phi(\phi_T(x)) \le c - \delta.$$

Proof: The proof is by contradiction. If the assertion were false then there would exist a sequence $x_\nu \in X$ such that $\Phi(x_\nu)$ and $\Phi(\phi_T(x_\nu))$ both converge to c. Since

$$\frac{d}{dt}\Phi(\phi_t(x_\nu)) = -\left\|\operatorname{grad}\Phi(\phi_t(x_\nu))\right\|^2,$$

we find that

$$\int_0^T \left\|\operatorname{grad}\Phi(\phi_t(x_\nu))\right\|^2 dt = \Phi(x_\nu) - \Phi(\phi_T(x_\nu)) \to 0.$$

Hence there must exist a sequence $t_\nu \in [0, T]$ such that

$$\lim_{\nu\to\infty} \operatorname{grad}\Phi(\phi_{t_\nu}(x_\nu)) = 0.$$

By the Palais–Smale condition the sequence $\phi_{t_\nu}(x_\nu)$ has a convergent subsequence and the limit point $x \in X$ of this subsequence satisfies $\operatorname{grad}\Phi(x) = 0$ and $\Phi(x) = c$. This contradicts our assumption that c be a regular value of Φ. □

Consider the space $\mathbb{R}^{2nN}$ of discrete loops in $\mathbb{R}^{2n}$ as a Hilbert space with inner product

$$\langle \mathbf{z}, \zeta\rangle_\tau = \tau \sum_{j=0}^{N-1} \langle z_j, \zeta_j\rangle.$$

This is a kind of discretized L^2 inner product and the corresponding norm is, of course, equivalent to the Euclidean norm. The gradient of the discrete symplectic action functional $\mathcal{A}_H^\tau$ with respect to this norm is given by

$$\operatorname{grad}\mathcal{A}_H^\tau(\mathbf{z}) = \zeta,$$

where

$$\xi_j = \frac{y_{j-1} - y_j}{\tau} - \frac{\partial V_\tau}{\partial x}(x_j, y_{j-1}), \qquad \eta_j = \frac{x_{j+1} - x_j}{\tau} - \frac{\partial V_\tau}{\partial y}(x_{j+1}, y_j).$$

(See Lemma 9.2.4.) If H has quadratic growth, then the gradient flow of $\mathcal{A}_H^\tau$ exists for all time. Furthermore, if H is the extended Hamiltonian function of Lemma 12.5.3, then the set of critical points of $\mathcal{A}_H^\tau$ is compact. We now prove that $\mathcal{A}_H^\tau$ satisfies the Palais–Smale condition.

Lemma 12.5.8 *Let H be any Hamiltonian which satisfies the quadratic growth condition* (12.5.1) *and suppose that there exist constants $c > 0$ and $\delta > 0$ such that*

$$|z| > c \qquad \Longrightarrow \qquad |\phi_H^1(z) - z| > \delta. \tag{12.5.6}$$

Then, for every sufficiently small $\tau > 0$, the functional $\mathcal{A}_H^\tau : \mathbb{R}^{2nN} \to \mathbb{R}$ satisfies the Palais–Smale condition.

Proof: We first prove that there exists a constant $\tau_0 > 0$ such that

$$\left|z_{j+1} - \phi_H^\tau(z_j)\right|^2 \le 2\tau^2\left(\left|\xi_{j+1}\right|^2 + \left|\eta_j\right|^2\right) \tag{12.5.7}$$

for $0 < \tau < \tau_0$ and $\mathbf{z} = \{z_j\}_{j\in\mathbb{Z}} \in \mathbb{R}^{2nN}$, where ξ_j and η_j denote the components of $\zeta = \operatorname{grad}\mathcal{A}_H^\tau(\mathbf{z})$. To see this, write

$$z'_{j+1} = (x'_{j+1}, y'_{j+1}) = \phi_H^\tau(z_j)$$

and note that

$$\begin{aligned} x'_{j+1} &= x_j + \tau\frac{\partial V_\tau}{\partial y}(x'_{j+1}, y_j), \qquad & x_{j+1} &= x_j + \tau\frac{\partial V_\tau}{\partial y}(x_{j+1}, y_j) + \tau\eta_j, \\ y'_{j+1} &= y_j - \tau\frac{\partial V_\tau}{\partial x}(x'_{j+1}, y_j), \qquad & y_{j+1} &= y_j - \tau\frac{\partial V_\tau}{\partial x}(x_{j+1}, y_j) - \tau\xi_{j+1}. \end{aligned}$$

With $\tau\cdot\sup\|\partial_y\partial_x V_\tau\| \le \alpha < 1$ and $\tau\cdot\sup\|\partial_x\partial_x V_\tau\| \le \alpha$ we obtain

$$\left|x_{j+1} - x'_{j+1}\right| \le \frac{\tau}{1-\alpha}\left|\eta_j\right|, \qquad \left|y_{j+1} - y'_{j+1}\right| \le \frac{\alpha\tau}{1-\alpha}\left|\eta_j\right| + \tau\left|\xi_{j+1}\right|.$$

This proves (12.5.7).

By Exercise 12.5.2, $C := \sup_{0\le t\le 1}\sup_{z\in\mathbb{R}^{2n}}\|d\phi_H^t(z)\| < \infty$. Hence

$$\begin{aligned} \left|z_0 - \phi_H^1(z_0)\right| &\le \sum_{j=0}^{N-1}\left|\phi_H^{1-(j+1)\tau}(z_{j+1}) - \phi_H^{1-j\tau}(z_j)\right| \\ &\le C\sum_{j=0}^{N-1}\left|z_{j+1} - \phi_H^\tau(z_j)\right| \\ &\le C\sqrt{N\sum_{j=0}^{N-1}\left|z_{j+1} - \phi_H^\tau(z_j)\right|^2} \\ &\le C\sqrt{2N\tau^2\sum_{j=0}^{N-1}\left|\zeta_j\right|^2} \\ &\le 2C\left\|\operatorname{grad}\mathcal{A}_H^\tau(\mathbf{z})\right\|_\tau. \end{aligned}$$

The penultimate inequality follows from (12.5.7) and the last inequality uses $N\tau = 1$. Thus we have proved that, for every sufficiently small number $\tau > 0$ and every $\mathbf{z} = \{z_j\}_{j\in\mathbb{Z}} \in \mathbb{R}^{2nN}$,

$$\sup_{j\in\mathbb{Z}}\left|\phi_H^1(z_j) - z_j\right| \le 2C\left\|\operatorname{grad}\mathcal{A}_H^\tau(\mathbf{z})\right\|_\tau. \tag{12.5.8}$$

Here we have used the invariance of the functional $\mathcal{A}_H^\tau$ under the obvious $\mathbb{Z}/N\mathbb{Z}$-action on $\mathbb{R}^{2nN}$. It follows from (12.5.8) and (12.5.6) that

$$\left\|\operatorname{grad}\mathcal{A}_H^\tau(\mathbf{z})\right\|_\tau \le \frac{\delta}{2C} \qquad \Longrightarrow \qquad \sup_{j\in\mathbb{Z}}|z_j| \le c.$$

This shows that $\mathcal{A}_H^\tau$ satisfies the Palais–Smale condition. □

A linking argument

In view of Lemma 12.5.3 and Corollary 12.5.6 we must prove that the discrete symplectic action has a critical point $\mathbf{z}$ such that $\mathcal{A}_H^\tau(\mathbf{z}) > 0$. The proof is based on a linking argument and this requires some preparation. The function $\mathcal{A}_H^\tau$ is a perturbation of the functional

$$\mathcal{A}^\tau(\mathbf{z}) = \sum_{j=0}^{N-1} \langle y_j, x_{j+1} - x_j \rangle = \frac{1}{2}\langle \mathbf{z}, L^\tau \mathbf{z} \rangle_\tau$$

which corresponds to $H = 0$. Here L^τ is the Hessian of $\mathcal{A}^\tau$ with respect to the inner product $\langle \cdot, \cdot \rangle_\tau$. It is a self-adjoint operator on $\mathbb{R}^{2nN}$ which sends $\zeta = (\zeta_j)_{j\in\mathbb{Z}}$ to the sequence $\zeta' = (\zeta'_j)_{j\in\mathbb{Z}}$, where

$$\xi'_{j+1} = \frac{\eta_j - \eta_{j+1}}{\tau}, \qquad \eta'_j = \frac{\xi_{j+1} - \xi_j}{\tau}.$$

(See the proof of Lemma 9.2.6.) There is an eigenspace decomposition

$$\mathbb{R}^{2nN} = E^- \cup E^0 \cup E^+$$

such that $E^0 = \ker L^\tau$, the eigenvalues of $L^\tau|_{E^+}$ are positive, and the eigenvalues of $L^\tau|_{E^-}$ are negative. The subspace $E^0 \cong \mathbb{R}^{2n}$ consists of the constant sequences.

Lemma 12.5.9 *The smallest positive eigenvalue of L^τ is less than or equal to 2π. For each eigenvalue there exists an eigenvector $\zeta = (\zeta_j)_{j\in\mathbb{Z}}$ such that $\zeta_j \in \mathbb{R}^2 \times 0 \subset \mathbb{R}^{2n}$ for all j. Moreover,*

$$\lambda \in \sigma(L^\tau) \qquad \Longleftrightarrow \qquad -\lambda \in \sigma(L^\tau).$$

Proof: The last assertion follows from the fact that

$$L^\tau(\xi, \eta) = \lambda(\xi, \eta) \quad \Longleftrightarrow \quad L^\tau(\xi, -\eta) = -\lambda(\xi, -\eta).$$

To prove the other assertions we examine the eigenvalues of L^τ. The equation $L^\tau(\xi, \eta) = \lambda(\xi, \eta)$ is equivalent to

$$\lambda\xi_{j+1} = \frac{\eta_j - \eta_{j+1}}{\tau}, \qquad \lambda\eta_j = \frac{\xi_{j+1} - \xi_j}{\tau}.$$

These equations can be rewritten in the form

$$\begin{pmatrix} \xi_{j+1} \\ \eta_{j+1} \end{pmatrix} = \begin{pmatrix} 1 & \lambda/N \\ -\lambda/N & 1 - (\lambda/N)^2 \end{pmatrix} \begin{pmatrix} \xi_j \\ \eta_j \end{pmatrix}.$$

Hence the eigenvalues of L^τ are the real numbers λ such that

$$\det\left(\mathbb{1} - A_N(\lambda)^N\right) = 0, \qquad A_N(\lambda) = \begin{pmatrix} 1 & \lambda/N \\ -\lambda/N & 1 - (\lambda/N)^2 \end{pmatrix}.$$

The eigenvectors have the form $\zeta = (\zeta_j)_{j\in\mathbb{Z}}$, where $\zeta_j = (\zeta_{j1}, \dots, \zeta_{jn}) \in \mathbb{R}^{2n}$ and

$$\zeta_{j\nu} = A_N(\lambda)^j \zeta_{0\nu}, \qquad A_N(\lambda)^N \zeta_{0\nu} = \zeta_{0\nu}.$$

Note that $\det(A_N(\lambda)) = 1$ and $\mathrm{trace}(A_N(\lambda)) = 2 - \lambda^2/N^2 \in [-2, 2]$ whenever $\lambda^2/N^2 \leq 4$. If $|\lambda| > 2N$ then the eigenvalues of $A_N(\lambda)$ have modulus not equal to 1 and hence λ is not an eigenvalue of L^τ. If $|\lambda| \le 2N$ then the eigenvalues of $A_N(\lambda)$ are given by $e^{\pm i\theta_N(\lambda)}$, where $\theta_N : [-2N, 2N] \to \mathbb{R}$ is the unique continuous function such that $\theta_N(0) = 0$ and

$$\cos\theta_N(\lambda) = \frac{1}{2}\mathrm{trace}(A_N(\lambda)) = 1 - \frac{\lambda^2}{2N^2}$$

for $-2N \le \lambda \le 2N$. The derivative of θ_N is given by

$$N\theta_N{}'(\lambda) = \frac{1}{\sqrt{1-\lambda^2/4N^2}} \ge 1,$$

and hence $N\theta_N(\lambda) \ge \lambda$ for $\lambda \ge 0$. Now the smallest positive eigenvalue of L^τ is the unique number $\lambda_N > 0$ such that $N\theta_N(\lambda_N) = 2\pi$. The previous inequality shows that $\lambda_N \le 2\pi$. This proves Lemma 12.5.9. □

Exercise 12.5.10 Prove that

$$\lim_{N\to\infty} A_N(\lambda)^N = \begin{pmatrix} \cos\lambda & \sin\lambda \\ -\sin\lambda & \cos\lambda \end{pmatrix}.$$

Hint: The sequence $(1 + \Lambda/N)^N$ converges to $\exp(\Lambda)$ for $\Lambda \in \mathbb{R}^{2\times 2}$, uniformly on compact subsets of $\mathbb{R}^{2\times 2}$. □

Remark 12.5.11 The operator L^τ is a discretization of the L^2-Hessian

$$L : W^{1,2}(\mathbb{R}/\mathbb{Z}, \mathbb{R}^{2n}) \to L^2(\mathbb{R}/\mathbb{Z}, \mathbb{R}^{2n})$$

of the unperturbed continuous action functional Φ corresponding to $H = 0$. This operator is given by

$$L(\xi, \eta) = (-\dot{\eta}, \dot{\xi}).$$

The eigenvalues of L are $2\pi k$ for $k \in \mathbb{Z}$ with eigenvectors

$$\zeta_k(t) = e^{-2\pi i k t}\zeta_k(0).$$

The eigenvalues of L^τ converge to those of L as $\tau = 1/N \to 0$. To see this note that $N\theta_N(\lambda)$ converges to λ, uniformly on compact subsets of $\mathbb{R}$. Now the eigenvalues of L^τ are the numbers $\lambda \in [-2N, 2N]$ with $N\theta_N(\lambda) \in 2\pi\mathbb{Z}$. These numbers converge to $2\pi\mathbb{Z}$. In particular, the sequence λ_N of smallest positive eigenvalues of $L^{1/N}$, which is defined by $N\theta_N(\lambda_N) = 2\pi$, converges to 2π. □

Lemma 12.5.12 *Assume that $H_0 : Z^{2n}(1) \to \mathbb{R}$ satisfies conditions* (I), (II), *and* (III) *on page 490. Then the Hamiltonian flow of H_0 has a nonconstant 1-periodic solution.*

Proof: By Remark 12.5.1, we may assume without loss of generality that the open set U in (III) is a neighbourhood of 0. Let $H : \mathbb{R}^{2n} \to \mathbb{R}$ be the extension of H_0, introduced in Lemma 12.5.3, and let $V_\tau : \mathbb{R}^{2n} \to \mathbb{R}$ be the corresponding generating function, introduced in Lemma 12.5.4 for $\tau = 1/N$ sufficiently small. We shall prove in three steps that the discrete symplectic action functional $\mathcal{A}_H^\tau$ has a critical point with $\mathcal{A}_H^\tau > 0$.

Step 1 *Define*

$$\Gamma := \left\{ \mathbf{z} \in E^+ \,|\, \|\mathbf{z}\|_\tau = \alpha \right\}.$$

If $\alpha > 0$ is sufficiently small then $\inf_\Gamma \mathcal{A}_H^\tau > 0$.

By Lemma 12.5.4, V_τ vanishes near zero. Hence $\mathcal{A}_H^\tau$ agrees with the unperturbed discrete symplectic action $\mathcal{A}^\tau$ in some neighbourhood of zero.

Step 2 *By Lemma 12.5.9, there exists an eigenvector $\zeta \in E^+$ of L^τ with eigenvalue $\lambda \le 2\pi$ and $\zeta_j \in \mathbb{R}^2 \times 0$ for all j. Define*

$$\Sigma = \left\{ \mathbf{z} + s\zeta \,|\, \mathbf{z} \in E^- \oplus E^0,\ \|\mathbf{z}\|_\tau \le T,\ 0 \le s \le T \right\}.$$

If T is sufficiently large then $\sup_{\partial\Sigma} \mathcal{A}_H^\tau \le 0$.

By construction, there exists a constant $C > 0$ such that

$$H(z_1, w) \ge (\pi + \varepsilon)|z_1|^2 + \frac{\pi}{2}|w|^2 - C.$$

Hence it follows from Lemma 12.5.4 (ii) that

$$V_\tau(z_1, w) \ge \left(\pi + \frac{\varepsilon}{2}\right)|z_1|^2 + \frac{\pi}{4}|w|^2 - 2C \tag{12.5.9}$$

for τ sufficiently small.

Now let $\mathbf{z} = \mathbf{z}^- + \mathbf{z}^0 \in E^- \oplus E^0$ and $s \in \mathbb{R}$. Then

$$\begin{aligned}
\mathcal{A}_H^\tau(\mathbf{z} + s\zeta) &= \mathcal{A}^\tau(\mathbf{z}^-) + s^2\mathcal{A}^\tau(\zeta) - \frac{1}{N}\sum_{j=0}^{N-1} V_\tau(x_{j+1} + s\xi_{j+1}, y_j + s\eta_j) \\
&\le \frac{\lambda s^2}{2}\|\zeta\|_\tau^2 - \left(\pi + \frac{\varepsilon}{2}\right)\|\mathbf{z}_1 + s\zeta\|_\tau^2 - \frac{\pi}{4}\|\mathbf{w}\|_\tau^2 + 2C \\
&\le \frac{\lambda s^2}{2}\|\zeta\|_\tau^2 - \left(\pi + \frac{\varepsilon}{2}\right)\|s\zeta\|_\tau^2 - \frac{\pi}{4}\|\mathbf{z}\|_\tau^2 + 2C \\
&\le -\frac{\varepsilon s^2}{2}\|\zeta\|_\tau^2 - \frac{\pi}{4}\|\mathbf{z}\|_\tau^2 + 2C.
\end{aligned}$$

In the second inequality we have used the fact that $\mathcal{A}^\tau(\mathbf{z}^-) \le 0$ and $\mathcal{A}^\tau(\zeta) = \lambda\|\zeta\|_\tau^2/2$. In the third inequality we have used the fact that $\mathbf{z}_1$ is perpendicular to ζ. In the last inequality we have used the fact that $\lambda \le 2\pi$. Now the boundary of Σ consists of those $\mathbf{z} + s\zeta \in \Sigma$ for which one of the following holds:

(i) $\|\mathbf{z}\|_\tau = T$ and $0 \le s \le T$,
(ii) $\|\mathbf{z}\|_\tau \le T$ and $s = T$,
(iii) $\|\mathbf{z}\|_\tau \le T$ and $s = 0$.

In the first two cases, it follows from the above estimate that $\mathcal{A}_H^\tau(\mathbf{z}+s\zeta) \le 0$. In the last case this follows from the fact that $\mathcal{A}^\tau \le 0$ on $E^- \oplus E^0$ and $V_\tau \ge 0$. Indeed, if V_τ were negative somewhere then, by (12.5.9), V_τ would have a negative critical value and hence, by Lemma 12.5.4, H would too. This is impossible by assumption.

Step 3 *The function $\mathcal{A}_H^\tau : \mathbb{R}^{2nN} \to \mathbb{R}$ has a positive critical value.*

For $t \ge 0$, denote by $\Sigma_t := \phi_t(\Sigma)$ the image of Σ under the gradient flow of $\mathcal{A}_H^\tau$. Define

$$c := \inf_{t\ge 0} \sup_{\Sigma_t} \mathcal{A}_H^\tau.$$

We will prove that $c > 0$ and c is a critical value of $\mathcal{A}_H^\tau$. By definition, $\partial\Sigma$ and Γ have linking number 1. (The linking number can be defined as the intersection number with Γ of any *ball* filling in the *sphere* $\partial\Sigma$.) Now the function $\mathcal{A}_H^\tau$ decreases along the gradient flow and hence it follows from Step 2 that

$$\sup_{\partial\Sigma_t} \mathcal{A}_H^\tau \le 0 \qquad \text{for all } t \ge 0.$$

By Step 1, this implies $\partial\Sigma_t \cap \Gamma = \emptyset$ for all $t \ge 0$ and so the linking number remains unchanged. Hence $\Sigma_t \cap \Gamma \ne \emptyset$ for every $t \ge 0$ and it follows again from Step 1 that

$$\sup_{\Sigma_t} \mathcal{A}_H^\tau \ge \inf_{\Gamma} \mathcal{A}_H^\tau > 0$$

for all $t \ge 0$ and hence $c > 0$.

Now suppose that c is not a critical value of $\mathcal{A}_H^\tau$. By Lemma 12.5.3 (iii), the Hamiltonian function H satisfies the condition (12.5.6) of Lemma 12.5.8. Hence $\mathcal{A}_H^\tau$ satisfies the Palais–Smale condition. Hence there exists a constant $\delta > 0$ such that the assertion of Lemma 12.5.7 holds with $T = 1$ and $\mathcal{A} = \mathcal{A}_H^\tau$. By definition of c, we may choose $t^* \ge 0$ such that

$$\sup_{\Sigma_{t^*}} \mathcal{A}_H^\tau \le c + \delta$$

and hence

$$\sup_{\Sigma_{t^*+1}} \mathcal{A}_H^\tau \le c - \delta.$$

But this contradicts the definition of c. Thus we have proved that c is a positive critical value of $\mathcal{A}_H^\tau$. By Corollary 12.5.6, this gives rise to a 1-periodic solution $z \in \mathrm{Per}(H)$ with $\mathcal{A}_H(z) > 0$. By Lemma 12.5.3, any such solution is nonconstant and is also a periodic solution of the Hamiltonian flow of H_0. This proves Lemma 12.5.12. □

Proof of Theorem 12.4.3: If $H \in \mathcal{H}(Z^{2n}(1))$ with $M = \sup H > \pi$, then $H_0 = M - H$ satisfies (I), (II), (III) on page 490. Hence, by Lemma 12.5.12, the Hamiltonian flow of H_0 has a nonconstant 1-periodic solution, and so has the Hamiltonian flow of H. This proves Theorem 12.4.3. □

In all the essential points our proof of Theorem 12.4.3 agrees with that given by Hofer and Zehnder [324]. This includes the extension of the Hamiltonian function, the Palais–Smale condition, and the linking argument. The only difference is that we use the discrete symplectic action functional $\mathcal{A}_H^\tau$ on $\mathbb{R}^{2nN}$, while Hofer and Zehnder work with the usual symplectic action functional $\mathcal{A}_H$ on the Sobolev space $H^{1/2}(S^1, \mathbb{R}^{2n})$. The advantage of our approach is that it does not require any infinite-dimensional functional analysis. Since our procedure for obtaining a critical point depends on the existence of a *global* generating function for symplectomorphisms close to the identity, it will only work in $\mathbb{R}^{2n}$ or other spaces (like the torus and cotangent bundles) which admit global symplectic coordinates. Apparently, similar restrictions apply to the variational approach of Hofer and Zehnder in the Sobolev space $H^{1/2}(S^1, \mathbb{R}^{2n})$ (cf. [325]). However, it may be possible to use our finite-dimensional approach to relate the Hofer–Zehnder capacity to Viterbo's in [646].

13

QUESTIONS OF EXISTENCE AND UNIQUENESS

This chapter discusses the fundamental existence and uniqueness questions in symplectic topology: which manifolds admit symplectic structures, and to what extent are they unique. There are partial answers for some classes of manifolds, particularly in dimension four, and many related open problems. The present chapter begins in Section 13.1 with a precise formulation of some relevant questions and continues in Section 13.2 by describing some related examples. Much of what we know is in dimension four, largely because of the existence of the powerful theory developed by Taubes and Seiberg–Witten in the 1990s. Section 13.3 outlines this theory, while Section 13.4 applies it in various explicit cases. This chapter does not offer many proofs, but only sketches a few arguments and points the reader towards the relevant references.

13.1 Existence and uniqueness of symplectic structures

Existence

Let M be a connected $2n$-dimensional manifold. The existence problem for symplectic forms can be stated as follows.

Which cohomology classes $a \in H^2(M;\mathbb{R})$ are represented by symplectic forms? Which homotopy classes of nondegenerate 2-forms contain a symplectic form?

For open manifolds (i.e. connected manifolds without boundary) these questions are completely answered by Gromov's h-principle [288, 190]. His theorem asserts that for every open manifold M and every cohomology class $a \in H^2(M;\mathbb{R})$, the inclusion of the space of symplectic forms representing the class a into the space of nondegenerate 2-forms on M is a homotopy equivalence. In particular, there exists a symplectic form in the cohomology class a if and only if M admits an almost complex structure and hence admits a nondegenerate 2-form (see Proposition 4.1.1 and Theorem 7.3.1). As we mentioned at the end of Section 7.3, the work [192] on symplectic cobordisms gives one a little control over boundary conditions.

For closed (i.e. compact without boundary) oriented $2n$-manifolds, necessary conditions for the existence of a symplectic form are that there is an almost complex structure and a cohomology class $a \in H^2(M;\mathbb{R})$ with $a^n > 0$. Whether these conditions are also sufficient is wide open in dimensions $2n \geq 6$. In dimension four, additional necessary conditions for existence arise from Taubes–Seiberg–Witten theory [606, 607]. These conditions show, for example, that an oriented connected sum of two 4-manifolds, each with $b^+ > 0$, cannot carry any symplectic form compatible with the orientation (see Proposition 13.3.13 in Section 13.3). The simplest example with this property is $\mathbb{C}P^2 \# \mathbb{C}P^2 \# \mathbb{C}P^2$, which has an almost complex structure by Exercise 4.1.11. One can consider analogous manifolds in dimension six (for example the connected sum of a closed symplectic 6-manifold with $\mathbb{C}P^3$ or with $S^3 \times S^3$), and it is an open question whether such a manifold carries a symplectic form. In dimension four, Taubes–Seiberg–Witten theory also gives constraints on the homotopy classes of almost complex structures. For example, every symplectic form on the 4-torus or on the K3-surface has first Chern class zero. On the one hand this can be viewed as a nonexistence theorem for symplectic forms in a given homotopy class of nondegenerate 2-forms; on the other hand, for the K3-surface, this gives rise to a uniqueness theorem which asserts that any two symplectic forms are homotopic as nondegenerate 2-forms.

Uniqueness

For a more precise discussion of the uniqueness problem in symplectic topology it is convenient to introduce the following **equivalence relations on the space of symplectic forms**. Throughout, M is a closed connected manifold. Consider the following statements for two symplectic forms ω_0, ω_1 on M.

(a) ω_0 and ω_1 are connected by a path of cohomologous symplectic forms.

(b) ω_0 and ω_1 are connected by a path of symplectic forms.

(c) ω_0 and ω_1 are connected by a path of nondegenerate 2-forms.

(d) ω_0 and ω_1 have the same first Chern class in $H^2(M;\mathbb{Z})$.

($\mathbf{A_0}$) There is a diffeomorphism ϕ of M, inducing the identity on cohomology, such that $\omega_0 = \phi^*\omega_1$.

($\mathbf{B_0}$) There is a diffeomorphism ϕ of M, inducing the identity on cohomology, such that ω_0 and $\phi^*\omega_1$ are connected by a path of symplectic forms.

($\mathbf{C_0}$) There is a diffeomorphism ϕ of M, inducing the identity on cohomology, such that ω_0 and $\phi^*\omega_1$ are connected by a path of nondegenerate 2-forms.

(A) There is a diffeomorphism ϕ of M such that $\omega_0 = \phi^*\omega_1$.

(B) There is a diffeomorphism ϕ of M such that ω_0 and $\phi^*\omega_1$ are connected by a path of symplectic forms.

(C) There is a diffeomorphism ϕ of M such that ω_0 and $\phi^*\omega_1$ are connected by a path of nondegenerate 2-forms.

(D) There is a diffeomorphism ϕ of M such that $c_1(\omega_0) = \phi^*c_1(\omega_1)$.

Two symplectic forms ω_0 and ω_1 are called **isotopic** if they are related by (a) (i.e. a path of cohomologous symplectic forms), they are called **homotopic** if they are related by (b) (i.e. a path of symplectic forms), they are called **diffeomorphic** or **symplectomorphic** if they are related by (A) (i.e. a diffeomorphism), and they are called **deformation equivalent** if they are related by (B) (i.e. a diffeomorphism, followed by a path of symplectic forms).[40]

For closed manifolds these equivalence relations are related as follows:

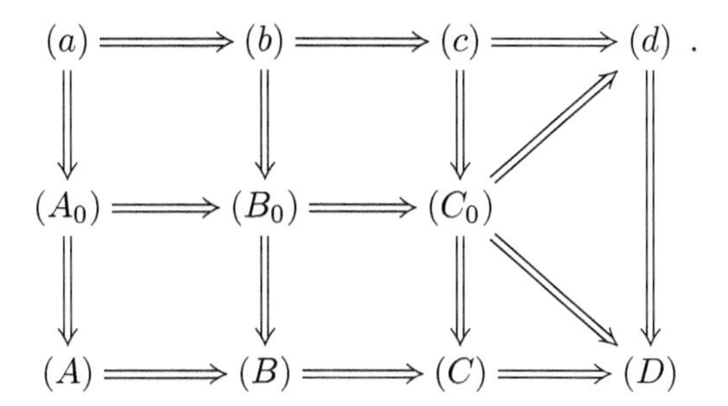

In particular, by Moser isotopy, assertion (a) holds if and only if there exists a diffeomorphism $\phi : M \to M$ isotopic to the identity such that

$$\phi^*\omega_1 = \omega_0.$$

Thus (a) implies (A_0). For (A_0) equivalent symplectic forms represent the same cohomology class and for (B_0) they have the same first Chern class and the same Gromov–Witten invariants. The relations (A_0) and (B_0) were used by Karshon–Kessler–Pinsonnault [357, 358, 359] in their study of blowups of $\mathbb{C}\mathrm{P}^2$.

The uniqueness problem in symplectic topology is the problem of understanding these equivalence relations. To formulate some precise questions, fix a closed oriented $2n$-manifold M. For $a \in H^2(M;\mathbb{R})$ with $a^n > 0$, denote by

$$\mathscr{S}_a := \left\{\omega \in \Omega^2(M) \,|\, d\omega = 0,\, \omega^n > 0,\, [\omega] = a\right\} \tag{13.1.1}$$

the set of symplectic forms representing the class a. For any nondegenerate 2-form $\rho \in \Omega^2(M)$ compatible with the orientation, denote by $c_1(\rho)$ the (common) first Chern class of the almost complex structures tamed by ρ. Denote by $\mathcal{C} \subset H^2(M;\mathbb{Z})$ the set of all first Chern classes associated to nondegenerate 2-forms compatible with the orientation, and by

$$\mathcal{C}_{\mathrm{symp}} := \bigcup_a \mathcal{C}_a \subset \mathcal{C}, \qquad \mathcal{C}_a := \{c_1(\omega) \,|\, \omega \in \mathscr{S}_a\}, \tag{13.1.2}$$

the set of all first Chern classes of symplectic forms compatible with the orientation. Here the union runs over all cohomology classes $a \in H^2(M;\mathbb{R})$ with $a^n > 0$. We also use the notation

$$\mathcal{K}^c := \left\{a \in H^2(M;\mathbb{R}) \,|\, a^n > 0 \text{ and } \exists\, \omega \in \mathscr{S}_a \text{ with } c_1(\omega) = c\right\} \tag{13.1.3}$$

[40] In the literature the term *'deformation equivalence'* is used with two different meanings, namely for the relations (b) and (B). The relation (b) is sometimes also called strong deformation equivalence, and the relation (B) is sometimes called weak deformation equivalence.

for the set of de Rham cohomology classes of symplectic forms, compatible with the orientation and with first Chern class c, and

$$\mathcal{K}_{\mathrm{symp}} := \{a \in H^2(M;\mathbb{R}) \,|\, a^n > 0,\, \mathscr{S}_a \neq \emptyset\} = \bigcup_{c \in \mathcal{C}_{\mathrm{symp}}} \mathcal{K}^c \qquad (13.1.4)$$

for the set of all de Rham cohomology classes of symplectic forms compatible with the orientation.

The reader is cautioned that $\mathcal{C}_{\mathrm{symp}}$ is a set of integral cohomology classes, and hence a subset of a $\mathbb{Z}$-module, while $\mathcal{K}_{\mathrm{symp}}$ is a set of de Rham cohomology classes and an open subset of a vector space. Clearly $\mathcal{K}_{\mathrm{symp}}$ is always the union of the symplectic cones $\mathcal{K}^c$ as c varies over $\mathcal{C}_{\mathrm{symp}}$. Proposition 13.3.11 shows that when M is a closed oriented smooth four-manifold, this union is *disjoint*. Thus in this case $\mathcal{K}_{\mathrm{symp}}$ has a chamber structure, which turns out to be particularly interesting in the case $b^+ = 1$; cf. Figure 13.4. Still in dimension four, we emphasize that often, but not always, the orientation of M is determined by the first Chern class via the Hirzebruch signature formula

$$c^2 = 2\chi + 3\sigma$$

(see Remark 4.1.10). The exceptions are four-manifolds with Euler characteristic zero, since reversing the orientation changes the signs of both σ and c^2.

Uniqueness questions

Here are some questions one can ask about the aforementioned equivalence relations. Throughout we consider the C^∞ topology on the space of symplectic forms and assume that $\mathscr{S}_a$ is nonempty, respectively, that M carries a symplectic form.

1. *Is the space $\mathscr{S}_a$ connected (and hence path-connected)?*

2. *Are any two symplectic forms in $\mathscr{S}_a$ diffeomorphic (by a diffeomorphism inducing the identity on cohomology)?*

3. *Are any two symplectic forms on M deformation equivalent?*

When $\#\mathcal{C}_a > 1$ the answer to Question 1 above is negative, because the class $c_1(\omega_t)$ does not change along a path ω_t of nondegenerate forms. In Proposition 13.3.11 we shall see that $\#\mathcal{C}_a = 1$ for every closed symplectic 4-manifold. In other words, the cohomology class of ω determines $c_1(\omega)$. This is somewhat surprising in view of the fact that there are four-manifolds that support symplectic forms whose first Chern classes are so different that even condition (D) does not hold (see McMullen–Taubes [476], Smith [592, 593], and Vidussi [641, 642]). In fact, Question 1 is completely open for closed symplectic four-manifolds: There is no known pair (M, a) (with M a closed four-manifold) with disconnected $\mathscr{S}_a$, nor is there one with $\mathscr{S}_a$ nonempty and connected. (As we will see in Section 13.4, for many examples this question is closely related to the properties of the diffeomorphism group.) A related existence problem is whether every cohomology class $c \in \mathcal{C}$ is the first Chern class of a symplectic form, or equivalently, whether $\mathcal{C}_{\rm symp}$ is equal to $\mathcal{C}$. As noted above, the answer to this question is *'no'* for the 4-torus (see Example 13.4.8). We will see in Example 13.4.4 that it is also *'no'* for blowups of $\mathbb{C}\mathrm{P}^2$ at nine or more points but is *'yes'* for the blowup of $\mathbb{C}\mathrm{P}^2$ at up to eight points. Most known results about the uniqueness problem require deep analytical techniques such as pseudoholomorphic curves, Floer homology, and, in dimension four, Taubes–Seiberg–Witten theory.

13.2 Examples

This section presents a miscellany of examples. We begin with some results on the extension or filling problem. Rather little is known here. Since by Gromov's Theorem 7.3.1 the complement $M \setminus B$ of a ball in an almost complex manifold always admits a symplectic structure; if one could understand when this extends to the whole of M we would have a good handle on the existence problem. Next we discuss symplectic structures on Euclidean space, in particular the existence of exotic structures that are not covered by a single Darboux chart. We end with a variety of examples of different symplectic structures on manifolds of dimension six and above.

Symplectic filling

In order to get a good geometric understanding of what a symplectic structure is, it is important to understand the relative version of the existence problem that has to do with extending forms. Simple examples show that it is not always possible to extend a symplectic form which is defined on an open subset U of M to a symplectic form on the whole of M, even when there is a suitable cohomology class and an extension by a nondegenerate form. For example, consider the unit ball $M = B^{2n}$ in $\mathbb{R}^{2n}$ with $n \geq 2$, and suppose that ω is a nondegenerate closed 2-form near the boundary ∂B^{2n}. A necessary condition for ω to extend to a symplectic form over B^{2n} is that ω must extend to a nondegenerate form over B^{2n}. Moreover, if λ is any 1-form near ∂B^{2n} such that $\omega = d\lambda$, there is a symplectic extension $\tilde{\omega}$ of ω only if

$$\int_{\partial B^{2n}} \lambda \wedge (d\lambda)^{n-1} > 0. \tag{13.2.1}$$

To see this, observe that because the de Rham cohomology group $H^2(B^{2n}, \partial B^{2n})$ vanishes, if $\tilde{\omega}$ exists we can extend λ to a 1-form $\tilde{\lambda}$ such that $d\tilde{\lambda} = \tilde{\omega}$. Stokes' theorem then implies that

$$0 < \int_{B^{2n}} \tilde{\omega}^n = \int_{B^{2n}} (d\tilde{\lambda})^n = \int_{\partial B^{2n}} \lambda \wedge (d\lambda)^{n-1}.$$

It follows as in the proof of Proposition 3.5.33 that the last integral above is independent of the choice of λ.

Exercise 13.2.1 Let $\iota : S^3 \to \mathbb{R}^4$ be a map of the form

$$(x_1, y_1, x_2, y_2) \mapsto (\theta(x_1, y_1), y_1, x_2, y_2).$$

Show that it is possible to choose the function $\theta : \mathbb{R}^2 \to \mathbb{R}$ so that ι is an immersion and so that the integral of $\lambda \wedge d\lambda$ over S^3 is negative (see Fig. 13.1). Show further that the 2-form $\omega = \iota^*\omega_0$ extends to a positively oriented nondegenerate form over B^4, but not to a symplectic form. □

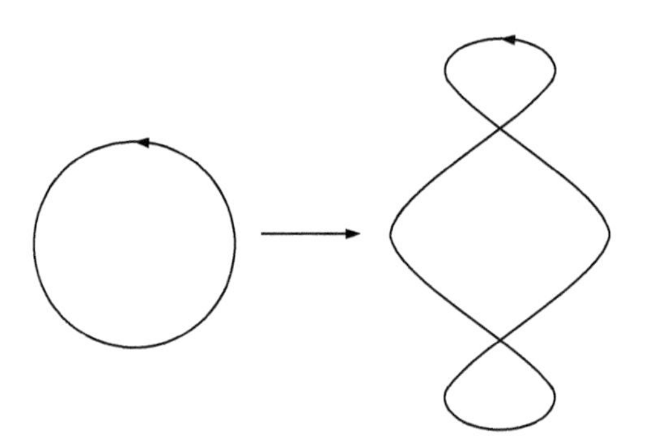

FIG. 13.1. Immersing S^3.

Here is a more subtle version of the above example proposed by Eliashberg.

Example 13.2.2 This example describes an immersion $\iota : S^3 \to \mathbb{R}^4$ such that the pullback $\iota^*\omega_0$ of the standard form on $\mathbb{R}^4$ is not fillable even though the positivity condition (13.2.1) is satisfied. Consider first the polydisc

$$P = B^2(\varepsilon) \times B^2(1)$$

with the symplectic form induced by the standard symplectic structure on $\mathbb{R}^4$. We first claim that if W is a symplectic 4-manifold with boundary $\partial W = \partial P$, then the Gromov width of W is bounded by

$$w_G(W) \le \pi\varepsilon^2. \tag{13.2.2}$$

To see this, note that for every $\delta > 0$ the manifold P can be symplectically embedded into a product $S \times \Sigma$ of two spheres with areas

$$\text{area}(S) = \pi(\varepsilon + \delta)^2, \qquad \text{area}(\Sigma) = \pi(1 + \delta)^2.$$

Joining W to the complement $S \times \Sigma - P$ gives rise to a symplectic 4-manifold X that contains a J-holomorphic sphere of zero self-intersection and area $\pi(\varepsilon + \delta)^2$. Hence it follows from Remark 12.1.3 that $w_G(W) \le w_G(X) \le \pi(\varepsilon + \delta)^2$. Since $\delta > 0$ can be chosen arbitrarily small, it follows that every symplectic filling W of ∂P satisfies (13.2.2).

Now we modify the boundary of P as follows. Geometrically we would like to cut a 4-ball of radius r (and hence of volume $\pi^2 r^4/2$) out of P, where

$$r^2/\sqrt{2} < \varepsilon < r.$$

In practice the hole is not contained in P, so that the boundary of the resulting space is immersed rather than embedded. To be precise, let $Q = B^4(r)$ and identify a small open 3-disc $S \subset \partial P$ with an open 3-disc $T \subset \partial Q$ by an orientation-preserving diffeomorphism $\phi : S \to T$ that extends to a symplectomorphism between some neighbourhoods of S and T, respectively. Now define

$$R = (\partial P \setminus S) \cup_\phi (\partial Q \setminus T),$$

where the points in $\partial_P S$ and $\partial_Q T$ are identified via ϕ. Smoothing the corners of R we obtain an immersion $\iota : S^3 \to \mathbb{R}^4$ such that the pullback form $\iota^*\omega_0$ does not admit a symplectic filling. If it did, then any such filling (W, ω) with $(\partial W, \omega) = (R, \iota^*\omega_0)$ could be extended, by adding the 4-ball Q, to obtain a filling (W', ω') of $(\partial P, \omega_0)$. By construction, this filling W' would contain a symplectically embedded ball of Gromov width πr^2 and hence would satisfy

$$w_G(W') \geq \pi r^2 > \pi \varepsilon^2$$

in contradiction to (13.2.2). This shows that $(R, \iota^*\omega_0)$ does not admit a symplectic filling, even though $\iota^*\omega_0$ satisfies (13.2.1) whenever $r^2/\sqrt{2} < \varepsilon$. □

Remark 4.5.2 (ix) describes another situation in which a filling does not exist. However, as we explain in Example 7.3.3, in many cases it is possible to fill if one allows one singular point.

Exotic structures on $\mathbb{R}^{2n}$

In order to bypass some of the questions about the structure at infinity, let us first ask the question: Is there any symplectic structure ω on $\mathbb{R}^{2n}$ which is essentially different from the standard one? By this we mean that the form ω cannot be obtained from the standard form by pulling it back via an embedding $\iota : \mathbb{R}^{2n} \to \mathbb{R}^{2n}$. Such a form will be called **exotic**. In [287] Gromov proved the existence of an exotic symplectic structure on $\mathbb{R}^{2n}$ by using the following result.

Theorem 13.2.3 (Gromov) *Consider a Lagrangian embedding $\iota : L \to \mathbb{R}^{2n}$ of a compact n-dimensional manifold L into $(\mathbb{R}^{2n}, \omega_0)$. Then the closed 1-form*

$$\iota^*\lambda_0 \in \Omega^1(L), \qquad \lambda_0 = \sum_{j=1}^n y_j dx_j,$$

represents a nontrivial cohomology class in $H^1(L; \mathbb{R})$.

Proof: See [287] and also [470, Theorem 9.2.1]. □

Recall that a Lagrangian submanifold L in an exact symplectic manifold $(M, -d\lambda)$ is called **exact** if λ restricts to an exact 1-form on L (see Definition 9.3.5). Thus, Theorem 13.2.3 asserts that there is no exact Lagrangian embedding of a compact manifold into $(\mathbb{R}^{2n}, \omega_0)$. This follows from Gromov's Theorem 11.3.10, which states that if L is an exact compact Lagrangian submanifold of T^*Y and if $\psi : T^*Y \to T^*Y$ is a Hamiltonian symplectomorphism, then $L \cap \psi(L) \neq \emptyset$. This result is stated in Section 11.3 for compact manifolds Y, but also applies to certain noncompact manifolds, and in particular to $Y = \mathbb{R}^n$. Since any compact submanifold of $T^*\mathbb{R}^n = \mathbb{R}^{2n}$ may be disjoined from itself by a Hamiltonian isotopy — for example, by a translation — there can be no compact exact Lagrangian submanifolds in $\mathbb{R}^{2n}$. The next example shows that there do exist exact Lagrangian immersions into $\mathbb{R}^{2n}$.

Example 13.2.4 Let S^n denote the sphere $\sum_{j=0}^n \xi_j^2 = 1$ in $\mathbb{R}^{n+1}$. The immersion

$$\iota : S^n \to \mathbb{R}^{2n} : (\xi_0, \ldots, \xi_n) \mapsto (\xi_1, \ldots, \xi_n, \xi_0\xi_1, \ldots, \xi_0\xi_n)$$

is Lagrangian since $\iota^*\omega_0 = \sum_{j=1}^n \xi_j d\xi_j \wedge d\xi_0 = 0$. Note that $\iota : S^n \to \mathbb{R}^{2n}$ is not an embedding since the north and the south pole are both mapped to the origin (see Fig. 13.2). □

Exercise 13.2.5 In the situation of Theorem 13.2.3, verify that the cohomology class $[\iota^*\lambda] \in H^1(L; \mathbb{R})$ is independent of the choice of the 1-form λ with $\omega_0 = -d\lambda$. □

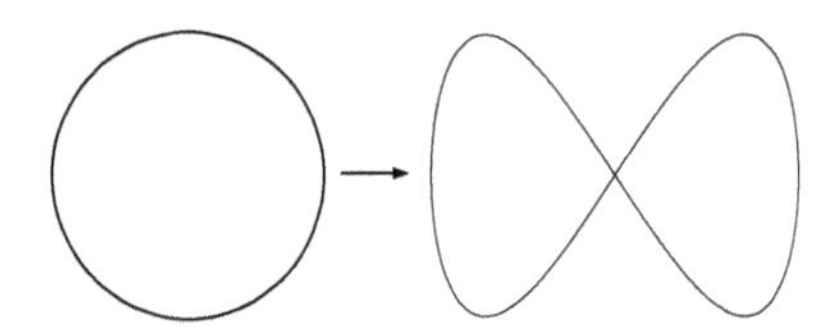

FIG. 13.2. The map $(x, y) \mapsto (y, xy)$.

In [287] Gromov used Theorem 7.3.1 and Theorem 13.2.3 to give a nonconstructive existence proof for exotic symplectic structures on $\mathbb{R}^{2n}$. Later Bates and Peschke [54] found a simple and explicit formula which we now describe.

Example 13.2.6 (Bates and Peschke) Consider the following 1-form on $\mathbb{R}^4$:

$$\sigma = \cos(r_1^2)(x_1 dy_1 - y_1 dx_1) + \cos(r_2^2)(x_2 dy_2 - y_2 dx_2),$$

where $r_1^2 := x_1^2 + y_1^2$ and $r_2^2 := x_2^2 + y_2^2$. Observe that σ vanishes on the embedded 2-torus

$$L = \{(x, y) \in \mathbb{R}^2 \times \mathbb{R}^2 \,|\, r_1 = r_2 = \sqrt{\pi/2}\}.$$

Furthermore, the 2-form

$$d\sigma = a(r_1)dx_1 \wedge dy_1 + a(r_2)dx_2 \wedge dy_2, \qquad a(r) = 2\cos r^2 - 2r^2 \sin r^2$$

is nondegenerate except at points where one of $a(r_1)$ or $a(r_2)$ vanishes. Since these functions cannot vanish simultaneously at points on the 3-sphere

$$S^3 = \{(x, y) \in \mathbb{R}^2 \times \mathbb{R}^2 \,|\, r_1^2 + r_2^2 = \pi\},$$

the 2-form $d\sigma$ does not vanish on S^3. This implies that there exists a 1-form μ such that $\mu \wedge d\sigma$ is a volume form on S^3. To be more explicit, we may take

$$\mu = a(r_1)(x_2dy_2 - y_2dx_2) + a(r_2)(x_1dy_1 - y_1dx_1).$$

It follows that the 2-form

$$\omega = -d(\sigma + f\mu), \qquad f(x, y) = r_1^2 + r_2^2 - \pi,$$

is nondegenerate in a neighbourhood of S^3. To see this, note that f vanishes on S^3 and

$$\omega \wedge \omega \;=\; \big(d\sigma + 2df \wedge \mu\big) \wedge d\sigma + f\big(f d\mu \wedge d\mu + 2d\mu \wedge df \wedge \mu + 2d\mu \wedge d\sigma\big).$$

A calculation shows that

$$\big(d\sigma + 2df \wedge \mu\big) \wedge d\sigma = 2\big(a(r_1)a(r_2) + 2r_1^2a(r_2)^2 + 2r_2^2a(r_1)^2\big)dx_1 \wedge dy_1 \wedge dx_2 \wedge dy_2,$$

which is nonzero at the points of S^3. Hence ω is nondegenerate near S^3. Note also that $L \subset S^3$ is an exact Lagrangian torus. In fact, the primitive $\sigma + f\mu$ is zero on L. Finally, we choose some point $x \in S^3$ which is not on L and identify $\mathbb{R}^4$ with $(S^3 \setminus \{x\}) \times (-\varepsilon, \varepsilon)$ using stereographic projection.

Thus we have constructed a 2-form $\omega = -d\lambda$ on $\mathbb{R}^4$ which admits an exact embedded Lagrangian torus

$$\iota : \mathbb{T}^2 \to \mathbb{R}^4, \qquad \iota^*\lambda = 0.$$

We claim that this form ω is exotic. For otherwise ω would have the form $\psi^*\omega_0$ for some embedding $\psi : \mathbb{R}^4 \hookrightarrow \mathbb{R}^4$. Then $\psi^*\lambda_0 - \lambda$ would be a closed 1-form and so there would exist a function $\theta : \mathbb{R}^4 \to \mathbb{R}$ such that

$$\psi^*\lambda_0 - \lambda = d\theta.$$

Hence the 1-form

$$(\psi \circ \iota)^*\lambda_0 = \iota^*(\lambda + d\theta) = d(\theta \circ \iota)$$

would be exact, which would mean that the embedding

$$\psi \circ \iota : \mathbb{T}^2 \to (\mathbb{R}^4, \omega_0)$$

is exact, in contradiction to Theorem 13.2.3.

Another explicit construction of an exotic structure on $\mathbb{R}^4$ may be found in Leung–Symington [415, §5]. We remark that it is an open question whether the symplectic form ω in the present example is the pullback of the standard symplectic form ω_0 by a *self-immersion* of $\mathbb{R}^4$. The image of L under such an immersion would have to be exact with at least one self-intersection. Such tori do exist; for example, one could take the product of n exact figure-8 immersions of the circle $S^1 \to \mathbb{R}^2 : (x, y) \to (y, xy)$, as in Example 13.2.4. □

In Example 13.2.6, as well as Gromov's example in [287], the symplectic manifold has no claim to any kind of completeness at infinity. It is a much deeper and more interesting problem to find symplectic manifolds that are convex at infinity (Definition 3.5.39) and diffeomorphic, but not symplectomorphic to $(\mathbb{R}^{2n}, \omega_0)$. In [496], Müller constructed a symplectic form on $\mathbb{R}^6$ which is exotic because it admits an embedded Lagrangian 3-sphere, and which admits a Liouville vector field X which is complete in the sense that it integrates to a family of diffeomorphisms; however, it is not convex at infinity. In Müller's example the symplectic structure is obtained from a nonstandard contact structure on $\mathbb{R}^5$ via symplectization (see Section 3.5). In [472], McDuff and Traynor proved that the manifold

$$E_s := \mathbb{R}^{2n} \setminus \left\{ z \,\middle|\, x_1 = 0, \sum_j (x_j^2 + y_j^2) \geq s^2 \right\}$$

is convex at infinity. This example arose in the discussion in Section 1.2 of the camel problem as the complement of the wall-with-hole. The manifold (E_s, ω_0) is not symplectomorphic to $(\mathbb{R}^{2n}, \omega_0)$, because a theorem of Gromov [287] implies that the space of embeddings of a symplectic ball of radius bigger than s into E_s is not path-connected, while the corresponding space of embeddings into $\mathbb{R}^{2n}$ is, by Exercise 7.1.27 (see Fig. 13.3). In [474, 475], Mark McLean constructed, for each $n \geq 4$, infinitely many pairwise nonsymplectomorphic finite type Weinstein manifolds (Definition 7.4.5) that are all diffeomorphic to $\mathbb{R}^{2n}$. Exotic symplectic structures were also constructed on many other Liouville manifolds (i.e. symplectic manifolds that are convex at infinity and admit global Liouville vector fields) by Abouzaid–Seidel [6] and others. The paper [6] is also an excellent reference for the history of this problem.

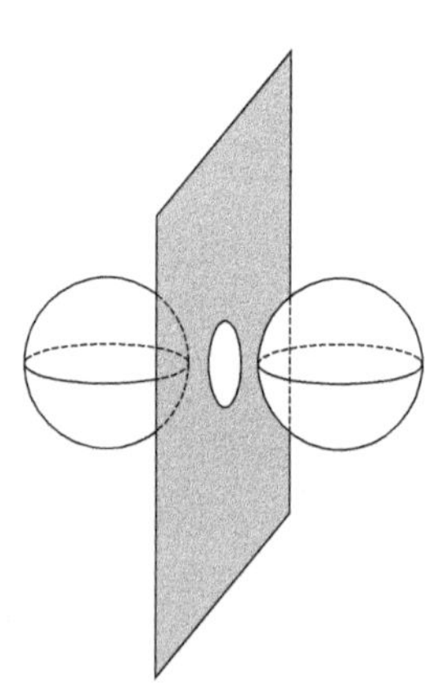

FIG. 13.3. Nonisotopic balls in E_s.

Here is one situation in which symplectic forms on Euclidean space are known to be isomorphic. It is a classical theorem in Riemannian geometry that a complete simply connected Riemannian manifold of nonpositive sectional curvature is always diffeomorphic to $\mathbb{R}^n$. McDuff [446] proved that this result extends to the symplectic category. The proof is elementary, involving the use of comparison theorems from Riemannian geometry.

Proposition 13.2.7 *A complete simply connected Kähler manifold of nonpositive sectional curvature is symplectomorphic to standard Euclidean space.*

Proof: See McDuff [446]. □

Example 13.2.8 Gromov [287] has proved that any symplectic structure on the connected sum $M = \mathbb{C}\mathrm{P}^3 \# \mathbb{T}^6$ is nonstandard on the submanifold $\mathbb{C}\mathrm{P}^2 \subset \mathbb{C}\mathrm{P}^3$. Details of the proof may be found in [451]. (One can also prove this by using part (ix) of Remark 4.5.2 in Chapter 4. Namely, if there were a symplectic form σ on $M = \mathbb{C}\mathrm{P}^3 \# T^6$ which had the standard $\mathbb{C}\mathrm{P}^2$ as a symplectic submanifold, then a neighbourhood of infinity in the open manifold $(\mathbb{C}\mathrm{P}^3 \# T^6) \setminus \mathbb{C}\mathrm{P}^2$ would be symplectomorphic to an open annulus $\{x \in \mathbb{R}^6 \,|\, \lambda_1 < |x| < \lambda_2\}$, and one could make it asymptotically flat by gluing on $\mathbb{R}^6 \setminus B^6(\lambda_2)$. This is impossible by part (ix) of Remark 4.5.2.) Note that this manifold M admits an almost complex structure and a suitable cohomology class $a \in H^2(M;\mathbb{R})$. It is an open question whether it admits a symplectic structure. □

Example 13.2.9 (A six-manifold) This example is due to McDuff [445] (see also [470, Theorem 9.7.4]). Here the set $\mathscr{S}_a$ is disconnected. However, by an arbitrarily small perturbation of the cohomology class a, the two known distinct connected components of $\mathscr{S}_a$ merge to a single connected component. McDuff's paper [445] also contains a variant of this construction in dimension eight, where two cohomologous symplectic forms are not even diffeomorphic.

Consider the manifold $M := \mathbb{T}^2 \times S^2 \times S^2$. Identify the 2-torus with the product of two circles. For $\theta \in S^1$ and $z \in S^2 \subset \mathbb{R}^3$, denote the rotation about the axis through z by the angle θ by $\phi_{z,\theta} : S^2 \to S^2$. Consider the diffeomorphism $\psi : M \to M$ defined by

$$\psi(\theta_1, \theta_2, z_1, z_2) := (\theta_1, \theta_2, z_1, \phi_{z_1,\theta_1}(z_2)), \qquad \theta_1, \theta_2 \in S^1, \qquad z_1, z_2 \in S^2.$$

It acts as the identity on cohomology. Let $\omega \in \Omega^2(M)$ be the product symplectic form, where both S^2 factors have the same area, and denote

$$a := [\omega] = [\psi^*\omega] \in H^2(M;\mathbb{R}).$$

McDuff's theorem asserts that ω and $\psi^*\omega$ can be joined by a path of symplectic forms, but not by a path of cohomologous symplectic forms. The proof that ω and $\psi^*\omega$ are not isotopic cannot be based on the Gromov–Witten invariants because these are invariant under deformation of symplectic forms (equivalence relation (B) on page 504). In other words, the evaluation maps from the relevant moduli spaces of holomorphic spheres are homologous. However, it turns out that they have different Hopf invariants. The argument breaks down for symplectic forms where the S^2 factors have different areas, because in that case the relevant moduli spaces are no longer compact. □

Example 13.2.10 A quite different example is provided by Ruan [546]. He finds two symplectic structures on a certain closed 6-manifold Z which are not deformation equivalent (i.e. related by a diffeomorphism followed by a path of symplectic forms). Observe first that a Kähler manifold carries a natural deformation class of symplectic forms: the set of Kähler forms on a complex manifold is convex, and so all such forms are homotopic (i.e. can be joined by a path of symplectic forms). Ruan considers pairs of homeomorphic but nondiffeomorphic Kähler surfaces (X, ω_X) and (Y, ω_Y) that become diffeomorphic when multiplied by S^2. For example, one can take X to be $\mathbb{C}P^2$ blown up at eight points, and Y to be the Barlow surface. Then $X \times S^2$ and $Y \times S^2$ can both be identified with the same manifold Z via diffeomorphisms

$$f_X : X \times S^2 \to Z, \qquad f_Y : Y \times S^2 \to Z.$$

Hence Z carries two symplectic forms

$$\tau_X = (f_X^{-1})^*(\omega_X \times \sigma), \qquad \tau_Y = (f_Y^{-1})^*(\omega_Y \times \sigma),$$

where σ is an area form on S^2. By using results on the classification of almost complex structures on 6-manifolds due to Wall, one can show that the diffeomorphisms f_X and f_Y can be chosen such that the almost complex structures associated to τ_X and τ_Y are homotopic. Hence the deformation classes of τ_X and τ_Y belong to the same path component of the space of nondegenerate (nonclosed) 2-forms.

Let us now consider the homology classes $c_{A,\tau}$ in Z that are defined by evaluation maps $\mathrm{ev} : \mathcal{M}(A, J) \times_{\mathrm{G}} S^2 \to Z$ (see Section 4.5). First suppose that J_Y is compatible with τ_Y and, via f_Y, diffeomorphic to a product structure on $Y \times S^2$. We claim that the moduli spaces $\mathcal{M}(A, J_Y)$ are empty for all classes A in the image of the homomorphism $H_2(Y) \to H_2(Z)$ induced by the map $Y \to Z : y \mapsto f_Y(y, \mathrm{pt})$. To see this, note that the projection of $Z \cong Y \times S^2$ onto Y is J_Y-holomorphic. Hence any J_Y-holomorphic curve must project to a holomorphic curve in Y, which cannot be constant by the assumption on A. Since the Barlow surface Y contains no nontrivial holomorphic spheres, $\mathcal{M}(A, J_Y)$ must be empty as claimed. Hence the classes c_{A,τ_Y} vanish for all such A. Moreover, they depend only on the deformation class of τ_Y by general Gromov–Witten theory.

On the other hand, these classes do not vanish when we look at the deformation class containing τ_X. To see this, consider $A \in H_2(Z)$ of the form

$$A = f_{X*}[\Sigma \times pt],$$

where $\Sigma \subset X$ is one of the exceptional divisors. It is not hard to see that this class A is in the image of the map $H_2(Y) \to H_2(Z)$ so that our previous argument applies. However, if J_X is the pushforward under f_X of a product complex structure on $X \times S^2$, it is easy to see that the moduli space $\mathcal{M}(A, J_X)/\mathrm{G}$ is diffeomorphic to S^2. Indeed, there is a 2-parameter family of J_X-holomorphic curves $f_X(\Sigma \times \{w\})$ in the class A parametrized by $w \in S^2$. Moreover, one can show that J_X satisfies the appropriate regularity condition. Hence the corresponding invariant c_{A,τ_X} equals the nonzero class $[\Sigma \times S^2]$. It follows that τ_X and τ_Y are not deformation equivalent.

Thus the manifold $Z \cong X \times S^2 \cong Y \times S^2$ carries two different deformation equivalence classes of symplectic forms. It was noted by Kaoru Ono that in this particular example one cannot choose representatives τ_1 and τ_2 of the two deformation equivalence classes which represent in the same cohomology class. However, Ruan also has more elaborate examples in which different deformation equivalence classes do have cohomologous representatives. □

Example 13.2.11 In [346], Ionel and Parker computed the Gromov invariants, and hence the Seiberg–Witten invariants, of products of the circle with three-dimensional mapping tori. As a result they were able to construct, for each integer $n \geq 2$, infinitely many symplectic four-manifolds that are homeomorphic, but not diffeomorphic, to the elliptic surface $E(n)$. The manifold $E(n)$ can be obtained as the fibre connected sum of n copies of the nine point blowup

$$E(1) := \mathbb{C}\mathrm{P}^2 \# 9 \overline{\mathbb{C}\mathrm{P}}^2$$

of the projective plane which admits the structure of a Lefschetz fibration over $\mathbb{C}\mathrm{P}^1$ with 12 singular fibres and the regular fibres diffeomorphic to the 2-torus. The construction of Ionel–Parker involves Gompf surgery and fibred knots. The same result was obtained by Fintushel–Stern [219] and, for $n = 2$ where $E(2)$ is the K3-surface, by Vidussi [641]. Taking the product with S^2 one obtains manifolds that are all diffeomorphic to $M := E(n) \times S^2$. Ionel–Parker [346] proved that this gives rise to infinitely many symplectic structures on $E(n) \times S^2$ that are pairwise not deformation equivalent. □

In [579], Seidel developed a new technique for the construction of deformation inequivalent symplectic structures in all dimensions, based on Fukaya categories, and explained many interesting examples.

Example 13.2.12 A longstanding question in symplectic topology is whether two closed smooth manifolds of the same dimension are diffeomorphic if and only if their cotangent bundles with their canonical symplectic structures are symplectomorphic. Interesting examples to consider are homotopy spheres of the same dimension; they have diffeomorphic cotangent bundles. In [3], Abouzaid proved that if Σ is a homotopy $(4k+1)$-sphere that does not bound a parallelizable manifold then $T^*\Sigma$ and T^*S^{4k+1} are not symplectomorphic. Such manifolds exist for many, but not all, $k \geq 2$. For example there are six diffeomorphism classes of such homotopy spheres in dimension 9, two in dimension 13, and fourteen in dimension 17, but none in dimension 61 (see Kervaire–Milnor [367] and Milnor [486]). When $T^*\Sigma$ and T^*S^{4k+1} are symplectomorphic, Abouzaid constructs the required parallelizable manifold with boundary Σ via a suitable moduli space of holomorphic discs. See also [173]. □

13.3 Taubes–Seiberg–Witten theory

The study of symplectic four-manifolds was revolutionized in 1994 and 1995 by Taubes–Seiberg–Witten theory. In [606, 607] Taubes proved that symplectic four-manifolds have nontrivial Seiberg–Witten invariants and that these share certain properties with the Seiberg–Witten invariants of Kähler surfaces. These relatively elementary results already have far reaching consequences in symplectic topology, explained below. In a series of papers [608, 609, 610, 611], he then proved a much deeper theorem which relates the Seiberg–Witten invariants to a variant of the Gromov–Witten invariant, called the **Gromov invariant**, which counts possibly disconnected embedded J-holomorphic curves representing a given homology class. These deeper results have many further consequences in four-dimensional symplectic topology, some of which will be discussed below. The proofs go far beyond the scope of the present book. We begin by discussing some relevant properties of spin^c structures and the Seiberg–Witten invariants of smooth four-manifolds, then discuss Taubes' elementary theorem about the Seiberg–Witten invariants of symplectic four-manifolds and formulate some of its consequences, before moving on to the Gromov invariants, Taubes' deeper theory about *'Seiberg–Witten equals Gromov'*, and some of the consequences of this deeper theory.

Spin^c structures

Let M be a closed oriented smooth four-manifold. Denote the set of (equivalence classes of) spin^c structures on M by $\mathrm{Spin}^c(M)$.[41] The tensor product with complex line bundles defines a free and transitive action

$$H^2(M;\mathbb{Z}) \times \mathrm{Spin}^c(M) \to \mathrm{Spin}^c(M) : (e, \Gamma) \mapsto \Gamma_e.$$

Each $\Gamma \in \mathrm{Spin}^c(M)$ has a first Chern class $c_1(\Gamma) \in H^2(M;\mathbb{Z})$ and

$$c_1(\Gamma_e) = c_1(\Gamma) + 2e. \tag{13.3.1}$$

Thus if $H^2(M;\mathbb{Z})$ has no 2-torsion (for example if $H_1(M;\mathbb{Z}) = 0$) the class e is determined by $c_1(\Gamma_e) - c_1(\Gamma)$ and the spin^c structure Γ by $c_1(\Gamma)$.

Remark 13.3.1 Every nondegenerate 2-form $\rho \in \Omega^2(M)$, compatible with the orientation, determines an equivalence class of spin^c structures $\Gamma_\rho \in \mathrm{Spin}^c(M)$,

[41] A spin^c-structure on an oriented Riemannian 4-manifold is a pair of rank-2 Hermitian vector bundles $W^\pm$, equipped with a bundle homomorphism $\gamma : TM \to \mathrm{Hom}(W^-, W^+)$ satisfying

$$\gamma(v)^*\gamma(v) = |v|^2 \mathbb{1}, \qquad \gamma(e_0)^*\gamma(e_1)\gamma(e_2)^*\gamma(e_3) = \mathbb{1}$$

for all $v \in TM$, and positive orthonormal frames e_0, e_1, e_2, e_3 of TM. The equivalence relation is given by bundle isomorphisms preserving γ and by deformations of the Riemannian metric and γ. The bundles $W^\pm$ have the same first Chern class $c_1(\gamma) := c_1(W^+) = c_1(W^-)$. □

associated to the ρ-compatible almost complex structures.[42] It depends only on the homotopy class of ρ and has the first Chern class $c_1(\Gamma_\rho) = c_1(\rho)$. The equivalence class of the tensor product of Γ_ρ with a Hermitian line bundle with first Chern class e is denoted by $\Gamma_{\rho,e}$. Thus $\Gamma_{\rho,0} = \Gamma_\rho$. □

Remark 13.3.2 (Involution) The space of spin^c structures on a closed oriented smooth four-manifold M carries a natural involution

$$\mathrm{Spin}^c(M) \to \mathrm{Spin}^c(M) : \Gamma \mapsto \overline{\Gamma}$$

obtained by reversing the complex structures on $W^{\pm}$. It satisfies $c_1(\overline{\Gamma}) = -c_1(\Gamma)$ and $\overline{\Gamma}_e = \Gamma_{-c-e}$, where $c := c_1(\Gamma)$, as well as

$$\overline{\Gamma}_{\rho,e} = \Gamma_{\rho,-c-e} = \Gamma_{-\rho,-e}, \qquad c := c_1(\rho), \tag{13.3.2}$$

for every $e \in H^2(M;\mathbb{Z})$ and every nondegenerate 2-form ρ. □

Remark 13.3.3 (Homological orientation) A Riemannian metric on M determines subspaces $H^{2,\pm}(M;\mathbb{R})$ of $H^2(M;\mathbb{R})$, consisting of the cohomology classes of all self-dual, respectively anti-self-dual, harmonic 2-forms. Their dimensions are independent of the Riemannian metric and are denoted by

$$b^+ := \dim H^{2,+}(M;\mathbb{R}), \qquad b^- := \dim H^{2,-}(M;\mathbb{R}).$$

Thus $b^+ + b^- = b_2$ is the second Betti-number and $b^+ - b^- = \sigma(M)$ is the **signature** of M. A **homological orientation of** M is the choice of an orientation of the real vector space

$$H^0(M;\mathbb{R}) \oplus H^1(M;\mathbb{R}) \oplus H^{2,+}(M;\mathbb{R})$$

for some, and hence every, Riemannian metric on M. Every nondegenerate 2-form ρ that is compatible with the orientation naturally determines such a homological orientation (see Donaldson [142]). For a Kähler surface (M, ω, J) the homological orientation is determined by the complex structure $\alpha \mapsto *(\omega \wedge \alpha) = -\alpha \circ J$ on the space of harmonic 1-forms and the complex structure

$$(f, \tau) \mapsto \left(\frac{\omega \wedge \tau}{\omega \wedge \omega}, -\iota(J)\tau - f\omega\right), \qquad \iota(J)\tau := \tfrac{1}{2}\big(\tau(J\cdot,\cdot) + \tau(\cdot, J\cdot)\big)$$

[42] Let J be an almost complex structure on M that is compatible with ρ and denote by $\langle\cdot,\cdot\rangle := \rho(\cdot, J\cdot)$ the associated Riemannian metric. The **canonical spin^c structure**

$$\gamma_{\rho,J} : TM \to \mathrm{Hom}(W_J^-, W_J^+)$$

is defined by $W_J^- := \Lambda^{0,1}T^*M$, $W_J^+ := \Lambda^{0,0}T^*M \oplus \Lambda^{0,2}T^*M$, and

$$\gamma_{\rho,J}(v)\alpha := \left(-2^{1/2}\iota(v)\alpha, -2^{-1/2}\big(\iota(Jv)\rho + i\iota(v)\rho\big) \wedge \alpha\right)$$

for $v \in T_pM$ and $\alpha \in \Lambda^{0,1}T_p^*M$. Thus $c_1(\gamma_{\rho,J}) = c_1(TM, J) = c_1(\rho)$. □

on the space of pairs (f, τ) consisting of a constant function $f : M \to \mathbb{R}$ and a self-dual harmonic 2-form τ. Equivalently, this orientation is determined by the complex orientation of $H^{0,0}(M) \oplus H^{0,1}(M) \oplus H^{0,2}(M)$ via the isomorphism

$$H^0(M) \oplus H^1(M) \oplus H^{2,+}(M) \to H^{0,0}(M) \oplus H^{0,1}(M) \oplus H^{0,2}(M)$$

that sends a triple (f, α, τ) to the triple

$$\left(f - i\frac{\omega \wedge \tau}{\omega \wedge \omega}, \tfrac{1}{2}(\alpha + i\alpha \circ J), \tfrac{1}{2}\left(\tau - \frac{\omega \wedge \tau}{\omega \wedge \omega}\omega + i\iota(J)\tau\right)\right).$$

Here the second term is $\alpha^{0,1}$ and the last term is $\tau^{0,2}$. This formula shows that

$$\mathfrak{o}_{-\omega} = (-1)^{\frac{\chi+\sigma}{4}}\mathfrak{o}_\omega. \tag{13.3.3}$$

The number $\chi + \sigma = 2 - 2b_1 + 2b^+$ is divisible by 4 for every closed Kähler surface (b_1 is even and b^+ is odd) and indeed for every closed almost complex 4-manifold. The formula (13.3.3) remains valid for any nondegenerate 2-form.

The motivation for this choice of the homological orientation arises from the fact that the Seiberg–Witten equations come in two flavours, a Dirac version for general smooth four-manifolds and a holomorphic version for Kähler surfaces and symplectic four-manifolds. The sign is chosen so that the canonical isomorphism between the moduli spaces is holomorphic in the Kähler case, and the Seiberg–Witten invariant associated to Γ_ω is plus one in the symplectic case. □

Remark 13.3.4 (Positive cone) If $b^+ = 1$ then the *positive cone*

$$\mathcal{P} := \left\{a \in H^2(M; \mathbb{R}) \,|\, a^2 > 0\right\}$$

has two connected components. The choice of a connected component $\mathbf{p} \subset \mathcal{P}$ is equivalent to the choice of an orientation of the one-dimensional vector space $H^{2,+}(M; \mathbb{R})$ (for some, and hence every, Riemannian metric on M). For every symplectic form ω that is compatible with the orientation, denote by $\mathbf{p}_\omega \subset \mathcal{P}$ the connected component containing the cohomology class $[\omega]$. □

The Seiberg–Witten invariants

The **Seiberg–Witten invariants** of a smooth oriented four-manifold M with $b^+ > 1$ take the form of a map

$$\mathrm{Spin^c}(M) \to \mathbb{Z} : \Gamma \mapsto \mathrm{SW}(M, \mathfrak{o}, \Gamma).$$

This map depends on the choice of a homological orientation $\mathfrak{o}$ of M (see Remark 13.3.3). Changing the homological orientation reverses the sign of the invariant. When $b^+ = 1$ the Seiberg–Witten invariant also depends on the choice of a connected component $\mathbf{p}$ of $\mathcal{P}$ (see Remark 13.3.4). The Seiberg–Witten invariant in either case can only be nonzero if

$$c^2 \geq 2\chi + 3\sigma, \tag{13.3.4}$$

where $c := c_1(\Gamma)$ and σ and χ are the signature and Euler characteristic of M. This is because, in the regular case, the dimension of the Seiberg–Witten moduli space $\mathcal{M}^{\mathrm{SW}}(M, \Gamma)$ is given by

$$\dim \mathcal{M}^{\mathrm{SW}}(M, \Gamma) = c_2(W^+) = \frac{c^2 - 2\chi - 3\sigma}{4}. \tag{13.3.5}$$

The number $\frac{1}{2}(\chi + \sigma) = 1 - b_1 + b^+$ is an integer and $\frac{1}{4}(c^2 - \sigma)$ is the real index of the Dirac operator and hence is an even integer.

Remark 13.3.5 The Seiberg–Witten invariants can only be nonzero when $\chi+\sigma$ is divisible by four, or equivalently when $b_1 - b^+$ is odd, or equivalently, when the dimension (13.3.5) of the Seiberg–Witten moduli space is even. The Hirzebruch signature formula (Remark 4.1.10) asserts that equality holds in (13.3.4) for the spin^c structures associated to nondegenerate 2-forms (Remark 13.3.1). □

Definition 13.3.6 *Let M be a closed oriented smooth four-manifold with*

$$b^+ > 1.$$

A cohomology class $c \in H^2(M; \mathbb{Z})$ is called a **(Seiberg–Witten) basic class** *if there is a spin^c structure Γ with $c_1(\Gamma) = c$ and nonzero Seiberg–Witten invariant. A closed oriented smooth four-manifold with $b^+ > 1$ is said to have* **(Seiberg–Witten) simple type** *if every basic class c satisfies $c^2 = 2\chi + 3\sigma$.* □

By a theorem of Seiberg and Witten there are only finitely many basic classes. All closed symplectic four-manifolds with $b^+ > 1$ have simple type (see Corollary 13.3.24 below). It is an open question whether there is any closed oriented smooth four-manifold with $b^+ > 1$ that does not have simple type.

Remark 13.3.7 (Symmetry) The Seiberg–Witten invariants of Γ and $\overline{\Gamma}$ (see Remark 13.3.2) are related by

$$\mathrm{SW}(M, \mathfrak{o}, \overline{\Gamma}) = (-1)^{\frac{\chi+\sigma}{4}} \mathrm{SW}(M, \mathfrak{o}, \Gamma) \tag{13.3.6}$$

when $b^+ > 1$, and by

$$\mathrm{SW}(M, \mathfrak{o}, \mathbf{p}, \overline{\Gamma}) = (-1)^{\frac{\chi+\sigma}{4}} \mathrm{SW}(M, \mathfrak{o}, -\mathbf{p}, \Gamma) \tag{13.3.7}$$

when $b^+ = 1$. (By Remark 13.3.5 the number $\frac{\chi+\sigma}{4}$ is an integer.) It follows that the basic classes occur in pairs $\pm c$. □

Remark 13.3.8 (Wall crossing) Let M be a closed oriented smooth four-manifold with $b^+ = 1$. Then reversing the connected component of the positive cone $\mathcal{P}$ is governed by the wall-crossing formula of Li–Liu [421]. If

$$H^1(M;\mathbb{R}) = 0$$

and $c_1(\Gamma)^2 \geq 2\chi + 3\sigma$, then

$$\mathrm{SW}(M,\mathfrak{o},\mathbf{p},\Gamma) - \mathrm{SW}(M,\mathfrak{o},-\mathbf{p},\Gamma) = 1.$$

Here the homological orientation $\mathfrak{o} = \mathfrak{o}_{\mathbf{p}}$ is determined by the standard orientation of $H^0(M;\mathbb{R})$ and the orientation of $H^{2,+}(M;\mathbb{R})$ induced by $-\mathbf{p}$. When M carries a Kähler form ω with $[\omega] \in \mathbf{p}$, this is the homological orientation induced by ω, as in Remark 13.3.3. We emphasize that the connected component $\mathbf{p} \subset \mathcal{P}$ is used for the choice of the perturbation of the Seiberg–Witten equations and the homological orientation $\mathfrak{o}$ is used to orient the Seiberg–Witten moduli spaces. The minus sign in the relation between $\mathbf{p}$ and $\mathfrak{o}_{\mathbf{p}}$ is a consequence of the various sign choices that are implicit in our discussion. □

Remark 13.3.9 (Vanishing) If M is a closed oriented Riemannian four-manifold with positive scalar curvature, then the unperturbed Seiberg–Witten equations on M have no solutions for any spin^c structure and hence the Seiberg–Witten invariants of M are zero. This can be viewed as a nonlinear analogue of Lichnerowicz' vanishing theorem for the Â-genus of manifolds with positive scalar curvature. A refinement of this observation asserts that the Seiberg–Witten invariants of an oriented connected sum of two smooth four-manifolds, each with $b^+ > 0$, must vanish. (Here we refer to the pointwise connected sum discussed in Exercise 7.1.4.) □

The Seiberg–Witten invariants of symplectic four-manifolds

The next theorem was proved by Taubes. For the notation Γ_ω and $\Gamma_{\omega,e}$ see Remark 13.3.1, for the notation $\mathbf{p}_\omega$ see Remark 13.3.4, and for the notation $\mathfrak{o}_\omega$ see Remark 13.3.3.

Theorem 13.3.10 (Taubes) *Let (M,ω) be a closed symplectic four-manifold with $b^+ \geq 2$ and define $a := [\omega] \in H^2(M;\mathbb{R})$. Then*

$$\mathrm{SW}(M,\mathfrak{o}_\omega,\Gamma_\omega) = 1, \tag{13.3.8}$$

$$\mathrm{SW}(M,\mathfrak{o}_\omega,\Gamma_{\omega,e}) \neq 0 \quad \Longrightarrow \quad a \cdot e \geq 0, \tag{13.3.9}$$

$$\mathrm{SW}(M,\mathfrak{o}_\omega,\Gamma_{\omega,e}) \neq 0 \quad \text{and} \quad a \cdot e = 0 \quad \Longrightarrow \quad e = 0 \tag{13.3.10}$$

for all $e \in H^2(M;\mathbb{Z})$. These assertions continue to hold in the case $b^+ = 1$, with $\mathrm{SW}(M,\mathfrak{o}_\omega,\Gamma_{\omega,e})$ replaced by $\mathrm{SW}(M,\mathfrak{o}_\omega,\mathbf{p}_\omega,\Gamma_{\omega,e})$

Proof: See Taubes [606, 607]. □

An immediate consequence of Theorem 13.3.10 is the uniqueness of the first Chern class for cohomologous symplectic forms in dimension four. It is due to Li–Liu [424, Proposition 4.1].

Proposition 13.3.11 *Let M be a closed smooth four-manifold. Two cohomologous symplectic forms on M have equivalent spin^c structures and hence have the same first Chern class.*

Proof: Let ω, ω' be cohomologous symplectic forms on M and $a := [\omega] = [\omega']$. Choose the orientation such that $a^2 > 0$ and choose e such that $\Gamma_{\omega'} = \Gamma_{\omega,e}$. Assume first that $b^+ > 1$. By (13.3.8), $\mathrm{SW}(M, \mathfrak{o}_{\omega'}, \Gamma_{\omega,e}) = \mathrm{SW}(M, \mathfrak{o}_{\omega'}, \Gamma_{\omega'}) = 1$ and so $\mathrm{SW}(M, \mathfrak{o}_\omega, \Gamma_{\omega,e}) \neq 0$. Hence $a \cdot e \geq 0$ by (13.3.9). Interchanging the roles of ω and ω' gives $a \cdot e \leq 0$, hence $a \cdot e = 0$, and hence $e = 0$ by (13.3.10). This shows that $\Gamma_{\omega'} = \Gamma_{\omega,e} = \Gamma_\omega$ and $c_1(\omega') = c_1(\omega) + 2e = c_1(\omega)$. The proof for $b^+ = 1$ is verbatim the same. Include $\mathbf{p}_\omega$, respectively $\mathbf{p}_{\omega'}$, as an argument of SW and use the fact that $\mathbf{p}_\omega = \mathbf{p}_{\omega'}$, because $[\omega] = [\omega']$. □

The next proposition is a special case of a result by Conolly–Lê–Ono [126]. It strengthens Proposition 13.3.11 in the simply connected case.

Proposition 13.3.12 *Let M be a simply connected closed smooth 4-manifold.*

(i) *Two symplectic forms on M with the same first Chern class and inducing the same orientation on M are homotopic as nondegenerate 2-forms.*

(ii) *Two cohomologous symplectic forms on M are homotopic as nondegenerate 2-forms.*

Proof: (Following Conolly–Lê–Ono [126].) Assume $b^+ = 1$. (The case $b^+ > 1$ is easier.) Fix an orientation of M and an integral lift $c \in H^2(M;\mathbb{Z})$ of the second Stiefel–Whitney class with $c^2 = 2\chi + 3\sigma$. Let $\mathscr{R}^c$ be the set of nondegenerate 2-forms on M with first Chern class c and compatible with the orientation. By Remark 4.1.12, $\mathscr{R}^c$ has precisely two connected components. Let $\omega, \omega' \in \mathscr{R}^c$ be symplectic forms. Then $c_1(\omega') = c_1(\omega) = c$. Since M is simply connected the equivalence class of a spin^c structure is uniquely determined by its first Chern class (see equation (13.3.1)) and hence $\Gamma_{\omega'} = \Gamma_\omega$. Moreover, by (13.3.8),

$$\mathrm{SW}(M, \mathfrak{o}_\omega, \mathbf{p}_\omega, \Gamma_\omega) = 1 = \mathrm{SW}(M, \mathfrak{o}_{\omega'}, \mathbf{p}_{\omega'}, \Gamma_{\omega'}). \tag{13.3.11}$$

The wall-crossing formula in Li–Liu [421] asserts that

$$\mathrm{SW}(M, \mathfrak{o}_\omega, \mathbf{p}_\omega, \Gamma_\omega) - \mathrm{SW}(M, \mathfrak{o}_\omega, -\mathbf{p}_\omega, \Gamma_\omega) = 1. \tag{13.3.12}$$

(See Remark 13.3.8.) Since $\mathfrak{o}_\omega = \pm\mathfrak{o}_{\omega'}$ and $\Gamma_\omega = \Gamma_{\omega'}$, it follows from equation (13.3.11) that

$$\mathrm{SW}(M, \mathfrak{o}_\omega, \mathbf{p}_\omega, \Gamma_\omega) = 1 = \pm\mathrm{SW}(M, \mathfrak{o}_\omega, \mathbf{p}_{\omega'}, \Gamma_\omega).$$

Hence the difference $\mathrm{SW}(M, \mathfrak{o}_\omega, \mathbf{p}_\omega, \Gamma_\omega) - \mathrm{SW}(M, \mathfrak{o}_\omega, \mathbf{p}_{\omega'}, \Gamma_\omega)$ is even, so that $\mathbf{p}_{\omega'} = \mathbf{p}_\omega$ by (13.3.12). Since $\Gamma_{\omega'} = \Gamma_\omega$ and $\mathbf{p}_{\omega'} = \mathbf{p}_\omega$ it follows from (13.3.11) that $\mathfrak{o}_{\omega'} = \mathfrak{o}_\omega$. In [142, Prop. 3.25] and [143, Lemma 6.4] Donaldson proved that there is a free involution on $\pi_0(\mathscr{R}^c)$, which reverses the homological orientation. But we have just seen that any two symplectic forms in $\mathscr{R}^c$ induce the same homological orientation. Hence they must belong to the same connected component of $\mathscr{R}^c$. This proves (i). Assertion (ii) follows from (i) and Proposition 13.3.11. □

Proposition 13.3.11 shows that for every closed oriented smooth 4-manifold, there is a chamber structure on the set

$$\mathcal{K}_{\mathrm{symp}} = \bigcup_{c \in \mathcal{C}_{\mathrm{symp}}} \mathcal{K}^c$$

of all de Rham cohomology classes of symplectic forms compatible with the orientation (see equation (13.1.4)). This chamber structure is examined in Section 13.4 below in some basic examples.

Proposition 13.3.13 *If a closed symplectic four-manifold (M, ω) is diffeomorphic to an oriented connected sum $X \# Y$ by an orientation-preserving diffeomorphism, then either $b^+(X) = 0$ or $b^+(Y) = 0$.*

Proof: If $b^+(X)$ and $b^+(Y)$ are both positive, then the Seiberg–Witten invariants of the oriented connected sum $X \# Y$ vanish by Remark 13.3.9. Hence it follows from Theorem 13.3.10 that $X \# Y$ does not carry a symplectic form that is compatible with the orientation. □

Example 13.3.14 It follows from Proposition 13.3.13 that the oriented connected sum of three copies of $\mathbb{C}\mathrm{P}^2$, which does have an almost complex structure (see Example 4.1.13), has no symplectic structure with either orientation. The oriented connected sum $\mathbb{T}^4 \# \mathbb{C}\mathrm{P}^2$ does carry a symplectic structure that is compatible with the reversed orientation, as the blowup of the four-torus. □

Exceptional spheres

The next lemma is of preparatory nature. It will be used to prove that smoothly embedded spheres with self-intersection number minus one (and first Chern number one in the case $b^+ = 1$) give rise to symplectically embedded spheres (see Corollary 13.3.27 below). Here we only show that they determine spin^c structures with nonzero Seiberg–Witten invariants via Theorem 13.3.10. This is closely related to the **blowup formula** for the Seiberg–Witten invariants which asserts, in the case $b^+ > 1$, that if $S \subset M$ is a smoothly embedded sphere with self-intersection number minus one, then every Seiberg–Witten basic class on M has the form $c \pm \mathrm{PD}(S)$, where $c \in H^2(M; \mathbb{Z})$ is Poincaré dual to an oriented 2-dimensional submanifold of $M \setminus S$. However, we shall not use the blowup formula in the proof and instead use a trick of Taubes.

Lemma 13.3.15 *Let (M, ω) be a closed symplectic four-manifold and suppose that $E, E' \in H_2(M; \mathbb{Z})$ are homology classes with self-intersection number -1 that can be represented by smoothly embedded spheres. Then the following holds.*

(i) $c_1(E)\omega(E) > 0$ *and*

$$\mathrm{SW}(M, \mathfrak{o}_\omega, \Gamma_{\omega, c_1(E)\mathrm{PD}(E)}) = 1 \qquad \text{if } b^+ > 1,$$
$$\mathrm{SW}(M, \mathfrak{o}_\omega, \mathbf{p}_\omega, \Gamma_{\omega, c_1(E)\mathrm{PD}(E)}) = 1 \qquad \text{if } b^+ = 1.$$

(ii) *If $c_1(E) = c_1(E') = 1$ then $E \cdot E' \geq -1$.*

Proof: The proof rests on the fact that as explained on page 538, there is an orientation-preserving diffeomorphism $\phi : M \to M$ with support in a neighbourhood of a smoothly embedded sphere representing the class E that acts on $H^2(M;\mathbb{Z})$ by reflection about the hyperplane orthogonal to $e := \mathrm{PD}(E)$, i.e.

$$\phi^* a = a + 2(a \cdot e)e. \tag{13.3.13}$$

When there is no 2-torsion in $H^2(M;\mathbb{Z})$ it follows that

$$\Gamma_{\phi^*\omega} = \phi^*\Gamma_\omega = \Gamma_{\omega,me}, \qquad m := c_1(E). \tag{13.3.14}$$

(All three spin^c structures have the same first Chern class.) It turns out that equation (13.3.14) remains valid even in the presence of 2-torsion in $H^2(M;\mathbb{Z})$. Moreover $m = c_1(E)$ is odd because $E \cdot E$ is odd. It follows from (13.3.13) that

$$\int_M \omega \wedge \phi^*\omega = \int_M \omega \wedge \omega + 2\omega(E)^2 > 0,$$

and hence $\mathbf{p}_{\phi^*\omega} = \mathbf{p}_\omega$ in the case $b^+ = 1$. Moreover, close examination of Remark 13.3.3 shows that ϕ preserves the homological orientation and hence $\mathfrak{o}_{\phi^*\omega} = \mathfrak{o}_\omega$. Thus it follows from equation (13.3.14) and Theorem 13.3.10 that $\mathrm{SW}(M, \mathfrak{o}_\omega, \mathbf{p}_\omega, \Gamma_{\omega,me}) = 1$ in the case $b^+ = 1$, and $\mathrm{SW}(M, \mathfrak{o}_\omega, \Gamma_{\omega,me}) = 1$ in the case $b^+ > 1$. Since $mE \neq 0$ it follows also from Theorem 13.3.10 that $m\omega(E) > 0$ and this proves (i). Now abbreviate $k := E \cdot E'$ and assume $c_1(E) = c_1(E') = 1$. Then it follows from equation (13.3.14) that

$$\phi^*\Gamma_{\omega,\mathrm{PD}(E')} = \Gamma_{\phi^*\omega,\phi^*\mathrm{PD}(E')} = \Gamma_{\phi^*\omega,\mathrm{PD}(E'+2kE)} = \Gamma_{\omega,\mathrm{PD}(E+E'+2kE)}.$$

Hence $\mathrm{SW}(M, \mathfrak{o}_\omega, \mathbf{p}_\omega, \Gamma_{\omega,\mathrm{PD}(E+E'+2kE)}) \neq 0$ and $\omega(E + E' + 2kE) \geq 0$ by Theorem 13.3.10. Interchange the roles of E and E' to obtain $\omega(E + E' + 2kE') \geq 0$. Add the two inequalities to obtain $2(k+1)\omega(E + E') \geq 0$. Since $\omega(E + E') > 0$ by (i) it follows that $k \geq -1$. This proves Lemma 13.3.15. □

Example 13.3.16 Let M be the 2-point blowup of $\mathbb{C}\mathrm{P}^2$ and let L, E_1, E_2 be the standard basis of $H_2(M;\mathbb{Z})$. Suppose that ω is a symplectic form with

$$c_1(\omega) = \mathrm{PD}(3L - E_1 - E_2).$$

Since M is a blowup of $\mathbb{C}\mathrm{P}^2$, the homology classes $E_1, E_2, L - E_1 - E_2$ are represented by smoothly embedded spheres with self-intersection number minus one and first Chern number one. Hence $\omega(E_1), \omega(E_2), \omega(L - E_1 - E_2)$ are positive by Lemma 13.3.15. This implies that

$$[\omega] = \mathrm{PD}(\lambda L - \lambda_1 E_1 - \lambda_2 E_2), \qquad \lambda_1, \lambda_2 > 0, \qquad \lambda_1 + \lambda_2 < \lambda.$$

Moreover, the homology class $E := 2L - 2E_1 - E_2$ is the image of E_2 under the reflection about the exceptional sphere in the class $L - E_1 - E_2$ and hence can be represented by a smoothly embedded sphere with self-intersection number minus one. It satisfies $c_1(E) = 3$ and $E \cdot (-E_1) = -2$. Thus the hypothesis $c_1(E) = c_1(E') = 1$ cannot be removed in part (ii) of Lemma 13.3.15. Moreover, $\mathrm{SW}(M, \mathfrak{o}_\omega, \mathbf{p}_\omega, \Gamma_{\omega,3\mathrm{PD}(E)}) = 1$ by part (i) of Lemma 13.3.15. □

The generalized adjunction inequality

The Seiberg–Witten invariants were used by Kronheimer–Mrowka [381], Morgan–Szabó–Taubes [494], and Ozsvath–Szabó [519] to settle the **symplectic Thom conjecture**, which asserts that every connected 2-dimensional symplectic submanifold of a closed symplectic 4-manifold minimizes the genus in its homology class.

Theorem 13.3.17 *Let M be a closed smooth four-manifold and let $\Sigma \subset M$ be an oriented connected 2-dimensional submanifold of genus g representing a nontorsion homology class $A \in H^2(M;\mathbb{Z})$.*

(i) *Assume $b^+ > 1$ and $A \cdot A \geq 0$ and let $c \in H^2(M;\mathbb{Z})$ be a Seiberg–Witten basic class. Then A, g, and c satisfy the* **generalized adjunction inequality**

$$2g - 2 \geq A \cdot A + |\langle c, A \rangle|. \tag{13.3.15}$$

(ii) *If ω is a symplectic form on M and $\omega(A) > 0$ then*

$$2g - 2 \geq A \cdot A - c_1(A). \tag{13.3.16}$$

Proof: Part (i) was proved by Kronheimer–Mrowka [381] and independently by Morgan–Szabó–Taubes [494]. For $A \cdot A \geq 0$, part (ii) was proved by Morgan–Szabó–Taubes [494] and for $A \cdot A < 0$ by Ozsvath–Szabó [519]. □

The nontorsion condition cannot be removed in (i) and neither can the condition $\omega(A) > 0$ in (ii). Contractible embedded spheres are counterexamples.

Corollary 13.3.18 (Symplectic Thom conjecture) *Every symplectic surface in a closed symplectic 4-manifold minimizes the genus in its homology class.*

Proof: The genus of a connected symplectic surface in the homology class A is given by $2g - 2 = A \cdot A - c_1(A)$ (see Example 4.4.5). Hence the assertion follows from part (ii) of Theorem 13.3.17. □

The original Thom conjecture asserts that every connected oriented 2-dimensional submanifold of $\mathbb{C}\mathrm{P}^2$ in the class $d[\mathbb{C}\mathrm{P}^1]$ has genus $g \geq \frac{(d-1)(d-2)}{2}$. This was confirmed by Kronheimer and Mrowka in [381]. An earlier version for Kähler surfaces with $b^+ > 1$ and nonnegative self-intersection was proved by Kronheimer and Mrowka in [379,380], using Donaldson theory. The general symplectic Thom conjecture was confirmed by Ozsvath and Szabó [519].

Remark 13.3.19 Theorem 13.3.17 can also be used to establish nonexistence results for symplectic forms. A case in point is the K3-surface with the orientation reversed. It satisfies $b^+ > 1$ and contains embedded spheres with positive self-intersection numbers. Hence its Seiberg–Witten invariants must vanish because otherwise these spheres would violate the generalized adjunction inequality (13.3.15). It follows that every symplectic form on the K3-surface is compatible with the standard orientation. Similarly, every closed oriented smooth 4-manifold M with $b^+ > 1$ that contains an oriented embedded surface Σ with $\Sigma \cdot \Sigma > 2g - 2 \geq 0$ has vanishing Seiberg–Witten invariants and hence does not support a symplectic structure compatible with the orientation. □

The Gromov invariants

The definition of the Gromov invariants goes far beyond the scope of this book and we content ourselves with outlining some of their key features. The overall idea is to assign to each integral homology class $A \in H_2(M;\mathbb{Z})$ in a closed symplectic four-manifold (M,ω) an integer $\mathrm{Gr}(A)$ that represents the algebraic count of (possibly disconnected) embedded J-holomorphic curves representing the class A. Here are several observations that might help in understanding the properties of this invariant and its close relation with Seiberg–Witten theory.

Fix a closed symplectic four-manifold (M,ω), an ω-compatible almost complex structure J, and a homology class $A \in H_2(M;\mathbb{Z})$. Let $C \subset M$ be a connected almost complex submanifold of real dimension two, representing the class A. (Such a submanifold can also be written as the image of a J-holomorphic curve $u : \Sigma \to M$ defined on a closed Riemann surface (Σ, j) and we will use the term *J-holomorphic curve* interchangeably for the map u modulo biholomorphic reparametrization and its image $C = u(\Sigma)$.) The genus $g = g(C)$, the first Chern number $c_1(T_CM) = c_1(A)$, and the self-intersection number $c_1(\nu_C) = A \cdot A$ are related by the adjunction formula

$$2g - 2 + c_1(A) = A \cdot A. \tag{13.3.17}$$

(See Example 4.4.5.) Now suppose our homology class A can be represented as a sum of homology classes $A_i \in H_2(M;\mathbb{Z})$ such that

$$A = \sum_i A_i, \qquad A_i \cdot A_i + c_1(A_i) \geq 0, \qquad A_i \cdot A_j = 0 \text{ for } i \neq j. \tag{13.3.18}$$

Denote by $\mathcal{M}_i$ the moduli space of all connected embedded J-holomorphic curves representing the class A_i. Their genus g_i is determined by the adjunction formula $2g_i - 2 + c_1(A_i) = A_i \cdot A_i$. Hence the number $A_i \cdot A_i + c_1(A_i) = 2g_i - 2 + 2c_1(A_i)$ is the dimension of $\mathcal{M}_i$ for a generic almost complex structure (see Remark 4.5.1). Thus the condition $A_i \cdot A_i + c_1(A_i) \geq 0$ is automatically satisfied when $\mathcal{M}_i \neq \emptyset$ for a generic J. Moreover, the condition $A_i \cdot A_j = 0$ guarantees that two connected J-holomorphic curves representing the classes A_i and A_j with $i \neq j$ either do not intersect or coincide. Assume for simplicity that $A_i \neq A_j$ for $i \neq j$, so they cannot coincide. Then each union $C := C_1 \cup \cdots \cup C_\ell$ of curves $C_i \in \mathcal{M}_i$ represents the class A. In the regular case the union of the products $\mathcal{M}_1 \times \cdots \times \mathcal{M}_\ell$ over all tuples A_i that satisfy (13.3.18) is a smooth moduli space $\mathcal{M}^{\mathrm{Gr}}(A,J)$, and

$$\dim \mathcal{M}^{\mathrm{Gr}}(A,J) = \sum_i \big(2g_i - 2 + 2c_1(A_i)\big) = A \cdot A + c_1(A). \tag{13.3.19}$$

This number agrees with the dimension of the Seiberg–Witten moduli space associated to the Spin^c structure $\Gamma_{\omega,e}$ with $e := \mathrm{PD}(A)$. Namely, $c_1(\Gamma_{\omega,e}) = c + 2e$, where $c := c_1(\omega)$ and $c^2 = 2\chi + 3\sigma$. Hence, by equation (13.3.5),

$$\dim \mathcal{M}^{\mathrm{SW}}(M, \Gamma_{\omega,e}) = \frac{(c+2e)^2 - 2\chi - 3\sigma}{4} = A \cdot A + c_1(A).$$

If $A \neq 0$ and $\mathcal{M}^{\mathrm{Gr}}(A, J) \neq \emptyset$ it also follows that $\omega(A) > 0$, and this corresponds to the assertions of Theorem 13.3.10. Now define

$$k := k(A) := \tfrac{1}{2}\big(A \cdot A + c_1(A)\big) = \tfrac{1}{2} \dim \mathcal{M}^{\mathrm{Gr}}(A, J) \tag{13.3.20}$$

and include k marked points, to obtain an oriented moduli space $\mathcal{M}_k^{\mathrm{Gr}}(A, J)$ of dimension $4k$ and an evaluation map $\mathrm{ev}_{A,k} : \mathcal{M}_k^{\mathrm{Gr}}(A, J) \to M^k$. If this moduli space is compact, the Gromov invariant is defined as the degree

$$\mathrm{Gr}(M, \omega, A) := \mathrm{Gr}(A) := \deg(\mathrm{ev}_{A,k}). \tag{13.3.21}$$

Thus $\mathrm{Gr}(A)$ is the algebraic number of (possibly disconnected) embedded J-holomorphic curves in the class A passing through k points in general position. For $A = 0$ the convention is $\mathrm{Gr}(0) := 1$. If $k(A) = 0$ the degree is understood as the number of elements of $\mathcal{M}^{\mathrm{Gr}}(A, J)$, counted with appropriate signs.

In general, however, the moduli space $\mathcal{M}^{\mathrm{Gr}}(A, J)$ will not be compact and one has to examine *Gromov* compactness in this setting, where holomorphic spheres can bubble off and surfaces of genus g can degenerate into nodal surfaces. Moreover, one does have to deal with the possibility that $A_i = A_j$ for $i \neq j$. Then $A_i \cdot A_i = 0$ and by (13.3.17), (13.3.18), this implies $2 - 2g_i = c_1(A_i) \geq 0$. Thus either $g_i = 0$ (embedded spheres with self–intersection zero) or $g_i = 1$ (embedded tori with self-intersection zero). The genus zero case is quite easy to handle because then $A_i \cdot A_i + c_1(A_i) = 2$, and the *bad case* $C_i = C_j$ for $i \neq j$ can be avoided by the choice of the regular value of the evaluation map. The genus one case with $A_i \cdot A_i = c_1(A_i) = 0$ is more serious. In this situation the *bad case* $C_i = C_j$ for $i \neq j$ cannot be avoided, and one does have to deal with multiple covers of embedded J-holomorphic tori with self-intersection numbers zero. This leads to subtleties in the choice of signs for the count in the Gromov invariant. Also exceptional spheres play a special role in the case $b^+ = 1$. Once the Gromov invariant has been defined, one can show that it depends only on the connected component of ω in the space of symplectic forms. In fact, as we shall see below, it is an invariant of the smooth structure of M. All these technical issues will not be discussed in the present overview and the reader is referred to Taubes [610] and McDuff [457].

Example 13.3.20 Consider the complex projective plane $(M, \omega) = (\mathbb{CP}^2, \omega_{\mathrm{FS}})$ and let $L := [\mathbb{CP}^1] \in H_2(\mathbb{CP}^2; \mathbb{Z})$ be the homology class of a line. Then $c_1(L) = 3$ and $L \cdot L = 1$. The dimension of the moduli space $\mathcal{M}^{\mathrm{Gr}}(L, J)$ is $2k(L) = 4$ and $\mathrm{Gr}(L)$ counts the number of lines through two points. Thus $\mathrm{Gr}(L) = 1$. A similar calculation shows that $k(2L) = 5$ and, since there is a unique conic through any five generic points in $\mathbb{CP}^2$, it follows that $\mathrm{Gr}(2L) = 1$. More generally, one obtains $k(dL) = \frac{d^2+3d}{2}$ and each holomorphic curve in $\mathbb{CP}^2$ of degree d is the zero set of a homogeneous polynomial in three variables of degree d, unique up to a scalar factor. The total number of coefficients is $\frac{(d+1)(d+2)}{2} = k(dL) + 1$. Up to scaling the coefficients are uniquely determined by specifying a generic set of $k(dL)$ points at which it vanishes, and so $\mathrm{Gr}(dL) = 1$ for every integer $d \geq 1$.

As another example, let $M \to B$ be a symplectic ruled surface, that is a symplectic S^2-bundle over a Riemann surface B, and let F be the fibre class. Then $F \cdot F = 0$ and $c_1(F) = 2$. Hence $k(F) = 1$. Since there is a unique fibre through each point of M it follows that $\mathrm{Gr}(F) = 1$.

As a third example, let M be the blowup of a Kähler surface and let E be the homology class of the exceptional divisor with its complex orientation. Then $E \cdot E = -1$ and $c_1(E) = 1$ so $E \cdot E + c_1(E) = 0$. There is a unique (regular) holomorphic sphere representing the class E and thus $\mathrm{Gr}(E) = 1$.

As a fourth example, let M be a K3-surface with an embedded rational curve in the class A. Then $A \cdot A = -2$ and $c_1(A) = 0$ so $A \cdot A + c_1(A) < 0$. In this case the holomorphic sphere will not persist under perturbation of J and the Gromov invariant $\mathrm{Gr}(A)$ vanishes. (The moduli space has 'dimension' minus two.) □

Example 13.3.20 shows that the Gromov invariant $\mathrm{Gr}(A)$ can be nontrivial for infinitely many homology classes $A \in H_2(M;\mathbb{Z})$. However, this can only happen when $b^+ = 1$. By Taubes' *'Seiberg–Witten equals Gromov'* theorem, symplectic four-manifolds with $b^+ > 1$ have only finitely many classes A with $\mathrm{Gr}(A) \neq 0$.

Seiberg–Witten equals Gromov

Throughout this subsection we denote by $K := -c_1(\omega) \in H^2(M;\mathbb{Z})$ the **canonical class** of a closed symplectic four-manifold (M, ω). It will sometimes be convenient to use the same letter K for its Poincaré dual in $H_2(M;\mathbb{Z})$.

Theorem 13.3.21 (Taubes) *Let (M, ω) be a closed symplectic four-manifold with $b^+ > 1$. Then $\mathrm{SW}(M, \mathfrak{o}_\omega, \Gamma_{\omega,\mathrm{PD}(A)}) = \mathrm{Gr}(M, \omega, A)$ for all $A \in H_2(M;\mathbb{Z})$.*

Proof: See Taubes [608, 609, 611]. □

Taubes' theorem was extended to symplectic four-manifolds with $b^+ = 1$ by Li and Liu [423]. In this case the correspondence between the Seiberg–Witten and the Gromov invariants is somewhat more subtle. One either has to modify the Gromov invariants by allowing for multiple covers of embedded J-holomorphic spheres with self-intersection number minus one (see McDuff [457]), or impose conditions on the homology classes A. (The class $A := 3E$ in Example 13.3.16 violates the condition (13.3.23) below.) To formulate the result, define

$$\begin{aligned} \mathcal{E} &:= \left\{ E \in H_2(M;\mathbb{Z}) \,\middle|\, \begin{array}{l} E \cdot E = -1 \text{ and } E \text{ can be represented} \\ \text{by a smoothly embedded sphere} \end{array} \right\}, \\ \mathcal{E}_\omega &:= \left\{ E \in H_2(M;\mathbb{Z}) \,\middle|\, \begin{array}{l} E \cdot E = -1 \text{ and } E \text{ can be represented} \\ \text{by a symplectically embedded sphere} \end{array} \right\}. \end{aligned} \tag{13.3.22}$$

Theorem 13.3.22 (Li–Liu) *Let (M, ω) be a closed symplectic 4-manifold with $b^+ = 1$ and choose a homology class $A \in H_2(M;\mathbb{Z})$ such that*

$$A \cdot E \geq -1 \qquad \forall\, E \in \mathcal{E}_\omega. \tag{13.3.23}$$

Then $\mathrm{SW}(M, \mathfrak{o}_\omega, \mathfrak{p}_\omega, \Gamma_{\omega,\mathrm{PD}(A)}) = \mathrm{Gr}(M, \omega, A)$.

Proof: See Li–Liu [423]. □

Corollary 13.3.23 (Taubes) *Let (M,ω) be a closed symplectic 4-manifold and let $A \in H_2(M;\mathbb{Z})$. If $b^+ > 1$ assume $\mathrm{SW}(M,\mathfrak{o}_\omega,\Gamma_{\omega,\mathrm{PD}(A)}) \neq 0$ and if $b^+ = 1$ assume $\mathrm{SW}(M,\mathfrak{o}_\omega,\mathfrak{p}_\omega,\Gamma_{\omega,\mathrm{PD}(A)}) \neq 0$ and $A \cdot E \geq -1$ for all $E \in \mathcal{E}_\omega$. Then A can be represented by a symplectic submanifold $C = C_1 \cup \cdots \cup C_\ell \subset M$ such that each connected component C_i satisfies*

$$K \cdot C_i \leq g(C_i) - 1 \leq C_i \cdot C_i \tag{13.3.24}$$

with equality in the case $b^+ > 1$. If $A = 0$ then $C = \emptyset$.

Proof: By assumption and Theorems 13.3.21 and 13.3.22, $\mathrm{Gr}(M,\omega,A) \neq 0$ and hence the homology class A can be represented by a symplectic submanifold $C = C_1 \cup \cdots \cup C_\ell$ of M, where the C_i denote the connected components of C. Here it is important to note that if one of the connected components of the J-holomorphic curves used in the definition of the Gromov invariant $\mathrm{Gr}(M,\omega,A)$ is a multiply covered torus with self-intersection number zero, one can use several parallel copies of this torus to obtain the required sympectic submanifold. The number $C_i \cdot C_i - K \cdot C_i$ is the dimension of the moduli space $\mathcal{M}^{\mathrm{Gr}}(A_i, J)$ associated to $A_i = [C_i]$, and hence is nonnegative for each i by construction of the Gromov invariant. Since $2g(C_i) - 2 = C_i \cdot C_i + K \cdot C_i$ by the adjunction formula (13.3.17), this proves (13.3.24).

Now assume $b^+ > 1$ and denote by $A_i := [C_i]$ the homology class of the connected component C_i. Then it follows from the definition of the Gromov invariant that $\mathrm{Gr}(M,\omega,A_i) \neq 0$. Moreover, $A_i \cdot A_i \geq -1$ by (13.3.24), and if $A_i \cdot A_i = -1$ then C_i is an exceptional J-holomorphic sphere with $K \cdot C_i = -1$ and so equality holds in (13.3.24). Hence assume $A_i \cdot A_i \geq 0$. Then it follows from the symmetry in Remark 13.3.7 and equation (13.3.2) that

$$\begin{aligned}
\mathrm{Gr}(M,\omega,K - A_i) &= \mathrm{SW}(M,\mathfrak{o}_\omega,\Gamma_{\omega,-c-\mathrm{PD}(A_i)}) \\
&= \mathrm{SW}(M,\mathfrak{o}_\omega,\overline{\Gamma}_{\omega,\mathrm{PD}(A_i)}) \\
&= (-1)^{\frac{\chi+\sigma}{4}}\mathrm{SW}(M,\mathfrak{o}_\omega,\Gamma_{\omega,\mathrm{PD}(A_i)}) \\
&= (-1)^{\frac{\chi+\sigma}{4}}\mathrm{Gr}(M,\omega,A_i) \neq 0
\end{aligned}$$

for every i. Hence the homology class $K - A_i$ can also be represented by a symplectic submanifold $C' = C_1' \cup \cdots \cup C_{\ell'}'$. At this point it is convenient to go back to the definition of the Gromov invariant, and choose J-holomorphic representatives, allowing for multiply covered tori. By positivity of intersections we then have $C_i \cdot C_j' \geq 0$ for all j with $C_j' \neq C_i$, and if $C_j' = C_i$ for some j we also have $C_i \cdot C_j' = C_i \cdot C_i \geq 0$ by assumption. Take the sum over all j to obtain

$$0 \leq \sum_j C_i \cdot C_j' = A_i \cdot (K - A_i) = K \cdot A_i - A_i \cdot A_i \leq 0$$

and hence $A_i \cdot A_i = K \cdot A_i$. This implies equality in (13.3.24) in the case $b^+ > 1$ and completes the proof of Corollary 13.3.23. □

Corollary 13.3.24 (Taubes) *Let (M, ω) be a closed symplectic four-manifold with $b^+ > 1$. Then the following holds.*

(i) *(M, ω) has Seiberg–Witten simple type.*

(ii) *The canonical class K is Poincaré dual to a symplectic submanifold of M.*

(iii) *$K \cdot [\omega] \geq 0$ with equality if and only if $K = 0$.*

(iv) *If (M, ω) is minimal then $K \cdot K \geq 0$.*

Proof: Let $c \in H^2(M; \mathbb{Z})$ be a Seiberg–Witten basic class. Then there exists a spin^c structure Γ with first Chern class $c_1(\Gamma) = c$ such that $\mathrm{SW}(M, \mathfrak{o}_\omega, \Gamma) \neq 0$. Choose $e \in H^2(M; \mathbb{Z})$ such that $\Gamma = \Gamma_{\omega,e}$ and define $A := \mathrm{PD}(e) \in H_2(M; \mathbb{Z})$. Then $\mathrm{Gr}(M, \omega, A) = \mathrm{SW}(M, \mathfrak{o}_\omega, \Gamma_{\omega,e}) \neq 0$ by Theorem 13.3.21 and $e^2 = K \cdot e$ because equality holds in (13.3.24). Since $c = 2e - K$ it follows that

$$c^2 = K^2 = 2\chi + 3\sigma.$$

This proves (i). Now Theorem 13.3.10 asserts that $\mathrm{SW}(M, \mathfrak{o}_\omega, \Gamma_\omega) = 1$. Hence it follows from Remark 13.3.7 and equation (13.3.2) with $c := c_1(\omega) = -K$ that

$$\mathrm{SW}(M, \mathfrak{o}_\omega, \Gamma_{\omega,-c}) = \mathrm{SW}(M, \mathfrak{o}_\omega, \overline{\Gamma}_\omega) = (-1)^{\frac{\chi+\sigma}{4}} \mathrm{SW}(M, \mathfrak{o}_\omega, \Gamma_\omega) = (-1)^{\frac{\chi+\sigma}{4}}.$$

Hence it follows from Corollary 13.3.23 that K is Poincaré dual to a symplectic submanifold $C \subset M$. This proves (ii). Moreover, $K \cdot [\omega] = \int_C \omega$ and this proves (iii). Part (iv) follows from Corollary 13.3.23 which asserts that the connected components C_i of C all satisfy $C_i \cdot C_i = K \cdot C_i = g(C_i) - 1$. If this number is negative then C_i is a symplectically embedded sphere with self-intersection number minus one, in contradiction to the assumption that (M, ω) is minimal. Hence $K \cdot C_i \geq 0$ for all i and hence $K \cdot K = \sum_i K \cdot C_i \geq 0$. □

Another proof of Corollary 13.3.24, using symplectic Lefschetz pencils instead of Taubes–Seiberg–Witten theory, was given by Donaldson and Smith in [162].

Corollary 13.3.25 (Taubes) *Let ω be any symplectic form on $\mathbb{C}\mathrm{P}^2$. Then there exists a symplectically embedded sphere $C \subset \mathbb{C}\mathrm{P}^2$ such that $C \cdot C = 1$.*

Proof: Since $\mathbb{C}\mathrm{P}^2$ admits a diffeomorphism that acts as minus the identity on $H_2(M; \mathbb{Z})$, we may assume without loss of generality that $\int_{\mathbb{C}\mathrm{P}^1} \omega > 0$. Then we have $\mathbf{p}_\omega = \mathbf{p}_{\omega_{\mathrm{FS}}}$ by Remark 13.3.1 and $\mathfrak{o}_\omega = \mathfrak{o}_{\omega_{\mathrm{FS}}}$ by Remark 13.3.3. Moreover, $c_1(\omega) = c_1(\omega_{\mathrm{FS}})$ and this implies $\Gamma_\omega = \Gamma_{\omega_{\mathrm{FS}}}$. Now let $L \in H_2(\mathbb{C}\mathrm{P}^2; \mathbb{Z})$ be the homology class of a line and denote $e := \mathrm{PD}(L)$. Then $\mathrm{Gr}(\mathbb{C}\mathrm{P}^2, \omega_{\mathrm{FS}}, L) = 1$ by Example 13.3.20 and hence $\mathrm{SW}(\mathbb{C}\mathrm{P}^2, \mathfrak{o}_{\omega_{\mathrm{FS}}}, \mathbf{p}_{\omega_{\mathrm{FS}}}, \Gamma_{\omega_{\mathrm{FS}},e}) = 1$ by Theorem 13.3.21. (This computation can also be derived from the wall-crossing formula in Remark 13.3.8.) Hence it follows from Theorem 13.3.21 for $(\mathbb{C}\mathrm{P}^2, \omega)$ that

$$\mathrm{Gr}(\mathbb{C}\mathrm{P}^2, \omega, L) = \mathrm{SW}(\mathbb{C}\mathrm{P}^2, \mathfrak{o}_\omega, \mathbf{p}_\omega, \Gamma_{\omega.e}) = \mathrm{SW}(\mathbb{C}\mathrm{P}^2, \mathfrak{o}_{\omega_{\mathrm{FS}}}, \mathbf{p}_{\omega_{\mathrm{FS}}}, \Gamma_{\omega_{\mathrm{FS}},e}) = 1.$$

Hence L is Poincaré dual to an ω-symplectic submanifold C. Since any two nonempty ω-symplectic submanifolds of $\mathbb{C}\mathrm{P}^2$ have positive intersection number, it follows that C is connected. Hence $2g(C) - 2 = L \cdot L + K \cdot L = -2$, by the adjunction formula (13.3.17), and so $g(C) = 0$. □

Corollary 13.3.26 (Taubes) *Any two cohomologous symplectic forms on* $\mathbb{C}\mathrm{P}^2$ *are diffeomorphic.*

Proof: Let ω_{FS} be the Fubini–Study form and ω any other symplectic form in the same cohomology class. By Corollary 13.3.25 the homology class $[\mathbb{C}\mathrm{P}^1]$ can be represented by an ω-symplectic sphere C. Hence a theorem of Gromov asserts that there exists a diffeomorphism $\phi : \mathbb{C}\mathrm{P}^2 \to \mathbb{C}\mathrm{P}^2$ such that $\phi(C) = \mathbb{C}\mathrm{P}^1$ and $\phi^*\omega_{\mathrm{FS}} = \omega$ (see [287] and [470, Theorem 9.4.1]). □

Recall that $\mathcal{E}$ denotes the set of all homology classes $E \in H_2(M;\mathbb{Z})$ that satisfy $E \cdot E = -1$ and can be represented by smoothly embedded spheres. For $b^+ > 1$ the next corollary is due to Taubes [608] and for $b^+ = 1$ to Li–Liu [423].

Corollary 13.3.27 *Let* (M, ω) *be a closed symplectic four-manifold.*

(i) *If* $E \in \mathcal{E}$ *satisfies* $c_1(E) = 1$ *then* E *can be represented by a symplectically embedded sphere.*

(ii) *If* $b^+ > 1$ *then every homology class* $E \in \mathcal{E}$ *satisfies* $c_1(E) = \pm 1$.

Proof: Assume first that $b^+ > 1$ and denote $m := c_1(E)$. Then it follows from part (i) of Lemma 13.3.15 that the homology class $A = mE$ satisfies the hypotheses of Corollary 13.3.23. Hence mE can be represented by a symplectic submanifold $C = C_1 \cup \cdots \cup C_\ell$ satisfying $K \cdot C_i = g(C_i) - 1 = C_i \cdot C_i$ for all i. Since $\sum_i C_i \cdot C_i = -m^2$ at least one of the C_i satisfies $C_i \cdot C_i = -1$. Choose the ordering such that $C_1 \cdot C_1 = -1$. Then $-1 = C_1 \cdot C = m(C_1 \cdot E)$ and so $m = \pm 1$. Assume $m = 1$. We must prove that $\ell = 1$. To see this, define

$$E_0 := E, \qquad E_1 := [C_1], \qquad A := -\sum_{i>1}[C_i], \qquad E_k := E_0 + kA.$$

Then $c_1(A) = 0$ and $E_0 \cdot A = A \cdot A = 0$, so $c_1(E_k) = 1$ and $E_j \cdot E_k = -1$ for all j, k. We prove by induction that $\pm E_k \in \mathcal{E}$ for every $k \in \mathbb{N}$. This holds by assumption for $k = 0, 1$. Assume $E_{k-1}, E_k \in \mathcal{E}$ for some $k \in \mathbb{N}$ and denote $e_k := \mathrm{PD}(E_k)$. Then there is a diffeomorphism $\phi_k : M \to M$ such that

$$\phi_k^* a = a + 2(a \cdot e_k)e_k$$

for $a \in H^2(M;\mathbb{Z})$. In particular,

$$\phi_k^* e_{k-1} = e_{k-1} - 2e_k = -e_{k+1},$$

so $\pm E_{k+1} \in \mathcal{E}$. Thus $E_k \in \mathcal{E}$ and $c_1(E_k) = 1$ for all $k \in \mathbb{Z}$. It follows that E_k can be represented by a symplectic submanifold, hence $\omega(E_k) = \omega(E_0) + k\omega(A) \geq 0$ for all $k \in \mathbb{N}$, hence $\omega(A) = 0$, and hence $\ell = 1$ as claimed. Thus we have proved (i) and (ii) under the assumption $b^+ > 1$.

Now assume $b^+ = 1$ and $c_1(E) = 1$. Then $\mathrm{SW}(M, \mathfrak{o}_\omega, \mathbf{p}_\omega, \Gamma_{\omega,\mathrm{PD}(E)}) = 1$ by part (i) of Lemma 13.3.15 and $E \cdot E' \geq -1$ for every $E' \in \mathcal{E}_\omega$ by part (ii) of Lemma 13.3.15. Hence the class $A = E$ satisfies the hypotheses of Corollary 13.3.23 and the proof proceeds as in the case $b^+ > 1$. This proves Corollary 13.3.27. □

Remark 13.3.28 (i) Corollary 13.3.25 was extended by Liu [431] to minimal closed symplectic 4-manifolds (M, ω) that satisfy the condition $K \cdot K < 0$. He proved that every such manifold contains a symplectically embedded sphere with nonnegative self-intersection. Here is a sketch. Since $K \cdot K < 0$ and (M, ω) is minimal, it follows from Corollary 13.3.24 that $b^+ = 1$. One can then use the wall-crossing formula in Remark 13.3.8 to compute the Seiberg–Witten invariants. In [431], Liu used this to show that there is a cohomology class $e \in H^2(M; \mathbb{Z})$ with $\mathrm{SW}(M, \Gamma_{\omega,e}) \neq 0$ and $e \cdot e + K \cdot e < 0$. Corollary 13.3.23 then asserts that e is Poincaré dual to a symplectic submanifold $C = C_1 \cup \cdots \cup C_\ell$ of M. Since $C \cdot C + K \cdot C < 0$ it follows that $C_i \cdot C_i + K \cdot C_i < 0$ for some i. By the adjunction formula (13.3.17) this implies $2g(C_i) - 2 = C_i \cdot C_i + K \cdot C_i < 0$. Hence $g(C_i) = 0$ and $C_i \cdot C_i + K \cdot C_i = -2$. Thus $C_i \cdot C_i \geq -1$ by (13.3.24), and hence $C_i \cdot C_i \geq 0$ because (M, ω) is minimal. By a theorem of McDuff [470, Theorem 9.4.1] the existence of a symplectic sphere with nonnegative self-intersection implies that (M, ω) is rational or ruled, and the condition $K \cdot K < 0$ shows that it is a ruled surface over a base of genus greater than one.

(ii) In [431], Liu also proved that a minimal closed symplectic four-manifold (M, ω) is rational or ruled if and only if $K \cdot [\omega] < 0$. The latter was independently proved by Ohta–Ono [511]. They also proved that the condition $K \cdot [\omega] < 0$ is equivalent to the existence of a metric with positive scalar curvature. These results can be used to prove that any symplectic 4-manifold that admits a symplectic submanifold C that is not an exceptional sphere and has $c_1(C) > 0$ is a blowup of a rational or ruled manifold, since it must satisfy $K \cdot [\omega] < 0$. This generalizes a well-known theorem in complex geometry.

(iii) In [422], Li and Liu proved that if $\pi : M \to \Sigma$ is a smooth S^2-bundle over a closed orientable surface, then any symplectic form ω on M is diffeomorphic to a symplectic form which is compatible with the given ruling π by a diffeomorphism that acts trivially on homology. □

Exercise 13.3.29 Let (M, ω) be a closed symplectic four-manifold with $b^+ > 1$, define $K := -\mathrm{PD}(c_1(\omega)) \in H_2(M; \mathbb{Z})$, and let $A \in H_2(M; \mathbb{Z})$. Prove the following.

(i) If $\mathrm{Gr}(A) \neq 0$ then $0 \leq \omega(A) \leq \omega(K)$ with equality on the left if and only if $A = 0$, and equality on the right if and only if $A = K$.

(ii) If (M, ω) is minimal and $\mathrm{Gr}(A) \neq 0$ then $A \cdot A \geq 0$.

(iii) If (M, ω) is minimal and $A \neq 0$, $A \cdot A = 0$, $\mathrm{Gr}(A) \neq 0$, then $c_1(A) = 0$ and A can be represented by a disjoint union of symplectically embedded tori with self-intersection number zero.

(iv) If $\mathrm{Gr}(A) \neq 0$ then $\mathrm{Gr}(K - A) \neq 0$ and $c := \mathrm{PD}(2A - K)$ is a basic class. □

Exercise 13.3.30 (i) Prove that $c = 0$ is the only Seiberg–Witten basic class of the 4-torus. Deduce that every symplectic form ω on the 4-torus satisfies $c_1(\omega) = 0$.

(ii) Prove that $c = 0$ is the only Seiberg–Witten basic class of the K3-surface (with its standard orientation). Deduce that every symplectic form ω on a K3-surface satisfies $c_1(\omega) = 0$. **Hint:** See Remark 13.3.19. □

More consequences for the uniqueness problem

Gromov's success in establishing that a symplectic form on $\mathbb{C}P^2$ admitting a symplectically embedded 2-sphere of self-intersection 1 is unique up to a scaling factor led to a series of efforts to extend this to ruled surfaces and to blowups. The uniqueness problem for ruled surfaces was finally solved by Lalonde–McDuff [392] by the process of inflation. We saw in Exercise 6.2.7 that any two symplectic forms on a ruled surface that are compatible with the same ruling are homotopic, i.e. they can be joined by a path of not necessarily cohomologous symplectic forms. The process of inflation allows one to convert this deformation into an isotopy. Here is the basic lemma. We have added some unnecessarily stringent conditions to avoid problems caused by multiply covered tori. A homology class $A \in H_2(M;\mathbb{Z})$ is called **primitive** if it is not a positive multiple of any other class in $H_2(M;\mathbb{Z})$.

Lemma 13.3.31 (Inflation lemma) *Let (M,ω) be a closed symplectic 4-manifold with $b^+ = 1$ and let $A \in H_2(M;\mathbb{Z})$ such that $A \cdot A \geq 0$ and $\mathrm{Gr}(A) \neq 0$. If $A^2 = 0$ assume also that A is a primitive class. Then, given any smooth family $\omega_t, 0 \leq t \leq 1$, of symplectic forms on M with $\omega_0 = \omega$, there is a family ρ_t of closed 2-forms on M, representing the class $\mathrm{PD}(A)$, such that the 2-forms*

$$\omega_t + \kappa(t)\rho_t, \quad 0 \leq t \leq 1,$$

are symplectic whenever $\kappa(t) \geq 0$.

Proof: See Lalonde–McDuff [392]. □

Thus one can change the cohomology class of a deformation by pushing it in a direction that is Poincaré dual to any class A with a nontrivial Gromov invariant. (See McDuff–Opshtein [468] for relative versions of this construction.) In Remark 13.3.28, we outlined an argument which shows that when $b^+ = 1$ the nontriviality of the wall-crossing numbers implies the nonvanishing of the Gromov invariants. Furthermore, the nontriviality of the wall-crossing numbers is a purely topological phenomenon. When $H^1(M;\mathbb{R}) = 0$, these numbers are always equal to 1 by Kronheimer–Mrowka [381] (see Remark 13.3.8). A result of Li–Liu [421] in the case $H^1(M;\mathbb{R}) \neq 0$ asserts that the numbers vanish identically if and only if $a \cup b = 0$ for all $a, b \in H^1(M;\mathbb{R})$. The following result was proved in McDuff [458] when M is simply connected or has $H^1(M;\mathbb{R})^2 \neq 0$, and was extended to the general case by Li–Liu [424, Proposition 4.11].

Theorem 13.3.32 *Let (M,ω) be a closed symplectic four-manifold with $b^+ = 1$. Then any path of symplectic forms on M with cohomologous endpoints is homotopic with fixed endpoints to an isotopy.*

Proof: See McDuff [458], Biran [65], and Li–Liu [424, Proposition 4.11]. □

The proof uses the fact that, by Taubes–Seiberg–Witten theory and the wall-crossing formula, there are enough nonvanishing Gromov invariants so that one can use Lemma 13.3.31 to force the cohomology class of the deformation to be constant. Combining this with our previous results we get the following theorem about the uniqueness problem.

Theorem 13.3.33 *Let M be $\mathbb{C}\mathrm{P}^2$ or a ruled surface.*

(i) *Any two cohomologous symplectic forms on M are diffeomorphic.*

(ii) *Let $\widetilde{M}$ be the blowup of M at k points, and suppose that $\widetilde{\omega}_0$ and $\widetilde{\omega}_1$ are two cohomologous symplectic forms on $\widetilde{M}$ that are obtained by blowup from symplectic forms ω_0 and ω_1 on M. Then $\widetilde{\omega}_0$ and $\widetilde{\omega}_1$ are diffeomorphic.*

Proof: We prove (i). For $M = \mathbb{C}\mathrm{P}^2$ this is the assertion of Corollary 13.3.26. If M is ruled then, by part (i) of Remark 13.3.28, we can assume both forms are compatible with the same ruling. Therefore, they are homotopic by Exercise 6.2.7 and isotopic by Theorem 13.3.32. Hence they are diffeomorphic. This proves (i).

We prove (ii). By (i), we may assume that $\omega_0 = \omega_1$. Then it follows from Theorem 7.1.23 that $\widetilde{\omega}_0$ and $\widetilde{\omega}_1$ are homotopic. Hence, by Theorem 13.3.32, they are isotopic, and hence diffeomorphic. □

One can generalize part (ii) of Theorem 13.3.33. For example, Li–Liu [422] prove that if (M, ω) is any blowup of a rational or ruled surface with canonical class $K = -c_1(\omega)$, then any two symplectic forms on M whose canonical class also equals K are deformation equivalent, i.e. related by a diffeomorphism followed by a path of symplectic forms.

Finally we remark that the inflation lemma implies that if A is any homology class with $\mathrm{Gr}(A) \neq 0$ there is a symplectic form on M whose cohomology class is arbitrarily close to $\mathrm{PD}(A)$. This is the approach used by Biran [66] to construct full fillings of $\mathbb{C}\mathrm{P}^2$ by k balls for $k \geq 10$.

13.4 Symplectic four-manifolds

This section is a collection of examples. In examining these examples we shall see that the uniqueness questions raised in Section 13.1 are intimately connected to the properties of the diffeomorphism group $\mathrm{Diff}(M)$. The subgroups

$$\begin{aligned}\mathrm{Diff}(M, a) &:= \{\phi \in \mathrm{Diff}(M) \,|\, \phi^* a = a\}, \\ \mathrm{Diff}_h(M) &:= \{\phi \in \mathrm{Diff}(M) \,|\, \phi_* = \mathrm{id} : H_*(M; \mathbb{Z}) \to H_*(M; \mathbb{Z})\},\end{aligned}$$

for $a \in H^2(M; \mathbb{R})$, will play an important role. The identity component $\mathrm{Diff}_0(M)$ of $\mathrm{Diff}(M)$ is contained in both subgroups. For a closed symplectic manifold (M, ω) denote by $\mathrm{Symp}_0(M, \omega)$ the identity component of the group $\mathrm{Symp}(M, \omega)$ of symplectomorphisms, by $\mathrm{Symp}_h(M, \omega) := \mathrm{Symp}(M, \omega) \cap \mathrm{Diff}_h(M)$ the group of symplectomorphisms that induce the identity on homology, and by $\mathscr{S}_\omega$ the connected component of ω in the space $\mathscr{S}_a$ of symplectic forms representing the cohomology class $a := [\omega]$. By Moser isotopy (Theorem 3.2.4), the map $\mathrm{Diff}_0(M) \to \mathscr{S}_\omega : \phi \mapsto \phi^* \omega$ is surjective and there is a fibration

$$\mathrm{Symp}(M, \omega) \cap \mathrm{Diff}_0(M) \hookrightarrow \mathrm{Diff}_0(M) \longrightarrow \mathscr{S}_\omega. \tag{13.4.1}$$

Hence there is a homeomorphism $\mathscr{S}_\omega \cong \mathrm{Diff}_0(M)/(\mathrm{Symp}(M, \omega) \cap \mathrm{Diff}_0(M))$.

Example 13.4.1 (The projective plane) Let $M = \mathbb{C}\mathrm{P}^2$, fix a cohomology class $a \in H^2(M;\mathbb{R})$ with $a^2 \neq 0$, and let ω be the Fubini–Study form in the class a. By a theorem of Taubes [608] every symplectic form in the class a is diffeomorphic to ω (Corollary 13.3.26). Hence the map $\mathrm{Diff}_h(M) \to \mathscr{S}_a : \phi \mapsto \phi^*\omega$ descends to a homeomorphism $\mathscr{S}_a \cong \mathrm{Diff}_h(M)/\mathrm{Symp}(M,\omega)$. A theorem of Gromov [287] asserts that $\mathrm{Symp}(M,\omega)$ retracts onto the isometry group $\mathrm{PU}(3)$ of $\mathbb{C}\mathrm{P}^2$ (Example 10.4.2) and hence is connected.

Thus $\mathscr{S}_a$ is connected if and only if every diffeomorphism of $\mathbb{C}\mathrm{P}^2$ that acts as the identity on cohomology is isotopic to the identity, i.e. if and only if $\mathrm{Diff}_h(\mathbb{C}\mathrm{P}^2)$ is connected. (This is an open problem.) □

Example 13.4.2 (The product $S^2 \times S^2$) The discussion of Example 13.4.1 carries over to $M := S^2 \times S^2$ as follows. First, every class $a \in H^2(M;\mathbb{R})$ with $a^2 \neq 0$ is represented by a symplectic form. Second, theorems of Gromov [287] and McDuff [449] assert that every symplectic form for which the homology classes $A := [S^2 \times \{\mathrm{pt}\}]$ and $B := [\{\mathrm{pt}\} \times S^2]$ (with either orientation) are represented by symplectically embedded spheres is diffeomorphic to a standard (i.e. product) form $\lambda_1 \mathrm{pr}_1^*\sigma \oplus \lambda_2 \mathrm{pr}_2^*\sigma$, where σ is an area form on S^2 (see [470, Theorem 9.4.7]). Third, Taubes' theorem establishes the existence of the required symplectic spheres. Fourth, a theorem of Gromov [287] asserts that the group $\mathrm{Symp}_h(M,\omega)$ retracts onto the isometry group $\mathrm{SO}(3)\times\mathrm{SO}(3)$ when A, B have the same area (see [470, Theorem 9.5.1]). Fifth, a theorem of Abreu and McDuff [8,12] asserts that the symplectomorphism group is connected when A, B have different areas. Thus in both cases

$$\mathscr{S}_a \cong \mathrm{Diff}(M,a)/\mathrm{Symp}(M,\omega) \cong \mathrm{Diff}_h(M)/\mathrm{Symp}_h(M,\omega)$$

and the group $\mathrm{Symp}_h(M,\omega)$ is connected.

This shows that $\mathscr{S}_a$ is connected if and only if $\mathrm{Diff}_h(S^2 \times S^2)$ is connected.

The above argument shows that $\mathrm{Symp}(M,\omega) \cap \mathrm{Diff}_0(M) = \mathrm{Symp}_0(M,\omega)$ for every symplectic form ω on $M = S^2 \times S^2$. Now consider the fibration

$$\mathrm{Symp}_0(M,\omega) \to \mathrm{Diff}_0(M) \to \mathscr{S}_\omega$$

in (13.4.1). Gromov showed in [287] that there is an element in $\pi_1(\mathrm{Diff}_0(M))$ with nontrivial image in $\pi_1(\mathscr{S}_\omega)$ when the spheres have equal area, but trivial image when $\langle a, A\rangle > \langle a, B\rangle$. It turned out that this loop is given by a circle action that fixes the points of the diagonal and antidiagonal spheres, and hence can be realized on the rational ruled surface $\mathbb{P}(L \oplus \mathbb{C})$ where $L \to \mathbb{C}\mathrm{P}^1$ has Chern class 2; cf. Example 5.1.11. Interchanging the roles of A, B we see that there are in fact two such loops in $\mathrm{Diff}_0(M)$ though at most one can be represented in $\mathrm{Symp}_0(M,\omega)$. These loops, together with their higher Samelson products with the elements in $\pi_3(\mathrm{Symp}_0(M,\omega))$ represented by the $\mathrm{SO}(3)$ factors, generate the elements of the rational homotopy of $\mathrm{Symp}_0(M,\omega)$ that disappear in $\mathrm{Diff}_0(M)$ (see Abreu–McDuff [8,12]). □

Example 13.4.3 (Ruled surfaces) Let M be an orientable smooth four-manifold that admits the structure of a fibration $S^2 \hookrightarrow M \xrightarrow{\pi} \Sigma$ over a closed orientable surface Σ of positive genus $g(\Sigma)$ with fibres diffeomorphic to the 2-sphere. Fix an orientation of M and an orientation of the fibres. Let $F \in H_2(M;\mathbb{Z})$ be the homology class of the fibre. Call a symplectic form ω on M **compatible with the fibration** if it restricts to a symplectic form on each fibre. Call it **compatible with the orientations** if its cohomology class a satisfies

$$a^2 > 0, \qquad \langle a, F\rangle > 0.$$

Here are some basic facts.

(i) By Taubes–Seiberg–Witten theory (Section 13.3) a class $a \in H^2(M;\mathbb{R})$ is represented by a symplectic form if and only if $a^2 \neq 0$ and $\langle a, F\rangle \neq 0$. Any such cohomology class is uniquely determined by the numbers a^2 and $\langle a, F\rangle$.

(ii) A theorem of McDuff [449] (see also [470, Theorem 9.4.1]) shows that every symplectic form on M that admits a symplectically embedded two-sphere in the class F or $-F$ is diffeomorphic to one that is compatible with the fibration. The existence of the required sphere follows from Taubes–Seiberg–Witten theory.

(iii) M admits an orientation-preserving diffeomorphism that preserves the fibration and reverses the orientation of the fibre, and also an orientation reversing diffeomorphism that preserves the fibration and the orientation of the fibre. Thus every symplectic form on M is diffeomorphic to one that is compatible with the fibration and orientations.

(iv) A theorem of Lalonde–McDuff [393, 394] asserts that any two symplectic forms ω_0, ω_1 on M that are compatible with the fibration and orientations can be joined by a path of symplectic forms. They also proved that the path can be chosen in the same cohomology class when $[\omega_0] = [\omega_1]$. Thus

$$\omega_0 \overset{(a)}{\sim} \omega_1 \qquad \Longleftrightarrow \qquad \omega_0 \overset{(b)}{\sim} \omega_1 \;\text{ and }\; [\omega_0] = [\omega_1]. \tag{13.4.2}$$

Here (a) and (b) denote the equivalence relations on page 504. *Thus on a ruled surface any two symplectic forms are deformation equivalent, and they are diffeomorphic if they represent the same cohomology class. In the latter case they are homotopic if and only if they are isotopic.*

(v) The properties of the fibration (13.4.1) are now more difficult to establish, and have only been worked out in any detail for the trivial bundle $M = \Sigma \times S^2$. Assume without loss of generality that $\langle a, F\rangle = 1$ and $\lambda := \langle a, [\Sigma \times \mathrm{pt}]\rangle > 0$. Under the assumption $\lambda > \lfloor g(\Sigma)/2 \rfloor$, Buse [88] proved that a symplectomorphism of (M,ω) is symplectically isotopic to the identity if and only if it is smoothly isotopic to the identity. In particular, the condition on λ is always satisfied when Σ is a torus or a sphere. At the time of writing (in 2016) it is an open question whether the result continues to hold for all values of λ. Buse's proof is based on earlier work by McDuff in [459]. It relies on the existence of suitable J-holomorphic curves and hence on the nonvanishing of certain Gromov invariants. This is why the condition on λ appears. □

Example 13.4.4 (Blowups of the projective plane) This example occupies the next five pages. Fix an integer $k \geq 0$ and denote by

$$M := M_k := \mathbb{C}\mathrm{P}^2 \# k\overline{\mathbb{C}\mathrm{P}}^2$$

the k-fold blowup of $\mathbb{C}\mathrm{P}^2$ with its complex orientation. Let $L \in H_2(M;\mathbb{Z})$ be the homology class of the line with self-intersection number $L \cdot L = 1$ and let $E_1, \dots, E_k \in H_2(M;\mathbb{Z})$ be the homology classes of the exceptional divisors with self-intersection numbers $E_i \cdot E_i = -1$. Then, by (4.4.3) with $2\chi + 3\sigma = 9 - k$, the set $\mathcal{C} \subset H^2(M;\mathbb{Z})$ of first Chern classes of almost complex structures on M that are compatible with the orientation is given by

$$\mathcal{C} = \left\{ c = \mathrm{PD}\left(nL - \sum_{i=1}^k n_i E_i\right) \,\middle|\, \begin{array}{l} n,\, n_1,\, \dots,\, n_k \text{ are odd,} \\ n^2 - \sum_{i=1}^k n_i^2 = 9 - k \end{array} \right\}.$$

The standard first Chern class is $c_{\mathrm{std}} = 3L - \sum_i E_i$. As in (13.3.22) denote by

$$\mathcal{E} \subset H_2(M;\mathbb{Z})$$

the set of all integral homology classes with square $E \cdot E = -1$ that are represented by smoothly embedded spheres. Wall [654, 655] proved that if $k \leq 9$ the set $\mathcal{E}$ has a purely numerical description:

$$\mathcal{E} = \left\{ E = mL + \sum_{i=1}^k m_i E_i \,\middle|\, E \cdot E = m^2 - \sum_{i=1}^k m_i^2 = -1 \right\} \quad \text{for } k \leq 9.$$

However, in general this is not true. For $c \in \mathcal{C}$ define

$$\mathcal{E}^c := \{E \in \mathcal{E} \,|\, \langle c, E\rangle = 1\}.$$

In [418] Tian-Jun Li proved that every smoothly embedded -1 sphere is homologous to a sphere that is symplectically embedded with respect to some symplectic form ω. Therefore its homology class E belongs to $\mathcal{E}^c$, where $c = c_1(\omega)$. Conversely, if $E \in \mathcal{E}^c$ then E is represented by a symplectically embedded sphere for every ω with $c_1(\omega) = c$ (see Corollary 13.3.27). Hence

$$\mathcal{E} = \bigcup_{c \in \mathcal{C}_{\mathrm{symp}}} \mathcal{E}^c.$$

This is *not* a disjoint union. Here are the main results about symplectic structures on M_k. The first three statements are proved in Li–Liu [422, 424] and Li–Li [416].

(I) *Fix a cohomology class* $a \in H^2(M;\mathbb{R})$ *such that* $a^2 > 0$. *Then* $\mathscr{S}_a \neq \emptyset$ *if and only if* $\langle a, E\rangle \neq 0$ *for all* $E \in \mathcal{E}$ *or, equivalently, there exists an element* $c \in \mathcal{C}_{\mathrm{symp}}$ *such that* $\langle a, E\rangle > 0$ *for all* $E \in \mathcal{E}^c$.

(II) *$\mathcal{C}_{\mathrm{symp}} = \mathcal{C}$ for $k \le 8$ and $\mathcal{C}_{\mathrm{symp}} \subsetneq \mathcal{C}$ for $k \ge 9$. Moreover, the diffeomorphism group acts transitively on $\mathcal{C}_{\mathrm{symp}}$.*

(III) *If ω_0, ω_1 are symplectic forms on M, then there exists a diffeomorphism ϕ of M such that ω_0 and $\phi^*\omega_1$ can be joined by a path of symplectic forms.*

Note 1. By (I) the hypersurfaces $\langle a, E\rangle = 0$ for $E \in \mathcal{E}$ divide the space $\{a^2 > 0\}$ into connected open cones $\mathcal{K}$, each of which contains symplectic forms (see Figure 13.4). Although it is not known whether the set of corresponding forms is connected, one can show that the cohomology classes of each component of this space of forms fill out the whole cone $\mathcal{K}$. Moreover, the map $\omega \mapsto c_1(\omega)$ sets up a bijection between the set of these cones $\mathcal{K}$ and $\mathcal{C}_{\mathrm{symp}}$. Indeed, $c_1(\omega)$ is determined on each cone by the fact that the walls of $\mathcal{K}$ are given by classes E that are represented by ω-exceptional spheres[43] and hence have $c_1(E) = 1$.

Here are some consequences for the symplectic uniqueness problem. Assertion (V) is Theorem 1.1 in McDuff [453]. For (IV) and (V) see also Karshon–Kessler–Pinsonnault [358, 3.11] and Karshon–Kessler [357, 1.5 and 1.6].

(IV) *Two symplectic forms ω_0, ω_1 on M have the same first Chern class if and only if there exists a diffeomorphism ϕ of M, inducing the identity on homology, such that ω_0 and $\phi^*\omega_1$ can be joined by a path of symplectic forms.*

(V) *Two symplectic forms ω_0, ω_1 on M represent the same cohomology class if and only if there exists a diffeomorphism ϕ of M, inducing the identity on homology, such that $\phi^*\omega_1 = \omega_0$.*

Thus two symplectic forms on M are cohomologous if and only if they are related by (A_0), the equivalence relations (B_0), (c), (d) on page 504 agree by (IV) and Proposition 13.3.12, and any two symplectic forms on M are deformation equivalent by (II) and (IV). The paper [358] gives an effective method to determine whether two classes in $\mathcal{K}^{c_{\mathrm{std}}}$ (see equation (13.1.3)) are diffeomorphic, which is based on understanding the action of the Cremona group (defined below).

It follows from (IV) and (V) that if $\mathrm{Diff}_h(M)$ is connected, then the space $\mathscr{S}_a$ is connected for every $a \in H^2(M;\mathbb{R})$ and the space of symplectic forms with given first Chern class $c \in \mathcal{C}_{\mathrm{symp}}$ is also connected.

The converse is an open question which requires further study of the symplectomorphism group of M for general k. The following note explains what is involved.

Note 2. Suppose $\mathscr{S}_a$ is connected, fix a symplectic form $\omega \in \mathscr{S}_a$ and choose a diffeomorphism $\psi \in \mathrm{Diff}_h(M)$. Then $\psi^*\omega$ is isotopic to ω. Thus there is a smooth isotopy ϕ_t such that $\phi_0 = \mathrm{id}$ and $\phi_1^*\psi^*\omega = \omega$, so $\psi \circ \phi_1 \in \mathrm{Symp}_h(M,\omega)$. To conclude that ψ is isotopic to the identity we would have to know that

[43] Every ω-exceptional sphere Z determines such a wall since, in view of the normal form of ω near Z described in Theorem 7.1.21 one can always homotop ω to make the size $\langle[\omega], E\rangle$ of such a sphere arbitrarily small. The converse follows from the second statement in (I) since each wall of $\mathcal{K}$ gives rise to one of its defining inequalities $\langle[\omega], E\rangle > 0$.

$\mathrm{Symp}_h(M,\omega) \subset \mathrm{Diff}_0(M)$. The group $\mathrm{Symp}_h(M,\omega)$ is connected for $\mathbb{C}\mathrm{P}^2$ and its blowups at up to four points (see Example 10.4.2). It is disconnected for the monotone blowup of $\mathbb{C}\mathrm{P}^2$ at five to eight points (see Seidel [575]). However, the known examples here are squares of generalized Dehn twists and hence are smoothly isotopic to the identity. *When $k \geq 5$ it is an open problem whether*

$$\mathrm{Symp}_h(M,\omega) \subset \mathrm{Diff}_0(M)$$

for the k-point blowup of $\mathbb{C}\mathrm{P}^2$.

Action of the diffeomorphism group on homology The proofs of (I) through (V) are based on understanding the action of the diffeomorphism group of M on homology. There are two kinds of diffeomorphisms of M that are relevant here, both acting by reflection on $H_2(M;\mathbb{Z})$.

• Reflection in a smoothly embedded -1 sphere in class E acts on homology by

$$A \mapsto A + 2(A \cdot E)E,$$

preserving the intersection form but not the first Chern class. In particular, the diffeomorphism sends E to $-E$.

• Reflection in a smoothly embedded -2 sphere in class R acts on homology by

$$A \mapsto A + (A \cdot R)R,$$

preserving the intersection form and also the first Chern class c if $c(R) = 0$. For example, with the standard first Chern class

$$c_{\mathrm{std}} = 3L - \sum_{i=1}^{k} E_i,$$

we can take $R = L - E_1 - E_2 - E_3$. If the symplectic form ω lies in a class a with $\langle a, R\rangle = 0$, then R can be realized as a Lagrangian sphere, in which case the reflection is simply a generalized Dehn twist in this sphere (see Exercise 6.3.7) and in particular preserves the symplectic form.

Reflections of the first kind act nontrivially on the set $\mathcal{C}$. It follows from the results of Wall [654, 655] that the reflections of the first kind generate the group of automorphisms of the intersection form for $k \leq 9$, and that this group acts transitively on $\mathcal{C}$ for $k \leq 8$. In contrast, reflections of the second kind permute the elements of the set $\mathcal{E}^c$ when $c(R) = 0$. In particular, Li–Li show in [416, Lemma 3.4] that when $k \geq 3$ there is a transitive action on $\mathcal{E}^{c_{\mathrm{std}}}$ by the group generated by reflection in the class $R = L - E_1 - E_2 - E_3$ and permutations of the E_i. (This is the **Cremona group.**) Thus $\mathcal{E}^{c_{\mathrm{std}}}$ is the orbit of E_1 under this action. These observations form the basis for the proof of (I) through (V).

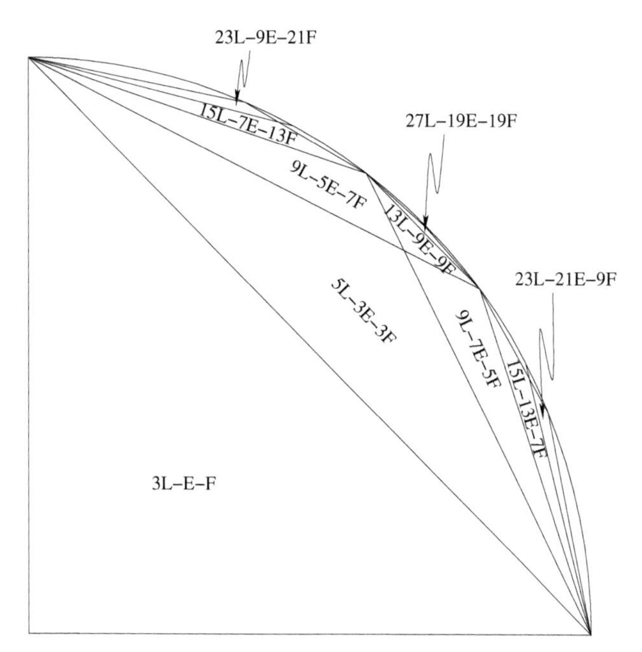

FIG. 13.4. The symplectic chamber structure on $H^2(\mathbb{C}\mathrm{P}^2\#2\overline{\mathbb{C}\mathrm{P}}^2;\mathbb{R})$ labelled by the Chern classes $c \in \mathcal{C}_{symp}$.

The case $k = 2$**.** Consider the two point blowup $M = M_2$ of $\mathbb{C}\mathrm{P}^2$. Since the diffeomorphism group acts transitively on $\mathcal{C}$ by the results of Wall [654, 655], it follows that for each $c \in \mathcal{C}$ the set $\mathcal{E}^c$ consists of three elements E_1^c, E_2^c, E_3^c and the numbering can be chosen such that

$$E_1^c \cdot E_3^c = E_2^c \cdot E_3^c = 1, \qquad E_1^c \cdot E_2^c = 0. \tag{13.4.3}$$

(For the standard first Chern class $c_{\rm std} := \mathrm{PD}(3L - E_1 - E_2) \in \mathcal{C}$ these can be chosen as $E_1^{c_{\rm std}} = E_1$, $E_2^{c_{\rm std}} = E_2$, $E_3^{c_{\rm std}} = L - E_1 - E_2$.) Corollary 13.3.27 asserts that the classes E_i^c are represented by symplectically embedded spheres for every symplectic form ω with first Chern class $c_1(\omega) = c$. Thus, if $a \in H^2(M;\mathbb{R})$ has a symplectic representative with first Chern class c, then $\langle a, E\rangle > 0$ for all $E \in \mathcal{E}^c$. The converse holds by the Nakai–Moishezon criterion (see Demazure [133] and McDuff–Polterovich [469] for a direct symplectic proof). Thus

$$\begin{aligned} \mathcal{K}^c &:= \left\{a \in H^2(M;\mathbb{R}) \,|\, \text{there is an } \omega \in \mathscr{S}_a \text{ such that } c_1(\omega) = c\right\} \\ &= \left\{a \in H^2(M;\mathbb{R}) \,|\, a^2 > 0,\ \langle a, E\rangle > 0 \; \forall\, E \in \mathcal{E}^c\right\}. \end{aligned} \tag{13.4.4}$$

(The condition $a^2 > 0$ in (13.4.4) is actually redundant for the two point blowup of $\mathbb{C}\mathrm{P}^2$.) A combinatorial argument shows that $a \in \mathcal{K}_{\rm symp}$ if and only if $a^2 > 0$ and $\langle a, E\rangle \neq 0$ for *all* $E \in \mathcal{E}$. These observations imply assertions (I) and (II) for $k = 2$. The cones $\mathcal{K}^c$ define the chamber structure on $H^2(M;\mathbb{R})$. This is illustrated in Figure 13.4, where the standard basis of $H_2(M;\mathbb{Z})$ is denoted by

L, E, F and the symplectic cohomology classes satisfy the normalization conditions $\langle a, L\rangle = 1$, $\langle a, E\rangle > 0$, $\langle a, F\rangle > 0$. The chamber structure for $k = 2$ is determined by lines connecting pairs of rational points on the unit circle. These straight lines are in one-to-one correspondence with the pairs $\pm E$ in $\mathcal{E}$ and reflections along the corresponding planes $\langle a, E\rangle = 0$ permute the chambers $\mathcal{K}^c$. Each chamber intersects the disc $\mathcal{K} \cap \{\langle a, L\rangle = 1\}$ in a triangle with sides of two kinds. One has both vertices on the boundary circle, while the other two have one common interior vertex.

To prove (III), (IV), and (V) one can start with an explicit complex model M for the 2-fold blowup of the projective plane with the standard first Chern class $c_{\mathrm{std}} = \mathrm{PD}(3L - E_1 - E_2)$ in the basis L, E_1, E_2 of $H_2(M; \mathbb{Z})$. Given any symplectic form ω on M with first Chern class c, there exist classes $E_1^c, E_2^c, E_3^c \in \mathcal{E}^c$ as in (13.4.3). By Taubes–Seiberg–Witten theory the classes $L^c := E_1^c + E_2^c + E_3^c$ and E_1^c, E_2^c are represented by disjoint embedded J-holomorphic spheres. (For E_i^c see Corollary 13.3.27.) These can then be used to construct a diffeomorphism $\phi : M \to M$ such that $\phi^*\omega$ belongs to the connected component of the symplectic forms obtained by the blowup construction of Theorem 7.1.21. The construction of ϕ is based on the Weinstein Neighbourhood Theorem 3.4.10, on symplectic blowing down as explained in Section 7.1, on a theorem of Gromov [286] (see also [470, Theorem 9.4.1]) to construct a symplectomorphism from the blown-down manifold to $\mathbb{C}\mathrm{P}^2$, then on the normalization of the diffeomorphism near the blowup points via a suitable Hamiltonian isotopy (see Proposition 7.1.22), and finally on the uniqueness result for blowup forms in Theorem 7.1.23. (For more details see Salamon [558].) When $c = c_{\mathrm{std}}$ one obtains $\phi \in \mathrm{Diff}_h(M)$. This proves (III) and (IV). Assertion (V) then follows from Theorem 13.3.32.

The case $1 \le k \le 8$**.** The story is much the same for all del Pezzo surfaces

$$\mathbb{C}\mathrm{P}^2 \# k \overline{\mathbb{C}\mathrm{P}}^2, \qquad 1 \le k \le 8.$$

In particular, the same argument as in the case $k = 2$ establishes condition (I) and equation (13.4.4) for $1 \le k \le 8$. When $3 \le k \le 8$ the condition $a^2 > 0$ in (I) is again redundant: see [565]. Moreover, the set $\mathcal{E}^c$ of exceptional classes is finite. Here is a list

$$\#\mathcal{E}^c = \begin{cases} 1, \text{ if } k = 1, \\ 3, \text{ if } k = 2, \\ 6, \text{ if } k = 3, \\ 10, \text{ if } k = 4, \\ 16, \text{ if } k = 5, \\ 27, \text{ if } k = 6, \\ 56, \text{ if } k = 7, \\ 240, \text{ if } k = 8. \end{cases}$$

(For $k = 6$ these are the 27 lines on a cubic in $\mathbb{C}\mathrm{P}^3$; cf. Example 4.4.3.)

The case $k \geq 9$. In this situation the diffeomorphism group no longer acts transitively on $\mathcal{C}$. Moreover, if $c \in \mathcal{C}_{\text{symp}}$, the set $\mathcal{E}^c$ is infinite. For $k = 9$ the **elliptic surface**

$$M := E(1) := \mathbb{C}\mathrm{P}^2 \# 9\overline{\mathbb{C}\mathrm{P}}^2. \tag{13.4.5}$$

admits the structure of a holomorphic Lefschetz fibration over $\mathbb{C}\mathrm{P}^1$ with elliptic curves as regular fibres and twelve singular fibres. (Choose two cubics in general position and blow up their nine points of intersection, as in Example 7.4.13.) In the standard basis the homology class of the fibre is

$$F := 3L - E_1 - E_2 - \cdots - E_9.$$

Its Poincaré dual $c_{\text{std}} = \mathrm{PD}(F) = c_1(TM, J_{\text{std}})$ is the first Chern class of the standard complex structure. For $k = 9$ the theory of Wall [654] is still applicable, and shows that for every $c \in \mathcal{C}$, there is a unique odd integer $m \geq 1$ such that c is diffeomorphic to mc_{std}. By Taubes' results, c is the first Chern class of a symplectic form if and only if $m = 1$. This amplifies the statement in (II) above that $\mathcal{C} \neq \mathcal{C}_{\text{symp}}$ when $k \geq 9$. For $k \geq 9$, Li–Liu proved in [424, Theorem 1] that $c \in \mathcal{C}_{\text{symp}}$ if and only if c is diffeomorphic to c_{std}. With this understood, one can use blowing down as outlined in the case $k = 2$ to prove assertions (III)-(V) and one only needs to verify equation (13.4.4) for the standard chamber. Note that now the condition $a^2 > 0$ in (I) is needed, since the first Chern class c_{std} satisfies $\langle c, E\rangle > 0$ for all $E \in \mathcal{E}_e$ while $c^2 = 0$.

This completes our discussion of blowups of the projective plane. For the case of ruled surfaces, see Holm–Kessler [328]. □

Example 13.4.5 (Recognizing blowups) We now discuss a question raised in Section 7.1 about symplectic forms on blowup manifolds which serves to illustrate the special nature of rational and ruled four-manifolds. Let $\pi_M : \widetilde{M} \to M$ be the one-point blowup of a symplectic 4-manifold (M, ω) with exceptional divisor in the class E, and define

$$a := [\omega] \in H^2(M),\ c := c_1(\omega),\ \widetilde{a} := \pi_M^* a,\ \widetilde{c} := \pi_M^* c,\ e := \mathrm{PD}(E) \in H^2(\widetilde{M}).$$

Assume that $\widetilde{\omega}$ is any symplectic form on $\widetilde{M}$ such that

$$[\widetilde{\omega}] = \widetilde{a} - \pi\lambda^2 e, \qquad c_1(\widetilde{\omega}) = \pi^* c - e. \tag{13.4.6}$$

On can ask the following questions (see the discussion after Corollary 7.1.24).

(a) *Is $(\widetilde{M}, \widetilde{\omega})$ symplectomorphic to a blowup of (M, ω) with weight $\pi\lambda^2$?*

(b) *Does (M, ω) admit a symplectic embedding of the ball $B(\lambda)$?*

(c) *Are any two cohomologus symplectic forms on $\widetilde{M}$ diffeomorphic?*

A positive answer to question (b) is necessary for addressing question (a), and question (c) is of course a separate issue that one can consider for any manifold. However, if questions (b) and (c) have positive answers, then so does question (a) by the blowup construction in Section 7.1. In general one could attempt to approach question (a) as follows. By Corollary 13.3.27, the manifold $(\widetilde{M}, \widetilde{\omega})$ contains a symplectically embedded sphere in the class E. Blowing this down as in Section 7.1, we get a symplectic manifold (M', ω') that contains a symplectically embedded ball $B(\lambda)$. However, there is no guarantee that M' is diffeomorphic to M since the action of the diffeomorphism group of $\widetilde{M}$ on the set of smoothly embedded spheres in the class E may well not be transitive. Further, even if M' is diffeomorphic to M it might not follow that (M', ω') and (M, ω) are symplectomorphic. Thus the answer to question (a) remains open in general.

We now explain why all three questions have positive answers for blowups of the projective plane and of ruled surfaces. Question (c) has a positive answer for all blowups of $\mathbb{C}\mathrm{P}^2$ by part (V) of Example 13.4.4 and for all blowups of ruled surfaces by a theorem of Li–Liu [422].

To address question (b), assume first that M is a $(k-1)$-point blowup of $\mathbb{C}\mathrm{P}^2$ with the standard basis $L, E_1, \dots, E_{k-1}$ of $H_2(M;\mathbb{Z})$. By part (II) of Example 13.4.4 we may assume that

$$c = \mathrm{PD}(3L - E_1 - \cdots - E_{k-1}).$$

Then L, E_i lift to classes in $H_2(\widetilde{M};\mathbb{Z})$, denoted by the same letters, and there is an additional exceptional class E_k, obtained by blowing up a point in M. By assumption, $\widetilde{\omega}$ is a symplectic form on $\widetilde{M}$ with

$$c_1(\widetilde{\omega}) = \mathrm{PD}(3L - E_1 - \cdots - E_k), \qquad \widetilde{\omega}(E_k) = \pi\lambda^2.$$

Hence the classes $E_1, \dots, E_k$ all have first Chern number one and thus, by Corollary 13.3.27, are represented by disjoint symplectically embedded spheres $S_1, \dots, S_k \subset \widetilde{M}$. (To see that they are disjoint, we use the fact that because 'Seiberg–Witten equals Gromov' these spheres can be assumed to be J-holomorphic for the same J, and hence are disjoint by the positivity of intersections of J-holomorphic curves. Alternatively, use Corollary 13.3.23 with $A = \sum_{i=1}^k E_i$.) The key issue to understand is why, after blowing these spheres down, we obtain a manifold diffeomorphic to $\mathbb{C}\mathrm{P}^2$. This follows from the fact that L can be represented by another symplectically embedded sphere S (with $S \cdot S = 1$) that is disjoint from the S_i and hence survives the blowing down construction. Since the blown down manifold is minimal, it follows from a theorem of Gromov–McDuff (see [470, Theorem 9.4.1]) that it is symplectomorphic to $\mathbb{C}\mathrm{P}^2$, with the appropriately scaled Fubini–Study form, and contains disjoint embedded balls $B_1, \dots, B_k$ of appropriate radii. Blowing up at $k-1$ points, the balls $B_1, \dots, B_{k-1}$ give rise to a symplectic form on M in the same homology class as ω, and hence symplectomorphic to ω. Now the ball B_k is symplectomorphic to $B(\lambda)$ by construction and embeds symplectically into (M, ω). This answers question (b) in the affirmative. Hence because (c) holds in this case, (a) also holds.

For blowups of $S^2 \times S^2$ the argument is essentially the same, except that the role of the sphere S is now played by a pair of symplectically embedded spheres with zero self-intersection that intersect transversally and positively in exactly one point. Then one can use [470, Theorem 9.4.7] to identify the blown down manifold with $S^2 \times S^2$. For blowups of ruled surfaces over a base of positive genus, the argument is again the same with the role of S played by an embedded symplectic sphere with self-intersection number zero in the homology class of the fibre. One can then use McDuff's theorem in [449] (see also [470, Theorem 9.4.1]) to identify the blown down manifold with the original ruled surface. □

Example 13.4.6 (Deformation equivalence) In [592, 593], Ivan Smith constructed a simply connected four-manifold X that admits two symplectic forms ω_0, ω_1 such that $c_1(\omega_0)$ is divisible by three and $c_1(\omega_1)$ is primitive. His four-manifold is obtained by forming a fibre connected sum of $\mathbb{T}^4$ with five copies of $E(1)$. The first Chern classes of ω_0 and ω_1 are not diffeomorphic, and hence ω_0 and ω_1 are not deformation equivalent. In fact, for each integer N he constructed a simply connected four-manifold X_N with at least N different deformation equivalence classes of symplectic forms, distinguished by the divisibility properties of their first Chern classes. Taking the product $X_N \times \mathbb{T}^{2n}$, one obtains a $(4+2n)$-manifold with N pairwise deformation inequivalent symplectic forms.

An earlier example of a symplectic four-manifold with two deformation inequivalent symplectic forms was constructed by McMullen–Taubes [476]. In their example the symplectic forms are distinguished by their Seiberg–Witten invariants (see also [642]). In [641] Vidussi constructed homotopy K3's with arbitrarily many deformation inequivalent symplectic forms. □

Example 13.4.7 (K3-surfaces) A closed complex surface is called a **complex K3-surface** if it is simply connected and has first Chern class zero. A closed oriented smooth four-manifold is called a **smooth K3-surface** if it is simply connected and admits an integrable almost complex structure that is compatible with the orientation and has first Chern class zero. Every K3-surface is spin and has Euler characteristic 24 and signature -16 [345, Prop 1.3.5]. Moreover, any two K3-surfaces are diffeomorphic [345, Thm 7.1.2], Explicit examples of K3-surfaces are quartics in $\mathbb{CP}^3$ (Example 4.4.3) and the fibre connected sum $E(2) = E(1)\#_{\mathbb{T}^2}E(1)$ of two elliptic surfaces as in (13.4.5) (Exercise 7.2.7). We now sketch the proof of the following.

Let M be a smooth K3-surface. Then every class $a \in H^2(M;\mathbb{R})$ with $a^2 > 0$ can be represented by a symplectic form with an associated hyperKähler structure.

To see this, denote by $\mathcal{J}_{\mathrm{int}}(M)$ the space of integrable almost complex structures on M and, for $J \in \mathcal{J}_{\mathrm{int}}(M)$, denote by $\mathcal{K}_J \subset H^2(M;\mathbb{R})$ the **Kähler cone**, i.e. the set of all cohomology classes of Kähler forms on (M, J). Every $J \in \mathcal{J}_{\mathrm{int}}(M)$ has first Chern class zero [241] and has a nonempty Kähler cone [345, Ch. 7, §3.2]. If $a_0 \in \mathcal{K}_J$, then Theorem 8.5.2 in [345] asserts that

$$\mathcal{K}_J = \left\{ a \in H^{1,1}_J(M;\mathbb{R}) \;\middle|\; \begin{array}{l} a^2 > 0,\ a \cdot a_0 > 0,\ \langle a, [C] \rangle > 0 \text{ for every} \\ \text{nonconstant } J\text{-holomorphic sphere } C \end{array} \right\}. \qquad (13.4.7)$$

Here $H_J^{1,1}(M;\mathbb{R}) \subset H^2(M;\mathbb{R})$ denotes the space of de Rham cohomology classes of real valued closed 2-forms $\rho \in \Omega^2(M)$ that satisfy $\rho(J\cdot, J\cdot) = \rho$ (see equation (4.3.8)). Further, because the intersection form on $H^{1,1}$ has signature $(1,19)$ (see below), the condition $a \cdot a_0 > 0$ picks out one of the two connected components of the positive cone $\{a \in H^{1,1} \,|\, a^2 > 0\}$. By a deep theorem of Yau, every Kähler class $a \in \mathcal{K}_J$ can be represented by a Kähler form ω such that the Kähler metric $g = \omega(\cdot, J\cdot)$ is Ricci flat and so determines a hyperKähler structure (see [345, Thm 7.3.6]). Thus it remains to show that for every cohomology class $a \in H^2(M;\mathbb{R})$ with $a^2 > 0$ there exists a complex structure $J \in \mathcal{J}_{\mathrm{int}}(M)$ such that $a \in \mathcal{K}_J$. To see this, note first that every complex K3 surface has geometric genus $p_g = h^{2,0} = \frac{1}{2}(b^+ - 1) = 1$ and hence carries a nonzero holomorphic 2-form, unique up to a nonzero constant factor. The holomorphic 2-form vanishes nowhere because M is simply connected and $c_1 = 0$, so the canonical bundle admits a holomorphic trivialization. The real and imaginary part of the holomorphic 2-form span a 2-dimensional subspace $\Lambda_J \subset H^2(M;\mathbb{R})$ which is **positive** in the sense that the restriction of the intersection pairing to Λ_J is positive definite. The complexification of Λ_J is the 2-dimensional complex subspace

$$H^{2,0}(X) \oplus H^{0,2}(X)$$

of $H^2(M;\mathbb{C})$. In particular, the Hodge numbers for every K3-surface are

$$h^{2,0} = h^{0,2} = 1, \qquad h^{1,1} = 20.$$

Conversely, it is shown in [345, Thm 7.4.1] that for every 2-dimensional positive linear subspace $\Lambda \subset H^2(M;\mathbb{R})$ there is a complex structure $J \in \mathcal{J}_{\mathrm{int}}(M)$ such that $\Lambda_J = \Lambda$. Given a cohomology class $a \in H^2(M;\mathbb{R})$ with $a^2 > 0$, choose a 2-dimensional positive linear subspace $\Lambda \subset H^2(M;\mathbb{R})$ that is perpendicular to a, but not perpendicular to any integral cohomology class in $H^2(M;\mathbb{Z}) \setminus \mathbb{R}a$. Then the associated complex structure J can only have J-holomorphic curves that are Poincaré dual to multiples of the class a, because the pullback of a holomorphic 2-form under every J-holomorphic curve vanishes so that the cohomology classes in Λ must evaluate to zero on the homology class of every J-holomorphic curve. Hence it follows from equation (13.4.7) that either $a \in \mathcal{K}_J$ or $a \in \mathcal{K}_{-J}$.

By Taubes' Theorem 13.3.10, every symplectic form on a smooth K3-surface is compatible with the standard orientation and has first Chern class zero (see Remark 13.3.19 and Exercise 13.3.30). Hence it follows from Proposition 13.3.12 that any two symplectic forms on a K3-surface are homotopic as nondegenerate 2-forms. However, *it is an open question whether any two symplectic forms on a smooth K3-surface are homotopic through symplectic forms.* Likewise, *it is an open question whether any two cohomologous symplectic forms on a smooth K3-surface are isotopic.* □

Example 13.4.8 (The four-torus) Every cohomology class $a \in H^2(\mathbb{T}^4;\mathbb{R})$ with $a^2 \neq 0$ is represented by a translation invariant symplectic form and, by Taubes' Theorem 13.3.10, every symplectic form on $\mathbb{T}^4$ has first Chern class zero (see Exercise 13.3.30). There are infinitely many homotopy classes of nondegenerate 2-forms with first Chern class zero and compatible with a fixed orientation. The set of such homotopy classes is in bijective correspondence to the set $\mathbb{Z}/2\mathbb{Z} \times H^3(\mathbb{T}^4;\mathbb{Z})$ (see Remark 4.1.12). The Conolly-Lê-Ono argument in [126], reproduced in the proof of Proposition 13.3.12, removes only half of these classes as candidates for containing a symplectic form. *It remains an open question whether any two symplectic forms on $\mathbb{T}^4$, that induce the same orientation, are homotopic as nondegenerate 2-forms.* □

Example 13.4.9 (The one-point blowup of the four-torus) A theorem by Latschev–McDuff–Schlenk [400] (see also Entov–Verbitsky [205]) asserts that a closed four-ball admits a symplectic embedding into the four-torus with any constant coefficient symplectic form if and only if the volume of the four-ball is smaller than the volume of the four-torus. This settles the existence problem for the one-point blowup of the four-torus $M := \mathbb{T}^4\#\overline{\mathbb{CP}}^2$. Namely, let $E \in H_2(M;\mathbb{Z})$ be the homology class of the exceptional divisor. Then a cohomology class $a \in H^2(M;\mathbb{R})$ can be represented by a symplectic form if and only if $a^2 \neq 0$ and $\langle a, E\rangle \neq 0$. Thus the symplectic cone of M is strictly bigger than the Kähler cone. The uniqueness problem for symplectic forms on M is still far from understood. □

Example 13.4.10 (Kähler surfaces of general type) Let (M,ω,J) be a minimal Kähler surface of general type with first Chern class $c := c_1(\omega)$ such that

$$b^+ > 1, \qquad c^2 > 0, \qquad c\cdot[\omega] < 0.$$

Then $\pm c$ are the only Seiberg–Witten basic classes. Hence, by Taubes' results in [606], every symplectic form ρ on M satisfies

$$c_1(\rho)\cdot[\rho] < 0, \qquad c_1(\rho) = \pm c.$$

Thus any two symplectic forms, compatible with the orientation, have the same first Chern class up to sign. *It is an open question whether every cohomology class a with $a^2 > 0$ and $a \cdot c < 0$ can be represented by a symplectic form with first Chern class c* (as conjectured by Tian-Jun Li). □

Example 13.4.11 (Topology of $\mathscr{S}_a$) It is also interesting to study the topology of the space $\mathscr{S}_a$ of symplectic forms in a given cohomology class beyond the question of connectedness. Thus let us consider the connected component $\mathscr{S}_\omega$ of $\mathscr{S}_a$ containing ω and the fibration

$$\mathrm{Symp}(M,\omega)\cap \mathrm{Diff}_0(M) \to \mathrm{Diff}_0(M) \to \mathscr{S}_\omega$$

in (13.4.1). The map $\mathrm{Diff}_0(M) \to \mathscr{S}_\omega : \psi \mapsto \psi^*\omega$ induces a homeomorphism

$$\mathscr{S}_\omega \cong \mathrm{Diff}_0(M)/(\mathrm{Symp}(M,\omega)\cap\mathrm{Diff}_0(M)).$$

If there is a symplectomorphism ϕ that is smoothly, but not symplectically, isotopic to the identity one obtains a loop

$$[0,1] \to \mathscr{S}_\omega : t \to \omega_t := \phi_t^* \omega, \qquad \omega_0 = \omega_1 = \omega,$$

that is not contractible in $\mathscr{S}_\omega$. Indeed, if $[0,1] \to \mathrm{Diff}(M) : t \mapsto \phi_t$ is a smooth isotopy from $\phi_0 = \mathrm{id}$ to $\phi_1 = \phi$, then by Moser isotopy the loop $t \mapsto \omega_t$ is contractible in $\mathscr{S}_\omega$ if and only if the path $t \mapsto \phi_t$ is isotopic, with fixed endpoints, to a path of symplectomorphisms. It is contractible in the space of all, not necessarily cohomologous, symplectic forms, whenever ϕ becomes symplectically isotopic to the identity by a deformation of the pair (ω, ϕ).

Let (M, ω) be a symplectic four-manifold satisfying the assumptions of Seidel's Theorems A or B in Example 10.4.3. Then M contains a Lagrangian sphere L and the square of the generalized Dehn twist τ_L along L is smoothly, but not symplectically, isotopic to the identity. Hence the space $\mathscr{S}_\omega$ is not simply connected. However, the nontrivial loops in $\mathscr{S}_\omega$ that arise from generalized Dehn twists *are* contractible in the space of nondegenerate 2-forms.

In [377], Kronheimer used Seiberg–Witten theory to prove the existence of symplectic four-manifolds (M, ω) such that $\mathscr{S}_\omega$ is not simply connected. In fact, he developed a method for constructing, for each integer $n \geq 1$, symplectic four-manifolds (M, ω) such that $\pi_{2n-1}(\mathscr{S}_\omega) \neq 0$. □

Most of the symplectic four-manifolds discussed in detail in the present section have a rather simple form, such as the torus, the complex projective plane, or ruled surfaces, as well as their blowups. The study of symplectic four-manifolds has become a vast subject of research over the last 20 to 30 years. The discovery of Seiberg–Witten theory has led to a large number of interesting constructions of *exotic* smooth four-manifolds of a given homeomorphism type, and many of the resulting examples carry symplectic structures (see the discussion of exotic K3-surfaces and other exotic elliptic surfaces in Example 13.2.11 and the construction of smooth four-manifolds with many deformation inequivalent symplectic forms mentioned in Example 13.4.6).

It was only in the early 1990s that examples of simply connected symplectic four-manifolds were discovered that do not admit any Kähler structures. The first examples of this type were constructed by Gompf and Mrowka [279]. Their manifolds were formed by taking the twisted connected sum of two Dolgachev surfaces along torus fibres. (A **Dolgachev surface** is the result of performing two logarithmic transformations on the fibres of the basic elliptic surface $E(1)$ of Example 7.1.7.) The proof that the resulting manifolds are not Kähler was based on a difficult calculation of their Donaldson invariants. The corresponding untwisted sums (i.e. those constructed with the obvious framing) are standard elliptic surfaces.

In the other direction the first examples of smooth four-manifolds that have nonzero Seiberg–Witten invariants but do not carry any symplectic forms were constructed by Szabó [604] using a very delicate analysis of the Seiberg–Witten invariants of a fibre sum. Later Fintushel and Stern [219] found a whole family of examples by taking the fibre connected sum of a K3-surface (Example 13.4.7) with a 4-manifold M_K that is constructed from $S^1 \times S^3$ by 0-surgery on $S^1 \times K$, where K is a knot in S^3. The Seiberg–Witten invariants of the resulting manifolds X_K can be assembled into a polynomial that turns out to coincide with the Alexander polynomial $A(K)$. It then follows from Taubes–Seiberg–Witten theory that X_K can only admit a symplectic structure when $A(K)$ is monic. Moreover, when K is a fibred knot its Alexander polynomial is monic and X_K does admit a symplectic structure [219]. On the other hand, any knot whose Alexander polynomial is not monic gives rise to a 4-manifold X_K, homeomorphic to the K3-surface, with nonzero Seiberg–Witten invariants but no symplectic structure.

This discussion shows that symplectic four-manifolds form a category that lies genuinely between Kähler surfaces and smooth four-manifolds with nonzero Seiberg–Witten invariants, and one might ask which properties symplectic four-manifolds share with Kähler surfaces. One question is whether every closed simply connected symplectic four-manifold satisfies the **Miyaoka–Yau inequality**

$$3\sigma \leq \chi. \tag{13.4.8}$$

Here σ is the signature and χ is the Euler characteristic. By the Hirzebruch signature formula (4.1.7), the Miyaoka–Yau inequality can be expressed in the form

$$c_1^2 \leq 3c_2.$$

By a classical theorem of Miyaoka [489] and Yau [691], this inequality holds for every Kähler surface of general type and every Kähler surface that admits an Einstein metric. A theorem of LeBrun [404, 405] asserts that the Miyaoka–Yau inequality holds for symplectic four-manifolds that admit Einstein metrics. The Miyaoka–Yau inequality is violated by ruled surfaces over a base of genus $g > 1$ with at most $g - 2$ points blown up. It is an open question whether there is any simply connected symplectic four-manifold that violates the Miyaoka–Yau inequality; further, the only known simply connected symplectic four-manifold that satisfies $3\sigma = \chi$ is the complex projective plane. For a study of symplectic 4-manifolds *near the Bogomolov–Miyaoka–Yau line* $3\sigma = \chi$, see the work of Akhmedov–Park–Urzúa [17] and Stipsicz [598, 599].

This line of enquiry is a part of the so-called **symplectic geography problem** that addresses the question of which pairs χ, σ (with $\chi + \sigma$ divisible by four) can be realized by symplectic four-manifolds, and of what can be said about symplectic four-manifolds with a given Euler characteristic and signature. It is often convenient to express this data in terms of the numbers

$$\chi_h := \frac{\chi + \sigma}{4}, \qquad c := 2\chi + 3\sigma.$$

Then the Miyaoka–Yau inequality takes the form $c \leq 9\chi_h$. For a beautiful exposition of the geography problem for general smooth four-manifolds and many related results and conjectures, see Fintushel–Stern [221].

Another direction is the work of Tian-Jun Li [419, 420] and others on carrying over the notion of Kodaira dimension from Kähler surfaces to symplectic four-manifolds (see page 174). The **Kodaira dimension** of a minimal closed symplectic four-manifold (M, ω) is defined by

$$\kappa(M,\omega) := \begin{cases} -\infty, & \text{if } K^2 < 0 \text{ or } K \cdot [\omega] < 0, \\ 0, & \text{if } K^2 = 0 \text{ and } K \cdot [\omega] = 0, \\ 1, & \text{if } K^2 = 0 \text{ and } K \cdot [\omega] > 0, \\ 2, & \text{if } K^2 > 0 \text{ and } K \cdot [\omega] > 0, \end{cases} \tag{13.4.9}$$

where $K := -c_1(\omega)$ denotes the canonical class. The **Kodaira dimension** of a nonminimal symplectic four-manifold is defined as the Kodaira dimension of any of its minimal models. Tian-Jun Li showed that the symplectic Kodaira dimension is independent of the choice of the minimal model and that it is a diffeomorphism invariant, while Ho–Li [311] showed that it is invariant under Luttinger surgery (Example 3.4.16). In Remark 13.3.28 we have seen that a closed symplectic four-manifold has Kodaira dimension $-\infty$ if and only if it is a blowup of the complex projective plane or of a ruled surface. Symplectic four-manifolds with Kodaira dimension zero were studied by Tian-Jun Li [419] and symplectic four-manifolds with Kodaira dimension one by Baldridge–Li [49]. Symplectic four-manifolds with Kodaira dimension two are the symplectic analogues of Kähler surfaces of general type (Example 13.4.10) and there are many simply connected nonKähler examples (see Fintushel–Stern [219]).

An interesting open problem is whether there are exotic smooth structures (carrying symplectic forms) on the projective plane, on the one-point blowup of the projective plane, or on the product of two 2-spheres. On higher blowups of the projective plane exotic smooth structures were discovered by many authors. For more details about this and many other open problems see the next and final chapter of this book.

14

OPEN PROBLEMS

In the final chapter of this book we discuss some open questions and conjectures that either have served as guiding lights or have emerged in the study of symplectic topology over the last quarter of a century. Since symplectic topology has grown into a vast area of research, it is impossible to be complete. Often, we mention only the most recent papers, since earlier relevant work can be discovered through their references. The choice of topics necessarily reflects our tastes and preferences: we have highlighted some of the open problems that seem to us to be both appealing in their own right and central to symplectic topology. Readers should be aware that this list will inevitably become out of date and hence at best can provide a snapshot of where the field is at the time of writing.

14.1 Symplectic structures

An in depth discussion of the existence and uniqueness problem for symplectic structures is contained in Section 13.1. Sections 13.2 and 13.4 explain what is known about these problems in various examples. Here we highlight some of the open questions that arise from this discussion. We begin with the fundamental existence problem.

Problem 1 (Existence of symplectic structures)
Is there an example of a closed oriented manifold M of dimension $2n \geq 6$, a nondegenerate 2-form ρ on M compatible with the orientation, and a cohomology class $a \in H^2(M;\mathbb{R})$ satisfying $a^n > 0$, such that ρ is not homotopic to a symplectic form representing the class a?

There are many such examples in dimension four (see Sections 13.3 and 13.4). However, the question is open in higher dimensions. In contrast, the uniqueness problem up to isotopy for cohomologous symplectic forms is completely open in dimension four, while counterexamples are known in higher dimensions (see Example 13.2.9). Recall the notation $\mathscr{S}_a$ for the space of symplectic forms representing the cohomology class $a \in H^2(M;\mathbb{R})$.

Problem 2 (Uniqueness in dimension four)
(a) *Is there a closed four-manifold M and a cohomology class $a \in H^2(M;\mathbb{R})$ such that $\mathscr{S}_a$ is disconnected?*
(b) *Is there a closed four-manifold M and a cohomology class $a \in H^2(M;\mathbb{R})$ such that $\mathscr{S}_a$ is nonempty and connected?*

Problem 3 (Projective plane uniqueness conjecture)
The space of symplectic forms on the complex projective plane representing a fixed cohomology class is contractible.

Problem 4 (HyperKähler surface uniqueness conjecture)
If M is a closed hyperKähler surface (i.e. a 4-torus or a K3-surface) and the cohomology class $a \in H^2(M;\mathbb{R})$ satisfies $a^2 > 0$, then $\mathscr{S}_a$ is connected.

The uniqueness conjecture for the four-torus is a longstanding open question. By Seiberg–Witten theory, every symplectic form on the four-torus or a K3-surface has first Chern class zero (see Example 13.4.7). In the case of a K3-surface it even follows that any two symplectic forms are homotopic as nondegenerate 2-forms. The analogous statement for two symplectic forms on the four-torus that induce the same orientation is an open problem. The **Donaldson geometric flow** as outlined in [149] is a conjectural approach to settle Problem 4. This geometric flow equation is valid, in principle, for all symplectic four-manifolds, and on $\mathbb{C}\mathrm{P}^2$ it has only one critical point [376]. This motivates the conjecture in Problem 3 which, if true, would imply that $\mathrm{Diff}_h(\mathbb{C}\mathrm{P}^2)$, the group of diffeomorphisms that act trivially on homology, retracts onto $\mathrm{PU}(3)$. However, one might expect that $\mathscr{S}_a$ is disconnected in some examples and in those cases something to go wrong with the analysis (such as failure of long time existence or lack of convergence).

Problem 5 (Donaldson four-six question)
Let (X, ω_X), (Y, ω_Y) be closed symplectic four-manifolds such that X and Y are homeomorphic. Are X and Y diffeomorphic if and only if the product manifolds

$$(X \times S^2, \omega_X \oplus \sigma), \qquad (Y \times S^2, \omega_Y \oplus \sigma)$$

are symplectically deformation equivalent?

Examples of nondeformation equivalent symplectic structures on the same symplectic four-manifold X are discussed in Example 13.4.6. In particular, the examples of Ivan Smith [592, 593] are distinguished by the divisibility properties of their first Chern classes and hence continue to be nondeformation equivalent on $X \times \mathbb{T}^2$. Hence the 2-sphere in Problem 5 cannot be replaced by the 2-torus.

Problem 5 is nontrivial in either direction. When X and Y *are not* diffeomorphic, but are homeomorphic so $X \times S^2$ and $Y \times S^2$ *are* diffeomorphic, the question suggests that the two symplectic structures on these six-manifolds should still *remember* the differences in the smooth structures on X and Y. At the time of writing the only known methods for distinguishing smooth structures on four-manifolds are the Donaldson invariants and the Seiberg–Witten invariants. By Taubes–Seiberg–Witten theory (see Section 13.3) the Seiberg–Witten invariants can also be interpreted as invariants of the symplectic structure, and it is plausible that they give rise to a method for distinguishing the symplectic structures on the products with the two-sphere. (See the related work of Herrera [308].) However, this observation does not get to the heart of the problem:

the Seiberg–Witten invariants may well not be strong enough to distinguish all pairs of smooth structures, and also we have very little understanding of six-dimensional symplectic manifolds and their symplectomorphisms to help us in going in the other direction. An example where X and Y *are not* diffeomorphic, $X \times S^2$ and $Y \times S^2$ *are* diffeomorphic, and the symplectic forms on $X \times S^2$ and $Y \times S^2$ can be distinguished by their Gromov–Witten invariants, was found by Ruan [546] (see Example 13.2.10). More examples along these lines were found by Ruan–Tian [550] and Ionel–Parker [346]. Also the examples of Vidussi [641, 642] furnish symplectic structures on homotopy K3's whose products with the two-sphere are not symplectically deformation equivalent. (See Example 13.2.11.)

Problem 6 (Donaldson almost complex structure question)
Let (M, ω) be a closed symplectic four-manifold and let J be an ω-tame almost complex structure on M. Does there exist a symplectic form that is compatible with J (and represents the same cohomology class as ω in the case $b^+ = 1$)?

This question has a positive answer for $M = \mathbb{C}\mathrm{P}^2$ by Taubes [618] and for $M = S^2 \times S^2$ by Li-Zhang [425] (see Remark 4.1.3). The next question is related to both to the work of Fintushel–Stern mentioned on page 547, and to Problem 5.

Problem 7 (Symplectic knot surgery)
Let X_K be obtained from a K3-surface X by surgery along a knot $K \subset S^3$.
(a) *If X_K admits a symplectic structure, does it follow that K is a fibred knot?*
(b) *If K and K' are fibred knots with the same Alexander polynomial, are $X_K \times S^2$ and $X_{K'} \times S^2$ symplectically deformation equivalent?*

The next question is related to work of Friedl–Vidussi, Ozsvath, Taubes.

Problem 8 (Symplectic mapping torus)
If M is a closed 3-manifold such that $M \times S^1$ admits a symplectic form, does it follow that M is a mapping torus?

Tian-Jun Li has extended the notion of the Kodaira dimension from Kähler surfaces to symplectic 4-manifolds (see page 548). In [420] he has formulated various conjectures about the behaviour of this invariant. Here is one of his conjectures related to Kähler surfaces of general type. It is motivated by an observation of Witten, which asserts that in this case the canonical class and its negative are the only basic classes (see Example 13.4.10).

Problem 9 (Tian-Jun Li's general type existence conjecture)
Let (M, ω, J) be a minimal Kähler surface of general type with first Chern class $c := c_1(\omega)$, so that

$$b^+ > 1, \qquad [\omega] \cdot c < 0, \qquad c \cdot c > 0.$$

Let $a \in H^2(M; \mathbb{R})$ be such that

$$a^2 > 0, \qquad a \cdot c < 0.$$

Then there exists a symplectic form on M in the class a.

The next question concerns the extent to which results known about symplectic forms on the blowup in four dimensions extend to higher dimensions.

Problem 10 (Blowup uniqueness)
Let $\pi_M : \widetilde{M} \to M$ be the one-point blowup of a symplectic $2n$-manifold (M, ω). Define $a := [\omega] \in H^2(M)$, $c := c_1(\omega)$, $\widetilde{a} := \pi_M^* a$, $\widetilde{c} := \pi_M^* c$, and let $e \in H^2(\widetilde{M})$ be the Poincaré dual of the class of the exceptional divisor. Fix a constant $r > 0$.
(a) *If $\widetilde{M}$ admits a symplectic form in the cohomology class $\widetilde{a} - \pi r^2 e$ and with first Chern class $\widetilde{c} - (n-1)e$, does it follow that there exists a symplectic embedding of the closed ball $B^{2n}(r)$ of radius r into M?*
(b) *Is every pair of blowup symplectic forms on $\widetilde{M}$ that arises from two normalized symplectic embeddings of the same ball into M isotopic?*

The answer to question (a) is positive for blowups of rational and ruled 4-manifolds by Example 13.4.5, while the answer to (b) is positive for all symplectic 4-manifolds with $b^+ = 1$ by Remark 7.1.28.

Problem 11 (Auroux uniqueness question)
Let (M_1, ω_1) and (M_2, ω_2) be closed symplectic four-manifolds such that the cohomology classes $[\omega_1]$ and $[\omega_2]$ have integral lifts. Suppose M_1 and M_2 have the same Euler characteristic, signature, and volume, and that $c_1(\omega_1) \cdot [\omega_1] = c_1(\omega_2) \cdot [\omega_2]$. Does there exist a collection of disjoint Lagrangian tori in M_1 such that Luttinger surgery along them gives rise to a manifold symplectomorphic to (M_2, ω_2)?

For Luttinger surgery see Example 3.4.16. A related result by Auroux [37] asserts that under the assumptions of Problem 11 there exist isomorphic Donaldson hypersurfaces $\Sigma_i \subset M_i$ (see Section 7.4) and a symplectic 4-manifold X with an embedded hypersurface $\Sigma \subset X$ diffeomorphic to Σ_i and with opposite normal bundle such that the fibre sums (see Section 7.2) $M_1 \#_\Sigma X$ and $M_2 \#_\Sigma X$ are symplectomorphic. Another related result by Auroux asserts that if (W_1, ω_1) and (W_2, ω_2) are 4-dimensional Weinstein manifolds (see Definition 7.4.5) with isomorphic contact boundaries and the same Euler characteristic and signature, then they are related by attaching Weinstein handles along the same Legendrian knots in $\partial W_1 \cong \partial W_2$ and subsequent deformation (see [41, Theorem 6.1]).

Problem 12 (Exotic projective plane)
Does there exist a closed symplectic four-manifold that is homeomorphic, but not diffeomorphic, to the complex projective plane, the one-point blowup of the complex projective plane, or the product of two 2-spheres?

The quest for exotic smooth structures on four-manifolds has a long history. For blowups of the complex projective plane it was shown by Donaldson [141], long before the advent of the Seiberg–Witten invariants, that there is an exotic smooth structure on the 9-point blowup of $\mathbb{CP}^2$ (that supports a symplectic form). This was gradually extended by several authors, including Stipsicz–Szabó [601] for the 6-point blowup, Park–Stipsicz–Szabó [524] and Fintushel–Stern [220] for the 5-point blowup, Baldridge–Kirk [47], Akhmedov–Park [15],

Fintushel–Park–Stern [218], and Fintushel–Stern [224] for the 3-point blowup, and Akhmedov–Park [16] and Fintushel–Stern [223] for the 2-point blowup of the complex projective plane. The holy grail in this quest would be a positive answer to Problem 12 for the complex projective plane. This would imply a negative answer to part (b) of the next question.

Problem 13 (Miyaoka–Yau inequality)
(a) *Is there any simply connected closed symplectic four-manifold whose Euler characteristic and signature satisfy* $3\sigma > \chi$*?*
(b) *Is every simply connected closed symplectic four-manifold whose Euler characteristic and signature satisfy* $3\sigma = \chi$ *diffeomorphic to* $\mathbb{C}\mathrm{P}^2$*?*

For a brief discussion of the Miyaoka–Yau inequality and some known results, see page 547. A more general question is which Chern numbers c_1^2 and c_2 are realized by simply connected closed symplectic four-manifolds and what can be deduced about the symplectic manifolds from the Chern numbers. This is the so-called *symplectic geography problem* and is briefly discussed on page 547.

14.2 Symplectomorphisms

Fix a closed symplectic manifold (M, ω) and consider the homomorphism

$$\pi_0(\mathrm{Symp}(M, \omega)) \to \pi_0(\mathrm{Diff}(M)). \tag{14.2.1}$$

A fundamental problem in symplectic topology is to understand the kernel and image of this homomorphism. The kernel is trivial if and only if every symplectomorphism that is smoothly isotopic to the identity is symplectically isotopic to the identity. One can also strengthen the question and ask whether the group $\mathrm{Symp}_h(M, \omega)$, of symplectomorphisms that act trivially on homology, is connected.[44]

There are known examples where this group is connected, even though it is an open question whether or not $\mathrm{Diff}_h(M)$ is connected, such as $\mathbb{C}\mathrm{P}^2$, the one-point blowup of $\mathbb{C}\mathrm{P}^2$, $S^2 \times S^2$ (Example 10.4.2), and the monotone k-point blowups of the projective plane for $2 \le k \le 4$ (see Evans [210] and Example 10.4.3). A natural conjecture is that the homomorphism (14.2.1) is injective for all ruled surfaces. This is known under some assumptions (Example 13.4.3) but is an open problem in general.

Problem 14 (Symplectic isotopy conjecture for ruled surfaces)
A symplectomorphism of a ruled surface is smoothly isotopic to the identity if and only if it is symplectically isotopic to the identity.

Here is another specific instance of this problem (see Example 10.4.1).

[44]For simplicity, we will restrict this discussion to simply connected manifolds. If $\pi_1(M) \neq 0$, the problem should be refined; one could ask, for example, whether every symplectomorphism that acts trivially on homology and on $\pi_1(M)$ is symplectically isotopic to the identity, or whether every symplectomorphism that is homotopic to the identity is symplectically isotopic to the identity.

Problem 15 (Symplectic isotopy conjecture for tori)
A symplectomorphism of $\mathbb{T}^{2n}$ with a translation invariant symplectic form induces the identity on homology if and only if it is symplectically isotopic to the identity. Equivalently, every exact symplectomorphism of $\mathbb{T}^{2n}$ is Hamiltonian.

There are many examples of simply connected symplectic 4-manifolds where the homomorphism (14.2.1) is not injective. This includes the monotone k-point blowups of the projective plane for $5 \le k \le 8$ by the work of Seidel (Example 10.4.3). The related question of whether $\mathrm{Symp}_h(M, \omega)$ is connected can also have negative answers in higher dimensions, as shown by unpublished work of Michael Callahan from the early 1990s. He proved that the symplectomorphism of the 6-dimensional moduli space of flat SO(3)-connections over a genus-2 Riemann surface that is induced by a Dehn twist along a separating loop is not symplectically isotopic to the identity. Some other explicit examples were found by Dimitroglou Rizell–Evans [138]. So far, all known constructions of nontrivial elements in the kernel of the homomorphism (14.2.1) are based on the use of iterated Dehn twists, around (families of) Lagrangian spheres or other manifolds with periodic geodesic flow. However, the full symplectic mapping class group $\pi_0(\mathrm{Symp}(M, \omega))$ can be very rich for some closed symplectic manifolds that do not contain any Lagrangian spheres. Examples are products of K3-surfaces with suitable symplectic forms, as explained to us by Ivan Smith.

Even if $\mathrm{Symp}_h(M, \omega)$ is not connected it is sometimes interesting to understand whether it is a subgroup of $\mathrm{Diff}_0(M)$. A specific instance of this is the following question, motivated by the discussion in Note 2 on page 537.

Problem 16 (Smooth isotopy for blowups of $\mathbb{C}\mathrm{P}^2$)
Let (M, ω) be the k-point blowup of the complex projective plane with $k \ge 5$ and any symplectic form. Is $\mathrm{Symp}_h(M, \omega) \subset \mathrm{Diff}_0(M)$?

Now consider the image of the homomorphism (14.2.1). The isotopy class of a diffeomorphism $\phi : M \to M$ belongs to this image if and only if ϕ is isotopic to a symplectomorphism. Necessary conditions are the following.

(S) *$\phi^*\omega - \omega$ is exact and $\omega, \phi^*\omega$ can be joined by a path of nondegenerate 2-forms.*

One can ask whether every diffeomorphism that satisfies (S) is smoothly isotopic to a symplectomorphism. For simply connected symplectic four-manifolds the second condition in (S) is redundant by Proposition 13.3.12. For $M = \mathbb{C}\mathrm{P}^2$ a positive answer is equivalent to connectivity of the space of cohomologous symplectic forms in Problem 3. For the standard symplectic form on the 6-manifold $M = \mathbb{T}^2 \times S^2 \times S^2$ there exists a diffeomorphism ϕ that satisfies (S) but is not isotopic to a symplectomorphism, by a theorem of McDuff (Example 13.2.9). In this example, ω and $\phi^*\omega$ can even be joined by a path of symplectic forms, with necessarily varying cohomology classes.

Similar questions are interesting for compactly supported symplectomorphisms of noncompact symplectic manifolds. For example, a theorem of Gromov [287] asserts that the group $\mathrm{Symp}_c(\mathbb{R}^4, \omega_0)$ is contractible (see also [470,

Theorem 9.5.2]). In fact, this continues to hold for the group of symplectomorphisms of $(\mathbb{R}^4, \omega_0)$ with support contained in the open four-ball. In higher dimensions even the question of connectivity is open. A first test case is the group of compactly supported diffeomorphisms of $\mathbb{R}^6$. Its connected components form a cyclic group of order 28 (corresponding to the 28 diffeomorphism types of homotopy 7-spheres). This leads to the following question.

Problem 17 (Symplectomorphism group of $\mathbb{R}^{2n}$)
(a) *Is every compactly supported symplectomorphism of* $(\mathbb{R}^{2n}, \omega_0)$ *smoothly isotopic to the identity (through an isotopy with uniform compact support)?*
(b) *Is every compactly supported symplectomorphism of* $(\mathbb{R}^{2n}, \omega_0)$ *symplectically isotopic to the identity (through an isotopy with uniform compact support)?*

Since the answer to question (b) in Problem 17 is positive for $2n = 4$, it follows from symplectic rigidity that the group

$$\mathrm{Symp}_{\mathrm{c},0}(\mathbb{R}^4, \omega_0) = \mathrm{Symp}_{\mathrm{c}}(\mathbb{R}^4, \omega_0)$$

is a C^0-closed subset of $\mathrm{Diff}_{\mathrm{c}}(\mathbb{R}^4)$. This is an open question for $\mathbb{R}^{2n}$ when $2n \geq 6$.

Problem 18 (C^0 closure of the identity component)
Let (M, ω) be a symplectic manifold without boundary. Is the identity component $\mathrm{Symp}_{\mathrm{c},0}(M, \omega)$ *of the group* $\mathrm{Symp}_{\mathrm{c}}(M, \omega)$ *of compactly supported symplectomorphisms a closed subset of* $\mathrm{Symp}_{\mathrm{c}}(M, \omega)$ *with respect to the C^0-topology?*

It is also interesting to investigate the more general question of what is known about the topology of the symplectomorphism group $\mathrm{Symp}(M, \omega)$ versus the topology of the diffeomorphism group of M. This question is only understood in very few cases (Examples 10.4.2 and 10.4.3). Via Moser isotopy it is closely related to the topology of the space of symplectic forms (Example 13.4.11).

The next conjecture is a restatement of Conjecture 10.2.18. For a discussion of this conjecture and some known results see Section 10.2.

Problem 19 (C^0 flux conjecture)
The group of compactly supported Hamiltonian symplectomorphisms is a C^0-closed subgroup of $\mathrm{Symp}_{\mathrm{c},0}(M, \omega)$.

Hofer geometry

There are various important questions that can be posed about the Hofer metric on the group of Hamiltonian symplectomorphisms. One central problem is whether the Hofer diameter is always infinite.

Problem 20 (Hofer diameter conjecture)
The group of Hamiltonian symplectomorphisms has infinite diameter in the Hofer metric for every nonempty closed symplectic manifold of positive dimension.

The Hofer diameter conjecture has been confirmed for $\mathbb{C}\mathrm{P}^n$ and for symplectically aspherical manifolds; cf. page 480. In general it is an open problem. Here we discuss the proof in Entov–Polterovich [198] for $M = S^2$, which is closely related to a question posed by Kapovich and Polterovich in 2006. Their question was posed for the 2-sphere but it makes sense for any closed manifold and is based on the following concept. Let $\mathcal{G}$ be a topological group equipped with a bi-invariant metric $\rho : \mathcal{G} \times \mathcal{G} \to [0, \infty)$. A group homomorphism $\mathbb{R} \to \mathcal{G} : t \mapsto \phi_t$ is called a **quasi-geodesic one parameter group** if there exist constants $\delta > 0$ and $c > 0$ such that

$$\delta|t| \leq \rho(\mathbb{1}, \phi_t) \leq c|t| \qquad \text{for all } t \in \mathbb{R}. \tag{14.2.2}$$

A quasi-geodesic one parameter group is called **quasi-dense** in $\mathcal{G}$ if

$$A(\{\phi_t\}) := \sup_{\psi \in \mathcal{G}} \inf_{t \in \mathbb{R}} \rho(\psi, \phi_t) < \infty. \tag{14.2.3}$$

One can prove that the finiteness of the number $A(\{\phi_t\})$ is independent of the choice of the quasi-geodesic one parameter group in $\mathcal{G}$. (Here is an argument explained to us by Polterovich: Let ϕ_t and ϕ'_t be two quasi-geodesic one parameter groups satisfying (14.2.2) and assume $A(\{\phi_t\}) < \infty$. Then there exists a constant A and a map $\mathbb{Z} \to \mathbb{Z} : j \mapsto k_j$ such that $\rho(\phi'_k, \phi_{j_k}) \leq A$ for all $k \in \mathbb{Z}$. It follows from the quasi-geodesic property that $\lim_{|k|\to\infty} |j_k| = \infty$. Moreover, it follows from (14.2.2) and the bi-invariance of the metric that, for all $k \in \mathbb{N}$, we have $\delta|j_{-k} + j_k| \leq \rho(\mathbb{1}, \phi_{j_k + j_{-k}}) \leq \rho(\phi_{-j_k}, \phi'_{-k}) + \rho(\phi'_{-k}, \phi_{j_{-k}}) \leq 2A$. Thus, reversing the orientation of $t \mapsto \phi'_t$ if necessary, we obtain $\lim_{k\to\infty} j_k = \infty$ and $\lim_{k\to\infty} j_{-k} = -\infty$. It then follows from (14.2.2) and the bi-invariance of the metric that $|j_{k+1} - j_k| \leq \rho(\phi_{j_k}, \phi_{j_{k+1}})/\delta \leq (2A + \rho(\mathbb{1}, \phi'_1))/\delta =: B$. This implies that every ϕ_j has distance at most $A + cB$ to some ϕ'_k and hence $A(\{\phi'_t\}) < \infty$.)

Now let (M, ω) be a closed symplectic manifold and consider the Hofer metric ρ on the group $\mathcal{G} = \mathrm{Ham}(M, \omega)$ of Hamiltonian symplectomorphisms. In the case $M = S^2$ it was shown by Entov–Polterovich in [198, Section 5.5] that the one parameter subgroup generated by a Hamiltonian function $H : S^2 \to \mathbb{R}$ is a quasi-geodesic whenever $\int_{S^2} H\sigma = 0$, H is a Morse function, zero is a regular value of H, and no connected component of $H^{-1}(0)$ divides the 2-sphere into two connected components of equal area. The existence of a quasi-geodesic one parameter subgroup shows that $\mathrm{Ham}(S^2)$ has infinite Hofer diameter.

Problem 21 (Kapovich–Polterovich)
Let $\mathbb{R} \to \mathrm{Ham}(S^2) : t \mapsto \phi_t$ be a 1-parameter subgroup generated by $H \in C^\infty(S^2)$ and assume that there is a constant $\delta > 0$ such that $\rho(\phi_t, \mathrm{id}) \geq \delta|t|$ for all $t \in \mathbb{R}$. Is the number

$$A(H) := \sup_{\psi \in \mathrm{Ham}(S^2)} \inf_{t \in \mathbb{R}} \rho(\psi, \phi_t)$$

finite or infinite?

As noted above, the finiteness of the number $A(H)$ in Problem 21 is independent of H. The question asks if some, and hence every, quasi-geodesic one parameter group in $\mathrm{Ham}(S^2)$ is quasi-dense with respect to the Hofer metric. A positive answer (i.e. finiteness of $A(H)$), would assert that $\mathrm{Ham}(S^2)$ is contained in a tube of finite Hofer distance around the one parameter subgroup $\{\phi_t \mid t \in \mathbb{R}\}$. The expected answer is that $A(H)$ is infinite. This is formulated as a much more general conjecture by Polterovich–Shelukhin [533]. They introduce the invariant

$$\mathrm{aut}(M,\omega) := \sup_{\psi \in \mathrm{Ham}(M,\omega)} \inf_{H \in C^\infty(M)} \rho(\psi, \phi_H). \tag{14.2.4}$$

Thus, $\mathrm{aut}(M,\omega)$ is the supremum over all $\psi \in \mathrm{Ham}(M,\omega)$ of the Hofer distance of ψ to the set of all Hamiltonian symplectomorphisms that are generated by time-independent Hamiltonian functions. Polterovich and Shelukhin conjecture that this invariant is always infinite. This would imply the Hofer diameter conjecture and show that the number $A(H)$ in Problem 21 is always infinite.

Problem 22 (Autonomous Hamiltonian conjecture)
The number $\mathrm{aut}(M,\omega)$ *is infinite for every closed symplectic manifold* (M,ω).

Polterovich–Shelukhin [533] verify this conjecture for Riemann surfaces of genus at least four and their products with symplectically aspherical manifolds. They also derive consequences of this result in Hamiltonian dynamics.

Calabi quasimorphisms

An important development since the mid 1990s, pursued by Entov, Polterovich, Py, and many others, was to bring the theory of quasimorphisms to bear on questions in symplectic topology and Hamiltonian dynamics. A map $\mu : \mathcal{G} \to \mathbb{R}$ on a group $\mathcal{G}$ is called a **quasimorphism** if there is a constant $c > 0$ such that

$$|\mu(\phi \circ \psi) - \mu(\phi) - \mu(\psi)| \leq c \tag{14.2.5}$$

for all $\phi, \psi \in \mathcal{G}$. A quasimorphisms $\mu : \mathcal{G} \to \mathbb{R}$ is called **homogeneous** if

$$\mu(\phi^k) = k\mu(\phi) \tag{14.2.6}$$

for all $\phi \in \mathcal{G}$ and all $k \in \mathbb{Z}$. The relevant case for the discussion at hand is when $\mathcal{G} = \mathrm{Ham}(M,\omega)$ is the group of Hamiltonian symplectomorphisms on a closed symplectic manifold. A homogeneous quasimorphism $\mu : \mathrm{Ham}(M,\omega) \to \mathbb{R}$ is called a **Calabi quasimorphism** if, for every displaceable open set $U \subset M$ and every compactly supported time-dependent Hamiltonian $\{H_t\}_{0 \leq t \leq 1}$, we have

$$\overline{\bigcup_{0 \leq t \leq 1} \mathrm{supp}(H_t)} \subset U \qquad \Longrightarrow \qquad \mu(\phi_H) = \int_0^1 \int_M H_t \omega^n \, dt. \tag{14.2.7}$$

In other words, if ϕ is generated by a time-dependent Hamiltonian function with compact support in a displaceable open set, then $\mu(\phi)$ agrees with the Calabi invariant of ϕ (see page 411). Although Calabi quasimorphisms are more naturally defined on the universal cover of $\mathrm{Ham}(M)$, they sometimes descend to

$\mathrm{Ham}(M)$. The existence of Calabi quasimorphisms on $\mathrm{Ham}(M)$ was established by Entov–Polterovich [198] for $M = \mathbb{C}\mathrm{P}^n$ and by Pierre Py [534,535] for all closed symplectic 2-manifolds of positive genus. For an extensive discussion of Calabi quasimorphisms and their applications in symplectic topology, see the papers [73, 198,199,200,201,202,203,204] by Entov, Polterovich, and their collaborators, and the book by Polterovich–Rosen [532]. In [203, Section 5.2] Entov, Polterovich, and Py posed the following questions.

Problem 23 (Quasimorphism question)
(a) *Does there exist a nonzero homogeneous quasimorphism* $\mu : \mathrm{Ham}(S^2) \to \mathbb{R}$ *that is continuous with respect to the* C^0*-topology on* $\mathrm{Ham}(S^2)$*?*
(b) *If yes, can it be made Lipschitz with respect to the Hofer metric?*

They show in [203] that the difference of two Calabi quasimorphisms on $\mathrm{Ham}(S^2)$ is C^0-continuous, although the Calabi homomorphism, and hence all Calabi quasimorphisms, are not C^0-continuous. They conclude that a negative answer to question (a) in Problem 23 would imply that the Calabi quasimorphism on $\mathrm{Ham}(S^2)$ constructed by Entov–Polterovich [198] is unique, while a positive answer to question (b) in Problem 23 would imply that $A(H) = \infty$ in Problem 21. For a full exposition of this circle of ideas, including a discussion of related developments and open questions, see the book by Polterovich–Rosen [532].

Finite subgroups of $\mathrm{Symp}(M, \omega)$

In [638], Polterovich observed that the group of Hamiltonian symplectomorphisms of a closed symplectically aspherical manifold has no nontrivial finite subgroups because there is a measure of the size of the iterates ϕ^k of an element $\phi \in \mathrm{Ham}(M) \setminus \{\mathrm{id}\}$ that grows at least linearly with k. This is not true for a manifold such as S^2 that supports an action of S^1. Now manifolds with a Hamiltonian circle action are symplectically uniruled, by a theorem of McDuff [462]. This is a condition on the genus zero Gromov–Witten invariants (see the discussion on page 563 below). Thus one might investigate connections between the properties of the Gromov–Witten invariants and the geometry of the group of Hamiltonian symplectomorphisms. One such question is the following.

Problem 24 (Finite groups of Hamiltonian symplectomorphisms)
Does there exist a closed symplectic manifold (M, ω) *with vanishing genus zero Gromov–Witten invariants such that* $\mathrm{Ham}(M, \omega)$ *has a nontrivial finite subgroup?*

For other interesting questions about the structure of finite subgroups of $\mathrm{Symp}(M, \omega)$, see Mundet-i-Riera [499] and the references therein.

14.3 Lagrangian submanifolds and cotangent bundles

Lagrangians are an important class of submanifolds of symplectic manifolds, and we concentrate here on questions about their existence and properties. Much of the current interest in understanding them is motivated by mirror symmetry, because Lagrangian manifolds (perhaps decorated with additional structure) form the objects of the Fukaya category, one side of the mirror symmetry correspondence. The influence of this powerful ansatz can be seen even within the rather narrowly focused questions discussed here: for example, the work of Vianna [640] discussed after Problem 25 was motivated by ideas from the mirror side. However, to discuss mirror symmetry itself is beyond the scope of this book.

Lagrangian tori

A longstanding open question which has attracted the attention of many researchers has been the *Audin conjecture*, which asserts that every Lagrangian torus in $\mathbb{R}^{2n}$ with the standard symplectic structure has minimal Maslov number two. After various partial results by Viterbo [645] (for $n = 2$), Buhovski [81], and Fukaya–Oh–Ohta–Ono [249, Theorem 6.4.35] (in the monotone case, cf. Definition 3.4.4), Damian [131], and others, the Audin conjecture was settled in 2014 by Cieliebak–Mohnke [117] (see Remark 3.4.6). Another proof of the Audin conjecture was outlined by Fukaya [247]. The standard torus $S^1 \times S^1$, also called the Clifford torus, and the Chekanov torus in Example 3.4.7 are, up to Hamiltonian isotopy, the only known examples of monotone Lagrangian tori with factor $\pi/2$ in $(\mathbb{R}^4, \omega_0)$. This leads to the following natural question.

Problem 25 (Monotone Lagrangian tori in $\mathbb{R}^4$)
Is every monotone Lagrangian torus in $(\mathbb{R}^4, \omega_0)$ Hamiltonian isotopic to either the Clifford torus or the Chekanov torus?

In [42], Auroux constructs infinitely many monotone Lagrangian tori in $\mathbb{R}^6$ (with factor $\pi/2$) that are pairwise not Hamiltonian isotopic and even not ambiently symplectomorphic, i.e. there is no symplectomorphism of $\mathbb{R}^6$ taking one to another. By taking products with S^1, he obtains infinitely many pairwise nonsymplectomorphic monotone Lagrangian tori in $(\mathbb{R}^{2n}, \omega_0)$ for every $n \geq 3$. After arbitrarily small nonmonotone Lagrangian perturbations, the tori constructed by Auroux become Hamiltonian isotopic to standard product tori. For many other interesting results about monotone Lagrangian tori in $\mathbb{R}^{2n}$ for $n \geq 4$, see the work of Dimitroglou Rizell–Evans [137].

After appropriate rescaling, the Clifford and Chekanov tori in $\mathbb{C}^2$ can also be embedded as monotone Lagrangian tori into $\mathbb{C}\mathrm{P}^2$. In [639], Vianna found a third monotone Lagrangian torus in $\mathbb{C}\mathrm{P}^2$, that is not Hamiltonian isotopic to the Clifford and Chekanov tori. The three monotone Lagrangian tori in $\mathbb{C}\mathrm{P}^2$ are distinguished by their relative Gromov–Witten invariants (the count of pseudoholomorphic discs with Maslov index two). In a subsequent paper [640], Vianna constructed infinitely many monotone Lagrangian tori in $\mathbb{C}\mathrm{P}^2$ that are pairwise not Hamiltonian isotopic. As explained in [639, §6.1], these tori can be considered as subsets of suitable affine charts in $\mathbb{C}\mathrm{P}^2$ and hence can also be embedded as monotone Lagrangian tori into $(\mathbb{C}^2, \omega_0)$. It is as yet unknown whether the resulting tori in $\mathbb{C}^2$ are Hamiltonian isotopic to the appropriately scaled Clifford or Chekanov torus.

In Problem 25 it makes sense to drop the monotonicity hypothesis and replace Hamiltonian isotopy by Lagrangian isotopy. In particular, the Chekanov torus is Lagrangian isotopic to the Clifford torus. This leads to the following question.

Problem 26 (Exotic nonmonotone Lagrangian tori in $\mathbb{R}^4$)
Does there exist a Lagrangian 2-torus in $(\mathbb{R}^4, \omega_0)$ that is not Lagrangian isotopic to the Clifford torus?

The next question concerns Lagrangian tori in symplectic 4-tori. Call a symplectic 4-torus $(\mathbb{T}^4, \omega)$ with a translation invariant symplectic form **irrational** if the integrals of ω over the six coordinate 2-tori are rationally independent.

Problem 27 (Irrational four-torus question)
Is every Lagrangian 2-torus in an irrational symplectic 4-torus displaceable by a Hamiltonian isotopy?

The above questions concentrate on the four-dimensional case, and have obvious analogues in higher dimensions. In particular, one might ask if every monotone torus in $\mathbb{C}\mathrm{P}^n$ (or more generally a product of projective spaces) is 'local', i.e. Hamiltonian isotopic to a torus in the affine part of the manifold, a result that is known in dimension four by [682].

Lagrangian knots

Another question concerns the existence of local Lagrangian knots. It goes back to a theorem by Eliashberg and Polterovich [195] which asserts that every Lagrangian embedding $\iota : \mathbb{R}^2 \to \mathbb{R}^4$ that is standard at infinity, is isotopic to the standard linear embedding by a compactly supported Hamiltonian isotopy of $\mathbb{R}^4$. The analogous question in higher dimensions is open.

Problem 28 (Local Lagrangian knots)
Fix an integer $n \geq 3$, let $\iota_0 : \mathbb{R}^n \to \mathbb{R}^{2n}$ be the standard Lagrangian embedding given by

$$\iota_0(x) := (x, 0) \qquad \text{for } x \in \mathbb{R}^n,$$

and let $\iota : \mathbb{R}^n \to \mathbb{R}^{2n}$ be a Lagrangian embedding that agrees with ι_0 outside of a compact set. Does it follow that ι is isotopic to ι_0 by a compactly supported (Hamiltonian) isotopy?

In a similar vein, Dimitroglou Rizell–Evans [137] proved that if $n \geq 5$ is an odd integer, then any two embedded Lagrangian tori in $(\mathbb{R}^{2n}, \omega_0)$ are smoothly isotopic. The questions discussed here can of course be extended to general symplectic $2n$-manifolds (M, ω) and one can ask whether any two homologous Lagrangian embeddings of a given n-manifold into (M, ω) are smoothly (or Lagrangian or Hamiltonian) isotopic. In this connection, Borman–Li–Wu [76] asserts that certain rational projective surfaces contain pairs of embedded Lagrangian real projective planes that are homologous but not smoothly isotopic.

Lagrangian spheres

Lagrangian spheres in smooth projective varieties (i.e. symplectic manifolds that are embedded as complex submanifolds in some projective space and inherit their symplectic structure from the Fubini–Study form) arise naturally as vanishing cycles when the algebraic variety in question is the fibre of a Lefschetz fibration. Thus the study of Lagrangian spheres plays a central role in both symplectic and algebraic geometry. For various interesting results about Lagrangian spheres in Milnor fibres, see the work of Seidel [577, 578] and Keating [361]. The following problem is a fundamental question in the subject and was pointed out to us by Ivan Smith.

Problem 29 (Donaldson's Lagrangian sphere question) *Is every Lagrangian sphere in a smooth algebraic variety Hamiltonian isotopic to a vanishing cycle of an algebraic degeneration?*

Cotangent bundles

Problem 30 (Nearby Lagrangian conjecture) *Let N be a closed smooth manifold. Then every exact Lagrangian submanifold $L \subset T^*N$ is Hamiltonian isotopic to the zero section.*

The nearby Lagrangian conjecture is a longstanding open problem in symplectic topology and there are quite a few partial results towards it. In particular the projection $L \to N$ along the fibres of T^*N is a homotopy equivalence by a result of Abouzaid–Kragh [5]. Constraints on exact Lagrangian submanifolds of cotangent bundles were found by Fukaya–Seidel–Smith [253, 254]. However, the nearby Lagrangian conjecture has only been confirmed for $N = S^1$, where it can be proved with elementary methods, and for $N = S^2$ by Richard Hind [309]. It is open for all other manifolds.

Problem 31 (Eliashberg cotangent bundle question) *Let N_0, N_1 be closed smooth manifolds that are homeomorphic. Are they diffeomorphic if and only if their cotangent bundles T^*N_0 and T^*N_1 (with their standard symplectic structures) are symplectomorphic?*

The first result in this direction is due to Abouzaid, who showed in [37] that if Σ is an exotic $(4k+1)$-dimensional sphere that does not bound a parallelizable manifold (and these exist), then $T^*\Sigma$ and T^*S^{4k+1} are *not* symplectomorphic. He does this by showing that every homotopy sphere that embeds as a Lagrangian in T^*S^{4k+1} must bound a parallelizable manifold, which he constructs directly out of certain perturbed J-holomorphic curves. (See Example 13.2.12.) This result has now been extended by Ekholm–Kragh–Smith [173].

Lagrangian Hofer geometry

A circle in a symplectic 2-sphere is called an **equator** if it divides the 2-sphere into two connected components of equal area. Any two equators are Hamiltonian isotopic (see part (v) of Exercise 3.4.5).

Problem 32 (Equator conjecture)
The space of equators on (S^2, σ) has infinite Hofer diameter.

This problem has been known to Lalonde and Polterovich since 1991; a positive answer would imply that the quantity $A(H)$ in Problem 21 is infinite. The analogous conjecture for *diameters of the disc* was confirmed by Khanevsky [368] in 2009. This question is, of course, a Lagrangian version of Problem 20, and can be asked for the orbit of any Lagrangian submanifold under the Hamiltonian group. Besides Ostover's observation in [518] and Usher's work in [635], very little is known about this question. In particular, to our knowledge there is no monotone Lagrangian in a closed manifold for which this orbit has been shown to have finite diameter.

14.4 Fano manifolds

A closed complex n-manifold M is called a **Fano manifold** if the top exterior power $L := \Lambda^{n,0}TM$ of its tangent bundle is **ample**, i.e. there is a $k \in \mathbb{N}$ such that $L^{\otimes k}$ is **very ample**, i.e. the holomorphic sections of $L^{\otimes k}$ determine an immersion into projective space. By pulling back the Fubini–Study form on projective space one obtains after scaling a Kähler form in the de Rham cohomology class $c_1(TM) \in H^2(M;\mathbb{R})$. Conversely, if (M, ω, J) is a closed Kähler manifold such that $[\omega] = c_1(\omega)$ then (M, J) is Fano by the Kodaira embedding theorem.

A theorem of Mori asserts that every Fano manifold is **uniruled**, i.e. admits a nonconstant holomorphic sphere (also called a *rational curve*) through every point. In fact he proved that every Fano manifold is **rationally connected**, i.e. for any two points in M there is a finite connected sequence of rational curves (also called a *genus zero stable map*) containing both points. This can be used to prove that every Fano manifold is simply connected. Mori's proofs of these theorems involve intrinsic algebraic geometric methods in characteristic p, and it would be interesting to find analytic proofs based on the theory of holomorphic curves. So far no one has succeeded in this endeavour.

Problem 33 (Mori theory question)
Is there an analytic proof of Mori's theorem that every closed Fano manifold is rationally connected?

In another direction one can ask if Mori's results carry over to nonKähler symplectic manifolds. The natural analogue of a Fano manifold in symplectic topology is a **monotone symplectic manifold**, i.e. a closed symplectic manifold (M, ω) such that $c_1(\omega) \in H^2(M;\mathbb{Z})$ is an integral lift of $[\omega] \in H^2(M;\mathbb{R})$. The above results do not all extend. Indeed, Fine–Panov [216] developed some ideas in Reznikov [538] to construct infinitely many examples in each dimension greater than or equal to twelve of nonsimply connected monotone symplectic manifolds. These manifolds cannot support any Kähler structure at all since their fundamental groups are hyperbolic. Further, since they fibre over hyperbolic manifolds, they are not rationally connected. However, one can still hope that a symplectic analogue of the uniruled condition might hold. This condition needs to be translated with some care. A first basic question would be whether every closed monotone symplectic manifold (M,ω) contains a symplectically embedded 2-sphere. For every symplectic manifold of dimension at least six, monotone or not, this question has a positive answer if and only if the cohomology class of ω does not vanish on $\pi_2(M)$. This follows from an h-principle for symplectic embeddings due to Gromov [288] (see also Eliashberg–Mishachev [190] and the discussion on page 327). Once such a symplectically embedded 2-sphere has been found, there is one through every point in M because the symplectomorphism group of (M,ω) acts transitively on M. Therefore a more interesting question is whether there is a nonconstant J-holomorphic sphere through every point for every ω-compatible almost complex structure. This question can be recast and strengthened in terms of the Gromov–Witten invariants. A closed symplectic manifold (M,ω) is called **symplectically uniruled** if there is a nonzero Gromov–Witten invariant with a point constraint. This means that there exists a nonzero class $A \in H_2(M;\mathbb{Z})$ and an integer k such that, for one and hence every regular ω-compatible almost complex structure J, the homology class in $H_*(M^k)$ represented by the evaluation map $\mathrm{ev}_{A,k,J}$ in (4.5.5) has a nonzero intersection pairing with a product homology class of the form $\mathrm{pt} \times c_2 \times \cdots \times c_k$. A class A with this property is called a **uniruling class**. With this definition it was proved by Hu–Li–Ruan [333] that every uniruled Kähler manifold is symplectically uniruled.

Problem 34 (Rational curves in monotone symplectic manifolds)
Is every closed monotone symplectic manifold (M,ω) symplectically uniruled?

As we discuss in more detail below, one way to approach such a problem might be to use the decomposition in Section 14.5.

There are many other questions one can ask about monotone symplectic manifolds. Here is a question pointed out to us by Paul Biran and Slava Kharlamov. If (M,ω) is a closed monotone symplectic manifold and $\tau : M \to M$ is an anti-symplectic involution so that its fixed point set $L := \mathrm{Fix}(\tau)$ is a Lagrangian submanifold, does there exist an upper bound, independent of τ, on the number of connected components of L that are tori? Apparently a reasonable guess would be that for $\dim M = 2n \geq 4$ an upper bound is given by $n+1-N$, where $N \in \mathbb{N}$ is the **minimal Chern number**, defined by $N\mathbb{Z} = \langle c_1(TX), H_2(X;\mathbb{Z})\rangle$. At least in complex dimensions $n = 2, 3$, there seems to be no known example with more than $n+1-N$ toral components.

14.5 Donaldson hypersurfaces

In the mid nineties Donaldson proved that for every closed symplectic manifold (M,ω) with an integral symplectic form the cohomology class $k[\omega]$ is Poincaré dual to a symplectic submanifold $Z \subset M$ for every sufficiently large integer k (see Section 7.4). The symplectic submanifolds constructed by Donaldson have very special properties that are not necessarily shared by all symplectic submanifolds in the same homology class. For example they are connected when M is connected and has dimension at least four, and they are simply connected when M is simply connected and has dimension at least six. Moreover, by an as yet unpublished theorem of Giroux, for k sufficiently large they can be chosen such that the complement $W := M \setminus Z$ is a Weinstein manifold (see Definition 7.4.5). On the other hand a theorem of McLean (also not yet published) asserts that certain closed symplectic manifolds contain symplectic submanifolds that are Poincaré dual to $k[\omega]$, but yet do not have a Weinstein complement. These submanifolds must therefore be different from those constructed by Donaldson. This leads to the question of how the term *Donaldson hypersurface* should be defined. As a temporary working definition we call a codimension two symplectic submanifold $Z \subset M$ of a closed integral symplectic manifold (M, ω) a **Donaldson hypersurface** if it is Poincaré dual to the cohomology class $k[\omega]$ for some integer $k > 0$.

By the theorems of Donaldson and Giroux, every closed integral symplectic manifold admits a decomposition into a Donaldson hypersurface $Z \subset M$ Poincaré dual to $k[\omega]$ and a Weinstein manifold $W := M \setminus Z$ for every sufficiently large integer $k > 0$. This decomposition is so far rather little understood but its properties seem likely to affect the symplectic topology of (M, ω) in very interesting ways. For example, (M, ω) is called **subcritical** if, for some k, it admits such a decomposition in which the Weinstein manifold W has the homotopy type of a cell complex with cells of dimension at most $n-1$. Moreover, there is the notion of a *flexible Weinstein manifold* introduced by Cieliebak–Eliashberg [111]. One can then compare the properties of closed integral symplectic manifolds that admit a Donaldson–Giroux decomposition with subcritical Weinstein part with those where the Weinstein part is flexible.

One can also ask whether these decompositions have consequences for the Gromov–Witten invariants, for example whether symplectic manifolds with flexible decompositions are uniruled. The work of Jian He [306] explains a new way to understand the Gromov–Witten invariants of subcritical manifolds in the Kähler case, giving a new proof that such a manifold is uniruled.

It is also important to gain a better understanding of the difference between Liouville and Weinstein domains. The symplectic form ω on a Liouville domain (W, ω, X) is exact, i.e. $\omega = \mathcal{L}_X \omega = d\iota(X)\omega$, and the corresponding Liouville vector field X points out along the boundary ∂W (see Definition 3.5.32), but a Weinstein domain also carries a function f that satisfies the stringent requirements in Definition 7.4.5: f must be a generalized Morse function and also such that the flow of X is gradient like, i.e. $df \cdot X \geq 0$. Neither of these conditions is well understood. Here is a sample question.

Problem 35 (Eliashberg's Weinstein manifold question)
Let (W, ω) be a noncompact connected symplectic manifold without boundary, let X be a global complete Liouville vector field on W, and let $f : W \to \mathbb{R}$ be a smooth function that is proper and bounded below such that $df \cdot X \geq 0$. Can the pair (X, f) be perturbed to a Weinstein structure as in Definition 7.4.5 (i.e. f is a generalized Morse function that satisfies (a) and (b))?

The next question was posed by Donaldson and seconded by Eliashberg.

Problem 36 (Hypersurface stability question)
Do Donaldson hypersurfaces 'stabilize' in dimension six? More precisely, is there a family of simply connected smooth four-manifolds $X_{b^+,b^-,\varepsilon}$ (parametrized by pairs of nonnegative integers $b^\pm$ and elements $\varepsilon \in \{0,1\}$, with Euler characteristic $2+b^++b^-$, signature b^+-b^-, and even/odd intersection form when $\varepsilon = 0/1$) having the following significance: If (M,ω) is a simply connected closed symplectic 6-manifold with integral symplectic form then, for k sufficiently large, the Donaldson hypersurface $X \subset M$ Poincaré dual to $k[\omega]$ can be constructed to be diffeomorphic to one of the manifolds $X_{b^+,b^-,\varepsilon}$?

This question is backed up by the observation that hyperplane sections of simply connected algebraic 3-folds are minimal Kähler surfaces of general type whenever they have sufficiently large degree. Hence they have very simple Seiberg–Witten invariants, namely, their only basic classes are $\pm K$ (see Example 13.4.10). Thus, when they have isomorphic intersection forms, their diffeomorphism types cannot be distinguished by any currently known method.

14.6 Contact geometry

The existence problem for contact structures was settled in 2014 by Borman–Eliashberg–Murphy [75]. A necessary condition for the existence of a co-oriented contact structure on a $(2n+1)$-dimensional manifold M is that the tangent bundle admits a rank-$2n$ subbundle $\xi \subset TM$, equipped with a nondegenerate 2-form ω, such that the rank-1 quotient bundle TM/ξ admits a trivialization, i.e. there exists a 1-form $\alpha \in \Omega^1(M)$ such that $\xi = \ker\alpha$. In other words, there exists a pair (α,ω) consisting of a nonvanishing 1-form $\alpha \in \Omega^1(M)$ and a nondegenerate 2-form ω on $\xi := \ker\alpha$. Such a pair (α,ω) is called an **almost contact structure**. They also introduced the notion of an *overtwisted contact structure* in higher dimensions (see Definition 3.5.4 for dimension three) and proved that the map $\alpha \mapsto (\alpha, d\alpha|_{\ker\alpha})$ from the space of overtwisted contact forms to the space of almost contact structures is a weak homotopy equivalence. In particular, the existence of an almost contact structure is equivalent to the existence of an overtwisted contact structure, and every homotopy class of almost contact structures contains a unique isotopy class of overtwisted contact structures. (See also the discussion on page 137 in Section 3.5.) Their theorem does not settle the existence and uniqueness problem for **tight** (i.e. not overtwisted) contact structures. An important problem is which homotopy classes of almost contact structures contain an isotopy class consisting of tight contact structures and, if so, how many such isotopy classes it contains. There are many results about this problem in dimension three, some of which are discussed in Section 3.5. For spheres of dimension $4m+1$ the problem was studied by Ustilovsky in the late 1990s. In [637] he found infinitely many isotopy classes of contact structures on the 5-sphere, that are all homotopic as contact structures and are tight because they are symplectically fillable. However the technical background for this result was not put in place until Gutt's thesis [302]; see also Bourgeois–Oancea [79]. A general question in higher dimensions is which closed contact manifolds are fillable by Liouville domains, respectively Weinstein domains (see Definitions 3.5.32, 3.5.38, and 7.4.5). Borman–Eliashberg–Murphy [75] proved that overtwisted contact structures are not symplectically fillable. Further, in [192] Eliashberg–Murphy construct many symplectic cobordisms, showing in particular that in all dimensions ≥ 5 there is a symplectic cobordism whose inner (i.e. concave) boundary is overtwisted while the outer (convex) boundary is tight.

Another central problem that has inspired many developments in symplectic and contact geometry is the Weinstein conjecture which has already been discussed in Section 1.2 for hypersurfaces of $(\mathbb{R}^{2n}, \omega_0)$.

Problem 37 (Weinstein conjecture)
The Reeb vector field of every contact form on a closed contact manifold has a closed orbit.

For contact hypersurfaces of Euclidean space the Weinstein conjecture was settled by Viterbo [643] in the late 1980s (see Theorem 12.4.6). It was confirmed by Hofer [317] for the 3-sphere and by Taubes [612] for all closed contact 3-manifolds. In [130], Cristofaro-Gardiner and Hutchings used embedded contact homology [337,340] to establish the existence of at least two closed Reeb orbits in dimension three. In [342], Hutchings and Taubes extended the Weinstein conjecture to **stable Hamiltonian structures** on 3-manifolds. In higher dimensions the general Weinstein conjecture is open.

Problem 38 (Arnold chord conjecture)
If M is a closed contact manifold and $L \subset M$ is a Legendrian submanifold then, for every contact form on M, there exists a Reeb orbit with endpoints on L.

In [343, 344] Hutchings and Taubes settled the Arnold chord conjecture in dimension three. Their work is based on Hutchings' embedded contact homology and a theorem of Taubes [613, 614, 615, 616, 617] which asserts that embedded contact homology is isomorphic to Seiberg–Witten Floer homology. The Arnold chord conjecture is open in higher dimensions.

An important notion in contact topology is that of a **positive loop of contactomorphisms**, i.e. a loop of contactmorphisms $\psi_t = \psi_{t+1}$ of a contact manifold (M, ξ) that is generated by a time-dependent family of positive Hamiltonian functions $H_t = H_{t+1} = \alpha(X_t) : M \to (0, \infty)$ (Lemma 3.5.14) for some, and hence every, contact form α. By a theorem of Eliashberg–Kim–Polterovich [189], the nonexistence of a positive loop is equivalent to the orderability of the identity component $\mathrm{Cont}_0(M, \xi)$ of the group of contactomorphisms, while the nonexistence of a contractible positive loop is equivalent to the orderability of the universal cover of $\mathrm{Cont}_0(M, \xi)$.

Problem 39 (Positive loop question)
Which closed contact manifolds (M, ξ) admit (contractible) positive loops of contactomorphisms?

Positive loops of contactomorphisms obviously exist on contact manifolds with periodic Reeb flows, such as pre-quantization circle bundles over symplectic manifolds (Example 3.5.11) and unit cotangent bundles of closed Riemannian manifolds with periodic geodesic flows (Example 3.5.7). These loops are all noncontractible [189]. Eliashberg–Kim–Polterovich [189] also found contractible positive loops of contactomorphisms for the standard contact structures on spheres (Example 3.5.9). The Eliashberg–Kim–Polterovich construction can be adapted to the regular level sets of plurisubharmonic Morse functions on symplectic $2n$-manifolds with critical points of indices at most $n-2$. These are apparently the only known existence results. Nonexistence is known in various cases such as the standard contact structure on $\mathbb{R}\mathrm{P}^{2n-1}$.

Another fundamental question in contact topology is whether contact manifolds are determined by their symplectizations (Definition 3.5.24); see Cieliebak–Eliashberg [111, p. 239].

Problem 40 (Symplectization question)
Do there exist closed contact 3-manifolds (M,ξ) and (M',ξ') that are not contactomorphic but have symplectomorphic symplectizations?

Examples of such contact manifolds in dimensions at least five were found by Sylvain Courte [129]. In dimension three this problem is open.

14.7 Continuous symplectic topology

Symplectic rigidity asserts that the group of symplectomorphisms is a closed subset of the group of all diffeomorphisms with respect to the C^0-topology (Theorem 12.2.1). This has prompted the development of continuous symplectic topology into an important and very active research area, that studies questions such as the relation between the C^0-topology and that given by the Hofer metric, and the persistence of symplectic phenomena under C^0 limits.

A rigidity theorem by Cardin–Viterbo [93] asserts that if F and G are smooth functions on a symplectic manifold that can be approximated in the C^0-topology by sequences of smooth functions F_n and G_n with vanishing Poisson bracket, then the Poisson bracket of F and G must also vanish. Their result was refined by Entov–Polterovich [201] who showed that the supremum norm of the Poisson bracket $\{F,G\}$ is a lower bound for the limit inferior of the supremum norms of $\{F_n,G_n\}$. Their result was in turn refined by Buhovski [82] when he established his *'2/3-law'* for the convergence rate. These results also led to the introduction of new invariants by Buhovsky–Entov–Polterovich [86] and they are closely related to the study of Calabi quasimorphisms on the universal cover of the group of Hamiltonian symplectomorphisms (see Section 14.2) and to the study of *symplectic quasi-states* on the space of functions on a symplectic manifold as well as to Floer homology (see Section 11.4).

Symplectic rigidity suggests that there is a notion of a symplectic homeomorphism. For example, one can define a **symplectic homeomorphism** between two open subsets $U,V\subset\mathbb{R}^{2n}$ as a homeomorphism $\phi:U\to V$ such that every point in U has an open neighbourhood such that the restriction of ϕ to this neighbourhood preserves a given symplectic capacity of all open sets (e.g. the Hofer–Zehnder capacity or the Gromov width). Another possible definition is that each point in U has a neighbourhood such that the restriction of ϕ to this neighbourhood can be approximated in the C^0-topology by a sequence of symplectomorphisms, and the same holds for ϕ^{-1}. As mentioned in the discussion at the end of Section 12.2, there are many open questions about the precise relation between the different definitions. However, they have in common that the symplectic homeomorphisms are the morphisms of a category (i.e. compositions and inverses of symplectic homeomorphisms are again symplectic homeomorphisms) and that the symplectic form is preserved by every smooth symplectic homeomorphism that is defined on a symplectic manifold. Given such choice one can define a **topological symplectic manifold** as a topological manifold equipped with an atlas whose transition maps are symplectic homeomorphisms. The following question was pointed out to us by Yasha Eliashberg and Leonid Polterovich. It has been known to them since the early 1990s.

Problem 41 (Topological symplectic four-sphere)
Does the 4*-sphere admit the structure of a topological symplectic manifold?*

A related and somewhat vaguely worded question is which symplectic properties are preserved by symplectic homeomorphisms. The notion of a symplectic homeomorphism seems to be considerably more flexible than that of a symplectomorphism. For example Buhovsky–Opshtein [84] construct symplectic homeomorphisms of Euclidean space $\mathbb{C}^3$ (in the sense of C^0 approximation by symplectomorphisms) that restrict to the map $(z, 0, 0) \mapsto (\frac{1}{2}z, 0, 0)$ for z in the open unit disc. On the other hand, Humilière–Leclercq–Seyfaddini [334] show that symplectic homeomorphisms are compatible with the construction in Proposition 5.4.5 of the symplectic reduction of a coisotropic manifold.

In dimension two a well-known folk theorem states that every area-preserving homeomorphism can be C^0-approximated by area-preserving diffeomorphisms. (For a short proof see Sikorav [587].) Nevertheless, there are still many open questions. The following question about area-preserving homeomorphisms has attracted much interest, partly because of its immediacy and partly because it can be tackled from many different points of view.

Problem 42 (Simplicity conjecture)
The group $\mathrm{Homeo}_c(\mathbb{D})$ *of compactly supported area preserving homeomorphisms of the open two-disc is not simple.*

There have been various attempts to settle this conjecture by constructing normal subgroups of $\mathrm{Homeo}_c(\mathbb{D})$; the difficulty has always been to show that these subgroups are proper. For example, in [413], Le Roux reduces this problem to a question about properties of *fragmentations* of elements $\phi \in \mathrm{Ham}_c(\mathbb{D})$ into products of symplectomorphisms with support in discs of small area. Oh–Müller [510] suggest a different approach in which the normal subgroup is the group of hameomorphisms. These are limits of symplectomorphisms under a combination of the Hofer metric topology and the C^0-topology (see Definition 12.3.17 in Section 12.3). The following problem describes an elementary example of an area-preserving homeomorphism that might not belong to this group. Note that every continuous function $f : (0,1] \to \mathbb{R}$ that equals zero near 1 determines a compactly supported area-preserving homeomorphism $\phi_f : \mathbb{D} \to \mathbb{D}$, defined by $\phi_f(0) := 0$ and

$$\phi_f(re^{i\theta}) := re^{i(\theta+f(r))} \tag{14.7.1}$$

for $0 < r \leq 1$ and $\theta \in \mathbb{R}$.

Problem 43 (Infinite twist conjecture)
If $\lim_{r\to 0} f(r) = \infty$ *then* ϕ_f *is not a hameomorphism.*

A positive answer to the infinite twist conjecture would also settle the simplicity conjecture in Problem 42. For further background see Le Roux [414].

14.8 Symplectic embeddings

The Gromov nonsqueezing Theorem 12.1.1 was the the first striking result about a nontrivial obstruction to the existence of a symplectic embedding and has led to many important developments in the subject. One can extend this question and ask in general when one symplectic manifold can be symplectically embedded into another of the same dimension. For example, one can ask when one $2n$-dimensional open ellipsoid can be embedded into another.

Problem 44 (Symplectic ellipsoid embeddings)
For $n \in \mathbb{N}$ *and* $a = (a_1, \ldots, a_n) \in \mathbb{R}^n$ *with* $0 < a_1 \leq a_2 \leq \cdots \leq a_n$ *define*

$$E(a) := \left\{ z = (z_1, \ldots, z_n) \in \mathbb{C}^n \,\middle|\, \sum_{i=1}^{n} \frac{\pi |z_i|^2}{a_i} < 1 \right\}$$

Under what conditions on a_i *and* b_i *does there exist a symplectic embedding of* $E(a)$ *into* $E(b)$ *(with respect to the standard symplectic structure* ω_0*)?*

For $n = 1$ the obvious answer is if and only if $a_1 \leq b_1$. For $n = 2$ an answer to the symplectic ellipsoid embedding problem was given by McDuff [463] in 2009. This answer was later reformulated by Hutchings [338, 339] and McDuff [467]. Define the ordered sequence

$$0 < c_1(a) \leq c_2(a) \leq c_3(a) \leq \cdots,$$

by ordering the set $\{n_1 a_1 + n_2 a_2 \,|\, n_i \in \mathbb{Z},\, n_i \geq 0,\, n_1 + n_2 > 0\}$ with multiplicities. Hutchings and McDuff proved that there is a symplectic embedding of $E(a)$ into $E(b)$ if and only if $c_k(a) \leq c_k(b)$ for all $k \in \mathbb{N}$. In Hutchings' work these numbers appear as invariants of embedded contact homology and he shows that the condition is necessary for the existence of a symplectic embedding. McDuff proved that the condition is sufficient. In dimensions $2n \geq 6$ the symplectic ellipsoid embedding problem is open in this general form. In [301] Guth constructed an embedding that shows that the obvious extension of the above condition to $n > 2$ is no longer satisfied by all ellipsoidal embeddings; however, so far no other substitute condition has been proposed. Of course, when $a_1 = a_2 = \cdots = a_n$ this becomes a question about the Gromov width of the ellipsoid $E(b)$ (see equation (12.1.2)), and the answer follows from Gromov's nonsqueezing theorem.

There are many open questions about the Gromov width, and more generally about symplectic packing in the sense of Remark 7.1.31. For example, it would be interesting to understand which manifolds are fully fillable by one ball, in the sense that they contain a symplectically embedded ball that occupies an arbitrarily large fraction of its volume. Another question concerns monotone manifolds with symplectic form normalized so that $[\omega] = c_1(\omega)$. In the four-dimensional case these manifolds are $S^2 \times S^2$ and k-point blowups of $\mathbb{C}P^2$ for $k \leq 8$, and they have integral Gromov width (the supremum of the numbers πr^2 such that there exists a symplectic embedding of the ball of radius r; see (12.1.2)). For example, the Gromov width of $\mathbb{C}P^2$ with the appropriately rescaled Fubini–Study form is 3, and for $S^2 \times S^2$ with the product form it is 2.

Problem 45 (Gromov width of monotone manifolds)
If (M, ω) is a closed symplectic manifold such that

$$[\omega] = c_1(\omega),$$

is its Gromov width at least one?

Here is a question of a different flavour.

Problem 46 (Gromov width of cohomologous symplectic forms)
Is there a closed manifold M with cohomologous symplectic forms ω_0, ω_1 such that (M, ω_0) and (M, ω_1) have different Gromov widths?

Related to the existence question is also a uniqueness problem, in this case whether two symplectic embeddings of a ball into a symplectic manifold are symplectically isotopic. By a theorem of McDuff [450], the space of symplectic embeddings of the closed ball $\overline{B}^4(r)$ into the open ball $B^4(1)$ is connected for every radius $0 < r < 1$. It is a remarkable fact that the analogous question is open in higher dimensions even for arbitrarily small radii.

Problem 47 (Ball isotopy question)
Let $n \geq 3$. Is there a constant $0 < \varepsilon < 1$ such that the space of symplectic embeddings of $\overline{B}^{2n}(\varepsilon)$ into $B^{2n}(1)$ is connected?

Problem 48 (Symplectic camel problem)
Is there any closed $2n$-dimensional symplectic manifold (M, ω) and a real number $r > 0$ such that the space of symplectic embeddings $\overline{B}^{2n}(r) \to M$ is disconnected?

For open manifolds the answer to Problem 48 is positive by Gromov's camel obstruction. Gromov's counterexamples are convex at infinity. (See the discussion on page 512 and Definition 3.5.39.) Returning to the existence problem, the following question about the existence of several disjoint symplectically embedded balls (the *packing problem*) was discussed in Remark 7.1.31.

Problem 49 (Ball packing)
For which integers $n \geq 3$ and $k \geq 2$ is

$$v_k(B^{2n}) := \sup \left\{ \frac{k\mathrm{Vol}(B^{2n}(\varepsilon))}{\mathrm{Vol}(B^{2n}(1))} \;\middle|\; \begin{array}{l} \textit{there exist symplectic embeddings} \\ \iota_i : \overline{B}^{2n}(\varepsilon) \to B^{2n}(1),\ i = 1, \dots, k, \\ \textit{with pairwise disjoint images} \end{array} \right\} < 1?$$

When $n = 2$ it is known that $v_k(B^4) < 1$ for $k = 2, 3, 5$ by a result of Gromov [287], that $v_k(B^4) < 1$ for $k = 6, 7, 8$ by a result of McDuff–Polterovich [469], and that $v_k(B^4) = 1$ for $k \geq 9$ by a result of Biran [66]. When $n \geq 3$ it is known that $v_k(B^{2n}) = 1$ for k sufficiently large by a result of Buse–Hind [89]. However, Gromov showed in [287] that the disjoint union of two open balls of radii r_1, r_2 embeds into $B^{2n}(1)$ if and only if $r_1^2 + r_2^2 \leq 1$. Hence

$$v_2(B^{2n}) = \frac{1}{2^{n-1}}$$

for all $n \geq 1$.

There are many interesting open problems about embedding other shapes. For example, one can consider the analogue of Problem 49 with domain (and perhaps also target) the polydisc

$$P(a_1, \dots, a_n) := B^2(r_1) \times \cdots \times B^2(r_n), \qquad \pi r_i^2 = a_i.$$

In general, polydiscs are harder to understand than ellipsoids, at least as domains. There are now two proofs that the closed polydisc $\overline{P(1,2)}$ embeds in the open ball $B^4(r)$ only if $\pi r^2 > 3$, one by Hind–Lisi [310] using finite energy foliations and one by Hutchings [341] that uses refined information about the chain complex of embedded contact homology.

14.9 Symplectic topology of Euclidean space

In the final section of this book we return to the symplectic topology of Euclidean space, which is where we started in Chapter 1. One open problem is whether every symplectic form on $\mathbb{R}^{2n}$ which is *standard at infinity* is diffeomorphic to the standard symplectic form.

Problem 50 (Standard-at-infinity)
Let $n \geq 3$ and let ω be a symplectic form on $\mathbb{R}^{2n}$ that agrees with ω_0 on the complement of a compact set. Is $(\mathbb{R}^{2n}, \omega)$ symplectomorphic to $(\mathbb{R}^{2n}, \omega_0)$?

The same question in dimension four has a positive answer by a celebrated theorem of Gromov [287]. A theorem of Floer–Eliashberg–McDuff [451] asserts that if (M, ω) is a symplectic manifold with $\pi_2(M) = 0$ that is symplectomorphic to $(\mathbb{R}^{2n}, \omega_0)$ outside of a compact set, then M is necessarily diffeomorphic to $\mathbb{R}^{2n}$. Mark McLean and others constructed many examples of symplectic structures on $\mathbb{R}^{2n}$ that are convex at infinity, but are not symplectomorphic to $(\mathbb{R}^{2n}, \omega_0)$ (see McLean [474, 475], Abouzaid–Seidel [6], and Seidel [576]). Problem 50 can be rephrased as the question of which symplectic $2n$-manifolds can have the standard $(2n-1)$-sphere as a contact boundary.

Problem 51 (Symplectic Hadamard question)
Let (M, ω) be a connected, simply connected, symplectic $2n$-manifold and let J be an ω-compatible almost complex structure such that the Riemannian metric $\langle \cdot, \cdot \rangle := \omega(\cdot, J\cdot)$ is complete and has nonpositive sectional curvature. Is (M, ω) symplectomorphic to $(\mathbb{R}^{2n}, \omega_0)$?

By Hadamard's theorem the manifold M in Problem 51 is diffeomorphic to $\mathbb{R}^{2n}$. By a theorem of McDuff [446] the question has a positive answer whenever J is integrable.

In [136], Dimitroglou Rizell constructed Lagrangian submanifolds $L \subset \mathbb{R}^{4k+2}$, diffeomorphic to $S^1 \times S^{2k}$, with infinite Gromov width (the supremum over all numbers πr^2 such that there is a symplectic embedding of $B^{2n}(r)$ into $\mathbb{R}^{2n}$ which sends the intersection of $B^{2n}(r)$ with a Lagrangian plane to L). His work builds on the ideas of Murphy [500] on *loose Legendrean knots* and of Eliashberg–Murphy [191] on *Lagrangian caps*. Such examples exist also in $\mathbb{R}^{2n}$ for any $n \geq 3$ but the question is open in $\mathbb{R}^4$.

Problem 52 (Lagrangian infinite width)
Is there a compact Lagrangian submanifold in $\mathbb{R}^4$ with infinite Gromov width?

This question brings us back to symplectic capacities as discussed in Section 12.1. Recall that a symplectic capacity is called *normalized* if the unit ball and the unit cylinder in $\mathbb{R}^{2n}$ have capacity π. A longstanding conjecture about symplectic capacities is the following.

Problem 53 (Convex capacity conjecture)
All normalized capacities agree on convex subsets of Euclidean space.

This conjecture, if true, would imply the following conjecture by Viterbo which relates the capacity of a general convex subset of Euclidean space to the capacity of the ball.

Problem 54 (Viterbo's symplectic isoperimetric conjecture)
All normalized capacities satisfy the inequality

$$\frac{c(\Sigma)}{\mathrm{Vol}(\Sigma)^{1/n}} \leq \frac{c(B)}{\mathrm{Vol}(B)^{1/n}} \tag{14.9.1}$$

for every compact convex set $\Sigma \subset \mathbb{R}^{2n}$ with nonempty interior.

It is known that the inequality (14.9.1) holds up to a constant factor which is independent of the dimension. An equivalent formulation is the inequality

$$\mathrm{Vol}(\Sigma) \geq \frac{c(\Sigma)^n}{n!}. \tag{14.9.2}$$

for every compact convex set $\Sigma \subset \mathbb{R}^{2n}$ with nonempty interior. In [46], Artstein, Karasev, and Ostrover proved that the Hofer–Zehnder capacity satisfies the inequality $c_{\mathrm{HZ}}(K \times K^*) \geq 4$ for every symmetric (i.e. $v \in K$ implies $-v \in K$) convex set $K \subset \mathbb{R}^n$ with nonempty interior. Assuming Viterbo's isoperimetric inequality for the Hofer–Zehnder capacity, one would then get

$$\mathrm{Vol}(K \times K^*) \geq \frac{4^n}{n!} \tag{14.9.3}$$

for every symmetric convex set $K \subset \mathbb{R}^n$ with nonempty interior. (Here K is the unit ball of a norm on $\mathbb{R}^n$ and $K^* \subset (\mathbb{R}^n)^*$ is the unit ball of the dual norm. It is known from Kuperberg [383] that $\mathrm{Vol}(K \times K^*) \geq \pi^n/n!$.) The inequality (14.9.3) is known as the **Mahler conjecture**. Its appearance in this context shows yet again the tight connection between the symplectic world and other ideas in geometry.

APPENDIX A

SMOOTH MAPS

This appendix clarifies the notion of smooth maps on manifolds with boundary used throughout the book. Section A.1 discusses smooth functions on manifolds with corners. Section A.2 discusses how to extend smooth (homotopies of) embeddings of a closed ball $B \subset \mathbb{R}^m$ into a manifold M to smooth (homotopies of) embeddings of $\mathbb{R}^m$ into M. This is used in Chapter 3, where the extension results are carried over to the symplectic setting. Section A.3 gives an explicit formula for a smooth function that is used in Section 7.1 for the construction of symplectic forms on blow-ups.

A.1 Smooth functions on manifolds with corners

Denote by

$$Q^m := \{x = (x_1, \dots, x_m) \in \mathbb{R}^m \mid x_i \geq 0 \text{ for } i = 1, \dots, m\}$$

the right upper quadrant in $\mathbb{R}^m$ and by

$$\mathbb{N}_0 := \mathbb{N} \cup \{0\}$$

the set of nonnegative integers.

Definition A.1.1 *Let $V \subset Q^m$ be an open set in the relative topology. A function $f : Q^m \cap V \to \mathbb{R}$ is called* **smooth** *if its restriction to* $\operatorname{int}(Q^m) \cap V$ *is smooth and all its partial derivatives extend to continuous functions on $Q^m \cap V$.*

In Milnor [484] and Guillemin–Pollack [293], a function $f : X \to \mathbb{R}$, defined on an arbitrary subset $X \subset \mathbb{R}^m$, is called smooth if, for every $x \in X$, there is an open neighbourhood $U \subset \mathbb{R}^m$ of x and a smooth function $F : U \to \mathbb{R}$ such that the restriction of F to $X \cap U$ agrees with f. The next theorem shows that, for $X = Q^m$, their definition of *smooth* agrees with that in Definition A.1.1.

Theorem A.1.2 *Let $f : Q^m \to \mathbb{R}$ be a compactly supported smooth function and let $U \subset \mathbb{R}^m$ be an open neighbourhood of Q^m. Then there is a compactly supported smooth function $F : \mathbb{R}^m \to \mathbb{R}$ such that*

$$F|_{Q^m} = f, \qquad \operatorname{supp}(F) \subset U.$$

The proof is based on the following lemma.

Lemma A.1.3 *Let $\delta > 0$ and let $a_k, c_k \in \mathbb{R}$ be sequences such that*

$$c_k \geq \delta^{-1} + |a_k| \tag{A.1.1}$$

for $k \in \mathbb{N}_0$. Let $\beta : \mathbb{R} \to [0,1]$ be a smooth cutoff function such that

$$\beta(x) = \begin{cases} 1, \text{for } |x| \leq 1/3, \\ 0, \text{for } |x| \geq 2/3, \end{cases} \tag{A.1.2}$$

and define $F : \mathbb{R} \to \mathbb{R}$ by

$$F(x) := \sum_{k=0}^{\infty} \frac{a_k \big(\beta(c_k x)x\big)^k}{k!} \qquad \text{for} \quad x \in \mathbb{R}. \tag{A.1.3}$$

Then F is smooth, vanishes for $|x| \geq \delta$, and satisfies $F^{(\ell)}(0) = a_\ell$ for $\ell \in \mathbb{N}_0$.

Proof: The right hand side in (A.1.3) vanishes for $|x| \geq \delta$. It converges absolutely and uniformly because $|x\beta(x)| \leq 1$ for all $x \in \mathbb{R}$ and hence, for $k \geq 1$,

$$|\beta(c_k x)x| \leq \frac{1}{c_k} \leq \delta \leq 1, \qquad |a_k \beta(c_k x)x| \leq \frac{|a_k|}{c_k} \leq 1.$$

This shows that F is continuous on $(-\infty, 0]$ and $F(0) = a_0$. Differentiate the right hand side of equation (A.1.3) term by term to obtain

$$F'(x) = \sum_{k=1}^{\infty} \frac{a_k \big(\beta(c_k x)x\big)^{k-1}}{(k-1)!} \big(\beta(c_k x) + \beta'(c_k x)c_k x\big). \tag{A.1.4}$$

Use the inequalities $|a_k \beta(c_k x)x| \leq 1$ and $|\beta'(c_k x)c_k x| \leq \max|\beta'|$ for all $x \in \mathbb{R}$ and $k \geq 2$ to show that the right hand side in equation (A.1.4) converges uniformly and absolutely on $\mathbb{R}$. Hence F is continuously differentiable and $F'(0) = a_1$.

To examine the higher derivatives of F, define $\beta_k : \mathbb{R} \to \mathbb{R}$ by

$$\beta_k(x) := \beta(c_k x)x$$

for $k \in \mathbb{N}_0$ and $x \in \mathbb{R}$. Then, for every $i \in \mathbb{N}$ there is a constant $C_i > 0$ such that

$$\left|\beta_k(x)^{i-1} \tfrac{d^i}{dx^i} \beta_k(x)\right| \leq C_i$$

for all $k \in \mathbb{N}_0$ and all $x \in \mathbb{R}$. Moreover, $F = \sum_{k=0}^{\infty} \frac{a_k \beta_k^k}{k!}$ and hence

$$F^{(\ell)} = G_\ell + \sum_{j=1}^{\ell} \sum_{k=\ell}^{\infty} \frac{a_k \beta_k^{k-\ell}}{(k-\ell+j-1)!} p_{j,k,\ell}, \tag{A.1.5}$$

where G_ℓ is a finite sum, satisfying $G_\ell(0) = a_\ell$, and $p_{j,k,\ell}$ is a polynomial in the functions $\beta_k^{i-1} \cdot \partial_x^i \beta_k$ for $i = 1, \ldots, j$. Thus the functions $p_{j,k,\ell}$ satisfy a uniform bound, independent of k, and hence the series converges absolutely and uniformly. This shows that F is smooth and $F^{(\ell)}(0) = a_\ell$ for all $\ell \in \mathbb{N}$. This proves Lemma A.1.3. $\square$

Proof of Theorem A.1.2: We prove the following by induction on n.

Claim. *Let $f : Q^m \to \mathbb{R}$ be a compactly supported smooth function, let $U \subset \mathbb{R}^m$ be an open neighbourhood of Q^m, and let $n \in \{1, \ldots, m\}$. Then there exists a compactly supported smooth function $F_n : \mathbb{R}^n \times Q^{m-n} \to \mathbb{R}$ such that*

$$F_n|_{Q^m} = f, \qquad \mathrm{supp}(F_n) \subset U.$$

Choose constants

$$c_k \geq \delta^{-1} + \sup_{(x_2,\ldots,x_m)\in Q^{m-1}} \left|\partial_1^k f(0, x_2, \ldots, x_k)\right|$$

for $k \in \mathbb{N}_0$ and define the function $F_1 : \mathbb{R} \times Q^{m-1} \to \mathbb{R}$ by $F_1|_{Q^m} := f$ and

$$F_1(x) := \sum_{k=0}^{\infty} \frac{\partial_1^k f(0, x_2, \ldots, x_m)}{k!} \big(\beta(c_k x_1) x_1\big)^k$$

for $x_1 < 0$ and $(x_2, \ldots, x_m) \in Q^{m-1}$. By Lemma A.1.3 this function is smooth. It is supported in U whenever $\delta > 0$ is chosen sufficiently small. Let $n \in \{2, \ldots, m\}$ and suppose F_{n-1} has been constructed. Choose constants

$$c_k \geq \delta^{-1} + \sup_{\substack{(x_1,\ldots,x_{n-1})\in\mathbb{R}^{n-1} \\ (x_{n+1},\ldots,x_m)\in Q^{m-n}}} \left|\partial_n^k F_{n-1}(x_1, \ldots, x_{n-1}, 0, x_n, \ldots, x_m)\right|$$

and define $F_n : \mathbb{R}^n \times Q^{m-n} \to \mathbb{R}$ by $F|_{\mathbb{R}^{n-1}\times Q^{m-n+1}} := F_{n-1}$ and

$$F_n(x) := \sum_{k=0}^{\infty} \frac{\partial_n^k F_{n-1}(x_1, \ldots, x_{n-1}, 0, x_n, \ldots, x_m)}{k!} \big(\beta(c_k x_n) x_n\big)^k$$

for $(x_1, \ldots, x_{n-1}) \in \mathbb{R}^{n-1}$, $x_n < 0$, and $(x_n, \ldots, x_m) \in Q^{m-n}$. By Lemma A.1.3, F_n satisfies the requirements of the claim and this proves Theorem A.1.2. □

We now briefly discuss the notion of a manifold with corners. In the following we will mostly be interested in smooth functions whose domain is either a closed ball B, i.e. a manifold with boundary, or is the product $B \times [0, 1]$, which is a manifold with corners of codimension 2. For an in-depth discussion of manifolds with corners and their tangent bundles, see Joyce [351] and the references cited therein.

Definition A.1.4 *A* **manifold with corners (of dimension** m**)** *is a second countable Hausdorff topological space M, equipped with an open cover $\{U_\alpha\}_\alpha$ and a collection of homeomorphisms $\phi_\alpha : U_\alpha \to \phi_\alpha(U_\alpha)$ (called* **coordinate charts***) onto open sets $\phi_\alpha(U_\alpha) \subset Q^m$ (in the relative topology induced by $\mathbb{R}^m$) such that the transition maps $\phi_{\beta\alpha} := \phi_\beta \circ \phi_\alpha^{-1} : \phi_\alpha(U_\alpha \cap U_\beta) \to \phi_\beta(U_\alpha \cap U_\beta)$ are smooth in the sense of Definition A.1.1. The collection of coordinate charts $\{U_\alpha, \phi_\alpha\}_\alpha$ is called an* **atlas**.

Here is the corresponding notion of smooth map.

Definition A.1.5 *Let M be an m-manifold with corners and with atlas*

$$\{U_\alpha, \phi_\alpha\}_\alpha.$$

A function $f : M \to \mathbb{R}$ is called **smooth** *if the composition*

$$f_\alpha := f \circ \phi_\alpha^{-1} : \phi_\alpha(U_\alpha) \to \mathbb{R}$$

is smooth in the sense of Definition A.1.1 for every α. A function $f : M \to \mathbb{R}^n$ is called **smooth** *if its coordinate functions $f_i : M \to \mathbb{R}$ are smooth for $i = 1, \dots, n$. A function $f : M \to N$ with values in a manifold N is called* **smooth** *if its composition with an embedding $\iota : N \to \mathbb{R}^n$ is smooth.*

For these to be sensible definitions, we need to know that the corner structure of M, i.e. the stratification of its boundary into pieces of different dimension, is preserved by the coordinate changes. This is not true in the topological category; for example, Q^2 is homeomorphic to the half space $Q^1 \times \mathbb{R}$. However, the next exercise shows that it is true in the smooth category.

Exercise A.1.6 Let M be an m-manifold with corners with atlas $\{U_\alpha, \phi_\alpha\}_\alpha$. Let $p \in M$, choose indices α, β such that $p \in U_\alpha \cap U_\beta$, and define the coordinate vectors $x, y \in Q^m$ by

$$x = (x_1, \dots, x_m) := \phi_\alpha(p), \qquad y = (y_1, \dots, y_m) := \phi_\beta(p).$$

Prove that $\#\{i \mid x_i = 0\} = \#\{j \mid y_j = 0\}$.

Exercise A.1.6 shows that there is a well-defined map

$$\nu : X \to \{1, \dots, m\}$$

which assigns to each point $p \in M$ the number of components of the vector $\phi_\alpha(p) \in Q^m$ that vanish for some, and hence every, index α such that $p \in U_\alpha$. For $k \in \{0, 1, \dots, m\}$ define the set

$$\partial^k M := \{p \in M \mid \nu(p) \geq k\}.$$

Thus

$$\partial^m M \subset \partial^{m-1} M \subset \cdots \subset \partial^2 M \subset \partial^1 M \subset \partial^0 M = M.$$

The **boundary of** M is the set

$$\partial M := \partial^1 M,$$

and the **interior of** M is the set

$$\mathrm{int}(M) := M \setminus \partial M.$$

Exercise A.1.7 Prove that $\partial^k M$ is an $(m - k)$-manifold with corners.

A.2 Extension

We now show that ball embeddings have smooth extensions. Let $B(\lambda) \subset \mathbb{R}^m$ denote the closed ball of radius λ.

Theorem A.2.1 (Smooth embedding extension) *Let M be a smooth manifold without boundary and fix a constant $\lambda > 0$.*

(i) *Every embedding $\psi : B(\lambda) \to M$ extends to an embedding $\Psi : \mathbb{R}^m \to M$ whose image has compact closure.*

(ii) *Let $\psi : [0,1] \times B(\lambda) \to M$ be a smooth map such that $\psi(t,\cdot) : B(\lambda) \to M$ is an embedding for every t. Then there is a smooth extension $\Psi : [0,1] \times \mathbb{R}^m \to M$ of ψ, whose image has compact closure, such that $\Psi(t,\cdot) : \mathbb{R}^m \to M$ is an embedding for every t.*

(iii) *Let $\psi : [0,1] \times B(\lambda) \to M$ be as in (ii) and let $\Psi_0, \Psi_1 : \mathbb{R}^m \to M$ be embeddings, whose images have compact closure, such that $\Psi_i|_{B(\lambda)} = \psi(i,\cdot)$ for $i = 0,1$. Then the extension Ψ in (ii) can be chosen such that $\Psi(i,\cdot) = \Psi_i$ for $i = 0,1$.*

The proof is based on the following lemma.

Lemma A.2.2 (Smooth extension) *Fix a constant $\lambda \geq 0$.*

(i) *Every smooth function $f : B(\lambda) \to \mathbb{R}$ extends to a compactly supported smooth function $F : \mathbb{R}^m \to \mathbb{R}$.*

(ii) *Every smooth function $f : [0,1] \times B(\lambda) \to \mathbb{R}$ extends to a compactly supported smooth function $F : [0,1] \times \mathbb{R}^m \to \mathbb{R}$.*

(iii) *Let $f : [0,1] \times B(\lambda) \to \mathbb{R}$ be a smooth function and let $F_0, F_1 : \mathbb{R}^m \to \mathbb{R}$ be compactly supported smooth functions such that $F_i|_{B(\lambda)} = f(i,\cdot)$ for $i = 0,1$. Then the extension F in (ii) can be chosen such that $F(i,\cdot) = F_i$ for $i = 0,1$.*

Proof: Assertion (i) is trivial for $\lambda = 0$ (take any smooth cutoff function). To prove it for $\lambda > 0$ choose a smooth cutoff function $\beta : [0,\infty) \to [0,1]$ that satisfies (A.1.2). For $x \in \partial B(\lambda)$, define $f_x : (-\infty, 0] \to \mathbb{R}$ by

$$f_x(s) := f(e^s x)$$

for $s \leq 0$. Choose a sequence of real numbers c_k such that $c_k \geq 1 + |f_x^{(k)}(0)|$ for all $k \in \mathbb{N}_0$ and all $x \in \partial B(\lambda)$. Define $F : \mathbb{R}^m \to \mathbb{R}$ by $F|_{B(\lambda)} := f$ and

$$F(e^s x) := \sum_{k=1}^{\infty} \frac{f_x^{(k)}(0)}{k!} \bigl(\beta(c_k s) s\bigr)^k$$

for $x \in \partial B(\lambda)$ and $s \geq 0$. Then F is smooth by Lemma A.1.3 and this proves (i). The same formula proves (ii) because of the smooth dependence of F on f. To prove (iii), assume that $G : [0,1] \times \mathbb{R}^m \to \mathbb{R}$ is any compactly supported smooth extension of f and let $F_0, F_1 : \mathbb{R}^m \to \mathbb{R}$ be compactly supported smooth functions such that $F_i|_{B(\lambda)} = f(i, \cdot)$ for $i = 0, 1$. Let $\beta : [0,1] \to [0,1]$ be a smooth cutoff function that satisfies (A.1.2). Then the function

$$F(t,x) := G(t,x) + \beta(t)\big(F_0(x) - G(0,x)\big) + \big(1 - \beta(t)\big)\big(F_1(x) - G(1,x)\big)$$

satisfies (iii). This proves Lemma A.2.2. □

Proof of Theorem A.2.1: Choose any embedding of M (or of an open subset of M that contains the image of f and has compact closure) into some Euclidean space $\mathbb{R}^n$. Let $U \subset \mathbb{R}^n$ be a tubular neighbourhood of the image of the embedding and denote by $\pi : U \to M$ the projection. Use part (i) of Lemma A.2.2 to construct any smooth extension of ψ to a function with values in $\mathbb{R}^n$ and compose it with π to obtain a smooth extension $\phi : B(\lambda + \varepsilon) \to M$ for some $\varepsilon > 0$. Shrinking ε, if necessary, we may assume without loss of generality that ϕ is an embedding. Now choose a diffeomorphism

$$\rho : (0, \infty) \to (0, \lambda + \varepsilon)$$

such that $\rho(r) = r$ for $0 < r \leq \lambda$. Then the map $\Psi : \mathbb{R}^m \to M$, defined by

$$\Psi(x) := \phi\left(\rho(|x|)\frac{x}{|x|}\right)$$

for $x \in \mathbb{R}^m \setminus \{0\}$ and $\Psi(0) := \phi(0)$, is an embedding and agrees with ψ on $B(\lambda)$. The same argument, using part (ii) of Lemma A.2.2, proves assertion (ii).

We prove (iii). The same argument as in the proof of (i) and (ii), using part (iii) of Lemma A.2.2, gives rise to a smooth map $\phi : \Omega \to M$, defined on an open neighbourhood $\Omega \subset [0,1] \times \mathbb{R}^m$ of $(\{(0,1)\} \times \mathbb{R}^m) \cup ([0,1] \times B(\lambda))$, such that $\phi|_{[0,1]\times B(\lambda)} = \psi$, $\phi(i, \cdot) = \Psi_i$ for $i = 0, 1$, the image of ϕ has compact closure, and the map

$$\phi_t := \phi(t, \cdot) : \Omega_t := \{x \in \mathbb{R}^m \,|\, (t,x) \in \Omega\} \to M$$

is a embedding for every t. Now choose a smooth function $(0,1) \to (0,\infty) : t \mapsto \varepsilon_t$ such that $B(\lambda + \varepsilon_t) \subset \Omega_t$ for all t, and ε_t tends to infinity for $t \to 0$ and $t \to 1$. Then choose a smooth function

$$\bigcup_{0 \leq t \leq 1} \{t\} \times (0, \lambda + \varepsilon_t) \to (0, \infty) : (t, r) \mapsto \rho_t(r)$$

such that $\rho_t : (0,\infty) \to (0, \lambda + \varepsilon_t)$ is a diffeomorphism satisfying $\rho_t|_{(0,\lambda]} = \mathrm{id}$ for all t, and $\rho_0 = \rho_1 = \mathrm{id}$. Then the function $\Psi : [0,1] \times \mathbb{R}^m \to M$, defined by

$$\Psi(t,x) := \phi_t\left(\rho_t(|x|)\frac{x}{|x|}\right), \qquad 0 \leq t \leq 1,$$

for $x \in \mathbb{R}^m \setminus \{0\}$ and $\Psi_t(0) := \phi_t(0)$, satisfies the requirements of (iii). This proves Theorem A.2.1. □

A.3 Construction of a smooth function

The following construction is used in Section 7.1.

Lemma A.3.1 *There exists a smooth function*

$$(0,\infty)^2 \times (0,1) \to \mathbb{R} : (r,\lambda,\varepsilon) \mapsto f_{\lambda,\varepsilon}(r)$$

such that

$$f_{\lambda,\varepsilon}(r) = \begin{cases} \sqrt{\lambda^2+r^2}, & \textit{for } r \le \varepsilon, \\ r, & \textit{for } r \ge \lambda+\varepsilon, \end{cases} \qquad f'_{\lambda,\varepsilon}(r) > 0 \tag{A.3.1}$$

for all $\varepsilon \in (0,1)$ *and all* $\lambda, r > 0$

Consider first the case $\lambda = 1$ and fix a constant $\varepsilon > 0$. Then it is a standard problem in analysis to construct a function $f_{1,\varepsilon} : (0,\infty) \to \mathbb{R}$ that satisfies (A.3.1). Moreover, the set of all such functions is a convex open set in the set of all smooth functions $f : (0,\infty) \to \mathbb{R}$ that satisfy the constraints $f(r) = \sqrt{1+\varepsilon^2}$ for $r \le \varepsilon$ and $f(r) = r$ for $r \ge 1+\varepsilon$. By varying $\varepsilon \in (0,1)$ and using this convexity, one can prove the existence of a smooth function $(r,\varepsilon) \mapsto f_{1,\varepsilon}(r)$ that satisfies (A.3.1) for $\lambda = 1$, and then, by varying λ similarly, for all $\lambda > 0$. Below we construct the functions $f_{\lambda,\varepsilon}$ explicitly.

Exercise A.3.2 Complete the details of the above proof.

Proof of Lemma A.3.1: The explicit construction of $f_{\lambda,\varepsilon}$ requires two preparatory steps.

Step 1. *There is a smooth function* $\mathbb{R} \times (4,\infty) \to \mathbb{R} : (t,c) \mapsto \alpha_c(t)$ *such that*

$$\alpha_c(t) = \begin{cases} 0, & \textit{for } t \notin [0,c], \\ t+1, & \textit{for } t \in [1,c-3], \end{cases} \qquad 0 \le \alpha_c(t) \le t+3, \qquad \int_0^c \alpha_c = \frac{c^2}{2}. \tag{A.3.2}$$

Let $\phi : [0,1] \to [0,2]$ and $\psi : [-1,1] \to [-1,1]$ be smooth functions such that

$$\phi(t) = \begin{cases} 0, & \text{for } t \le 1/2, \\ t+1, & \text{for } t \ge 3/4, \end{cases} \qquad \psi(-t) = -\psi(t), \qquad \psi|_{[1/2,1]} \equiv 1, \tag{A.3.3}$$

and define

$$\alpha_c(t) := \begin{cases} \phi(t), & \text{for } 0 \le t \le 1, \\ t+1, & \text{for } 1 \le t \le c-3, \\ c - \phi(c-2-t), & \text{for } c-3 \le t \le c-2, \\ \frac{c}{2}(1-\psi(t+1-c)), & \text{for } c-2 \le t \le c. \end{cases} \tag{A.3.4}$$

The map $(c,t) \mapsto \alpha_c(t)$ is well-defined and smooth and satisfies the first two equations in (A.3.2). Moreover,

$$\int_1^{c-3} \alpha_c = \frac{c^2}{2} - 2c, \qquad \int_{c-3}^{c-2} \alpha_c = c - \int_0^1 \phi, \qquad \int_{c-2}^{c} \alpha_c = c,$$

and hence α_c also satisfies the integral condition in (A.3.2). This proves Step 1.

Step 2. *There exists a smooth function*

$$(0,\infty)\times(0,1)\to\mathbb{R}:(r,\varepsilon)\mapsto\beta_\varepsilon(r)$$

such that

$$\beta_\varepsilon(r)=\begin{cases}1+r^2, & \text{for } r\le\varepsilon,\\ r^2, & \text{for } r\ge 1+\varepsilon,\end{cases}\qquad \beta_\varepsilon'(r)>0 \tag{A.3.5}$$

for all $\varepsilon\in(0,1)$ *and all* $r>0$.

Let α_c be as in Step 1 and define

$$\beta_\varepsilon(r):=1+r^2-\frac{\varepsilon}{2}\int_0^r\alpha_{4/\varepsilon}\left(4\frac{s-\varepsilon}{\varepsilon}\right)ds \tag{A.3.6}$$

for $0<\varepsilon<1$ and $r>0$. If $0<r\le\varepsilon$ then the integral vanishes and so $\beta_\varepsilon(r)=1+r^2$. Moreover, since $\alpha_{4/\varepsilon}(t)\le t+3$ for every $t\in\mathbb{R}$, we have

$$\beta_\varepsilon'(r)=2r-\frac{\varepsilon}{2}\alpha_{4/\varepsilon}\left(4\frac{r-\varepsilon}{\varepsilon}\right)\ge 2r-\frac{\varepsilon}{2}\left(\frac{4r}{\varepsilon}-1\right)=\frac{\varepsilon}{2}.$$

This shows that $\beta_\varepsilon'(r)>0$ for $\varepsilon\le r\le 1+\varepsilon$ and $\beta_\varepsilon'(r)=2r$ for $r\ge 1+\varepsilon$. Moreover, it follows from equation (A.3.2) in Step 1 that

$$\begin{aligned}\beta_\varepsilon(1+\varepsilon)&=1+(1+\varepsilon)^2-\frac{\varepsilon}{2}\int_\varepsilon^{1+\varepsilon}\alpha_{4/\varepsilon}\left(4\frac{s-\varepsilon}{\varepsilon}\right)ds\\&=1+(1+\varepsilon)^2-\frac{\varepsilon^2}{8}\int_0^{4/\varepsilon}\alpha_{4/\varepsilon}(t)\,dt\\&=(1+\varepsilon)^2.\end{aligned}$$

Hence $\beta_\varepsilon(r)=r^2$ for $r\ge 1+\varepsilon$. This proves Step 2.

Let β_ε be as in Step 2. Then the function

$$f_{\lambda,\varepsilon}(r):=\lambda\sqrt{\beta_{\varepsilon/\lambda}(r/\lambda)}$$

satisfies the requirements of Lemma A.3.1. □

REFERENCES

[1] Abbondandolo, A., Majer, P. (2014). A nonsqueezing theorem for convex images of the Hilbert ball, hrrp://arxiv.org/abs/1405.3200

[2] Abouzaid, M. (2012). Nearby Lagrangians with vanishing Maslov class are homotopy equivalent. *Inventiones Mathematicae*, **189**, no. 2, 25–313. hrrp://arxiv.org/abs/1005.0358

[3] Abouzaid, M. (2012). Framed bordism and Lagrangian embeddings of exotic spheres. *Annals of Mathematics*, **175**, 71–185. http://arxiv.org/abs/0812.4781

[4] Abouzaid, M. (2014). Family Floer homology and mirror symmetry. http://arxiv.org/abs/1404.2659

[5] Abouzaid, M., Kragh, T. (2016). On the immersion classes of nearby Lagrangians. *Journal of Topology*, **9** no. 1, 232–244. http://arxiv.org/abs/1305.6810

[6] Abouzaid, M., Seidel, P. (2010). Altering symplectic manifold by homologous recombination. http://arxiv.org/abs/1007.3281v3

[7] Abraham, R., Marsden, J. (1978). *Foundations of Mechanics*. Second edition. Addison-Wesley, Reading.

[8] Abreu, M. (1998). Topology of symplectomorphism groups of $S^2 \times S^2$. *Inventiones Mathematicae*, **131**, 1–24.

[9] Abreu, M. (1998). Kähler geometry of toric varieties and extremal metrics. *International Journal of Mathematics*, **9**, 641–651. http://arxiv.org/abs/dg-ga/9711014

[10] Abreu, M., Granja, G., Kitchloo, N. (2005). Moment maps, symplectomorphism groups and compatible complex structures. *Journal of Symplectic Geometry*, **3**, no. 4, 655–670.

[11] Abreu, M., Macarini, L. (2013). Remarks on Lagrangian intersections in toric manifolds. *Transactions of the American Mathematical Society*, **365**, no. 7, 3851–3875. http://arxiv.org/abs/1105.0640

[12] Abreu, M., McDuff, D. (2000). Topology of symplectomorphism groups of rational ruled surfaces. *Journal of the American Mathematical Society*, **13**, 971–1009.

[13] Aebischer, B. *et al.* (1994). *Symplectic Geometry*. Progress in Mathematics **124**, Birkhäuser, Basel.

[14] Ahara, K., Hattori, A. (1991). 4-dimensional symplectic S^1 manifolds admitting moment map. *Journal of the Faculty of Science, University of Tokyo, Section 1A, Mathematics*, **38**, 251–298.

[15] Akhmedov, A., Park, B.D. (2008). Exotic smooth structures on small 4-manifolds. *Inventiones Mathematicae*, **173**, 209–223.

[16] Akhmedov, A., Park, B.D. (2010). Exotic smooth structures on small 4-manifolds with odd signatures. *Inventiones Mathematicae*, **181**, 577–603.

[17] Akhmedov, A., Park, B.D., Urzúa, G. (2010). Spin symplectic 4-manifolds near Bogomolov-Miyaoka-Yau line. *Journal of Gökova Geometry Topology*, **4**, 55–66. http://gokovagt.org/journal/2010/jggt10-akhmparkurzu.pdf

[18] Akveld, M., Salamon, D. (2001). Loops of Lagrangian submanifolds and pseudoholomorphic disks, *Geometric and Functional Analysis*, **11**, 609–650. http://www.math.ethz.ch/~salamon/PREPRINTS/meike.pdf

[19] Albers, P., Bramham, B., Wendl, C. (2010). On non-separating contact hypersurfaces in symplectic 4-manifolds. *Algebraic & Geometric Topology*, **10**, 697–737.

[20] Alkoumi, A., Schlenk, F. (2015). Shortest closed billiard orbits on convex tables. *Manuscripta Mathematica*, **147**, no. 3-4, 365–380. http://arxiv.org/abs/math/1408.5255

[21] Alvarez, J.-C. (1995). The symplectic geometry of spaces of geodesics. PhD thesis, Rutgers.

[22] Anjos, S., Granja, G. (2004). Homotopy decomposition of a group of symplectomorphisms of $S^2 \times S^2$. *Topology*, **43**, 599–618.

[23] Arnold, V.I. (1978). *Mathematical Methods in Classical Mechanics.* Springer-Verlag, Berlin.

[24] Arnold, V.I. (1986). First steps in symplectic topology. *Russian Mathematical Surveys*, **41**, 1–21.

[25] Arnold, V.I., Givental, A. (1990). *Symplectic Topology. Dynamical Systems* **IV**. Springer-Verlag, Berlin.

[26] Atiyah, M.F. (1982). Convexity and commuting Hamiltonians. *Bulletin of the London Mathematical Society*, **14**, 1–15.

[27] Atiyah, M.F., Bott, R. (1982). The Yang Mills equations over Riemann surfaces. *Philosophical Transactions of the Royal Society of London* A, **308**, 523–615.

[28] Atiyah, M.F., Bott, R. (1984). The moment map and equivariant cohomology. *Topology*, **23**, 1–28.

[29] Audin, M. (1988). Fibrés normaux d'immersions en dimension Moité, points doubles d'immersions Lagrangiennes et plongements totalement réels. *Commentarii Mathematici Helvetici*, **63**, 593–623.

[30] Audin, M. (1990). Hamiltoniens periodiques sur les variétés symplectiques compactes de dimension 4. In *Géométrie symplectique et mechanique, Proceedings (1988)* (ed. C. Albert). Lecture Notes in Mathematics **1416**, Springer-Verlag, Berlin, pp. 1–25.

[31] Audin, M. (1991). Examples de variétés presques complexes. *L'Enseignement Mathématique*, **37**, 175–190.

[32] Audin, M. (1991). *The Topology of Torus Actions on Symplectic Manifolds*. Birkhäuser, Basel.

[33] Audin, M., Damian, M. (2010). *Théorie de Morse et Homologie de Floer*. Savior Actuels, EDP Sciences, CNRS Éditions, Paris. English translation: Universitext, Springer-Verlag, Berlin.

[34] Audin, M., Lafontaine, F. (ed.) (1994). *Holomorphic Curves in Symplectic Geometry*. Progress in Mathematics **117**, Birkhäuser, Basel.

[35] Audin, M., Lalonde, F., Polterovich, L. (1994). Symplectic rigidity: Lagrangian submanifolds. In [34], pp. 271–321.

[36] Auroux, D. (1997). Asymptotically holomorphic families of symplectic submanifolds. *Geometric and Functional Analysis*, **7**, 971–995.

[37] Auroux, D. (2005). A stable classification of Lefschetz fibrations. *Geometry & Topology*, **9**, 203–217. http://arxiv.org/abs/math/0412120

[38] Auroux, D. (2007). Mirror symmetry and T-duality in the complement of the anticanonical divisor. *Journal of Gökova Geometry Topology*, **1**, 51–91. http://arxiv.org/abs/0706.3207

[39] Auroux, D. (2009). Special Lagrangian fibrations, wall-crossing, and mirror symmetry. *Surveys in Differential Geometry*, Volume XIII, edited by H.D. Cao and S.T. Yau, International Press, 2009, pp 1–47. http://arxiv.org/abs/0902.1595v1

[40] Auroux, D. (2005). Some open questions about symplectic 4-manifolds, singular plane curves, and braid group factorizations. *Proceedings of the 4th European Congress of Mathematics*, Stockholm, 27 June – 2 July 2004, EMS 2005, pp 23–40. http://arxiv.org/abs/math/0410119

[41] Auroux, D. (2015). Factorizations in $\mathrm{SL}(2,\mathbb{Z})$ and simple examples of inequivalent Stein fillings. *Journal of Symplectic Geometry*, **13** no. 2, 261–277. http://arxiv.org/abs/1311.0847.

[42] Auroux, D. (2015). Infinitely many monotone Lagrangian tori in $\mathbb{R}^6$. *Inventiones Mathematicae*, **201** no. 3, 909–924. http://arxiv.org/abs/1407.3725.

[43] Auroux, D., Katzarkov, L. (2008). A degree doubling formula for braid monodromies and Lefschetz pencils. *Pure and Applied Mathematics Quarterly*, **4**, no. 2, part 1, 237–318.

[44] Auroux, D., Donaldson, S.K., Katzarkov, L. (2003). Luttinger surgery along Lagrangian tori and non-isotropy for singular symplectic plane curves. *Mathematische Annalen*, **326**, 185–203. http://arxiv.org/abs/math/0206005

[45] Auroux, D., Donaldson, S.K., Katzarkov, L. (2005). Singular Lefschetz pencils. *Geometry & Topology*, **9**, 1043–1114. http://arxiv.org/abs/math/0410332
[46] Artstein–Avidan, K., Karasev, R., Ostrover, Ya. (2014). From symplectic measurements to the Mahler conjecture. *Duke Mathematical Journal*, **163** no. 11, 2003–2022. http://arxiv.org/abs/1303.4197
[47] Baldridge, S., Kirk, P. (2008). A symplectic manifold homeomorphic but not diffeomorphic to $\mathbb{C}\mathrm{P}^2\#3\overline{\mathbb{C}\mathrm{P}}^2$. *Geometry & Topology*, **12**, 919–940.
[48] Baldridge, S., Kirk, P. (2013). Coisotropic Luttinger surgery and some new symplectic 6-manifolds Calabi-Yau 6-manifolds. *Indiana Univ. Math. J.*, **62** no. 5, 1457–1471.
[49] Baldridge, S., Li, T.J. (2005). Geography of symplectic 4-manifolds with Kodaira dimension one. *Algebraic & Geometric Topology*, **5**, 355–368. http://arxiv.org/abs/math/0505030
[50] Banyaga, A. (1978). Sur la structure du groupe des difféomorphismes qui préservent une forme symplectique. *Commentarii Mathematici Helvetici*, **53**, 174–227.
[51] Banyaga, A. (1997). *The Structure of Classical Diffeomorphism Groups.* Mathematics and its Applications, **400**. Kluwer, Dordrecht.
[52] Barge, J., Ghys, É. (1988), Surfaces et cohomologie borné, *Inventiones Mathematicae*, **92**, 509–526.
[53] Barth, W., Peters, C., Van de Ven, A. (1984). *Compact Complex Surfaces.* Springer-Verlag, Berlin.
[54] Bates, L., Peschke, G. (1990). A remarkable symplectic structure. *Journal of Differential Geometry*, **32**, 533–538.
[55] Bates, S. (1994). Symplectic end invariants and C^0 symplectic topology. Ph.D. thesis, University of California, Berkeley.
[56] Bates, S. (1994). Entropy dimension of action spectra. Preprint.
[57] Bates, S. (1994). On certain ω-convex hypersurfaces of $\mathbb{R}^{2n}$. Preprint.
[58] Bates, S. (1995). Some simple continuity properties of symplectic capacities. In [319], pp. 185–195.
[59] Bates, S., Weinstein, A. (1993). *Lectures on Geometric Quantization.* Berkeley Lecture Notes.
[60] Bennequin, D. (1983). Entrelacements et équations de Pfaff. *Astérisque*, **107–8**, 87–161.
[61] Berline, M., Vergne, M. (1983). Zeros d'un champ de vecteurs et classes caracteristiques equivariantes. *Duke Mathematical Journal*, **50**, 539–549.
[62] Bers, L., John, F., Schechter, M. (1964). *Partial Differential Equations.* Interscience, New York.
[63] Bhupal, M. (1998). Legendrian intersections in the 1-jet bundle. Ph.D. thesis, University of Warwick.
[64] Bialy, M., Polterovich, L. (1994). Geodesics of Hofer's metric on the group of Hamiltonian diffeomorphisms. *Duke Mathematical Journal*, **76**, 273–92
[65] Biran, P. (1996). Connectedness of spaces of symplectic embeddings. *International Mathematics Research Notices*, **7** 487–491.
[66] Biran, P. (1997). Symplectic packing in dimension 4. *Geometric and Functional Analysis*, **7**, 420–437.
[67] Biran, P. (1999). A stability property of symplectic packing. *Inventiones Mathematicae*, **136** (1999), 123–155.
[68] Biran, P. (2001). Lagrangian barriers and symplectic embeddings. *Geometric and Functional Analysis*, **11**, 407–469.
[69] Biran, P. (2006). Lagrangian Non-Intersections. *Geometric and Functional Analysis*, **16**, 279–326.
[70] Biran, P., Cieliebak, K. (2001). Symplectic topology on subcritical manifolds. *Commentarii Mathematici Helvetici*, **76**, 712–753.
[71] Biran, P., Cornea, O. (2009). Rigidity and uniruling for Lagrangian submanifolds. *Geometry & Topology* **13**, 2881–2989. http://arxiv.org/abs/0808.2440
[72] Biran, P., Khanevsky, M. (2013). A Floer–Gysin exact sequence for Lagrangian submanifolds. *Commentarii Mathematici Helvetici*, **88**, 899–952.

[73] Biran, P., Entov, M., Polterovich, L. (2004). Calabi quasimorphisms of the symplectic ball. *Communications in Contemporary Mathematics*, **6**, 793–802.

[74] Birkhoff, G.D. (1925). An extension of Poincaré's last geometric theorem. *Acta Mathematica*, **47**, 297–311.

[75] Borman, M. S., Eliashberg, Ya., Murphy, E. (2015). Existence and classification of overtwisted contact structures in all dimensions. *Acta Mathematica*, **215** no. 2, 281–361. Preprint. `http://arxiv.org/abs/1404.6157`

[76] Borman, M. S., Li, T.-J., Wu, W. (2014). Spherical Lagrangians via ball packings and symplectic cutting. *Selecta Mathematica*, **20**, 261–283. `http://arxiv.org/abs/1211.5952`

[77] Bott, R. (1982). Lectures on Morse theory, old and new. *Bulletin of the American Mathematical Society*, **7**, 331–358.

[78] Bott, R., Tu, L. (1982). *Differential Forms in Algebraic Topology.* Graduate Texts in Mathematics, **82**. Springer-Verlag, Berlin.

[79] Bourgeois, F., Oancea, A. (2012). S^1-equivariant symplectic homology and linearized contact homology. Preprint. `http://arxiv.org/abs/1212.3731`

[80] Brown, M., Neumann, W.D. (1977). Proof of the Poincaré Birkhoff fixed point theorem. *Michigan Mathematical Journal*, **24**, 21–31.

[81] Buhovski, L. (2010). The Maslov class of Lagrangian tori and quantum products in Floer cohomology. *Journal of Topology and Analysis*, **2**, 57–75. `http://arxiv.org/abs/math/0608063`

[82] Buhovski, L. (2010). The 2/3 convergence rate for the Poisson bracket. Preprint. *Geometric and Functional Analysis*, **19** (2010), 1620-1649.

[83] Buhovski, L. (2013). Towards the C^0 flux conjecture. Preprint. `http://arxiv.org/pdf/1309.1325.pdf`

[84] Buhovski, L., Opshtein, E. (2014). Some quantitative results in C^0 symplectic topology. *Algebraic & Geometric Topology*, **14** no. 6, 3493–3508. `http://arxiv.org/pdf/1404.0875.pdf`

[85] Buhovski, L., Seyfaddini, S. (2013). Uniqueness of generating Hamiltonians for continuous Hamiltonian flows. *Journal of Symplectic Geometry*, **11**, 1–161.

[86] Buhovsky, L., Entov, M., Polterovich, L. (2012). Poisson brackets and symplectic invariants. *Selecta Mathematica*, **12**, 89-157.

[87] Burago, D., Ivanov, S., Polterovich, L. (2008). Conjugation-invariant norms on groups of geometric origin. *Groups of diffeomorphisms*, 221–250 *Advanced Studies in Pure Mathematics* **52**, Math. Soc. Japan, Tokyo.

[88] Buse, O. (2011). Negative inflation and stability in symplectomorphism groups of ruled surfaces. *Journal of Symplectic Geometry*, **9** no. 2, 147-160.

[89] Buse, O., Hind, R. (2013). Ellipsoidal embeddings and symplectic packing stability, *Compositio Mathematicae*, **149**, 889–902.

[90] Buse, O., Hind, R., Opshtein, E. (2016). Packing stability for symplectic 4-manifolds. *Transactions of the American Mathematical Society*, **368** no. 11, 8209–8222. `http://arxiv.org/abs/1404.4183`

[91] Calabi, E. (1958). Construction and properties of some 6-dimensional almost complex manifolds. *Transactions of the American Mathematical Society*, **87**, 407–438.

[92] Calabi, E. (1970). On the group of automorphisms of a symplectic manifold. In *Problems in analysis* (ed. R. Gunning), pp. 1–26. Princeton University Press.

[93] Cardin, F., Viterbo C. (2008). Commuting Hamiltonians and Hamilton-Jacobi multi-time equations. *Duke Mathematical Journal*, **144**, 235–284.

[94] Catanese, F., Tian, G. eds. (2008). *Symplectic 4-manifolds and Algebraic Surfaces.* CIME Summer school, Centraro, Italy 2003, Lecture Notes in Mathematics **1938**, Springer-Verlag, Berlin.

[95] Candelas, P., Ossa, X.C. de la (1990). Moduli space of Calabi–Yau manifolds. University of Texas Report UTTG-07.

[96] Chaperon, M. (1983). Quelques questions de Géométrie symplectique. *Seminaire Bourbaki. Astérisque*, **105–6**, 231–250.

[97] Chaperon, M. (1984). Une idée du type 'géodesiques brisées'. *Comptes Rendues de l'Academie des Sciences*, **298**, 293–296.

[98] Chaperon, M. (1996). On generating families. In [319], pp. 283–296.

[99] Chekanov, Yu. (1996). Lagrangian tori in symplectic vector spaces and global symplectomorphisms. *Mathematische Zeitschrift* **335**, 547–559.

[100] Chekanov, Yu. (1996). Hofer's symplectic energy and Lagrangian intersections. In [620], pp. 296–306.

[101] Chekanov, Yu. (1996). Critical points of quasi-functions and generating families of Legendrian manifolds. *Functional Analysis and its Applications*, **30**, 118–128.

[102] Chekanov, Yu. (2000). Invariant Finsler metrics on the space of Lagrangian embeddings. *Mathematische Zeitschrift*, **234**, (2000), 605–619.

[103] Chekanov, Yu., Schlenk, F. (2010). Notes on monotone Lagrangian twist tori, *Electron. Res. Announc. Math. Sci.* **17**, 104–121.

[104] Chen, B., Li, An-Min, Wang, Bai-Ling. (2013). Virtual neighbourhood technique for pseudo-holomorphic spheres. Preprint. `http://arxiv.org/abs/1306.3276`

[105] Chen, X.X., Donaldson, S.K., Sun, S. (2014). Kähler-Einstein metrics and stability. *International Mathematics Research Notices*, no 8, 2119–2125. `http://arxiv.org/abs/1210.7494`.

[106] Chen, X.X., Donaldson, S.K., Sun, S. (2015). Kähler-Einstein metrics on Fano manifolds, I: approximation of metrics with cone singularities. *Journal of the American Mathematical Society*, **28** no. 1, 183–197. `http://arxiv.org/abs/1211.4566`

[107] Chen, X.X., Donaldson, S.K., Sun, S. (2015). Kähler-Einstein metrics on Fano manifolds, II: limits with cone angle less than 2π. *Journal of the American Mathematical Society*, **28** no. 1, 199–234. `http://arxiv.org/abs/1212.4714`

[108] Chen, X.X., Donaldson, S.K., Sun, S. (2015). Kähler-Einstein metrics on Fano manifolds, III: limits as cone angle approaches 2π and completion of the main proof. *Journal of the American Mathematical Society*, **28** no. 1, 235–278. `http://arxiv.org/abs/1302.0282`

[109] Cho, Y., Hwang, T., Suh, Dong Yup. (2015). Semifree Hamiltonian circle actions on 6-dimensional symplectic manifolds with non-isolated fixed point set. *Journal of Symplectic Geometry*, **13**, no. 4, 963–1000. `http://arxiv.org/abs/1005.0193`

[110] Cieliebak, K. (1997). Symplectic boundaries: creating and destroying closed characteristics. *Geometric and Functional Analysis*, **7**, 269–321.

[111] Cieliebak, K., Eliashberg, Y. (2012). *From Stein to Weinstein and Back. Symplectic Geometry of Affine Complex Manifolds.* American Mathematical Society Colloquium Publications **59**, American Mathematical Society , Providence, RI.

[112] Cieliebak, K., Floer, A., Hofer, H. (1995). Symplectic homology II: a general construction. *Mathematische Zeitschrift*, **218**, 103–122.

[113] Cieliebak, K. Floer, A., Hofer, H., Wysocki, K. (1996). Applications of symplectic homology II: stability of the action spectrum. *Mathematische Zeitschrift*, **223**, 27–45.

[114] Cieliebak, K., Gaio, A.R., Salamon, D.A. (2000). J-holomorphic curves, moment maps, and invariants of Hamiltonian group actions. *International Mathematics Research Notices*, **10**, 831–882.

[115] Cieliebak, K., Gaio, A.R., Mundet i Riera, I., Salamon, D.A. (2002). The symplectic vortex equations and invariants of Hamiltonian group actions. *Journal of Symplectic Geometry*, **1**, 543–645.

[116] Cieliebak, K., Hofer, H., Latschev, J., Schlenk, F. (2007). Quantitative symplectic geometry. *Dynamics, Ergodic Theory, Geometry MSRI*, **54** (2007), 1–44. `http://arxiv.org/abs/0506191`

[117] Cieliebak, K., Mohnke, K. (2014). Punctured holomorphic curves and Lagrangian embeddings. Preprint. `http://arxiv.org/abs/1411.1870v1`

[118] Cieliebak, K., Mundet-i-Riera, I., Salamon, D.A. (2003). Equivariant moduli problems, branched manifolds, and the Euler class. *Topology* **42** (2003), 641–700.

[119] Clemens, H., Kollar, J., Mori, S. (1989). Higher-dimensional complex geometry. *Astérisque*, **166**, 144 pages.

[120] Colin, V., Ghiggini, P., Honda, K. (2012). The equivalence of Heegaard Floer homology

and embedded contact homology via open book decompositions I. Preprint, 2012. http://arxiv.org/abs/1208.1074.

[121] Colin, V., Ghiggini, P., Honda, K. (2012). The equivalence of Heegaard Floer homology and embedded contact homology via open book decompositions II. Preprint, 2012. http://arxiv.org/abs/1208.1077.

[122] Colin, V., Ghiggini, P., Honda, K. (2012). The equivalence of Heegaard Floer homology and embedded contact homology via open book decompositions III: from hat to plus. Preprint. http://arxiv.org/abs/1208.1526.

[123] Conley, C. (1978). *Isolated Invariant Sets and the Morse Index*. CBMS Notes, **38**, American Mathematical Society, Providence, RI.

[124] Conley, C., Zehnder, E. (1983). The Birkhoff–Lewis fixed point theorem and a conjecture of V.I. Arnold. *Inventiones Mathematicae*, **73**, 33–49.

[125] Conley, C., Zehnder, E. (1984). Morse type index theory for flows and periodic solutions for Hamiltonian systems. *Communications on Pure and Applied Mathematics*, **37**, 207–253.

[126] Conolly, F., Lê, H.V., Ono, K. (1997). Almost complex structures which are compatible with Kähler or symplectic structures. *Annals of Global Analysis and Geometry*, **15**, 325–334.

[127] Cordero, L., Fernández, M., Gray, A. (1986). Symplectic manifolds with no Kähler structure. *Topology*, **25**, 375–80.

[128] Courant, R., Hilbert, D. (1962). *Methods of Mathematical Physics II: Partial Differential Equations*. Interscience, New York.

[129] Courte, S. (2014). Contact manifolds with symplectomorphic symplectizations. *Geometry & Topology*, **18** no. 1, 1–15. http://arxiv.org/abs/1212.5618

[130] Cristofaro-Gardiner, D., Hutchings, M. (2016). From one Reeb orbit to two. *Journal of Differential Geometry*, **102** no. 1, 25–36. http://arxiv.org/abs/1202.4839.

[131] Damian, M. (2012). Floer homology on the universal cover, a proof of Audin's conjecture, and other constraints on Lagrangian submanifolds. *Commentarii Mathematici Helvetici* **87**, 433–463. http://arxiv.org/abs/1006.3398

[132] Delzant, T. (1988). Hamiltoniens périodiques et image convexe de l'application moment. *Bulletin de la Société Mathmatique de France*, **116**, 315–339.

[133] Demazure, M. (1980). Surfaces de del Pezzo II–V. *Séminar sur les singularités de surfaces (1976–1977)*, Lecture Notes in Mathematics **777**, Springer-Verlag, Berlin.

[134] De Silva, V. (1998). Products in symplectic Floer homology of Lagrangian intersections. PhD Thesis, University of Oxford, 1998.

[135] De Silva, V., Robbin, J., Salamon D. (2014). *Combinatorial Floer Homology.* Memoirs of the American Mathematical Society. http://arxiv.org/abs/1205.0533

[136] Dimitroglou Rizell, G. (2013). Exact Lagrangian caps and non-uniruled Lagrangian caps. Preprint. http://arxiv.org/abs/1306.4667v5

[137] Dimitroglou Rizell, G., Evans, J. (2012). Unlinking and unknottedness of monotone Lagrangian submanifolds. Preprint. http://arxiv.org/abs/1211.6633

[138] Dimitroglou Rizell, G., Evans, J. (2014). Exotic spheres and the topology of symplectomorphism groups. Preprint. http://arxiv.org/abs/1407.3173

[139] Donaldson, S.K. (1983). A new proof of a theorem of Narasimhan and Seshadri. *Journal of Differential Geometry*, **18**, 269–277.

[140] Donaldson, S.K. (1985). Anti-self-dual Yang-Mills connections on complex algebraic surfaces and stable vector bundles. *Proceedings of the London Mathematical Society*, **50**, 1–26.

[141] Donaldson, S.K. (1987). Irrationality and the h-cobordism conjecture. *Journal of Differential Geometry*, **26**, 141–168.

[142] Donaldson, S.K. (1987). The Orientation of Yang–Mills Moduli Spaces and Four-Manifold Topology. *Journal of Differential Geometry*, **26**, 397–428.

[143] Donaldson, S.K. (1990). Polynomial Invariants of Smooth Four-Manifolds. *Topology*, **29**, 257–315.

[144] Donaldson, S.K. (1990). Yang–Mills invariants of four manifolds. In *Geometry of low-dimensional manifolds*, Vol. 1: *Gauge theory and algebraic surfaces* (ed. S.K. Donaldson and C.B. Thomas), London Mathematical Society Lecture Notes **150**, Cambridge University Press, pp. 5–40.
[145] Donaldson, S.K. (1992). Boundary value problems for Yang-Mills fields. *Journal of Differential Geometry*, **28**, 89–122.
[146] Donaldson, S.K. (1996). Symplectic submanifolds and almost complex geometry. *Journal of Differential Geometry*, **44**, 666–705.
[147] Donaldson, S.K. (1999). Symmetric spaces, Kähler geometry, and Hamiltonian dynamics. *American Mathematical Society Translations - Series 2*, **196**, 13–33.
[148] Donaldson, S.K. (1999). Lefschetz pencils on symplectic manifolds. *Journal of Differential Geometry*, **53**, 205–236.
[149] Donaldson, S.K. (1999). Moment Maps and Diffeomorphisms. *Asian Journal of Mathematics*, **3**, 1–16. *Surveys in Differential Geometry*, Volume **VII**, *Papers dedicated to Atiyah, Bott, Hirzebruch, and Singer*, edited by S.-T. Yau, International Press, 2000, pp. 107–127. `http://bogomolov-lab.ru/G-sem/AJM-3-1-001-016.pdf`
[150] Donaldson, S.K. (2002). *Floer Homology Groups in Yang–Mills Theory*. Cambridge Tracts in Mathematics **147**, Cambridge University Press 2002.
[151] Donaldson, S.K. (2002). Conjectures in Kähler geometry. *Strings and Geometry*, Proceedings of the Clay Mathematics Institute 2002 Summer School on Strings and Geometry, Isaac Newton Institute, Cambridge, United Kingdom, March 24–April 20, 2002. *Clay Mathematical Proceedings* **3**, American Mathematical Society 2004, pp 71–78. `http://www2.imperial.ac.uk/~skdona/CLAY.PDF`
[152] Donaldson, S.K. (2002). Scalar curvature and stability of toric varieties. *Journal of Differential Geometry*, **62** (2002), 289–349.
[153] Donaldson, S.K. (2003). Moment maps in differential geometry. *Surveys in Differential Geometry*, Volume **VIII**. *Lectures on geometry and topology, held in honor of Calabi, Lawson, Siu, and Uhlenbeck*, edited by S.-T. Yau, International Press, 2003, pp 171–189. `http://www2.imperial.ac.uk/~skdona/donaldson-ams.ps`
[154] Donaldson, S.K. (2004). Lefschetz pencils and the mapping class group. *Problems on Mapping Class Groups and Related Topics*, edited by Benson Farb, Proceedings of symposia in pure mathematics **74**, American Mathematical Society, 2006, pp 151–163. `http://www2.imperial.ac.uk/~skdona/MCGROUP.PDF`
[155] Donaldson, S.K. (2005). What is a pseudoholomorphic curve? *Notices of the American Mathematical Society*, October 2005, 1026–1027. `http://www.ams.org/notices/200509/what-is.pdf`
[156] Donaldson, S.K. (2006). Two-forms on four-manifolds and elliptic equations. *Inspired by S. S. Chern*, edited by Phillip A. Griffiths, Nankai Tracts Mathematics **11**, World Scientific, 153–172. `http://arxiv.org/abs/math/0607083`
[157] Donaldson, S.K. (2009). Lie algebra theory without algebra. In *"Algebra, Arithmetic, and Geometry, In honour of Yu. I. Manin"*, edited by Yuri Tschinkel and Yuri Zarhin, Progress in Math. **269**, Birkhäuser, 249–266. `http://arxiv.org/abs/math/0702016`
[158] Donaldson, S.K. (2014). The Ding functional, Berndtsson convexity, and moment maps. Preprint, October 2014.
[159] Donaldson, S.K., Fine, J. (2006). Toric anti-self-dual 4-manifolds via complex geometry. *Mathematische Annalen*, **336**, 281–309. `http://arxiv.org/abs/math/0602423`
[160] Donaldson, S.K., Kronheimer, P. (1990). *The Geometry of Four-Manifolds*. Clarendon Press, Oxford.
[161] Donaldson, S.K., Segal, E. (2011). Gauge theory in higher dimensions II. *Surveys in differential geometry*, Volume **XVI**, Geometry of special holonomy and related topics, pp. 1–41. MR2893675
[162] Donaldson, S.K., Smith, I. (2003). Lefschetz pencils and the canonical class for symplectic 4-manifolds. *Topology*, **42**, 743–785.
[163] Donaldson, S.K., Thomas, R.P. (1998). Gauge theory in higher dimensions. *The Geometric Universe (Oxford, 1996)* Oxford University Press, 31–47.

[164] Dostoglou, S., Salamon, D. (1993). Instanton homology and symplectic fixed points. *Symplectic Geometry*, edited by D. Salamon, LMS Lecture Notes Series **192**, Cambridge University Press, Cambridge, pp. 57–93.

[165] Dostoglou, S., Salamon, D. (1994). Cauchy–Riemann operators, self-duality, and the spectral flow. *First European Congress of Mathematics, Volume I, Invited Lectures (Part 1)*, edited by Joseph et al, Birkhäuser, Progress in Mathematics **119**, 1994, pp. 511–545.

[166] Dostoglou, S., Salamon, D. (1994). Self-dual instantons and holomorphic curves, *Annals of Mathematics* **139** (1994), 581–640. Corrigendum, *Annals of Mathematics* **165** (2007), 665–673.

[167] Dubrovin, B. (2014). Gromov–Witten invariants and integrable hierarchies of topological type. in *Topology, Geometry, Integrable Systems, and Mathematical Physics,* pp. 141?171, American Mathematical Society Translations Series 2, **234**, American Mathematical Society, Providence, RI, 1999. http://arxiv.org/abs/1312.0799v2

[168] Duistermaat, J.J., Heckman, G.J. (1982). On the variation in the cohomology of the symplectic form of the reduced phase space. *Inventiones Mathematicae*, **69**, 259–269.

[169] Duncan, David L. (2013). Higher rank instanton cohomology and the quilted Atiyah–Floer conjecture. Preprint. http://arxiv.org/abs/1311.5609v2

[170] Ekeland, I. (1989). *Convexity Methods in Hamiltonian Mechanics.* Ergebnisse Math **19**, Spinger-Verlag, New York.

[171] Ekeland, I., Hofer, H. (1989). Symplectic topology and Hamiltonian dynamics. *Mathematische Zeitschrift*, **200**, 355–378.

[172] Ekeland, I., Hofer, H. (1990). Symplectic topology and Hamiltonian dynamics II. *Mathematische Zeitschrift.*, **203**, 553–569.

[173] Ekholm, T., Kragh, T., Smith, I. (2016). Lagrangian exotic spheres. *Journal of Topology and Analysis*, **8** no. 3, 375–397. http://arxiv.org/abs/1503.00473v2

[174] Eliashberg, Y. (1979). Estimates on the number of fixed points of area preserving transformations. Syktyvkar University Preprint.

[175] Eliashberg, Y. (1982). Rigidity of symplectic and contact structures. Abstracts of reports to the 7th Leningrad International Topology Conference.

[176] Eliashberg, Y. (1987). A theorem on the structure of wave fronts and its applications in symplectic topology. *Functional Analysis and Applications*, **21**, 65–72.

[177] Eliashberg, Y. (1989). Classification of overtwisted contact structures on 3-manifolds. *Inventiones Mathematicae*, **98**, 623–637.

[178] Eliashberg, Y. (1990). Filling by holomorphic discs and its applications. *Geometry of Low-Dimensional Manifolds, Part 2*, (Durham, 1989), London Math. Soc. Lecture Note Series, Volume **151**, Cambridge University Press, 1990, pp 45–67.

[179] Eliashberg, Y. (1990). Topological characterization of Stein manifolds of dimension > 2. *International Journal of Mathematics*, **1**, 29–47.

[180] Eliashberg, Y. (1990). New invariants of open symplectic and contact manifolds. *Journal of the American Mathematical Society*, **4** , 513–520.

[181] Eliashberg, Y. (1991). On symplectic manifolds with some contact properties. *Journal of Differential Geometry*, **33**, 233–238.

[182] Eliashberg, Y. (1992). Contact manifolds twenty years since J. Martinet's work. *Annales de l'Institut Fourier*, **42**, 165–192.

[183] Eliashberg, Y. (1993). Classification of contact structures on $\mathbb{R}^3$. *International Mathematics Research Notices*, 87–91.

[184] Eliashberg, Y. (1996). Unique holomorphically fillable contact structure on the 3-torus, *International Mathematics Research Notices*, no. 2, 77-82.

[185] Eliashberg, Y. (2004). A few remarks about symplectic filling. *Geomerty & Topology* **8**, 277–293.

[186] Eliashberg, Y., Givental, A., Hofer, H. (2000). Introduction to symplectic field theory. *Geometric and Functional Analysis Special Volume*, Part **II**, 560–673. http://arxiv.org/abs/math/0010059

[187] Eliashberg, Y., Gromov, M. (1991). Convex symplectic manifolds. In *Proceeding of the Symposium on Pure Mathematics*, **52**, part 2, 135–162.

[188] Eliashberg, Y., Hofer, H. (1992). Towards the definition of symplectic boundary. *Geometric and Functional Analysis*, **2**, 211–220.

[189] Eliashberg, Y., Kim, S.S., Polterovich, L. (2006). Geometry of contact transformations: orderability versus squeezing. *Geometry & Topology* **10**, 1635–1747, and **13** (2009), 1175–1176.

[190] Eliashberg, Y., Mishachev, N. (2002). *Introduction to the h-Principle.* Graduate studies in Mathematics, **48**, American Mathematical Society, Providence, RI.

[191] Eliashberg, Y., Murphy, E. (2013). Lagrangian caps. *Geometric and Functional Analysis*, **23** no. 5, 1483–1514. `http://arxiv.org/abs/1303.0586`

[192] Eliashberg, Y., Murphy, E. (2015). Making cobordisms symplectic. Preprint. `http://arxiv.org/abs/1504.06312`

[193] Eliashberg, Y., Polterovich, L. (1993). Bi-invariant metrics on the group of Hamiltonian diffeomorphisms. *International Journal of Mathematics*, **4**, 727–738.

[194] Eliashberg, Y., Polterovich, L. (1994). New applications of Luttinger's surgery. *Commentarii Mathematicae Helvetica*, **69**, 512–522.

[195] Eliashberg, Y., Polterovich, L. (1996). Local Lagrangian 2-knots are trivial. *Annals of Mathematics* **144**, 61–76.

[196] Eliashberg, Y., Polterovich, L. (1997). The problem of Lagrangian knots in symplectic four-manifolds. *Geometric Topology, Proceedings of the 1993 Georgia International Topology Conference*, edited by William H. Kazez, Studies in Advanced Mathematics, American Mathematical Society/International Press, pp 313–327.

[197] Eliashberg, Y., Thurston, W. (1998). *Confoliations.* University Lecture Series, 13. Amer, Math. Soc.

[198] Entov, M., Polterovich, L. (2003). Calabi quasimorphism and quantum homology. *International Mathematics Research Notices*, 1635–1676. `http://arxiv.org/abs/math/0205247`

[199] Entov, M., Polterovich, L. (2006). Quasi-states and symplectic intersections. *Comm. Math. Helv.* **81** (2006), 75–99.

[200] Entov, M., Polterovich L. (2009). Rigid subsets of symplectic manifolds, *Compositio Mathematica* **145** (2009), 773–826.

[201] Entov, M., Polterovich, L. (2010). C^0-rigidity of Poisson brackets. *Proceedings of the Joint Summer Research Conference on Symplectic Topology and Measure-Preserving Dynamical Systems* (eds. A. Fathi, Y.-G. Oh and C. Viterbo), Contemporary Mathematics **512**, American Mathematical Society, pp. 25–32.

[202] Entov, M., Polterovich, L. (2009). C^0-rigidity of the double Poisson bracket. *International Mathematical Research Notices*, 1134–1158.

[203] Entov, M., Polterovich, L., Py, P. (2012). On continuity of quasi-morphisms for symplectic maps, (with an appendix by Michael Khanevsky). *Progress in Math.* **296** (2012), 169–197. `http://arxiv.org/abs/0904.1397`

[204] Entov, M., Polterovich, L., Zapolsky, F. (2007). Quasi-morphisms and the Poisson bracket. *Pure and Applied Mathematics Quarterly* **3** (2007), 1037–1055.

[205] Entov, M., Verbitsky, M. (2016). Unobstructed symplectic packing for tori and hyperkähler manifolds. *Journal of Topology and Analysis*, **8** no. 4, 589–626. `http://arxiv.org/abs/1412.7183`

[206] Epstein, D.B.A. (1966). Curves on 2-manifolds and isotopies, *Acta Math.*, **115**, 83–107.

[207] Epstein, D.B.A. (1970). The simplicity of certain groups of homeomorphisms. *Compositiones Mathematicae*, **22**, 165–173.

[208] Etnyre, J.B. (2004). Planar open book decompositions and contact structures. *International Mathematics Research Notices*, 4255–4267.

[209] Etnyre, J.B., Honda, K. (2002). On symplectic cobordisms. *Mathematische Annalen* **323**, 31–39.

[210] Evans, J.D. (2011). Symplectic mapping class groups of some Stein and rational surfaces, *J. Symplectic Geom.* **9**, no. 1, 45–82.

[211] Fathi, A. (1980). Transformations et homéomorphismes préservent la mesure. Systèmes dynamiques minimaux. Thesis, Orsay.

[212] Fernández, M., Gotay, M., Gray, A. (1988). Compact parallelizable 4-dimensional symplectic and complex manifolds. *Proceedings of the American Mathematical Society*, **103**, 1209–1212.

[213] Fine, J. (2011). A gauge theoretic approach to the anti-self-dual Einstein equations. Preprint, 2011. http://arxiv.org/abs/1111.5005

[214] Fine, J. (2014). The Hamiltonian geometry of the space of unitary connections with symplectic curvature. *Journal of Symplectic Geometry*, **12**, no. 1, 105–123. http://arxiv.org/abs/1101.2420

[215] Fine, J., Krasnov, K., Panov, D. (2014). A gauge theoretic approach to Einstein 4-manifolds. *New York Journal of Mathematics* **20**, 293–323. http://arxiv.org/abs/1312.2831

[216] Fine, J., Panov, D. (2010). Hyperbolic Geometry and non-Kähler manifolds with trivial canonical bundle, *Geometry & Topology* **14**, 1723–1763. http://arxiv.org/abs/0905.3237

[217] Fine, J., Panov, D. (2013), The diversity of symplectic Calabi–Yau six manifolds. *Journal of Topology*, **6**, 644–658. http://arxiv.org/abs/1108.5944

[218] Fintushel, R., Park, B.D., Stern, R. (2007). Reverse engineering small 4-manifolds *Algebraic & Geometric Topology* **7**, 2103–2116.

[219] Fintushel, R., Stern, R. (1998). Knots, links and 4-manifolds. *Inventiones Mathematicae*, **134** (1998), 363–400. http://arxiv.org/abs/dg-ga/9612014

[220] Fintushel, R., Stern, R. (2006). Double node neighborhoods and families of simply connected 4-manifolds with $b^+ = 1$. *Journal of the American Mathematical Society*, **19**, 171–180.

[221] Fintushel, R., Stern, R. (2006). Six lectures on symplectic four-manifolds. *Low Dimensional Topology*, edited by Tomasz Mrowka and Peter Ozsvath, IAS/Park City Mathematics Series, Volume **15**, pp 265–313. http://arxiv.org/abs/math/0610700

[222] Fintushel, R., Stern, R. (2008). Surgery on nullhomologous tori and simply connected 4-manifolds with $b^+ = 1$. *Journal of Topology* **1**, 1–15.

[223] Fintushel, R., Stern, R. (2011). Pinwheels and nullhomologous surgery on 4-manifolds with $b^+ = 1$. *Algebraic & Geometric Topology* **11**, 1649–1699.

[224] Fintushel, R., Stern, R. (2012). Surgery on nullhomologous tori. *Geometry & Topology* **12**, 61–81. http://arxiv.org/abs/1111.4509

[225] Floer, A. (1987). A refinement of the Conley index and an application to the stability of hyperbolic invariant sets. *Ergodic Theory and Dynamical Systems*, **7**, 93–103.

[226] Floer, A. (1988). Morse theory for Lagrangian intersections. *Journal of Differential Geometry*, **28**, 513–547.

[227] Floer, A. (1988). A relative Morse index for the symplectic action. *Communications on Pure and Applied Mathematics*, **41**, 393–407.

[228] Floer, A. (1988). The unregularized gradient flow of the symplectic action. *Communications on Pure and Applied Mathematics*, **41**, 775–813.

[229] Floer, A. (1989). Witten's complex and infinite dimensional Morse theory. *Journal of Differential Geometry*, **30**, 207–221.

[230] Floer, A. (1989). Symplectic fixed points and holomorphic spheres. *Communications in Mathematical Physics*, **120**, 575–611.

[231] Floer, A. (1988). An instanton invarant for 3-manifolds. *Communications in Mathematical Physics*, **118**, (1988), 215–240.

[232] Floer, A. (1990). Instanton homology, surgery, and knots. *Geometry of Low-Dimensional Manifolds, Volume 1: Gauge Theory and Algebraic Surfaces*, Proceedings of the Durham Symposium, July 1989. Edited by S.K. Donaldson and C.B. Thomas. London Mathematical Soiety, Lecture Notes Series **150**, 1990, pp. 97–114.

[233] Floer, A. (1995). Instanton homology and Dehn surgery. *The Floer Memorial Volume*, edited by H. Hofer, C.H. Taubes, A. Weinstein, E.Zehnder. Progress in Mathematics, Volume **133**, 1995, pp. 77–97.

[234] Floer, A., Hofer, H. (1994). Symplectic homology I: Open sets in $\mathbb{C}^n$. *Mathematische Zeitschrift*, **215**, 37–88.

[235] Floer, A., Hofer, H., Salamon, D. (1995). Transversality in elliptic Morse theory for the symplectic action, *Duke Mathematical Journal*, **80**, 251–292.
[236] Floer, A., Hofer, H., Wysocki, K. (1994). Applications of symplectic homology I. *Mathematische Zeitschrift*, **217**, 577–606.
[237] Fortune, B., Weinstein, A. (1985). A symplectic fixed point theorem for complex projective space. *Bulletin of the American Mathematical Society*, **12**, 128–130.
[238] Frankel, T. (1959). Fixed points and torsion on Kähler manifolds. *Annals of Mathematics* **70**, 1–8.
[239] Franks, J. (1992). Geodesics on S^2 and periodic points of annulus diffeomorphisms. *Inventiones Mathematicae*, **108**, 403–418.
[240] Frauenfelder, U. (2004). The Arnold-Givental conjecture and moment Floer homology. *International Mathematics Research Notices*, 2179–2269.
[241] Friedman, R., Morgan, J. (1994). *Smooth four-manifolds and complex surfaces.* Ergebnisse der Mathematik und ihrer Grenzgebiete, Volume **27**, Springer-Verlag, Berlin.
[242] Fueter, R. (1932). Analytische Funktionen einer Quaternionenvariablen. *Commentarii Mathematici Helvetici*, **4** (1932), 9–20.
[243] Fueter, R. (1937). Die Theorie der regulären Funktionen einer Quaternionenvariablen. *Compte Rendus (ICM Oslo 1936)* **1** (1937), 75–91.
[244] Fujiki, A. (1990). Moduli spaces of polarized algebraic varieties and Kähler metrics. *Sûgaku* **42** (1990), 231–243. English Translation: *Sûgaku Expositions* **5** (1992), 173–191.
[245] Fukaya, K. (1993). Morse homotopy, A_∞-categories, and Floer homologies. *Proceedings of the GARC workshop in Geometry and Topology*, edited by H.J. Kim, Seoul National University, pp. 1–102.
[246] Fukaya, K. (1997). Floer homology for 3-manifolds with boundary I, Preprint. `http://www.kusm.kyoto-u.ac.jp/~fukaya/fukaya.html`.
[247] Fukaya, K. (2006). Application of Floer homology of Lagrangian submani- folds to symplectic topology. *Morse theoretic methods in nonlinear analysis and in symplectic topology*, edited by P. Biran, O. Cornea and F. Lalonde, Springer-Verlag, Berlin. pp. 231–276.
[248] Fukaya, K., Oh, Y.-G. (1998). Zero-loop open strings in the cotangent bundle and Morse homotopy. *Asian Journal of Mathematics*, **1**, 96–180.
[249] Fukaya, K., Oh, Y.-G., Ohta, H., Ono, K. (2009). *Lagrangian Intersection Floer Theory: Anomaly and Obstruction.* AMS/IP Studies in Advanced Mathematics, vol. **46**, American Mathematical Society, Providence, RI.
[250] Fukaya, K., Oh, Y.-G., Ohta, H., Ono, K. (2010). Lagrangian Floer theory and mirror symmetry on compact toric manifolds, I. *Duke Mathematical Journal*, **151**, no. 1, 23-174.
[251] Fukaya, K., Oh, Y.-G., Ohta, H., Ono, K. (2012). Technical details on Kuranishi structures and virtual fundamental chain. `http://arxiv.org/abs/math/1209.4410`
[252] Fukaya, K., Ono, K. (1999). Arnold conjecture and Gromov–Witten invariants. *Topology*, **38**, 933–1048.
[253] Fukaya, K., Seidel, P., Smith, I. (2008). Exact Lagrangian submanifolds in simply connected cotangent bundles. *Inventiones Mathematicae*, **172**, (2008), 1–27.
[254] Fukaya, K., Seidel, P., Smith, I. (2009). The symplectic geometry of cotangent bundles from a categorical viewpoint, in *Homological Mirror Symmetry*, Lecture Notes in Physics, vol **757**, Springer-Verlag, Berlin, pp. 1–26.
[255] Futaki, A. (1983). An obstruction to the existence of Einstein-Kähler metrics. *Inventiones Mathematicae*, **73**, 437–443.
[256] Gadbled, A. (2013). On exotic monotone tori in $\mathbb{C}\mathrm{P}^2$ and $S^2 \times S^2$. *Journal of Symplectic Geometry*, **11**, 343–361. `http://arxiv.org/abs/1103.3487v1`
[257] Gambaudo, J.-M., Ghys, E. (1997). Enlacements asymptotiques. *Topology* **36** no. 6, 1355–1379.
[258] Geiges, H. (1991). Contact structures on 1-connected 5-manifolds. *Mathematika*, **38**, 303–311.
[259] Geiges, H. (1992). Symplectic structures on T^2-bundles over T^2. *Duke Mathematical Journal*, **67**, no. 3, 539–555.

[260] Geiges, H. (1994). Symplectic manifolds with disconnected boundaries of contact type. *International Mathematics Research Notices*, **1**, 23–30.

[261] Geiges, H. (2008). *An Introduction to Contact Topology.* Cambridge Studies in Advanced Mathematics **109**, CUP, Cambridge, UK.

[262] Georgoulas, V., Robbin, J., Salamon, D. (2013). The moment-weight inequality and the Hilbert–Mumford criterion. Preprint. `http://arxiv.org/abs/1311.0410` `http://www.math.ethz.ch/~salamon/PREPRINTS/momentweight.pdf`

[263] Ginzburg, V.L. (1995). An embedding $S^{2n-1} \to \mathbb{R}^{2n}$, $2n-1 \geq 7$, whose Hamiltonian flow has no periodic trajectories. *International Mathematics Research Notices*, **2**, 83–98.

[264] Ginzburg, V.L. (1997). A smooth counterexample to the Hamiltonian Seifert conjecture in $\mathbb{R}^6$. *International Mathematics Research Notices*, No 13, 641–650.

[265] Ginzburg, V.L. (1999). Hamiltonian dynamical systems without periodic orbits. *Northern California Symplectic Geometry Seminar*, American Mathematical Society Translations Series 2, **196**, American Mathematical Society, Providence, RI, 1999, pp. 35–48.

[266] Ginzburg, V.L., Gurel, B. (2003). A C^2-smooth counterexample to the Hamiltonian Seifert conjecture in $\mathbb{R}^4$. *Annals of Mathematics* **158** (2003), 953–976.

[267] Ginzburg, V.L., Hein, D. (2012). Hyperkähler Arnold conjecture and its generalizations. *International Journal of Mathematics* **23**, 1250077. `http://www.worldscientific.com/doi/pdf/10.1142/S0129167X12500772`

[268] Ginzburg, V.L., Hein, D. (2013). The Arnold conjecture for Clifford symplectic pencils. *Israel Journal of Mathematics* **196**, 95–112.

[269] Giroux, E. (1991). Convexité en topologie de contact. *Commentarii Mathematici Helvetici*, **66**, 637–677.

[270] Giroux, E. (2001). Structure de contact sur les variétés fibrées en cercles au-dessus d'une surface. *Commentarii Mathematici Helvetici*, **76** no. 2, 218–262. `http://arxiv.org/abs/math/9911235`

[271] Giroux, E. (2002). Geometrie de contacte: déla dimension trois ver les dimensions supérieures. In 'Proceedings of the International Congress of Mathematicians', **vol II** (Beijing 2002), Higher Education Press, Beijing, 405–414.

[272] Giroux, E., Goodman, N. (2006). On the stable equivalence of open books in three manifolds, *Geometry and Topology* **10**, 97–114.

[273] Givental, A. (1987). Periodic maps in symplectic topology. *Functional Analysis and its Applications* **21**, 271–283.

[274] Givental, A. (1995). A fixed point theorem for toric manifolds. In reference [319] below, pp. 445–483.

[275] Goldman, W. (1984). The symplectic nature of fundamental groups of surfaces. *Advances in Mathematics*, **54**, 200–225.

[276] Golé, C. (1994). Periodic orbits for Hamiltonian systems on cotangent bundles. *Transactions of the American Mathematical Society*, **343**, 327–347.

[277] Gompf, R. (1994). Some new symplectic manifolds. *Turkish Mathematical Journal*, **18**, 7–15.

[278] Gompf, R. (1995). A new construction of symplectic manifolds. *Annals of Mathematics*, **142**, 527–595.

[279] Gompf, R., Mrowka, T. (1993). Irreducible 4-manifolds need not be complex. *Annals of Mathematics*, **138**, 61–111.

[280] Gompf, R., Stipcicz, T. (1999). *4-manifolds and Kirby Calculus.* Graduate Studies in Mathematics, 20. American Mathematical Society, Providence, RI.

[281] Gonzalez, E., Woodward, C. (2012). Quantum cohomology and toric minimal model programs. `http://arxiv.org/abs/math/1207.3253`

[282] Gotay, M., Lashof, R., Sniatycki, J., Weinstein, A. (1983). Closed forms on symplectic fiber bundles. *Commentarii Mathematici Helvetici*, **58**, 617–621.

[283] Gottlieb, D. H. (1965). A certain subgroup of the fundamental group. *American Journal of Mathematics*, **87**, 840–856.

[284] Greene, R., Shiohama, K. (1979). Diffeomorphisms and volume-preserving embeddings of non-compact manifolds. *Transactions of the American Mathematical Society*, **225**,

403–414.
[285] Griffiths, P., Harris, J. (1978). *Principles of Algebraic Geometry.* Wiley, New York.
[286] Gromov, M. (1969). Stable mappings of foliations into manifolds. *Mathematics of the USSR - Isvestija*, **33**, 671–694.
[287] Gromov, M. (1985). Pseudo holomorphic curves in symplectic manifolds. *Inventiones Mathematicae*, **82**, 307–347.
[288] Gromov, M. (1986). *Partial Differential Relations.* Springer-Verlag, Berlin.
[289] Gromov, M. (1987). Soft and hard symplectic geometry. In *Proceedings of the ICM at Berkeley 1986*, Vol. **1**, American Mathematical Society, Providence, RI, pp. 81–98.
[290] Gromov, M. (1988). Dimension, nonlinear spectra and width. in *Geometric Aspects of Functional Analysis (1986/87)*, 132–184, Springer Lecture Notes in Mathematics 1317, Springer-Verlag, Berlin.
[291] Gualtieri, M. (2011). Generalized complex geometry. *Annals of Mathematics*, **174**, 75–123.
[292] Guillemin, V. Lerman, E., Sternberg, S. (1996). *Symplectic Fibrations and Multiplicity Diagrams.* Cambridge University Press, Cambridge, UK.
[293] Guillemin, V., Pollack, V. (1974). *Differential Topology.* Prentice-Hall.
[294] Guillemin, V., Sjamaar, R. (2005). *Convexity Properties of Hamiltonian Group Actions.* CRM Monograph Series **26**, American Mathematical Society, Providence, R!.
[295] Guillemin, V., Sternberg, S. (1982). Convexity properties of the moment map. *Inventiones Mathematicae*, **67**, 491–514.
[296] Guillemin, V., Sternberg, S. (1982). Geometric quantization and multiplicities of group representations. *Inventiones Mathematicae*, **67**, 515–538.
[297] Guillemin, V., Sternberg, S. (1984). *Symplectic Techniques in Physics.* Cambridge University Press, Cambridge, UK.
[298] Guillemin, V., Sternberg, S. (1989). Birational equivalence in the symplectic category. *Inventiones Mathematicae*, **97**, 485–522.
[299] Guillemin, V., Sternberg, S. (1990). *Geometric Asymptotics.* American Mathematical Society Mathematical Surveys and Monographs **14**, American Mathematical Society, Providence, RI.
[300] Guillermou, S., Kashiwara, M., Schapira, P. (2012) Sheaf quantization of Hamiltonian isotopies and applications to nondisplacebility problems. *Duke Mathematical Journal*, **161**, no. 2, 201–245.
[301] Guth, L. (2008). Symplectic embeddings of polydiscs. *Inventiones Mathematicae*, **172**, 477–489. http://arxiv.org/abs/0709.1957
[302] Gutt, J. (2014). On the minimal number of periodic Reeb orbits on contact manifolds. PhD thesis, Université Libre de Bruxelles and Université de Strasbourg.
[303] Haefliger, A. (1971). Lectures on a theorem of Gromov. In *Liverpool Singularities Conference.* Lecture Notes in Mathematics **209**, pp. 128–141. Springer-Verlag, Berlin.
[304] Hano, J.-I. (1957). On Kählerian homogeneous spaces of unimodular Lie groups. *American Journal of Mathematics* **79**, 885–900.
[305] Haydys, A., Walpuski, T. (2015). A compactness theorem for the Seiberg-Witten equation with multiple spinors in dimension three. *Geometric and Functional Analysis* **25**, no. 6, 1799–1821. https://arxiv.org/abs/1406.5683
[306] He, Jian. (2013). Correlators and descendants of subcritical Stein manifolds. *International Journal of Mathematics*, **24** no. 2, 1350004. http://arxiv.org/abs/0810.4174v3
[307] Herman, M. (1994). Private communication.
[308] Herrera, H. (2002). Gromov invariants of S^2-bundles over 4-manifolds. Proceedings of the 1999 Georgia Topology conference (Athens, GA). *Topology and its Applications*, **124** no. 2, 327–345.
[309] Hind, R. (2012). Lagrangian unknottedness in Stein surfaces. *Asian Journal of Mathematics*, **16**, 1–36.
[310] Hind, R., Lisi, S. (2015). Symplectic embeddings of polydiscs. *Selecta Mathematica (New Series)* **21** no. 3, 1099–1120. http://arxiv.org/abs/01304.3065
[311] Ho, C.-I., Li, T.-J. (2012). Luttinger surgery and Kodaira dimension. *Asian Journal of Mathematics*, **16**, 299–318. http://arxiv.org/abs/1108.0479

[312] Hofer, H. (1985). Lagrangian embeddings and critical point theory. *Annales de l'Institut Henri Poincaré – analyse nonlinéaire* **2**, 407–462.
[313] Hofer, H. (1988). Ljusternik–Schnirelman theory for Lagrangian intersections. *Annales de l'Institut Henri Poincaré – analyse nonlinéaire* **5**, 465–99.
[314] Hofer, H. (1990). On the topological properties of symplectic maps. *Proceedings of the Royal Society of Edinburgh* **115**, 25–38.
[315] Hofer, H. (1992). Symplectic capacities. In *Geometry of Low-Dimensional Manifolds*, Vol 2 (ed. S. Donaldson and C.B. Thomas), London Mathematical Society Lecture Notes **150**. Cambridge University Press, pp. 15–34.
[316] Hofer, H. (1993). Estimates for the energy of a symplectic map. *Commentarii Mathematici Helvetici* **68**, 48–72.
[317] Hofer, H. (1993). Pseudoholomorphic curves in symplectizations with applications to the Weinstein conjecture in dimension 3. *Inventiones Mathematicae*, **114**, 515–563.
[318] Hofer, H., Salamon, D. (1995). Floer homology and Novikov rings. In [319], pp. 483–524.
[319] Hofer, H., Taubes, C.H., Weinstein, A., Zehnder, E. (ed.) (1995). *The Floer Memorial Volume*. Birkhäuser, Basel.
[320] Hofer, H., Wysocki, K., Zehnder, E. (2007). A general Fredholm theory I, A splicing based differential geometry, *Journal of the European Mathematical Society*, **9**, 841–876.
[321] Hofer, H., Wysocki, K., and Zehnder, E. (2009). A general Fredholm theory II, Implicit Function theorems, *Geometric and Functional Analysis*, **19**, 206–293.
[322] Hofer, H., Wysocki, K., Zehnder, E. (2009). A general Fredholm theory III, Fredholm functors and polyfolds, *Geometry & Topology*, **13**, 2279–2387.
[323] Hofer, H., Wysocki, K., Zehnder, E. (2012). *Applications of Polyfold theory I: The Polyfolds of Gromov–Witten theory*, `http://arxiv.org/abs/1107.2097`
[324] Hofer, H., Zehnder, E. (1990). A new capacity for symplectic manifolds. In *Analysis et cetera* (ed. P.H. Rabinowitz and E. Zehnder), Academic Press, New York, pp. 405–429.
[325] Hofer, H., Zehnder, E. (1994). *Symplectic Capacities*. Birkhäuser, Basel.
[326] Hohloch, S., Noetzel, G., Salamon, D.A. (2009). Hypercontact structures and Floer homology. *Geometry & Topology*, **13**, 2543–2617.
[327] Hohloch, S., Noetzel, G., Salamon, D.A. (2009). Floer homology groups in hyperkähler geometry. *New Perspectives and Challenges in Symplectic Field Theory*, eds. M. Abreu, F. Lalonde, L. Polterovich. CRM, Volume **49**, American Mathematical Society, 2009, pp. 251–261.
[328] Holm, T., Kessler, L. (2015). Circle actions on symplectic four-manifolds. `https://arxiv.org/abs/1507.05972`
[329] Honda, K. (2000). On the classification of tight contact structures, I. *Geometry and Topology*, **4**, 309–368, and erratum *Geometry and Topology*, **5** (2001), 925–938.
[330] Honda, K. (2000). On the classification of tight contact structures, II. *Journal of Differential Geometry*, **55**, 1–191.
[331] Hörmander, L. (1971). Fourier integral operators I. *Acta Mathematica*, **127**, 79–183.
[332] Horn, A. (1954). Doubly stochastic matrices and the diagonal of a rotation matrix. *American Journal of Mathematics* **76**, 620–630.
[333] Hu, J., Li, T.-J., Ruan, Yongbin. (2008). Birational cobordism invariance of uniruled symplectic manifolds. *Inventiones Mathematicae*, **172**, no. 2, 231–275.
[334] Humiliére, V. Leclercq, R., Seyfaddini, S. (2014) Reduction of symplectic homeomorphisms, Preprint. `http://arxiv.org/abs/1407.6330`
[335] Husemoller, D. (1974). *Fibre Bundles*. Springer-Verlag, New York.
[336] Hutchings, M. (2010). Taubes' proof of the Weinstein conjecture in dimension three. *Bulletin of the American Mathematical Society*, **47**, 73–125.
[337] Hutchings, M. (2010). Embedded contact homology and its applications. in *Proceedings of the 2010 International Congress of Mathematicians*, Volume **II**, pp. 1022–1041, Hindustan Book Agency, New Delhi.
[338] Hutchings, M. (2011). Quantitative embedded contact homology. *Journal of Differential Geometry*, **88**, 231–266.

[339] Hutchings, M. (2011). Recent progress on symplectic embedding problems in four dimensions. *Proceedings of the National Academy of Sciences of the USA* **108**, 8093–8099.
[340] Hutchings, M. (2014). Lecture notes on embedded contact homology. in *Contact and Symplectic Topology*, 389–484, Bolyai Society Mathematical Studies, 26, János Bolyai Mathematical Society, Budapest, Hungary. http://arxiv.org/abs/1303.5789
[341] Hutchings, M. (2016). Beyond ECH capacities. *Geometry & Topology*, **20** no. 2, 1085–1126. http://arxiv.org/abs/1409.1352
[342] Hutchings, M., Taubes, C. (2009). The Weinstein conjecture for stable Hamiltonian structures. *Geometry & Topology*, **13**, 901–941.
[343] Hutchings, M., Taubes, C. (2011). Proof of the Arnold chord conjecture in three dimensions I. *Mathematical Research Letters*, **18**, 295–313.
[344] Hutchings, M., Taubes, C. (2013). Proof of the Arnold chord conjecture in three dimensions II. *Geometry & Topology* **17**, 2601–2688.
[345] Huybrechts, D. (2015). *Lectures on K3 surfaces.* http://www.math.uni-bonn.de/people/huybrech/K3Global.pdf
[346] Ionel, E.N., Parker, T.H. (1999). Gromov invariants and symplectic maps. *Mathematische Annalen* **314** (1999), 127–158. http://arxiv.org/abs/dg-ga/9703013
[347] Jeffrey, L., Kirwan, F. (1995). Localization for nonabelian group actions. *Topology* **34**, 291–327.
[348] Jeffrey, L., Weitsman, J. (1994). Toric structures on the moduli space of flat connections on a Riemann surface: volumes and the moment map. *Advances in Mathematics* **106**, 151–168.
[349] Joyce, D. (2000) *Compact Manifolds with Special Holonomy.* Oxford Mathematical Monographs. Oxford University Press, Oxford, UK.
[350] Joyce, D. (2006) *Riemannian Holonomy Groups in Calibrated Geometry.* Oxford Graduate Texts in Mathematics **12**. Oxford University Press, Oxford, UK.
[351] Joyce, D. (2012). On manifolds with corners, in *Advances in geometric analysis* pp. 225?258, Advanced Lectures in Mathematics **21**, International Press, Somerville, MA. http://arxiv.org/abs/0910.3518
[352] Joyce, D. (2014). A new definition of Kuranishi space, Preprint. http://arxiv.org/abs/1407.6908
[353] Kanda, Y. (1997). The classification of tight contact structures on the 3-torus. *Communications in Analysis and Geometry* bf 5, 413–438.
[354] Karshon, Y. (1994). Periodic Hamiltonian flows on four dimensional manifolds. In [620], pp. 43–47.
[355] Karshon, Y. (1999). Periodic Hamiltonian flows on four dimensional manifolds. *Memoirs of the American Mathematical Society*, **672**.
[356] Karshon, Y., Bjorndahl, C. (2010). Revisiting Tietze–Nakajima: local and global convexity for maps, *Canadian Journal of Mathematics*, **62**, 975–993. http://arxiv.org/abs/0701745
[357] Karshon, Y., Kessler, L. (2014). Distinguishing symplectic blowups of the complex projective plane. http://arxiv.org/abs/1407.5312
[358] Karshon, Y., Kessler, L., Pinsonnault, M. (2012). Symplectic blowups of the complex projective plane and counting toric actions (21 December 2012). http://www.math.toronto.edu/karshon/.
[359] Karshon, Y., Kessler, L., Pinsonnault, M. (2014). Counting toric actions on symplectic four-manifolds. http://arxiv.org/abs/1409.6061
[360] Kasper, B. (1990). Examples of symplectic structures on fiber bundles. Ph.D. thesis, SUNY, Stony Brook.
[361] Keating, A. (2015). Lagrangian tori in four-dimensional Milnor fibers, *Geometric and Functional Analysis*, **25** no. 6, 1822–1901. http://arxiv.org/abs/math/1405.0744
[362] Kedra, J. (2005). Evaluation fibrations and the topology of symplectomorphisms. *Proceedings of the American Mathematical Society*, **133**, 305–312. http://arxiv.org/abs/math/0305325

[363] Kedra, J., Kotschick, D., Morita, S. (2006). Crossed Flux homomorphisms and vanishing theorems for flux groups. *Geometric and Functional Analysis*, **16**, 1246–1273.
[364] Kedra, J., McDuff, D. (2005). Homotopy properties of Hamiltonian group actions, *Geometry & Topology*, **9**, 121–162.
[365] Kempf, G. (1978). Instability in invariant theory. *Annals of Mathematics*, **108**, 299–317.
[366] Kempf, G., Ness, L. (1978). The length of vectors in representation spaces. Springer Lecture Notes **732**, *Algebraic Geometry, Proceedings*, Copenhagen, 1978, pp. 233–244.
[367] Kervaire, M.A., Milnor, J. (1963). *Annals of Mathematics*, **77**, 504–537.
[368] Khanevsky, M. (2009). Hofer's metric on the space of diameters. *Journal of Topology and Analysis*, **1** (2009), 407–416.
[369] Kirwan, F. (1984). *Cohomology of Quotients in Symplectic and Algebraic Geometry*. Mathematics Notes, **31**. Princeton University Press, Princeton.
[370] Kirwan, F. (1984). Convexity properties of the moment mapping III. *Inventiones Mathematicae*, **77**, 547–552.
[371] Kobayashi, S., Nomizu, K. (1963). *Differential Geometry*. Wiley-Interscience, New York.
[372] Kodaira, K. (1954). On Kähler varietes of restricted type. *Annals of Mathematics*, **60**, 28–48.
[373] Kontsevich, M., Manin, Yu. (1994). Gromov-Witten classes, quantum cohomology, and enumerative geometry. *Comm. Math. Phys.*, **164**, 525–562.
[374] Kostant, B. (1973). On convexity, the Weyl group, and the Iwasawa decomposition. *Annales de Science de l'École Normale Supérieure*, **6**, 413–455.
[375] Krom, R. (2015). The Donaldson geometric flow is a local smooth semiflow. Preprint, December 2015. `http://arxiv.org/abs/1512.09199`
[376] Krom, R., and Salamon, D.A. (2015). The Donaldson geometric flow for symplectic four-manifolds. Preprint, December 2015. `http://arxiv.org/abs/1512.09198`
[377] Kronheimer, P.B. (1997). Some nontrivial families of symplectic structures. `http://www.math.harvard.edu/~kronheim/papers.html`
[378] Kronheimer, P., Mrowka, T. (1994). Recurrence relations and asymptotics for four-manifold invariants. *Bulletin of the American Mathematical Society*, **30**, 215–221.
[379] Kronheimer, P., Mrowka, T. (1993). Gauge theory for embedded surfaces I. *Topology*, **32**, 773–826.
[380] Kronheimer, P., Mrowka, T. (1995). Gauge theory for embedded surfaces II. *Topology*, **34**, 37–97.
[381] Kronheimer, P., and Mrowka, T. (1994). The genus of embedded surfaces in the projective plane, *Mathematical Research Letters*, **1**, 797–808.
[382] Kronheimer, P., Mrowka, T. (2007). *Monopoles and Three-Manifolds*. New Mathematical Mongraphs **10**, Cambridge University Press, 2007.
[383] Kuperberg, G. (2009). From the Mahler conjecture to Gauss linking integrals. *Geometric and Functional Analysis*, **18**, 870–892. `http://arxiv.org/abs/math/0610904`
[384] Kutluhan, C., Lee, Y.-J., Taubes, C.H. (2010). HF=HM I: Heegaard Floer homology and Seiberg–Witten Floer homology. `http://arxiv.org/abs/1007.1979`
[385] Kutluhan, C., Lee, Y.-J., Taubes, C.H. (2010). HF=HM II: Reeb orbits and holomorphic curves for the ech/Heegaard-Floer correspondence. `http://arxiv.org/abs/1008.1595`
[386] Kutluhan, C., Lee, Y.-J., Taubes, C.H. (2010). HF=HM III: Holomorphic curves and the differential for the ECH/Heegaard Floer correspondence. `http://arxiv.org/abs/1010.3456`
[387] Kutluhan, C., Lee, Y.-J., Taubes, C.H. (2011). HF=HM IV: The Seiberg-Witten Floer homology and ech correspondence. `http://arxiv.org/abs/1107.2297`
[388] Kutluhan, C., Lee, Y.-J., Taubes, C.H. (2012). HF=HM V: Seiberg–Witten–Floer homology and handle addition. `http://arxiv.org/abs/1204.0115`
[389] Lalonde, F. (1994). Isotopy of symplectic balls, Gromov's radius, and structure of ruled symplectic manifolds. *Mathematische Annalen*, **300**, 273–296.
[390] Lalonde, F., McDuff, D. (1995). The geometry of symplectic energy. *Annals of Mathematics*, **141**, 349–371.

[391] Lalonde, F., McDuff, D. (1995). Hofer's L^∞ geometry: geodesics and stability, I, II. *Inventiones Mathematicae*, **122**, 1–33, 35–69.

[392] Lalonde, F., McDuff, D. (1995). Local nonsqueezing theorems and stability. *Geometric and Functional Analysis*, **5**, 364–386.

[393] Lalonde, F., McDuff, D. (1996). The classification of ruled symplectic 4-manifolds, *Mathematical Research Letters*, **3**, 769–778.

[394] Lalonde, F., McDuff, D. (1996). J-curves and the classification of rational and ruled symplectic 4-manifolds. In [620], pp. 3–42.

[395] Lalonde, F., McDuff, D. (1997). Positive paths in the linear symplectic group. In *The Arnold–Gelfand mathematical seminars, Geometry and Singularity theory*, edited by Arnold, Gelfand *et al*, Birkhäuser, Basel, pp. 361–388.

[396] Lalonde, F., McDuff, D. (2003). Symplectic structures on fiber bundles. *Topology* **42**, 309–347, with erratum in *Topology* **44**, (2005), 1301-1303.

[397] Lalonde, F., McDuff, D., Polterovich, L. (1997). On the Flux conjectures. Preprint DG/9706015, in the Proceedings of the CRM Workshop on Geometry, Topology and Dynamics, Montreal 1995, CRM Special Series published by the American Mathematical Society, Providence, RI.

[398] Lalonde, F., McDuff, D., Polterovich, L. (1999). Topological rigidity of Hamiltonian loops and quantum homology. *Inventiones Mathematicae*, **135**, 369–385. `http://arxiv.org/abs/dg-ga/9710017`

[399] Lanzat, S. (2013). Quasi-morphisms and symplectic quasi states for convex symplectic manifolds. *International Mathematics Research Notices*, 5321–5365. `http://arxiv.org/abs/1110.1555`

[400] Latschev, J., McDuff, D., Schlenk, F. (2013). The Gromov-width of four-dimensional tori. *Geometry & Topology*, **17**, 2813–2853. `http://arxiv.org/abs/1111.6566v2`

[401] Latschev, J., Wendl, C. (2011). Algebraic Torsion in Contact Manifolds (with an appendix by M. Hutchings). *Geometric and Functional Analysis*, **21**, 1144–1195.

[402] Laudenbach, F., Sikorav, J.P. (1985). Persistance d'intersection avec la section nulle au cours d'une isotopie hamiltonienne dans un fibré cotangent. *Inventiones Mathematicae*, **82**, 349–357.

[403] Lawson, H.B., Michelsohn, M.-L. (1989). *Spin Geometry.* Princeton Mathematical Series **58**, Princeton University Press, Princeton.

[404] LeBrun, C. (1995). Einstein metrics and Mostow rigidity. *Mathematical Research Letters*, **2**, 1–8.

[405] LeBrun, C. (2009). Einstein metrics, complex surfaces, and symplectic 4-manifolds. *Mathematical Proceedings of the Cambridge Philosophical Society*, **147**, 1–8. `http://arxiv.org/abs/0803.3743`

[406] Lekili, Y. (2009). Wrinkled fibrations on near symplectic manifolds. *Geometry & Topology*, **13**, 277–318. `http://arxiv.org/abs/0712.2202`

[407] Lerman, E. (1988). How fat is a fat bundle? *Letters in Mathematical Physics*, **15**, 335–339.

[408] Lerman, E. (1995). Symplectic cuts. *Mathematical Research Letters*, **2**, 247–258.

[409] Lerman, E. (2005). Gradient flow of the norm squared of the moment map. *L'Enseignement Mathémathique* **51**, 117–127.

[410] Lerman, L., Meinrenken, E., Tolman, S., Woodward, C. (1998). Non-abelian convexity by symplectic cuts. *Topology*, **37**, 245–259. `http://arxiv.org/abs/dg-ga/9603015`

[411] Lerman, E., Sjamaar, R. (1991). Stratified symplectic spaces and reduction. *Annals of Mathematics*, **134**, 375–422.

[412] Lerman, E., Montgomery, R., Sjamaar, R. (1993). Examples of singular reduction. In *Symplectic Geometry* (ed. D. Salamon), London Mathematical Society Lecture Note Series, **192**. Cambridge University Press, Cambridge, UK, pp. 127–155.

[413] Le Roux, F. (2010). Simplicity of the group of compactly supported area preserving homeomorphisms of the open disc and fragmentation of symplectic diffeomorphisms. *Journal of Symplectic Geometry*, **8**, 73–93. `http://arxiv.org/abs/0901.2428`

[414] Le Roux, F. (2010). Six questions, a proposition and two pictures on Hofer distance for Hamiltonian diffeomorphisms on surfaces. *Contemporary Mathematics* , **516**, 33–40.
[415] Leung, C., Symington, M. (2010). Almost toric symplectic four-manifolds, *Journal of Symplectic Geometry*, **8**, no. 2, 143–187.
[416] Li, B.-H., Li, T.-J. (2002). Symplectic genus, minimal genus, and diffeomorphisms. *Asian Journal of Mathematics*, **26**, 123–144. `http://arxiv.org/abs/math/0108227`
[417] Li, J., Li, T.-J., Wu, W. (2015). On the symplectic mapping class group of small rational manifolds. *Michigan Mathematics Journal*, **64**, no. 2, 319–333. `http://arxiv.org/abs/1310.7329v2`
[418] Li, T.-J. (1999). Smoothly embedded spheres in symplectic four-manifolds. *Proceedings of the American Mathematical Society*, **127**, 609–613.
[419] Li, T.-J. (2006). Symplectic 4-manifolds with Kodaira dimension zero. *Journal of Differential Geometry*, **74**, 177–352.
[420] Li, T.-J. (2008). The space of symplectic structures on closed 4-manifolds. *Proceedings of the 3rd ICCM 2004*, AMS/IP *Studies in Advanced Mathematics* **42** (2008), 259–277. `http://arxiv.org/abs/arXiv:0805.2931`
[421] Li, T.-J., Liu, A. (1995). General wall crossing formula. *Mathematical Research Letters*, **2**, 797–810.
[422] Li, T.-J., Liu, A. (1995). Symplectic structure on ruled surfaces and generalized adjunction formula. *Mathematical Research Letters*, **2**, 453–471.
[423] Li, T.-J., Liu, A. (1999). On the equivalence between SW and GT in the case $b^+ = 1$. *International Mathematical Research Notices*, 335–345.
[424] Li, T.-J., Liu, A. (2001). Uniqueness of symplectic canonical class, surface cone and symplectic cone of manifolds with $b^1 = 1$. *Journal of Differential Geometry*, **58**, 331–370.
[425] Li, T.-J., Zhang, W. (2009). Comparing tamed and compatible symplectic cones and cohomological properties of almost complex manifolds. *Communications in Analysis and Geometry*, **17**, 651–683. `http://arxiv.org/abs/0708.2520`
[426] Li, Ping, Liu, Kefeng (2011). Some remarks on circle action on manifolds. *Mathematical Research Letters*, **18**, 437–446. `http://arxiv.org/abs/1008.4826`
[427] Li, T.-J., Ruan, Y. (2009) Symplectic birational geometry. *New Perspectives and Challenges in Symplectic Field Theory*, eds. M. Abreu, F. Lalonde, L. Polterovich. CRM, Volume **49**, American Mathematical Society, 2009, pp. 307–326.
[428] Li, J., Tian, G. (1998). Virtual moduli cycles and Gromov–Witten invariants for general symplectic manifolds. *Topics in Symplectic 4-manifolds (Irvine CA 1996)*, International Press, Cambridge, MA (1998), 47–83.
[429] Li-Bland, D., Weinstein, A. (2014). Selective categories and linear canonical relations. *Symmetry, Integrability and Geometry: Methods and Applications* **10**, 100, 31 pages. `http://arxiv.org/abs/1401.7302`
[430] Lipyanskiy, M. (2014). Gromov–Uhlenbeck compactness. Preprint. `http://arxiv.org/abs/1409.1129`
[431] Liu, A. (1995). Some new applications of general wall crossing formula. *Mathematical Research Letters*, **3**, 569–586.
[432] Liu, G., Tian, G. (1998). Floer homology and Arnold conjecture. *Journal of Differential Geometry*, **49**, 1–74.
[433] Long, Y. (2002). *Index Theory for Symplectic Paths with Applications.* Progrress in Mathematics **207**, Birkhäuser, Basel.
[434] Lupton, G., Oprea, J. (1995). Cohomologically symplectic spaces, toral actions and the Gottlieb group. *Transactions of the American Mathematical Society* **347**, 261–288.
[435] Luttinger, K. (1995). Lagrangian tori in $\mathbb{R}^4$. *Journal of Differential Geometry*, **42**, 220–228.
[436] Mabuchi, T. (1986). K-energy maps integrating Futaki invariants. *Tohoku Mathematics Journal* **38**, 575–593.
[437] Marsden, J. (1992). *Lectures on Mechanics.* Cambridge University Press, Cambridge, UK.

[438] Marsden, J., Weinstein, A. (1974). Reduction of symplectic manifolds with symmetry. *Reports on Mathematical Physics* **5**, 121–130.
[439] Martin, S.K. (1994). Cohomology rings of symplectic quotients. Ph.D. thesis. University of Oxford.
[440] Massot, P., Niederkrüger, K., Wendl, C. (2013). Weak and strong fillability of higher dimensional contact manifolds. *Inventiones Mathematicae*, **192**, 287–373.
[441] Mather, J.N. (1985). More Denjoy minimal sets for area preserving diffeomorphisms. *Commentarii Mathematici Helvetici* **60**, 508–557.
[442] McCarthy, J., Wolfson, J. (1994). Symplectic normal connect sum. *Topology* **33**, 729–764.
[443] McDuff, D. (1984). Symplectic diffeomorphisms and the flux homomorphism. *Inventiones Mathematicae*, **77**, 353–366.
[444] McDuff, D. (1984). Examples of simply connected symplectic non Kaehlerian manifolds. *Journal of Differential Geometry*, **20**, 267–727.
[445] McDuff, D. (1987). Examples of symplectic structures. *Inventiones Mathematicae*, **89**, 13–36.
[446] McDuff, D. (1988). The symplectic structure of Kähler manifolds of non-positive curvature. *Journal of Differential Geometry*, **28**, 467–475.
[447] McDuff, D. (1988). The moment map for circle actions on symplectic manifolds. *Journal of Geometrical Physics*, **5**, 149–160.
[448] McDuff, D. (1990). Elliptic methods in symplectic geometry. *Bulletin of the American Mathematical Society*, **23**, 311–358.
[449] McDuff, D. (1990). Rational and ruled symplectic 4-manifolds. *Journal of the American Mathematical Society*, **3**, 679–712; erratum **5** (1992), 987–988.
[450] McDuff, D. (1991). Blowing up and symplectic embeddings in dimension 4. *Topology*, **30**, 409–421.
[451] McDuff, D. (1991). Symplectic manifolds with contact type boundary. *Inventiones Mathematicae*, **103**, 651–671.
[452] McDuff, D. (1991). Immersed spheres in symplectic 4-manifolds. *Annales de l'Institut Fourier*, **42**, 369–392.
[453] McDuff, D. (1993). Remarks on the uniqueness of symplectic blowing-up. In *Symplectic Geometry* (ed. D. Salamon), London Mathematical Society Lecture Note Series, **192**. Cambridge University Press, Cambridge, UK, pp. 157–168.
[454] McDuff, D. (1994). Notes on ruled symplectic 4-manifolds. *Transactions of the American Mathematical Society*, **345**, 623–639.
[455] McDuff, D. (1994). Singularities and positivity of intersections of J-holomorphic curves. In [34], pp. 191–216.
[456] McDuff, D. (1995). An irrational ruled symplectic manifold. In [319], pp. 545–554.
[457] McDuff, D. (1997). Lectures on Gromov invariants for symplectic 4-manifolds. In *Gauge Theory and Symplectic Geometry* (ed. Hurtubise and Lalonde), NATO ASI series, C-488, Kluwer, Dordrecht, pp. 175–210.
[458] McDuff, D. (1998). From symplectic deformation to isotopy. *Topics in Symplectic Four-Manifolds*, edited by R.J. Stern, International Press Lecture Series, Volume **1**, International Press, Cambridge, MA, pp. 85–100.
[459] McDuff, D. (2000). Almost complex structures on $S^2 \times S^2$, *Duke Mathematical Journal*, **101**, 135–177.
[460] McDuff, D. (2002). Geometric variants of the Hofer norm. *Journal of Symplectic Geometry*, **1**, 197–252.
[461] McDuff, D. (2005). Enlarging the Hamiltonian group. *Journal of Symplectic Geometry*, **3**, 481–530.
[462] McDuff, D. (2009). Hamiltonian S^1-manifolds are uniruled. *Duke Mathematical Journal*, **146**, 449–507.
[463] McDuff, D. (2009). Symplectic embeddings of 4-dimensional ellipsoids. *Journal of Topology*, **2**, no. 3, 589–623.
[464] McDuff, D. (2010). Monodromy in Hamiltonian Floer theory. *Commentarii Mathematicii Helvetici*, **85**, (2010), 95–133.

[465] McDuff, D. (2010). Loops in the Hamiltonian group: a survey. in *Symplectic Topology and Measure Preserving Dynamical Systems*, 127–148, *Contemporary Mathematics* **512**, American Mathematical Society, Providence, RI.
[466] McDuff, D. (2011). The topology of toric symplectic manifolds. *Geometry & Topology*, **15**, 145–190. http://arxiv.org/abs/1004.3227
[467] McDuff, D. (2011). The Hofer conjecture on embedding symplectic ellipsoids. *Journal of Differential Geometry*, **88**, 519–532.
[468] McDuff, D., Opshtein, E. (2015). Nongeneric J-holomorphic curves and singular inflation. *Algebraic & Geometric Topology*, **15**, no. 1, 231–286. http://arxiv.org/abs/1309.6425
[469] McDuff, D., Polterovich, L. (1994). Symplectic packings and algebraic geometry. *Inventiones Mathematicae*, **115**, 405–429.
[470] McDuff, D., Salamon, D. (2012). *J-holomorphic Curves and Symplectic Topology, Second Edition.* Colloquium Publications, Volume **52**, American Mathematical Society, Providence, RI.
[471] McDuff, D., Schlenk, F. (2012). The embedding capacity of 4-dimensional symplectic ellipsoids. *Annals of Mathematics*, **175**, 1191–1282.
[472] McDuff, D., Traynor, L. (1993). The 4-dimensional symplectic camel and related results. In *Symplectic Geometry* (ed. D. Salamon), pp. 169–182. London Mathematical Society Lecture Note Series, **192**. Cambridge University Press, Cambridge, UK.
[473] McDuff, D., Wehrheim, K. (2015). The fundamental class of smooth Kuranishi atlases with trivial isotropy. to appear in *Journal of Topology and Analysis.* Preprint. http://arxiv.org/abs/1508.01560
[474] McLean, M. (2009). Lefschetz fibrations and symplectic homology. *Geometry & Topology* **13**, 1877–1944. http://arxiv.org/abs/0709.1639
[475] McLean, M. (2012). The growth rate of symplectic homology and affine varieties. *Geometric and Functional Analysis*, **22**, no. 2, 369?442. http://arxiv.org/abs/1109.4466
[476] McMullen, T.C., Taubes, C.H. (1999). 4-manifolds with inequivalent symplectic forms and 3-manifolds with inequivalent fibrations. *Mathematical Research Letters*, **6**, 681–696.
[477] Meckert, C. (1982). Forme de contacte sur la somme connexe de deux variétés de contacte. *Annales de l' Institut Fourier (Grenoble)*, **32**, 251–260.
[478] Meyer, K.R., Hall, G.R. (1992). *Introduction to Hamiltonian Dynamical Systems and the N-body Problem.* Applied Mathematics Series, **90**. Springer-Verlag, Berlin.
[479] Micallef, M., White, B. (1994). The structure of branch points in minimal surfaces and in pseudoholomorphic curves. *Annals of Mathematics*, **139**, 35–85.
[480] Milinković, D., Oh, Y.-G. (1997). Floer homology as the stable Morse homology. *Journal of the Korean Mathematical Society* **34**, 1065–1087.
[481] Milinković, D. (2001). Geodesics on the space of Lagrangian submanifolds in cotangent bundles. *Proceedings of the American Mathematical Society*, **129**, 1843–1851.
[482] Milnor, J. (1958). On simply connected 4-manifolds. In *Symposium Internacionale de Topologia Algebraica, (International symposium on algebraic topology)*, pp. 122–128, Universidad Nacional Autónoma de México and UNESCO, Mexico City
[483] Milnor, J. (1964). *Morse Theory.* Annals of Mathematics Studies, **51**. Princeton University Press, Princeton, NJ.
[484] Milnor, J. (1997). *Topology from the Differentiable Viewpoint.* Based on notes by D. Weaver. Revised reprint of the 1965 original. *Princeton Landmarks in Mathematics*, Princeton University Press, Princeton, NJ.
[485] Milnor, J. (1965). *Lectures on the h-Cobordism Theorem.* Princeton Universty Press, Princeton, NJ.
[486] Milnor, J. (2011). Differential Topology Forty-six Years Later. *Notices of the American Mathematical Society.* **58**, no. 6, 804– 809.
http://www.ams.org/notices/201106/rtx110600804p.pdf
[487] Milnor, J., Husemoller, D. (1970). *Symmetric Bilinear Forms.* Springer-Verlag, Berlin.
[488] Milnor, J., Stasheff, J.D. (1974). *Characteristic Classes.* Annals of Mathematics Studies, **76**. Princeton University Press.

[489] Miyaoka, Y. (1977). On the Chern numbers of surfaces of general type. *Inventiones Mathematicae*, **42**, 225–237.
[490] Moser, J.K. (1965). On the volume elements on manifolds. *Transactions of the American Mathematical Society* **120**, 280–296.
[491] Moser, J.K. (1966). A rapidly convergent iteration method and non-linear partial differential differential equations I & II. *Annali della Scuola Normale Superiore di Pisa* **20**, 265–315; 499–535.
[492] Moser, J.K. (1981). *Integrable Hamiltonian Systems and Spectral Theory.* Lezioni Fermiane, Scuola Normale Superiore, Pisa.
[493] Moser, J.K. (1986). Monotone twist maps and the calculus of variations. *Ergodic Theory and Dynamical Systems* **6**, 401–414.
[494] Morgan, J., Szabó, Z., Taubes, C.H. (1996). A product formula for the Seiberg-Witten invariants and the generalized Thom conjecture. *Journal of Differential Geometry*, **44**, 706–788.
[495] Morita, S., Kotschick, D. (2005). Signatures of foliated surface bundles and the symplectomorphism groups of surfaces. *Topology* **44**, 131–149.
[496] Müller, M.-P. (1990). Une structure symplectique sur $\mathbb{R}^6$ avec une sphere lagrangienne plongeé et un champ de Liouville complet. *Commentarii Mathematici Helvetici*, **65**, 623–663.
[497] Mumford, D., Fogarty, J., Kirwan, F. (1994). *Geometric Invariant Theory.* 3rd edition, Ergebnisse der Mathematik, Springer-Verlag, New York.
[498] I. Mundet-i-Riera (2003). Hamiltonian Gromov–Witten invariants. *Topology* **42**, 525–553.
[499] I. Mundet-i-Riera (2015). The symplectomorphism groups of $T^2 \times S^2$ are Jordan. Preprint. `http://arxiv.org/abs/1502.02420`
[500] Murphy, E. (2012). Loose Legendrian embeddings in high dimensional contact manifolds. `http://arxiv.org/abs/1201.2245`
[501] Narasimhan, M.S., Ramanan, S. (1961). Existence of universal connections. *American Journal of Mathematics* **83**, 563–572.
[502] Narasimhan, M.S., Seshadri, C.S. (1965). Stable and unitary vector bundles on compact Riemann surfaces. *Annals of Mathematics*, **82**, 540–567.
[503] Ness, L. (1978). Mumford's numerical function and stable projective hypersurfaces. Springer Lecture Notes **732**, *Algebraic Geometry, Proceedings*, Copenhagen, 1978, pp. 417–454.
[504] Ness, L. (1984). A stratification of the null cone by the moment map. *American Journal of Mathematics*, **106**, 1281–1329.
[505] Oakley, J., Usher, M. (2016). On certain Lagrangian submanifolds of $S^2 \times S^2$ and $\mathbb{C}P^n$. *Algebraic & Geometric Topology*, **16** no. 1, 149?209. `http://arxiv.org/abs/1311.5152`
[506] Oh, Y.-G. (1993). Floer cohomology of Lagrangian intersections discs and pseudoholomorphic discs I, II. *Communications in Pure and Applied Mathematics*, **46**, 949–994, 995–1012. III. Arnold–Givental conjecture. In [319], pp. 555-574.
[507] Oh, Y.-G. (1996). Relative Floer and quantum cohomology and the symplectic topology of Lagrangian submanifolds. In [620], pp. 201–267.
[508] Oh, Y.-G. (1997). Gromov–Floer theory and disjunction energy of compact Lagrangian embeddings. *Mathematical Research Letters*, **4**, 985–1005.
[509] Oh, Y.-G. (2005). Spectral invariants, analysis of the Floer moduli space and Geometry of the Hamiltonian Diffeomorphism group. *Duke Mathematical Journal*, **130**, no. 2, 199–295.
[510] Oh, Y.-G., Müller, S. (2007). The group of Hamiltonian homeomorphisms and C^0 symplectic topology. *Journal of Symplectic Geometry*, **5**, (2007), 167–220.
[511] Ohta, H., Ono, K. (1996). Notes on symplectic 4-manifolds with $b_2^+ = 1$, II. *International Journal of Mathematics*, **7**, 755–770.
`http://www.worldscientific.com/doi/pdf/10.1142/S0129167X96000402`
[512] Okounkov, A., Pandharipande, R. (2006). Gromov-Witten theory, Hurwitz theory, and completed cycles. *Annals of Mathematics*, **163**, 517–560.

[513] Ono, K. (1984). Some remarks on group actions in symplectic geometry. *Journal of the Faculty of Science of the University of Tokyo, Section 1A, Mathematics*, **35**, 431–437.

[514] Ono, K. (1992). Obstruction to circle group action preserving symplectic structure. *Hokkaido Mathematical Journal* **21**, 99–102.

[515] Ono, K. (1995). The Arnold conjecture for weakly monotone symplectic manifolds. *Inventiones Mathematicae*, **119**, 519–537.

[516] Ono, K. (2006). Floer–Novikov cohomology and the flux conjecture. *Geometric and Functional Analysis*, **16**, 981–1020.

[517] Oprea, J., Rudyak, Y.B. (1999). On the Lusternick–Schnirelmann category of symplectic manifolds and the Arnold conjecture. *Mathematische Zeitschrift* **230**, 673–678. `http://arxiv.org/abs/dg-ga/9708007`

[518] Ostrover, Ya. (2003). A comparison of Hofer's metrics on Hamiltonian diffeomorphisms and Lagrangian submanifolds. *Communications in Contemporary Mathematics*, **5** (2003), no. 5., 803–811.

[519] Ozsváth, P., Szabó, Z. (2000). The symplectic Thom conjecture. *Annals of Mathematics*, **151**, 93–124.

[520] Ozsváth, P., Szabó, Z. (2004). Holomorphic disks and topological invariants for closed three-manifolds. *Annals of Mathematics*, **159**, 1027–1158.

[521] Ozsváth, P., Szabó, Z. (2004). Holomorphic disks and three-manifold invariants: properties and applications. *Annals of Mathematics*, **159**, 1059–1245.

[522] Palais, R. (1960). On the local triviality for the restriction map for embeddings. *Commentarii Mathematici Helvetici*, **34**, 306–312.

[523] Pardon, J. (2016) An algebraic approach to fundamental cycles on moduli spaces of J-holomorphic curves. *Geometry & Topology*, **20**, no. 2, 779–1034. `http://arxiv.org/abs/math/1309.2370`

[524] Park, J., Stipsicz, A., Szabó, Z. (2005). Exotic smooth structures on $\mathbb{C}\mathrm{P}^2\#5\overline{\mathbb{C}\mathrm{P}}^2$. *Mathematical Research Letters*, **12**, 701–712. `http://arxiv.org/abs/math/0412216`

[525] Piunikhin, S., Salamon, D., Schwarz, M. (1996). Symplectic Floer-Donaldson theory and quantum cohomology. *Contact and Symplectic Geometry*, edited by C.B. Thomas, Publications of the Newton Institute, Cambridge University Press 1996, 171–200.

[526] Poenaru, V. (1971). Homotopy theory and differentiable singularities. In *Manifolds – Amsterdam 1970*. Lecture Notes in Mathematics, **197**, pp. 106–132. Springer-Verlag, Berlin.

[527] Polterovich, L. (1993). Symplectic displacement energy for Lagrangian submanifolds. *Ergodic Theory and Dynamical Systems* **13**, 357–367.

[528] Polterovich, L. (1996). Gromov's K-area and symplectic rigidity. *Geometric and Functional Analysis* **6**, 726–739.

[529] Polterovich, L. (1998). Symplectic aspects of the first eigenvalue. *Journal für die Riene und Angewandte Mathematik [Crelle's Journal]*, **502**, 1–17.

[530] Polterovich, L. (1998). Hofer's diameter and Lagrangian intersections. *International Mathematical Research Notices*, **4**, 217–223.

[531] Polterovich, L. (1998). Geometry on the group of Hamiltonian diffeomorphisms. *Proceedings of the ICM*, Berlin, 1998, Vol. II, pp. 401–410.

[532] Polterovich, L., Rosen, D. (2014). *Function Theory on Symplectic Manifolds.* CRM Monograph Series, Volume **34**, American Mathematical Society.

[533] Polterovich, L., Shelukhin, E. (2014). *Autonomous Hamiltonian flows, Hofer's geometry and persistence modules.* `http://arxiv.org/abs/1412.8277`

[534] Py, P. (2006). Quasi-morphismes de Calabi et graphe de Reeb sur la tore. *Comptes Rendus Mathématique. Acadḿie des Sciences. Paris*, **343**, 323–328.

[535] Py, P. (2006). Quasi-morphismes et invariant de Calabi. *Annales Scientifiques de l'École Normale Supérieure, Quatrième Série*, **39**, 177–195.

[536] Rabinowitz, P.H. (1978). Periodic solutions of Hamiltonian systems. *Communications in Pure and Applied Mathematics* **31**, 157–184.

[537] Ratiu, T, Weinstein, A., Zung, N.T. (eds.) (2011). *Lectures on Poisson Geometry* Summer School, Trieste 2005, *Geometry & Topology Monographs* **17**, Geometry & Topology

Publications, Coventry.

[538] Reznikov, A.G. (1993). Symplectic twistor spaces. *Annals of Global Analysis and Geometry,* **11** (2), 109–118.

[539] Robbin, J.W., Salamon, D.A. (1988). Dynamical systems, shape theory, and the Conley index. *Ergodic Theory and Dynamical Systems,* **8**, 375–393.

[540] Robbin, J.W., Salamon, D.A. (1993). Phase functions and path integrals. In *Symplectic Geometry* (ed. D. Salamon), pp. 203–226. London Mathematical Society Lecture Note Series, **192**. Cambridge University Press, Cambridge, UK.

[541] Robbin, J.W., Salamon, D.A. (1993). The Maslov index for paths. *Topology,* **32**, 827–844.

[542] Robbin, J.W., Salamon, D.A. (1995). The spectral flow and the Maslov index. *Bulletin of the London Mathematical Society,* **27**, 1-33.

[543] Robbin, J.W., Salamon, D.A. (1996). Path integrals on phase space and the metaplectic representation. *Mathematische Zeitschrift* **221**, 307–335.

[544] Robbin, J.W., Salamon, D.A. (2013). *Introduction to Differential Geometry.* `http://www.math.ethz.ch/~salamon/PREPRINTS/diffgeo.pdf`

[545] Ruan, Y. (1993). Symplectic topology and extremal rays. *Geometric and Functional Analysis,* **3**, 395–430.

[546] Ruan, Y. (1994). Symplectic topology on algebraic 3-folds. *Journal of Differential Geometry,* **39**, 215–227.

[547] Ruan, Y. (1996). Topological sigma model and Donaldson type invariants in Gromov theory. *Duke Mathematical Journal,* **83**, 461–500

[548] Ruan, Y. (1999). Virtual neighborhoods and pseudoholomorphic curves, *Turkish Journal of Mathematics,* **23**, 161–231.

[549] Ruan, Y., Tian, G. (1995). A mathematical theory of quantum cohomology. *Journal of Differential Geometry,* **42**, 259–367.

[550] Ruan, Y., Tian, G. (1997). Higher genus symplectic invariants and sigma model coupled with gravity. *Inventiones Mathematicae,* **130**, 455–516.

[551] Rudyak, Y.B. (1999). On analytical applications of stable homotopy (the Arnold conjecture, critical points). *Mathematische Zeitschrift,* **230**, 659–672. `http://arxiv.org/abs/dg-ga/9708008`

[552] Salamon, D.A. (1985). Connected simple systems and the Conley index for isolated invariant sets. *Transactions of the American Mathematical Society* **291**, 1–41.

[553] Salamon, D.A. (1990). Morse theory, the Conley index and Floer homology. *Bulletin of the London Mathematical Society,* **22**, 113–140.

[554] Salamon, D.A. (1996) Removable singularities and a vanishing theorem for Seiberg–Witten invariants. *Turkish Journal of Mathematics,* **20**, 61–73.

[555] Salamon, D.A. (1997). Lectures on Floer homology. In *Symplectic Geometry and Topology,* IAS/Park City Mathematics Series, Vol. **7**, edited by Y. Eliashberg and L. Traynor, American Mathematical Society, Providence, RI, 1999, pp. 145–229.

[556] Salamon, D.A. (1999). *Spin Geometry and Seiberg–Witten Invariants.* unpublished manuscript. `http://www.math.ethz.ch/~salamon/PREPRINTS/witsei.pdf`

[557] Salamon, D.A. (2004). The Kolmogorov-Arnold-Moser theorem. *Mathematical Physics Electronic Journal* **10** (2004), paper 3. `http://www.ma.utexas.edu/mpej/Vol/10/3.pdf`

[558] Salamon, D.A. (2013). Uniqueness of symplectic structures. *Acta Mathematica Vietnamica* **38**, 123–144.

[559] Salamon, D.A. (2013). The three dimensional Fueter equation and divergence-free frames. *Abhandlungen aus dem Mathematischen Seminar der Universitat Hamburg* **82**, 1–28. `http://lanl.arxiv.org/abs/1202.4165`

[560] Salamon, D.A., Walpuski, T. (2017). Notes on the Octonions. *Proceedings of the 23rd Gökova Geometry-Topology Conference 2016,* edited by S. Akbulut, D. Auroux, and T. Önder. `http://arxiv.org/abs/1005.2820`

[561] Salamon, D.A., Wehrheim, K. (2008). Instanton Floer homology with Lagrangian boundary conditions. *Geometry & Topology,* **12**, (2008), 747–918.

[562] Salamon, D.A., Zehnder, E. (1989). KAM theory in configuration space. *Commentarii Mathematici Helvetici,* **64**, 84–132.

[563] Salamon, D.A., Zehnder, E. (1992). Morse theory for periodic solutions of Hamiltonian systems and the Maslov index. *Communications in Pure and Applied Mathematics* **45**, 1303–1360.

[564] Sandon, S. (2010). An integer valued bi-invariant metric on the group of contactomorphisms of $R^{2n} \times S^1$. *Journal of Topology and Analysis*, **2**, 327–339.

[565] Schlenk, F. (2015). Symplectic embedding problems, in preparation.

[566] Schur, L. (1923). Über eine Klasse von Mittelbildungen mit Anwendungen auf die Determinantentheorie. *Sitzungsberichte der Berliner Mathematischen Gesellschaft* **22**, 9–20.

[567] Schwarz, M. (1995). Cohomology operations from S^1-cobordisms in Floer theory. PhD thesis, ETH-Zürich, 1995.

[568] Schwarz, M. (2000). On the action spectrum for closed symplectially aspherical manifolds, *Pacific Journal of Mathematics* **193**, (2000), 419–461.

[569] Seidel, P. (1997). π_1 of symplectic automorphism groups and invertibles in quantum cohomology rings. *Geometric and Functional Analysis*, **7**, 1046–1095.

[570] Seidel, P. (1997). Floer homology and the symplectic isotopy problem. Ph.D. thesis, Oxford.

[571] Seidel, P. (1999). On the group of symplectic automorphisms of $\mathbb{C}\mathrm{P}^m \times \mathbb{C}\mathrm{P}^n$. *Northern California Symplectic Geometry Seminar*, American Mathematical Society, Providence RI. *American Mathematical Society Translations Series 2*, **196**, pp. 237–250.

[572] Seidel, P. (2002). Fukaya categories and deformations. *Proceedings of the International Congress of Mathematicians (Beijing 2002)*, Vol **II**, Higher Education Press, 2002, Beijing, pp. 351–360.

[573] Seidel, P. (2003). A long exact sequence in symplectic Floer homology. *Topology*, **42**, 1003–1063.

[574] Seidel, P. (2008). *Fukaya Categories and Picard–Lefschetz Theory.* Zurich Lectures in Advanced Mathematics, European Mathematical Society, Zürich.

[575] Seidel, P. (2008). Lectures on four-dimensional Dehn twists. *Symplectic Four-Manifolds and Algebraic Surfaces*, Lecture Notes in Mathematics **1938**, Springer-Verlag, Berlin, pp. 231–267.

[576] Seidel, P. (2011). Simple examples of distinct Liouville type symplectic structures. *Journal of Topology and Analysis*, **3** no. 1, 1?5. `http://arxiv.org/abs/1011.0394v2`

[577] Seidel, P. (2012). Lagrangian homology spheres in (A_m) Milnor fibres via $\mathbb{C}^*$-equivariant A_∞ modules. *Geometry & Topology* **16**, 2343–2389. `http://arxiv.org/abs/1202.1955`

[578] Seidel, P. (2014). Disjoinable Lagrangian spheres and dilations. *Inventiones Mathematicae*, **197**, no. 2, 299–359. `http://arxiv.org/abs/1307.4819`

[579] Seidel, P. (2014). *Abstract Analogues of Flux as Symplectic Invariants.* Mémoires de la Société Mathématiques de France, **137**, 135 pages. `http://arxiv.org/abs/1108.1415`

[580] Shelukhin, E. (2014). The Action homomorphism, quasimorphisms and moment maps on the space of almost complex structures. *Commentarii Mathematici Helvetici*, **89** no. 1, 69–123. `http://arxiv.org/abs/1105.5814`

[581] Shelukhin, E. (2015). Enlacements asymptotiques revisités. *Annales Mathématiques du Québec*, **39**, no. 2, 205–208. `http://arxiv.org/abs/1411.1458`

[582] Siebert, B. (1999). Symplectic Gromov–Witten invariants, in *New Trends in Algebraic Geometry, (Warwick 1996)*, London Math Society Lecture Notes **264**, Cambridge University Press, Cambridge, pp. 375–424.

[583] Siegel, C.L. (1964). *Symplectic Geometry*. Academic Press, New York.

[584] Siegel, C.L., Moser, J.K. (1971). *Lectures on Celestial Mechanics.* Springer-Verlag.

[585] Sikorav, J.-C. (1989). Rigidité symplectique dans le cotangent de $\mathbb{T}^n$. *Duke Mathematical Journal*, **59**, 227–231.

[586] Sikorav, J.-C. (1990). *Systemes Hamiltoniens et topologie symplectique.* University of Pisa, ETS Editrice, Pisa.

[587] Sikorav, J.-C. (1991). Quelques propriétés des plongements Lagrangiens. *Bulletin de la Société Mathématique de France* **46**, 151–167.

[588] Sikorav, J. (2007). Approximation of a volume-preserving homeomorphism by a volume-preserving diffeomorphism.

http://www.umpa.ens-lyon.fr/~symplexe/Documents/volume-preserving-approximation-2.pdf

[589] Singer, I.M., Sternberg, S. (1965). The infinite groups of Lie and Cartan I. *Journal d'Analyse Mathématique* **15**, 1–114.

[590] Sjamaar, R. (1990). Singular orbit spaces in Riemannian and symplectic geometry. Ph.D. thesis, University of Utrecht.

[591] Smale, S. (1959). Diffeomorphisms of the 2-sphere. *Proceedings of the American Mathematical Society* **10**, 621–626.

[592] Smith, I. (2000). On moduli spaces of symplectic forms. *Mathematical Research Letters,* **7**, 779–788.

[593] Smith, I. (2001). Torus fibrations on symplectic 4-manifolds. *Turkish Journal of Mathematics*, **25**, 69–95.

[594] Smith, I., Thomas. R.P., Yau, S.-T. (2002). Symplectic conifold transitions. *Journal of Differential Geometry*, **62**, 209–242. http://arxiv.org/abs/math.SG/0209319

[595] Solomon, J.P. (2013). The Calabi homomorphism, Lagrangian paths and special Lagrangians. *Mathematische Annalen* **357**, 1389–1424.
http://arxiv.org/abs/1209.4737

[596] Souriou, J.M. (1997). *Structure of Dynamical Systems: A Symplectic View of Physics.* Progress in Mathematics, Volume **149**, Birkhäuser, Basel.

[597] Sternberg, S. (1977). Minimal coupling and the symplectic mechanics of a classical particle in the presence of a Yang–Mills field. *Proceedings of the National Academy of Sciences of the USA*, **74**, 5253–54.

[598] Stipsicz, A. (1998). Simply connected 4-manifolds near the Bogomolov–Miyaoka–Yau line. *Mathematical Research Letters*, **5**, 723–730.

[599] Stipsicz, A. (1999). Simply connected symplectic 4-manifolds with positive signature. *Proceedings of 6th Gökova Geometry-Topology Conference*, pp. 145–150.

[600] Strominger, A., Yau, S.-T., Zaslow, E. (1996). Mirror symmetry is T-duality. *Nuclear Physics B*, **479**, 243–259.

[601] Stipsicz, A., Szabó, Z. (2005). Exotic smooth structures on $\mathbb{CP}^2\#6\overline{\mathbb{CP}}^2$. *Geometry & Topology*, **9**, 813–832 http://arxiv.org/abs/math/0411258

[602] Struwe, M. (1990). Existence of periodic orbits of Hamiltonian systems on almost every energy surface. *Boletin do Sociedade Brasiliense de Matematicas*, **20**, 49–58.

[603] Symington, M. (1998). Symplectic rational blowdowns. *Journal of Differential Geometry,* **50**, 505–518.
http://arxiv.org/abs/math/9802079v1

[604] Szabó, Z. (1998). Simply-connected irreducible 4-manifolds with no symplectic structures. *Inventiones Mathematicae*, **123**, 457–466.

[605] Tamarkin, D. (2008). Microlocal condition for non-displaceability. Preprint.
http://arxiv.org/abs/math/0809.1584

[606] Taubes, C.H. (1994). The Seiberg–Witten invariants and symplectic forms. *Mathematical Research Letters*, **1**, 809–822.

[607] Taubes, C.H. (1995). More constraints on symplectic forms from Seiberg–Witten invariants. *Mathematical Research Letters*, **2**, 9–14.

[608] Taubes, C.H. (1995). The Seiberg–Witten and the Gromov invariants. *Mathematical Research Letters*, **2**, 221–238.

[609] Taubes, C.H. (1996). SW$\Longrightarrow$ Gr: From the Seiberg–Witten equations to pseudo-holomorphic curves. *Journal of the American Mathematical Society*, **9**, 845–918.

[610] Taubes, C.H. (1996). Counting pseudoholomorphic submanifolds in dimension four. *Journal of Differential Geometry*, **44**, 818–893.

[611] Taubes, C.H. (2000). *Seiberg-Witten and Gromov Invariants for Symplectic Four-Manifolds.* First International Press Lecture Series 2, International Press, Boston, MA.

[612] Taubes, C.H. (2007). The Seiberg-Witten equations and the Weinstein conjecture. *Geometry & Topology,***11**, 2117–2202.

[613] Taubes, C.H. (2010). Embedded contact homology and Seiberg–Witten Floer homology I. *Geometry & Topology*, **14**, 2497–2581.

[614] Taubes, C.H. (2010). Embedded contact homology and Seiberg–Witten Floer homology II. *Geometry & Topology*, **14**, 2583–2720.

[615] Taubes, C.H. (2010). Embedded contact homology and Seiberg–Witten Floer homology III. *Geometry & Topology*, **14**, 2721–2817.

[616] Taubes, C.H. (2010). Embedded contact homology and Seiberg–Witten Floer homology IV. *Geometry & Topology*, **14**, 2819–2960.

[617] Taubes, C.H. (2010). Embedded contact homology and Seiberg–Witten Floer homology V. *Geometry & Topology*, **14**, 2961–3000.

[618] Taubes, C.H. (2011). Tamed to compatible: Symplectic forms via moduli space integration. *Journal of Symplectic Geometry*, **9**, 161–250. `http://arxiv.org/abs/0910.5440`

[619] Théret, D. (1995). Thèse de Doctorat, Univerité Denis Diderot (Paris 7).

[620] Thomas, C. (ed.) (1996). *Symplectic and Contact Geometry*. Publications of the Newton Institute, Volume **8**. Cambridge University Press, Cambridge, UK.

[621] Thomas, R.P. (2001). Moment maps, monodromy and mirror manifolds. *Symplectic Geometry and Mirror Symmetry* (Seoul, 2000), World Scientific Publications, River Edge, NJ, pp. 467–498. `http://arxiv.org/abs/math/0104196`

[622] Thomas, R.P. (2006). Notes on GIT and symplectic reduction for bundles and varieties. *Surveys in Differential Geometry* **10**, *A tribute to S.-S. Chern*, edited by Shing-Tung Yau, International Press, Somerville, MA. `http://arxiv.org/abs/math.AG/0512411`

[623] Thurston, W. (c. 1972). On the structure of the group of volume-preserving diffeomorphisms. Unpublished.

[624] Thurston, W. (1976). Some simple examples of symplectic manifolds. *Proceedings of the American Mathematical Society* **55**, 467–468.

[625] Thurston, W. (1976). Existence of codimension 1 foliations. *Annals of Mathematics*, **104**, 249–268.

[626] Tian, G. (1997) Kähler-Einstein metrics with positive scalar curvature. *Inventiones Mathematicae*, **130**, 1–37.

[627] Tischler, D. (1977). Closed 2-forms and an embedding theorem for symplectic manifolds. *Journal of Differential Geometry*, **12**, 229–235.

[628] Tolman, S. (2010). On a symplectic generalization of Petrie's conjecture. *Transactions of the American Mathematical Society*, **362**, 3963–3996. `http://arxiv.org/abs/0903.4918`

[629] Tolman, S. (2015). Non-Hamiltonian actions with isolated fixed points. Preprint, October 2015. `http://arxiv.org/abs/1510.02829`

[630] Tolman, S., Weitsman, J. (2000). On semi free circle actions with isolated fixed points, *Topology*, **39** no. 2, 299–310.

[631] Torres, R., Yazinski, J. (2015). Geography of symplectic 4- and 6-manifolds. *Topology Proceedings*, **46**, 87?115. `http://arxiv.org/abs/1107.3954`

[632] Uhlenbeck, K., Yau, S.-T. (1986). On the existence of Hermitian Yang-Mills connections in stable vector bundles. *Communications on Pure and Applied Mathematics* **39**, 257–293.

[633] Usher, M. (2012). Many closed symplectic manifolds have infinite Hofer-Zehnder capacity. *Transactions of the American Mathematical Society*, **364** (2012), 5913–5943. `http://arxiv.org/abs/1101.4986`

[634] Usher, M. (2013). Hofer's metric and boundary depth. *Annales de Science de l'École Normale Supérieure*, **46**, 57–128. `http://arxiv.org/abs/1107.4599`

[635] Usher, M. (2015). Observations on the Hofer distance between closed sets. *Mathematical Research Letters*, **22** no. 6, 180–1820. `http://arxiv.org/abs/1409.2577`

[636] Ustilovsky, I. (1996). Conjugate points on geodesics of Hofer's metric. *Differential Geometry and its Applications* **6**, 327–342.

[637] Ustilovsky, I. (1999). Infinitely many contact structures on S^{4m+1}. *International Mathematics Research Notices*, **14**, 781–791.

[638] Van de Ven, A. (1966). On the Chern numbers of complex and almost complex manifolds. *Proceedings of the National Academy of Sciences of the USA* **55**, 1624–1627.

[639] Vianna, R. (2014). On exotic Lagrangian tori in $\mathbb{CP}^2$. *Geometry & Topology*, **18** no. 4, 2419–2476. `http://arxiv.org/abs/1305.7512`

[640] Vianna, R. (2016). Infinitely many exotic monotone Lagrangian tori in $\mathbb{C}P^2$. *Journal of Topology*, **9** no. 2, 53–551. `http://arxiv.org/abs/1409.2850`
[641] Vidussi, S. (2001). Homotopy K3's with several symplectic structures. *Geometry & Topology*, **5**, 267–285. `http://arxiv.org/abs/math/0103158`
[642] Vidussi, S. (2001). Smooth structure of some symplectic surfaces. *Michigan Mathematics Journal*, **49**, 325–330. `http://arxiv.org/abs/math/0103216`
[643] Viterbo, C. (1987). A proof of the Weinstein conjecture in $\mathbb{R}^{2n}$. *Annales de l' Institut Henri Poincaré – Analyse nonlinéaire* **4**, 337–357.
[644] Viterbo, C. (1989). Capacités symplectiques et applications. Séminaire Bourbaki 1988–89, exposé n° **714**, *Astérisque 177–178*, 345–362.
[645] Viterbo, C. (1990). A new obstruction to embedding Lagrangian tori. *Inventiones Mathematicae*, **100**, 301–320.
[646] Viterbo, C. (1992). Symplectic topology as the geometry of generating functions. *Mathematische Annalen* **292**, 685–710.
[647] Viterbo, C. (1999). Functors and computations in Floer homology with applications, I *Geometric and Functional Analysis*, **9** no. 5., 985–1033. part II is a preprint, Orsay.
[648] Viterbo, C. (2000). Metric and isoperimetric problems in symplectic geometry. *Journal of the American Mathematical Society*, **13** 411-431.
[649] Viterbo, C. (2006). On the uniqueness of generating Hamiltonian for continuous limits of Hamiltonians flows. *International Mathematics Research Notices*, Article ID 34028, Erratum ID 38784.
[650] Voisin, C. (2004). On the homotopy types of Kähler compact and complex projective manifolds. *Inventiones Mathematicae*, **157**, no. 2, 329–343.
[651] Voisin, C. (2006). On the homotopy types of Kähler manifolds and the birational Kodaira problem. *Journal of Differential Geometry*, **72**, 43–71.
[652] Wall, C.T.C. (1962). On the orthogonal groups of unimodular quadratic forms. *Mathematische Annalen*, **147**, 328–338.
[653] Wall, C.T.C. (1963). On the orthogonal groups of unimodular quadratic forms II. *Journal für die Riene und Angewandte Mathematik [Crelle's Journal]*, **213**, 122–136.
[654] Wall, C.T.C. (1964). Diffeomorphisms of 4-Manifolds. *Journal of the London Mathematical Society*, **39**, 131–140.
[655] Wall, C.T.C. (1964). On simply connected 4-Manifolds. *Journal of the London Mathematical Society*, **39**, 141–149.
[656] Walpuski, T. (2012). G2-instantons, associative submanifolds and Fueter sections. To appear in *Communications in Analysis and Geometry*. `http://lanl.arxiv.org/abs/1205.5350`
[657] Walpuski, T. (2013). G2-instantons on generalized Kummer constructions. *Geometry & Topology*, **17** no. 4, 2345–2388. `http://lanl.arxiv.org/abs/1109.6609`
[658] Walpuski, T. (2014). Spin(7)-instantons, Cayley submanifolds and Fueter sections. To appear in *Communications in Mathematical Physics*. `https://arxiv.org/abs/1409.6705`
[659] Walpuski, T. (2015). A compactness theorem for Fueter sections. Preprint, July 2015. `https://arxiv.org/abs/1507.03258`
[660] Wehrheim, K. (2004). Banach space valued Cauchy-Riemann equations with totally real boundary conditions. *Communications in Contemporary Mathematics* , **6**, 601–635.
[661] Wehrheim, K. (2005). Anti-self-dual instantons with Lagrangian boundary conditions I: Elliptic theory. *Communications in Mathematical Physics*, **254**, 45–89.
[662] Wehrheim, K. (2005). Anti-self-dual instantons with Lagrangian boundary conditions II: Bubbling. *Communications in Mathematical Physics*, **258**, 275–315.
[663] Wehrheim, K. (2005). Lagrangian boundary conditions for anti-self-dual connections and the Atiyah-Floer conjecture. *Journal of Symplectic Geometry*, **3**, 703–747.
[664] Wehrheim, K., Woodward, C.T. (2010). Quilted Floer cohomology. *Geometry & Topology* **14**, 833–902.
[665] Wehrheim, K., Woodward, C.T. (2012). Quilted Floer trajectories with constant components. *Geometry & Topology* **16**, 127–154.

[666] Wehrheim, K., Woodward, C.T. (2010). Functoriality for Lagrangian correspondences in Floer theory. *Quantum Topology*, **1**, 129–170.
[667] Wehrheim, K., Woodward, C.T. (2012). Floer cohomology and geometric composition of Lagrangian correspondences. *Advances in Mathematics* **230**, 833–902.
[668] Wehrheim, K., Woodward, C.T. (2015). Pseudoholomorphic quilts. *Journal of Symplectic Geometry*, **13**, no. 4, 849?904
[669] Weinstein, A. (1971). Symplectic manifolds and their Lagrangian submanifolds. *Advances in Mathematics* **6**, 329–346.
[670] Weinstein, A. (1972). The invariance of Poincaré's generating function for canonical transformations. *Inventiones Mathematicae*, **16**, 202–213.
[671] Weinstein, A. (1977). *Lectures on Symplectic Manifolds*. CBMS Conference Series, **29**. American Mathematical Society, Providence, RI.
[672] Weinstein, A. (1978). A universal phase space for particles in Yang-Mills fields. *Letters in Mathematical Physics*, **2**, 417–420.
[673] Weinstein, A. (1979). On the hypotheses of Rabinowitz's periodic orbit theorems. *Journal of Differential Equations*, **33**, 353–358.
[674] Weinstein, A. (1980). Fat bundles and symplectic manifolds. *Advances in Mathematics*, **37**, 239–50.
[675] Weinstein, A. (1981). Symplectic geometry. *Bulletin of the American Mathematical Society*, **5**, 1–13.
[676] Weinstein, A. (1983). The local structure of Poisson manifolds. *Journal of Differential Geometry*, **18**, 523–557.
[677] Weinstein, A. (1983). Sophus Lie and symplectic geometry. *Expositiones Mathematicae*, **1**, 95–96.
[678] Weinstein, A. (1991). Contact surgery and symplectic handlebodies. *Hokkaido Mathematical Journal*, **20**, 241–251.
[679] Weinstein, A. (1995). The symplectic structure on moduli space. In [319], pp. 627–626.
[680] Weinstein, A. (2010). Symplectic Categories. *Portugaliae Mathematica*, **67** no. 2, 261–278. `http://arxiv.org/abs/0911.4133`
[681] Weinstein, A. (2011). A note on the Wehrheim-Woodward category. *Journal of Geometric Mechanics*, **3**, 507–515. `http://arxiv.org/abs/1012.0105`
[682] Welschinger, J.-Y. (2007). Effective classes and Lagrangian tori in symplectic four-manifolds. *Journal of Symplectic Geometry*, **5**, no. 1, 9–18.
[683] Wendl, C. (2010). Strongly fillable contact manifolds and J-holomorphic foliations. *Duke Mathematical Journal*, **151**, 337–384.
[684] Wendl, C. (2010). *Lectures on lolomorphic curves in symplectic and contact topology.* `http://arxiv.org/abs/1011.1690`
[685] Wendl, C. (2013). Non-exact symplectic cobordisms between contact 3-manifolds. *Journal of Differential Geometry*, **95**, 121–182.
[686] Wendl, C. (2013). A hierarchy of local symplectic filling obstructions for contact 3-manifolds *Duke Mathematical Journal*, **162**, 2197–2283.
[687] Wendl, C. (2014). Contact hypersurfaces in uniruled symplectic manifolds always separate. *Journal of the London Mathematical Society*, (2) **89** no. 3, 832–852. `http://arxiv.org/abs/1202.4685`
[688] Witten, E. (1982). Supersymmetry and Morse theory. *Journal of Differential Geometry*, **17**, 661–692.
[689] Woodhouse, N.M.J. (1980). *Geometric Quantization.* Oxford University Press.
[690] Woodward, C.T. (2011). Moment maps and geometric invariant theory. `http://arxiv.org/abs/0912.1132v6`
[691] Yau, S.-T. (1977). Calabi's conjecture and some new results in algebraic geometry. *Proceedings of the National Academy of Sciences of the USA* , **74**, 1789–1799.
[692] Yau, S.-T. (1987). Open problems in geometry. *L'Enseignement Mathémathique* **33**, 109–158.
[693] Zehnder, E. (2010). *Lectures on Dynamical Systems.* EMS Textbooks in Mathematics, European Mathematical Society, Zürich.

INDEX